PRINCIPLES OF ANATOMY AND PHYSIOLOGY

Tenth Edition

Volume 4

The Maintenance and Continuity of the Human Body

Gerard J. Tortora

Bergen Community College

Sandra Reynolds Grabowski

Purdue University

John Wiley & Sons, Inc.

Senior Editor	Bonnie Roesch
Marketing Manager	Clay Stone
Developmental Editor	Ellen Ford
Production Director	Pamela Kennedy
Associate Production Manager	Kelly Tavares
Text and Cover Designer	Karin Gerdes Kincheloe
Art Coordinator	Claudia Durrell
Photo Editor	Hilary Newman
Chapter Opener Illustrations	Keith Kasnot

Cover photos: Photo of woman: ©PhotoDisc, Inc.
Photo of man: ©Ray Massey/Stone
Illustration and photo credits follow the Glossary.

This book was typeset by Progressive Information Technologies. It was printed and bound by Von Hoffmann Press, Inc. The cover was also printed by Von Hoffman Press, Inc.

The paper in this book was manufactured by a mill whose forest management programs include sustained yield harvesting of its timberlands. Sustained yield harvesting principles ensure that the number of trees cut each year does not exceed the amount of new growth.

This book is printed on acid-free paper. ∞

USA ISBN: 0-471-22934-2

Printed in the United States of America.
10 9 8 7 6 5 4 3 2 1

About The Cover

In choosing the cover illustration for the tenth edition of *Principles of Anatomy and Physiology,* we wanted to reflect the underlying principle in the textbook as well as our approach to presentation of the content. In a word, we wanted to suggest the idea of balance. Successfully practicing yoga requires finding the balance between the physical, mental, and spiritual aspects of the human body. Achieving this balance produces flexibility and strength, as demonstrated by the figures on the cover.

Similarly, balance has always been at the heart of the study of anatomy and physiology. Homeostasis, the underlying principle discussed throughout this textbook, is the state of equilibrium of the internal environment of the human body. During your study of anatomy and physiology you will learn about the structures and functions needed to regulate and maintain this balanced state and its importance in promoting optimal well being.

Balance also applies to the way we present the content in the tenth edition of *Principles of Anatomy & Physiology*. We offer a balanced coverage of structure and function; between normal and abnormal anatomy and physiology; between concepts and applications; between narrative and illustrations; between print and media; and among supplemental materials developed to meet the variety of teaching and learning styles. We hope that this balanced approach will provide you with the flexibility and strength to succeed in this course and in your future careers.

About the Authors

Gerard J. Tortora is Professor of Biology and former Coordinator at Bergen Community College in Paramus, New Jersey, where he teaches human anatomy and physiology as well as microbiology. He received his bachelor's degree in biology from Fairleigh Dickinson University and his master's degree in science education from Montclair State College. He is a member of many professional organizations, such as the Human Anatomy and Physiology Society (HAPS), the American Society of Microbiology (ASM), American Association for the Advancement of Science (AAAS), National Education Association (NEA), and the Metropolitan Association of College and University Biologists (MACUB).

Above all, Jerry is devoted to his students and their aspirations. In recognition of this commitment, Jerry was the recipient of MACUB's 1992 President's Memorial Award. In 1996, he received a National Institute for Staff and Organizational Development (NISOD) excellence award from the University of Texas and was selected to represent Bergen Community College in a campaign to increase awareness of the contributions of community colleges to higher education.

Jerry is the author of several best-selling science textbooks and laboratory manuals, a calling that often requires an additional 40 hours per week beyond his teaching responsibilities. Nevertheless, he still makes time for four or five weekly aerobic workouts that include biking and running. He also enjoys attending college basketball and professional hockey games and performances at the Metropolitan Opera House.

To my children, Lynne, Gerard, Kenneth, Anthony, and Andrew, who make it all worthwhile. — *G.J.T.*

Sandra Reynolds Grabowski is an instructor in the Department of Biological Sciences at Purdue University in West Lafayette, Indiana. For 25 years, she has taught human anatomy and physiology to students in a wide range of academic programs. In 1992 students selected her as one of the top ten teachers in the School of Science at Purdue.

Sandy received her BS in biology and her PhD in neurophysiology from Purdue. She is a founding member of the Human Anatomy and Physiology Society (HAPS) and has served HAPS as President and as editor of *HAPS News*. In addition, she is a member of the American Anatomy Association, the Association for Women in Science (AWIS), the National Science Teachers Association (NSTA), and the Society for College Science Teachers (SCST).

To students around the world, whose questions and comments continue to inspire the fine-tuning of this textbook. — *S.R.G.*

Preface

An anatomy and physiology course can be the gateway to a gratifying career in a host of health-related professions. As active teachers of the course, we recognize both the rewards and challenges in providing a strong foundation for understanding the complexities of the human body to an increasingly diverse population of students. Building on the unprecedented success of previous editions, the tenth edition of *Principles of Anatomy and Physiology* continues to offer a balanced presentation of content under the umbrella of our primary and unifying theme of homeostasis, supported by relevant discussions of disruptions to homeostasis. In addition, years of student feedback have convinced us that readers learn anatomy and physiology more readily when they remain mindful of the relationship between structure and function. As a writing team—an anatomist and a physiologist—our very different specializations offer practical advantages in fine-tuning the balance between anatomy and physiology.

Most importantly, our students continue to remind us of their needs for—and of the power of—simplicity, directness, and clarity. To meet these needs each chapter has been written and revised to include:

- clear, compelling, and up-to-date discussions of anatomy and physiology
- expertly executed and generously sized art
- classroom-tested pedagogy
- outstanding student study support.

As we revised the content for this edition we kept our focus on these important criteria for success in the anatomy and physiology classroom and have refined or added new elements to enhance the teaching and learning process.

Homeostasis: A Unifying Theme

The dynamic physiological constancy known as homeostasis is the prime theme in *Principles of Anatomy and Physiology*. We immediately introduce this unifying concept in Chapter 1 and describe how various feedback mechanisms work to maintain physiological processes within the narrow range that is compatible with life. Homeostatic mechanisms are discussed throughout the book, and homeostatic processes are clarified and reinforced through our well-received series of homeostasis feedback illustrations.

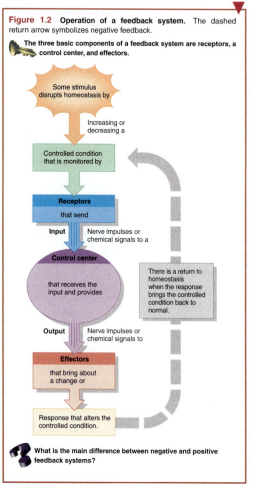

Figure 1.2 Operation of a feedback system. The dashed return arrow symbolizes negative feedback.

🎺 The three basic components of a feedback system are receptors, a control center, and effectors.

What is the main difference between negative and positive feedback systems?

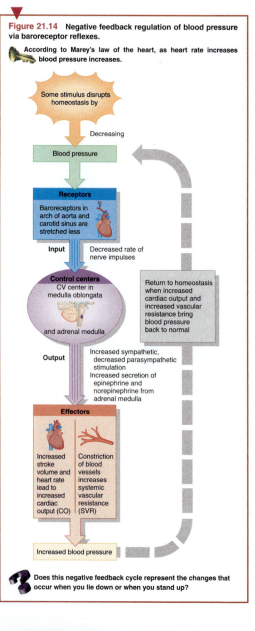

Figure 21.14 Negative feedback regulation of blood pressure via baroreceptor reflexes.

🎺 According to Marey's law of the heart, as heart rate increases blood pressure increases.

Does this negative feedback cycle represent the changes that occur when you lie down or when you stand up?

New to this edition are ten *Focus on Homeostasis* pages, one each for the integumentary, skeletal, muscular, nervous, endocrine, cardiovascular, lymphatic and immune, respiratory, digestive, and urinary systems. Incorporating both graphic and narrative elements, these pages explain, clearly and succinctly, how the system under consideration contributes to the homeostasis of each of the other body systems. Use of this feature will enhance student understanding of the links between body systems and how interactions among systems contribute to the homeostasis of the body as a whole

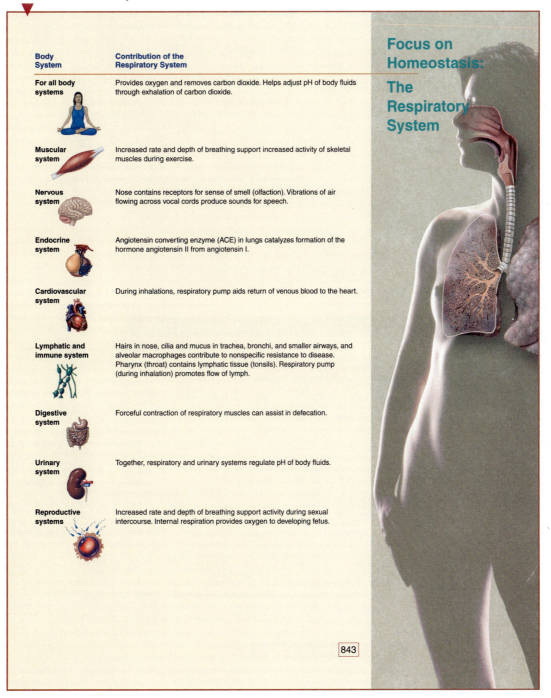

Body System	Contribution of the Respiratory System
For all body systems	Provides oxygen and removes carbon dioxide. Helps adjust pH of body fluids through exhalation of carbon dioxide.
Muscular system	Increased rate and depth of breathing support increased activity of skeletal muscles during exercise.
Nervous system	Nose contains receptors for sense of smell (olfaction). Vibrations of air flowing across vocal cords produce sounds for speech.
Endocrine system	Angiotensin converting enzyme (ACE) in lungs catalyzes formation of the hormone angiotensin II from angiotensin I.
Cardiovascular system	During inhalations, respiratory pump aids return of venous blood to the heart.
Lymphatic and immune system	Hairs in nose, cilia and mucus in trachea, bronchi, and smaller airways, and alveolar macrophages contribute to nonspecific resistance to disease. Pharynx (throat) contains lymphatic tissue (tonsils). Respiratory pump (during inhalation) promotes flow of lymph.
Digestive system	Forceful contraction of respiratory muscles can assist in defecation.
Urinary system	Together, respiratory and urinary systems regulate pH of body fluids.
Reproductive systems	Increased rate and depth of breathing support activity during sexual intercourse. Internal respiration provides oxygen to developing fetus.

Focus on Homeostasis:

The Respiratory System

843

In addition, we believe students can better understand normal physiological processes by examining situations in which diseases or disorders impair those processes. We offer three features that highlight these disruptions to homeostasis.

A perennial favorite among students, the intriguing **Clinical Applications** in every chapter explore the clinical, professional, or everyday relevance of a particular anatomical structure or its related function. Many are new to this edition, and all have been reviewed for accuracy and relevance. Each application directly follows the discussion to which it relates.

Aspirin and Thrombolytic Agents

In patients with heart and blood vessel disease, the events of hemostasis may occur even without external injury to a blood vessel. At low doses, **aspirin** inhibits vasoconstriction and platelet aggregation by blocking synthesis of thromboxane A2. It also reduces the chance of thrombus formation. Due to these effects, aspirin reduces the risk of transient ischemic attacks (TIA), strokes, myocardial infarction, and blockage of peripheral arteries.

Thrombolytic agents are chemical substances that are injected into the body to dissolve blood clots that have already formed to restore circulation. They either directly or indirectly activate plasminogen. The first thrombolytic agent, approved in 1982 for dissolving clots in the coronary arteries of the heart, was **streptokinase,** which is produced by streptococcal bacteria. A genetically engineered version of human **tissue plasminogen activator (t-PA)** is now used to treat both heart attacks and brain attacks (strokes) that are caused by blood clots. ■

The **Disorders: Homeostatic Imbalances** sections at the end of most chapters include concise discussions of major diseases and disorders that illustrate departures from normal homeostasis. They provide answers to many questions that students ask about medical problems.

DISORDERS: HOMEOSTATIC IMBALANCES

Disorders of epithelial tissues are mainly specific to individual organs, for example, peptic ulcer disease (PUD), which erodes the epithelial lining of the stomach or small intestine. For this reason, epithelial disorders are described together with the relevant body system throughout the text. The most prevalent disorders of connective tissues are **autoimmune diseases**—diseases in which antibodies produced by the immune system fail to distinguish what is foreign from what is self and attack the body's own tissues. The most common autoimmune disorder is rheumatoid arthritis, which attacks the synovial membranes of joints. Because connective tissue is one of the most abundant and widely distributed of the four main types of tissues, its disorders often affect multiple body systems. Common disorders of muscle tissue and nervous tissue are described at the ends of Chapters 10 and 12, respectively.

Sjögren's Syndrome

Sjögren's syndrome (SHŌ-grenz) is a common autoimmune disorder that causes inflammation and destruction of exocrine glands, especially the lacrimal (tear) glands and salivary glands. Signs include dryness of the mucous membranes in the eyes and mouth and salivary gland enlargement. Systemic effects include arthritis, difficulty swallowing, pancreatitis (inflammation of the pancreas), pleuritis (inflammation of the pleurae of the lungs), and migraine headaches. The disorder affects more females than males by a ratio of 9 to 1. About 20% of older adults experience some signs of Sjögren's. Treatment is supportive, including using artificial tears to moisten the eyes, sipping fluids, chewing sugarless gum, and using a saliva substitute to moisten the mouth.

Systemic Lupus Erythematosus

Systemic lupus erythematosus (er-i-thē-ma-TŌ-sus), **SLE,** or simply lupus, is a chronic inflammatory disease of connective tissue occurring mostly in nonwhite women during their childbearing years. It is an autoimmune disease that can cause tissue damage in every body system. The disease, which can range from a mild condition in most patients to a rapidly fatal disease, is marked by periods of exacerbation and remission. The prevalence of SLE is about 1 in 2000 persons, with females more likely to be afflicted than males by a ratio of 8 or 9 to 1.

Although the cause of SLE is not known, genetic, environmental, and hormonal factors are implicated. The genetic component is suggested by studies of twins and family history. Environmental factors include viruses, bacteria, chemicals, drugs, exposure to excessive sunlight, and emotional stress. With regard to hormones, sex hormones, such as estrogens, may trigger SLE.

Signs and symptoms of SLE include painful joints, low-grade fever, fatigue, mouth ulcers, weight loss, enlarged lymph nodes and spleen, sensitivity to sunlight, rapid loss of large amounts of scalp hair, and anorexia. A distinguishing feature of lupus is an eruption across the bridge of the nose and cheeks called a "butterfly rash." Other skin lesions may occur, including blistering and ulceration. The erosive nature of some SLE skin lesions was thought to resemble the damage inflicted by the bite of a wolf—thus, the term *lupus* (= wolf). The most serious complications of the disease involve inflammation of the kidneys, liver, spleen, lungs, heart, brain, and gastrointestinal tract. Because there is no cure for SLE, treatment is supportive, including anti-inflammatory drugs, such as aspirin, and immunosuppressive drugs.

MEDICAL TERMINOLOGY

Vocabulary-building glossaries of selected **Medical Terminology** and conditions also appear at the end of appropriate chapters. This feature has been expanded and updated for this edition.

Atrophy (AT-rō-fē; *a-* = without; *-trophy* = nourishment) A decrease in the size of cells, with a subsequent decrease in the size of the affected tissue or organ.

Hypertrophy (hī-PER-trō-fē; *hyper-* = above or excessive) Increase in the size of a tissue because its cells enlarge without undergoing cell division.

Tissue rejection An immune response of the body directed at foreign proteins in a transplanted tissue or organ; immunosuppressive drugs, such as cyclosporine, have largely overcome tissue rejection in heart-, kidney-, and liver-transplant patients.

Tissue transplantation The replacement of a diseased or injured tissue or organ. The most successful transplants involve use of a person's own tissues or those from an identical twin.

Xenotransplantation (zen'-ō-trans-plan-TĀ-shun; *xeno-* = strange, foreign) The replacement of a diseased or injured tissue or organ with cells or tissues from an animal. Porcine (from pigs) and bovine (from cows) heart valves are used for some heart-valve replacement surgeries.

Organization, Special Topics, and Content Improvements

The book follows the same unit and topic sequence as its nine earlier editions. It is divided into five principal sections: Unit 1, "Organization of the Human Body," provides an understanding of the structural and functional levels of the body, from molecules to organ systems. Unit 2, "Principles of Support and Movement," analyzes the anatomy and physiology of bones, joints, and muscles. Unit 3, "Control Systems of the Human Body," emphasizes the importance of neural communication in the immediate maintenance of homeostasis, the role of sensory receptors in providing information about the internal and external environment, and the significance of hormones in maintaining long-term homeostasis. Unit 4, "Maintenance of the Human Body," explains how body systems function to maintain homeostasis on a moment-to-moment basis through the processes of circulation, respiration, digestion, cellular metabolism, urinary functions, and buffer systems. Unit 5, "Continuity," covers the anatomy and physiology of the reproductive systems, development, and the basic concepts of genetics and inheritance.

DEVELOPMENT OF THE HEART

▶ **OBJECTIVE**

• **Describe the development of the heart.**

The *heart*, a derivative of **mesoderm**, begins to develop early in the third week after fertilization. In the ventral region of the embryo inferior to the foregut, the heart develops from a group of mesodermal cells called the **cardiogenic area.** The next step in its development is the formation of a pair of tubes, the **endocardial tubes,** which develop from the cardiogenic area (Figure 20.18). These tubes then unite to form a common tube, referred to as the **primitive heart tube.** Next, the primitive heart tube develops into five distinct regions: (1) **truncus arteriosus,** (2) **bulbus cordis,** (3) **ventricle,** (4) **atrium,** and (5) **sinus venosus.** Because the bulbus cordis and ventricle grow most rapidly, and because the heart enlarges more rapidly than its superior and inferior attachments, the heart first assumes a U-shape and then an S-shape. The flexures of the heart reorient the regions so that the atrium and sinus venosus eventually come to lie superior to the bulbus cordis, ventricle, and truncus arteriosus. Contractions

Development We often tell our students that they can better appreciate the "logic" of human anatomy by becoming aware of how various structures developed in the first place. As in previous editions, illustrated discussions of development are found near the conclusion of most body system chapters. Placing this coverage at the end of chapters enables students to master the anatomical terminology they need before attempting to learn about embryonic and fetal structures. The fetus icon designates the start of each development discussion.

AGING AND THE IMMUNE SYSTEM

▶ **OBJECTIVE**

• **Describe the effects of aging on the immune system.**

With advancing age, most people become more susceptible to all types of infections and malignancies. Their response to vaccines is decreased, and they tend to produce more autoantibodies (antibodies against their body's own molecules). In addition, the immune system exhibits lowered levels of function. For example, T cells become less responsive to antigens, and fewer T cells respond to infections. This may result from age-related atrophy of the thymus or decreased production of thymic hormones. Because the T cell population decreases with age, B cells are also less responsive. Consequently, antibody levels do not increase as rapidly in response to a challenge by an antigen, resulting in increased susceptibility to various infections. It is for this key reason that elderly individuals are encouraged to get influenza (flu) vaccinations each year.

Aging Students need to be reminded from time to time that anatomy and physiology is not static. As the body ages, its structures and related functions subtly change. Moreover, aging is a professionally relevant topic for the majority of this book's readers, who will go on to careers in *health-related* fields in which the average age of the client population is steadily advancing. For these reasons, age-related changes in anatomy and physiology are discussed at the end of sixteen chapters.

EXERCISE AND BONE TISSUE

▶ **OBJECTIVE**

• **Describe how exercise and mechanical stress affect bone tissue.**

Within limits, bone has the ability to alter its strength in response to changes in mechanical stress. When placed under stress, bone tissue adapts by becoming stronger through increased deposition of mineral salts and production of collagen fibers. Another effect of stress is to increase the production of calcitonin, a hormone that inhibits bone resorption. Without mechanical stress, bone does not remodel normally because bone resorption outstrips bone formation. Removal of mechanical stress weakens bone through demineralization (loss of bone minerals) and decreased numbers of collagen fibers.

The main mechanical stresses on bone are those that result from the pull of skeletal muscles and the pull of gravity. If a person is bedridden or has a fractured bone in a cast, the strength of the unstressed bones diminishes. Astronauts subjected to the microgravity of space also lose bone mass. In both cases, bone loss can be dramatic—as much as 1% per week. Bones of athletes, which are repetitively and highly stressed, become notably thicker and stronger than those of nonathletes. Weight-bearing activities, such as walking or moderate weight lifting, help build and retain bone mass. Adolescents and young adults should engage in regular weight-bearing exercise prior to the closure of the epiphyseal plates to help build total mass prior to its inevitable reduction with aging. Even elderly people can strengthen their bones by engaging in weight-bearing exercise.

Exercise Physical exercise can produce favorable changes in some anatomical structures and enhance many physiological functions, most notably those associated with the muscular, skeletal, and cardiovascular systems. This information is especially relevant to readers embarking on careers in physical education, sports training, and dance. Hence, key chapters include brief discussions of exercise-related considerations, which are signaled by a distinctive running shoe icon.

Every chapter in the tenth edition of *Principles of Anatomy and Physiology* incorporates a host of improvements to both the text and the art, many suggested by reviewers, educators, and students. In addition, most chapters offer new Clinical Applications. Here are some of the more noteworthy changes.

Chapter 2 / The Chemical Level of Organization
A new Clinical Application on "Saturated Fats, Cholesterol, and Atherosclerosis." Some sections are more concise, and the complete discussion of oxidation–reduction reactions is now in chapter 25.

Chapter 3 / The Cellular Level of Organization
Description of a newly discovered cellular organelle, the proteasome. New Clinical Applications on Medical Uses of Isotonic, Hypertonic, and Hypotonic Solutions; Proteasomes and Disease; Genomics; and Antisense Therapy. Transport through the Golgi complex, tumor-suppressor genes, and causes of cancer updated and revised.

Chapter 4 / The Tissue Level of Organization
New illustrations of cell junctions and types of multicellular exocrine glands; several new photomicrographs. Revisions in sections on cell junctions, the basement membrane, and ground substance of connective tissue.

Chapter 5 / The Integumentary System
New section on types of hairs. New Clinical Applications on Photosensitivity, Hair Removal, and Hair and Hormones; New illustrations of the rule of nines, types of epidermal cells, epidermal wound healing, and deep wound healing. Structure of epidermis and development of the integumentary system revised.

Chapter 6 / The Skeletal System: Bone Tissue
New illustrations of the histology of bone, endochondral ossification, and fracture repair.

Chapter 8 / The Skeletal System: The Appendicular Skeleton
New illustration of the hip bone. Addition of Hip Fracture to the disorders section at the end of the chapter.

Chapter 9 / Joints
New Clinical Applications on Rotator Cuff Injuries, Separated Shoulder, Tennis Elbow, Little League Elbow, Dislocated Radius, Swollen Knee, Dislocated Knee, and Lyme Disease.

Chapter 10 / Muscle Tissue
New coverage of exercise and skeletal muscles, fibromyalgia, and benefits of stretching. New illustrations on microscopic organization of skeletal muscle tissue, the neuromuscular junction, and development of the muscular system. Proteins in skeletal muscle and uses of Botox revised.

Chapter 11 / The Muscular System
New illustration of a newly discovered skeletal muscle, the sphenomandibularis muscle (see page 325).

Chapter 12 / Nervous Tissue
New Medical Terminology list. New illustrations of the structure of a neuron, myelinated and unmyelinated axons, and signal transmission at synapses. Updated definitions of sensory neuron, interneuron, and motor neuron.

Chapter 13 / The Spinal Cord and Spinal Nerves
New Medical Terminology list. New illustration of the connective tissue coverings of a spinal nerve.

Chapter 14 / The Brain and Cranial Nerves
New illustrations of the parts of the brain, protective coverings of the brain, pathways of cerebrospinal fluid flow, thalamus, development of the brain, seven figures on cranial nerves. Revised discussions of the thalamus and cranial nerves.

Chapter 15 / Sensory, Motor and Integrative Systems
New illustrations of muscle spindles, tendon organs, somatic sensory pathways, and motor pathways. New Clinical Applications on Phantom Limb Sensation, Amyotrophic Lateral Sclerosis, Damage to Basal Ganglia, and Damage to the Cerebellum. Section on Muscle Spindles rewritten to simplify terminology but more clearly explain function; sections on somatic motor pathways, memory, and sleep updated.

Chapter 16 / The Special Senses
New illustrations of taste buds, effects of tastants on taste neurons, and development of the eyes and ears. New Clinical Applications on Hyposmia, Macular Degeneration, Taste Aversion, Presbyopia, and Color and Night Blindness. New section on Development of the Eyes and Ears. Expanded Medical Terminology list.

Chapter 17 / The Autonomic Nervous System
New Table comparing sympathetic and parasympathetic divisions of the autonomic nervous system; new illustrations of sympathetic and parasympathetic divisions.

Chapter 18 / The Endocrine System
New illustrations of the location of endocrine glands, thyroid gland, parathyroid glands, adrenal glands, pancreas, and development of the endocrine system. New Clinical Applications on Administering Hormones, Oxytocin and Childbirth, Congenital Adrenal Hyperplasia, and Posttraumatic Stress Disorder. Photographs of people with endocrine disorders, including an interesting case of a twin with giantism (see page 625). New Medical Terminology list.

Chapter 19 / The Blood
New description of red bone marrow. New Clinical Applications on Iron Overload and Tissue Damage, Induced Polycythemia in Athletes, and Aspirin and Thrombolytic Agents.

Chapter 20 / The Heart
New section on ATP production in cardiac muscle fibers. New Clinical Applications on Heart Valve Disorders, Reperfusion Damage, Ectopic Pacemaker, Regeneration of Cardiac Muscle Tissue, Artificial Hearts. New illustrations of the serous pericardium, structure of the heart, coronary circulation, pacemaker potentials, route of action potential depolarization and repolarization, and development of the heart. Arrhythmias added to disorders section. Expanded Medical Terminology list.

Chapter 21 / Blood Vessels and Hemodynamics
New illustrations of the structure of blood vessels, types of capillaries, action of skeletal muscles, ANS innervation of the heart,

circulatory routes, principal branches of the aorta, ascending aorta, branches of the brachiocephalic trunk in the neck, cerebral arterial circle, branches of the abdominal aorta, arteries of the pelvis and lower limbs, pulmonary circulation, fetal circulation, and development of blood vessels and blood cells.

Chapter 22 / The Lymphatic and Immune System and Resistance to Disease

New section on stress and immunity. New illustrations of lymphatic tissue, routes for lymph drainage, lymph node structure, development of the lymphatic system, phagocytosis, and stages of inflammation. New Clinical Applications on Ruptured Spleen, Edema and Lymph Flow, and Microbial Evasion of Phagocytosis. Autoimmune diseases, severe combined immunodeficiency disease, SLE, and lymphomas added to the disorders section. Descriptions of thymus and lymph nodes revised.

Chapter 23 / The Respiratory System

New illustrations of the structures of the respiratory system, location of peripheral chemoreceptors that help regulate respiration, and development of the bronchial tubes and lungs. New Clinical Applications on Laryngitis, Cancer of the Larynx, and Nebulization. New discussion and explanation of ventilation-perfusion coupling, nitrogen narcosis, and decompression sickness. Addition of ARDS to disorders section. Updated percentages of carbon dioxide dissolved in blood plasma and carried by bicarbonate ions and hemoglobin.

Chapter 24 / The Digestive System

New illustrations of organs of the gastrointestinal tract, peritoneal folds, salivary glands, histology of the stomach, secretion of HCl by stomach cells, histology of the small intestine, anatomy of the large intestine, and histology of the large intestine. New Clinical Applications on Root Canal Therapy, Jaundice, Occult Blood, and Absorption of Alcohol.

Chapter 25 / Metabolism

New section on energy homeostasis. New Clinical Application on Emotional Eating and new Medical Terminology list. Updated sections on regulation of food intake and obesity.

Chapter 26 / The Urinary System

New illustrations of a sagittal section through the kidney, blood supply of the kidneys, anatomy of the ureter, urinary bladder, and urethra, and development of the urinary system. More concise sections on reabsorption and secretion. New Clinical Applications on Nephroptosis and Loss of Plasma Proteins. Addition of Urinary Bladder Cancer to the disorders section.

Chapter 27 / Fluid, Electrolyte, and Acid–Base Homeostasis

New Summary table of factors that maintain body water balance. Discussion and related figure describing how kidneys contribute to acid–base balance by secreting hydrogen ions moved from Chapter 26 to this chapter.

Chapter 28 / The Reproductive Systems

New illustrations to compare mitosis and meiosis and to show development of internal reproductive organs. New table reviewing oogenesis and development of ovarian follicles; new explanation of mittelschmerz and the role of leptin in puberty. New Clinical Applications on Uterine Prolapse, Episiotomy, and Female Athlete Triad. Addition of genital warts and premenstrual dysphoric disorder to the disorders section.

Chapter 29 / Development and Inheritance

New illustrations of fertilization, cleavage, formation of primary germ layers, gastrulation, development of the notochordal process, neurulation, development of chorionic villi, embryo folding, and development of pharyngeal arches. Beautiful new photographs of embryonic and fetal development. New Clinical Applications on Anencephaly and Stem Cell Research. Addition of Trinucleotide Repeat Diseases to the disorders section. Updated and revised sections on embryonic and fetal development, noninvasive prenatal tests, and inheritance.

Enhancement to the Illustration Program

New Design A textbook with beautiful illustrations or photographs on most pages requires a carefully crafted and functional design. The new design for the tenth edition has been transformed to assist students in making the most of the text's many features and outstanding art. A larger trim size provides even more space for the already large and highly praised illustrations and accommodates the larger font size for easier reading. Each page is carefully laid out to place related text, figures, and tables near one another, minimizing the need for page turning while reading a topic. New to this edition is the red print used to indicate the first mention of a figure or table. Not only is the reader alerted to refer to the figure or table, but the color print also serves as a place locator for easy return to the narrative.

Distinctive icons incorporated throughout the chapters signal special features and make them easy to find during review. These include the **key** with Key Concepts Statements; the **ques-**tion mark with the applicable questions that enhance every figure; the **stethoscope** indicating a clinical application within the chapter narrative; the **fetus icon** announcing the developmental anatomy section; the **running shoe** highlighting content relevant to exercise and the icons that indicate the study outline and distinctive types of **chapter-ending questions.**

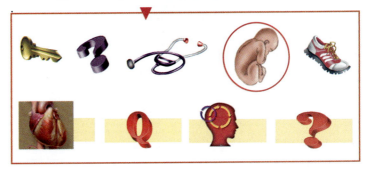

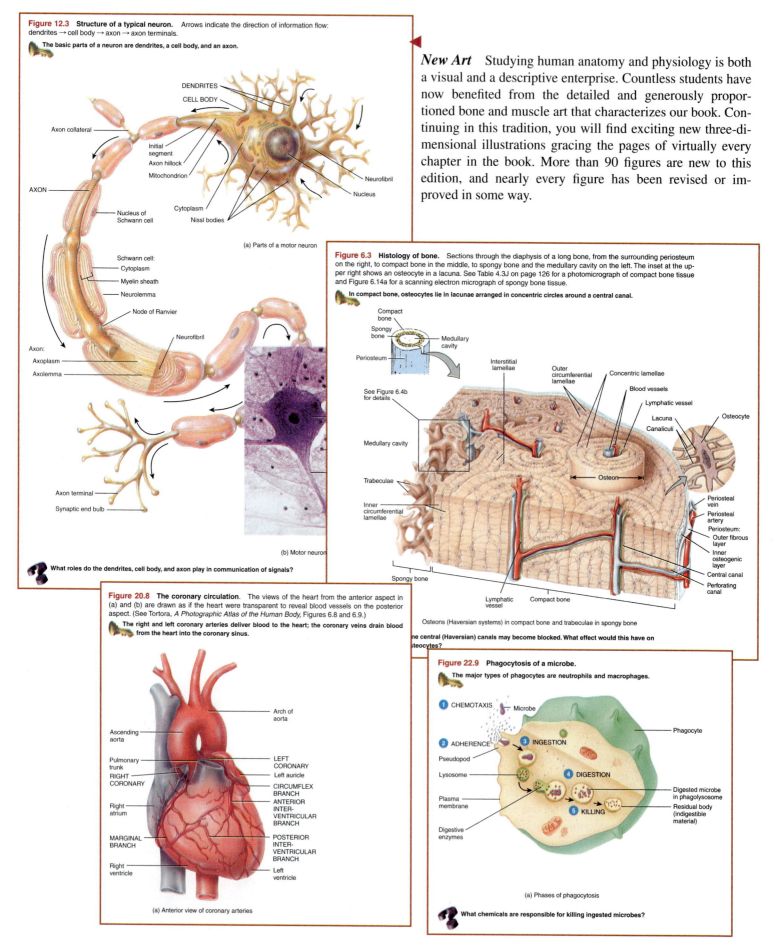

Figure 12.3 Structure of a typical neuron. Arrows indicate the direction of information flow: dendrites → cell body → axon → axon terminals.

The basic parts of a neuron are dendrites, a cell body, and an axon.

DENDRITES
CELL BODY
Axon collateral
Initial segment
Axon hillock
Mitochondrion
AXON
Neurofibril
Nucleus
Cytoplasm
Nissl bodies
Nucleus of Schwann cell
Schwann cell:
Cytoplasm
Myelin sheath
Neurolemma
Node of Ranvier
Neurofibril
Axon:
Axoplasm
Axolemma

(a) Parts of a motor neuron

Axon terminal
Synaptic end bulb

(b) Motor neuron

What roles do the dendrites, cell body, and axon play in communication of signals?

New Art Studying human anatomy and physiology is both a visual and a descriptive enterprise. Countless students have now benefited from the detailed and generously proportioned bone and muscle art that characterizes our book. Continuing in this tradition, you will find exciting new three-dimensional illustrations gracing the pages of virtually every chapter in the book. More than 90 figures are new to this edition, and nearly every figure has been revised or improved in some way.

Figure 6.3 Histology of bone. Sections through the diaphysis of a long bone, from the surrounding periosteum on the right, to compact bone in the middle, to spongy bone and the medullary cavity on the left. The inset at the upper right shows an osteocyte in a lacuna. See Table 4.3J on page 126 for a photomicrograph of compact bone tissue and Figure 6.14a for a scanning electron micrograph of spongy bone tissue.

In compact bone, osteocytes lie in lacunae arranged in concentric circles around a central canal.

Compact bone
Spongy bone
Medullary cavity
Periosteum

Interstitial lamellae
Outer circumferential lamellae
Concentric lamellae
Blood vessels
Lymphatic vessel
Lacuna
Osteocyte
Canaliculi
See Figure 6.4b for details
Medullary cavity
Osteon
Periosteal vein
Periosteal artery
Periosteum:
Outer fibrous layer
Inner osteogenic layer
Central canal
Perforating canal
Trabeculae
Inner circumferential lamellae
Spongy bone
Lymphatic vessel
Compact bone

Osteons (Haversian systems) in compact bone and trabeculae in spongy bone

ne central (Haversian) canals may become blocked. What effect would this have on steocytes?

Figure 20.8 The coronary circulation. The views of the heart from the anterior aspect in (a) and (b) are drawn as if the heart were transparent to reveal blood vessels on the posterior aspect. (See Tortora, *A Photographic Atlas of the Human Body,* Figures 6.8 and 6.9.)

The right and left coronary arteries deliver blood to the heart; the coronary veins drain blood from the heart into the coronary sinus.

Arch of aorta
Ascending aorta
Pulmonary trunk
RIGHT CORONARY
Right atrium
MARGINAL BRANCH
Right ventricle
LEFT CORONARY
Left auricle
CIRCUMFLEX BRANCH
ANTERIOR INTER-VENTRICULAR BRANCH
POSTERIOR INTER-VENTRICULAR BRANCH
Left ventricle

(a) Anterior view of coronary arteries

Figure 22.9 Phagocytosis of a microbe.

The major types of phagocytes are neutrophils and macrophages.

1 CHEMOTAXIS
Microbe
2 ADHERENCE
3 INGESTION
Pseudopod
Lysosome
4 DIGESTION
Plasma membrane
5 KILLING
Digestive enzymes
Phagocyte
Digested microbe in phagolysosome
Residual body (indigestible material)

(a) Phases of phagocytosis

What chemicals are responsible for killing ingested microbes?

New Histology-Based Art As part of our goal for continuous improvement, many of the anatomical illustrations based on histological preparations have been revised and redrawn for this edition.

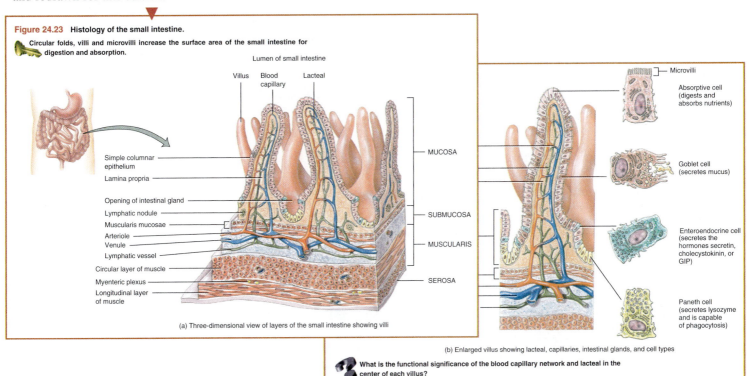

Figure 24.23 Histology of the small intestine.

Circular folds, villi and microvilli increase the surface area of the small intestine for digestion and absorption.

(a) Three-dimensional view of layers of the small intestine showing villi

(b) Enlarged villus showing lacteal, capillaries, intestinal glands, and cell types

What is the functional significance of the blood capillary network and lacteal in the center of each villus?

New Photomicrographs Dr. Michael Ross of the University of Florida has again provided us with more beautiful, customized photomicrographs of various tissues of the body. We have always considered Dr. Ross' photos among the best available and their inclusion in this edition greatly enhances the illustration program.

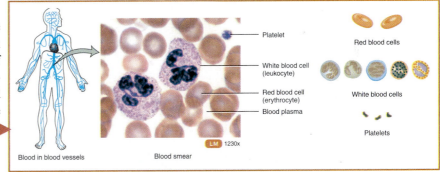

Cadaver Photos As before, we provide an assortment of large, clear cadaver photos at strategic points in many chapters. What is more, many anatomy illustrations are keyed to the large cadaver photos available in a companion text, *A Photographic Atlas of the Human Body*, by Gerard J. Tortora.

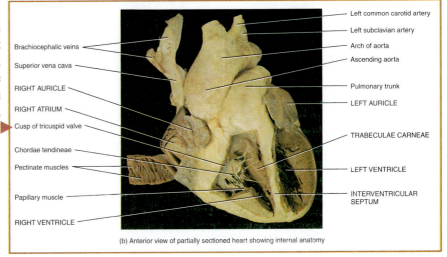

(b) Anterior view of partially sectioned heart showing internal anatomy

Helpful Orientation Diagrams Students sometimes need help figuring out the plane of view of anatomy illustrations — descriptions alone do not always suffice. An orientation diagram that depicts and explains the perspective of the view represented in the figure accompanies every major anatomy illustration. There are three types of diagrams: (1) planes used to indicate where certain sections are made when a part of the body is cut; (2) diagrams containing a directional arrow and the word "View" to indicate the direction from which the body part is viewed, and (3) diagrams with arrows leading from or to them that direct attention to enlarged and detailed parts of illustrations.

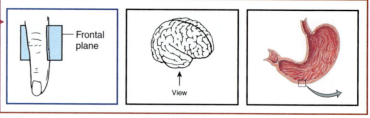

Key Concept Statements This art-related feature summarizes an idea that is discussed in the text and demonstrated in a figure. Each Key Concept Statement is positioned adjacent to its figure and is denoted by a distinctive key icon.

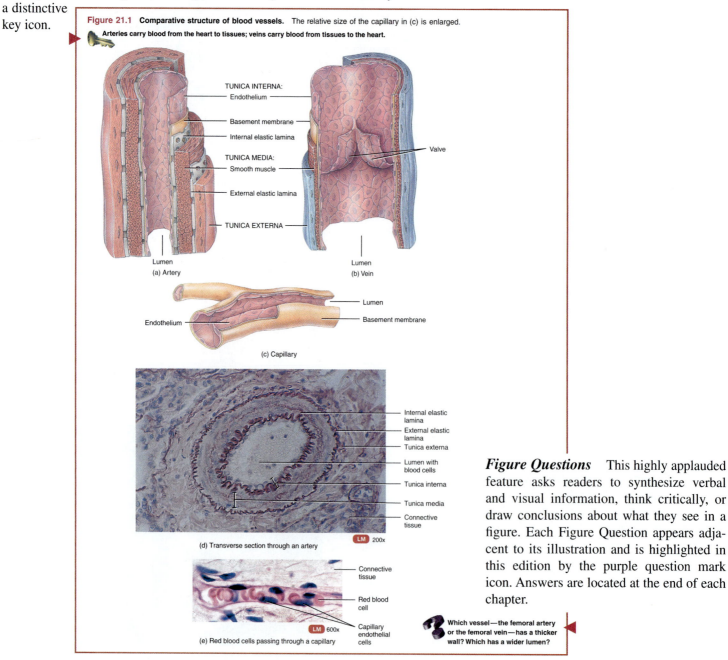

Figure 21.1 **Comparative structure of blood vessels.** The relative size of the capillary in (c) is enlarged.

Arteries carry blood from the heart to tissues; veins carry blood from tissues to the heart.

TUNICA INTERNA:
- Endothelium
- Basement membrane
- Internal elastic lamina

TUNICA MEDIA:
- Smooth muscle
- External elastic lamina

TUNICA EXTERNA

Valve

Lumen
(a) Artery

Lumen
(b) Vein

Lumen

Basement membrane

Endothelium

(c) Capillary

Internal elastic lamina
External elastic lamina
Tunica externa
Lumen with blood cells
Tunica interna
Tunica media
Connective tissue

LM 200x

(d) Transverse section through an artery

Connective tissue
Red blood cell
Capillary endothelial cells

LM 600x

(e) Red blood cells passing through a capillary

Figure Questions This highly applauded feature asks readers to synthesize verbal and visual information, think critically, or draw conclusions about what they see in a figure. Each Figure Question appears adjacent to its illustration and is highlighted in this edition by the purple question mark icon. Answers are located at the end of each chapter.

Which vessel—the femoral artery or the femoral vein—has a thicker wall? Which has a wider lumen?

Correlation of Sequential Processes Correlation of sequential processes in text and art is achieved through the use of numbered lists in the narrative that correspond to numbered segments in the accompanying figure. This approach is used extensively throughout the book to lend clarity to the flow of complex processes.

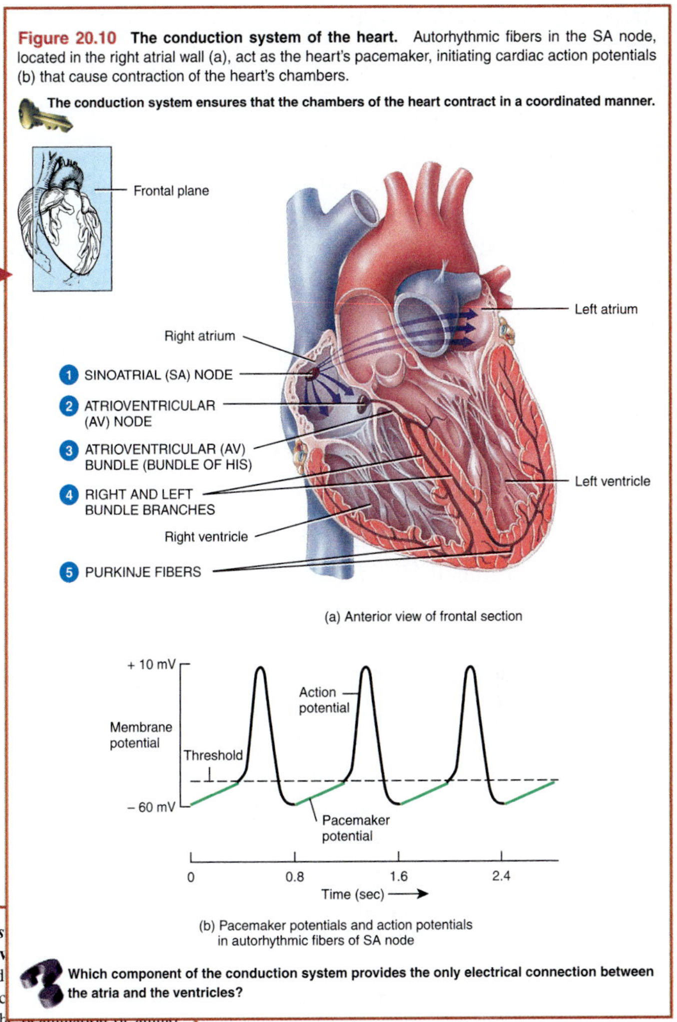

Functions Overview This feature succinctly lists the functions of an anatomical structure or body system depicted within a figure. The juxtaposition of text and art further reinforces the connection between structure and function.

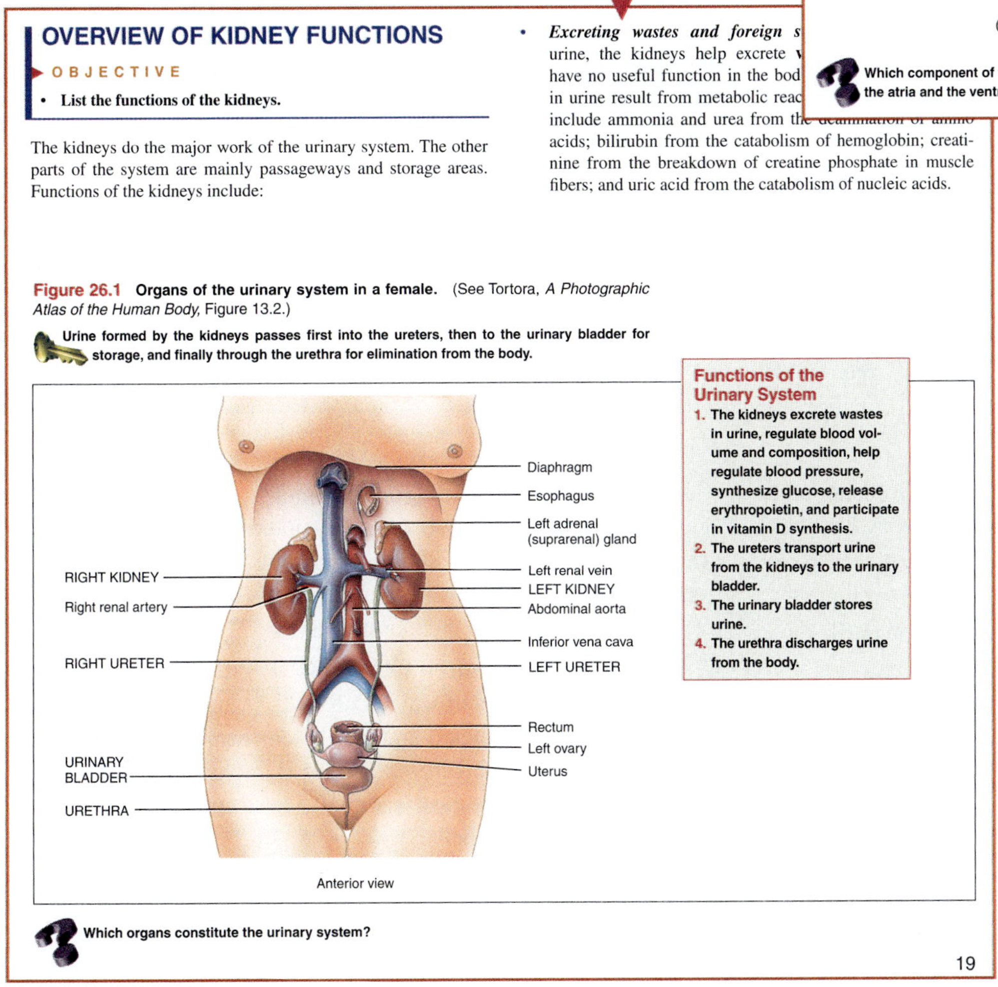

Figure 20.10 The conduction system of the heart. Autorhythmic fibers in the SA node, located in the right atrial wall (a), act as the heart's pacemaker, initiating cardiac action potentials (b) that cause contraction of the heart's chambers.

The conduction system ensures that the chambers of the heart contract in a coordinated manner.

Frontal plane

Right atrium
Left atrium

1 SINOATRIAL (SA) NODE
2 ATRIOVENTRICULAR (AV) NODE
3 ATRIOVENTRICULAR (AV) BUNDLE (BUNDLE OF HIS)
4 RIGHT AND LEFT BUNDLE BRANCHES

Left ventricle
Right ventricle

5 PURKINJE FIBERS

(a) Anterior view of frontal section

+ 10 mV

Action potential

Membrane potential
Threshold

− 60 mV

Pacemaker potential

0   0.8   1.6   2.4
Time (sec) ⟶

(b) Pacemaker potentials and action potentials in autorhythmic fibers of SA node

Which component of the conduction system provides the only electrical connection between the atria and the ventricles?

OVERVIEW OF KIDNEY FUNCTIONS

▶ **OBJECTIVE**

• **List the functions of the kidneys.**

The kidneys do the major work of the urinary system. The other parts of the system are mainly passageways and storage areas. Functions of the kidneys include:

• *Excreting wastes and foreign s* urine, the kidneys help excrete v have no useful function in the bod in urine result from metabolic reac include ammonia and urea from the deamination of amino acids; bilirubin from the catabolism of hemoglobin; creatinine from the breakdown of creatine phosphate in muscle fibers; and uric acid from the catabolism of nucleic acids.

Figure 26.1 Organs of the urinary system in a female. (See Tortora, *A Photographic Atlas of the Human Body,* Figure 13.2.)

Urine formed by the kidneys passes first into the ureters, then to the urinary bladder for storage, and finally through the urethra for elimination from the body.

Diaphragm
Esophagus
Left adrenal (suprarenal) gland
RIGHT KIDNEY
Left renal vein
LEFT KIDNEY
Right renal artery
Abdominal aorta
Inferior vena cava
RIGHT URETER
LEFT URETER
Rectum
Left ovary
URINARY BLADDER
Uterus
URETHRA

Anterior view

Functions of the Urinary System
1. The kidneys excrete wastes in urine, regulate blood volume and composition, help regulate blood pressure, synthesize glucose, release erythropoietin, and participate in vitamin D synthesis.
2. The ureters transport urine from the kidneys to the urinary bladder.
3. The urinary bladder stores urine.
4. The urethra discharges urine from the body.

Which organs constitute the urinary system?

19

Hallmark Features

The tenth edition of *Principles of Anatomy and Physiology* builds on the legacy of thoughtfully designed and class-tested pedagogical features that provide a complete learning system for students as they navigate their way through the text and course. All have been revised to reflect the enhancements to the text.

Chapter-opening Pages   Each chapter now begins with a chapter-opening page that includes a beautiful new piece of art related to the body system under consideration. A perspective on the chapter's content from either a recent student of the course or a practitioner in a health related field introduces current users to the relevance of the material or particularly interesting portions of the chapter. Suggestions for energizing their study with dynamic multimedia activities on the Foundations CD included with the text and specially developed internet activities on the companion website are highlighted.

Helpful Exhibits   Students of anatomy and physiology need extra help learning the many structures that constitute certain body systems—most notably skeletal muscles, articulations, blood vessels, and nerves. As in previous editions, the chapters that present these topics are organized around **Exhibits,** each of which consists of an overview, a tabular summary of the relevant anatomy, and an associated suite of illustrations or photographs. Each Exhibit is prefaced by an Objective and closes with a Checkpoint activity. We trust you will agree that our spaciously designed Exhibits are ideal study vehicles for learning anatomically complex body systems.

Student Objectives and Checkpoints   **Objectives** are found at the beginning of major sections throughout each chapter. Complementing this format, **Checkpoint** questions appear at strategic intervals within chapters to give students the chance to validate their understanding as they read.

Tools for Mastering Vocabulary   Students—even the best ones—generally find it difficult at first to read and pronounce anatomical and physiological terms. Moreover, as teachers we are sympathetic to the needs of the growing ranks of college students who speak English as a second language. For these reasons, we have endeavored to ensure that this book has a strong and helpful vocabulary component. The key terms in every chapter are emphasized by use of **boldface type.** We include **pronunciation guides** when major, or especially hard-to-pronounce, structures and functions are introduced in the discussions, Tables, or Exhibits. **Word roots** citing the Greek or Latin derivations of anatomical terms are offered as an additional aid. As a further service to readers, we provide a list of **Medical Terminology** at the conclusion of most chapters, and a comprehensive **Glossary** at the back of the book. Also included at the end of the book is a list of the basic building blocks of medical terminology—**Combining Forms, Word Roots, Prefixes, and Suffixes.**

Study Outline   As always, readers will benefit from the popular end-of-chapter **Study Outline** that is page referenced to the chapter discussions.

End-of-chapter Questions   The **Self-Quiz Questions** are written in a variety of styles that are calculated to appeal to readers' different testing preferences. **Critical Thinking Questions** challenge readers to apply concepts to real-life situations. The style of these questions ought to make students smile on occasion as well as think! Answers to the Self-Quiz and Critical Thinking Questions are located in Appendix E.

Complete Teaching And Learning Package

Continuing the tradition of providing a complete teaching and learning package, the tenth edition of *Principles of Anatomy and Physiology* is available with a host of carefully planned supplementary materials that will help you and your students attain the maximal benefit from our textbook. Please contact your Wiley sales representative for additional information about any of these resources.

Media

Dynamic CD-ROMs and an enhanced Companion Website have been developed to complement the tenth edition of *Principles of Anatomy and Physiology.* These are fully described on the inside front cover and include:

• **Student Companion CD-ROM,** provided free with the purchase of a new textbook

• **Dedicated Book Companion Website** free to students who purchase a new textbook

• **Interactions: Exploring the Functions of the Human Body.** The first CD in this dramatic new series—Foundations— is provided free to adopters of the text and to students purchasing a new text.

For Instructors

• **Full-color Overhead Transparencies** This set of full-color acetates includes over 650 figures, including histology micrographs, from the text — complete with figure numbers and captions. Transparencies have enlarged labels for effective use as overhead projections in the classroom. (0-471- 25425-8)

• **Instructor's Resource CD-ROM** For lecture presentation purposes, this cross-platform CD-ROM includes all of the line art from the text in labeled and unlabeled formats. In addition, a pre-designed set of Powerpoint Slides is included with text and images. The slides are easily edited to customize your presentations to your specific needs. (0-471- 25350-2)

- **Professor's Resource Manual** by Lee Famiano of Cuyahoga CC. This on-line manual provides instructors with tools to enhance their lectures. Key features include: Lecture Outlines, What's New and Different in this Chapter, Critical Thinking Questions, Teaching Tips, and a guide to other educational resources. The entire manual is found on the text's dedicated companion website and is fully downloadable. The electronic format allows you to customize lecture outlines or activities for delivery either in print or electronically to your students. (0-471-25355-3)

- **Printed Test Bank** by P. James Nielsen of Western Illinois University. A testbank of nearly 3,000 questions, many new to the tenth edition, is available. A variety of formats—multiple choice, short answer, matching and essay—are provided to accommodate different testing preferences. (0-471-25421-5)

- **Computerized Test Bank** An electronic version of the printed test bank is available on a cross-platform CD-ROM. (0-471-25146-1)

- **Interactions: Exploring the Functions of the Human Body**

- **WebCT** or **Blackboard** Course Management Systems with content prepared by Sharon Simpson are available.

- **Faculty Resource Network** – New from Wiley is a support structure to help instructors implement the dynamic new media that supports this text into their classrooms, laboratories, or on-line courses. Consult with your Wiley representative for details about this program.

For Students

- **Student Companion CD-ROM**
- **Interactions: Exploring the Functions of the Human Body**
- **Book Companion Website**
- **Learning Guide (0-471-43447-7)** by Kathleen Schmidt Prezbindowski, College of Mount St. Joseph. Designed specifically to fit the needs of students with different learning styles, this well-received guide helps students to more closely examine important concepts through a variety of activities and exercises. The 29 chapters in the *Learning Guide* parallel those of the textbook and include many activities, quizzes and tests for review and study.

- **Illustrated Notebook (0-471-25150-X)** A true companion to the text, this unique notebook is a tool for organized note taking in class and for review during study. Following the sequence in the textbook, each left-handed page displays an unlabeled black and white copy of every text figure. Students can fill in the labels during lecture or lab at the instructor's directions and take additional notes on the lined right-handed pages.

- **Anatomy and Physiology: A Companion Coloring Book (0-471-39515-3)** This helpful study aid features over 500 striking, original illustrations that, by actively coloring them, give students a clear understanding of key anatomical structures and physiological processes.

For the Laboratory

- **A Brief Atlas of the Human Skeleton** This brief photographic review of the human skeleton is provided free with every new copy of the text.

- **Laboratory Manual for Anatomy and Physiology** by Connie Allen and Valerie Harper, Edison Community College **(0-471-39464-5)** This new laboratory manual presents material covered in the 2-semester undergraduate anatomy & physiology laboratory course in a clear and concise way, while maintaining a student-friendly tone. The manual is very interactive and contains activities and experiments that enhance students' ability to both visualize anatomical structures and understand physiological topics.

- **Cat Dissection Manual** by Connie Allen and Valerie Harper, Edison Community College **(0-471-26457-1)** This manual includes photographs and illustrations of the cat along with guidelines for dissection. It is available independently as well as bundled with the main manual depending upon your an adoption needs.

- **Fetal Pig Dissection Manual** by Connie Allen and Valerie Harper, Edison Community College **(0-471-26458-X)** This manual includes photographs and illustrations of the fetal pig along with guidelines for dissection. It is available independently as well as bundled with the main manual depending upon your an adoption needs.

- **Photographic Atlas of the Human Body with Selected Cat, Sheep, and Cow Dissections** by Gerard Tortora **(0-471-37487-3)** This four-colored atlas is designed to support both study and laboratory experiences. Organized by body systems, the clearly labeled photographs provide a stunning visual reference to gross anatomy. Histological micrographs are also included. Many of the illustrations within *Principles of Anatomy and Physiology, 10e* are cross-referenced to this atlas.

Like each of our students, this book has a life of its own. The structure, content, and production values of *Principles of Anatomy and Physiology* are shaped as much by its relationship with educators and readers as by the vision that gave birth to the book ten editions ago. Today you, our readers, are the "heart" of this book. We invite you to continue the tradition of sending in your suggestions to us so that we can include them in the eleventh edition.

Gerard J. Tortora
Department of Science and Health, S229
Bergen Community College
400 Paramus Road
Paramus, NJ 07652

Sandra Reynolds Grabowski
Department of Biological Sciences
1392 Lilly Hall of Life Sciences
Purdue University
West Lafayette, IN 47907-1392
email: Sgrabows@bilbo.bio.purdue.edu

Acknowledgements

For the tenth edition of *Principles of Anatomy and Physiology*, we have enjoyed the opportunity of collaborating with a group of dedicated and talented professionals. Accordingly, we would like to recognize and thank the members of our book team, who often worked evenings and weekends, as well as days, to bring this book to you. At John Wiley & Sons, Inc., our Editor Bonnie Roesch again illuminates the path toward ever better books with her creative ideas and dedication. Bonnie was our valued editor during the seventh and eighth editions—welcome back and thanks! Karin Kincheloe is the Wiley designer whose vision for the tenth edition is a larger, more colorful, more user-friendly new style. Moreover, Karin laid out each page of the book to achieve the best possible placement of text, figures, and other elements. Both instructors and students will appreciate and benefit from the pedagogically effective and visually pleasing design elements that augment the content changes made to this edition. Danke sehr! Claudia Durrell, our Art Coordinator, has collaborated on this text since its seventh edition. Her artistic ability, organizational skills, attention to detail, and understanding of our illustration preferences greatly enhance the visual appeal and style of the figures. She remains a cornerstone of our projects—thank you, Claudia, for all your contributions. Kelly Tavares, Senior Production Editor, demonstrated her untiring expertise during each step of the production process. She coordinated all aspects of actually making and manufacturing the book. Kelly also was "on press" as the book was being printed to ensure the highest possible quality. Muito obrigada, Kelly for all the extra hours you spent to implement book-improving changes! Ellen Ford, Developmental Editor, shepherded the manuscript and electronic files during the revision process. She also contacted many reviewers for their comments and sought out former students to obtain their perspectives on learning anatomy and physiology for the new chapter-opening pages. Hillary Newman, Photo Editor, provided us with all of the photos we requested and did it with efficiency, accuracy, and professionalism. Mary O'Sullivan, Assistant Editor, coordinated the development of many of the supplements that support this text. We are most appreciative! Wiley Editorial Assistants Justin Bow and Kelli Coaxum helped with various aspects of the project and took care of many details. Thanks to all of you!

Jerri K. Lindsey, Tarrant County Junior College, Northeast and Caryl Tickner, Stark State College wrote the end-of-chapter Self-Quiz Questions. Joan Barber of Delaware Technical and Community College contributed the Critical Thinking Questions. Thanks to all three for writing questions that students will appreciate. And a special thank you to Kathleen Prezbindowski, who has authored the *Learning Guide* for many editions. The high quality of her study activities ensures student success.

Outstanding illustrations and photographs have always been a signature feature of *Principles of Anatomy and Physiology*. Respected scientific and medical illustrators on our team of exceptional artists include Mollie Borman, Leonard Dank, Sharon Ellis, Wendy Hiller Gee, Jean Jackson, Keith Kasnot, Lauren Keswick, Steve Oh, Lynn O'Kelley, Hilda Muinos, Tomo Narashima, Nadine Sokol, and Kevin Somerville. Mark Nielsen of the University of Utah provided many of the cadaver photos that appear in this edition. Artists at Imagineering created the amazing computer graphics images and provided all labeling of figures.

Reviewers

We are extremely grateful to our colleagues who reviewed the manuscript and offered insightful suggestions for improvement. The contributions of all these people, who generously provided their time and expertise to help us maintain the book's accuracy and clarity, are acknowledged in the list that follows.

Patricia Ahanotu, Georgia Perimeter College
Cynthia L. Beck, George Mason University
Clinton L. Benjamin, Lower Columbia College
Anna Berkovitz, Purdue University
Charles J. Biggers, University of Memphis
Mark Bloom, Tyler Junior College
Michele Boiani, University of Pennsylvania
Bruce M. Carlson, University of Michigan
Barbara Janson Cohen, Delaware County Community College
Matthew Jarvis Cohen, Delaware County Community College
Marcia Carol Coss, George Mason University
Victor P. Eroschenko, University of Idaho
Lorraine Findlay, Nassau Community College
Candice Francis, Palomar College
Christina Gan, Rogue Community College
Gregory Garman, Centralia College

Alan Gillen, Pensacola Christian College

Chaya Gopalan, St. Louis Community College

Janet Haynes, Long Island University

Clare Hayes, Metropolitan State College of Denver

James Junker, Campbell University

Gerald Karp, University of Florida

William Kleinelp, Middlesex County College

John Langdon, University of Indianapolis

John Lepri, University of North Carolina at Greensboro

Jerri K. Lindsey, Tarrant County College

Mary Katherine K. Lockwood, University of
    New Hampshire

Jennifer Lundmark, California State University, Sacramento

Paul Malven, Purdue University

G. K. Maravelas, Bristol Community College

Jane Marks, Paradise Valley Community College

Lee Meserve, Bowling Green State University

Javanika Mody, Anne Arundel Community College

Robert L. Moskowitz, Community College of Philadelphia

Shigihiro Nakajima, University of Illinois at Chicago

Jerry D. Norton, Georgia State University

Justicia Opoku, University of Maryland

Weston Opitz, Kansas Wesleyan University

Joann Otto, Purdue University

David Parker, Northern Virginia Community College

Karla Pouillon, Everett Community College

Linda Powell, Community College of Philadelphia

C. Lee Rocket, Bowling Green State University

Esmond J. Sanders, University of Alberta

Louisa Schmid, Tyler Junior College

Hans Schöler, University of Pennsylvania

Charles Sinclair, University of Indianapolis

Dianne Snyder, Augusta State University

Dennis Strete, McLennan Community College

Eric Sun, Macon State College

Antonietto Tan, Worcester State College

Jim Van Brunt, Rogue Community College

Jyoti Wagle, Houston Community College

Curt Walker, Dixie State College

DeLoris M. Wenzel, The University of Georgia

David Westmoreland, US Air Force Academy

Frederick E. Williams, University of Toledo

In addition, every chapter was read and reviewed by either an Anatomy & Physiology student or a health care professional now working in his or her chosen career. Their comments on the relevance and interest of the chapter topics is presented on each chapter opening page. We appreciate their time and effort, and especially their enthusiasm for the subject matter. We think that enthusiasm is catching! Thanks go out to the following:

Tamatha Adkins, RN

Molly Causby, Georgia Perimeter College

Margaret Chambers, George Mason University

RoNell Coco, Clark College

John Curra, Respiratory Technologist

Stephanie Hall Ford, Barry University

Helen Hart Ford, Radiological Technologist

Elizabeth Garrison, Modesto Junior College

Emily Gordon, Edison Community College

Kim Green, Edison Community College

Mike Grosse, Licensed Physical Therapy Assistant

Caroline Guerra, Broward Community College

Jill Haan, Cardiographic Technician

Dorie Hart, RN

Lisa P. Hubbard, Macon State College

Mark Johnson, Edison Community College

Wendy Lawrence, Troy State University

Candice Machado, Modesto Junior College

Susan Mahoney, Physical Therapist

Christine McGrellis, Mohawk Valley Community College

Jacqueline Opera, Medical Laboratory Technologist

Denise Pacheco, Modesto Junior College

Joan Petrokovsky, Licensed Massage Therapist

Toni Sheridan, Pharmacist

Barbara Simone, RN, FNP

Tiffany Smith, Stark State College of Technology

Sabrina von Brueckwitz, Edison Community College

To the Student

Your book has a variety of special features that will make your time studying anatomy and physiology a more rewarding experience. These have been developed based on feedback from students – like you – who have used previous editions of the text. A review of the preface will give you insight, both visually and in narrative, to all of the text's distinctive features.

Our experience in the classroom has taught us that student's appreciate a hint – both visually and verbally – at the beginning of each chapter about what to expect from its contents. Each chapter of your book begins with a stunning illustration depicting the system or main content being covered in the chapter. In addition, a short introduction to the chapter contents – usually written by a student who has recently completed the course, but occasionally by a practitioner in an allied health field – offers you an insight into some of the most intriguing or relevant aspects of the chapter. Links to activities on your **Foundations** CD, and special web-based activities called **Insights and Explorations** are suggested to make your study time worthwhile and interesting.

A Student's View

My medical coding classes stress that it is routine for coders to review lab work to find abnormalities that tie into the text of the patient's medical record. Only then can proper codes be assigned so the bill can be processed and sent to third party payers such as a health insurance company or Medicare. So, the chapter on Blood caught my attention because I've been thinking about patient blood work and the lab slips that report the findings: the actual numbers, the normal ranges, what the highs or lows in the various categories could mean.

It is the physician, of course, who interprets the results of the blood work and decides how it all fits into the picture of his patient's health and care. He documents this in the patient's medical record. An example might be a finding of anemia and then a blood transfusion as treatment. As a coder in a hospital looking at that patient's medical record, I could see the doctor's diagnosis of anemia and could verify that diagnosis by seeing the lab slips of the blood work and finding progress notes and orders concerning the transfusion.

Coders also need to know how to recognize results from blood work that show infection or sepsis. Table 19.2, on page 645, gives a quick summary of the significance of the information found in a complete blood count (CBS): high or low "counts" of the white blood cells (WBC).

Christine McBrellis
Mohawk Valley Community College

CHAPTER NINETEEN

The Cardiovascular System: The Blood

BLOOD AND HOMEOSTASIS
Blood contributes to homeostasis by transporting oxygen, carbon dioxide, nutrients, and hormones to and from your body's cells. It helps regulate body pH and temperature, and provides protection against disease through phagocytosis and the production of antibodies.

ENERGIZE YOUR STUDY
FOUNDATIONS CD
Anatomy Overview
• The Cardiovascular System
• Connective Tissue: Blood
Animations
• Systems Contributions to Homeostasis
• The Case of the Man with the Yellow Eyes
• The Case of the Worried Smoker
Concepts and Connections and Exercises reinforce your understanding

www.wiley.com/college/apcentral
INSIGHTS AND EXPLORATIONS
It seems that professional athletes just keep getting stronger and faster. Many of us might think that this improved performance is due to better training. It might surprise you to find out that recombinant DNA technology helps some athletes perform better. These athletes take recombinant human erythropoietin (rHuEpo) to increase the number of red blood cells in their bodies. This is called induced polycythemia and is believed to have played a role in the deaths of some professional cyclists in the 1990s. What is the normal concentration of red blood cells in a person? Is it different for an athlete? What about athletes who train at high altitudes? Come explore the answers to these questions and more in this web-based activity about polycythemia.

As you begin each narrative section of the chapter, be sure to take note of the **objectives** at the beginning of the section to help you focus on what is important as you read it.

FUNCTIONS AND PROPERTIES OF BLOOD

OBJECTIVES
• Describe the functions of blood.
• Describe the physical characteristics and principal components of blood.

Most cells of a multicellular organism cannot move around to obtain oxygen and nutrients or eliminate carbon dioxide and other wastes. Instead, these needs are met by two fluids: blood and interstitial fluid. **Blood** is a connective tissue composed of a liquid matrix ca

At the end of the section, take time to try and answer the **Checkpoint** questions placed there. If you can, then you are ready to move on to the next section. If you experience difficulty answering the questions, you may want to re-read the section before continuing.

CHECKPOINT
1. In what ways is blood plasma similar to and different from interstitial fluid?
2. How many kilograms or pounds of blood is present in your body?
3. How does the volume of blood plasma in your body compare to the volume of fluid in a two-liter bottle of Coke?
4. What are the major solutes in blood plasma? What does each do?
5. What is the significance of lower-than-normal or higher-than-normal hematocrit?

Studying the figures (illustrations that include artwork and photographs) in this book is as important as reading the text. To get the most out of the visual parts of this book, use the tools we have added to the figures to help you understand the concepts being presented. Start by reading the **legend**, which explains what the figure is about. Next, study the **key concept statement**, which reveals a basic idea portrayed in the figure. Added to many figures you will also find an **orientation diagram** to help you understand the perspective from which you are viewing a particular piece of anatomical art. Finally, at the bottom of each figure you will find a **figure question**. If you try to answer these questions as you go along, they will serve as self-checks to help you understand the material. Often it will be possible to answer a question by examining the figure itself. Other questions will encourage you to integrate the knowledge you've gained by carefully reading the text associated with the figure. Still other questions may prompt you to think critically about the topic at hand or predict a consequence in advance of its description in the text.

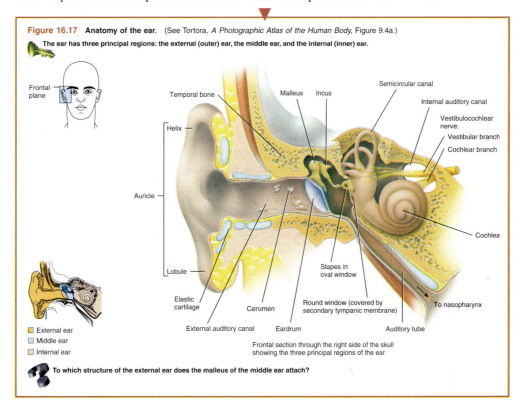

Figure 16.17   Anatomy of the ear.   (See Tortora, *A Photographic Atlas of the Human Body,* Figure 9.4a.)

The ear has three principal regions: the external (outer) ear, the middle ear, and the internal (inner) ear.

Frontal plane

Temporal bone

Malleus   Incus

Semicircular canal

Internal auditory canal

Vestibulocochlear nerve:
Vestibular branch
Cochlear branch

Helix

Auricle

Cochlea

Lobule

Stapes in oval window

To nasopharynx

Elastic cartilage

Cerumen

Round window (covered by secondary tympanic membrane)

External auditory canal   Eardrum

Auditory tube

☐ External ear
☐ Middle ear
☐ Internal ear

Frontal section through the right side of the skull showing the three principal regions of the ear

To which structure of the external ear does the malleus of the middle ear attach?

At the end of each chapter are other resources that you will find useful. **The Study Outline** is a concise statement of important topics discussed in the chapter. Page numbers are listed next to key concepts so you can easily refer to the specific passages in the text for clarification or amplification.

STUDY OUTLINE

SPINAL CORD ANATOMY
(p. 420)

1. The spinal cord is protected by the vertebral column, the meninges, cerebrospinal fluid, and denticulate ligaments.
2. The three meninges are coverings that run continuously around the spinal cord and brain. They are the dura mater, arachnoid mater, and pia mater.
3. The spinal cord begins as a continuation of the medulla oblongata and ends at about the second lumbar vertebra in an adult.
4. The spinal cord contains cervical and lumbar enlargements that serve as points of origin for nerves to the limbs.
5. The tapered inferior portion of the spinal cord is the conus medullaris, from which arise the filum terminale and cauda equina.
6. Spinal nerves connect to each segment of the spinal cord by two

roots. The posterior or dorsal root contains sensory axons, and the anterior or ventral root contains motor neuron axons.
7. The anterior median fissure and the posterior median sulcus partially divide the spinal cord into right and left sides.
8. The gray matter in the spinal cord is divided into horns, and the white matter into columns. In the center of the spinal cord is the central canal, which runs the length of the spinal cord.
9. Parts of the spinal cord observed in transverse section are the gray commissure; central canal; anterior, posterior, and lateral gray horns; and anterior, posterior, and lateral white columns, which contain ascending and descending tracts. Each part has specific functions.
10. The spinal cord conveys sensory and motor information by way of ascending and descending tracts, respectively.

The **Self-quiz Questions** are designed to help you evaluate your understanding of the chapter contents. **Critical Thinking Questions** are word problems that allow you to apply the concepts you have studied in the chapter to specific situations.

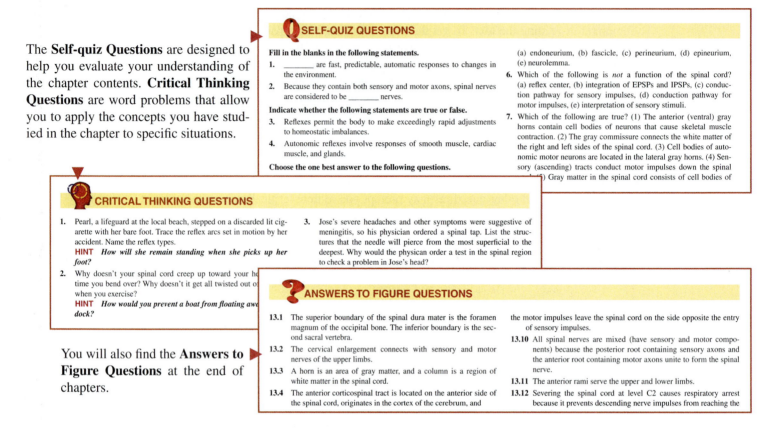

SELF-QUIZ QUESTIONS

Fill in the blanks in the following statements.

1. _____ are fast, predictable, automatic responses to changes in the environment.

2. Because they contain both sensory and motor axons, spinal nerves are considered to be _____ nerves.

Indicate whether the following statements are true or false.

3. Reflexes permit the body to make exceedingly rapid adjustments to homeostatic imbalances.

4. Autonomic reflexes involve responses of smooth muscle, cardiac muscle, and glands.

Choose the one best answer to the following questions.

(a) endoneurium, (b) fascicle, (c) perineurium, (d) epineurium, (e) neurolemma.

6. Which of the following is *not* a function of the spinal cord? (a) reflex center, (b) integration of EPSPs and IPSPs, (c) conduction pathway for sensory impulses, (d) conduction pathway for motor impulses, (e) interpretation of sensory stimuli.

7. Which of the following are true? (1) The anterior (ventral) gray horns contain cell bodies of neurons that cause skeletal muscle contraction. (2) The gray commissure connects the white matter of the right and left sides of the spinal cord. (3) Cell bodies of autonomic motor neurons are located in the lateral gray horns. (4) Sensory (ascending) tracts conduct motor impulses down the spinal [cord. (5)] Gray matter in the spinal cord consists of cell bodies of

CRITICAL THINKING QUESTIONS

1. Pearl, a lifeguard at the local beach, stepped on a discarded lit cigarette with her bare foot. Trace the reflex arcs set in motion by her accident. Name the reflex types.
   HINT *How will she remain standing when she picks up her foot?*

2. Why doesn't your spinal cord creep up toward your he[ad] time you bend over? Why doesn't it get all twisted out o[f] when you exercise?
   HINT *How would you prevent a boat from floating awa[y] dock?*

3. Jose's severe headaches and other symptoms were suggestive of meningitis, so his physician ordered a spinal tap. List the structures that the needle will pierce from the most superficial to the deepest. Why would the physican order a test in the spinal region to check a problem in Jose's head?

You will also find the **Answers to Figure Questions** at the end of chapters.

ANSWERS TO FIGURE QUESTIONS

13.1 The superior boundary of the spinal dura mater is the foramen magnum of the occipital bone. The inferior boundary is the second sacral vertebra.

13.2 The cervical enlargement connects with sensory and motor nerves of the upper limbs.

13.3 A horn is an area of gray matter, and a column is a region of white matter in the spinal cord.

13.4 The anterior corticospinal tract is located on the anterior side of the spinal cord, originates in the cortex of the cerebrum, and

the motor impulses leave the spinal cord on the side opposite the entry of sensory impulses.

13.10 All spinal nerves are mixed (have sensory and motor components) because the posterior root containing sensory axons and the anterior root containing motor axons unite to form the spinal nerve.

13.11 The anterior rami serve the upper and lower limbs.

13.12 Severing the spinal cord at level C2 causes respiratory arrest because it prevents descending nerve impulses from reaching the

---

Learning the language of anatomy and physiology can be one the more challenging aspects of taking this course. Throughout the text we have included pronunciations, and sometimes, Word Roots, for many terms that may be new to you. These appear in parentheses immediately following the new words, and the pronunciations are repeated in the glossary at the back of the book. You companion CD includes the complete glossary and offers you the opportunity to hear the words pronounced. In addition, the Companion Website offers you a review of these terms by chapter, pronounces them for you, and allows you the opportunity to create flash cards or quiz yourself on the many new terms.

Look at the words carefully and say them out loud several times. Learning to pronounce a new word will help you remember it and make it a useful part of your medical vocabulary. Take a few minutes to review the following pronunciation key, so it will be familiar to you when you encounter new words. The key is repeated at the beginning of the Glossary, as well.

Pronounciation Key

1. The most strongly accented syllable appears in capital letters, for example, bilateral (bī-LAT-er-al) and diagnosis (dī-ag-NŌ-sis).

2. If there is a secondary accent, it is noted by a prime ('), for example, constitution (kon'-sti-TOO-shun) and physiology (fiz'-ē-OL-ō-jē). Any additional secondary accents are also noted by a prime, for example, decarboxylation (dē'-kar-bok'-si-LĀ-shun).

3. Vowels marked by a line above the letter are pronounced with the long sound, as in the following common words:
   ā as in māke   ō as in pōle
   ē as in bē   ū as in cute
   ī as in īvy

4. Vowels not marked by a line above the letter are pronounced with the short sound, as in the following words:
   a as in above or at   o as in not
   e as in bet   u as in bud
   i as in sip

5. Other vowel sounds are indicated as follows:
   oy as in oil
   oo as in root

6. Consonant sounds are pronounced as in the following words:
   b as in bat   m as in mother
   ch as in chair   n as in no
   d as in dog   p as in pick
   f as in father   r as in rib
   g as in get   s as in so
   h as in hat   t as in tea
   j as in jump   v as in very
   k as in can   w as in welcome
   ks as in tax   z as in zero
   kw as in quit   zh as in lesion
   l as in let

Brief Table of Contents

Contents

Clinical Applications

Chapter 17 The Autonomic Nervous System 565

Clinical Applications

FOCUS ON HOMEOSTASIS: THE NERVOUS SYSTEM 582

Chapter 18 The Endocrine System 586

Clinical Applications

FOCUS ON HOMEOSTASIS: THE ENDOCRINE SYSTEM 624

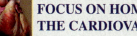

The Cardiovascular System: The Blood

A Student's View

My medical coding classes stress that it is routine for coders to review lab work to find abnormalities that tie into the text of the patient's medical record. Only then can proper codes be assigned so the bill can be processed and sent to third party payers such as a health insurance company or Medicare. So, the chapter on Blood caught my attention because I've been thinking about patient blood work and the lab slips that report the findings: the actual numbers, the normal ranges, what the highs or lows in the various categories could mean.

It is the physician, of course, who interprets the results of the blood work and decides how it all fits into the picture of his patient's health and care. He documents this in the patient's medical record. An example might be a finding of anemia and then a blood transfusion as treatment. As a coder in a hospital looking at that patient's medical record, I could see the doctor's diagnosis of anemia and could verify that diagnosis by seeing the lab slips of the blood work and finding progress notes and orders concerning the transfusion.

Coders also need to know how to recognize results from blood work that show infection or sepsis. Table 19.2, on page 645, gives a quick summary of the significance of the information found in a complete blood count (CBS): high or low "counts" of the white blood cells (WBC).

Christine McGrellis
Mohawk Valley Community College

BLOOD AND HOMEOSTASIS

Blood contributes to homeostasis by transporting oxygen, carbon dioxide, nutrients, and hormones to and from your body's cells. It helps regulate body pH and temperature, and provides protection against disease through phagocytosis and the production of antibodies.

ENERGIZE YOUR STUDY

FOUNDATIONS CD

Anatomy Overview
  • The Cardiovascular System
  • Connective Tissue: Blood

Animations
  • Systems Contributions to Homeostasis
  • The Case of the Man with the Yellow Eyes
  • The Case of the Worried Smoker

Concepts and Connections and Exercises reinforce your understanding

www.wiley.com/college/apcentral

INSIGHTS AND EXPLORATIONS

It seems that professional athletes just keep getting stronger and faster. Many of us might think that this improved performance is due to better training. It might surprise you to find out that recombinant DNA technology helps some athletes perform better. These athletes take recombinant human erythropoietin (rHuEpo) to increase the number of red blood cells in their bodies. This is called induced polycythemia and is believed to have played a role in the deaths of some professional cyclists in the 1990s. What is the normal concentration of red blood cells in a person? Is it different for an athlete? What about athletes who train at high altitudes? Come explore the answers to these questions and more in this web-based activity about polycythemia.

The **cardiovascular system** (*cardio-* = heart; *vascular* = blood vessels) consists of three interrelated components: blood, the heart, and blood vessels. The focus of this chapter is blood; the next two chapters will examine the heart and blood vessels, respectively. Blood transports various substances, helps regulate several life processes, and affords protection against disease. For all of its similarities in origin, composition, and functions, blood is as unique from one person to another as are skin, bone, and hair. Health-care professionals routinely examine and analyze its differences through various blood tests when trying to determine the cause of different diseases. The branch of science concerned with the study of blood, blood-forming tissues, and the disorders associated with them is **hematology** (hēm-a-TOL-ō -jē ; *hema-* or *hemato-* = blood; *-logy* = study of).

FUNCTIONS AND PROPERTIES OF BLOOD

▶ OBJECTIVES

- **Describe the functions of blood.**
- **Describe the physical characteristics and principal components of blood.**

Most cells of a multicellular organism cannot move around to obtain oxygen and nutrients or eliminate carbon dioxide and other wastes. Instead, these needs are met by two fluids: blood and interstitial fluid. **Blood** is a connective tissue composed of a liquid matrix called **plasma** that dissolves and suspends various cells and cell fragments. **Interstitial fluid** is the fluid that bathes body cells (see Figure 27.1 on page 992). Blood transports oxygen from the lungs and nutrients from the gastrointestinal tract to body cells. The oxygen and nutrients diffuse from the blood into the interstitial fluid and then into body cells. Carbon dioxide and other wastes move in the reverse direction, from body cells into the interstitial fluid and then into the blood.

Functions of Blood

Blood, the only liquid connective tissue, has three general functions:

1. *Transportation.* Blood transports oxygen from the lungs to the cells of the body and carbon dioxide from the body cells to the lungs for exhalation. It carries nutrients from the gastrointestinal tract to body cells and hormones from endocrine glands to other body cells. Blood also transports heat and waste products to the lungs, kidneys, and skin for elimination from the body.

2. *Regulation.* Circulating blood helps maintain homeostasis in all body fluids. Blood helps regulate pH through buffers. It also helps adjust body temperature through the heat-absorbing and coolant properties of the water (see page 39) in plasma and its variable rate of flow through the skin, where excess heat can be lost from the blood to the environment. Blood osmotic pressure also influences the water content of cells, mainly through interactions of dissolved ions and proteins.

3. *Protection.* Blood can clot, which protects against its excessive loss from the cardiovascular system after an injury. In addition, white blood cells protect against disease by carrying on phagocytosis. Several types of blood proteins, antibodies, such as interferons, and complement, help protect against disease in a variety of ways.

Physical Characteristics of Blood

Blood is denser and more viscous than water and feels slightly sticky. The temperature of blood is 38°C (100.4°F), about 1°C higher than oral or rectal body temperature, and it has a slightly alkaline pH ranging from 7.35 to 7.45. Blood constitutes about 20% of extracellular fluid, amounting to 8% of the total body mass. The blood volume is 5 to 6 liters (1.5 gal) in an average-sized adult male and 4 to 5 liters (1.2 gal) in an average-sized adult female. Several hormonal negative feedback systems ensure that blood volume and osmotic pressure remain relatively constant. Especially important systems are those involving aldosterone, antidiuretic hormone, and atrial natriuretic peptide, which regulate how much water is excreted in the urine (see pages 970–972).

Withdrawing Blood

Blood samples for laboratory testing may be obtained in several ways. The most common procedure is **venipuncture,** withdrawal of blood from a vein using a hypodermic needle and syringe. A tourniquet is wrapped around the arm above the venipuncture site, which causes blood to accumulate in the vein. This increased blood volume makes the vein stand out. Clenching the fist further causes it to stand out, making the venipuncture more successful. A common site for venipuncture is the median cubital vein anterior to the elbow (see Figure 21.26b on page 743). Another method of withdrawing blood is through a **finger** or **heel stick.** Diabetic patients who monitor their daily blood sugar typically perform a finger stick, and it is often used for drawing blood from infants and children. In an **arterial stick,** blood is withdrawn from an artery, for example, when it is important to know the level of oxygen in oxygenated blood. ■

Components of Blood

Whole blood has two components: (1) blood plasma, a watery liquid matrix that contains dissolved substances, and (2) formed elements, which are cells and cell fragments. If a sample of blood is centrifuged (spun) in a small glass tube, the cells sink to the bottom of the tube while the lighter-weight plasma forms a layer on top (Figure 19.1a). Blood is about 45% formed

Figure 19.1   Components of blood in a normal adult.

Blood is a connective tissue that consists of blood plasma (liquid) plus formed elements (red blood cells, white blood cells, and platelets).

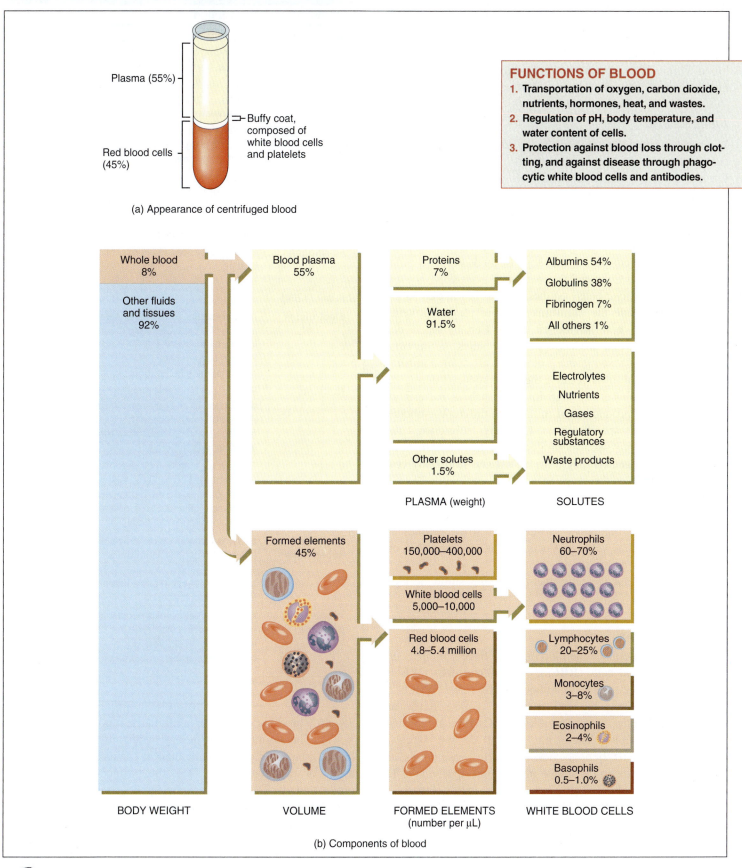

(a) Appearance of centrifuged blood

FUNCTIONS OF BLOOD

1. Transportation of oxygen, carbon dioxide, nutrients, hormones, heat, and wastes.
2. Regulation of pH, body temperature, and water content of cells.
3. Protection against blood loss through clotting, and against disease through phagocytic white blood cells and antibodies.

(b) Components of blood

 What is the approximate volume of blood in your body? (*Hint*: Each liter of blood has a mass of one kilogram.)

elements and 55% plasma. Normally, more than 99% of the formed elements are red-colored red blood cells (RBCs). Pale, colorless white blood cells (WBCs) and platelets occupy less than 1% of total blood volume. They form a very thin **buffy coat** layer between the packed RBCs and plasma in centrifuged blood. Figure 19.1b shows the composition of blood plasma and the numbers of the various types of formed elements in blood.

Blood Plasma

When the formed elements are removed from blood, a straw-colored liquid called **blood plasma** (or simply **plasma**) is left. Plasma is about 91.5% water and 8.5% solutes, most of which (7% by weight) are proteins. Some of the proteins in plasma are also found elsewhere in the body, but those confined to blood are called **plasma proteins.** Among other functions, these proteins play a role in maintaining proper blood osmotic pressure, which is an important factor in the exchange of fluids across capillary walls (discussed in Chapter 21).

Hepatocytes (liver cells) synthesize most of the plasma proteins, which include the **albumins** (54% of plasma proteins), **globulins** (38%), and **fibrinogen** (7%). Certain blood cells develop into cells that produce gamma globulins, an important type of globulin. These plasma proteins are also called **antibodies** or **immunoglobulins** because they are produced during certain immune responses. Foreign invaders such as bacteria and viruses stimulate production of millions of different antibodies. An antibody binds specifically to the foreign substance, called an **antigen,** that provoked its production. When an antigen and antibody come together, they form an **antigen–antibody complex,** which disables the invading antigen (see Figure 22.18 on page 791).

Besides proteins, other solutes in plasma include electrolytes, nutrients, regulatory substances such as enzymes and hormones, gases, and waste products such as urea, uric acid, creatinine, ammonia, and bilirubin.

Table 19.1 describes the chemical composition of blood plasma.

Formed Elements

The **formed elements** of the blood include three principal components: **red blood cells (RBCs), white blood cells (WBCs),** and **platelets** (Figure 19.2). Although RBCs and WBCs are whole cells, platelets are cell fragments. Unlike RBCs and platelets, which have just a few roles, WBCs have a number of specialized functions. There are several distinct types of WBCs—neutrophils, lymphocytes, monocytes, eosinophils, and basophils—each having a unique microscopic appearance. The roles of each type of WBC are discussed later in this chapter.

The percentage of total blood volume occupied by RBCs is called the **hematocrit** (he-MAT-ō-krit). For example, a hematocrit of 40 means that 40% of the volume of blood is composed of RBCs. The normal range of hematocrit for adult females is

Table 19.1	Substances in Blood Plasma
Constituent	**Description**
Water (91.5%)	Liquid portion of blood. Acts as solvent and suspending medium for components of blood; absorbs, transports, and releases heat.
Proteins (7.0)	Exert colloid osmotic pressure, which helps maintain water balance between blood and tissues and regulates blood volume.
Albumins	Smallest and most numerous plasma proteins; produced by liver. Function as transport proteins for several steroid hormones and for fatty acids.
Globulins	Produced by liver and by plasma cells, which develop from B lymphocytes. Antibodies (immunoglobulins) help attack viruses and bacteria. Alpha and beta globulins transport iron, lipids, and fat-soluble vitamins.
Fibrinogen	Produced by liver. Plays essential role in blood clotting.
Other Solutes (1.5%)	
Electrolytes	Inorganic salts. Positively charged ions (cations) include Na^+, K^+, Ca^{2+}, Mg^{2+}; negatively charged ions (anions) include Cl^-, HPO_4^{2-}, SO_4^{2-}, and HCO_3^-. Help maintain osmotic pressure and play essential roles in the function of cells.
Nutrients	Products of digestion pass into blood for distribution to all body cells. Include amino acids (from proteins), glucose (from carbohydrates), fatty acids and glycerol (from triglycerides), vitamins, and minerals.
Gases	Oxygen (O_2), carbon dioxide (CO_2), and nitrogen (N_2). Whereas more O_2 is associated with hemoglobin inside red blood cells, more CO_2 is dissolved in plasma. N_2 is present but has no known function in the body.
Regulatory substances	Enzymes, produced by body cells, catalyze chemical reactions. Hormones, produced by endocrine glands, regulate metabolism, growth, and development in body. Vitamins are cofactors for enzymatic reactions.
Waste products	Most are breakdown products of protein metabolism and are carried by blood to organs of excretion. Include urea, uric acid, creatine, creatinine, bilirubin, and ammonia.

38–46% (average = 42); for adult males, it is 40–54% (average = 47). The hormone testosterone, which is present in much higher concentration in males than in females, stimulates synthesis of erythropoietin (EPO), the hormone that stimulates production of RBCs. Thus, testosterone contributes to higher hematocrits in males. Lower values in women during their reproductive years also may be due to excessive loss of blood during menstruation. A significant drop in hematocrit indicates *anemia,* a lower-than-normal number of RBCs. In *polycythemia* the per-

Figure 19.2 Scanning electron micrograph of the formed elements of blood.

 The formed elements of blood are red blood cells (RBCs), white blood cells (WBCs), and platelets.

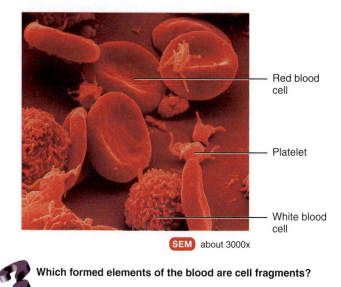

Red blood cell

Platelet

White blood cell

SEM about 3000x

Which formed elements of the blood are cell fragments?

centage of RBCs is abnormally high, and the hematocrit may be 65% or higher. Causes of polycythemia include abnormal increases in RBC production, tissue hypoxia, dehydration, and blood doping or use of EPO by athletes.

Induced Polycythemia in Athletes

Delivery of oxygen to muscles is a limiting factor in muscular feats. As a result, increasing the oxygen-carrying capacity of the blood enhances athletic performance, especially in endurance events. Because red blood cells are the main transport vehicle for oxygen, athletes have tried several means of increasing their hematocrit, causing **induced polycythemia,** to gain a competitive edge. Training at higher altitudes produces a natural increase in hematocrit because the lower amount of oxygen in inhaled air triggers release of more EPO. In past years, some athletes performed *blood doping* to elevate their hematocrit. In this procedure, blood cells are removed from the body and stored for a month or so, during which time the hematocrit returns to normal. Then, a few days before an athletic event, the RBCs are reinjected to increase hematocrit. More recently, athletes have enhanced their RBC production by injecting *Epoetin alfa (Procrit* or *Epogen),* a drug that is used to treat anemia. Epoetin alfa is identical to naturally occurring EPO because it is manufactured by recombinant DNA technology using the human gene that codes for erythropoietin. Practices that increase hematocrit are dangerous, however, because they increase the workload of the heart. The increased numbers of RBCs raise the viscosity of the blood, which increases the resistance to blood

flow and makes the blood more difficult for the heart to pump. Increased viscosity also contributes to high blood pressure and increased risk of stroke. During the 1980s, at least 15 competitive cyclists died from heart attacks or strokes linked to suspected use of Epoetin alfa. Although the International Olympics Committee bans Epoetin alfa use, enforcement is difficult because the drug is identical to naturally occurring EPO. ■

▶ **CHECKPOINT**

1. In what ways is blood plasma similar to and different from interstitial fluid?

2. How many kilograms or pounds of blood is present in your body?

3. How does the volume of blood plasma in your body compare to the volume of fluid in a two-liter bottle of Coke?

4. What are the major solutes in blood plasma? What does each do?

5. What is the significance of lower-than-normal or higher-than-normal hematocrit?

FORMATION OF BLOOD CELLS

▶ **OBJECTIVE**

• **Explain the origin of blood cells.**

Although some lymphocytes have a lifetime measured in years, most formed elements of the blood are continually dying and being replaced within hours, days, or weeks. Negative feedback systems regulate the total number of RBCs and platelets in circulation, and their numbers normally remain steady. The abundance of the different types of WBCs, however, varies in response to challenges by invading pathogens and other foreign antigens.

The process by which the formed elements of blood develop is called **hemopoiesis** (hē-mō-poy-Ē-sis; *-poiesis* = making) or *hematopoiesis.* Before birth, hemopoiesis first occurs in the yolk sac of an embryo and later in the liver, spleen, thymus, and lymph nodes of a fetus. In the last three months before birth, red bone marrow becomes the primary site of hemopoiesis and continues as the main source of blood cells after birth and throughout life.

Red bone marrow is a highly vascularized connective tissue located in the microscopic spaces between trabeculae of spongy bone tissue. It is present chiefly in bones of the axial skeleton, pectoral and pelvic girdles, and the proximal epiphyses of the humerus and femur (see Figure 6.1a on page 165). About 0.05−0.1% of red bone marrow cells are derived from mesenchymal cells called **pluripotent stem cells** (ploo-RIP-ō-tent; *pluri-* = several). Pluripotent stem cells are cells that have the capacity to develop into several different types of cells (Figure 19.3).

Stem cells in red bone marrow reproduce themselves, proliferate, and differentiate into cells that give rise to blood cells,

Figure 19.3 **Origin, development, and structure of blood cells.** Some of the generations of some cell lines have been omitted.

🔑 Blood cell production is called hemopoiesis and occurs only in red bone marrow after birth.

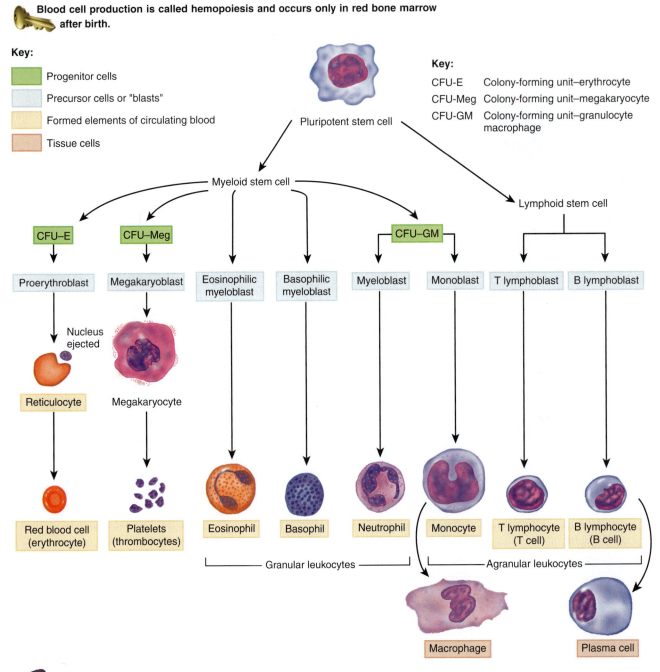

Key:

🟩 Progenitor cells

⬜ Precursor cells or "blasts"

🟨 Formed elements of circulating blood

🟧 Tissue cells

Key:

CFU-E   Colony-forming unit–erythrocyte

CFU-Meg   Colony-forming unit–megakaryocyte

CFU-GM   Colony-forming unit–granulocyte macrophage

Pluripotent stem cell

Myeloid stem cell — Lymphoid stem cell

CFU–E   CFU–Meg   CFU–GM

Proerythroblast   Megakaryoblast   Eosinophilic myeloblast   Basophilic myeloblast   Myeloblast   Monoblast   T lymphoblast   B lymphoblast

Nucleus ejected

Reticulocyte   Megakaryocyte

Red blood cell (erythrocyte)   Platelets (thrombocytes)   Eosinophil   Basophil   Neutrophil   Monocyte   T lymphocyte (T cell)   B lymphocyte (B cell)

└─ Granular leukocytes ─┘   └─ Agranular leukocytes ─┘

Macrophage   Plasma cell

From which connective tissue cells do pluripotent stem cells develop?

macrophages, reticular cells, mast cells, and adipocytes. The reticular cells produce reticular fibers, which form the stroma (framework) that supports red bone marrow cells. Blood from nutrient and metaphyseal arteries (see Figure 6.5 on page 169) enters a bone and passes into the enlarged and leaky capillaries, called *sinuses,* that surround red bone marrow cells and fibers. After blood cells form, they enter the sinuses and other blood vessels and leave the bone through nutrient and periosteal veins

(see Figure 6.5 on page 169). With the exception of lymphocytes, formed elements do not divide once they leave red bone marrow.

Pluripotent stem cells produce two further types of stem cells, called *myeloid stem cells* and *lymphoid stem cells.* Myeloid stem cells begin their development in red bone marrow and give rise to red blood cells, platelets, monocytes, neutrophils, eosinophils, and basophils. Lymphoid stem cells begin their development in red bone marrow but complete it in lymphatic tissues; they give rise to lymphocytes. Although the various stem cells have distinctive cell identity markers in their plasma membranes, they cannot be distinguished histologically and resemble lymphocytes.

During hemopoiesis, the myeloid stem cells differentiate into **progenitor cells** (pro-JEN-i-tor). Progenitor cells are no longer capable of reproducing themselves and are committed to giving rise to more specific elements of blood. Some progenitor cells are known as *colony-forming units (CFUs)*. After the CFU designation is an abbreviation that designates the mature elements in blood that they will produce: CFU–E ultimately produces erythrocytes (red blood cells), CFU–Meg produces megakaryocytes, the source of platelets, and CFU–GM ultimately produces granulocytes (specifically, neutrophils) and monocytes. Progenitor cells, like stem cells, resemble lymphocytes and cannot be distinguished by their microscopic appearance alone. Other myeloid stem cells develop directly into cells called precursor cells (described next). Lymphoid stem cells differentiate into T lymphoblasts and B lymphoblasts, which ultimately develop into B lymphocytes (B cells) and T lymphocytes (T cells), respectively.

In the next generation, the cells are called **precursor cells,** also known as **blasts.** Over several cell divisions they develop into the actual formed elements of blood. For example, monoblasts develop into monocytes, eosinophilic myeloblasts develop into eosinophils, and so on. Precursor cells have recognizable microscopic appearances.

Several hormones called **hemopoietic growth factors** regulate the differentiation and proliferation of particular progenitor cells. **Erythropoietin** (e-rith′-rō-POY-e-tin) or **EPO** increases the number of red blood cell precursors. The main producers of EPO are cells in the kidneys that lie between the kidney tubules (peritubular interstitial cells). With renal failure, EPO release slows and RBC production is inadequate. **Thrombopoietin** (throm′-bō-POY-e-tin) or **TPO** is a hormone produced by the liver that stimulates formation of platelets (thrombocytes) from megakaryocytes. Several different cytokines regulate development of different blood cell types. **Cytokines** are small glycoproteins that are produced by red bone marrow cells, leukocytes, macrophages, fibroblasts, and endothelial cells. They typically act as local hormones (autocrines or paracrines). Cytokines stimulate proliferation of progenitor cells in red bone marrow and regulate the activities of cells involved in nonspecific defenses (such as phagocytes) and immune responses (such as B cells and T cells). Two important families of cytokines that stimulate white blood cell formation are **colony-stimulating factors (CSFs)** and **interleukins.**

Medical Uses of Hemopoietic Growth Factors

Hemopoietic growth factors made available through recombinant DNA technology hold tremendous potential for medical uses when a person's natural ability to form new blood cells is diminished or defective. Recombinant erythropoietin (Epoetin alfa) is very effective in treating the diminished red blood cell production that accompanies end-stage kidney disease. Granulocyte-macrophage colony-stimulating factor and granulocyte CSF are given to stimulate white blood cell formation in cancer patients who are undergoing chemotherapy, which kills some of their red bone marrow cells as well as the cancer cells because both cell types are undergoing mitosis. Thrombopoietin shows great promise for preventing platelet depletion during chemotherapy. CSFs and thrombopoietin also improve the outcome of patients who receive bone marrow transplants. ■

▶ **C H E C K P O I N T**

6. Which hemopoietic growth factors regulate differentiation and proliferation of CFU–E and formation of platelets from megakaryocytes?

RED BLOOD CELLS

▶ **O B J E C T I V E**

- **Describe the structure, functions, life cycle, and production of red blood cells.**

Red blood cells (RBCs) or **erythrocytes** (e-RITH-rō-sīts; *erythro-* = red; *-cyte* = cell) contain the oxygen-carrying protein **hemoglobin,** which is a pigment that gives whole blood its red color. A healthy adult male has about 5.4 million red blood cells per microliter (μL) of blood,* and a healthy adult female has about 4.8 million. (One drop of blood is about 50 μL.) To maintain normal numbers of RBCs, new mature cells must enter the circulation at the astonishing rate of at least 2 million per second, a pace that balances the equally high rate of RBC destruction.

RBC Anatomy

RBCs are biconcave discs with a diameter of 7–8 μm (Figure 19.4a). Mature red blood cells have a simple structure. Their plasma membrane is both strong and flexible, which allows them to deform without rupturing as they squeeze through narrow capillaries. As you will see later, certain glycolipids in the plasma membrane of RBCs are antigens that account for the various blood groups such as the ABO and Rh groups. RBCs lack a nucleus and other organelles and can neither reproduce nor carry on extensive metabolic activities. The cytosol of RBCs contains hemoglobin molecules, which were synthesized before loss of the nucleus during RBC production and which constitute about 33% of the cell's weight.

---

*1 μL = 1 mm^3 = 10^{-6} liter.

Figure 19.4 The shapes of a red blood cell (RBC) and a hemoglobin molecule, and the structure of a heme group. In (b), note that each of the four polypeptide chains (blue) of a hemoglobin molecule has one heme group (gold), which contains an iron ion Fe^{2+} (red).

🔑 **The iron portion of a heme group binds oxygen for transport by hemoglobin.**

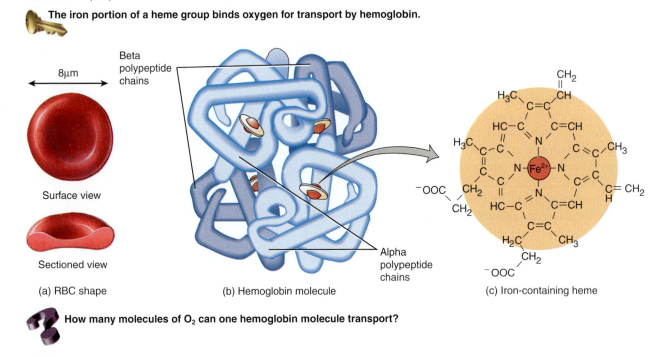

(a) RBC shape   (b) Hemoglobin molecule   (c) Iron-containing heme

❓ **How many molecules of O_2 can one hemoglobin molecule transport?**

RBC Physiology

Red blood cells are highly specialized for their oxygen transport function. Because mature RBCs have no nucleus, all their internal space is available for oxygen transport. Moreover, RBCs lack mitochondria and generate ATP anaerobically (without oxygen). Hence, they do not use up any of the oxygen they transport. Even the shape of an RBC facilitates its function. A biconcave disc has a much greater surface area for its volume than, say, a sphere or a cube. This shape provides a large surface area for the diffusion of gas molecules into and out of the RBC.

Each RBC contains about 280 million hemoglobin molecules and each hemoglobin can carry up to four oxygen molecules. A hemoglobin molecule consists of a protein called **globin,** composed of four polypeptide chains (two alpha and two beta chains), plus four nonprotein pigments called **hemes** (Figure 19.4b). One ringlike heme binds to each polypeptide chain. At the center of the heme ring is an iron ion (Fe^{2+}) that can combine reversibly with one oxygen molecule (Figure 19.4c). The oxygen picked up in the lungs is transported bound to the iron of the heme group. As blood flows through tissue capillaries, the iron-oxygen reaction reverses. Hemoglobin releases oxygen, which diffuses first into the interstitial fluid and then into cells.

Hemoglobin also transports about 13% of the total carbon dioxide, a waste product of metabolism. Blood flowing through tissue capillaries picks up carbon dioxide, some of which combines with amino acids in the globin part of hemoglobin. As blood flows through the lungs, the carbon dioxide is released from hemoglobin and then exhaled.

In addition to its key role in transporting oxygen and carbon dioxide, hemoglobin also plays a role in regulation of blood flow and blood pressure. The gaseous hormone **nitric oxide (NO),** produced by the endothelial cells that line blood vessels, binds to hemoglobin. Under some circumstances, hemoglobin releases NO. The released NO causes *vasodilation,* an increase in blood vessel diameter that occurs when the smooth muscle in the vessel wall relaxes. Vasodilation improves blood flow and enhances oxygen delivery to cells near the site of NO release.

RBC Life Cycle

Red blood cells live only about 120 days because of the wear and tear their plasma membranes undergo as they squeeze through blood capillaries. Without a nucleus and other organelles, RBCs cannot synthesize new components to replace damaged ones. The plasma membrane becomes more fragile with age, and the cells are more likely to burst, especially as they squeeze through narrow channels in the spleen. Ruptured red blood cells are removed from circulation and destroyed by fixed phagocytic macrophages in the spleen and liver, and the breakdown products are recycled, as follows (Figure 19.5):

1 Macrophages in the spleen, liver, or red bone marrow phagocytize ruptured and worn-out red blood cells.

2 The globin and heme portions of hemoglobin are split apart.

3 Globin is broken down into amino acids, which can be reused to synthesize other proteins.

Figure 19.5 Formation and destruction of red blood cells, and the recycling of hemoglobin components. RBCs circulate for about 120 days after leaving red bone marrow before they are phagocytized by macrophages.

The rate of RBC formation by red bone marrow equals the rate of RBC destruction by macrophages.

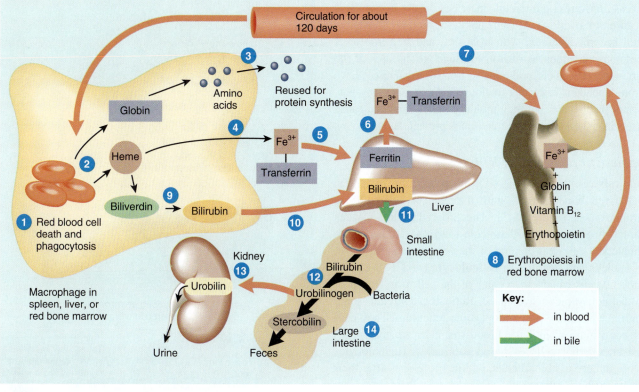

 What is the function of transferrin?

4 Iron removed from the heme portion is in the form of Fe^{3+}, which associates with the plasma protein **transferrin** (trans-FER-in; *trans-* = across; *ferr-* = iron), a transporter for Fe^{3+} in the bloodstream.

5 In muscle fibers, liver cells, and macrophages of the spleen and liver, Fe^{3+} detaches from transferrin and attaches to an iron-storage protein called **ferritin.**

6 Upon release from a storage site or absorption from the gastrointestinal tract, Fe^{3+} reattaches to transferrin.

7 The Fe^{3+}-transferrin complex is then carried to bone marrow, where RBC precursor cells take it up through receptor-mediated endocytosis (see Figure 3.13 on page 72) for use in hemoglobin synthesis. Iron is needed for the heme portion of the hemoglobin molecule, whereas amino acids are needed for the globin portion. Vitamin B_{12} is also needed for synthesis of hemoglobin.

8 Erythropoiesis in red bone marrow results in the production of red blood cells, which enter the circulation.

9 When iron is removed from heme, the non-iron portion of heme is converted to **biliverdin** (bil'-i-VER-din), a green pigment, and then into **bilirubin** (bil'-i-ROO-bin), a yellow-orange pigment.

10 Bilirubin enters the blood and is transported to the liver.

11 Within the liver, bilirubin is secreted by liver cells into bile, which passes into the small intestine and then into the large intestine.

12 In the large intestine, bacteria convert bilirubin into **urobilinogen** (ūr-ō-bī-LIN-ō-jen).

13 Some urobilinogen is absorbed back into the blood, converted to a yellow pigment called **urobilin** (ūr-ō-BĪ-lin), and excreted in urine.

14 Most urobilinogen is eliminated in feces in the form of a brown pigment called **stercobilin** (ster'-kō-BĪ-lin), which gives feces its characteristic color.

Iron Overload and Tissue Damage

Because free iron ions (Fe^{2+} and Fe^{3+}) bind to and damage molecules in cells or in the blood, transferrin and ferritin act as protective "protein escorts" during transport and storage of iron ions. As a result, plasma contains virtually no free iron. Moreover, only small amounts are available inside body cells for use in synthesis of iron-containing molecules such as the cytochrome pigments needed for ATP production in mitochondria (see Figure 25.7 on page 915). In cases of **iron overload,** the amount of iron present in the body builds up. Because we have no method for eliminating excess iron, any condition that increases dietary iron absorption can cause iron overload. At some point, the proteins transferrin and ferritin become saturated with iron ions, and free iron level rises. Common consequences of iron overload are diseases of the liver, heart, pancreatic islets, and gonads. Iron overload also allows certain iron-dependent microbes to flourish. Such microbes normally are not pathogenic, but they multiply rapidly and can cause lethal effects in a short time when free iron is present. ■

Erythropoiesis: Production of RBCs

Erythropoiesis (e-rith′-rō-poy-Ē-sis), the production of RBCs, starts in the red bone marrow with a precursor cell called a proerythroblast (see Figure 19.3). The proerythroblast divides several times, producing cells that begin to synthesize hemoglobin. Ultimately, a cell near the end of the development sequence ejects its nucleus and becomes a **reticulocyte** (re-TIK-ū-lō-sīt). Loss of the nucleus causes the center of the cell to indent, producing a distinctive biconcave shape. Reticulocytes retain some mitochondria, ribosomes, and endoplasmic reticulum. They pass from red bone marrow into the bloodstream by squeezing between the endothelial cells of blood capillaries. Reticulocytes develop into mature red blood cells, within 1 to 2 days after their release from red bone marrow.

Normally, erythropoiesis and red blood cell destruction proceed at roughly the same pace. If the oxygen-carrying capacity of the blood falls because erythropoiesis is not keeping up with RBC destruction, a negative feedback system steps up RBC production (Figure 19.6). The controlled condition is the amount of oxygen delivered to body tissues. Cellular oxygen deficiency, called **hypoxia** (hī-POKS-ē-a), may occur if too little oxygen enters the blood. For example, the lower oxygen content of air at high altitudes reduces the amount of oxygen in the blood. Oxygen delivery may also fall due to anemia, which has many causes: Lack of iron, lack of certain amino acids, and lack of vitamin B_{12} are but a few (see page 654). Circulatory problems that reduce blood flow to tissues may also reduce oxygen delivery. Whatever the cause, hypoxia stimulates the kidneys to step up the release of erythropoietin. This hormone circulates through the blood to the red bone marrow, where it speeds the development of proerythroblasts into reticulocytes. When the number of circulating RBCs increases, more oxygen can be delivered to body tissues.

Premature newborns often exhibit anemia, due in part to inadequate production of erythropoietin. During the first weeks after birth, the liver, not the kidneys, produces most EPO. Because the liver is less sensitive than the kidneys to hypoxia, newborns have a smaller EPO response to anemia than do adults.

Figure 19.6 **Negative feedback regulation of erythropoiesis (red blood cell formation).** Lower oxygen content of air at high altitudes, anemia, and circulatory problems may reduce oxygen delivery to body tissues.

🔑 **The main stimulus for erythropoiesis is hypoxia, a decrease in the oxygen-carrying capacity of the blood.**

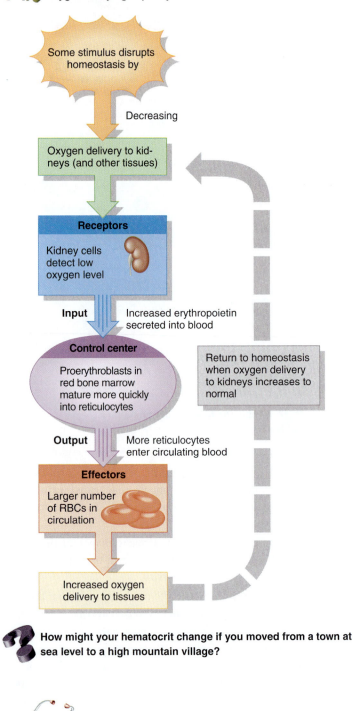

❓ **How might your hematocrit change if you moved from a town at sea level to a high mountain village?**

🩺 **Reticulocyte Count**

The rate of erythropoiesis is measured by a **reticulocyte count.** Normally, a little less than 1% of the oldest RBCs are replaced by newcomer reticulocytes on any given day. It then takes 1 to 2 days for the reticulocytes to lose the last vestiges of endoplasmic reticulum and become mature RBCs. Thus, reticulocytes account for about 0.5–1.5% of all RBCs in a normal blood sample. A low "retic" count in a person who is anemic might indicate shortage of erythropoietin or an inability of the red bone marrow to respond to EPO, perhaps because of a nutritional deficiency or leukemia. A high "retic" count might indicate a good red bone marrow response to previous blood loss or to iron therapy in someone who had been iron deficient. It could also point to illegal use of Epoetin alfa by an athlete. ■

▶ **CHECKPOINT**

7. How large are RBCs and what is their microscopic appearance?
8. How is hemoglobin recycled?
9. What is erythropoiesis? How does erythropoiesis affect hematocrit? What factors speed up and slow down erythropoiesis?

WHITE BLOOD CELLS

▶ **OBJECTIVE**

- Describe the structure, functions, and production of white blood cells.

Types of WBCs

Unlike red blood cells, white blood cells or **leukocytes** (LOO-kō-sīts; *leuko-* = white) have a nucleus and do not contain hemoglobin (Figure 19.7). WBCs are classified as either granular or agranular, depending on whether they contain conspicuous chemical-filled cytoplasmic vesicles (originally called granules) that are made visible by staining. *Granular leukocytes* include

neutrophils, eosinophils, and basophils; *agranular leukocytes* include lymphocytes and monocytes. As shown in Figure 19.3, monocytes and granular leukocytes develop from a myeloid stem cell. In contrast, lymphocytes develop from a lymphoid stem cell.

Granular Leukocytes

After staining, each of the three types of granular leukocytes displays conspicuous granules with distinctive coloration that can be recognized under a light microscope. The large, uniform-sized granules of an **eosinophil** (ē-ō-SIN-ō-fil) are *eosinophilic* (= eosin-loving)—they stain red-orange with acidic dyes (Figure 19.7a). The granules usually do not cover or obscure the nucleus, which most often has two or three lobes connected by either a thin strand or a thick strand. The round, variable-sized granules of a **basophil** (BĀ-sō-fil) are *basophilic* (= basic loving)—they stain blue-purple with basic dyes (Figure 19.7b). The granules commonly obscure the nucleus, which has two lobes. The granules of a **neutrophil** (NOO-trō-fil) are smaller, evenly distributed, and pale lilac in color (Figure 19.7c); the nucleus has two to five lobes, connected by very thin strands of chromatin. As the cells age, the number of nuclear lobes increases. Because older neutrophils thus have several differently shaped nuclear lobes, they are often called *polymorphonuclear leukocytes (PMNs)*, polymorphs, or "polys." Younger neutrophils are often called *bands* because their nucleus is more rod-shaped.

Agranular Leukocytes

Even though so-called agranular leukocytes possess cytoplasmic granules, the granules are not visible under a light microscope because of their small size and poor staining qualities.

The nucleus of a **lymphocyte** (LIM-fō-sīt) is round or slightly indented and stains dark. The cytoplasm stains sky blue and forms a rim around the nucleus. The larger the cell, the more cytoplasm is visible. Lymphocytes are classified as large or small based on cell diameter: 6–9 μm in small lymphocytes and 10–14 μm in large lymphocytes (Figure 19.7d). (Although the functional significance of the size difference between small and large lymphocytes is unclear, the distinction is still useful clinically because an increase in the number of large lymphocytes has diagnostic significance in acute viral infections and in some immunodeficiency diseases.)

Monocytes (MON-ō-sīts) are 12–20 μm in diameter (Figure 19.7e). The nucleus of a monocyte is usually kidney shaped or horseshoe shaped, and the cytoplasm is blue-gray and has a foamy appearance. The color and appearance are due to very fine *azurophilic granules* (az'-ū-rō-FIL-ik; *azur-* = blue; *-philic* = loving), which are lysosomes. The blood is merely a conduit for monocytes, which migrate from the blood into the tissues, where they enlarge and differentiate into **macrophages** (= large eaters). Some become **fixed macrophages,** which means they reside in a particular tissue; examples are alveolar macrophages in the lungs, macrophages in the spleen, and stellate reticuloendothelial (Kupffer) cells in the liver. Others become **wandering**

Figure 19.7 Types of white blood cells.

🔑 The shape of their nuclei and the staining properties of their cytoplasmic granules distinguish white blood cells from one another.

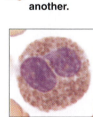
(a) Eosinophil

(b) Basophil

(c) Neutrophil

(d) Small lymphocyte

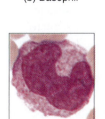
(e) Monocyte

LM all 1420x

❓ Which WBCs are called granular leukocytes? Why?

macrophages, which roam the tissues and gather at sites of infection or inflammation.

White blood cells and other nucleated body cells have proteins, called *major histocompatibility (MHC) antigens,* protruding from their plasma membrane into the extracellular fluid. These "cell identity markers" are unique for each person (except identical twins). Although RBCs possess blood group antigens, they lack the MHC antigens.

Functions of WBCs

In a healthy body, some WBCs, especially lymphocytes, can live for several months or years, but most live only a few days. During a period of infection, phagocytic WBCs may live only a few hours. WBCs are far less numerous than red blood cells, about 5000–10,000 cells per μL of blood. RBCs therefore outnumber white blood cells by about 700:1. **Leukocytosis** (loo′-kō-sī-TŌ-sis), an increase in the number of WBCs above 10,000/μm, is a normal, protective response to stresses such as invading microbes, strenuous exercise, anesthesia, and surgery. An abnormally low level of white blood cells (below 5000/μL) is termed **leukopenia** (loo′-kō-PĒ-nē-a). It is never beneficial and may be caused by radiation, shock, and certain chemotherapeutic agents.

The skin and mucous membranes of the body are continuously exposed to microbes and their toxins. Some of these microbes can invade deeper tissues to cause disease. Once pathogens enter the body, the general function of white blood cells is to combat them by phagocytosis or immune responses. To accomplish these tasks, many WBCs leave the bloodstream and collect at points of pathogen invasion or inflammation. Once granulocytes and monocytes leave the bloodstream to fight injury or infection, they never return to it. Lymphocytes, on the other hand, continually recirculate—from blood to interstitial spaces of tissues to lymphatic fluid and back to blood. Only 2% of the total lymphocyte population is circulating in the blood at any given time; the rest are in lymphatic fluid and organs such as the skin, lungs, lymph nodes, and spleen.

WBCs leave the bloodstream by a process termed **emigration** (em′-i-GRĀ-shun; *e-* = out; *migra-* = wander) in which they roll along the endothelium, stick to it, and then squeeze between endothelial cells (Figure 19.8). The precise signals that stimulate emigration through a particular blood vessel vary for the different types of WBCs. Molecules known as **adhesion molecules** help WBCs stick to the endothelium. For example, endothelial cells display adhesion molecules called *selectins* in response to nearby injury and inflammation. Selectins stick to carbohydrates on the surface of neutrophils, causing them to slow down and roll along the endothelial surface. On the neutrophil surface are other adhesion molecules called *integrins,* which tether neutrophils to the endothelium and assist their movement through the blood vessel wall and into the interstitial fluid of the injured tissue.

Neutrophils and macrophages are active in **phagocytosis;** they can ingest bacteria and dispose of dead matter (see Figure 3.14 on page 73). Several different chemicals released by

Figure 19.8   Emigration of white blood cells.

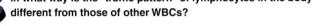 Adhesion molecules (selectins and integrins) assist the emigration of WBCs from the bloodstream into interstitial fluid.

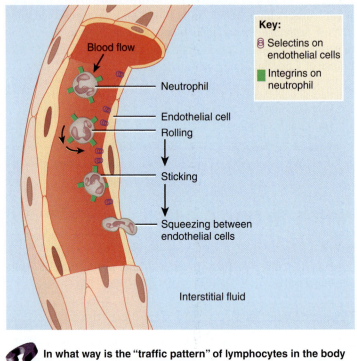

In what way is the "traffic pattern" of lymphocytes in the body different from those of other WBCs?

microbes and inflamed tissues attract phagocytes, a phenomenon called **chemotaxis.** Among the substances that provide stimuli for chemotaxis are toxins produced by microbes; kinins, which are specialized products of damaged tissues; and some of the colony-stimulating factors. The CSFs also enhance the phagocytic activity of neutrophils and macrophages.

Among WBCs, neutrophils respond most quickly to tissue destruction by bacteria. After engulfing a pathogen during phagocytosis, a neutrophil unleashes several chemicals to destroy the pathogen. These chemicals include the enzyme **lysozyme,** which destroys certain bacteria, and **strong oxidants,** such as the superoxide anion (O_2^-), hydrogen peroxide (H_2O_2), and the hypochlorite anion (OCl^-), which is similar to household bleach. Neutrophils also contain **defensins,** proteins that exhibit a broad range of antibiotic activity against bacteria and fungi. Within a neutrophil, vesicles containing defensins merge with phagosomes containing microbes. Defensins form peptide "spears" that poke holes in microbe membranes; the resulting loss of cellular contents kills the invader.

Monocytes take longer to reach a site of infection than do neutrophils, but they arrive in larger numbers and destroy more microbes. Upon arrival they enlarge and differentiate into wandering macrophages, which clean up cellular debris and microbes by phagocytosis after an infection.

Eosinophils leave the capillaries and enter tissue fluid. They are believed to release enzymes, such as histaminase, that combat the effects of histamine and other mediators of inflammation in allergic reactions. Eosinophils also phagocytize antigen–antibody complexes and are effective against certain parasitic worms. A high eosinophil count often indicates an allergic condition or a parasitic infection.

At sites of inflammation, basophils leave capillaries, enter tissues, and liberate heparin, histamine, and serotonin. These substances intensify the inflammatory reaction and are involved in hypersensitivity (allergic) reactions. Basophils are similar in function to mast cells, connective tissue cells that originate from pluripotent stem cells in red bone marrow. Like basophils, mast cells liberate mediators in inflammation, including heparin, histamine, and proteases. Mast cells are widely dispersed in the body, particularly in connective tissues of the skin and mucous membranes of the respiratory and gastrointestinal tracts.

Lymphocytes are the major soldiers in immune system battles (described in detail in Chapter 22). Three main types of lymphocytes are B cells, T cells, and natural killer cells. B cells are particularly effective in destroying bacteria and inactivating their toxins. T cells attack viruses, fungi, transplanted cells, cancer cells, and some bacteria. Immune responses carried out by B cells and T cells help combat infection and provide protection against some diseases. T cells also are responsible for transfusion reactions, allergies, and rejection of transplanted organs. Natural killer cells attack a wide variety of infectious microbes and certain spontaneously arising tumor cells.

An increase in the number of circulating WBCs usually indicates inflammation or infection. A physician may order a **differential white blood cell count** to detect infection or inflammation, determine the effects of possible poisoning by chemicals or drugs, monitor blood disorders (for example, leukemia) and effects of chemotherapy, or detect allergic reactions and parasitic infections. Because each type of white blood cell plays a different role, determining the *percentage* of each type in the blood assists in diagnosing the condition. Table 19.2 lists the significance of both elevated and depressed WBC counts.

Bone Marrow Transplant

A **bone marrow transplant** is the intravenous transfer of red bone marrow from a healthy donor to a recipient, with the goal of establishing normal hemopoiesis and consequently normal blood cell counts in the recipient. In patients with cancer or certain genetic diseases, the defective red bone marrow first must be destroyed by high doses of chemotherapy and whole body radiation just before the transplant takes place. The donor marrow must be very closely matched to that of the recipient, or rejection occurs. When a transplant is successful, stem cells from the donor marrow reseed and grow in the recipient's red bone marrow cavities. Bone marrow transplants have been used to treat aplastic anemia, certain types of leukemia, severe combined immunodeficiency disease (SCID), Hodgkin's disease, non-Hodgkin's lymphoma, multiple myeloma, thalassemia, sickle-cell disease, breast cancer, ovarian cancer, testicular cancer, and hemolytic anemia. ■

Table 19.2	Significance of High and Low White Blood Cell Counts	
WBC Type	**High Count May Indicate**	**Low Count May Indicate**
Neutrophils	Bacterial infection, burns, stress, inflammation.	Radiation exposure, drug toxicity, vitamin B_{12} deficiency, and systemic lupus erythematosus (SLE).
Lymphocytes	Viral infections, some leukemias.	Prolonged illness, immunosuppression, and treatment with cortisol.
Monocytes	Viral or fungal infections, tuberculosis, some leukemias, other chronic diseases.	Bone marrow suppression, treatment with cortisol.
Eosinophils	Allergic reactions, parasitic infections, autoimmune diseases.	Drug toxicity, stress.
Basophils	Allergic reactions, leukemias, cancers, hypothyroidism.	Pregnancy, ovulation, stress, and hyperthyroidism.

▶ **CHECKPOINT**

10. What is the importance of emigration, chemotaxis, and phagocytosis in fighting bacterial invaders?
11. How are leukocytosis and leukopenia different?
12. What is a differential white blood cell count?
13. The first letter of each word in the phrase "*N*ever *l*et *m*onkeys *e*at *b*ananas!" gives the order of WBCs from most numerous to least numerous. Which WBCs do n, l, m, e, and b stand for?
14. What functions do B cells, T cells, and natural killer cells perform?

PLATELETS

▶ **OBJECTIVE**

• **Describe the structure, function, and origin of platelets.**

Besides the immature cell types that develop into erythrocytes and leukocytes, hemopoietic stem cells also differentiate into cells that produce platelets. Under the influence of the hormone **thrombopoietin,** myeloid stem cells develop into megakaryocyte-colony-forming cells that, in turn, develop into precursor cells called megakaryoblasts (see Figure 19.3). Megakaryoblasts transform into megakaryocytes, huge cells that splinter into 2000 to 3000 fragments. Each fragment, enclosed by a piece of the plasma membrane, is a **platelet (thrombocyte).** Platelets break off from the megakaryocytes in red bone marrow and then enter the blood circulation. Between 150,000 and 400,000 platelets are present in each μL of blood. They are disc-shaped, 2–4 μm in diameter, and have many vesicles but no nucleus.

Platelets help stop blood loss from damaged blood vessels by forming a platelet plug. Their granules also contain chemicals that, once released, promote blood clotting. Platelets have a short life span, normally just 5 to 9 days. Aged and dead platelets are removed by fixed macrophages in the spleen and liver.

Table 19.3 summarizes the formed elements in blood.

Complete Blood Count

A **complete blood count** (CBC) is a valuable test that screens for anemia and various infections. Usually included are counts of RBCs, WBCs, and platelets per μL of whole blood; hematocrit; and differential white blood cell count. The amount of hemoglobin in grams per milliliter of blood also is determined.

Table 19.3   Summary of Formed Elements in Blood

Name and Appearance	Number	Characteristics*	Functions
Red blood cells (RBCs) or **erythrocytes**	4.8 million/μL in females; 5.4 million/μL in males.	7–8 μm diameter, biconcave discs, without a nucleus; live for about 120 days.	Hemoglobin within RBCs transports most of the oxygen and part of the carbon dioxide in the blood.
White blood cells (WBCs) or **leukocytes**	5000–10,000/μL.	Most live for a few hours to a few days.[†]	Combat pathogens and other foreign substances that enter the body.
Granular leukocytes			
Neutrophils	60–70% of all WBCs.	10–12 μm diameter; nucleus has 2–5 lobes connected by thin strands of chromatin; cytoplasm has very fine, pale lilac granules.	Phagocytosis. Destruction of bacteria with lysozyme, defensins, and strong oxidants, such as superoxide anion, hydrogen peroxide, and hypochlorite anion.
Eosinophils	2–4% of all WBCs.	10–12 μm diameter; nucleus has 2 or 3 lobes; large, red-orange granules fill the cytoplasm.	Combat the effects of histamine in allergic reactions, phagocytize antigen–antibody complexes, and destroy certain parasitic worms.
Basophils	0.5–1% of all WBCs.	8–10 μm diameter; nucleus has 2 lobes; large cytoplasmic granules appear deep blue-purple.	Liberate heparin, histamine, and serotonin in allergic reactions that intensify the overall inflammatory response.
Agranular leukocytes			
Lymphocytes (T cells, B cells, and natural killer cells)	20–25% of all WBCs.	Small lymphocytes are 6–9 μm in diameter; large lymphocytes are 10–14 μm in diameter; nucleus is round or slightly indented; cytoplasm forms a rim around the nucleus that looks sky blue; the larger the cell, the more cytoplasm is visible.	Mediate immune responses, including antigen–antibody reactions. B cells develop into plasma cells, which secrete antibodies. T cells attack invading viruses, cancer cells, and transplanted tissue cells. Natural killer cells attack a wide variety of infectious microbes and certain spontaneously arising tumor cells.
Monocytes	3–8% of all WBCs.	12–20 μm diameter; nucleus is kidney shaped or horseshoe shaped; cytoplasm is blue-gray and has foamy appearance.	Phagocytosis (after transforming into fixed or wandering macrophages).
Platelets (thrombocytes)	150,000–400,000/μL.	2–4 μm diameter cell fragments that live for 5–9 days; contain many vesicles but no nucleus.	Form platelet plug in hemostasis; release chemicals that promote vascular spasm and blood clotting.

*Colors are those seen when using Wright's stain.
[†]Some lymphocytes, called T and B memory cells, can live for many years once they are established.

Normal hemoglobin ranges are: infants, 14–20 g/100 mL of blood; adult females, 12–16 g/100 mL of blood; and adult males, 13.5–18 g/100 mL of blood. ■

▶ **CHECKPOINT**

15. How do RBCs, WBCs, and platelets compare with respect to size, number per μL of blood, and life span?

HEMOSTASIS

▶ **OBJECTIVES**

- **Describe the three mechanisms that contribute to hemostasis.**
- **Identify the stages of blood clotting and explain the various factors that promote and inhibit blood clotting.**

Hemostasis (hē-mō-STĀ-sis) is a sequence of responses that stops bleeding. When blood vessels are damaged or ruptured, the hemostatic response must be quick, localized to the region of damage, and carefully controlled. Three mechanisms reduce blood loss: (1) vascular spasm, (2) platelet plug formation, and (3) blood clotting (coagulation). When successful, hemostasis prevents **hemorrhage** (HEM-o-rij; -*rhage* = burst forth), the loss of a large amount of blood from the vessels. Hemostatic mechanisms can prevent hemorrhage from smaller blood vessels, but extensive hemorrhage from larger vessels usually requires medical intervention.

Vascular Spasm

When arteries or arterioles are damaged, the circularly arranged smooth muscle in their walls contracts immediately. Such a **vascular spasm** reduces blood loss for several minutes to several hours, during which time the other hemostatic mechanisms go into operation. The spasm is probably caused by damage to the smooth muscle, by substances released from activated platelets, and by reflexes initiated by pain receptors.

Platelet Plug Formation

Considering their small size, platelets store an impressive array of chemicals. Within many vesicles are clotting factors, ADP, ATP, Ca^{2+}, and serotonin. Also present are enzymes that produce thromboxane A2, a prostaglandin; *fibrin-stabilizing factor,* which helps to strengthen a blood clot; lysosomes; some mitochondria; membrane systems that take up and store calcium and provide channels for release of the contents of granules; and glycogen. Also within platelets is **platelet-derived growth factor (PDGF),** a hormone that can cause proliferation of vascular endothelial cells, vascular smooth muscle fibers, and fibroblasts to help repair damaged blood vessel walls.

Platelet plug formation occurs as follows (Figure 19.9):

1 Initially, platelets contact and stick to parts of a damaged blood vessel, such as collagen fibers of the connective tissue underlying the damaged endothelial cells. This process is called **platelet adhesion.**

Figure 19.9 Platelet plug formation.

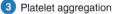 A platelet plug can stop blood loss completely if the hole in a blood vessel is small enough.

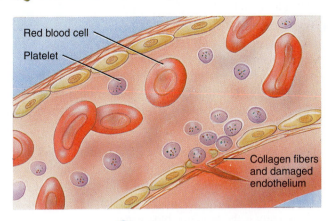

Red blood cell
Platelet
Collagen fibers and damaged endothelium

1 Platelet adhesion

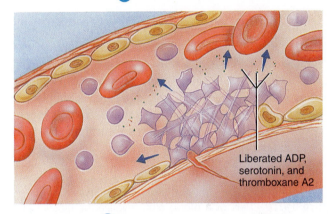

Liberated ADP, serotonin, and thromboxane A2

2 Platelet release reaction

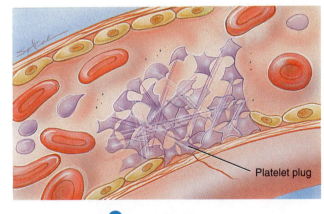

Platelet plug

3 Platelet aggregation

 Besides platelet plug formation, which two mechanisms also contribute to hemostasis?

2 Due to adhesion, the platelets become activated, and their characteristics change dramatically. They extend many projections that enable them to contact and interact with one another, and they begin to liberate the contents of their vesicles. This phase is called the **platelet release reaction.** Liberated ADP and thromboxane A2 play a major role by activating nearby platelets. Serotonin and thromboxane A2 function as vasoconstrictors, causing and sustaining contraction of vascular smooth muscle, which decreases blood flow through the injured vessel.

3 The release of ADP makes other platelets in the area sticky, and the stickiness of the newly recruited and activated platelets causes them to adhere to the originally activated platelets. This gathering of platelets is called **platelet aggregation.** Eventually, the accumulation and attachment of large numbers of platelets form a mass called a **platelet plug.**

A platelet plug is very effective in preventing blood loss in a small vessel. Although initially the platelet plug is loose, it becomes quite tight when reinforced by fibrin threads formed during clotting (see Figure 19.10). A platelet plug can stop blood loss completely if the hole in a blood vessel is not too large.

Blood Clotting

Normally, blood remains in its liquid form as long as it stays within its vessels. If it is drawn from the body, however, it thickens and forms a gel. Eventually, the gel separates from the liquid. The straw-colored liquid, called **serum,** is simply plasma minus the clotting proteins. The gel is called a **clot.** It consists of a network of insoluble protein fibers called fibrin in which the formed elements of blood are trapped (Figure 19.10).

Figure 19.10 Part of a blood clot. Notice the platelet and red blood cells entrapped in fibrin threads.

A blood clot is a gel that contains formed elements of the blood entangled in fibrin threads.

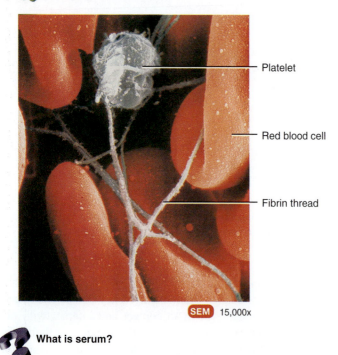

— Platelet

— Red blood cell

— Fibrin thread

SEM 15,000x

? What is serum?

The process of gel formation, called **clotting** or **coagulation,** is a series of chemical reactions that culminates in formation of fibrin threads. If blood clots too easily, the result can be **thrombosis**—clotting in an undamaged blood vessel. If the blood takes too long to clot, hemorrhage can occur.

Clotting involves several substances known as **clotting (coagulation) factors.** These factors include calcium ions (Ca^{2+}), several inactive enzymes that are synthesized by hepatocytes (liver cells) and released into the bloodstream, and various molecules associated with platelets or released by damaged tissues. Most clotting factors are identified by Roman numerals that indicate the order of their discovery, and not necessarily the order of their participation in the clotting process.

Clotting is a complex cascade of enzymatic reactions in which each clotting factor activates many molecules of the next one in a fixed sequence. Finally, a large quantity of product (the insoluble protein fibrin) is formed. Clotting can be divided into three stages (Figure 19.11):

1 First, several reactions of the clotting sequence end in formation of prothrombinase. Two pathways, called the extrinsic pathway (Figure 19.11a) and the intrinsic pathway (Figure 19.11b), lead to prothrombinase. Once prothrombinase is formed, the steps involved in the next two stages of clotting are the same for both the extrinsic and intrinsic pathways, and together these two stages are referred to as the common pathway.

2 Prothrombinase converts prothrombin (a plasma protein formed by the liver) into the enzyme thrombin.

3 Thrombin converts soluble fibrinogen (another plasma protein formed by the liver) into insoluble fibrin. Fibrin forms the threads of the clot.

The Extrinsic Pathway

The **extrinsic pathway** of blood clotting has fewer steps than the intrinsic pathway and occurs rapidly—within a matter of seconds if trauma is severe. It is so named because a tissue protein called **tissue factor (TF),** also known as **thromboplastin,** leaks into the blood from cells *outside (extrinsic to)* blood vessels and initiates the formation of prothrombinase. TF is a complex mixture of lipoproteins and phospholipids released from the surfaces of damaged cells. In the presence of Ca^{2+}, TF begins a sequence of reactions that ultimately activates clotting factor X (Figure 19.11a). Once factor X is activated, it combines with factor V in the presence of Ca^{2+} to form the active enzyme prothrombinase, completing the extrinsic pathway.

The Intrinsic Pathway

The **intrinsic pathway** of blood clotting is more complex than the extrinsic pathway, and it occurs more slowly, usually requiring several minutes. The intrinsic pathway is so-named because its activators are either in direct contact with blood or contained *within (intrinsic to)* the blood; outside tissue damage is not needed. If endothelial cells become roughened or damaged,

Figure 20.3 Structure of the heart: surface features.

Sulci and grooves that contain blood vessels and fat and mark the boundaries between the various chambers.

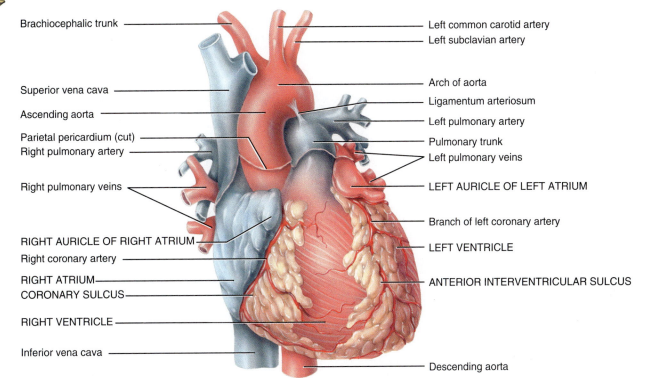

Brachiocephalic trunk
Left common carotid artery
Left subclavian artery
Superior vena cava
Arch of aorta
Ascending aorta
Ligamentum arteriosum
Left pulmonary artery
Parietal pericardium (cut)
Right pulmonary artery
Pulmonary trunk
Left pulmonary veins
Right pulmonary veins
LEFT AURICLE OF LEFT ATRIUM
RIGHT AURICLE OF RIGHT ATRIUM
Branch of left coronary artery
Right coronary artery
LEFT VENTRICLE
RIGHT ATRIUM
CORONARY SULCUS
ANTERIOR INTERVENTRICULAR SULCUS
RIGHT VENTRICLE
Inferior vena cava
Descending aorta

(a) Anterior external view showing surface features

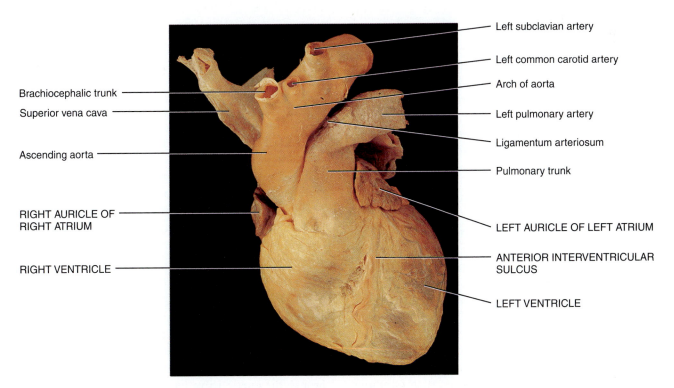

Left subclavian artery
Left common carotid artery
Arch of aorta
Brachiocephalic trunk
Superior vena cava
Left pulmonary artery
Ligamentum arteriosum
Ascending aorta
Pulmonary trunk
RIGHT AURICLE OF
RIGHT ATRIUM
LEFT AURICLE OF LEFT ATRIUM
RIGHT VENTRICLE
ANTERIOR INTERVENTRICULAR
SULCUS
LEFT VENTRICLE

(b) Anterior external view showing surface features

(continues)

(Figure 20.3 continued)

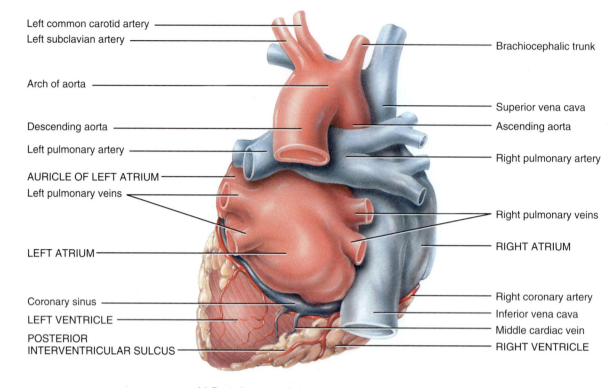

Left common carotid artery
Left subclavian artery
Arch of aorta
Descending aorta
Left pulmonary artery
AURICLE OF LEFT ATRIUM
Left pulmonary veins
LEFT ATRIUM
Coronary sinus
LEFT VENTRICLE
POSTERIOR INTERVENTRICULAR SULCUS

Brachiocephalic trunk
Superior vena cava
Ascending aorta
Right pulmonary artery
Right pulmonary veins
RIGHT ATRIUM
Right coronary artery
Inferior vena cava
Middle cardiac vein
RIGHT VENTRICLE

(c) Posterior external view showing surface features

The coronary sulcus forms a boundary between which chambers of the heart?

the remnant of the foramen ovale, an opening in the inter-atrial septum of the fetal heart that normally closes soon after birth (see Figure 21.31 on page 754). Blood passes from the right atrium into the right ventricle through a valve called the **tricuspid valve** (trī-KUS-pid; *tri-* = three; *cuspid* = point) because it consists of three leaflets or cusps (Figure 20.4a). The valves of the heart are composed of dense connective tissue covered by endocardium.

Right Ventricle

The **right ventricle** forms most of the anterior surface of the heart. The inside of the right ventricle contains a series of ridges formed by raised bundles of cardiac muscle fibers called **trabeculae carneae** (tra-BEK-ū-lē KAR-nē-ē; *trabec-ulae* = little beams; *carneae* = fleshy). Some of the trabeculae carneae convey part of the conduction system of the heart (see page 674). The cusps of the tricuspid valve are connected to tendonlike cords, the **chordae tendineae** (KOR-dē ten-DIN-ē-ē; *chord-* = cord; *tend-* = tendon), which, in turn, are connected to cone-shaped trabeculae carneae called **papillary muscles** (*papill-* = nipple). The right ventricle is separated from the left ventricle by a partition called the **interventricular septum.** Blood passes from the right ventricle through the

pulmonary valve into a large artery called the *pulmonary trunk,* which divides into right and left *pulmonary arteries.*

Left Atrium

The **left atrium** forms most of the base of the heart. It receives blood from the lungs through four *pulmonary veins.* Like the right atrium, the inside of the left atrium has a smooth posterior wall. Because pectinate muscles are confined to the auricle of the left atrium, the anterior wall of the left atrium also is smooth. Blood passes from the left atrium into the left ventricle through the **bicuspid (mitral) valve** (*bi-* = two), which has two cusps. The term *mitral* refers to its resemblance to a bishop's miter (hat), which is two-sided.

Left Ventricle

The **left ventricle** forms the apex of the heart. Like the right ventricle, the left ventricle also contains trabeculae carneae and has chordae tendinae that anchor the cusps of the bicuspid valve to papillary muscles. Blood passes from the left ventricle through the **aortic valve** into the largest artery of the body, the *ascending aorta* (*aorte* = to suspend, because the aorta once was believed to lift up the heart). Some of the blood in the aorta flows into the *coronary arteries,* which branch from the

Figure 20.4 Structure of the heart: internal anatomy.

Blood flows into the right atrium through the superior vena cava, inferior vena cava, and coronary sinus and into the left atrium through four pulmonary veins.

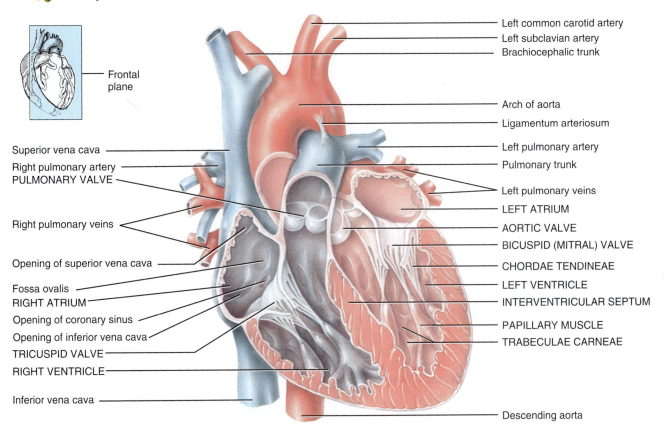

Frontal plane

Left common carotid artery
Left subclavian artery
Brachiocephalic trunk

Arch of aorta
Ligamentum arteriosum

Superior vena cava
Right pulmonary artery
PULMONARY VALVE

Left pulmonary artery
Pulmonary trunk
Left pulmonary veins
LEFT ATRIUM
AORTIC VALVE
BICUSPID (MITRAL) VALVE
CHORDAE TENDINEAE
LEFT VENTRICLE
INTERVENTRICULAR SEPTUM
PAPILLARY MUSCLE
TRABECULAE CARNEAE

Right pulmonary veins

Opening of superior vena cava

Fossa ovalis
RIGHT ATRIUM
Opening of coronary sinus
Opening of inferior vena cava
TRICUSPID VALVE
RIGHT VENTRICLE

Inferior vena cava

Descending aorta

(a) Anterior view of frontal section showing internal anatomy

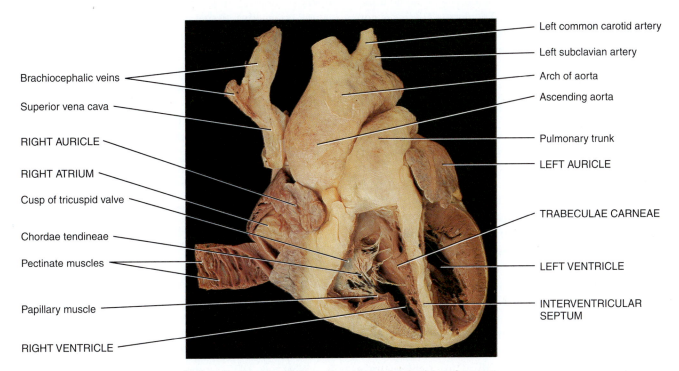

Brachiocephalic veins

Superior vena cava

RIGHT AURICLE

RIGHT ATRIUM

Cusp of tricuspid valve

Chordae tendineae

Pectinate muscles

Papillary muscle

RIGHT VENTRICLE

Left common carotid artery
Left subclavian artery
Arch of aorta
Ascending aorta

Pulmonary trunk

LEFT AURICLE

TRABECULAE CARNEAE

LEFT VENTRICLE

INTERVENTRICULAR SEPTUM

(b) Anterior view of partially sectioned heart showing internal anatomy

(continues)

(Figure 20.4 continued)

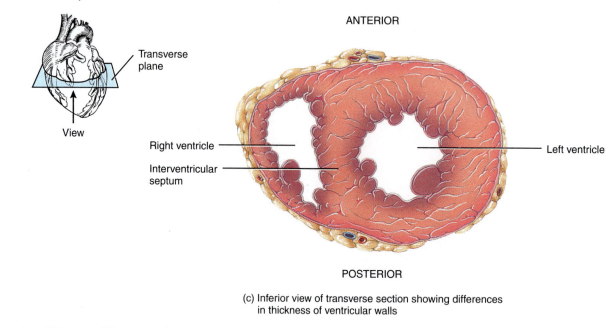

(c) Inferior view of transverse section showing differences in thickness of ventricular walls

How does thickness of the myocardium relate to the workload of a cardiac chamber?

ascending aorta and carry blood to the heart wall. The remainder of the blood passes into the *arch of the aorta* and *descending aorta* (*thoracic aorta* and *abdominal aorta*). Branches of the arch of the aorta and descending aorta carry blood throughout the body.

During fetal life, a temporary blood vessel, called the *ductus arteriosus,* shunts blood from the pulmonary trunk into the aorta. Hence, only a small amount of blood enters the nonfunctioning fetal lungs (see Figure 21.31). The ductus arteriosus normally closes shortly after birth, leaving a remnant known as the **ligamentum arteriosum,** which connects the arch of the aorta and pulmonary trunk (Figure 20.4a).

Myocardial Thickness and Function

The thickness of the myocardium of the four chambers varies according to each chamber's function. The thin-walled atria deliver blood into the adjacent ventricles. Because the ventricles pump blood greater distances, their walls are thicker (Figure 20.4a). Although the right and left ventricles act as two separate pumps that simultaneously eject equal volumes of blood, the right side has a much smaller workload. It pumps blood a short distance to the lungs at lower pressure and the resistance to blood flow is small. The left ventricle pumps blood great distances to all other parts of the body at higher pressure, and the resistance to blood flow is larger. Therefore, the left ventricle works much harder than the right ventricle to maintain the same rate of blood flow. The anatomy of the two ventricles confirms this functional difference—the muscular wall of the left ventricle is considerably thicker than the wall of the right ventricle (Figure 20.4c). Note also that the perimeter of the lumen (space)

of the left ventricle is roughly circular, whereas that of the right ventricle is somewhat crescent shaped.

Fibrous Skeleton of the Heart

In addition to cardiac muscle tissue, the heart wall also contains dense connective tissue that forms the **fibrous skeleton of the heart** (Figure 20.5). Essentially, the fibrous skeleton consists of dense connective tissue rings that surround the valves of the heart, fuse with one another, and merge with the interventricular septum. Four fibrous rings support the four valves of the heart and are fused to each other. The fibrous skeleton forms the foundation to which the heart valves attach and prevents overstretching of the valves as blood passes through them. It also serves as a point of insertion for bundles of cardiac muscle fibers and acts as an electrical insulator between the atria and ventricles.

▶ **C H E C K P O I N T**

1. Where are the apex and base of the heart?

2. Define each of the following external features of the heart: auricle, coronary sulcus, anterior interventricular sulcus, and posterior interventricular sulcus.

3. What are the characteristic internal features of each chamber of the heart?

4. Which blood vessels deliver blood to the right and left atria?

5. What is the relationship between wall thickness and function for the chambers of the heart?

6. What type of tissue composes the fibrous skeleton of the heart? What functions does this tissue perform?

Figure 20.5 Fibrous skeleton of the heart (depicted in blue).

 Fibrous rings support the four valves of the heart and are fused to each other.

Pulmonary valve —

Left coronary artery —

Aortic valve —

LEFT FIBROUS TRIGONE —

RIGHT FIBROUS TRIGONE —

Bicuspid valve —

LEFT ATRIOVENTRICULAR FIBROUS RING —

— PULMONARY FIBROUS RING

— CONUS TENDON

— AORTIC FIBROUS RING

— Right coronary artery

— Tricuspid valve

— RIGHT ATRIOVENTRICULAR FIBROUS RING

Superior view (the atria have been removed)

In what two ways does the fibrous skeleton contribute to the functioning of heart valves?

HEART VALVES AND CIRCULATION OF BLOOD

OBJECTIVES

- Describe the structure and function of the valves of the heart.
- Describe the flow of blood through the chambers of the heart and through the systemic and pulmonary circulations.
- Discuss the coronary circulation.

As each chamber of the heart contracts, it pushes a volume of blood into a ventricle or out of the heart into an artery. Valves open and close in response to *pressure changes* as the heart contracts and relaxes. Each of the four valves helps ensure the one-

way flow of blood by opening to let blood through and then closing to prevent the backflow of blood.

Operation of the Atrioventricular Valves

Because they are located between an atrium and a ventricle, the tricuspid and bicuspid valves are termed **atrioventricular (AV) valves.** When an AV valve is open, the pointed ends of the cusps project into the ventricle. Blood moves from the atria into the ventricles through open AV valves when atrial pressure is higher than ventricular pressure (Figure 20.6a, c). At this time, the papillary muscles are relaxed, and the chordae tendineae are slack. When the ventricles contract, the pressure of the blood drives the

Figure 20.6 Responses of the valves to the pumping of the heart.

Heart valves prevent the backflow of blood.

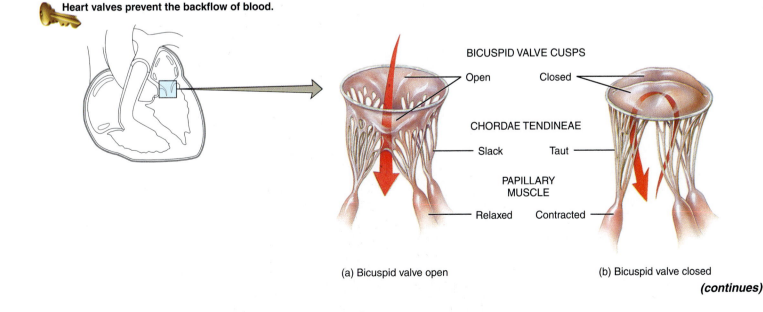

BICUSPID VALVE CUSPS

Open   Closed

CHORDAE TENDINEAE

Slack   Taut

PAPILLARY MUSCLE

Relaxed   Contracted

(a) Bicuspid valve open

(b) Bicuspid valve closed

(continues)

(Figure 20.6 continued)

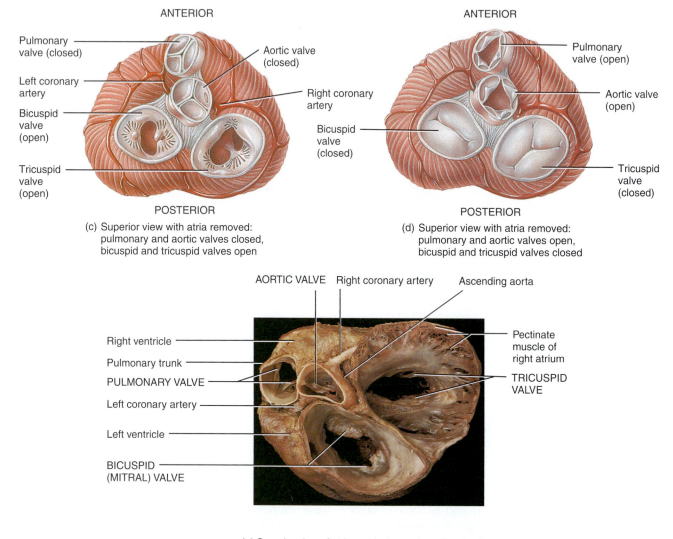

(c) Superior view with atria removed: pulmonary and aortic valves closed, bicuspid and tricuspid valves open

(d) Superior view with atria removed: pulmonary and aortic valves open, bicuspid and tricuspid valves closed

(e) Superior view of atrioventricular and semilunar valves

How do papillary muscles prevent AV valve cusps from everting or swinging upward into the atria?

cusps upward until their edges meet and close the opening (Figure 20.6b, d). At the same time, the papillary muscles are also contracting, which pulls·on and tightens the chordae tendineae. This prevents the valve cusps from everting (being forced to open in the opposite direction into the atria due to the high ventricular pressure). If the AV valves or chordae tendineae are damaged, blood may regurgitate (flow back) into the atria when the ventricles contract.

Operation of the Semilunar Valves

The aortic and pulmonary valves are known as the **semilunar (SL) valves** (*semi-* = half; *lunar* = moon-shaped) because they are made up of three crescent moon–shaped cusps (Figure 20.6c). Each cusp attaches to the artery wall by its convex outer margin. The SL valves allow ejection of blood from the heart into arteries but prevent backflow of blood into the ventricles. The free borders of the cusps project into the lumen of the

artery. When the ventricles contract, pressure builds up within the chambers. The semilunar valves open when pressure in the ventricles exceeds the pressure in the arteries, permitting ejection of blood from the ventricles into the pulmonary trunk and aorta (Figure 20.6d). As the ventricles relax, blood starts to flow back toward the heart. As the back-flowing blood fills the cusps, the semilunar valves close (Figure 20.6c).

Surprisingly perhaps, valves do not guard the entrance of the venae cavae into the right atrium or the pulmonary veins into the left atrium. As the atria contract, a small amount of blood does flow backward into these vessels. However, as the atrial muscle contracts, it compresses and nearly collapses the venous entry points thus minimizing backflow of blood.

Heart Valve Disorders

When heart valves operate normally, they open fully and close completely at the proper times. Failure of a heart valve to open fully is known as **stenosis** (= a narrowing) whereas failure of a valve to close completely is termed **insufficiency** or **incompetence.** In **mitral stenosis,** scar formation or a congenital defect causes narrowing of the mitral valve. **Mitral insufficiency** allows backflow of blood from the left ventricle into the left atrium. One cause is **mitral valve prolapse (MVP),** in which one or both cusps of the mitral valve protrude into the left atrium during ventricular contraction. Although a small volume of blood may flow back into the left atrium during ventricular contraction, mitral valve prolapse does not always pose a serious threat. Mitral insufficiency also can be due to a damaged mitral valve or ruptured chordae tendineae. In **aortic stenosis** the aortic valve is narrowed, and in **aortic insufficiency** there is backflow of blood from the aorta into the left ventricle.

Certain infectious diseases can damage or destroy the heart valves. One example is **rheumatic fever,** an acute systemic inflammatory disease that usually occurs after a streptococcal infection of the throat. The bacteria trigger an immune response in which antibodies that are produced to destroy the bacteria attack and inflame the connective tissues in joints, heart valves, and other organs. Even though rheumatic fever may weaken the entire heart wall, most often it damages the bicuspid (mitral) and aortic valves. ■

Systemic and Pulmonary Circulations

With each beat, the heart pumps blood into two closed circuits—the **systemic circulation** and the **pulmonary circulation** (*pulmon-* = lung). The two circuits are arranged in series: the output of one becomes the input of the other, as would happen if you attached two garden hoses together. The left side of the heart is the pump for the systemic circulation; it receives bright red, oxygen-rich blood from the lungs. The left ventricle ejects blood into the *aorta* (Figure 20.7a). From the aorta, the blood divides into separate streams, entering progressively smaller *systemic arteries* that carry it to all organs throughout the body—except for the air sacs (alveoli) of the lungs, which are supplied by the pulmonary circulation. In systemic tissues, arteries give rise to smaller-diameter *arterioles,* which finally lead into extensive beds of *systemic capillaries.* Exchange of nutrients and gases occurs across the thin capillary walls. Blood unloads O_2 (oxygen) and picks up CO_2 (carbon dioxide). In most cases, blood flows through only one capillary and then enters a *systemic venule.* Venules carry deoxygenated (oxygen-poor) blood away from tissues and merge to form larger *systemic veins.* Ultimately the blood flows back to the right atrium.

Figure 20.7   Systemic and pulmonary circulations.   Throughout this book, blood vessels that carry oxygenated blood (which looks bright red) are colored red, whereas those that carry deoxygenated blood (which looks dark red) are colored blue.

The left side of the heart pumps oxygenated blood into the systemic circulation to all tissues of the body except the air sacs (alveoli) of the lungs. The right side of the heart pumps deoxygenated blood into the pulmonary circulation to the air sacs (alveoli) of the lungs.

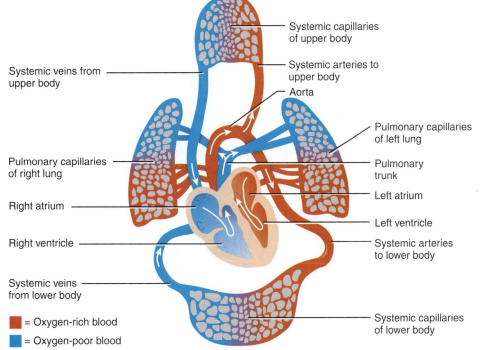

= Oxygen-rich blood

= Oxygen-poor blood

(a) Systemic and pulmonary circulations

(continues)

(Figure 20.7 continued)

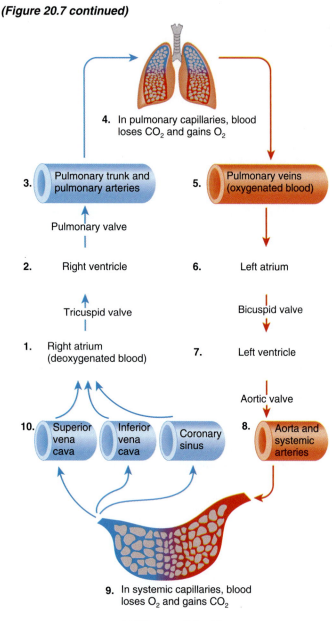

4. In pulmonary capillaries, blood loses CO_2 and gains O_2

3. Pulmonary trunk and pulmonary arteries

5. Pulmonary veins (oxygenated blood)

Pulmonary valve

2. Right ventricle

6. Left atrium

Tricuspid valve

Bicuspid valve

1. Right atrium (deoxygenated blood)

7. Left ventricle

Aortic valve

10. Superior vena cava | Inferior vena cava | Coronary sinus

8. Aorta and systemic arteries

9. In systemic capillaries, blood loses O_2 and gains CO_2

(b) Diagram of blood flow

In part (b), which numbers constitute the pulmonary circulation? Which constitute the systemic circulation?

The right side of the heart is the pump for the pulmonary circulation; it receives all the dark red, deoxygenated blood returning from the systemic circulation. Blood ejected from the right ventricle flows into the *pulmonary trunk,* which branches into *pulmonary arteries* that carry blood to the right and left lungs. In pulmonary capillaries, then, blood unloads CO_2, which is exhaled, and picks up O_2. The freshly oxygenated blood then flows into pulmonary veins and returns to the left atrium. Figure 20.7b reviews the route of blood flow through the chambers and valves of the heart and the pulmonary and systemic circulations.

Coronary Circulation

Nutrients could not diffuse quickly enough from blood in the chambers of the heart to supply all the layers of cells that make up the heart wall. For this reason, the myocardium has its own blood vessels, the **coronary circulation** (*coron-* = crown). The **coronary arteries** branch from the ascending aorta and encircle the heart like a crown encircles the head (Figure 20.8a). While the heart is contracting, little blood flows in the coronary arteries because they are squeezed shut. When the heart relaxes, however, the high pressure of blood in the aorta propels blood through the coronary arteries, into capillaries, and then into **coronary veins** (Figure 20.8b).

Coronary Arteries

Two coronary arteries, the right and left coronary arteries, branch from the ascending aorta and supply oxygenated blood to the myocardium (Figure 20.8a). The **left coronary artery** passes inferior to the left auricle and divides into the anterior interventricular and circumflex branches. The **anterior interventricular branch** or **left anterior descending (LAD) artery** extends along the anterior interventricular sulcus and supplies oxygenated blood to the walls of both ventricles. The **circumflex branch** lies in the coronary sulcus and distributes oxygenated blood to the walls of the left ventricle and left atrium.

The **right coronary artery** gives off small atrial branches that supply the right atrium. It continues inferior to the right auricle and divides into the posterior interventricular and marginal branches. The **posterior interventricular branch** follows the posterior interventricular sulcus and supplies the walls of the two ventricles with oxygenated blood. The **right marginal branch** in the coronary sulcus carries oxygenated blood to the myocardium of the right ventricle.

Most parts of the body receive blood from branches of more than one artery, and where two or more arteries supply the same region, they usually connect. These connections, called **anastomoses** (a-nas′-tō-MŌ-sēs), provide alternate routes for blood to reach a particular organ or tissue. The myocardium contains many anastomoses that connect branches of a given coronary artery or extend between branches of different coronary arteries. They provide detours for arterial blood if a main route becomes obstructed. Thus, heart muscle may receive sufficient oxygen even if one of its coronary arteries is partially blocked.

Coronary Veins

After blood passes through the coronary arteries, it flows into capillaries, where it delivers oxygen and nutrients and collects carbon dioxide and wastes, and then into veins. The deoxygenated blood then drains into a large vascular sinus on the posterior surface of the heart, called the **coronary sinus** (Figure 20.8b), which empties into the right atrium. A vascular sinus is a thin-walled vein that has no smooth muscle to alter its diameter. The main tributaries carrying blood into the coronary sinus are the **great cardiac vein,** which drains the anterior aspect of the heart, and the **middle cardiac vein,** which drains the posterior aspect of the heart.

Reperfusion Damage

When blockage of a coronary artery deprives the heart muscle of oxygen, **reperfusion,** reestablishing the blood flow, may damage the tissue further. This surprising effect is due to the formation of oxygen **free radicals** from the reintroduced oxygen. Free radicals are electrically charged molecules that have an unpaired electron (see Figure 2.3b on page 30). Such molecules are unstable and highly reactive. They cause chain reactions that lead to cellular damage and death. To counter the effects of oxygen free radicals, body cells produce enzymes that convert free radicals to less reactive substances. Two such enzymes are *superoxide dismutase* and *catalase.* In addition, some nutrients, such as vitamin E, vitamin C, beta-carotene, and selenium are antioxidants, which remove oxygen free radicals. Drugs that lessen reperfusion damage after a heart attack or stroke are being developed. ■

▶ **CHECKPOINT**

7. What causes the heart valves to open and to close?

8. In correct sequence, which heart chambers, heart valves, and blood vessels would a drop of blood encounter as it flows out of the right atrium until it reaches the aorta?

9. Which arteries deliver oxygenated blood to the myocardium of the left and right ventricles?

Figure 20.8 **The coronary circulation.** The views of the heart from the anterior aspect in (a) and (b) are drawn as if the heart were transparent to reveal blood vessels on the posterior aspect. (See Tortora, *A Photographic Atlas of the Human Body,* Figures 6.8 and 6.9.)

🔑 **The right and left coronary arteries deliver blood to the heart; the coronary veins drain blood from the heart into the coronary sinus.**

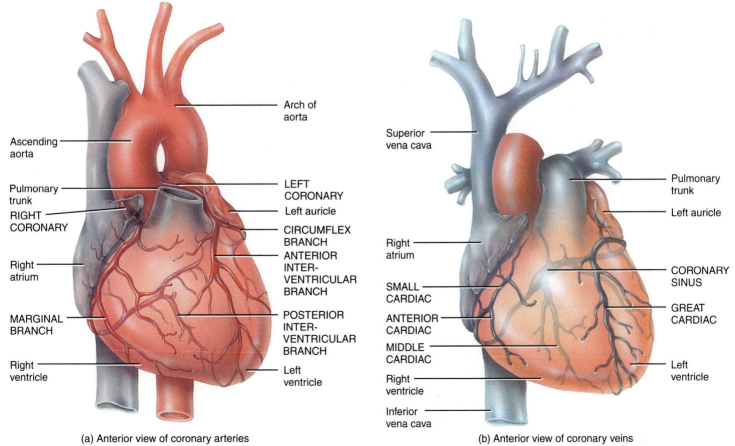

(a) Anterior view of coronary arteries

(b) Anterior view of coronary veins

(continues)

(Figure 20.8 continued)

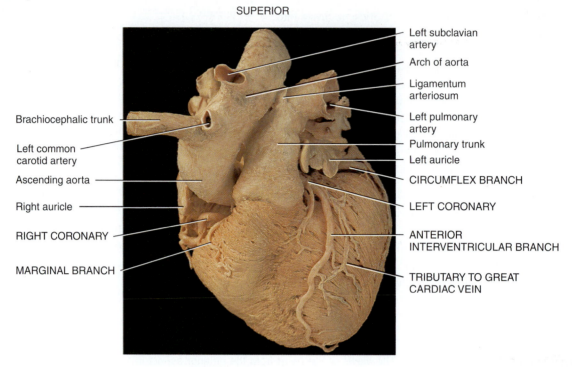

SUPERIOR

Left subclavian artery

Arch of aorta

Ligamentum arteriosum

Left pulmonary artery

Pulmonary trunk

Left auricle

CIRCUMFLEX BRANCH

LEFT CORONARY

ANTERIOR INTERVENTRICULAR BRANCH

TRIBUTARY TO GREAT CARDIAC VEIN

Brachiocephalic trunk

Left common carotid artery

Ascending aorta

Right auricle

RIGHT CORONARY

MARGINAL BRANCH

INFERIOR

(c) Anterior view of coronary blood vessels

Which coronary blood vessel delivers oxygenated blood to the left atrium and left ventricle?

CARDIAC MUSCLE TISSUE AND THE CARDIAC CONDUCTION SYSTEM

▶ **OBJECTIVES**

- **Describe the structural and functional characteristics of cardiac muscle tissue and the conduction system of the heart.**
- **Describe how an action potential occurs in cardiac contractile fibers.**
- **Describe the electrical events of a normal electrocardiogram (ECG).**

Histology of Cardiac Muscle Tissue

Compared with skeletal muscle fibers, cardiac muscle fibers are shorter in length and less circular in transverse section (Figure 20.9). They also exhibit branching, which gives individual cardiac muscle fibers a "stair-step" appearance (see Table 4.4B). A typical cardiac muscle fiber is 50–100 μm long and has a diameter of about 14 μm. Usually one centrally located nucleus is present, although an occasional cell may have two nuclei. The ends of cardiac muscle fibers connect to neighboring fibers by irregular transverse thickenings of the sarcolemma called **intercalated discs** (in-TER-kā-lāt-ed; *intercalat-* = to insert between). The discs contain **desmosomes,** which hold the fibers together, and

gap junctions, which allow muscle action potentials to conduct from one muscle fiber to its neighbors.

Mitochondria are larger and more numerous in cardiac than in skeletal muscle fibers. In a cardiac muscle fiber, they take up 25% of the cytosolic space as compared with only 2% of the cytosolic space in a skeletal muscle fiber. Cardiac muscle fibers have the same arrangement of actin and myosin, and the same bands, zones, and Z discs, as skeletal muscle fibers. The transverse tubules of cardiac muscle are wider but less abundant than those of skeletal muscle; the one transverse tubule per sarcomere is located at the Z disc. The sarcoplasmic reticulum of cardiac muscle fibers is somewhat smaller than the SR of skeletal muscle fibers. As a result, cardiac muscle has a smaller intracellular reserve of Ca^{2+}.

Regeneration of Heart Cells

The heart of a heart attack survivor often has regions of infarcted (dead) cardiac muscle tissue that typically are replaced with noncontractile fibrous scar tissue over time. Our inability to repair damage from a heart attack has been attributed to a lack of stem cells in cardiac muscle and to the absence of mitosis in mature cardiac muscle fibers. A recent study of heart transplant recipients by American and Italian scientists, however, provides evidence for significant replacement of heart cells. The researchers

Figure 20.9 Histology of cardiac muscle tissue. (See Table 4.4B for a light microscopic view of cardiac muscle.)

Cardiac muscle fibers connect to neighboring fibers by intercalated discs, which contain desmosomes and gap junctions.

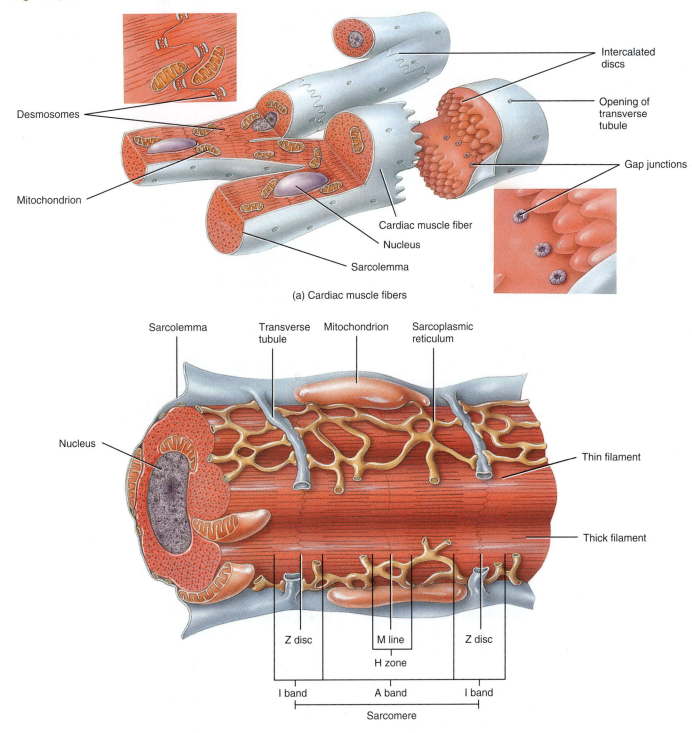

(a) Cardiac muscle fibers

(b) Arrangement of components in a cardiac muscle fiber

 What are the functions of intercalated discs in cardiac muscle fibers?

studied men who had received a heart from a female, and then looked for the presence of a Y chromosome in heart cells. (All female cells except gametes have two X chromosomes and lack the Y chromosome.) Several years after the transplant surgery, between 7% and 16% of the heart cells in the transplanted tissue, including cardiac muscle fibers and endothelial cells in coronary arterioles and capillaries, had been replaced by the recipient's own cells, as evidenced by the presence of a Y chromosome. The study also revealed cells with some of the characteristics of stem cells in both transplanted hearts and control hearts. Evidently, stem cells can migrate from the blood into the heart and differentiate into functional muscle and endothelial cells. The hope is that researchers can learn how to "turn on" such regeneration of heart cells to treat people with heart failure or cardiomyopathy (diseased heart). ■

Autorhythmic Fibers: The Conduction System

An inherent and rhythmical electrical activity is the reason for the heart's continuous beating. The source of this electrical activity is a network of specialized cardiac muscle fibers. These cells are called **autorhythmic fibers** (*auto-* = self) because they are self-excitable. Autorhythmic fibers repeatedly generate action potentials that trigger heart contractions. Autorhythmic fibers continue to stimulate a heart to beat even after it is removed from the body—for example, to be transplanted into another person—although all of its nerves have been cut. (Note: Surgeons do not attempt to reattach heart nerves during heart transplant operations. For this reason, it has been said that heart surgeons are better "plumbers" than they are "electricians.")

During embryonic development, about 1% of the cardiac muscle fibers become autorhythmic fibers, which have two important functions.

1. They act as a **pacemaker,** setting the rhythm of electrical excitation that causes contraction of the heart.

2. They form the **conduction system,** a network of specialized cardiac muscle fibers that provide a path for each cycle of cardiac excitation to progress through the heart. The conduction system ensures that cardiac chambers become stimulated to contract in a coordinated manner, which makes the heart an effective pump.

Cardiac action potentials propagate through the conduction system in the following sequence (Figure 20.10a):

1 Cardiac excitation normally begins in the **sinoatrial (SA) node,** located in the right atrial wall just inferior to the opening of the superior vena cava. SA node cells do not have a stable resting potential. Rather, they repeatedly depolarize to threshold spontaneously. The spontaneous depolarization is a **pacemaker potential.** When the pacemaker potential reaches threshold, it triggers an action potential (Figure 20.10b). Each action potential from the SA node propagates throughout both atria via gap junctions in the intercalated discs of atrial muscle fibers. Following the action potential, the atria contract.

2 By conducting along atrial muscle fibers, the action potential reaches the **atrioventricular (AV) node,** located in the septum between the two atria, just anterior to the opening of the coronary sinus (Figure 20.10a).

3 From the AV node, the action potential enters the **atrioventricular (AV) bundle** (also known as the **bundle of His**). This bundle is the only site where action potentials can conduct from the atria to the ventricles. (Elsewhere, the fibrous skeleton of the heart electrically insulates the atria from the ventricles.)

4 After propagating along the AV bundle, the action potential enters both the **right** and **left bundle branches.** The bundle branches extend through the interventricular septum toward the apex of the heart.

5 Finally, the large-diameter **Purkinje fibers** rapidly conduct the action potential from the apex of the heart upward to the remainder of the ventricular myocardium. Then, the ventricles contract, pushing the blood upward toward the semilunar valves.

On their own, autorhythmic fibers in the SA node initiate an action potential about every 0.6 second, or 100 times per minute. Normally, this rate is faster than that of any other autorhythmic fibers. So, action potentials from the SA node spread through the conduction system and stimulate other areas before the other areas are able to generate an action potential at their own, slower rate. Thus, the SA node is the normal pacemaker of the heart. Nerve impulses from the autonomic nervous system (ANS) and blood-borne hormones (such as epinephrine) *modify the timing and strength* of each heartbeat, but they *do not establish the fundamental rhythm.* In a person at rest, for example, acetylcholine released by the parasympathetic division of the ANS slows SA node pacing to about 75 action potentials per minute or one every 0.8 sec (Figure 20.10b).

If the SA node becomes damaged or diseased, the slower AV node can pick up the pacemaking task. Its rate of spontaneous depolarization is 40 to 60 times per minute. If the activity of both nodes is suppressed, the heartbeat may still be maintained by autorhythmic fibers in the ventricles—the AV bundle, a bundle branch, or Purkinje fibers. However, the pacing rate is so slow, only 20 to 35 beats per minute, that blood flow to the brain is inadequate. When this condition occurs, normal heart rhythm can be restored and maintained by surgically implanting an **artificial pacemaker,** a device that sends out small electrical currents to stimulate the heart to contract. Many of the newer pacemakers, called activity-adjusted pacemakers, automatically speed up the heartbeat during exercise.

Ectopic Pacemaker

Sometimes, a site other than the SA node becomes the pacemaker because it develops abnormal self-excitability. Such a site is called an **ectopic pacemaker** (ek-TOP-ik = displaced). An

Figure 20.10 The conduction system of the heart. Autorhythmic fibers in the SA node, located in the right atrial wall (a), act as the heart's pacemaker, initiating cardiac action potentials (b) that cause contraction of the heart's chambers.

🔑 **The conduction system ensures that the chambers of the heart contract in a coordinated manner.**

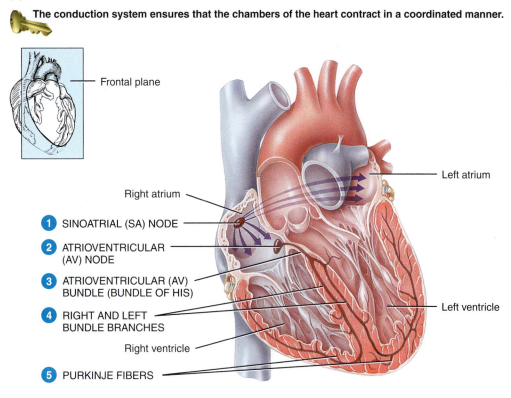

Frontal plane

Right atrium

Left atrium

① SINOATRIAL (SA) NODE

② ATRIOVENTRICULAR (AV) NODE

③ ATRIOVENTRICULAR (AV) BUNDLE (BUNDLE OF HIS)

④ RIGHT AND LEFT BUNDLE BRANCHES

Left ventricle

Right ventricle

⑤ PURKINJE FIBERS

(a) Anterior view of frontal section

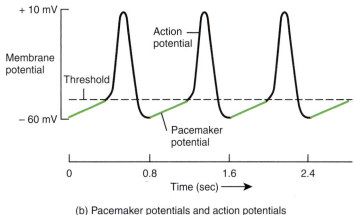

+ 10 mV

Membrane potential

Threshold

Action potential

− 60 mV

Pacemaker potential

0   0.8   1.6   2.4

Time (sec) ⟶

(b) Pacemaker potentials and action potentials in autorhythmic fibers of SA node

❓ **Which component of the conduction system provides the only electrical connection between the atria and the ventricles?**

ectopic pacemaker may operate only occasionally, producing extra beats, or it may pace the heart for some period. Triggers of ectopic activity include caffeine and nicotine, electrolyte imbalances, hypoxia, and toxic reactions to drugs such as digitalis. Students recording their own electrocardiograms (described shortly) in an anatomy and physiology lab may see an occasional ectopic beat if they have consumed too much coffee, tea, or Coke!

Action Potential and Contraction of Contractile Fibers

The action potential initiated by the SA node travels along the conduction system and spreads out to excite the "working" atrial and ventricular muscle fibers, which are called **contractile fibers.** An action potential occurs in a contractile fiber as follows (Figure 20.11):

① *Depolarization.* Unlike autorhythmic fibers, contractile fibers have a stable resting membrane potential that is close to −90 mV. When a contractile fiber is brought to threshold by an action potential in neighboring fibers, its **voltage-gated fast Na⁺ channels** open. These sodium ion channels are "fast" because they open very rapidly in response to a threshold-level depolarization. Opening of these channels allows Na⁺ inflow because the cytosol is electrically more negative than interstitial fluid and Na⁺ concentration is higher in interstitial fluid. Inflow of Na⁺ down the electrochemical gradient produces a **rapid depolarization.** Within a few milliseconds, the fast Na⁺ channels automatically inactivate and Na⁺ inflow decreases.

② *Plateau.* The next phase of an action potential in a contractile fiber is the **plateau,** a period of maintained depolarization. It is due in part to opening of **voltage-gated slow Ca²⁺ channels** in the sarcolemma and sarcoplasmic reticulum membrane. To trigger contraction, some calcium ions pass through the sarcolemma from the interstitial fluid (which has a higher Ca²⁺ concentration) into the cytosol. The small inflow of Ca²⁺ from interstitial fluid causes much more Ca²⁺ to pour out of the sarcoplasmic reticulum within the fiber. At the same time, the membrane permeability to potassium ions decreases due to closing of some K⁺ channels. For about 0.25 sec the membrane is depolarized, close to 0 mV, as a small outflow of K⁺ just balances the inflow of Ca²⁺. By comparison, depolarization in a neuron or skeletal muscle fiber is much briefer, about 1 msec (0.001 sec), because it lacks a plateau phase.

③ *Repolarization.* The recovery of the resting membrane potential during the **repolarization** phase of a cardiac action potential resembles that in other excitable cells. After a delay (which is particularly prolonged in cardiac muscle), **voltage-gated K⁺ channels** open, thereby increasing the membrane permeability to potassium ions. Outflow of K⁺ restores the negative resting membrane potential (−90 mV). At the same time, the calcium channels are closing, which also contributes to repolarization.

The mechanism of contraction is similar in cardiac and skeletal muscle: The electrical activity (action potential) leads to the mechanical response (contraction) after a short delay. As Ca²⁺ concentration rises inside a contractile fiber, Ca²⁺ binds to the regulatory protein troponin, which allows the actin and myosin filaments to begin sliding past one another, and tension starts to develop. Substances that alter the movement of Ca²⁺ through slow Ca²⁺ channels influence the strength of heart contractions. Epinephrine, for example, increases contraction force by enhancing Ca²⁺ flow into the cytosol.

In muscle, the **refractory period** is the time interval during which a second contraction cannot be triggered. The refractory period of a cardiac muscle fiber lasts longer than the contraction itself (Figure 20.11). As a result, another contraction cannot begin until relaxation is well underway. For this reason, tetanus (maintained contraction) cannot occur in cardiac muscle as it can in skeletal muscle. The advantage is apparent if you consider how the ventricles work. Their pumping function depends on alternating contraction (when they eject blood) and relaxation (when they refill). If heart muscle could undergo tetanus, blood flow would cease.

Figure 20.11 **Action potential in a ventricular contractile fiber.** The resting membrane potential is about −90 mV.

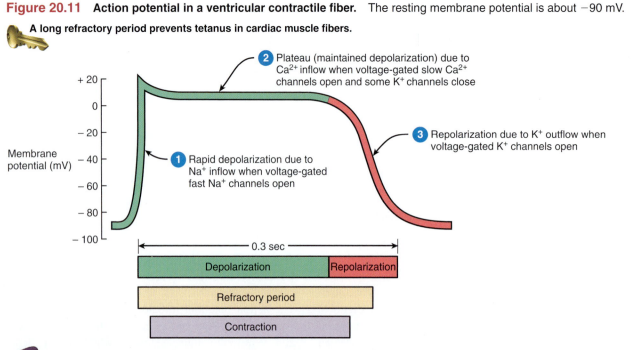

A long refractory period prevents tetanus in cardiac muscle fibers.

② Plateau (maintained depolarization) due to Ca²⁺ inflow when voltage-gated slow Ca²⁺ channels open and some K⁺ channels close

① Rapid depolarization due to Na⁺ inflow when voltage-gated fast Na⁺ channels open

③ Repolarization due to K⁺ outflow when voltage-gated K⁺ channels open

Membrane potential (mV)

0.3 sec

Depolarization | Repolarization

Refractory period

Contraction

 How does the duration of an action potential in a ventricular contractile fiber compare with that in a skeletal muscle fiber?

ATP Production in Cardiac Muscle

In contrast to skeletal muscle, cardiac muscle produces little of the ATP it needs by anaerobic cellular respiration (see Figure 10.13 on page 289). Instead, it relies almost exclusively on aerobic cellular respiration in its numerous mitochondria. The needed oxygen diffuses from blood in the coronary circulation and is released from myoglobin inside cardiac muscle fibers. Cardiac muscle fibers use several fuels to power mitochondrial ATP production. In a person at rest, the heart's ATP comes mainly from oxidation of fatty acids (60%) and glucose (35%), with smaller contributions from lactic acid, amino acids, and ketone bodies. During exercise, the heart's use of lactic acid, produced by actively contracting skeletal muscles, rises.

Like skeletal muscle, cardiac muscle also produces some ATP from creatine phosphate. One sign that a myocardial infarction (heart attack, see page 691) has occurred is the presence in blood of creatine kinase (CK), the enzyme that catalyzes transfer of a phosphate group from creatine phosphate to ADP to make ATP. Normally, CK and other enzymes are confined within cells. Injured or dying cardiac or skeletal muscle fibers release CK into the blood.

Electrocardiogram

As action potentials propagate through the heart, they generate electrical currents that can be detected at the surface of the body. An **electrocardiogram** (e-lek′-trō-KAR-dē-ō-gram), abbreviated either **ECG** or **EKG** (from the German word *Elektrokardiogram*) is a recording of these electrical signals. The ECG is a composite record of action potentials produced by all the heart muscle fibers during each heartbeat. The instrument used to record the changes is an **electrocardiograph.**

In clinical practice, electrodes are positioned on the arms and legs (limb leads) and at six positions on the chest (chest leads) to record the ECG. The electrocardiograph amplifies the heart's electrical signals and produces 12 different tracings from different combinations of limb and chest leads. Each limb and chest electrode records slightly different electrical activity because it is in a different position relative to the heart. By comparing these records with one another and with normal records, it is possible to determine (1) if the conducting pathway is abnormal, (2) if the heart is enlarged, and (3) if certain regions of the heart are damaged.

In a typical Lead II record (right arm to left leg), three clearly recognizable waves appear with each heartbeat (Figure 20.12). The first, called the **P wave,** is a small upward deflection on the ECG. The P wave represents **atrial depolarization,** which spreads from the SA node through contractile fibers in both atria. The second wave, called the **QRS complex,** begins as a downward deflection, continues as a large, upright, triangular wave, and ends as a downward wave. The QRS complex represents **rapid ventricular depolarization,** as the action potential spreads through ventricular contractile fibers. The third wave is a dome-shaped upward deflection called the **T wave.** It indicates

Figure 20.12 Normal electrocardiogram or ECG (Lead II). P wave = atrial depolarization; QRS complex = onset of ventricular depolarization; T wave = ventricular repolarization.

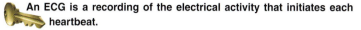

🔑 **An ECG is a recording of the electrical activity that initiates each heartbeat.**

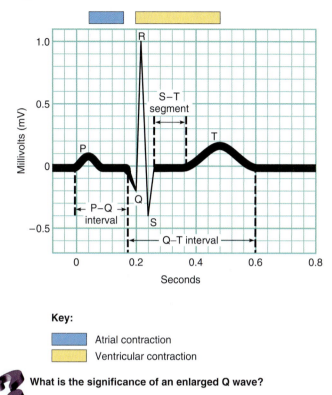

Key:

🟦 Atrial contraction

🟨 Ventricular contraction

❓ **What is the significance of an enlarged Q wave?**

ventricular repolarization and occurs just as the ventricles are starting to relax. The T wave is smaller and wider than the QRS complex because repolarization occurs more slowly than depolarization. During the plateau period of steady depolarization, the ECG tracing is flat.

In reading an ECG, the size of the waves can provide clues to abnormalities. Larger P waves, for example, indicate enlargement of an atrium; an enlarged Q wave may indicate a myocardial infarction; and an enlarged R wave generally indicates enlarged ventricles. The T wave is flatter than normal when the heart muscle is receiving insufficient oxygen—as, for example, in coronary artery disease. The T wave may be elevated in hyperkalemia (high blood K^+ level).

Sometimes it is helpful to evaluate the heart's response to the stress of physical exercise. Such a test is called a **stress electrocardiogram,** or **stress test.** Although narrowed coronary arteries may carry adequate oxygenated blood while a person is at rest, they will not be able to meet the heart's increased need for oxygen during strenuous exercise. This situation creates changes that can be seen on an electrocardiogram.

Analysis of an ECG also involves measuring the time spans between waves, which are called **intervals** or **segments.** For example, the **P-Q interval** is the time from the beginning of the P wave to the beginning of the QRS complex. It represents the conduction time from the beginning of atrial excitation to the beginning of ventricular excitation. Put another way, the P-Q interval is the time required for the action potential to travel through the atria, atrioventricular node, and the remaining fibers of the conduction system. In coronary artery disease and rheumatic fever, scar tissue may form in the heart. As the action potential detours around scar tissue, the P-Q interval lengthens.

The **S-T segment** begins at the end of the S wave and ends at the beginning of the T wave. It represents the time when the ventricular contractile fibers are depolarized during the plateau phase of the action potential. The S-T segment is elevated (above the baseline) in acute myocardial infarction and depressed (below the baseline) when the heart muscle receives insufficient oxygen. The **Q-T interval** extends from the start of the QRS complex to the end of the T wave. It is the time from the beginning of ventricular depolarization to the end of ventricular repolarization. The Q-T interval may be lengthened by myocardial damage, myocardial ischemia (decreased blood flow), or conduction abnormalities.

Correlation of ECG Waves with Atrial and Ventricular Systole

As we have seen, the atria and ventricles depolarize and then contract at different times because the conduction system routes cardiac action potentials along a specific pathway. The term **systole** (SIS-tō-lē = contraction) refers to the phase of contraction; the phase of relaxation is **diastole** (dī-AS-tō-lē = dilation or expansion). The ECG waves predict the timing of atrial and ventricular systole and diastole. At a heart rate of 75 beats per minute, the timing is as follows (Figure 20.13):

1 A cardiac action potential arises in the SA node. It propagates throughout the atrial muscle and down to the AV node in about 0.03 sec. As the atrial contractile fibers depolarize, the P wave appears in the ECG.

2 After the P wave begins, the atria contract (atrial systole). Conduction of the action potential slows at the AV node because the fibers there have much smaller diameters and fewer gap junctions. (Traffic slows in a similar way where a four-lane highway narrows to one lane in a construction zone!) The resulting 0.1-sec delay gives the atria time to contract, thus adding to the volume of blood in the ventricles, before ventricular diastole begins.

3 The action potential propagates rapidly again after entering the AV bundle. About 0.2 sec after onset of the P wave, it has propagated through the bundle branches, Purkinje fibers, and the entire ventricular myocardium. Depolarization progresses down the septum, upward from the apex, and outward from the endocardial surface, producing the QRS complex. At the same time, atrial repolarization is occurring, but it is not usually evident in an ECG because the larger QRS complex masks it. Atrial systole follows atrial repolarization.

4 Contraction of ventricular contractile fibers, that is ventricular systole, begins shortly after the QRS appears and continues during the S-T segment. As contraction proceeds from the apex toward the base of the heart, blood is squeezed upward toward the semilunar valves.

5 Repolarization of ventricular contractile fibers begins at the apex and spreads throughout the ventricular myocardium. This produces the T wave in the ECG about 0.4 sec after the P wave onset.

6 Shortly after the T wave begins, the ventricles start to relax (ventricular diastole). By 0.6 sec, ventricular repolarization is complete and ventricular contractile fibers are relaxed.

During the next 0.2 sec, contractile fibers in both the atria and ventricles are relaxed. At 0.8 sec, the P wave appears again in the ECG, the atria begin to contract, and the cycle repeats. As you can see, events in the heart occur in cycles that repeat for as long as you live. Next, we will see how the pressure changes associated with relaxation and contraction of the heart chambers allow the heart to alternately fill with blood and then eject blood into the aorta and pulmonary trunk.

▶ **CHECKPOINT**

10. How do cardiac muscle fibers differ structurally and functionally from skeletal muscle fibers?

11. In what ways are autorhythmic fibers similar to and different from contractile fibers?

12. What happens during each of the three phases of an action potential in ventricular contractile fibers?

13. In what ways are ECGs helpful in diagnosing cardiac problems?

14. How does each ECG wave, interval, and segment relate to contraction (systole) and relaxation (diastole) of the atria and ventricles?

THE CARDIAC CYCLE

▶ **OBJECTIVES**

• **Describe the pressure and volume changes that occur during a cardiac cycle.**

• **Relate the timing of heart sounds to the ECG waves and pressure changes during systole and diastole.**

A single **cardiac cycle** includes all the events associated with one heartbeat. Thus, a cardiac cycle consists of systole and diastole of the atria plus systole and diastole of the ventricles.

Pressure and Volume Changes During the Cardiac Cycle

In each cardiac cycle, the atria and ventricles alternately contract and relax, forcing blood from areas of higher pressure to areas of lower pressure. As a chamber of the heart contracts, blood pressure within it increases. Figure 20.14 shows the relation between the heart's electrical signals (ECG) and changes in atrial

Figure 20.13 **Timing and route of action potential depolarization and repolarization through the conduction system and myocardium.** Green indicates depolarization and red indicates repolarization. Arrows show the direction of propagation of the action potential.

🔑 **Depolarization causes contraction and repolarization causes relaxation of cardiac muscle fibers.**

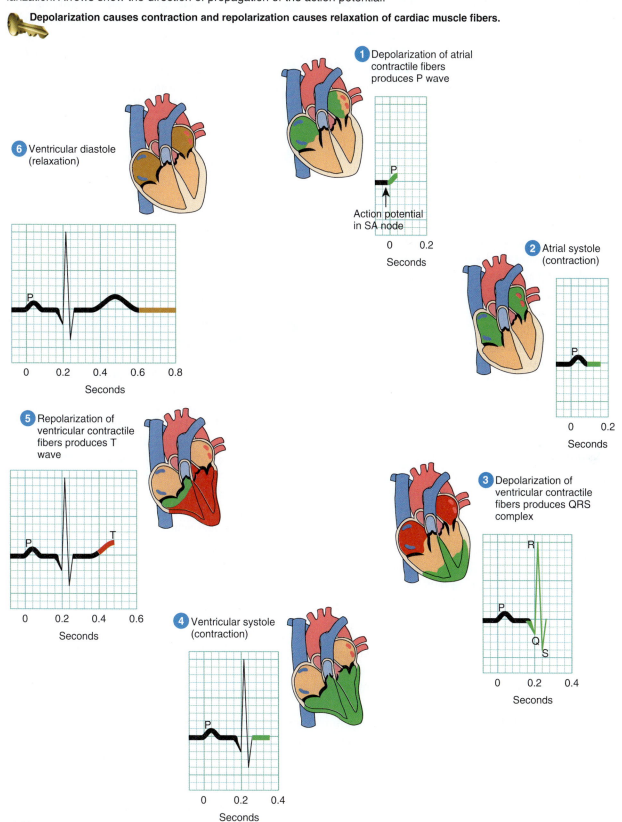

 Where in the conduction system do action potentials propagate most slowly?

pressure, ventricular pressure, aortic pressure, and ventricular volume during the cardiac cycle. The pressures given in Figure 20.14 apply to the left side of the heart; pressures on the right side are considerably lower. Each ventricle, however, expels the same volume of blood per beat, and the same pattern exists for both pumping chambers. When heart rate is 75 beats/min, a cardiac cycle lasts 0.8 sec. To examine and correlate the events taking place during a cardiac cycle, we will begin with atrial systole.

Atrial Systole

During **atrial systole,** which lasts about 0.1 sec, the atria are contracting. At the same time, the ventricles are relaxed.

1 Depolarization of the SA node causes atrial depolarization, marked by the P wave in the ECG.

2 Atrial depolarization causes atrial systole. As the atria contract, they exert pressure on the blood within, which forces blood through the open AV valves into the ventricles.

3 Atrial systole contributes a final 25 mL of blood to the volume already in each ventricle (about 105 mL). The end of atrial systole is also the end of ventricular diastole (relaxation). Thus, each ventricle contains about 130 mL at the end of its relaxation period (diastole). This blood volume is the **end-diastolic volume (EDV).**

4 The QRS complex in the ECG marks the onset of ventricular depolarization.

Ventricular Systole

During **ventricular systole,** which lasts about 0.3 sec, the ventricles are contracting. At the same time, the atria are relaxed, in **atrial diastole.**

5 Ventricular depolarization causes ventricular systole. As ventricular systole begins, pressure rises inside the ventricles and pushes blood up against the atrioventricular (AV) valves, forcing them shut. For about 0.05 seconds, both the SL (semilunar) and AV valves are closed. This is the period of **isovolumetric contraction** (*iso-* = same). During this interval, cardiac muscle fibers are contracting and exerting force but are not yet shortening. Thus, the muscle contraction is isometric (same length). Moreover, because all four valves are closed, ventricular volume remains the same (isovolumic).

6 Continued contraction of the ventricles causes pressure inside the chambers to rise sharply. When left ventricular pressure surpasses aortic pressure at about 80 millimeters of mercury (mmHg) and right ventricular pressure rises above the pressure in the pulmonary trunk (about 20 mmHg), both SL valves open. At this point, ejection of blood from the heart begins. The period when the SL valves are open is **ventricular ejection** and lasts for about 0.25 sec. The pressure in the left ventricle continues to rise to about 120 mmHg,

whereas the pressure in the right ventricle climbs to about 25–30 mmHg.

7 The left ventricle ejects about 70 mL of blood into the aorta and the right ventricle ejects the same volume of blood into the pulmonary trunk. The volume remaining in each ventricle at the end of systole, about 60 mL, is the **end-systolic volume (ESV). Stroke volume,** the volume ejected per beat from each ventricle, equals end-diastolic volume minus end-systolic volume: SV = EDV − ESV. At rest, the stroke volume is about 130 mL – 60 mL = 70 mL (a little more than 2 oz).

8 The T wave in the ECG marks the onset of ventricular repolarization.

Relaxation Period

During the **relaxation period,** which lasts about 0.4 sec, the atria and the ventricles are both relaxed. As the heart beats faster and faster, the relaxation period becomes shorter and shorter, whereas the durations of atrial systole and ventricular systole shorten only slightly.

9 Ventricular repolarization causes **ventricular diastole.** As the ventricles relax, pressure within the chambers falls, and blood in the aorta and pulmonary trunk begins to flow backward toward the regions of lower pressure in the ventricles. Backflowing blood catches in the valve cusps and closes the SL valves. The aortic valve closes at a pressure of about 100 mmHg. Rebound of blood off the closed cusps of the aortic valve produces the **dicrotic wave** on the aortic pressure curve. After the SL valves close, there is a brief interval when ventricular blood volume does not change because all four valves are closed. This is the period of **isovolumetric relaxation.**

10 As the ventricles continue to relax, the pressure falls quickly. When ventricular pressure drops below atrial pressure, the AV valves open, and **ventricular filling** begins. The major part of ventricular filling occurs just after the AV valves open. Blood that has been flowing into and building up in the atria during ventricular systole then rushes rapidly into the ventricles. At the end of the relaxation period, the ventricles are about three-quarters full. The P wave appears in the ECG, signaling the start of another cardiac cycle.

Heart Sounds

Auscultation (aws-kul-TĀ-shun; *ausculta-* = listening) is the act of listening to sounds within the body, and it is usually done with a stethoscope. The sound of the heartbeat comes primarily from blood turbulence caused by the closing of the heart valves. Smoothly flowing blood is silent. Recall the sounds made by white-water rapids or a waterfall as compared with the silence of a smoothly flowing river. During each cardiac cycle, there are four **heart sounds,** but in a normal heart only the first and second heart sounds (S1 and S2) are loud enough to be heard by

Figure 20.14 Cardiac cycle. (a) ECG. (b) Changes in left atrial pressure (green line), left ventricular pressure (blue line), and aortic pressure (red line) as they relate to the opening and closing of heart valves. (c) Changes in left ventricular volume. (d) Phases of the cardiac cycle.

A cardiac cycle is composed of all the events associated with one heartbeat.

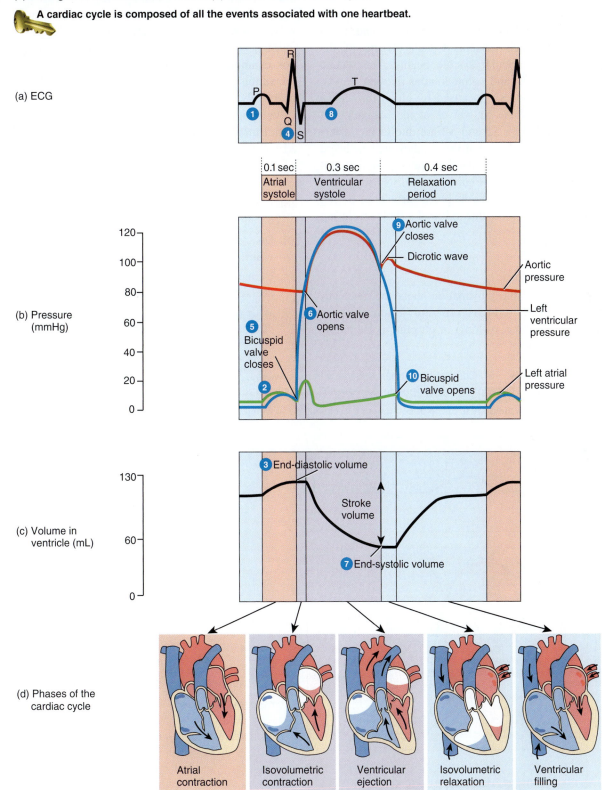

(a) ECG

0.1 sec | 0.3 sec | 0.4 sec
Atrial systole | Ventricular systole | Relaxation period

(b) Pressure (mmHg)

9 Aortic valve closes
Dicrotic wave
Aortic pressure
Left ventricular pressure
6 Aortic valve opens
5 Bicuspid valve closes
10 Bicuspid valve opens
Left atrial pressure
2

(c) Volume in ventricle (mL)

3 End-diastolic volume
Stroke volume
7 End-systolic volume

(d) Phases of the cardiac cycle

Atrial contraction | Isovolumetric contraction | Ventricular ejection | Isovolumetric relaxation | Ventricular filling

How much blood remains in each ventricle at the end of ventricular diastole in a resting person? What is this volume called?

listening through a stethoscope. Figure 20.15a shows the timing of heart sounds relative to other events in the cardiac cycle and where each sound is best heard.

The first sound (S1), which can be described as a **lubb** sound, is louder and a bit longer than the second sound. S1 is caused by blood turbulence associated with closure of the AV valves soon after ventricular systole begins. The second sound (S2), which is shorter and not as loud as the first, can be described as a **dupp** sound. S2 is caused by blood turbulence associated with closure of the SL valves at the beginning of ventricular diastole. Although S1 and S2 are due to blood turbulence associated with the closure of valves, they are best heard at the surface of the chest in locations that are slightly different from the locations of the valves (Figure 20.15b). Normally not loud enough to be heard, S3 is due to blood turbulence during rapid ventricular filling, and S4 is due to blood turbulence during atrial systole.

Heart Murmurs

Heart sounds provide valuable information about the mechanical operation of the heart. A **heart murmur** is an abnormal sound consisting of a clicking, rushing, or gurgling noise that is heard before, between, or after the normal heart sounds, or that may mask the normal heart sounds. Although some heart murmurs are "innocent," meaning they are not associated with a significant heart problem, most often a murmur indicates a valve disorder. When a heart valve exhibits stenosis, the heart murmur is heard while the valve should be fully open but is not. For example, mitral stenosis produces a murmur during the relaxation period, between S2 and the next S1. An incompetent heart valve, by contrast, causes a murmur to appear when the valve should be fully closed but is not. So, a murmur due to mitral incompetence occurs during ventricular systole, between S1 and S2. ■

▶ **CHECKPOINT**

15. Why must left ventricular pressure be greater than aortic pressure during ventricular ejection?

16. Does more blood flow through the coronary arteries during ventricular diastole or ventricular systole?

17. During which two periods of the cardiac cycle do the heart muscle fibers exhibit isometric contractions?

18. What events produce the four normal heart sounds? Which ones can usually be heard through a stethoscope?

Figure 20.15 Heart sounds. (a) Correlation of heart sounds with ECG waves, pressure changes, and valve opening and closing. (b) Location of valves (purple) and auscultation sites (red) for heart sounds.

Listening to sounds within the body is called auscultation; it is usually done with a stethoscope.

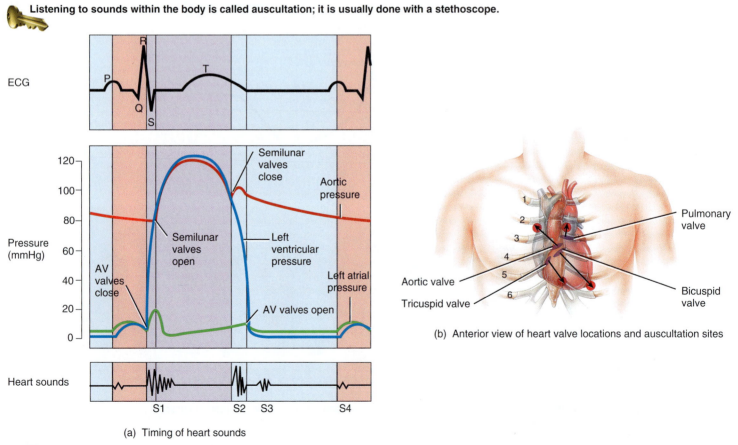

(a) Timing of heart sounds

(b) Anterior view of heart valve locations and auscultation sites

Which heart sound is related to blood turbulence associated with closure of the AV valves?

CARDIAC OUTPUT

• **Define cardiac output, and describe the factors that affect it.**

Although the heart has autorhythmic fibers that enable it to beat independently, its operation is governed by events occurring in the rest of the body. All body cells must receive a certain amount of oxygenated blood each minute to maintain health and life. When cells are metabolically active, as during exercise, they take up even more oxygen from the blood. During rest periods, cellular metabolic need is reduced, and the workload of the heart decreases.

Cardiac output (CO) is the volume of blood ejected from the left ventricle (or the right ventricle) into the aorta (or pulmonary trunk) each minute. Cardiac output equals the **stroke volume (SV),** the volume of blood ejected by the ventricle during each contraction, multiplied by the **heart rate (HR),** the number of heartbeats per minute:

$$\underset{\text{(mL/min)}}{CO} = \underset{\text{(mL/beat)}}{SV} \times \underset{\text{(beats/min)}}{HR}$$

In a typical resting adult male, stroke volume averages 70 mL/beat, and heart rate is about 75 beats/min. Thus, average cardiac output is

$$
\begin{aligned}
CO &= 70 \text{ mL/beat} \times 75 \text{ beats/min} \\
&= 5250 \text{ mL/min} \\
&= 5.25 \text{ L/min}
\end{aligned}
$$

This volume is close to the total blood volume, which is about 5 liters in a typical adult male. Thus, the entire blood volume flows through the pulmonary and systemic circulations each minute. When body tissues use more or less oxygen, cardiac output changes to meet the need. Factors that increase stroke volume or heart rate normally increase CO. During mild exercise, for example, stroke volume may increase to 100 mL/beat, and heart rate to 100 beats/min. Cardiac output then would be 10 L/min. During intense (but still not maximal) exercise, the heart rate may accelerate to 150 beats/min, and stroke volume may rise to 130 mL/beat, providing a cardiac output of 19.5 L/min.

Cardiac reserve is the difference between a person's maximum cardiac output and cardiac output at rest. A cardiac reserve that is four or five times the resting value is average. Top endurance athletes may have a cardiac reserve seven or eight times their resting CO. People with severe heart disease may have little or no cardiac reserve, which limits their ability to carry out even the simple tasks of daily living.

Regulation of Stroke Volume

A healthy heart will pump out all the blood that entered its chambers during the previous diastole. In other words, if more blood returns to the heart during diastole, then more blood is ejected during the next systole. At rest, the stroke volume is 50–60% of the end-diastolic volume because 40–50% remains

in the ventricles after each contraction (end-systolic volume). Three factors regulate stroke volume and ensure that the left and right ventricles pump equal volumes of blood: (1) **preload,** the degree of stretch on the heart before it contracts; (2) **contractility,** the forcefulness of contraction of individual ventricular muscle fibers; and (3) **afterload,** the pressure that must be exceeded before ejection of blood from the ventricles can occur.

Preload: Effect of Stretching

A greater preload (stretch) on cardiac muscle fibers prior to contraction increases their force of contraction. Within limits, the more the heart fills with blood during diastole, the greater the force of contraction during systole. This relationship is known as the **Frank–Starling law of the heart.** The preload is proportional to the volume of blood that fills the ventricles at the end of diastole, the end-diastolic volume (EDV). Normally, the greater the EDV, the more forceful the next contraction.

Two key factors determine EDV: (1) the duration of ventricular diastole and (2) **venous return,** the volume of blood returning to the right ventricle. When heart rate increases, the duration of diastole is shorter. Less filling time means a smaller EDV, and the ventricles may contract before they are adequately filled. By contrast, when venous return increases, a greater volume of blood flows into the ventricles, and the EDV is increased.

When heart rate exceeds about 160 beats/min, stroke volume usually declines due to the short filling time. At such rapid heart rates, EDV is less, and the preload is lower. People who have slow resting heart rates, in contrast, usually have large resting stroke volumes because filling time is prolonged and preload is larger.

The Frank–Starling law of the heart equalizes the output of the right and left ventricles and keeps the same volume of blood flowing to both the systemic and pulmonary circulations. If the left side of the heart pumps a little more blood than the right side, for example, the volume of blood returning to the right ventricle (venous return) increases. With increased EDV, then, the right ventricle contracts more forcefully on the next beat, and the two sides are again in balance.

Contractility

The second factor that influences stroke volume is myocardial **contractility,** which is the strength of contraction at any given preload. Substances that increase contractility are **positive inotropic agents,** whereas those that decrease contractility are **negative inotropic agents.** Thus, for a constant preload, the stroke volume increases when a positive inotropic substance is present. Positive inotropic agents often promote Ca^{2+} inflow during cardiac action potentials, which strengthens the force of the next contraction. Stimulation of the sympathetic division of the autonomic nervous system (ANS), hormones such as epinephrine and norepinephrine, increased Ca^{2+} level in the interstitial fluid, and the drug digitalis all have positive inotropic effects. In contrast, inhibition of the sympathetic division of the ANS, anoxia, acidosis, some anesthetics (for example, halothane), and increased K^+ level in the interstitial fluid have negative inotropic effects. *Calcium channel blockers* are drugs that can have a negative inotropic effect by reducing Ca^{2+} inflow, thereby decreasing the strength of the heartbeat.

Afterload

Ejection of blood from the heart begins when pressure in the right ventricle exceeds the pressure in the pulmonary trunk (about 20 mmHg), and when the pressure in the left ventricle exceeds the pressure in the aorta (about 80 mmHg). At that point, the higher pressure in the ventricles causes blood to push the semilunar valves open. The pressure that must be overcome before a semilunar valve can open is termed the **afterload.** At any given preload, an increase in afterload causes stroke volume to decrease, and more blood remains in the ventricles at the end of systole. Conditions that can increase afterload include hypertension (elevated blood pressure) and narrowing of arteries by atherosclerosis (see page 447).

Congestive Heart Failure

In **congestive heart failure (CHF),** the heart is a failing pump. Causes of CHF include coronary artery disease (see page 000), congenital defects, long-term high blood pressure (which increases the afterload), myocardial infarctions (regions of dead heart tissue due to a previous heart attack), and valve disorders. As the pump becomes less effective, more blood remains in the ventricles at the end of each cycle, and gradually the end-diastolic volume (preload) increases. Initially, increased preload may promote increased force of contraction (the Frank–Starling law of the heart), but as the preload increases further, the heart is overstretched and contracts less forcefully. The result is a potentially lethal positive feedback loop: Less-effective pumping leads to even lower pumping capability.

Often, one side of the heart starts to fail before the other. If the left ventricle fails first, it can't pump out all the blood it receives. As a result, blood backs up in the lungs and causes *pulmonary edema,* fluid accumulation in the lungs that can suffocate an untreated person. If the right ventricle fails first, blood backs up in the systemic veins and over time, the kidneys cause an increase in blood volume. In this case, the resulting *peripheral edema* usually is most noticeable in the feet and ankles. ■

Regulation of Heart Rate

As we have seen, cardiac output depends on heart rate as well as stroke volume. Adjustments in heart rate are important in the short-term control of cardiac output and blood pressure. The sinoatrial (SA) node initiates contraction and, if left to itself, would set a constant heart rate of about 100 beats/min. However, tissues require a different volume of blood flow under different conditions. During exercise, for example, cardiac output rises to supply working tissues with increased amounts of oxygen and nutrients. Stroke volume may fall if the ventricular myocardium

is damaged or if blood volume is reduced by bleeding. In these cases, homeostatic mechanisms act to maintain adequate cardiac output by increasing the heart rate and contractility. Among the several factors that contribute to regulation of heart rate, the most important are the autonomic nervous system and hormones released by the adrenal medullae (epinephrine and norepinephrine).

Autonomic Regulation of Heart Rate

Nervous system regulation of the heart originates in the **cardiovascular center** in the medulla oblongata. This region of the brain stem receives input from a variety of sensory receptors and from higher brain centers, such as the limbic system and cerebral cortex. The cardiovascular center then directs appropriate output by increasing or decreasing the frequency of nerve impulses in both the sympathetic and parasympathetic branches of the ANS (Figure 20.16).

Even before physical activity begins, especially in competitive situations, heart rate may climb. This anticipatory increase occurs because the limbic system sends nerve impulses to the cardiovascular center in the medulla. Then, as physical activity begins, **proprioceptors** that are monitoring the position of limbs and muscles send an increased frequency of nerve impulses to the cardiovascular center. Proprioceptor input is a major stimulus for the quick rise in heart rate that occurs at the onset of physical activity. Other sensory receptors that provide input to the cardiovascular center include **chemoreceptors,** which monitor chemical changes in the blood, and **baroreceptors,** which monitor stretching caused by blood pressure in major arteries and veins. Important baroreceptors located in the arch of the aorta and in the carotid arteries (see Figure 21.13 on page 711) detect changes in blood pressure. They provide input to the cardiovascular center when blood pressure changes. Baroreceptor reflexes, which are also important for the regulation of blood pressure, are discussed in detail in Chapter 21. Here we focus on the innervation of the heart by the sympathetic and parasympathetic branches of the ANS.

Sympathetic neurons extend from the medulla oblongata into the spinal cord. From the thoracic region of the spinal cord, sympathetic **cardiac accelerator nerves** extend out to the SA node, AV node, and most portions of the myocardium. Impulses in the cardiac accelerator nerves trigger the release of norepinephrine, which binds to beta-1 receptors on cardiac muscle fibers. This interaction has two separate effects: (1) In SA (and AV) node fibers, norepinephrine speeds the rate of spontaneous depolarization so that these pacemakers fire impulses more rapidly and heart rate increases. (2) In contractile fibers throughout the atria and ventricles, norepinephrine enhances Ca^{2+} entry through the voltage-gated slow Ca^{2+} channels, thereby increasing contractility. The result is a greater ejection of blood during systole. With a moderate increase in heart rate, stroke volume does not decline because the increased contractility offsets the decreased preload. With maximal sympathetic stimulation, however, heart rate may reach 200 beats/min in a 20-year-old person. At such a high heart rate, stroke volume is lower than at rest due

Figure 20.16 Nervous system control of the heart.

The cardiovascular center in the medulla oblongata controls both sympathetic and parasympathetic nerves that innervate the heart.

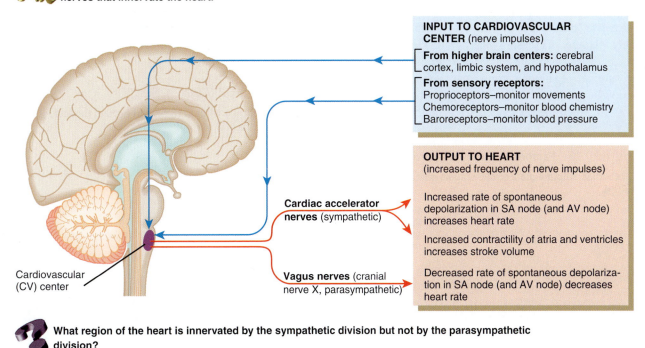

INPUT TO CARDIOVASCULAR CENTER (nerve impulses)

From higher brain centers: cerebral cortex, limbic system, and hypothalamus

From sensory receptors:
Proprioceptors–monitor movements
Chemoreceptors–monitor blood chemistry
Baroreceptors–monitor blood pressure

OUTPUT TO HEART
(increased frequency of nerve impulses)

Increased rate of spontaneous depolarization in SA node (and AV node) increases heart rate

Increased contractility of atria and ventricles increases stroke volume

Decreased rate of spontaneous depolarization in SA node (and AV node) decreases heart rate

Cardiac accelerator nerves (sympathetic)

Vagus nerves (cranial nerve X, parasympathetic)

Cardiovascular (CV) center

What region of the heart is innervated by the sympathetic division but not by the parasympathetic division?

to the very short filling time. The maximal heart rate declines with age; as a rule, subtracting one's age from 220 provides a good estimate of maximal heart rate in beats per minute.

Parasympathetic nerve impulses reach the heart via the right and left **vagus (X) nerves.** Vagal axons terminate in the SA node, AV node, and atrial myocardium. They release acetylcholine, which decreases heart rate by slowing the rate of spontaneous depolarization in autorhythmic fibers. As only a few vagal fibers innervate ventricular muscle, changes in parasympathetic activity have little effect on contractility of the ventricles.

A continually shifting balance exists between sympathetic and parasympathetic stimulation of the heart. At rest, parasympathetic stimulation predominates. The resting heart rate—about 75 beats/min—is usually lower than the autorhythmic rate of the SA node (about 100 beats/min). With maximal stimulation by the parasympathetic division, the heart can slow to 20 or 30 beats/min, or can even stop momentarily.

Chemical Regulation of Heart Rate

Certain chemicals influence both the basic physiology of cardiac muscle and the heart rate. For example, hypoxia (lowered oxygen level), acidosis (low pH), and alkalosis (high pH) all depress cardiac activity. Several hormones and cations have major effects on the heart:

1. Hormones. Epinephrine and norepinephrine (from the adrenal medullae) enhance the heart's pumping effectiveness. These hormones affect cardiac muscle fibers in much the same way as does norepinephrine released by cardiac accelerator nerves—they increase both heart rate and contractility. Exercise, stress, and excitement cause the adrenal medullae to release more hormones. Thyroid hormones also enhance cardiac contractility and increase heart rate. One sign of hyperthyroidism (excessive thyroid hormone) is **tachycardia,** an elevated resting heart rate.

2. Cations. Given that differences between intracellular and extracellular concentrations of several cations (for example, Na^+ and K^+) are crucial for the production of action potentials in all nerve and muscle fibers, it is not surprising that ionic imbalances can quickly compromise the pumping effectiveness of the heart. In particular, the relative concentrations of three cations—K^+, Ca^{2+}, and Na^+—have a large effect on cardiac function. Elevated blood levels of K^+ or Na^+ decrease heart rate and contractility. Excess Na^+ blocks Ca^{2+} inflow during cardiac action potentials, thereby decreasing the force of contraction, whereas excess K^+ blocks generation of action potentials. A moderate increase in interstitial (and thus intracellular) Ca^{2+} level speeds heart rate and strengthens the heartbeat.

Other Factors in Heart Rate Regulation

Age, gender, physical fitness, and body temperature also influence resting heart rate. A newborn baby is likely to have a resting heart rate over 120 beats/min; the rate then declines through childhood. Senior citizens may develop a more rapid heartbeat, however. Adult females often have slightly higher resting heart rates than adult males, although regular exercise tends to bring resting heart rate down in both sexes. A physically fit person may even exhibit **bradycardia,** a resting heart rate under 60 beats/min. This is a beneficial effect of endurance-type training because a slowly beating heart is more energy efficient than one that beats more rapidly.

Increased body temperature, as occurs during a fever or strenuous exercise, causes the SA node to discharge impulses more quickly, thereby increasing heart rate. Decreased body temperature decreases heart rate and strength of contraction.

During surgical repair of certain heart abnormalities, it is helpful to slow a patient's heart rate by **hypothermia** (hī-pō-THER-mē-a), in which the person's body is deliberately cooled to a low core temperature. Hypothermia also slows metabolism, which reduces the oxygen needs of the tissues, allowing the heart and brain to withstand short periods of interrupted or reduced blood flow during the procedure.

Figure 20.17 summarizes the factors that can increase stroke volume and heart rate to achieve an increase in cardiac output.

▶ **CHECKPOINT**

19. How is cardiac output calculated?

20. Define stroke volume (SV), and explain the factors that regulate it.

Figure 20.17   Factors that increase cardiac output.

🔑 **Cardiac output equals stroke volume multiplied by heart rate.**

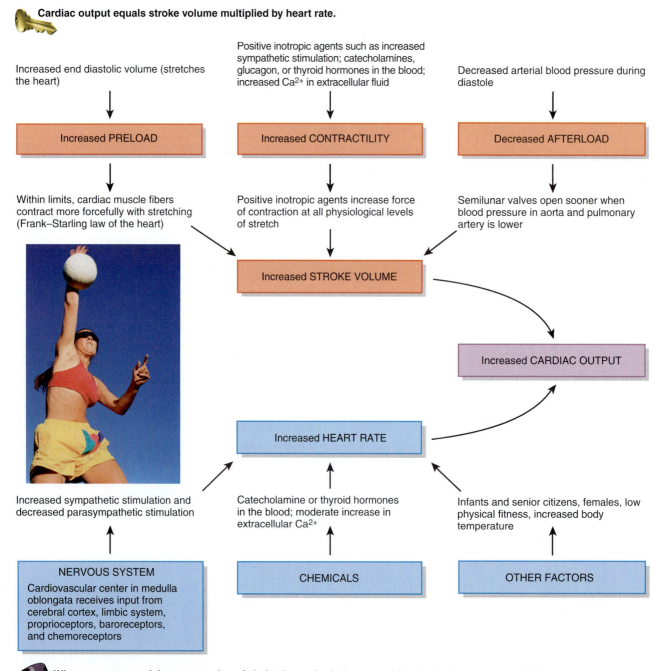

When you are exercising, contraction of skeletal muscles helps return blood to the heart more rapidly.
Would this effect tend to increase or decrease stroke volume?

21. What is the Frank–Starling law of the heart? What is its significance?

22. Define cardiac reserve. How does it change with training or with heart failure?

23. How do the sympathetic and parasympathetic divisions of the autonomic nervous system adjust heart rate?

EXERCISE AND THE HEART

▶ O B J E C T I V E

• **Explain the relationship between exercise and the heart.**

No matter what a person's level of fitness, it can be improved at any age with regular exercise. Of the various types of exercise, some are more effective than others for improving the health of the cardiovascular system. **Aerobics,** any activity that works large body muscles for at least 20 minutes, elevates cardiac output and accelerates metabolic rate. Three to five such sessions a week are usually recommended for improving the health of the cardiovascular system. Brisk walking, running, bicycling, cross-country skiing, and swimming are examples of aerobic activities.

Sustained exercise increases the oxygen demand of the muscles. Whether the demand is met depends mainly on the adequacy of cardiac output and proper functioning of the respiratory system. After several weeks of training, a healthy person increases maximal cardiac output, thereby increasing the maximal rate of oxygen delivery to the tissues. Oxygen delivery also rises because skeletal muscles develop more capillary networks in response to long-term training.

During strenuous activity, a well-trained athlete can achieve a cardiac output double that of a sedentary person, in part because training causes hypertrophy (enlargement) of the heart. Even though the heart of a well-trained athlete is larger, *resting* cardiac output is about the same as in a healthy untrained person, because stroke volume is increased while heart rate is decreased. The resting heart rate of a trained athlete often is only 40 to 60 beats per minute (resting bradycardia).

Help for Failing Hearts

As the heart fails, a person has decreasing ability to exercise or even to move around. A variety of surgical techniques and medical devices exist to aid a failing heart. For some patients, even a 10% increase in the volume of blood ejected from the ventricles can mean the difference between being bedridden and having limited mobility. **Heart transplants** are common today and produce good results, but the availability of donor hearts is very limited. For instance, there are 50 potential candidates for each of the 2500 donor hearts available each year in the United States. Another approach is to use **cardiac assist devices** and surgical procedures that augment heart function without removing the heart. Table 20.1 describes several of these.

Finally, scientists continue to develop and refine **artificial hearts,** mechanical devices that completely replace the functions of the natural heart. During the 1980s several patients received a Jarvik-7 artificial heart, which used an external power source to drive an internal pump by compressed air. In

Table 20.1	Cardiac Assist Devices and Procedures
Device	**Description**
Intra-aortic balloon pump (IABP)	A 40-mL polyurethane balloon mounted on a catheter is inserted into an artery in the groin and threaded into the thoracic aorta. An external pump inflates the balloon with gas at the beginning of ventricular diastole. As the balloon inflates, it pushes blood both backward toward the heart, which improves coronary blood flow, and forward toward peripheral tissues. The balloon then is rapidly deflated just before the next ventricular systole, making it easier for the left ventricle to eject blood. Because the balloon is inflated between heartbeats, this technique is called intra-aortic balloon counterpulsation.
Hemopump	This propeller-like pump is threaded through an artery in the groin and then into the left ventricle. There, the blades of the pump whirl at about 25,000 revolutions per minute, pulling blood out of the left ventricle and pushing it into the aorta.
Left ventricular assist device (LVAD)	The LVAD is a completely portable assist device. It is implanted within the abdomen and powered by a battery pack worn in a shoulder holster. The LVAD is connected to the patient's weakened left ventricle and pumps blood into the aorta. The pumping rate increases automatically during exercise.
Cardiomyoplasty	A large piece of the patient's own skeletal muscle (left latissimus dorsi) is partially freed from its connective tissue attachments and wrapped around the heart, leaving the blood and nerve supply intact. An implanted pacemaker stimulates the skeletal muscle's motor neurons to cause contraction 10–20 times per minute, in synchrony with some of the heartbeats.
Skeletal muscle assist device	A piece of the patient's own skeletal muscle is used to fashion a pouch that is inserted between the heart and the aorta, functioning as a booster heart. A pacemaker stimulates the muscle's motor neurons to elicit contraction.

1990, the U.S. Food and Drug Administration (FDA) banned use of the device because persistent problems with blood clotting caused strokes, and the chest tube for compressed air led to infections. More than a decade later, in July 2001, the first person received a fully self-contained artificial heart, the AbioCor Implantable Replacement Heart. Made of titanium, plastic, and epoxy, the 2-pound AbioCor heart is powered by a battery pack worn outside but with no wires penetrating through the skin. It alternately pumps blood from the left and then the right side of the heart. Because a permanent opening through the chest is not needed, the risk of infection should be much lower than with the Jarvik-7. Without the artificial heart, the first recipient was given little chance to survive more than a month due to his congestive heart failure, kidney disease, and diabetes. After surgery, he lived for 151 days, recovering enough to give several interviews and to enjoy a fishing trip. However, internal bleeding and organ failure unrelated to the AbioCor heart caused his death. By the end of 2001 several more people had received an AbioCor heart as part of the clinical trials needed to test any new device. Their progress provides encouragement that AbioCor hearts will be available on the market within the next few years. ■

DEVELOPMENT OF THE HEART

▶ **OBJECTIVE**

• **Describe the development of the heart.**

The *heart,* a derivative of **mesoderm,** begins to develop early in the third week after fertilization. In the ventral region of the embryo inferior to the foregut, the heart develops from a group of mesodermal cells called the **cardiogenic area.** The next step in its development is the formation of a pair of tubes, the **endocardial tubes,** which develop from the cardiogenic area (Figure 20.18). These tubes then unite to form a common tube, referred to as the **primitive heart tube.** Next, the primitive heart tube develops into five distinct regions: (1) **truncus arteriosus,** (2) **bulbus cordis,** (3) **ventricle,** (4) **atrium,** and (5) **sinus venosus.** Because the bulbus cordis and ventricle grow most rapidly, and because the heart enlarges more rapidly than its superior and inferior attachments, the heart first assumes a U-shape and then an S-shape. The flexures of the heart reorient the regions so that the atrium and sinus venosus eventually come to lie superior to the bulbus cordis, ventricle, and truncus arteriosus. Contractions

Figure 20.18   Development of the heart.   Arrows within the structures indicate the direction of blood flow.

🔑 The heart begins its development from a group of mesodermal cells called the cardiogenic area during the third week after fertilization.

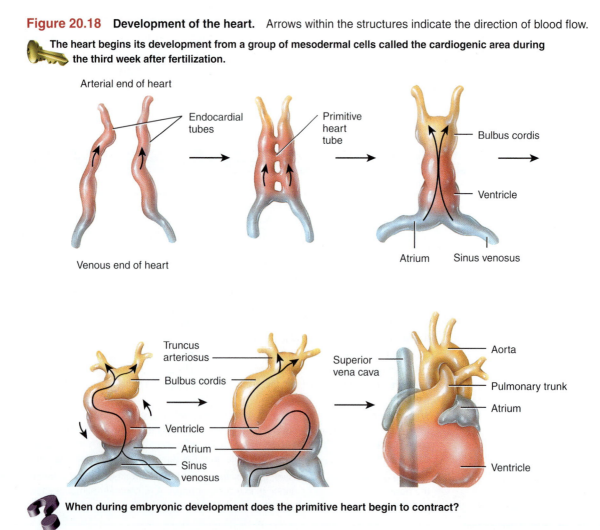

When during embryonic development does the primitive heart begin to contract?

of the primitive heart begin by the end of the third week of development; they originate in the sinus venosus and force blood through the tubular heart.

At about the seventh week of development, a partition called the **interatrial septum** forms in the atrial region. This septum divides the atrial region into a *right atrium* and *left atrium*. The opening in the partition is the **foramen ovale,** which normally closes at birth or soon thereafter and later forms a depression called the *fossa ovalis* (see Figure 20.4a). An **interventricular septum** also develops; it partitions the ventricular region into a *right ventri-*cle and *left ventricle.* The bulbus cordis and truncus arteriosus divide into two vessels, the *aorta* (arising from the left ventricle) and the *pulmonary trunk* (arising from the right ventricle). The great veins of the heart, the *superior vena cava* and the *inferior vena cava,* develop from the venous end of the primitive heart tube.

► **CHECKPOINT**

24. What are some of the cardiovascular benefits of regular exercise?

25. What structures develop from the bulbus cordis and truncus arteriosus?

DISORDERS: HOMEOSTATIC IMBALANCES

Coronary Artery Disease

In **coronary artery disease (CAD),** atherosclerotic plaques (described shortly) in coronary arteries reduce blood flow to the myocardium. CAD is the number-one killer of men and women in all ethnic groups, causing 750,000 deaths in the United States every year. Some individuals have no signs or symptoms; others experience angina pectoris (chest pain), and still others suffer a heart attack.

Risk Factors for CAD

People who possess combinations of certain risk factors are more likely to develop CAD. *Risk factors* are characteristics, symptoms, or signs present in a disease-free person that are statistically associated with a greater chance of developing a disease. Some risk factors for developing CAD can be minimized by lifestyle changes or by drugs. Among such risk factors for CAD are high blood cholesterol level, high blood pressure, cigarette smoking, obesity, diabetes mellitus, "type A" personality, and sedentary lifestyle. It has been found, for example, that moderate exercise and reducing fat in the diet can not only stop the progression of CAD, but actually reverse the process. Other risk factors are beyond our control. These include genetic predisposition (family history of CAD at an early age), age, and gender. Adult males are more likely than adult females to develop CAD until age 70 when the risks become roughly equal.

Development of Atherosclerotic Plaques

Although the following discussion of atherosclerosis applies to coronary arteries, the process can also occur in arteries outside the heart. Thickening of the walls of arteries and loss of elasticity are the main characteristics of a group of diseases called **arteriosclerosis** (ar-tē-rē-ō-skle-RŌ-sis; *sclero-* = hardening). One form of arteriosclerosis is **atherosclerosis** (ath-er-ō-skle-RŌ-sis), a progressive disease characterized by the formation of lesions called **atherosclerotic plaques** in the walls of large and medium-sized arteries (Figure 20.19). Atherosclerosis is initiated by one or more unknown factors that cause damage to the endothelial lining of the arterial wall. Factors that may initiate the process include high circulating LDL levels, cytomegalovirus (a common herpes virus), prolonged high blood pressure, carbon monoxide in cigarette smoke, and high blood glucose levels in diabetes mellitus. Atherosclerosis is thought to begin when one of these factors injures the endothelium of an artery, promoting the aggregation of platelets and attracting phagocytes.

At the injury site, cholesterol and triglycerides collect in the inner layer of the arterial wall. Contact with platelets, lipids, and other components of blood stimulates smooth muscle cells in the arterial wall to proliferate abnormally. As an atherosclerotic plaque develops and enlarges, it progressively obstructs blood flow. An additional danger is that a plaque may provide a roughened surface that attracts platelets,

Figure 20.19 Photomicrographs of a transverse section of (a) a normal artery and (b) an artery partially obstructed by an atherosclerotic plaque.

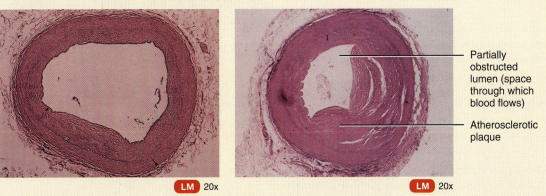

LM 20x
(a) Normal artery

LM 20x
(b) Obstructed artery

Partially obstructed lumen (space through which blood flows)

Atherosclerotic plaque

initiating clot formation, and further obstructing blood flow. Moreover, a thrombus or piece of a thrombus may dislodge to become an embolus and obstruct blood flow in other vessels.

Diagnosis of CAD

Cardiac catheterization (kath′-e-ter-i-ZĀ-shun) is an invasive procedure used to visualize the heart's coronary arteries, chambers, valves, and great vessels. It may also be used to measure pressure in the heart and blood vessels; to assess cardiac output; to measure the flow of blood through the heart and blood vessels; and to identify the location of septal and valvular defects. The basic procedure involves inserting a long, flexible, radiopaque **catheter** (plastic tube) into a peripheral vein (for right heart catheterization) or a peripheral artery (for left heart catheterization) and guiding it under fluoroscopy (x-ray observation).

Cardiac angiography (an′-jē-OG-ra-fē) is another invasive procedure in which a cardiac catheter is used to inject a radiopaque contrast medium into blood vessels or heart chambers. The procedure may be used to visualize coronary arteries, the aorta, pulmonary blood vessels, and the ventricles to assess structural abnormalities in blood vessels (such as atherosclerotic plaques and emboli), ventricular volume, wall thickness, and wall motion. Angiography can also be used to inject clot-dissolving drugs, such as streptokinase or tissue plasminogen activator (t-PA), into a coronary artery to dissolve an obstructing thrombus.

Treatment of CAD

Treatment options for CAD include drugs (nitroglycerine, beta blockers, cholesterol-lowering drugs, and clot-dissolving agents) and various surgical and nonsurgical procedures designed to increase the blood supply to or decrease the metabolic demands of the heart.

Coronary artery bypass grafting (CABG) is a surgical procedure in which a blood vessel from another part of the body is attached ("grafted") to a coronary artery to bypass an area of blockage. A piece of the grafted blood vessel is sutured between the aorta and the unblocked portion of the coronary artery (Figure 20.20a).

A nonsurgical procedure used to treat CAD is termed **percutaneous transluminal coronary angioplasty (PTCA)** (*percutaneous* = through the skin; *trans-* = across; *lumen* = an opening or channel in a tube; *angio-* = blood vessel; *-plasty* = to mold or to shape). In this procedure, a balloon catheter is inserted into an artery of an arm or leg and gently guided into a coronary artery (Figure 20.20b). Then, while dye is released, angiograms are taken to locate the plaques. Next, the catheter is advanced to the point of obstruction, and a balloonlike device is inflated with air to squash the plaque against the blood vessel wall. Because 30–50% of PTCA-opened arteries fail due to restenosis (renarrowing) within six months after the procedure is done, a stent may be inserted via a catheter. A *stent* is a stainless steel device, resembling a spring coil. It is permanently placed in an artery to keep the artery open (Figure 20.20c).

Myocardial Ischemia and Infarction

Partial obstruction of blood flow in the coronary arteries may cause **myocardial ischemia** (is-KĒ-mē-a; *ische-* = to obstruct; *-emia* = in the blood), a condition of reduced blood flow to the myocardium. Usually, ischemia causes **hypoxia** (reduced oxygen supply), which may weaken cells without killing them. **Angina pectoris** (an-JĪ-na, or AN-ji-na, PEK-to-ris), literally meaning "strangled chest," is a severe pain

Figure 20.20 Three procedures for reestablishing blood flow in occluded coronary arteries.

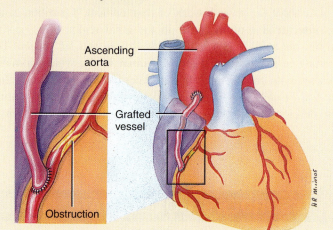

(a) Coronary artery bypass grafting (CABG)

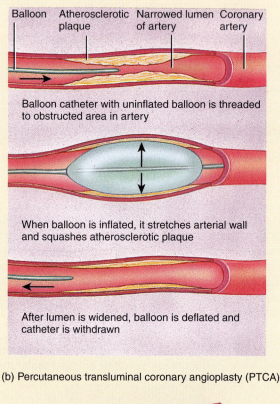

Balloon catheter with uninflated balloon is threaded to obstructed area in artery

When balloon is inflated, it stretches arterial wall and squashes atherosclerotic plaque

After lumen is widened, balloon is deflated and catheter is withdrawn

(b) Percutaneous transluminal coronary angioplasty (PTCA)

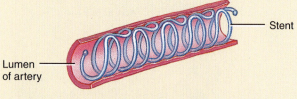

(c) Stent in an artery

that usually accompanies myocardial ischemia. Typically, sufferers describe it as a tightness or squeezing sensation, as though the chest were in a vise. Angina pectoris often occurs during exertion, when the heart demands more oxygen, and then disappears with rest. The pain associated with angina pectoris is often referred to the neck, chin, or down the left arm to the elbow. In some people, ischemic episodes occur without producing pain. This condition is known as **silent myocardial ischemia** and is particularly dangerous because the person has no forewarning of an impending heart attack.

A complete obstruction to blood flow in a coronary artery may result in a **myocardial infarction** (in-FARK-shun), or **MI,** commonly called a heart attack. *Infarction* means the death of an area of tissue because of interrupted blood supply. Because the heart tissue distal to the obstruction dies and is replaced by noncontractile scar tissue, the heart muscle loses some of its strength. The aftereffects depend partly on the size and location of the infarcted (dead) area. Besides killing normal heart tissue, an infarction may disrupt the conduction system of the heart and cause sudden death by triggering ventricular fibrillation. Treatment for a myocardial infarction may involve injection of a thrombolytic (clot-dissolving) agent such as streptokinase or t-PA, plus heparin (an anticoagulant), or performing coronary angioplasty or coronary artery bypass grafting. Fortunately, heart muscle can remain alive in a resting person if it receives as little as 10–15% of its normal blood supply. Furthermore, the extensive anastomoses of coronary blood vessels are a significant factor in helping some people survive myocardial infarctions.

Congenital Heart Defects

A defect that is present at birth, and usually before, is called a **congenital defect.** Many such defects are not serious and may go unnoticed for a lifetime. Others are life threatening and must be surgically repaired. Among the several congenital defects that affect the heart are the following:

*   **Coarctation** (kō′-ark-TĀ-shun) **of the aorta.** In this condition, a segment of the aorta is too narrow, and thus the flow of oxygenated blood to the body is reduced, the left ventricle is forced to pump harder, and high blood pressure develops.
*   **Patent ductus arteriosus.** In some babies, the ductus arteriosus, a temporary blood vessel between the aorta and the pulmonary trunk, remains open rather than closing shortly after birth. As a result, aortic blood flows into the lower-pressure pulmonary trunk, thus increasing the pulmonary trunk blood pressure and overworking both ventricles.
*   **Septal defect.** A septal defect is an opening in the septum that separates the interior of the heart into left and right sides. In an **atrial septal defect** the fetal foramen ovale between the two atria fails to close after birth. A **ventricular septal defect** is caused by incomplete development of the interventricular septum. In such cases, oxygenated blood flows directly from the left ventricle into the right ventricle, where it mixes with deoxygenated blood.
*   **Tetralogy of Fallot** (tet-RAL-ō-jē of fal-Ō). This condition is a combination of four developmental defects: an interventricular septal defect, an aorta that emerges from both ventricles instead of from the left ventricle only, a stenosed pulmonary valve, and an enlarged right ventricle. Deoxygenated blood from the right ventricle enters the left ventricle through the interventricular septum, mixes with oxygenated blood, and is pumped into the systemic circulation. Moreover, because the aorta emerges from the right ventricle and the pulmonary trunk is stenosed, very little blood reaches the pulmonary circulation. This causes cyanosis, the bluish discoloration most easily seen in nail beds and mucous membranes when the level of deoxygenated hemoglobin is high. For this reason, tetralogy of Fallot is one of the conditions that cause a "blue baby."

Arrhythmias

The usual rhythm of heartbeats, established by the SA node, is **normal sinus rhythm.** An **arrhythmia** (a-RITH-mē-a) or **dysrhythmia** is an irregularity in heart rhythm resulting from a defect in the conduction system of the heart. Arrhythmias can be caused by caffeine, nicotine, alcohol, and certain other drugs; anxiety; hyperthyroidism; potassium deficiency; and certain heart diseases.

Heart Block

One serious arrhythmia is **heart block,** in which propagation of action potentials through the conduction system is either slowed or blocked at some point. The most common site of blockage is the atrioventricular node, a condition called **atrioventricular (AV) block.** In *first-degree AV block,* the P-Q interval is prolonged, usually because conduction through the AV node is slower than normal. In *second-degree AV block,* some of the action potentials from the SA node are not conducted through the AV node. The result is "dropped" beats because excitation doesn't always reach the ventricles. In *third-degree (complete) AV block,* no SA node action potentials get through the AV node. Autorhythmic fibers in the atria and ventricles pace the upper and lower chambers separately. With complete AV block, the ventricular contraction rate is less than 40 beats/min.

Flutter and Fibrillation

Two other abnormal rhythms are **flutter** and **fibrillation. Atrial flutter** consists of rapid atrial contractions (240–360 beats/min) accompanied by a second-degree AV block. Flutter may result from rheumatic heart disease, coronary artery disease, or certain congenital heart diseases. In **atrial fibrillation** contraction of atrial fibers is asynchronous (not in unison) so that atrial pumping ceases altogether. Atrial fibrillation may occur in myocardial infarction, acute and chronic rheumatic heart disease, and hyperthyroidism. In an otherwise strong heart, atrial fibrillation reduces the pumping effectiveness of the heart by 20–30%.

Ventricular fibrillation is the most deadly arrhythmia, in which contractions of ventricular muscle fibers are completely asynchronous. Ventricular pumping stops, blood ejection ceases, and circulatory failure and death occur unless the situation is quickly corrected. A procedure known as **defibrillation,** in which a strong, brief electrical current is passed through the heart, often can stop ventricular fibrillation. The electric shock is applied via large paddle-shaped electrodes pressed against the skin of the chest. Patients who face a high risk of dying from heart rhythm disorders now can receive an **automatic implantable cardioverter defibrillator (AICD),** an implanted device that monitors their heart rhythm and delivers a small shock directly to the heart when a life-threatening rhythm disturbance occurs. Thousands of patients around the world have AICDs, including Dick Cheney, Vice President of the United States, who received a combination pacemaker-defibrillator in 2001.

Ventricular Premature Contraction

Another form of arrhythmia arises when an ectopic focus, a region of the heart other than the conduction system, becomes more excitable than normal and causes an occasional abnormal action potential to occur. As a wave of depolarization spreads outward from the ectopic focus, it causes a **ventricular premature contraction.** The contrac-tion occurs early in diastole before the SA node is normally scheduled to discharge its action potential. Ventricular premature contractions may be relatively benign and may be caused by emotional stress, excessive intake of stimulants such as caffeine or nicotine, and lack of sleep. In other cases, the premature beats may reflect an underlying pathology.

MEDICAL TERMINOLOGY

Angiocardiography (an′-jē-ō-kar′-dē-OG-ra-fē; *angio-* = vessel; *cardio-* = heart) X-ray examination of the heart and great blood vessels after injection of a radiopaque dye into the bloodstream.

Cardiac arrest (KAR-dē-ak a-REST) A clinical term meaning cessation of an effective heartbeat. The heart may be completely stopped or in ventricular fibrillation.

Cardiomegaly (kar′-dē-ō-MEG-a-lē; *mega* = large) Heart enlargement.

Cor pulmonale (CP) (kor pul-mōn-ALE; *cor-* = heart; *pulmon-* = lung) A term referring to right ventricular hypertrophy from disorders that bring about hypertension (high blood pressure) in the pulmonary circulation.

Ejection fraction The fraction of the end-diastolic volume (EDV) that is ejected during an average heartbeat. Equal to stroke volume (SV) divided by EDV.

Palpitation (pal′-pi-TĀ-shun) A fluttering of the heart or an abnormal rate or rhythm of the heart.

Paroxysmal tachycardia (par′-ok-SIZ-mal tak′-e-KAR-dē-a; *tachy-* = swift or fast) A period of rapid heartbeats that begins and ends suddenly.

Sudden cardiac death The unexpected cessation of circulation and breathing due to an underlying heart disease such as ischemia, myocardial infarction, or a disturbance in cardiac rhythm.

STUDY OUTLINE

ANATOMY OF THE HEART (p. 660)

1. The heart is located in the mediastinum; about two-thirds of its mass is to the left of the midline.
2. The heart is shaped like a cone lying on its side; its apex is the pointed, inferior part, whereas its base is the broad, superior part.
3. The pericardium is the membrane that surrounds and protects the heart; it consists of an outer fibrous layer and an inner serous pericardium, which is composed of a parietal and a visceral layer.
4. Between the parietal and visceral layers of the serous pericardium is the pericardial cavity, a potential space filled with a few milliliters of pericardial fluid that reduces friction between the two membranes.
5. Three layers make up the wall of the heart: epicardium (visceral layer of the serous pericardium), myocardium, and endocardium.
6. The epicardium consists of mesothelium and connective tissue, the myocardium is composed of cardiac muscle tissue, and the endocardium consists of endothelium and connective tissue.
7. The heart chambers include two superior chambers, the right and left atria, and two inferior chambers, the right and left ventricles.
8. External features of the heart include the auricles (flaps on each atrium that slightly increase their volume), the coronary sulcus between the atria and ventricles, and the anterior and posterior sulci between the ventricles on the anterior and posterior surfaces of the heart, respectively.
9. The right atrium receives blood from the superior vena cava, inferior vena cava, and coronary sinus. It is separated from the left atrium by the interatrial septum, which contains the fossa ovalis. Blood exits the right atrium through the tricuspid valve.

10. The right ventricle receives blood from the right atrium. It is separated from the left ventricle by the interventricular septum and pumps blood through the pulmonary valve into the pulmonary trunk.
11. Oxygenated blood enters the left atrium from the pulmonary veins and exits through the bicuspid (mitral) valve.
12. The left ventricle pumps oxygenated blood through the aortic valve into the aorta.
13. The thickness of the myocardium of the four chambers varies according to the chamber's function. The left ventricle, with the highest workload, has the thickest wall.
14. The fibrous skeleton of the heart is dense connective tissue that surrounds and supports the valves of the heart.

HEART VALVES AND CIRCULATION OF BLOOD (p. 667)

1. Heart valves prevent backflow of blood within the heart. The atrioventricular (AV) valves, which lie between atria and ventricles, are the tricuspid valve on the right side of the heart and the bicuspid (mitral) valve on the left. The semilunar (SL) valves are the aortic valve, at the entrance to the aorta, and the pulmonary valve, at the entrance to the pulmonary trunk.
2. The left side of the heart is the pump for the systemic circulation, the circulation of blood throughout the body except for the air sacs of the lungs. The left ventricle ejects blood into the aorta, and blood then flows into systemic arteries, arterioles, capillaries, venules, and veins, which carry it back to the right atrium.
3. The right side of the heart is the pump for pulmonary circulation, the circulation of blood through the lungs. The right ventricle ejects blood into the pulmonary trunk, and blood then flows into

pulmonary arteries, pulmonary capillaries, and pulmonary veins, which carry it back to the left atrium.

4. The coronary circulation provides blood flow to the myocardium. The main arteries of the coronary circulation are left and right coronary arteries; the main veins are the cardiac vein and the coronary sinus.

CARDIAC MUSCLE TISSUE AND THE CARDIAC CONDUCTION SYSTEM (p. 672)

1. Cardiac muscle fibers usually contain a single centrally located nucleus. Compared to skeletal muscle fibers, cardiac muscle fibers have more and larger mitochondria, slightly smaller sarcoplasmic reticulum, and wider transverse tubules, which are located at Z discs.
2. Cardiac muscle fibers are connected via end-to-end intercalated discs. Desmosomes in the discs provide strength and gap junctions allow muscle action potentials to conduct from one muscle fiber to its neighbors.
3. Autorhythmic fibers form the conduction system, cardiac muscle fibers that spontaneously depolarize and generate action potentials.
4. Components of the conduction system are the sinoatrial (SA) node (pacemaker), atrioventricular (AV) node, atrioventricular (AV) bundle (bundle of His), bundle branches, and Purkinje fibers.
5. Phases of an action potential in a ventricular contractile fiber include rapid depolarization, a long plateau, and repolarization.
6. Cardiac muscle tissue has a long refractory period, which prevents tetanus.
7. The record of electrical changes during each cardiac cycle is called an electrocardiogram (ECG). A normal ECG consists of a P wave (atrial depolarization), a QRS complex (onset of ventricular depolarization), and a T wave (ventricular repolarization).
8. The P–Q interval represents the conduction time from the beginning of atrial excitation to the beginning of ventricular excitation. The S–T segment represents the time when ventricular contractile fibers are fully depolarized.

THE CARDIAC CYCLE (p. 678)

1. A cardiac cycle consists of the systole (contraction) and diastole (relaxation) of both atria, plus the systole and diastole of both ventricles. With an average heartbeat of 75 beats/min, a complete cardiac cycle requires 0.8 seconds.

2. The phases of the cardiac cycle are (a) atrial systole, (b) ventricular systole, and (c) relaxation period.
3. S1, the first heart sound (lubb), is caused by blood turbulence associated with the closing of the atrioventricular valves. S2, the second sound (dupp), is caused by blood turbulence associated with the closing of semilunar valves.

CARDIAC OUTPUT (P. 683)

1. Cardiac output (CO) is the amount of blood ejected per minute by the left ventricle into the aorta (or by the right ventricle into the pulmonary trunk). It is calculated as follows: CO (mL/min) = stroke volume (SV) in mL/beat × heart rate (HR) in beats per minute.
2. Stroke volume (SV) is the amount of blood ejected by a ventricle during each systole.
3. Cardiac reserve is the difference between a person's maximum cardiac output and his or her cardiac output at rest.
4. Stroke volume is related to preload (stretch on the heart before it contracts), contractility (forcefulness of contraction), and afterload (pressure that must be exceeded before ventricular ejection can begin).
5. According to the Frank–Starling law of the heart, a greater preload (end-diastolic volume) stretching cardiac muscle fibers just before they contract increases their force of contraction until the stretching becomes excessive.
6. Nervous control of the cardiovascular system stems from the cardiovascular center in the medulla oblongata.
7. Sympathetic impulses increase heart rate and force of contraction; parasympathetic impulses decrease heart rate.
8. Heart rate is affected by hormones (epinephrine, norepinephrine, thyroid hormones), ions (Na^+, K^+, Ca^{2+}), age, gender, physical fitness, and body temperature.

EXERCISE AND THE HEART (p. 687)

1. Sustained exercise increases oxygen demand on muscles.
2. Among the benefits of aerobic exercise are increased cardiac output, decreased blood pressure, weight control, and increased fibrinolytic activity.

DEVELOPMENT OF THE HEART (p. 688)

1. The heart develops from mesoderm.
2. The endocardial tubes develop into the four-chambered heart and great vessels of the heart.

Q SELF-QUIZ QUESTIONS

Fill in the blanks in the following statements.

1. The chamber of the heart with the thickest myocardium is the _____.
2. The phase of heart contraction is called _____; the phase of heart relaxation is called _____.
3. Cardiac output equals _____ multiplied by heart rate.

Indicate whether the following statements are true or false.

4. In auscultation, the lubb represents closing of the semilunar valves and the dupp represents closing of the atrioventricular valves.
5. Positive inotropic agents decrease myocardial contractility.

6. The Frank–Starling law of the heart equalizes the output of the right and left ventricles and keeps the same volume of blood flowing to both the systemic and pulmonary circulations.

Choose the one best answer to the following questions.

7. Which of the following is the correct route of blood through the heart from the systemic circulation to the pulmonary circulation and back to the systemic circulation? (a) right atrium, tricuspid valve, right ventricle, pulmonary semilunar valve, left atrium, mitral valve, left ventricle, aortic semilunar valve, (b) left atrium, tricuspid valve, left ventricle, pulmonary semilunar valve, right atrium, mitral valve, right

ventricle, aortic semilunar valve, (c) left atrium, pulmonary semilunar valve, right atrium, tricuspid valve, left ventricle, aortic semilunar valve, right ventricle, mitral valve, (d) left ventricle, mitral valve, left atrium, pulmonary semilunar valve, right ventricle, tricuspid valve, right atrium, aortic semilunar valve, (e) right atrium, mitral valve, right ventricle, pulmonary semilunar valve, left atrium, tricuspid valve, left ventricle, aortic semilunar valve.

8. Which of the following represents the correct pathway for conduction of an action potential through the heart? (a) AV node, AV bundle, SA node, Purkinje fibers, bundle branches, (b) AV node, bundle branches, AV bundle, SA node, Purkinje fibers, (c) SA node, AV node, AV bundle, bundle branches, Purkinje fibers, (d) SA node, AV bundle, bundle branches, AV node, Purkinje fibers, (e) SA node, AV node, Purkinje fibers, bundle branches, AV bundle.

9. A softball player is found to have a resting cardiac ouput of 5.0 liters per minute and a heart rate of 50 beats per minute. What is her stroke volume? (a) 10 mL; (b) 100 mL; (c) 1000 mL; (d) 6000 mL; (e) The information given is insufficient to calculate stroke volume.

10. When would a murmur due to stenosis of a semilunar valve be heard? (a) between S1 and S2, when the valve should be fully open but is not; (b) between S1 and S2, when the valve should be fully closed but is not; (c) between S2 and S1, when the valve should be fully open but is not; (d) between S2 and S1, when the valve should be fully closed but is not.

11. Which of the following are *true*? (1) ANS regulation of heart rate originates in the cardiovascular center of the medulla oblongata. (2) Proprioceptor input is a major stimulus that accounts for the rapid rise in the heart rate at the onset of physical activity. (3) The vagus nerves release norepinephrine, causing the heart rate to increase. (4) Hormones from the adrenal medulla and the thyroid gland can increase the heart rate. (5) Hypothermia increases the heart rate. (a) 1, 2, 3, and 4, (b) 1, 2, and 4, (c) 2, 3, 4, and 5, (d) 3, 5, and 6, (e) 1, 2, 4, and 5

12. Match the following:
_____ (a) collects oxygenated blood from the pulmonary circulation
_____ (b) pumps deoxygenated blood to the lungs for oxygenation
_____ (c) their contraction pulls on and tightens the chordae tendineae, preventing the valve cusps from everting
_____ (d) increase blood-holding capacity of the atria
_____ (e) tendonlike cords connected to the atrioventricular valve cusps which, along with the papillary muscles, prevent valve eversion
_____ (f) pumps oxygenated blood to all body cells, except the air sacs of the lungs
_____ (g) collects deoxygenated blood from the systemic circulation
_____ (h) prevents backflow of blood from the right ventricle into the right atrium
_____ (i) left atrioventricular valve

(1) right atrium
(2) right ventricle
(3) left atrium
(4) left ventricle
(5) tricuspid valve
(6) bicuspid (mitral) valve
(7) chordae tendineae
(8) auricles
(9) papillary muscles

13. Match the following:
_____ (a) cardiac muscle tissue
_____ (b) blood vessels that pierce the heart muscle and supply blood to the cardiac muscle fibers
_____ (c) separate the upper and lower heart chambers, preventing backflow of blood from the inferior chambers back into the superior chambers
_____ (d) endothelial cells lining the interior of the heart; are continuous with the endothelium of the blood vessels
_____ (e) inner visceral layer of the serous pericardium; adheres tightly to the surface of the heart
_____ (f) the remnant of the foramen ovale, an opening in the interatrial septum of the fetal heart
_____ (g) outer layer of the serous pericardium; is fused to the fibrous pericardium
_____ (h) the gap junction and desmosome connections between individual cardiac muscle fibers
_____ (i) the superficial dense irregular connective tissue covering of the heart
_____ (j) prevent backflow of blood from the arteries into the inferior heart chambers

(1) fibrous pericardium
(2) parietal pericardium
(3) epicardium
(4) myocardium
(5) endocardium
(6) atrioventricular valves
(7) semilunar valves
(8) intercalated discs
(9) fossa ovalis
(10) coronary circulation

14. Match the following:
_____ (a) amount of blood contained in the ventricles at the end of ventricular relaxation
_____ (b) period of time when cardiac muscle fibers are contracting and exerting force but not shortening
_____ (c) amount of blood ejected per beat by each ventricle
_____ (d) amount of blood remaining in the ventricles following ventricular contraction
_____ (e) difference between a person's maximum cardiac output and cardiac output at rest
_____ (f) period of time when semilunar valves are open and blood flows out of the inferior chambers
_____ (g) period when all four valves are closed and ventricular blood volume does not change

(1) cardiac reserve
(2) stroke volume
(3) end-diastolic volume (EDV)
(4) isovolumetric relaxation
(5) end-systolic volume (ESV)
(6) ventricular ejection
(7) isovolumetric contraction

15. Match the following:

____ (a) indicates ventricular repolarization
____ (b) represents the time from the beginning of ventricular depolarization to the end of ventricular repolarization
____ (c) represents atrial depolarization
____ (d) represents the time when the ventricular contractile fibers are fully depolarized; occurs during the plateau phase of the action potential
____ (e) represents the onset of ventricular depolarization
____ (f) represents the conduction time from the beginning of atrial excitation to the beginning of ventricular excitation

(1) P wave
(2) QRS complex
(3) T wave
(4) P–Q interval
(5) S–T segment
(6) Q–T interval

CRITICAL THINKING QUESTIONS

1. Arian, a member of the college track team, volunteered for a study that the exercise physiology class was conducting on the cardiovascular system. His resting heart rate was measured at 55 beats per minute and he had a normal cardiac output (CO). After strenuous exercise, his cardiac reserve was calculated to be six times the resting value. Arian is in excellent physical condition. Predict Arian's resting CO and stroke volume. What would his CO be during strenuous exercise?

HINT *At rest, each ventricle pumps a volume of blood per minute that is about equal to the body's total blood volume.*

2. At family gatherings, Aunt Mercy likes to tell the story of when she had rheumatic fever as a child and how her mother treated her with home remedies. Now Aunt Mercy complains of a "weak heart." Could the childhood illness be related to her heart problem?

HINT *The immune system may mistakenly attack healthy body tissue along with the bacteria.*

3. Mr. Perkins is a large, 62-year-old man with a weakness for sweets and fried foods. His idea of exercise is to get up to change the channel while watching sports on television. Lately, he's been troubled by chest pains when he walks up stairs. His doctor told him to quit smoking and scheduled a cardiac angiography for next week. What is involved in performing this procedure? Why did the doctor order this test?

HINT *Mr Perkins' stress electrocardiogram revealed changes that suggest myocardial ischemia.*

ANSWERS TO FIGURE QUESTIONS

20.1 The mediastinum is the mass of tissue that extends from the sternum to the vertebral column between the lungs.

20.2 The visceral layer of the serous pericardium (epicardium) is both a part of the pericardium and a part of the heart wall.

20.3 The coronary sulcus forms a boundary between the atria and ventricles.

20.4 The greater the workload of a heart chamber, the thicker its myocardium.

20.5 The fibrous skeleton attaches to the heart valves and prevents overstretching of the valves as blood passes through them.

20.6 The papillary muscles contract, which pulls on the chordae tendineae and prevents valve cusps from everting.

20.7 Numbers 2 (right ventricle) through 6 depict the pulmonary circulation, whereas numbers 7 (left ventricle) through 10 and 1 (right atrium) depict the systemic circulation.

20.8 The circumflex artery delivers oxygenated blood to the left atrium and left ventricle.

20.9 The intercalated discs hold the cardiac muscle fibers together and enable action potentials to propagate from one muscle fiber to another.

20.10 The only electrical connection between the atria and the ventricles is the atrioventricular bundle.

20.11 The duration of an action potential is much longer in a ventricular contractile fiber (0.3 sec = 300 msec) than in a skeletal muscle fiber (1–2 msec).

20.12 An enlarged Q wave may indicate a myocardial infarction (heart attack).

20.13 Action potentials propagate most slowly through the AV node.

20.14 The amount of blood in each ventricle at the end of ventricular diastole—the end-diastolic volume—is about 130 mL in a resting person.

20.15 The first sound (S1), or lubb, is associated with closure of the AV valves.

20.16 The ventricular myocardium receives innervation from the sympathetic division only.

20.17 The skeletal muscle "pump" increases stroke volume by increasing preload (end-diastolic volume).

20.18 The heart begins to contract by the end of the third week of development.

The Cardiovascular System:

Blood Vessels

and Hemodynamics

BLOOD VESSELS HEMO-DYNAMICS AND HOMEOSTASIS

Blood vessels contribute to homeostasis by providing the structures for the flow of blood to and from the heart and the exchange of nutrients and wastes in tissues. They also play an important role in adjusting the velocity and volume of blood flow.

ENERGIZE YOUR STUDY

FOUNDATIONS CD

Anatomy Overview
• The Cardiovascular System

Animations
• Systems Contributions to Homeostasis
• Negative Feedback Control of Blood Pressure
• The Case of the Man with a Pain in His Jaw
• The Case of the Girl with the Fruity Breath

Concepts and Connections and Exercises reinforce your understanding

www.wiley.com/college/apcentral

INSIGHTS AND EXPLORATIONS

Heart bypass surgery has almost become commonplace in the United States. What is not so widely known is that many of these individuals must have bypass surgery a second time and suffer serious side effects from the surgery. Scientists are busy studying how blood vessels grow (angiogenesis) in order to some day use them in victims of heart attacks. Some are trying to grow blood vessels in the laboratory, while others are studying how time-release capsules containing growth factors can aid in the growth of blood vessels in the heart. Come explore more on the web about how blood vessels grow and how scientists hope this will help patients with heart disease.

A Practitioner's View

After studying anatomy and physiology, it is so neat to witness the real life events and have an understanding of the science behind it. Students headed for a health career will encounter events like the ones I describe on a daily basis.

As a Cardiographic Technician, one of the topics in this chapter that I find interesting and relevant to my job is syncope. Doctors order a test called a Tilt Table test for patients with a history of syncopal episodes (passing out), trying to recreate the event in a controlled environment to determine if the problem is cardiac in nature. This test is performed with the patient secured on a bed while heart rate, heart rhythm, and blood pressure are closely monitored as the bed is tilted upright to an angle between 60 and 80 degrees. The rapid change in vital signs that occurs in the seconds preceding the syncopal episode and also the rapid recovery once the patient is returned to a horizontal position are very interesting. Their blood pressure drops dramatically, sometimes completely inaudibly, and their pulse decreases to almost nothing for a brief moment. One of the Clinical Applications in this chapter gives possible explanations for the physiology of why this happens.

In cases where it does not appear that the patient is affected by the change in bed position, the physician might do a carotid sinus massage, also covered in one of the Clinical Applications. It was amazing to me the first time I saw this done because the patient's heart rhythm changed dramatically on the rhythm strips just by the doctor palpating and massaging the patient's neck in a specific spot. It was just like a textbook description of the events that occur when the carotid sinus baroreceptors are stretched.

Jill Haan, Cardiographic Technician
Home Hospital

The cardiovascular system contributes to homeostasis of other body systems by transporting and distributing blood throughout the body to deliver materials such as oxygen, nutrients, and hormones and to carry away wastes. This transport occurs via blood vessels, which form a closed system of tubes that carries blood away from the heart, transports it to the tissues of the body, and then returns it to the heart. Chapters 19 and 20 described the composition and functions of blood and the structure and function of the heart. In this chapter, the focus is on the structure and functions of the various types of blood vessels; on **hemodynamics** (hē-mō-dī-NAM-iks; *hemo-* = blood; *dynamics* = power), the forces involved in circulating blood throughout the body; and on the blood vessels that constitute the major circulatory routes.

STRUCTURE AND FUNCTION OF BLOOD VESSELS

▶ **O B J E C T I V E**

- **Contrast the structure and function of arteries, arterioles, capillaries, venules, and veins.**

The five main types of blood vessels are arteries, arterioles, capillaries, venules, and veins. **Arteries** (AR-ter-ēz) carry blood *away from the heart* to other organs. Large, elastic arteries leave the heart and divide into medium-sized, muscular arteries that branch out into the various regions of the body. Medium-sized arteries then divide into small arteries, which, in turn, divide into still smaller arteries called **arterioles** (ar-TER-ē-ōls). As the arterioles enter a tissue, they branch into myriad tiny vessels called **capillaries** (KAP-i-lar'-ēz = hairlike). The thin walls of capillaries allow exchange of substances between the blood and body tissues. Groups of capillaries within a tissue reunite to form small veins called **venules** (VEN-ūls). These, in turn, merge to form progressively larger blood vessels called veins. **Veins** (VĀNZ) are the blood vessels that convey blood from the tissues *back to the heart.* Because blood vessels require oxygen (O$_2$) and nutrients just like other tissues of the body, larger blood vessels are served by their own blood vessels, called **vasa vasorum** (literally, vasculature of vessels), located within their walls.

Arteries

In ancient times, **arteries** (*ar-* = air; *ter-* = to carry) were found empty at death and thus were thought to contain only air. The wall of an artery has three coats or tunics: (1) tunica interna, (2) tunica media, and (3) tunica externa (Figure 21.1). The innermost coat, the **tunica interna** or **intima,** contains a lining of simple squamous epithelium called *endothelium, a basement membrane,* and a layer of elastic tissue called the *internal elastic lamina.* The endothelium is a continuous layer of cells that line the inner surface of the entire cardiovascular system (the heart and all blood vessels). Normally, endothelium is the only tissue that makes contact with blood. The tunica interna is closest to the **lumen,** the hollow center through which blood flows. The middle coat, or **tunica media,** is usually the thickest layer. It consists of elastic fibers and smooth muscle fibers (cells) that extend circularly (in rings) around the lumen. Due to their plentiful elastic fibers, arteries normally have high *compliance,* which means that their walls easily stretch or expand without tearing in response to a small increase in pressure. The outer coat, the **tunica externa,** is composed mainly of elastic and collagen fibers. In muscular arteries (described shortly), an *external elastic lamina* composed of elastic tissue separates the tunica externa from the tunica media.

Sympathetic neurons of the autonomic nervous system innervate vascular smooth muscle. An increase in sympathetic stimulation typically stimulates the smooth muscle to contract, squeezing the vessel wall and narrowing the lumen. Such a decrease in the diameter of the lumen of a blood vessel is called **vasoconstriction.** In contrast, smooth muscle fibers relax when sympathetic stimulation decreases or when certain chemicals, such as nitric oxide, K$^+$, H$^+$, and lactic acid, are present. The resulting increase in lumen diameter is called **vasodilation.** Additionally, when an artery or arteriole is damaged, its smooth muscle contracts, producing vascular spasm of the vessel. Such a vasospasm limits blood flow through the damaged vessel and helps reduce blood loss if the vessel is small.

Elastic Arteries

The largest-diameter arteries, termed **elastic arteries** because the tunica media contains a high proportion of elastic fibers, have walls that are relatively thin in proportion to their overall diameter. Elastic arteries perform an important function: They help propel blood onward while the ventricles are relaxing. As blood is ejected from the heart into elastic arteries, their highly elastic walls stretch, accommodating the surge of blood. By stretching, the elastic fibers store mechanical energy for a short time, functioning as a **pressure reservoir** (Figure 21.2a on page 699). Then, the elastic fibers recoil and convert stored (potential) energy in the vessel into kinetic energy of the blood. Thus, blood continues to move through the arteries even while the ventricles are relaxed (Figure 21.2b). Because they conduct blood from the heart to medium-sized, more-muscular arteries, elastic arteries also are called *conducting arteries.* The aorta and the brachiocephalic, common carotid, subclavian, vertebral, pulmonary, and common iliac arteries are elastic arteries (see Figure 21.18).

Figure 21.1 **Comparative structure of blood vessels.** The relative size of the capillary in (c) is enlarged.

🔑 Arteries carry blood from the heart to tissues; veins carry blood from tissues to the heart.

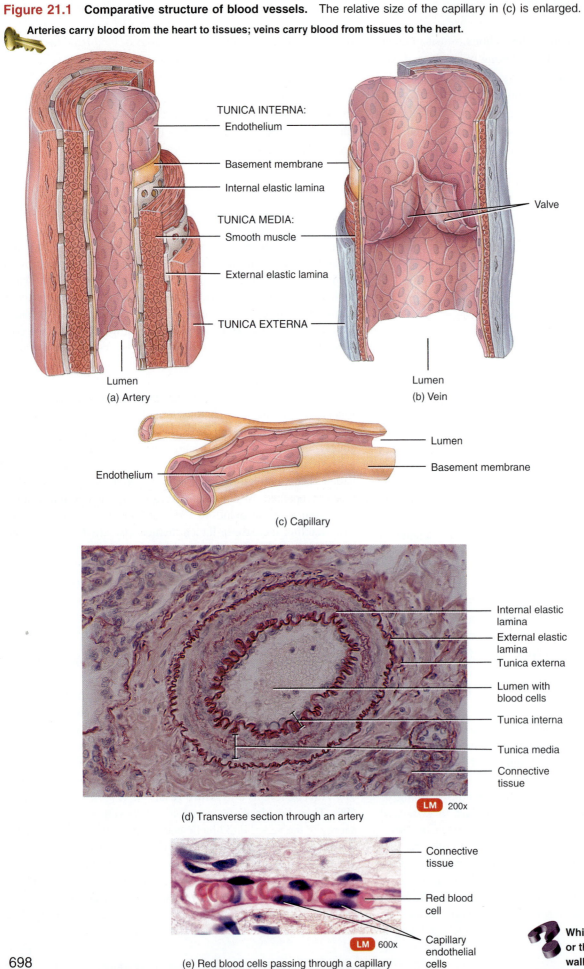

TUNICA INTERNA:
Endothelium

Basement membrane

Internal elastic lamina

TUNICA MEDIA:
Smooth muscle

External elastic lamina

TUNICA EXTERNA

Valve

Lumen
(a) Artery

Lumen
(b) Vein

Lumen

Basement membrane

Endothelium

(c) Capillary

Internal elastic lamina
External elastic lamina
Tunica externa
Lumen with blood cells
Tunica interna
Tunica media
Connective tissue

LM 200x

(d) Transverse section through an artery

Connective tissue
Red blood cell
Capillary endothelial cells

LM 600x

(e) Red blood cells passing through a capillary

Which vessel—the femoral artery or the femoral vein—has a thicker wall? Which has a wider lumen?

Figure 21.2 Pressure reservoir function of elastic arteries.

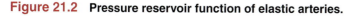

Recoil of elastic arteries keeps blood flowing during ventricular relaxation (diastole).

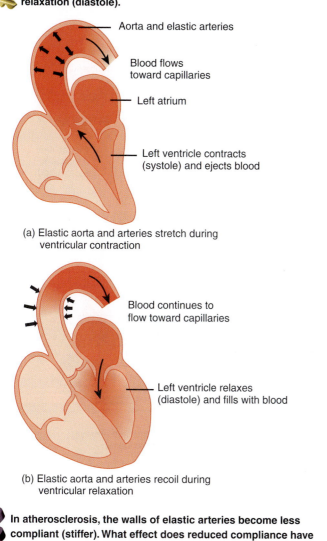

(a) Elastic aorta and arteries stretch during ventricular contraction

(b) Elastic aorta and arteries recoil during ventricular relaxation

In atherosclerosis, the walls of elastic arteries become less compliant (stiffer). What effect does reduced compliance have on the pressure reservoir function of arteries?

Muscular Arteries

Medium-sized arteries are called **muscular arteries** because their tunica media contains more smooth muscle and fewer elastic fibers than elastic arteries. Thus, muscular arteries are capable of greater vasoconstriction and vasodilation to adjust the rate of blood flow. The large amount of smooth muscle makes the walls of muscular arteries relatively thick. Muscular arteries also are called *distributing arteries* because they distribute blood to various parts of the body. Examples include the brachial artery in the arm and radial artery in the forearm (see Figure 21.18).

Arterioles

An **arteriole** (= small artery) is a very small (almost microscopic) artery that delivers blood to capillaries (Figure 21.3). Arterioles near the arteries from which they branch have a tunica interna like that of arteries, a tunica media composed of smooth

muscle and very few elastic fibers, and a tunica externa composed mostly of elastic and collagen fibers. In the smallest-diameter arterioles, which are closest to capillaries, the tunics consist of little more than a ring of endothelial cells surrounded by a few scattered smooth muscle fibers.

Arterioles play a key role in regulating blood flow from arteries into capillaries by regulating **resistance,** the opposition to blood flow. In a blood vessel, resistance is due mainly to friction between blood and the inner walls of blood vessels. When the blood vessel diameter is smaller, the friction is greater. Because contraction and relaxation of the smooth muscle in arteriole walls can change their diameter, arterioles are known as *resistance vessels.* Contraction of arteriolar smooth muscle causes vasoconstriction, which increases resistance and decreases blood flow into capillaries supplied by that arteriole. By contrast, relaxation of arteriolar smooth muscle causes vasodilation, which decreases resistance and increases blood flow into capillaries. A change in diameter of arterioles can also significantly affect blood pressure.

Capillaries

Capillaries are microscopic vessels that connect arterioles to venules (Figure 21.3). The flow of blood from arterioles to venules through capillaries is called the **microcirculation.** Capillaries are found near almost every cell in the body, but their number varies with the metabolic activity of the tissue they serve. Body tissues with high metabolic requirements, such as muscles, the liver, the kidneys, and the nervous system, use more O_2 and nutrients and thus have extensive capillary networks. Tissues with lower metabolic requirements, such as tendons and ligaments, contain fewer capillaries. Capillaries are absent in a few tissues, for instance, all covering and lining epithelia, the cornea and lens of the eye, and cartilage.

Capillaries are known as *exchange vessels* because their prime function is the exchange of nutrients and wastes between the blood and tissue cells through the interstitial fluid. The structure of capillaries is well suited to this function. Capillary walls are composed of only a single layer of endothelial cells and a basement membrane (see Figure 21.1e). They have no tunica media or tunica externa. Thus, a substance in the blood must pass through just one cell layer to reach the interstitial fluid and tissue cells. Exchange of materials occurs only through the walls of capillaries and the beginning of venules; the walls of arteries, arterioles, most venules, and veins present too thick a barrier. Capillaries form extensive branching networks that increase the surface area available for rapid exchange of materials. In most tissues, blood flows through only a small part of the capillary network when metabolic needs are low. However, when a tissue is active, such as contracting muscle, the entire capillary network fills with blood.

A **metarteriole** (*met-* = beyond) is a vessel that emerges from an arteriole and supplies a group of 10–100 capillaries that constitute a **capillary bed** (Figure 21.3a). The proximal end of

Figure 21.3 Arteriole, capillaries, and venule. Precapillary sphincters regulate the flow of blood through capillary beds.

🔑 **In capillaries, nutrients, gases, and wastes are exchanged between the blood and interstitial fluid.**

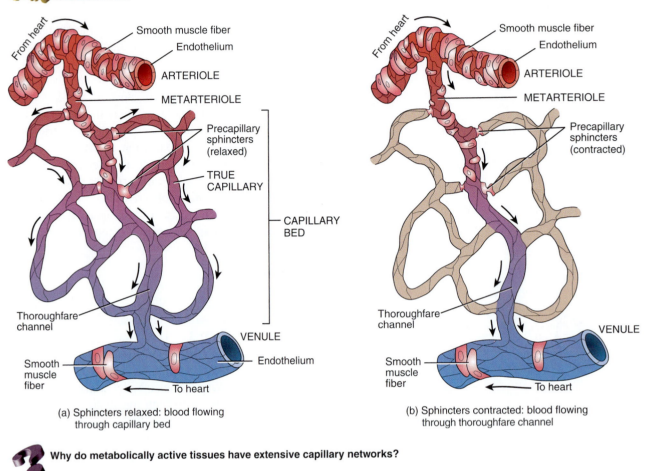

(a) Sphincters relaxed: blood flowing through capillary bed

(b) Sphincters contracted: blood flowing through thoroughfare channel

❓ **Why do metabolically active tissues have extensive capillary networks?**

a metarteriole is surrounded by scattered smooth muscle fibers whose contraction and relaxation help regulate blood flow through the capillary bed. The distal end of a metarteriole, which empties into a venule, has no smooth muscle fibers and is called a **thoroughfare channel.** Blood flowing through a thoroughfare channel bypasses the capillary bed.

True capillaries emerge from arterioles or metarterioles. At their sites of origin, a ring of smooth muscle fibers called a **precapillary sphincter** controls the flow of blood into a true capillary. When the precapillary sphincters are relaxed (open), blood flows into the capillary bed (Figure 21.3b); when precapillary sphincters contract (close or partially close), blood flow through the capillary bed ceases or decreases (Figure 21.3b). Typically, blood flows intermittently through a capillary bed due to alternating contraction and relaxation of the smooth muscle of metarterioles and the precapillary sphincters. This intermittent contraction and relaxation, which may occur 5 to 10 times per minute, is called **vasomotion.** In part, vasomotion is due to chemicals released by the endothelial cells; nitric oxide is one

example. At any given time, blood flows through only about 25% of a capillary bed.

The body contains three different types of capillaries: continuous capillaries, fenestrated capillaries, and sinusoids (Figure 21.4). Many capillaries are **continuous capillaries,** in which the plasma membranes of endothelial cells form a continuous tube that is interrupted only by **intercellular clefts,** which are gaps between neighboring endothelial cells (Figure 21.4a). Continuous capillaries are found in skeletal and smooth muscle, connective tissues, and the lungs.

Other capillaries of the body are **fenestrated capillaries** (*fenestr-* = window). The plasma membranes of the endothelial cells in these capillaries have many **fenestrations,** small pores (holes) ranging from 70 to 100 nm in diameter (Figure 21.4b). Fenestrated capillaries are found in the kidneys, villi of the small intestine, choroid plexuses of the ventricles in the brain, ciliary processes of the eyes, and endocrine glands.

Sinusoids are wider and more winding than other capillaries. Their endothelial cells may have unusually large fenestrations. In

Figure 21.4   Types of capillaries.

Capillaries are microscopic blood vessels that connect arterioles and venules.

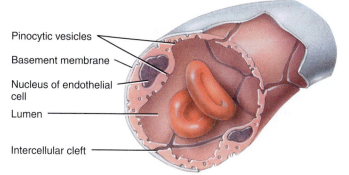

Pinocytic vesicles
Basement membrane
Nucleus of endothelial cell
Lumen
Intercellular cleft

(a) Continuous capillary formed by endothelial cells

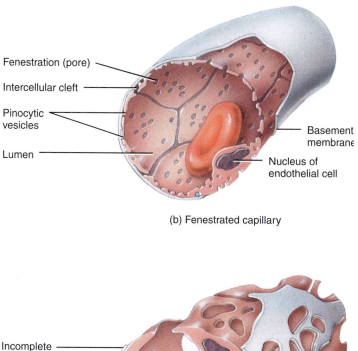

Fenestration (pore)
Intercellular cleft
Pinocytic vesicles
Lumen

Basement membrane
Nucleus of endothelial cell

(b) Fenestrated capillary

Incomplete basement membrane
Lumen

Nucleus of endothelial cell

Intercellular cleft

(c) Sinusoid

How do materials move across capillary walls?

addition to an incomplete or absent basement membrane (Figure 21.4c), sinusoids have very large intercellular clefts that allow proteins and in some cases even blood cells to pass from a tissue into the bloodstream. For example, newly formed blood cells enter the bloodstream through the sinusoids of red bone marrow. In addition, sinusoids contain specialized lining cells that are adapted to the function of the tissue. Sinusoids in the liver, for example, contain phagocytic cells that remove bacteria and other debris from the blood. The spleen, anterior pituitary, and parathyroid glands also have sinusoids.

Venules

When several capillaries unite, they form small veins called **venules** (= little veins). Venules collect blood from capillaries and drain into veins. The smallest venules, those closest to the capillaries, consist of a tunica interna of endothelium and a tunica media that has only a few scattered smooth muscle fibers and fibroblasts. Like capillaries, the walls of the smallest venules are very porous and are the site where many phagocytic white blood cells emigrate from the bloodstream into an inflamed or infected tissue. As venules become larger and converge to form veins, they contain the tunica externa characteristic of veins.

Veins

Although **veins** are composed of essentially the same three coats as arteries, the relative thicknesses of the layers are different. The tunica interna of veins is thinner than that of arteries; the tunica media of veins is much thinner than in arteries, with relatively little smooth muscle and elastic fibers. The tunica externa of veins is the thickest layer and consists of collagen and elastic fibers; the tunica externa of the inferior vena cava also contains longitudinal fibers of smooth muscle. Veins lack the external or internal elastic laminae found in arteries (see Figure 21.1b). Still, veins are distensible enough to adapt to variations in the volume and pressure of blood passing through them, although they are not designed to withstand high pressure. Furthermore, the lumen of a vein is larger than that of a comparable artery, and veins often appear collapsed (flattened) when sectioned.

Many veins, especially those in the limbs, also feature **valves**, which are thin folds of tunica interna that form flaplike cusps. The valve cusps project into the lumen, pointing toward the heart (Figure 21.5). Because the low blood pressure in veins allows blood returning to the heart to slow and even back up, the valves aid in venous return by preventing the backflow of blood.

A **vascular (venous) sinus** is a vein with a thin endothelial wall that has no smooth muscle to alter its diameter. In a vascular sinus, the surrounding dense connective tissue replaces the tunica media and tunica externa in providing support. For example, dural venous sinuses, which are supported by the dura mater, convey deoxygenated blood from the brain to the heart. Another example of a vascular sinus is the coronary sinus of the heart (see Figure 20.3c on page 664).

Figure 21.5   Venous valves.

🔑 Valves in veins allow blood to flow in one direction only—toward the heart.

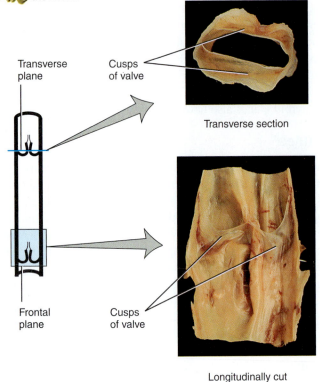

Transverse section

Longitudinally cut

Photograph of a valve in a vein

❓ Why are valves more important in arm veins and leg veins than in neck veins?

🩺 Varicose Veins

Leaky venous valves can cause veins to become dilated and twisted in appearance, a condition called **varicose veins** (*varic-* = a swollen vein). The condition may occur in the veins of almost any body part, but it is most common in the esophagus and in superficial veins of the lower limbs. The valvular defect may be congenital or may result from mechanical stress (prolonged standing or pregnancy) or aging. The leaking venous valves allow the backflow of blood, which causes pooling of blood. This, in turn, creates pressure that distends the vein and allows fluid to leak into the surrounding tissue. As a result, the affected vein and the tissue around it may become inflamed and painfully tender. Veins close to the surface, especially the saphenous vein, are highly susceptible to varicosities, whereas deeper veins are not as vulnerable because surrounding skeletal muscles prevent their walls from stretching excessively. Varicose veins in the anal canal are termed *hemorrhoids*. ■

Anastomoses

Most tissues of the body receive blood from more than one artery. The union of the branches of two or more arteries supply-ing the same body region is called an **anastomosis** (a-nas-tō-MŌ-sis = connecting; plural is **anastomoses**). Anastomoses between arteries provide alternate routes for blood to reach a tissue or organ. If blood flow stops for a short time when normal movements compress a vessel, or if a vessel is blocked by disease, injury, or surgery, then circulation to a part of the body is not necessarily stopped. The alternate route of blood flow to a body part through an anastomosis is known as **collateral circulation.** Anastomoses may also occur between veins and between arterioles and venules. Arteries that do not anastomose are known as **end arteries.** Obstruction of an end artery interrupts the blood supply to a whole segment of an organ, producing necrosis (death) of that segment. Alternate blood routes may also be provided by nonanastomosing vessels that supply the same region of the body.

Blood Distribution

The largest portion of your blood volume at rest—about 60%—is in systemic veins and venules (Figure 21.6). Systemic capillaries hold only about 5% of the blood volume, and arteries and arterioles about 15%. Because systemic veins and venules contain a large percentage of the blood volume, they function as **blood reservoirs** from which blood can be diverted quickly if the need arises. For example, during increased muscular activity, the cardiovascular center in the brain stem sends a larger number of sympathetic impulses to veins. The result is *venoconstriction,* constriction of veins, which reduces the volume of blood in

Figure 21.6   Blood distribution in the cardiovascular system at rest.

🔑 Because systemic veins and venules contain more than half the total blood volume, they are called blood reservoirs.

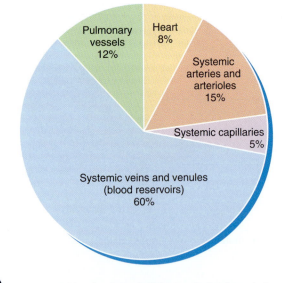

Pulmonary vessels 12%

Heart 8%

Systemic arteries and arterioles 15%

Systemic capillaries 5%

Systemic veins and venules (blood reservoirs) 60%

❓ If your total blood volume is 5 liters, what volume is in your venules and veins right now? In your capillaries?

reservoirs and allows a greater blood volume to flow to skeletal muscles, where it is needed most. A similar mechanism operates in cases of hemorrhage, when blood volume and pressure decrease; in this case, venoconstriction helps counteract the drop in blood pressure. Among the principal blood reservoirs are the veins of the abdominal organs (especially the liver and spleen) and the veins of the skin.

▶ **CHECKPOINT**

1. What is the function of elastic fibers and smooth muscle in the tunica media of arteries?

2. How are elastic arteries and muscular arteries different?

3. What structural features of capillaries allow the exchange of materials between blood and body cells?

4. Distinguish between pressure reservoirs and blood reservoirs. Why is each important?

5. What is the relationship between anastomoses and collateral circulation?

CAPILLARY EXCHANGE

▶ **OBJECTIVE**

- **Discuss the pressures that cause movement of fluids between capillaries and interstitial spaces.**

The mission of the entire cardiovascular system is to keep blood flowing through capillaries to allow **capillary exchange,** the movement of substances between blood plasma and interstitial fluid. The 5% of the blood in systemic capillaries at any given time is continually exchanging materials with interstitial fluid. Substances enter and leave capillaries by three basic mechanisms: diffusion, transcytosis, and bulk flow.

Diffusion

The most important method of capillary exchange is simple diffusion. Many substances, such as oxygen (O_2), carbon dioxide (CO_2), glucose, amino acids, and hormones, enter and leave capillaries by simple diffusion. Because O_2 and nutrients normally are present in higher concentrations in blood, they diffuse down their concentration gradients into interstitial fluid and then into body cells. CO_2 and other wastes released by body cells are present in higher concentrations in interstitial fluid, so they diffuse into blood.

All plasma solutes except proteins pass easily across most capillary walls. Except in sinusoids, the gaps between capillary endothelial cells are usually too small to allow proteins through. Thus, most plasma proteins remain in the blood. Smaller water-soluble substances, such as glucose and amino acids, pass through either fenestrations or intercellular clefts (see Figure 21.4). Lipid-soluble materials, such as O_2, CO_2, and steroid hormones, may pass directly through the lipid bilayer of

endothelial cell plasma membranes. The gaps between the endothelial cells in liver sinusoids are so large that they do allow proteins to enter the bloodstream. In fact, hepatocytes (liver cells) synthesize and release many proteins, such as fibrinogen (the main clotting protein) and albumin, which then diffuse into the sinusoids through the large intercellular clefts.

Capillaries in most areas of the brain, by contrast, are very "tight;" they allow only a few substances to enter and leave. The endothelial cells of most brain capillaries are nonfenestrated and are sealed together by tight junctions. The resulting blockade to movement of materials into and out of brain capillaries is known as the *blood–brain barrier* (see page 452). In brain areas that lack the blood–brain barrier, for example, the hypothalamus, pineal gland, and pituitary gland, materials undergo capillary exchange more freely.

Transcytosis

A small quantity of material crosses capillary walls by **transcytosis** (*trans-* = across; *cyt-* = cell; *-osis* = process). In this process, substances in blood plasma become enclosed within tiny pinocytic vesicles that first enter endothelial cells by endocytosis, then move across the cell and exit on the other side by exocytosis. This method of transport is important mainly for large, lipid-insoluble molecules that cannot cross capillary walls in any other way. For example, the hormone insulin (a small protein) enters the bloodstream by transcytosis, and certain antibodies (also proteins) pass from the maternal circulation into the fetal circulation by transcytosis.

Bulk Flow: Filtration and Reabsorption

Bulk flow is a passive process by which *large* numbers of ions, molecules, or particles in a fluid move together in the same direction. The substances move in unison, and they move at rates far greater than can be accounted for by diffusion alone. Bulk flow occurs from an area of higher pressure to an area of lower pressure, and it continues as long as a pressure difference exists. Whereas diffusion is more important for *solute exchange* between plasma and interstitial fluid, bulk flow is more important for regulation of the *relative volumes of blood and interstitial fluid.* Pressure-driven movement of fluid and solutes *from* blood capillaries *into* interstitial fluid is termed **filtration.** Pressure-driven movement *from* interstitial fluid *into* blood capillaries is called **reabsorption.**

Two pressures promote filtration: **blood hydrostatic pressure (BHP),** the pressure generated by the pumping action of the heart, and **interstitial fluid osmotic pressure.** The main pressure promoting reabsorption of fluid is **blood colloid osmotic pressure.** The balance of these pressures, called **net filtration pressure (NFP),** determines whether blood volume and interstitial fluid remain steady or change. Overall, the volume of fluid and solutes reabsorbed normally is almost as large as the volume filtered. This near equilibrium is known as **Starling's law of**

the capillaries. Let's see how these hydrostatic and osmotic pressures balance.

Within vessels, the hydrostatic pressure is due to the pressure that water in blood plasma exerts against blood vessel walls. The blood hydrostatic pressure (BHP) is about 35 millimeters of mercury (mmHg) at the arterial end of a capillary, and about 16 mmHg at the capillary's venous end (Figure 21.7). BHP "pushes" fluid out of capillaries into interstitial fluid. The opposing pressure of the interstitial fluid, called **interstitial fluid hydrostatic pressure (IFHP),** "pushes" fluid from interstitial spaces back into capillaries. However, IFHP is close to zero.

(IFHP is difficult to measure, and its reported values vary from small positive values to small negative values.) For our discussion we assume that IFHP equals 0 mmHg all along the capillaries.

The difference in osmotic pressure across a capillary wall is due almost entirely to the presence in blood of plasma proteins, which are too large to pass through either fenestrations or gaps between endothelial cells. **Blood colloid osmotic pressure (BCOP)** is a force caused by the colloidal suspension of these large proteins in plasma that averages 26 mmHg in most capillaries. The effect of BCOP is to "pull" fluid from interstitial

Figure 21.7 Dynamics of capillary exchange (Starling's law of the capillaries). Excess filtered fluid drains into lymphatic capillaries.

 Blood hydrostatic pressure pushes fluid out of capillaries (filtration), whereas blood colloid osmotic pressure pulls fluid into capillaries (reabsorption).

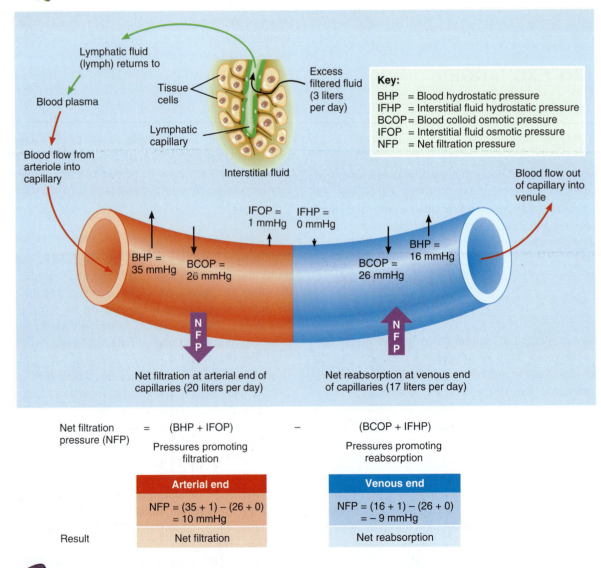

	Arterial end
	NFP = (35 + 1) − (26 + 0) = 10 mmHg
Result	Net filtration

	Venous end
	NFP = (16 + 1) − (26 + 0) = − 9 mmHg
	Net reabsorption

Net filtration pressure (NFP) = (BHP + IFOP) − (BCOP + IFHP)

Pressures promoting filtration — Pressures promoting reabsorption

? A person who has liver failure cannot synthesize the normal amount of plasma proteins. How does a deficit of plasma proteins affect blood colloid osmotic pressure, and what is the effect on capillary filtration and reabsorption?

spaces into capillaries. Opposing BCOP is **interstitial fluid osmotic pressure (IFOP),** which "pulls" fluid out of capillaries into interstitial fluid. Normally, IFOP is very small—0.1–5 mmHg—because only tiny amounts of protein are present in interstitial fluid. The small amount of protein that leaks from blood plasma into interstitial fluid does not accumulate there because it enters lymphatic fluid and is returned to the blood. For discussion, we can use a value of 1 mmHg for IFOP.

Whether fluids leave or enter capillaries depends on the balance of pressures. If the pressures that push fluid out of capillaries exceed the pressures that pull fluid into capillaries, fluid will move from capillaries into interstitial spaces (filtration). If, however, the pressures that push fluid out of interstitial spaces into capillaries exceed the pressures that pull fluid out of capillaries, then fluid will move from interstitial spaces into capillaries (reabsorption).

The net filtration pressure (NFP), which indicates the direction of fluid movement, is calculated as follows:

$$NFP = (BHP + IFOP) - (BCOP + IFHP)$$

Pressures that promote filtration   Pressures that promote reabsorption

At the arterial end of a capillary,

$$NFP = (35 + 1) \text{ mmHg} - (26 + 0) \text{ mmHg}$$
$$= 36 - 26 \text{ mmHg} = 10 \text{ mmHg}$$

Thus, at the arterial end of a capillary, there is a *net outward pressure* of 10 mmHg, and fluid moves out of the capillary into interstitial spaces (filtration).

At the venous end of a capillary,

$$NFP = (16 + 1) \text{ mmHg} - (26 + 0) \text{ mmHg}$$
$$= 17 - 26 \text{ mmHg} = -9 \text{ mmHg}$$

At the venous end of a capillary, the negative value (−9 mmHg) represents a *net inward pressure,* and fluid moves into the capillary from tissue spaces (reabsorption).

On average, about 85% of the fluid filtered out of capillaries is reabsorbed. The excess filtered fluid and the few plasma proteins that do escape from blood into interstitial fluid enter lymphatic capillaries (see Figure 22.2 on page 767). As lymph drains into the subclavian veins in the upper thorax (see Figure 22.3 on page 768), these materials return to the blood. Every day about 20 liters of fluid filter out of capillaries in tissues throughout the body. Of this fluid, 17 liters are reabsorbed and 3 liters enter lymphatic capillaries. (Excluded from these numbers are the 180 liters of fluid that are filtered from capillaries in the kidneys and the 178 liters that are reabsorbed during the formation of urine each day.)

Edema

If filtration greatly exceeds reabsorption, the result is **edema** (= swelling), an abnormal increase in interstitial fluid volume.

Edema is not usually detectable in tissues until interstitial fluid volume has risen to 30% above normal. Edema can result from either excess filtration or inadequate reabsorption.

Two situations may cause excess filtration:

- *Increased capillary blood pressure* causes more fluid to be filtered from capillaries.

- *Increased permeability of capillaries* raises interstitial fluid osmotic pressure by allowing some plasma proteins to escape. Such leakiness may be caused by the destructive effects of chemical, bacterial, thermal, or mechanical agents on capillary walls.

One situation commonly causes inadequate reabsorption:

- *Decreased concentration of plasma proteins* lowers the blood colloid osmotic pressure. Inadequate synthesis or loss of plasma proteins is associated with liver disease, burns, malnutrition, and kidney disease. ■

▶ **CHECKPOINT**

6. How can substances enter and leave blood plasma?

7. How do hydrostatic and osmotic pressures determine fluid movement across the walls of capillaries?

8. Write an equation to describe the pressures involved in Starling's law of the capillaries.

HEMODYNAMICS: FACTORS AFFECTING BLOOD FLOW

▶ **OBJECTIVES**

- **Explain the factors that regulate the volume of blood flow.**
- **Explain how blood pressure changes throughout the cardiovascular system.**
- **Describe the factors that determine mean arterial blood pressure and systemic vascular resistance.**
- **Describe the relation between cross-sectional area and velocity of blood flow.**

Blood flow is the volume of blood that flows through any tissue in a given time period (in mL/min). Total blood flow is cardiac output (CO), the volume of blood that circulates through systemic (or pulmonary) blood vessels each minute. In Chapter 20 we saw that cardiac output depends on heart rate and stroke volume: Cardiac output (CO) = heart rate (HR) × stroke volume (SV). How the cardiac output becomes distributed into circulatory routes that serve various body tissues depends on two more factors: (1) the *pressure difference* that drives the blood flow through a tissue and (2) the *resistance* to blood flow in specific blood vessels. Blood flows from regions of higher to lower pressure; the greater the pressure difference, the greater the blood flow. The higher the resistance, by contrast, the smaller the blood flow.

Blood Pressure

As you have just learned, blood flows from regions of higher pressure to regions of lower pressure; the greater the pressure difference, the greater the blood flow. Contraction of the ventricles generates **blood pressure (BP),** the hydrostatic pressure exerted by blood on the walls of a blood vessel. BP is highest in the aorta and large systemic arteries, where in a resting, young adult, BP rises to about 120 mmHg during systole (ventricular contraction) and drops to about 80 mmHg during diastole (ventricular relaxation). **Systolic blood pressure** is the highest pressure attained in arteries during systole, and **diastolic blood pressure** is the lowest arterial pressure during diastole (Figure 21.8). As blood leaves the aorta and flows through the systemic circulation, its pressure falls progressively as the distance from the left ventricle increases. Blood pressure decreases to about 35 mmHg as blood passes from systemic arteries through systemic arterioles and into capillaries, where the pressure fluctuations disappear. At the venous end of capillaries, blood pressure has dropped to about 16 mmHg. Blood pressure continues to drop as blood enters systemic venules and then veins because these vessels are farthest from the left ventricle. Finally, blood pressure reaches 0 mmHg as blood flows into the right ventricle.

Mean arterial blood pressure (MABP), the average pressure in arteries, is roughly one-third of the way between the diastolic and systolic pressures. It can be estimated as follows:

MABP = diastolic BP + 1/3 (systolic BP − diastolic BP)

Thus, in a person whose BP is 120/80 mmHg, MABP is about 93 mmHg.

We have already seen that cardiac output equals heart rate multiplied by stroke volume. Another way to calculate cardiac output is to divide mean arterial blood pressure (MABP) by resistance (R): CO = MABP ÷ R. By rearranging the terms of this equation, you can see that MABP = CO × R. If cardiac output rises due to an increase in stroke volume or heart rate, then the mean arterial blood pressure rises so long as resistance remains steady. Likewise, a decrease in cardiac output causes a decrease in blood pressure if resistance does not change.

Blood pressure also depends on the total volume of blood in the cardiovascular system. The normal volume of blood in an adult is about 5 liters (5.3 qt). Any decrease in this volume, as from hemorrhage, decreases the amount of blood that is circulated through the arteries each minute. A modest decrease can be compensated for by homeostatic mechanisms that help maintain blood pressure (described on page 709), but if the decrease in blood volume is greater than 10% of the total, blood pressure drops. Conversely, anything that increases blood volume, such as water retention in the body, tends to increase blood pressure.

Resistance

As noted earlier, **vascular resistance** is the opposition to blood flow due to friction between blood and the walls of blood ves-

Figure 21.8 Blood pressures in various parts of the cardiovascular system. The dashed line is the mean (average) blood pressure in the aorta, arteries, and arterioles.

Blood pressure rises and falls with each heartbeat in blood vessels leading to capillaries.

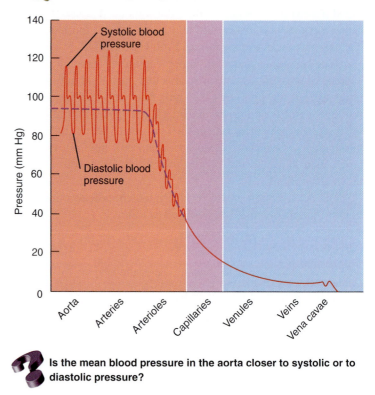

Is the mean blood pressure in the aorta closer to systolic or to diastolic pressure?

sels. Vascular resistance depends on (1) size of the blood vessel lumen, (2) blood viscosity, and (3) total blood vessel length.

1. *Size of the lumen.* The smaller the lumen of a blood vessel, the greater its resistance to blood flow. Resistance is inversely proportional to the fourth power of the diameter (d) of the blood vessel's lumen (R ∝ 1/d⁴). The smaller the diameter of the blood vessel, the greater the resistance it offers to blood flow. For example, if the diameter of a blood vessel decreases by one-half, its resistance to blood flow increases 16 times. Vasoconstriction narrows the lumen whereas vasodilation widens it. Normally, moment-to-moment fluctuations in blood flow through a given tissue are due to vasoconstriction and vasodilation of the tissue's arterioles. As arterioles dilate, resistance decreases, and blood pressure falls. As arterioles constrict, resistance increases, and blood pressure rises.

2. *Blood viscosity.* The viscosity (thickness) of blood depends mostly on the ratio of red blood cells to plasma (fluid) volume, and to a smaller extent on the concentration of proteins in plasma. The higher the blood's viscosity, the higher the resistance. Any condition that increases the viscosity of blood, such as dehydration or polycythemia (an unusually high number of red blood cells), thus increases blood pressure. A depletion of

plasma proteins or red blood cells, due to anemia or hemorrhage, decreases viscosity and thus decreases blood pressure.

3. ***Total blood vessel length.*** Resistance to blood flow through a vessel is directly proportional to the length of the blood vessel. The longer a blood vessel, the greater the resistance. Obese people often have hypertension (elevated blood pressure) due to an increase in total blood vessel length caused by the additional blood vessels in their adipose tissue. An estimated 650 km (about 400 miles) of additional blood vessels develop for each extra kilogram (2.2 lb) of fat.

Systemic vascular resistance (SVR), also known as *total peripheral resistance (TPR),* refers to all the vascular resistances offered by systemic blood vessels. The diameters of arteries and veins are large, and thus their resistance is very small because most of the blood does not come into physical contact with the walls of the blood vessel. The smallest vessels—arterioles, capillaries, and venules—contribute the most resistance. A major function of arterioles is to control SVR—and therefore blood pressure and blood flow to particular tissues—by changing their diameters. Arterioles need vasodilate or vasoconstrict only slightly to have a large effect on SVR. The main center for regulation of SVR is the vasomotor center in the brain stem (described shortly).

Venous Return

Venous return, the volume of blood flowing back to the heart through the systemic veins, occurs due to the pressure generated by contractions of the heart's left ventricle. The pressure difference from venules (averaging about 16 mmHg) to the right ventricle (0 mmHg), although small, normally is sufficient to cause venous return to the heart. If pressure increases in the right atrium or ventricle, venous return will decrease. One cause of increased pressure in the right atrium is an incompetent (leaky) tricuspid valve, which lets blood regurgitate (flow backward) as the ventricles contract. The result is decreased venous return and buildup of blood on the venous side of the systemic circulation.

When you stand, the pressure pushing blood up the veins in your lower limbs is barely enough to overcome the force of gravity pushing it back down. Besides the heart, two other mechanisms "pump" blood from the lower body back to the heart: (1) the skeletal muscle pump, and (2) the respiratory pump. Both pumps depend on the presence of valves in veins.

The **skeletal muscle pump** operates as follows (Figure 21.9):

1 While you are standing at rest, both the venous valve closer to the heart (proximal valve) and the one farther from the heart (distal valve) in this part of the leg are open, and blood flows upward toward the heart.

2 Contraction of leg muscles, such as when you stand on tiptoes or take a step, compresses the vein. The compression pushes blood through the proximal valve, an action called *milking.* At the same time, the distal valve in the uncom-

pressed segment of the vein closes as some blood is pushed against it. People who are immobilized through injury or disease lack these contractions of leg muscles. As a result, their venous return is slower and they may develop circulation problems.

3 Just after muscle relaxation, pressure falls in the previously compressed section of vein, which causes the proximal valve to close. The distal valve now opens because blood pressure in the foot is higher than in the leg, and the vein fills with blood from the foot.

The **respiratory pump** is also based on alternating compression and decompression of veins. During inhalation, the diaphragm moves downward, which causes a decrease in pressure in the thoracic cavity and an increase in pressure in the abdominal cavity. As a result, abdominal veins are compressed, and a greater volume of blood moves from the compressed abdominal veins into the decompressed thoracic veins and then into the right atrium. When the pressures reverse during exhalation, the valves in the veins prevent backflow of blood from the thoracic veins to the abdominal veins.

Figure 21.9 Action of the skeletal muscle pump in returning blood to the heart. 1 At rest, both proximal and distal venous valves are open and blood flows toward the heart. **2** Contraction of leg muscles pushes blood through the proximal valve while closing the distal valve. **3** As the leg muscles relax, the proximal valve closes and the distal valve opens. When the vein fills with blood from the foot, the proximal valve reopens.

🔑 **Milking refers to skeletal muscle contractions that drive venous blood toward the heart.**

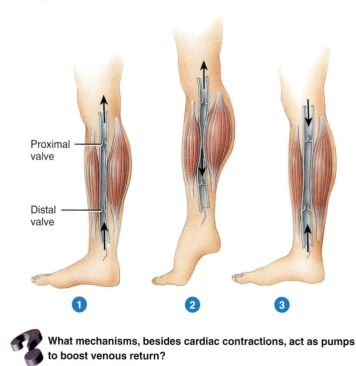

Proximal valve

Distal valve

1　**2**　**3**

❓ **What mechanisms, besides cardiac contractions, act as pumps to boost venous return?**

Figure 21.10 summarizes the factors that increase blood pressure through increasing cardiac output or systemic vascular resistance.

Velocity of Blood Flow

Earlier we saw that blood flow is the *volume* of blood that flows through any tissue in a given time period (in mL/min). The speed or *velocity* of blood flow (in cm/sec) is inversely related to the cross-sectional area. Velocity is slowest where the total cross-sectional area is greatest (Figure 21.11). Each time an artery branches, the total cross-sectional area of all its branches is greater than the cross-sectional area of the original vessel, and blood flow is slower in the branches. Conversely, when branches combine as venules unite to form veins, the total cross-sectional area becomes smaller and flow becomes faster. In an adult, the cross-sectional area of the aorta is only 3–5 cm², and the average velocity of the blood there is 40 cm/sec. In capillaries, the total cross-sectional area is 4500–6000 cm², and the velocity of blood flow is less than 0.1 cm/sec. In the two venae cavae combined, the cross-sectional area is about 14 cm², and the velocity

is about 15 cm/sec. Thus, the velocity of blood flow decreases as blood flows from the aorta to arteries to arterioles to capillaries, and increases as it leaves capillaries and returns to the heart. The relatively slow rate of flow through capillaries aids the exchange of materials between blood and interstitial fluid.

Circulation time is the time required for a drop of blood to pass from the right atrium, through the pulmonary circulation, back to the left atrium, through the systemic circulation down to the foot, and back again to the right atrium. In a resting person, circulation time normally is about 1 minute.

Syncope

Syncope (SIN-kō-pē), or fainting, is a sudden, temporary loss of consciousness that is not due to head trauma, followed by spontaneous recovery. It is most commonly due to cerebral ischemia, lack of sufficient blood flow to the brain. Syncope may occur for several reasons:

- *Vasodepressor syncope* is due to sudden emotional stress or real, threatened, or fantasized injury.

Figure 21.10 Summary of factors that increase blood pressure. Changes noted within green boxes increase cardiac output, whereas changes noted within blue boxes increase systemic vascular resistance.

🔑 **Increases in cardiac output and increases in systemic vascular resistance will increase mean arterial blood pressure.**

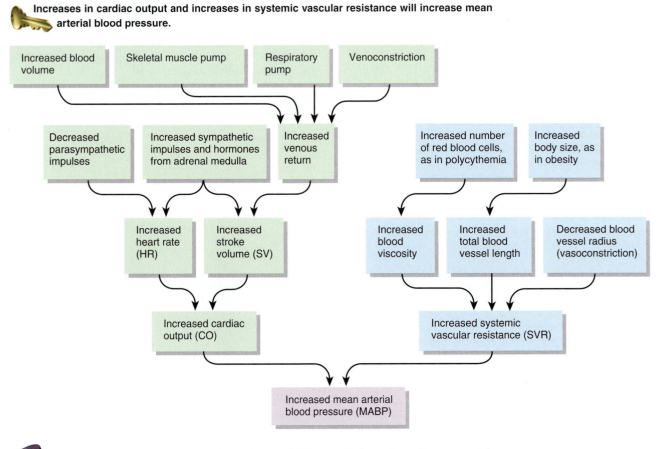

Which type of blood vessel exerts the major control of systemic vascular resistance, and how does it achieve this?

Figure 21.11 **Relationship between velocity (speed) of blood flow and total cross-sectional area in different types of blood vessels.**

 Velocity of blood flow is slowest in the capillaries because they have the largest total cross-sectional area.

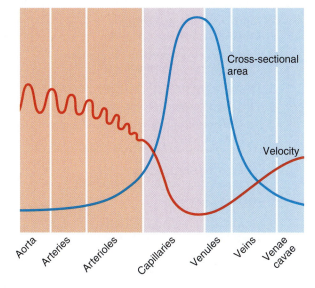

In which blood vessels is the velocity of flow fastest?

- *Situational syncope* is caused by pressure stress associated with urination, defecation, or severe coughing.
- *Drug-induced syncope* may be caused by drugs such as anti-hypertensives, diuretics, vasodilators, and tranquilizers.
- *Orthostatic hypotension,* an excessive decrease in blood pressure that occurs upon standing up, may cause fainting. ■

▶ C H E C K P O I N T

9. Explain how blood pressure and resistance determine volume of blood flow.

10. What is systemic vascular resistance and what factors contribute to it?

11. How is the return of venous blood to the heart accomplished?

12. Why is the velocity of blood flow faster in arteries and veins than in capillaries?

CONTROL OF BLOOD PRESSURE AND BLOOD FLOW

▶ O B J E C T I V E

- **Describe how blood pressure is regulated.**

Several interconnected negative feedback systems control blood pressure by adjusting heart rate, stroke volume, systemic vascular resistance, and blood volume. Some systems allow rapid adjustments to cope with sudden changes, such as the drop in blood pressure in the brain that occurs when you get out of bed; others act more slowly to provide long-term regulation of blood pressure. The body may also require adjustments to the distribution of blood flow. During exercise, for example, a greater percentage of the total blood flow is diverted to skeletal muscles.

Role of the Cardiovascular Center

In Chapter 20, we noted how the **cardiovascular (CV) center** in the medulla oblongata helps regulate heart rate and stroke volume. The CV center also controls neural, hormonal, and local negative feedback systems that regulate blood pressure and blood flow to specific tissues. Groups of neurons scattered within the CV center regulate heart rate, contractility (force of contraction) of the ventricles, and blood vessel diameter. Some of the CV center's neurons stimulate the heart (cardiostimulatory center), whereas others inhibit the heart (cardioinhibitory center). Still others control blood vessel diameter (vasomotor center), either by causing constriction (vasoconstrictor center) or dilation (vasodilator center). Because these clusters of neurons communicate with one another, function together, and are not clearly separated anatomically, we discuss them here as a group.

The cardiovascular center receives input both from higher brain regions and from sensory receptors (Figure 21.12). Nerve impulses descend from the cerebral cortex, limbic system, and hypothalamus to affect the cardiovascular center. For example, even before you start to run a race, your heart rate may increase due to nerve impulses conveyed from the limbic system to the CV center. If your body temperature rises during a race, the hypothalamus sends nerve impulses to the CV center. The resulting vasodilation of skin blood vessels allows heat to dissipate more rapidly from the surface of the skin. The three main types of sensory receptors that provide input to the cardiovascular center are proprioceptors, baroreceptors, and chemoreceptors. *Proprioceptors* monitor movements of joints and muscles and provide input to the cardiovascular center during physical activity. Their activity accounts for the rapid increase in heart rate at the beginning of exercise. *Baroreceptors* monitor changes in pressure and stretch in the walls of blood vessels, and *chemoreceptors* monitor the concentration of various chemicals in the blood.

Output from the cardiovascular center flows along sympathetic and parasympathetic neurons of the ANS (Figure 21.12). Sympathetic impulses reach the heart via the **cardiac accelerator nerves.** An increase in sympathetic stimulation increases heart rate and contractility, whereas a decrease in sympathetic stimulation decreases heart rate and contractility. Parasympathetic stimulation, conveyed along the **vagus nerves (cranial nerve X),** decreases heart rate. Thus, opposing sympathetic (stimulatory) and parasympathetic (inhibitory) influences control the heart.

The cardiovascular center also continually sends impulses to smooth muscle in blood vessel walls via **vasomotor nerves.**

Figure 21.12   Location and function of the cardiovascular (CV) center in the medulla oblongata. The CV center receives input from higher brain centers, proprioceptors, baroreceptors, and chemoreceptors. Then, it provides output to the sympathetic and parasympathetic divisions of the autonomic nervous system (ANS).

🔑 **The cardiovascular center is the main region for nervous system regulation of the heart and blood vessels.**

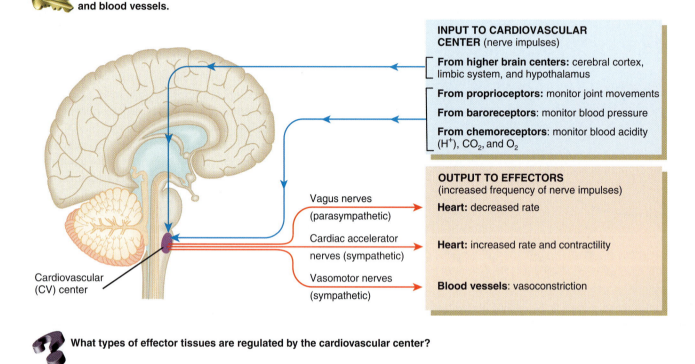

INPUT TO CARDIOVASCULAR CENTER (nerve impulses)

From higher brain centers: cerebral cortex, limbic system, and hypothalamus

From proprioceptors: monitor joint movements

From baroreceptors: monitor blood pressure

From chemoreceptors: monitor blood acidity (H^+), CO_2, and O_2

OUTPUT TO EFFECTORS (increased frequency of nerve impulses)

Heart: decreased rate

Heart: increased rate and contractility

Blood vessels: vasoconstriction

Vagus nerves (parasympathetic)

Cardiac accelerator nerves (sympathetic)

Vasomotor nerves (sympathetic)

Cardiovascular (CV) center

What types of effector tissues are regulated by the cardiovascular center?

These sympathetic neurons exit the spinal cord through all thoracic and the first one or two lumbar spinal nerves and then pass into the sympathetic trunk ganglia (see Figure 17.2 on page 569). From there, impulses propagate along sympathetic neurons that innervate blood vessels in viscera and peripheral areas. The vasomotor region of the cardiovascular center continually sends impulses over these routes to arterioles throughout the body, but especially to those in the skin and abdominal viscera. The result is a moderate state of tonic contraction or vasoconstriction, called **vasomotor tone,** that sets the resting level of systemic vascular resistance. Sympathetic stimulation of most veins causes constriction that moves blood out of venous blood reservoirs and increases blood pressure.

Neural Regulation of Blood Pressure

The nervous system regulates blood pressure via negative feedback loops that occur as two types of reflexes: baroreceptor reflexes and chemoreceptor reflexes.

Baroreceptor Reflexes

Baroreceptors, pressure-sensitive sensory receptors, are located in the aorta, internal carotid arteries (arteries in the neck that supply blood to the brain), and other large arteries in the neck and chest. They send impulses to the cardiovascular center to help regulate blood pressure. The two most important **baroreceptor reflexes** are the carotid sinus reflex and the aortic reflex.

Baroreceptors in the wall of the carotid sinuses initiate the **carotid sinus reflex,** which helps regulate blood pressure in the brain. The **carotid sinuses** are small widenings of the right and left internal carotid arteries just above the point where they branch from the common carotid arteries (Figure 21.13). Blood pressure stretches the wall of the carotid sinus, which stimulates the baroreceptors. Nerve impulses propagate from the carotid sinus baroreceptors over sensory axons in the glossopharyngeal nerves (cranial nerve IX) to the cardiovascular center in the medulla oblongata. Baroreceptors in the wall of the ascending aorta and arch of the aorta initiate the **aortic reflex,** which regulates systemic blood pressure. Nerve impulses from aortic baroreceptors reach the cardiovascular center via sensory axons of the vagus nerves (cranial nerve X).

When blood pressure falls, the baroreceptors are stretched less, and they send nerve impulses at a slower rate to the cardiovascular center (Figure 21.14 on page 712). In response, the CV center decreases parasympathetic stimulation of the heart via motor axons of the vagus nerves and increases sympathetic stimulation of the heart via cardiac accelerator nerves. Another consequence of increased sympathetic stimulation is increased epinephrine and

Figure 21.13 **ANS innervation of the heart and the baroreceptor reflexes that help regulate blood pressure.**

🔑 **Baroreceptors are pressure-sensitive neurons that monitor stretching.**

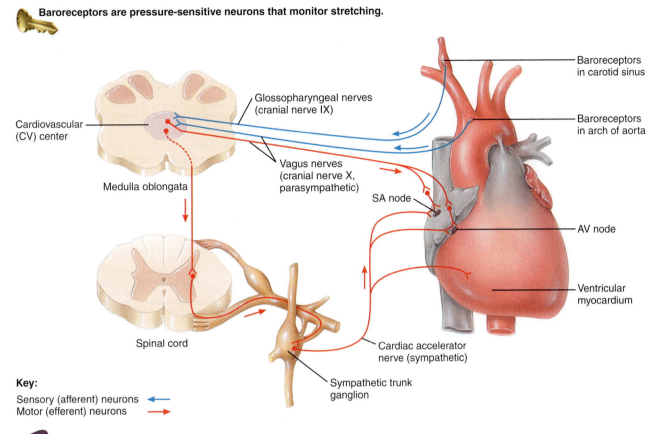

Key:
Sensory (afferent) neurons ⟵
Motor (efferent) neurons ⟶

❓ **Which cranial nerves conduct impulses to the cardiovascular center from baroreceptors in the carotid sinuses and the arch of the aorta?**

norepinephrine secretion by the adrenal medulla. As the heart beats faster and more forcefully, and as systemic vascular resistance increases, cardiac output and systemic vascular resistance rise, and blood pressure increases to the normal level.

Conversely, when an increase in pressure is detected, the baroreceptors send impulses at a faster rate. The CV center responds by increasing parasympathetic stimulation and decreasing sympathetic stimulation. The resulting decreases in heart rate and force of contraction reduce the cardiac output. Moreover, the cardiovascular center also slows the rate at which it sends sympathetic impulses along vasomotor neurons that normally cause vasoconstriction. The resulting vasodilation lowers systemic vascular resistance. Decreased cardiac output and decreased systemic vascular resistance both lower systemic arterial blood pressure.

Moving from a prone (lying down) to an erect position decreases blood pressure and blood flow in the head and upper part of the body. The baroreceptor reflexes, however, quickly counteract the drop in pressure. Sometimes these reflexes operate more slowly than normal, especially in the elderly, in which case a person can faint due to reduced brain blood flow upon standing up too quickly.

Carotid Sinus Massage and Carotid Sinus Syncope

Because the carotid sinus is close to the anterior surface of the neck, it is possible to stimulate the baroreceptors there by putting pressure on the neck. Physicians sometimes use **carotid sinus massage,** which involves carefully massaging the neck over the carotid sinus, to slow heart rate in a person who has paroxysmal supraventricular tachycardia, a type of tachycardia that originates in the atria. Anything that stretches or puts pressure on the carotid sinus, such as hyperextension of the head, tight collars, or carrying heavy shoulder loads, may also slow heart rate and can cause **carotid sinus syncope,** fainting due to inappropriate stimulation of the carotid sinus baroreceptors. ■

Chemoreceptor Reflexes

Chemoreceptors, sensory receptors that monitor the chemical composition of blood, are located close to the baroreceptors of the carotid sinus and arch of the aorta in small structures called **carotid bodies** and **aortic bodies,** respectively. These chemoreceptors detect changes in blood level of O_2, CO_2, and H^+. *Hypoxia* (lowered O_2 availability), *acidosis* (an increase in H^+

Figure 21.14 Negative feedback regulation of blood pressure via baroreceptor reflexes.

According to Marey's law of the heart, as heart rate increases blood pressure increases.

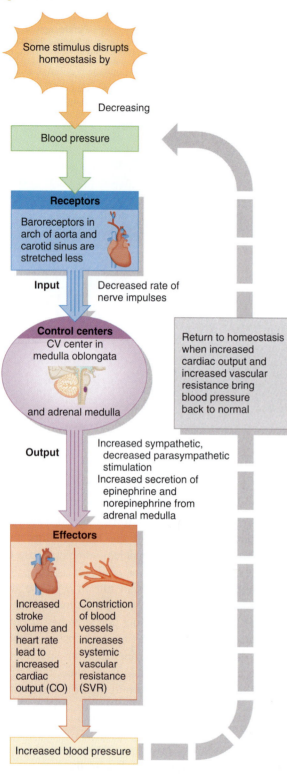

Some stimulus disrupts homeostasis by

Decreasing

Blood pressure

Receptors

Baroreceptors in arch of aorta and carotid sinus are stretched less

Input Decreased rate of nerve impulses

Control centers
CV center in medulla oblongata

and adrenal medulla

Return to homeostasis when increased cardiac output and increased vascular resistance bring blood pressure back to normal

Output Increased sympathetic, decreased parasympathetic stimulation

Increased secretion of epinephrine and norepinephrine from adrenal medulla

Effectors

Increased stroke volume and heart rate lead to increased cardiac output (CO)

Constriction of blood vessels increases systemic vascular resistance (SVR)

Increased blood pressure

Does this negative feedback cycle represent the changes that occur when you lie down or when you stand up?

concentration), or *hypercapnia* (excess CO_2) stimulates the chemoreceptors to send impulses to the cardiovascular center. In response, the CV center increases sympathetic stimulation to arterioles and veins, producing vasoconstriction and an increase in blood pressure. These chemoreceptors also provide input to the respiratory center in the brain stem to adjust the rate of breathing.

Hormonal Regulation of Blood Pressure

Several hormones help regulate blood pressure and blood flow by altering cardiac output, changing systemic vascular resistance, or adjusting the total blood volume:

1. ***Renin–angiotensin–aldosterone (RAA) system.*** When blood volume falls or blood flow to the kidneys decreases, juxtaglomerular cells in the kidneys secrete **renin** into the bloodstream. In sequence, renin and angiotensin converting enzyme (ACE) act on their substrates to produce the active hormone **angiotensin II,** which raises blood pressure in two ways. First, angiotensin II is a potent vasoconstrictor; it raises blood pressure by increasing systemic vascular resistance. Second, it stimulates secretion of **aldosterone,** which increases reabsorption of sodium ions (Na^+) and water by the kidneys. The water reabsorption increases total blood volume, which increases blood pressure.

2. ***Epinephrine and norepinephrine.*** In response to sympathetic stimulation, the adrenal medulla releases epinephrine and norepinephrine. These hormones increase cardiac output by increasing the rate and force of heart contractions. They also cause vasoconstriction of arterioles and veins in the skin and abdominal organs and vasodilation of arterioles in cardiac and skeletal muscle, which helps increase blood flow to muscle during exercise.

3. ***Antidiuretic hormone (ADH).*** ADH is produced by the hypothalamus and released from the posterior pituitary in response to dehydration or decreased blood volume. Among other actions, ADH causes vasoconstriction, which increases blood pressure. For this reason ADH is also called **vasopressin.**

4. ***Atrial natriuretic peptide (ANP).*** Released by cells in the atria of the heart, ANP lowers blood pressure by causing vasodilation and by promoting the loss of salt and water in the urine, which reduces blood volume.

Table 21.1 summarizes the regulation of blood pressure by hormones.

Autoregulation of Blood Pressure

In each capillary bed, local changes can regulate vasomotion. When vasodilators produce local dilation of arterioles and relaxation of precapillary sphincters, blood flow into capillary networks is increased, which restores O_2 level to normal. Vasoconstrictors have the opposite effect. The ability of a tissue to

Table 21.1 Blood Pressure Regulation by Hormones

Factor Influencing Blood Pressure	Hormone	Effect on Blood Pressure
Cardiac Output		
Increased heart rate and contractility	Norepinephrine Epinephrine	Increase
Systemic Vascular Resistance		
Vasoconstriction	Angiotensin II	Increase
	Antidiuretic hormone (vasopressin)	
	Norepinephrine*	
	Epinephrine*	
Vasodilation	Atrial natriuretic peptide	Decrease
	Epinephrine†	
	Nitric oxide	
Blood Volume		
Blood volume increase	Aldosterone Antidiuretic hormone	Increase
Blood volume decrease	Atrial natriuretic peptide	Decrease

*Acts at α_1 receptors in arterioles of abdomen and skin.
†Acts at β_2 receptors in arterioles of cardiac and skeletal muscle; norepinephrine has a much smaller vasodilating effect.

automatically adjust its blood flow to match its metabolic demands is called **autoregulation.** In tissues such as the heart and skeletal muscle, where the demand for O_2 and nutrients and for the removal of wastes can increase as much as tenfold during physical activity, autoregulation is an important contributor to increased blood flow through the tissue. Autoregulation also controls regional blood flow in the brain; blood distribution to various parts of the brain changes dramatically for different mental and physical activities. During a conversation, for example, blood flow increases to the motor speech areas when one is talking and increases to the auditory areas when one is listening.

Two general types of stimuli cause autoregulatory changes in blood flow:

1. *Physical changes.* Warming promotes vasodilation, whereas cooling causes vasoconstriction. Furthermore, smooth muscle in arteriole walls exhibits a **myogenic response**—it contracts more forcefully when it is stretched and relaxes when stretching lessens. If, for example, blood flow through an arteriole decreases, then stretching of the arteriole walls decreases. As a result, the smooth muscle relaxes and produces vasodilation, which increases blood flow.

2. *Vasodilating and vasoconstricting chemicals.* Several types of cells—including white blood cells, platelets, smooth muscle fibers, macrophages, and endothelial cells—release a wide variety of chemicals that alter blood-vessel diameter. Vasodilating chemicals released by metabolically active tissue cells include K^+, H^+, lactic acid (lactate), and adenosine (from ATP). Another important vasodilator released by endothelial cells is nitric oxide (NO), named endothelium-derived relaxation factor (EDRF) before its chemical identity was known. Tissue trauma or inflammation causes release of vasodilating kinins and histamine. Vasoconstrictors include thromboxane A2, superoxide radicals, serotonin (from platelets), and endothelins (from endothelial cells).

An important difference between the pulmonary and systemic circulations is their autoregulatory response to changes in O_2 level. The walls of blood vessels in the systemic circulation *dilate* in response to low O_2. With vasodilation, O_2 delivery increases, which restores the normal O_2 level. By contrast, the walls of blood vessels in the pulmonary circulation *constrict* in response to low levels of O_2. This response ensures that blood mostly bypasses those alveoli (air sacs) in the lungs that are poorly ventilated by fresh air. Thus, most blood flows to better-ventilated areas of the lung.

▶ **CHECKPOINT**

13. What are the principal inputs to and outputs from the cardiovascular center?

14. How do the carotid sinus and the aortic reflexes operate?

15. What is the role of chemoreceptors in the regulation of blood pressure?

16. How do hormones regulate blood pressure?

17. What is autoregulation and how does it differ in the systemic and pulmonary circulations?

CHECKING CIRCULATION

▶ **OBJECTIVE**

• **Define pulse, and define systolic, diastolic, and pulse pressures.**

Pulse

The alternate expansion and recoil of elastic arteries after each systole of the left ventricle create a traveling pressure wave that is called the **pulse.** The pulse is strongest in the arteries closest to the heart, becomes weaker in the arterioles, and disappears altogether in the capillaries. The pulse may be felt in any artery that lies near the surface of the body and runs over a bone or other firm structure. Table 21.2 depicts some common pulse points.

The pulse rate normally is the same as the heart rate, about 70 to 80 beats per minute at rest. **Tachycardia** (tak′-i-KAR-dē-a; *tachy-* = fast) is a rapid resting heart or pulse rate over 100

Table 21.2 Pulse Points

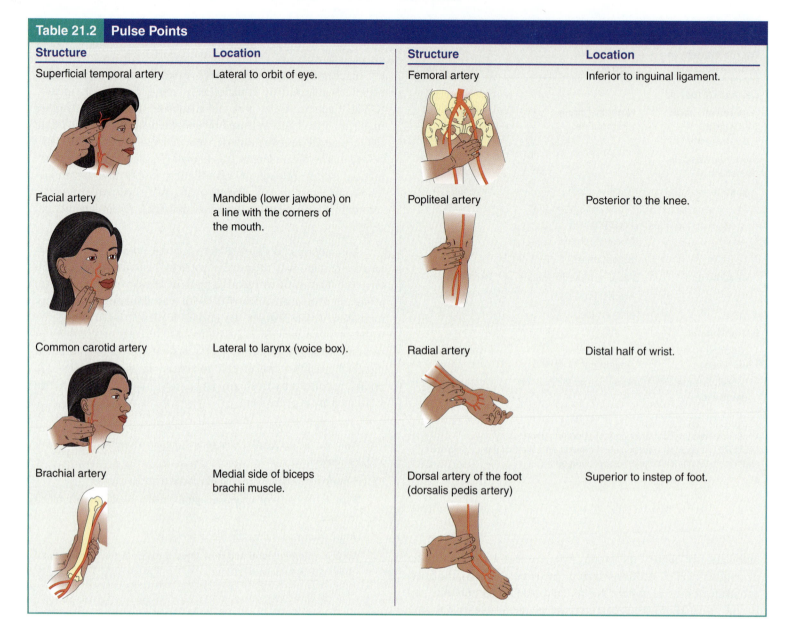

Structure	Location	Structure	Location
Superficial temporal artery	Lateral to orbit of eye.	Femoral artery	Inferior to inguinal ligament.
Facial artery	Mandible (lower jawbone) on a line with the corners of the mouth.	Popliteal artery	Posterior to the knee.
Common carotid artery	Lateral to larynx (voice box).	Radial artery	Distal half of wrist.
Brachial artery	Medial side of biceps brachii muscle.	Dorsal artery of the foot (dorsalis pedis artery)	Superior to instep of foot.

beats/min. **Bradycardia** (brād′-i-KAR-dē-a; *brady-* = slow) is a slow resting heart or pulse rate under 60 beats/min. Endurance-trained athletes normally exhibit bradycardia.

Measuring Blood Pressure

In clinical use, the term **blood pressure** usually refers to the pressure in arteries generated by the left ventricle during systole and the pressure remaining in the arteries when the ventricle is in diastole. Blood pressure is usually measured in the brachial artery in the left arm (see Figure 21.20a). The device used to measure blood pressure is a **sphygmomanometer** (sfig′-mō-ma-NOM-e-ter; *sphygmo-* = pulse; *manometer* = instrument used to measure pressure). When the pressure cuff is inflated above the blood pressure attained during systole, the artery is com-

pressed so that blood flow stops. The technician places a stethoscope below the cuff, and slowly deflates the cuff. When the cuff is deflated enough to allow the artery to open, a spurt of blood passes through, resulting in the first sound heard through the stethoscope. This sound corresponds to **systolic blood pressure (SBP),** the force of blood pressure on arterial walls just after ventricular contraction (Figure 21.15). As the cuff is deflated further, the sounds suddenly become faint. This level, called the **diastolic blood pressure (DBP),** represents the force exerted by the blood remaining in arteries during ventricular relaxation. At pressures below diastolic blood pressure, sounds disappear altogether. The various sounds that are heard while taking blood pressure are called **Korotkoff sounds** (kō-ROT-kof).

The normal blood pressure of a young adult male is 120 mmHg systolic and 80 mmHg diastolic, reported as "120 over

Figure 21.15   Relationship of blood pressure changes to cuff pressure.

As the cuff is deflated, sounds first occur at the systolic blood pressure; the sounds suddenly become faint at the diastolic blood pressure.

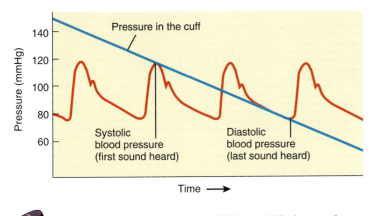

If a blood pressure is reported as "142 over 95," what are the diastolic, systolic, and pulse pressures? Does this person have hypertension as defined on page 758?

80" and written as 120/80. In young adult females, the pressures are 8 to 10 mmHg less. People who exercise regularly and are in good physical condition may have even lower blood pressures. Thus, blood pressure slightly lower than 120/80 may be a sign of good health and fitness.

The difference between systolic and diastolic pressure is called **pulse pressure.** This pressure, normally about 40 mmHg, provides information about the condition of the cardiovascular system. For example, conditions such as atherosclerosis and patent (open) ductus arteriosus greatly increase pulse pressure. The normal ratio of systolic pressure to diastolic pressure to pulse pressure is about 3:2:1.

▶ **CHECKPOINT**

18. Where may the pulse be felt?

19. What do tachycardia and bradycardia mean?

20. How are systolic and diastolic blood pressures measured with a sphygmomanometer?

SHOCK AND HOMEOSTASIS

▶ **OBJECTIVE**

• **Define shock, and describe the four types of shock.**

Shock is a failure of the cardiovascular system to deliver enough O_2 and nutrients to meet cellular metabolic needs. The causes of shock are many and varied, but all are characterized by inadequate blood flow to body tissues. With inadequate oxygen delivery, cells switch from aerobic to anaerobic production of ATP, and lactic acid accumulates in body fluids. If shock persists, cells and organs become damaged, and cells may die unless proper treatment begins quickly.

Types of Shock

Shock can be of four different types: (1) **hypovolemic shock** (hī-pō-vō-LĒ-mik; *hypo-* = low; *-volemic* = volume) due to decreased blood volume, (2) **cardiogenic shock** due to poor heart function, (3) **vascular shock** due to inappropriate vasodilation, and (4) **obstructive shock** due to obstruction of blood flow.

A common cause of hypovolemic shock is acute (sudden) hemorrhage. The blood loss may be external as occurs in trauma, or internal, as in rupture of an aortic aneurysm. Loss of body fluids through excessive sweating, diarrhea, or vomiting also can cause hypovolemic shock. Other conditions—for instance, diabetes mellitus—may cause excessive loss of fluid in the urine. Sometimes, hypovolemic shock is due to inadequate intake of fluid. Whatever the cause, when the volume of body fluids falls, venous return to the heart declines, filling of the heart lessens, stroke volume decreases, and cardiac output decreases.

In cardiogenic shock, the heart fails to pump adequately, most often because of a myocardial infarction (heart attack). Other causes of cardiogenic shock include poor perfusion of the heart (ischemia), heart valve problems, excessive preload or afterload, impaired contractility of heart muscle fibers, and arrhythmias.

Even with normal blood volume and cardiac output, shock may occur if blood pressure drops due to a decrease in systemic vascular resistance. A variety of conditions can cause inappropriate dilation of arterioles or venules. In *anaphylactic shock,* a severe allergic reaction—for example, to a bee sting—releases histamine and other mediators that cause vasodilation. In *neurogenic shock,* vasodilation may occur following trauma to the head that causes malfunction of the cardiovascular center in the medulla. Shock stemming from certain bacterial toxins that produce vasodilation is termed *septic shock.* In the United States, septic shock causes more than 100,000 deaths per year and is the most common cause of death in hospital critical care units.

Obstructive shock occurs when blood flow through a portion of the circulation is blocked. The most common cause is *pulmonary embolism,* a blood clot lodged in a blood vessel of the lungs.

Homeostatic Responses to Shock

The major mechanisms of compensation in shock are *negative feedback systems* that work to return cardiac output and arterial blood pressure to normal. When shock is mild, compensation by homeostatic mechanisms prevents serious damage. In an otherwise healthy person, compensatory mechanisms can maintain adequate blood flow and blood pressure despite an acute blood loss of as much as 10% of total volume. Figure 21.16 shows several of the negative feedback systems that respond to hypovolemic shock.

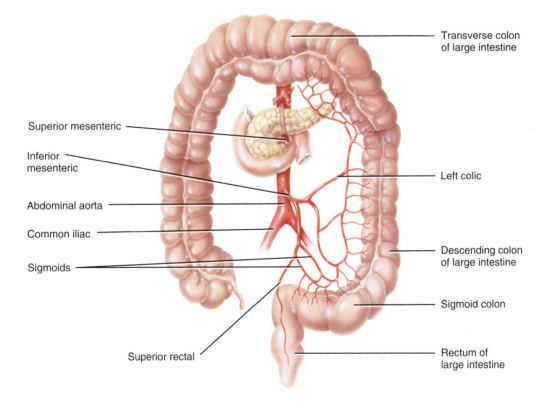

Transverse colon
of large intestine

Superior mesenteric

Inferior
mesenteric

Abdominal aorta

Common iliac

Sigmoids

Left colic

Descending colon
of large intestine

Sigmoid colon

Superior rectal

Rectum of
large intestine

(c) Anterior view of inferior mesenteric artery and its branches

 Where does the abdominal aorta begin?

Exhibit 21.6 Arteries of the Pelvis and Lower Limbs (Figure 21.23)

▶ **O B J E C T I V E**

• Identify the two major branches of the common iliac arteries.

The abdominal aorta ends by dividing into the right and left **common iliac arteries.** These, in turn, divide into the **internal** and **external iliac arteries.** In sequence, the external iliacs become the **femoral arteries**

in the thighs, the **popliteal arteries** posterior to the knee, and the **anterior** and **posterior tibial arteries** in the legs.

▶ **C H E C K P O I N T**

What general regions do the internal and external iliac arteries supply?

Branch	Description and Region Supplied
Common iliac arteries (IL-ē-ak = pertaining to the ilium)	At about the level of the fourth lumbar vertebra, the abdominal aorta divides into the right and left **common iliac arteries,** the terminal branches of the abdominal aorta. Each passes inferiorly about 5 cm (2 in.) and gives rise to two branches: internal iliac and external iliac arteries. The general distribution of the common iliac arteries is to the pelvis, external genitals, and lower limbs.
Internal iliac arteries	The **internal iliac (hypogastric) arteries** are the primary arteries of the pelvis. They begin at the bifurcation of the common iliac arteries anterior to the sacroiliac joint at the level of the lumbosacral intervertebral disc. They pass posteromedially as they descend in the pelvis and divide into anterior and posterior divisions. The general distribution of the internal iliac arteries is to the pelvis, buttocks, external genitals, and thigh.
External iliac arteries	The **external iliac arteries** are larger than the internal iliac arteries. Like the internal iliac arteries, they begin at the bifurcation of the common iliac arteries. They descend along the medial border of the psoas major muscles following the pelvic brim, pass posterior to the midportion of the inguinal ligaments, and become the femoral arteries. The general distribution of the external iliac arteries is to the lower limbs. Specifically, branches of the external iliac arteries supply the muscles of the anterior abdominal wall, the cremaster muscle in males and the round ligament of the uterus in females, and the lower limbs.
Femoral arteries (FEM-o-ral = pertaining to the thigh)	The **femoral arteries** descend along the anteromedial aspects of the thighs to the junction of the middle and lower third of the thighs. Here they pass through an opening in the tendon of the adductor magnus muscle, where they emerge posterior to the femurs as the popliteal arteries. A pulse may be felt in the femoral artery just inferior to the inguinal ligament. The general distribution of the femoral arteries is to the lower abdominal wall, groin, external genitals, and muscles of the thigh.
Popliteal arteries (pop'-li-TĒ-al = posterior surface of the knee)	The **popliteal arteries** are the continuation of the femoral arteries through the popliteal fossa (space behind the knee). They descend to the inferior border of the popliteus muscles, where they divide into the anterior and posterior tibial arteries. A pulse may be detected in the popliteal arteries. In addition to supplying the adductor magnus and hamstring muscles and the skin on the posterior aspect of the legs, branches of the popliteal arteries also supply the gastrocnemius, soleus, and plantaris muscles of the calf, knee joint, femur, patella, and fibula.
Anterior tibial arteries (TIB-ē-al = pertaining to the shin bone)	The **anterior tibial arteries** descend from the bifurcation of the popliteal arteries. They are smaller than the posterior tibial arteries. The anterior tibial arteries descend through the anterior muscular compartment of the leg. They pass through the interosseous membrane that connects the tibia and fibula, lateral to the tibia. The anterior tibial arteries supply the knee joints, anterior compartment muscles of the legs, skin over the anterior aspects of the legs, and ankle joints. At the ankles, the anterior tibial arteries become the **dorsal arteries of the foot (dorsalis pedis arteries),** also arteries from which a pulse may be detected. The dorsalis pedis arteries supply the muscles, skin, and joints on the dorsal aspects of the feet. On the dorsum of the feet, the dorsalis pedis arteries give off a transverse branch at the first (medial) cuneiform bone called the **arcuate arteries** (arcuat- = bowed) that run laterally over the bases of the metatarsals. From the arcuate arteries branch **dorsal metatarsal arteries,** which supply the feet. The dorsal metatarsal arteries terminate by dividing into the **dorsal digital arteries,** which supply the toes.
Posterior tibial arteries	The **posterior tibial arteries,** the direct continuations of the popliteal arteries, descend from the bifurcation of the popliteal arteries. They pass down the posterior muscular compartment of the legs posterior to the medial malleolus of the tibia. They terminate by dividing into the medial and lateral plantar arteries. Their general distribution is to the muscles, bones, and joints of the leg and foot. Major branches of the posterior tibial arteries are the **fibular (peroneal) arteries,** which supply the fibularis, soleus, tibialis posterior, and flexor hallucis muscles. They also supply the fibula, tarsus, and lateral aspect of the heel. The bifurcation of the posterior tibial arteries into the medial and lateral plantar arteries occurs deep to the flexor retinaculum on the medial side of the feet. The **medial plantar arteries** (PLAN-tar = sole of foot) supply the abductor hallucis and flexor digitorum brevis muscles and the toes. The **lateral plantar arteries** unite with a branch of the dorsalis pedis arteries to form the **plantar arch.** The arch begins at the base of the fifth metatarsal and extends medially across the metacarpals. As the arch crosses the foot, it gives off **plantar metatarsal arteries,** which supply the feet. These terminate by dividing into **plantar digital arteries,** which supply the toes.

SCHEME OF DISTRIBUTION

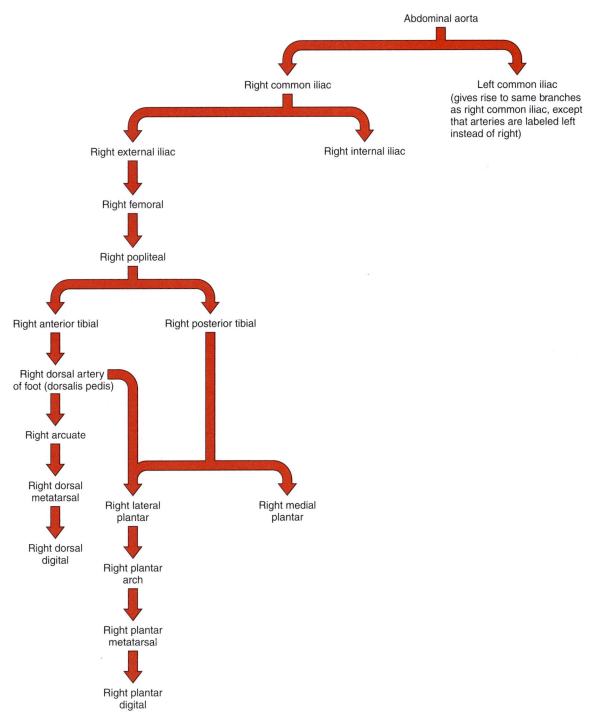

Abdominal aorta

Right common iliac

Left common iliac
(gives rise to same branches
as right common iliac, except
that arteries are labeled left
instead of right)

Right external iliac

Right internal iliac

Right femoral

Right popliteal

Right anterior tibial

Right posterior tibial

Right dorsal artery
of foot (dorsalis pedis)

Right arcuate

Right dorsal
metatarsal

Right lateral
plantar

Right medial
plantar

Right dorsal
digital

Right plantar
arch

Right plantar
metatarsal

Right plantar
digital

(continues)

Exhibit 21.6   Arteries of the Pelvis and Lower Limbs (Figure 21.23) (continued)

Figure 21.23   Arteries of the pelvis and right lower limb.

The internal iliac arteries carry most of the blood supply to the pelvic viscera and wall.

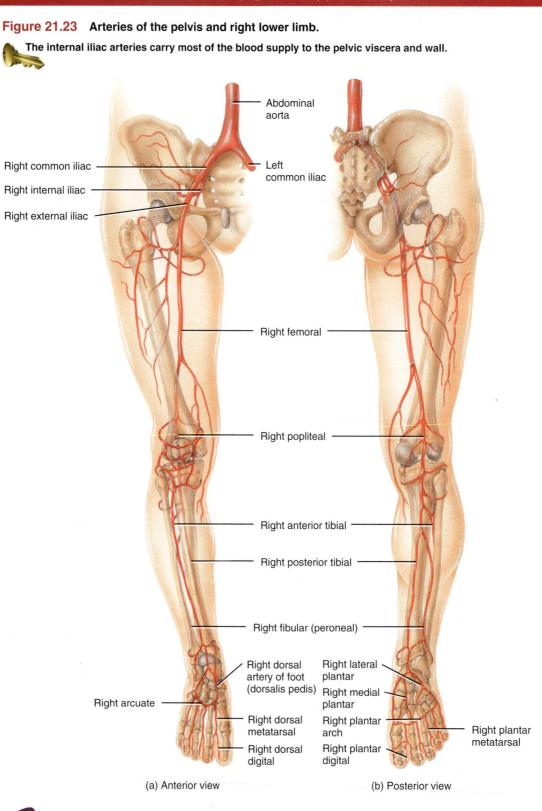

(a) Anterior view

(b) Posterior view

At what point does the abdominal aorta divide into the common iliac arteries?

Exhibit 21.7   Veins of the Systemic Circulation (Figure 21.24)

▶ **OBJECTIVE**

• Identify the three systemic veins that return deoxygenated blood to the heart.

Whereas arteries distribute blood to various parts of the body, veins drain blood away from them. For the most part, arteries are deep, whereas veins may be superficial or deep. Superficial veins are located just beneath the skin and can easily be seen. Because there are no large superficial arteries, the names of superficial veins do not correspond to those of arteries. Superficial veins are clinically important as sites for withdrawing blood or giving injections. Deep veins generally travel alongside arteries and usually bear the same name. Arteries usually follow definite pathways; veins are more difficult to follow because they connect in irregular networks in which many tributaries merge to form a large vein. Although only one systemic artery, the aorta, takes oxygenated blood away from the heart (left ventricle), three systemic veins, the **coronary sinus, superior vena cava,** and **inferior vena cava,** return deoxygenated blood to the heart (right atrium). The coronary sinus receives blood from the cardiac veins; the superior vena cava receives blood from other veins superior to the diaphragm, except the air sacs (alveoli) of the lungs; the inferior vena cava receives blood from veins inferior to the diaphragm.

▶ **CHECKPOINT**

What are the three tributaries of the coronary sinus?

Vein	Description and Region Supplied
Coronary sinus (KOR-ō-nar-ē; *corona* = crown)	The **coronary sinus** is the main vein of the heart; it receives almost all venous blood from the myocardium. It is located in the coronary sulcus (see Figure 20.3c) and opens into the right atrium between the orifice of the inferior vena cava and the tricuspid valve. It is a wide venous channel into which three veins drain. It receives the **great cardiac vein** (in the anterior interventricular sulcus) into its left end, and the **middle cardiac vein** (in the posterior interventricular sulcus) and the **small cardiac vein** into its right end. Several **anterior cardiac veins** drain directly into the right atrium.
Superior vena cava (SVC) (VĒ-na CĀ-va; *vena* = vein; *cava* = cavelike)	The **SVC** is about 7.5 cm. (3 in.) long and 2 cm (1 in.) in diameter and empties its blood into the superior part of the right atrium. It begins posterior to the right first costal cartilage by the union of the right and left brachiocephalic veins and ends at the level of the right third costal cartilage, where it enters the right atrium. The SVC drains the head, neck, chest, and upper limbs.
Inferior vena cava (IVC)	The **IVC** is the largest vein in the body, about 3.5 cm (1.4 in.) in diameter. It begins anterior to the fifth lumbar vertebra by the union of the common iliac veins, ascends behind the peritoneum to the right of the midline, pierces the costal tendon of the diaphragm at the level of the eighth thoracic vertebra, and enters the inferior part of the right atrium. The IVC drains the abdomen, pelvis, and lower limbs. The inferior vena cava is commonly compressed during the later stages of pregnancy by the enlarging uterus, producing edema of the ankles and feet and temporary varicose veins.

(continues)

Figure 21.24   **Principal veins.**

Deoxygenated blood returns to the heart via the superior and inferior venae cavae and the coronary sinus.

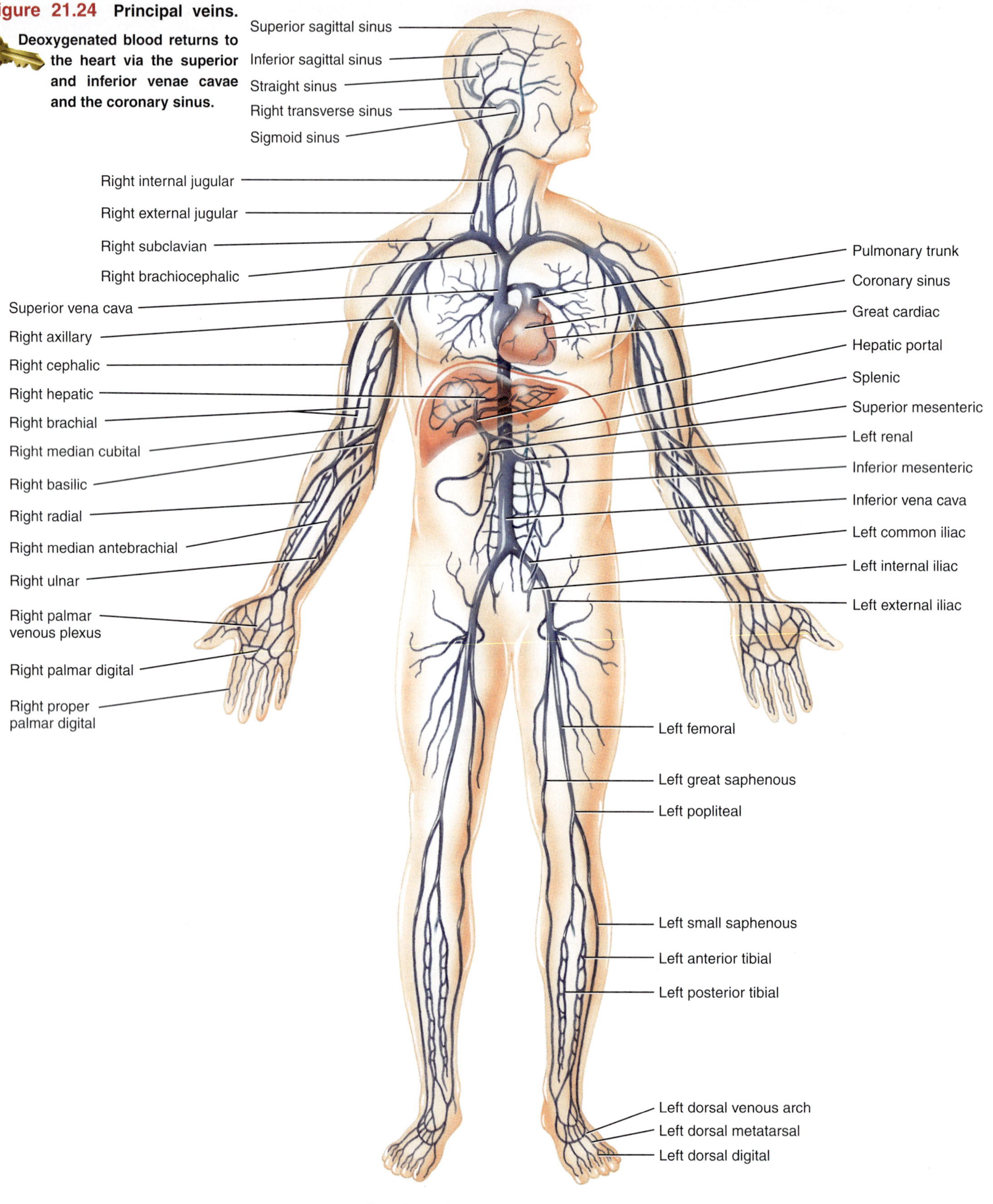

Superior sagittal sinus

Inferior sagittal sinus

Straight sinus

Right transverse sinus

Sigmoid sinus

Right internal jugular

Right external jugular

Right subclavian

Right brachiocephalic

Superior vena cava

Right axillary

Right cephalic

Right hepatic

Right brachial

Right median cubital

Right basilic

Right radial

Right median antebrachial

Right ulnar

Right palmar venous plexus

Right palmar digital

Right proper palmar digital

Pulmonary trunk

Coronary sinus

Great cardiac

Hepatic portal

Splenic

Superior mesenteric

Left renal

Inferior mesenteric

Inferior vena cava

Left common iliac

Left internal iliac

Left external iliac

Left femoral

Left great saphenous

Left popliteal

Left small saphenous

Left anterior tibial

Left posterior tibial

Left dorsal venous arch

Left dorsal metatarsal

Left dorsal digital

Overall anterior view of the principal veins

 Which general regions of the body are drained by the superior vena cava and the inferior vena cava?

Exhibit 21.8   Veins of the Head and Neck (Figure 21.25)

▶ **OBJECTIVE**

• Identify the three major veins that drain blood from the head.

Most blood draining from the head passes into three pairs of veins: the **internal jugular, external jugular,** and **vertebral veins.** Within the brain, all veins drain into dural venous sinuses and then into the internal jugular veins. **Dural venous sinuses** are endothelial-lined venous channels between layers of the cranial dura mater.

▶ **CHECKPOINT**

Which general areas are drained by the internal jugular, external jugular, and vertebral veins?

Vein	Description and Region Drained
Internal jugular veins (JUG-ū-lar; *jugular* = throat)	The flow of blood from the dural venous sinuses into the internal jugular veins is as follows (Figure 21.25): The **superior sagittal sinus** (SAJ-i-tal = straight) begins at the frontal bone, where it receives a vein from the nasal cavity, and passes posteriorly to the occipital bone. Along its course, it receives blood from the superior, medial, and lateral aspects of the cerebral hemispheres, meninges, and cranial bones. The superior sagittal sinus usually turns to the right and drains into the right transverse sinus.
	The **inferior sagittal sinus** is much smaller than the superior sagittal sinus; it begins posterior to the attachment of the falx cerebri and receives the great cerebral vein to become the straight sinus. The great cerebral vein drains the deeper parts of the brain. Along its course the inferior sagittal sinus also receives tributaries from the superior and medial aspects of the cerebral hemispheres.
	The **straight sinus** runs in the tentorium cerebelli and is formed by the union of the inferior sagittal sinus and the great cerebral vein. The straight sinus also receives blood from the cerebellum and usually drains into the left transverse sinus.
	The **transverse sinuses** begin near the occipital bone, pass laterally and anteriorly, and become the sigmoid sinuses near the temporal bone. The transverse sinuses receive blood from the cerebrum, cerebellum, and cranial bones.
	The **sigmoid sinuses** (SIG-moyd = S-shaped) are located along the temporal bone. They pass through the jugular foramina, where they terminate in the internal jugular veins. The sigmoid sinuses drain the transverse sinuses.
	The **cavernous sinuses** (KAV-er-nus = cavelike) are located on either side of the sphenoid bone. They receive blood from the ophthalmic veins from the orbits, and from the cerebral veins from the cerebral hemispheres. They ultimately empty into the transverse sinuses and internal jugular veins. The cavernous sinuses are unique because they have nerves and a major blood vessel passing through them on their way to the orbit and face. The oculomotor nerve (cranial nerve III), trochlear nerve (cranial nerve IV), and ophthalmic and maxillary branches of the trigeminal nerve (cranial nerve V), as well as the internal carotid arteries pass through the cavernous sinuses.
	The right and left **internal jugular veins** pass inferiorly on either side of the neck lateral to the internal carotid and common carotid arteries. They then unite with the subclavian veins posterior to the clavicles at the sternoclavicular joints to form the right and left **brachiocephalic veins** (brā-kē-ō-se-FAL-ik; *brachio-* = arm; *cephal-* = head). From here blood flows into the superior vena cava. The general structures drained by the internal jugular veins are the brain (through the dural venous sinuses), face, and neck.
External jugular veins	The right and left **external jugular veins** begin in the parotid glands near the angle of the mandible. They are superficial veins that descend through the neck across the sternocleidomastoid muscles. They terminate at a point opposite the middle of the clavicle, where they empty into the subclavian veins. The general structures drained by the external jugular veins are external to the cranium, such as the scalp and superficial and deep regions of the face.
Vertebral veins (VER-te-bral; *vertebra* = vertebrae)	The right and left **vertebral veins** originate inferior to the occipital condyles. They descend through successive transverse foramina of the first six cervical vertebrae and emerge from the foramina of the skin cervical vertebra to enter the brachiocephalic veins in the root of the neck. The vertebral veins drain deep structures in the neck such as the cervical vertebrae, cervical spinal cord, and some neck muscles.

(continues)

Exhibit 21.8   Veins of the Head and Neck (Figure 21.25) *(continued)*

SCHEME OF DRAINAGE

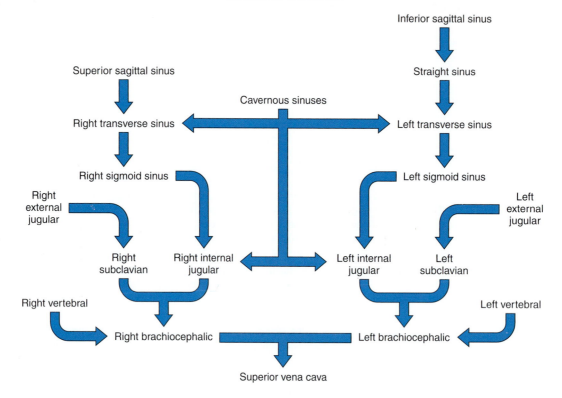

Figure 21.25   Principal veins of the head and neck.

Blood draining from the head passes into the internal jugular, external jugular, and vertebral veins.

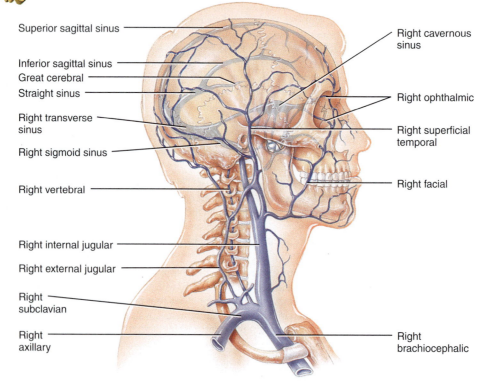

Right lateral view

Into which veins in the neck does all venous blood in the brain drain?

Exhibit 21.9 Veins of the Upper Limbs (Figure 21.26)

► **OBJECTIVE**

• Identify the principal veins that drain the upper limbs.

Both superficial and deep veins return blood from the upper limbs to the heart. **Superficial veins** are located just deep to the skin and are often visible. They anastomose extensively with one another and with deep veins, and they do not accompany arteries. Superficial veins are larger than deep veins and return most of the blood from the upper limbs. **Deep veins** are located deep in the body. They usually accompany arteries and have the same names as the corresponding arteries. Both superficial and deep veins have valves, which are more numerous in the deep veins.

► **CHECKPOINT**

Where do the cephalic, basilic, median antebrachial, radial, and ulnar veins originate?

Vein	Description and Region Drained
Superficial	
Cephalic veins (se-FAL-ik = pertaining to the head)	The principal superficial veins that drain the upper limbs are the cephalic and basilic veins. They originate in the hand and convey blood from the smaller superficial veins into the axillary veins. The **cephalic veins** begin on the lateral aspect of the **dorsal venous networks of the hands (dorsal venous arches),** networks of veins on the dorsum of the hands formed by the **dorsal metacarpal veins** (Figure 21.26a). These veins, in turn, drain the **dorsal digital veins,** which pass along the sides of the fingers. Following their formation from the dorsal venous networks of the hands, the cephalic veins arch around the radial side of the forearms to the anterior surface and ascend through the entire limbs along the anterolateral surface. The cephalic veins end where they join the axillary veins, just inferior to the clavicles. **Accessory cephalic veins** originate either from a venous plexus on the dorsum of the forearms or from the medial aspects of the dorsal venous networks of the hands and unite with the cephalic veins just inferior to the elbow. The cephalic veins drain blood from the lateral aspect of the upper limbs.
Basilic veins (ba-SIL-ik = royal, of prime importance)	The **basilic veins** begin on the medial aspects of the dorsal venous networks of the hands and ascend along the posteromedial surface of the forearm and anteromedial surface of the arm (Figure 21.26b). They drain blood from the medial aspects of the upper limbs. Anterior to the elbow, the basilic veins are connected to the cephalic veins by the **median cubital veins** (*cubital* = pertaining to the elbow), which drain the forearm. If veins must be punctured for an injection, transfusion, or removal of a blood sample, the medial cubital veins are preferred. After receiving the median cubital veins, the basilic veins continue ascending until they reach the middle of the arm. There they penetrate the tissues deeply and run alongside the brachial arteries until they join the brachial veins. As the basilic and brachial veins merge in the axillary area, they form the axillary veins.
Median antebrachial veins (an'-tē-BRĀ-kē-al; *ante-* = before, in front of; *brachi-* =arm)	The **median antebrachial veins (median veins of the forearm)** begin in the **palmar venous plexuses,** networks of veins on the palms. The plexuses drain the **palmar digital veins** in the fingers. The median antebrachial veins ascend anteriorly in the forearms to join the basilic or median cubital veins, sometimes both. They drain the palms and forearms.
Deep	
Radial veins (RĀ-dē-al = pertaining to the radius)	The paired **radial veins** begin at the **deep palmar venous arches** (Figure 21.26c). These arches drain the **palmar metacarpal veins** in the palms. The radial veins drain the lateral aspects of the forearms and pass alongside the radial arteries. Just inferior to the elbow joint, the radial veins unite with the ulnar veins to form the brachial veins.
Ulnar veins (UL-nar = pertaining to the ulna)	The paired **ulnar veins,** which are larger than the radial veins, begin at the **superficial palmar venous arches.** These arches drain the **common palmar digital veins** and the **proper palmar digital veins** in the fingers. The ulnar veins drain the medial aspect of the forearms, pass alongside the ulnar arteries, and join with the radial veins to form the brachial veins.
Brachial veins (BRĀ-kē-al; *brachi-* = arm)	The paired **brachial veins** accompany the brachial arteries. They drain the forearms, elbow joints, arms, and humerus. They pass superiorly and join with the basilic veins to form the axillary veins.
Axillary veins (AK-sil-ār-ē; *axilla* = armpit)	The **axillary veins** ascend to the outer borders of the first ribs, where they become the subclavian veins. The axillary veins receive tributaries that correspond to the branches of the axillary arteries. The axillary veins drain the arms, axillas, and superolateral chest wall.
Subclavian veins (sub-KLĀ-vē-an; *sub-* = under; *clavian* = pertaining to the clavicle)	The **subclavian veins** are continuations of the axillary veins that terminate at the sternal end of the clavicles, where they unite with the internal jugular veins to form the brachiocephalic veins. The subclavian veins drain the arms, neck, and thoracic wall. The thoracic duct of the lymphatic system delivers lymph into the left subclavian vein at the junction with the internal jugular. The right lymphatic duct delivers lymph into the right subclavian vein at the corresponding junction (see Figure 22.3a).

(continues)

Exhibit 21.9   **Veins of the Upper Limbs (Figure 21.26)** *(continued)*

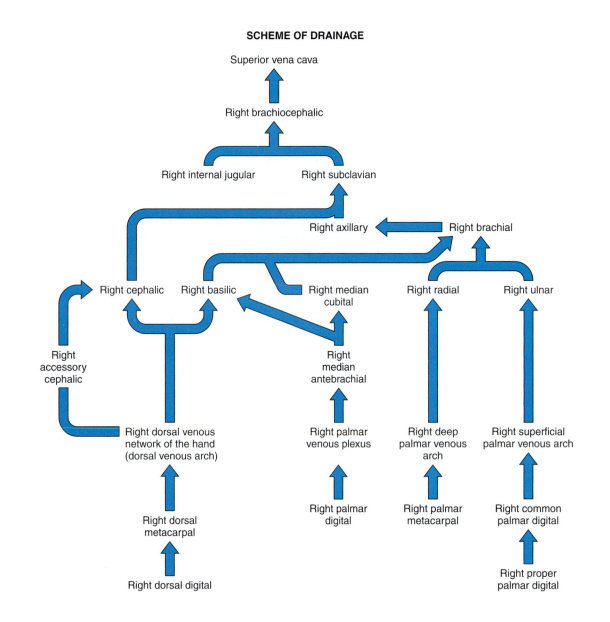

SCHEME OF DRAINAGE

Figure 21.26   **Principal veins of the right upper limb.**

🔑 Deep veins usually accompany arteries that have similar names.

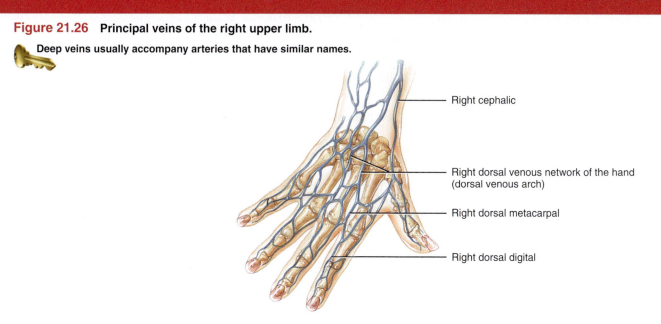

(a) Posterior view of superficial veins of the hand

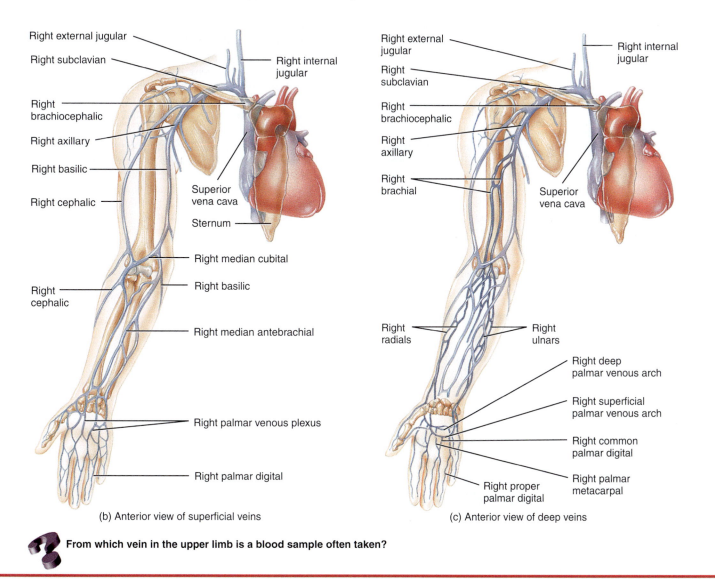

(b) Anterior view of superficial veins

(c) Anterior view of deep veins

❓ **From which vein in the upper limb is a blood sample often taken?**

Exhibit 21.10 **Veins of the Thorax (Figure 21.27)**

▶ OBJECTIVE

• Identify the components of the azygos system of veins.

Although the brachiocephalic veins drain some portions of the thorax, most thoracic structures are drained by a network of veins, called the **azygos system,** that runs on either side of the vertebral column. The system consists of three veins—the **azygos, hemiazygos,** and **accessory hemiazygos veins**—that show considerable variation in origin,

course, tributaries, anastomoses, and termination. Ultimately they empty into the superior vena cava.

▶ CHECKPOINT

What is the importance of the azygos system relative to the inferior vena cava?

Vein	Description and Region Drained
Brachiocephalic vein (brā′-kē-ō-se-FAL-ik; *brachio-* = arm; *cephalic* = pertaining to the head)	The right and left **brachiocephalic veins,** formed by the union of the subclavian and internal jugular veins, drain blood from the head, neck, upper limbs, mammary glands, and superior thorax. The brachiocephalic veins unite to form the superior vena cava. Because the superior vena cava is to the right of the body's midline, the left brachiocephalic vein is longer than the right. The right brachiocephalic vein is anterior and to the right of the brachiocephalic trunk. The left brachiocephalic vein is anterior to the brachiocephalic trunk, the left common carotid and left subclavian arteries, the trachea, the left vagus nerve (cranial nerve X), and phrenic nerve.
Azygos system (az-ī-gos = unpaired)	The **azygos system,** besides collecting blood from the thorax and abdominal wall, may serve as a bypass for the inferior vena cava that drains blood from the lower body. Several small veins directly link the azygos system with the inferior vena cava. Large veins that drain the lower limbs and abdomen conduct blood into the azygos system. If the inferior vena cava or hepatic portal vein becomes obstructed, the azygos system can return blood from the lower body to the superior vena cava.
Azygos vein	The **azygos vein** is anterior to the vertebral column, slightly to the right of the midline. It usually begins at the junction of the right ascending lumbar and right subcostal veins near the diaphragm. At the level of the fourth thoracic vertebra, it arches over the root of the right lung to end in the superior vena cava. Generally, the azygos vein drains the right side of the thoracic wall, thoracic viscera, and abdominal wall. Specifically, the azygos vein receives blood from most of the **right posterior intercostal, hemiazygos, accessory hemiazygos, esophageal, mediastinal, pericardial,** and **bronchial veins.**
Hemiazygos vein (HEM-ē-az-ī-gos; *hemi-* = half)	The **hemiazygos vein** is anterior to the vertebral column and slightly to the left of the midline. It often begins at the junction of the left ascending lumbar and left subcostal veins. It terminates by joining the azygos vein at about the level of the ninth thoracic vertebra. Generally, the hemiazygos vein drains the left side of the thoracic wall, thoracic viscera, and abdominal wall. Specifically, the hemiazygos vein receives blood from the ninth through eleventh **left posterior intercostal, esophageal, mediastinal,** and sometimes the **accessory hemiazygos veins.**
Accessory hemiazygos vein	The **accessory hemiazygos vein** is also anterior to the vertebral column and to the left of the midline. It begins at the fourth or fifth intercostal space and descends from the fifth to the eighth thoracic vertebra or ends in the hemiazygos vein. It terminates by joining the azygos vein at about the level of the eighth thoracic vertebra. The accessory hemiazygos vein drains the left side of the thoracic wall. It receives blood from the fourth through eighth **left posterior intercostal veins** (the first through third left posterior intercostal veins open into the left brachiocephalic vein), **left bronchial,** and **mediastinal veins.**

SCHEME OF DRAINAGE

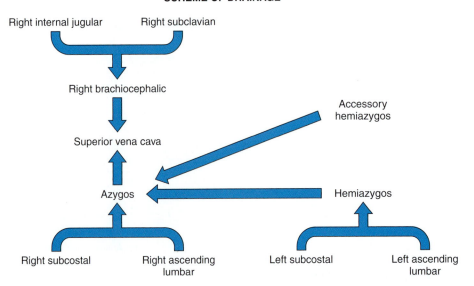

Figure 21.27 Principal veins of the thorax, abdomen, and pelvis.

Most thoracic structures are drained by the azygos system of veins.

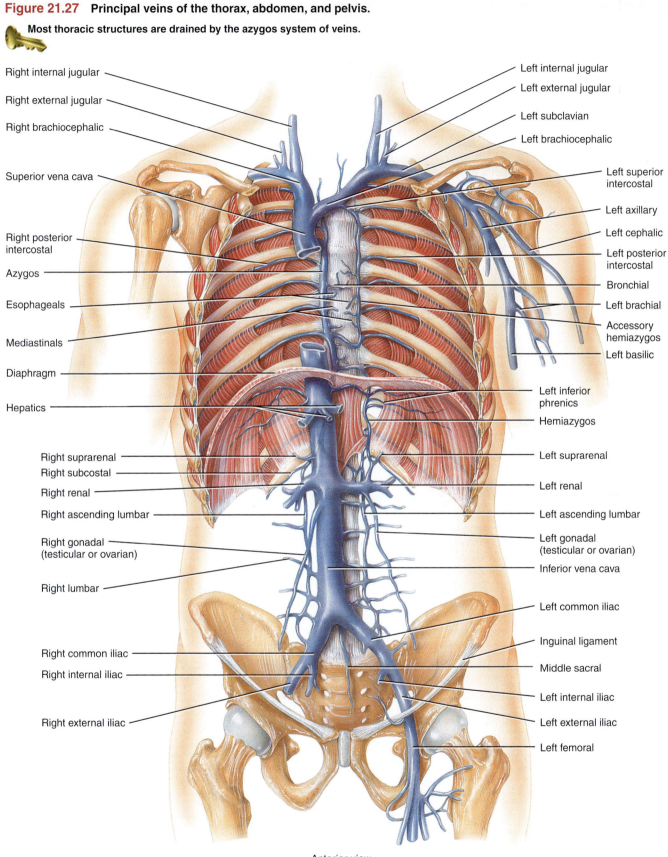

Right internal jugular

Right external jugular

Right brachiocephalic

Superior vena cava

Right posterior intercostal

Azygos

Esophageals

Mediastinals

Diaphragm

Hepatics

Right suprarenal

Right subcostal

Right renal

Right ascending lumbar

Right gonadal (testicular or ovarian)

Right lumbar

Right common iliac

Right internal iliac

Right external iliac

Left internal jugular

Left external jugular

Left subclavian

Left brachiocephalic

Left superior intercostal

Left axillary

Left cephalic

Left posterior intercostal

Bronchial

Left brachial

Accessory hemiazygos

Left basilic

Left inferior phrenics

Hemiazygos

Left suprarenal

Left renal

Left ascending lumbar

Left gonadal (testicular or ovarian)

Inferior vena cava

Left common iliac

Inguinal ligament

Middle sacral

Left internal iliac

Left external iliac

Left femoral

Anterior view

Which vein returns blood from the abdominopelvic viscera to the heart?

Exhibit 21.11   Veins of the Abdomen and Pelvis (Figure 21.27)

▶ OBJECTIVE

• Identify the principal veins that drain the abdomen and pelvis.

Blood from the abdominal and pelvic viscera and abdominal wall returns to the heart via the **inferior vena cava.** Many small veins enter the inferior vena cava. Most carry return flow from parietal branches of the abdominal aorta, and their names correspond to the names of the arteries.

The inferior vena cava does not receive veins directly from the gastrointestinal tract, spleen, pancreas, and gallbladder. These organs pass their blood into a common vein, the **hepatic portal vein,** which delivers the blood to the liver. The superior mesenteric and splenic veins unite to form the hepatic portal vein (see Figure 21.29). This special flow of venous blood is called **hepatic portal circulation,** which is described shortly. After passing through the liver for processing, blood drains into the hepatic veins, which empty into the inferior vena cava.

▶ CHECKPOINT

What structures do the lumbar, gonadal, renal, suprarenal, inferior phrenic, and hepatic veins drain?

Vein	Description and Region Drained
Inferior vena cava (VĒ-na CĀ-va; *vena* = vein; *cava* = cavelike)	The two common iliac veins that drain the lower limbs, pelvis, and abdomen unite to form the **inferior vena cava** (see Figure 21.27). The inferior vena cava extends superiorly through the abdomen and thorax to the right atrium.
Common iliac veins (IL-ē-ak = pertaining to the ilium)	The **common iliac veins** are formed by the union of the internal and external iliac veins anterior to the sacroiliac joint and represent the distal continuation of the inferior vena cava at their bifurcation. The right common iliac vein is much shorter than the left and is also more vertical. Generally, the common iliac veins drain the pelvis, external genitals, and lower limbs.
Internal iliac veins	The **internal iliac veins** begin near the superior portion of the greater sciatic notch and run medial to their corresponding arteries. Generally, the veins drain the thigh, buttocks, external genitals, and pelvis.
External iliac veins	The **external iliac veins** are companions of the internal iliac arteries and begin at the inguinal ligaments as continuations of the femoral veins. They end anterior to the sacroiliac joint where they join with the internal iliac veins to form the common iliac veins. The external iliac veins drain the lower limbs, cremaster muscle in males, and the abdominal wall.
Lumbar veins (LUM-bar = pertaining to the loin)	A series of parallel **lumbar veins,** usually four on each side, drain blood from both sides of the posterior abdominal wall, vertebral canal, spinal cord, and meninges. The lumbar veins run horizontally with the lumbar arteries. The lumbar veins connect at right angles with the right and left **ascending lumbar veins,** which form the origin of the corresponding azygos or hemiazygos vein. The lumbar veins drain blood into the ascending lumbars and then run to the inferior vena cava, where they release the remainder of the flow.
Gonadal veins (gō-NAD-al; *gono* = seed) [**testicular** (tes-TIK-ū-lar) or **ovarian** (ō-VAR-ē-an)]	The **gonadal veins** ascend with the gonadal arteries along the posterior abdominal wall. In the male, the gonadal veins are called the testicular veins. The **testicular veins** drain the testes (the left testicular vein empties into the left renal vein, and the right testicular vein drains into the inferior vena cava). In the female, the gonadal veins are called the ovarian veins. The **ovarian veins** drain the ovaries. The left ovarian vein empties into the left renal vein, and the right ovarian vein drains into the inferior vena cava.
Renal veins (RĒ-nal = pertaining to the kidney)	The **renal veins** are large and pass anterior to the renal arteries. The left renal vein is longer than the right renal vein and passes anterior to the abdominal aorta. It receives the left testicular (or ovarian), left inferior phrenic, and usually left suprarenal veins. The right renal vein empties into the inferior vena cava posterior to the duodenum. The renal veins drain the kidneys.
Suprarenal veins (soo′-pra-RĒ-nal; *supra* = above)	The **suprarenal veins** drain the adrenal (suprarenal) glands (the left suprarenal vein empties into the left renal vein, and the right suprarenal vein empties into the inferior vena cava).
Inferior phrenic veins (FREN- ik = pertaining to the diaphragm)	The **inferior phrenic veins** drain the diaphragm (the left inferior phrenic vein usually sends one tributary to the left suprarenal vein, which empties into the left renal vein, and another tributary that empties into the inferior vena cava; the right inferior phrenic vein empties into the inferior vena cava).
Hepatic veins (he-PAT-ik = pertaining to the liver)	The **hepatic veins** drain the liver.

SCHEME OF DRAINAGE

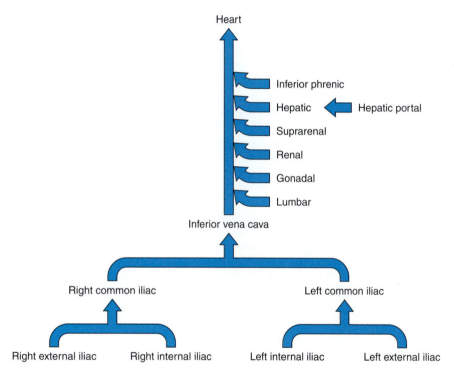

Exhibit 21.12   Veins of the Lower Limbs (Figure 21.28)

▶ **OBJECTIVE**

• Identify the principal superficial and deep veins that drain the lower limbs.

As with the upper limbs, blood from the lower limbs is drained by both **superficial** and **deep veins.** The superficial veins often anastomose with each other and with deep veins along their length. Deep veins, for the most part, have the same names as corresponding arteries. All veins of the lower limbs have valves, which are more numerous than in veins of the upper limbs.

▶ **CHECKPOINT**

Why are the great saphenous veins clinically important?

Vein	Description and Region Drained
Superficial veins	
Great saphenous veins (sa-FĒ-nus; *saphen-* = clearly visible)	The **great (long) saphenous veins,** the longest veins in the body, ascend from the foot to the groin in the subcutaneous layer. They begin at the medial end of the dorsal venous arches of the foot. The **dorsal venous arches** (VĒ-nus) are networks of veins on the dorsum of the foot formed by the **dorsal digital veins,** which collect blood from the toes, and then unite in pairs to form the **dorsal metatarsal veins,** which parallel the metatarsals. As the dorsal metatarsal veins approach the foot, they combine to form the dorsal venous arches. The great saphenous veins pass anterior to the medial malleolus of the tibia and then superiorly along the medial aspect of the leg and thigh just deep to the skin. They receive tributaries from superficial tissues and connect with the deep veins as well. They empty into the femoral veins at the groin. Generally, the great saphenous veins drain mainly the medial side of the leg and thigh, the groin, external genitals, and abdominal wall. Along their length, the great saphenous veins have from 10 to 20 valves, with more located in the leg than the thigh. These veins are more likely to be subject to varicosities than other veins in the lower limbs because they must support a long column of blood and are not well supported by skeletal muscles. The great saphenous veins are often used for prolonged administration of intravenous fluids. This is particularly important in very young children and in patients of any age who are in shock and whose veins are collapsed. The great saphenous veins are also often used as a source of vascular grafts, especially for coronary bypass surgery. In the procedure, the vein is removed and then reversed so that the valves do not obstruct the flow of blood.
Small saphenous veins	The **small (short) saphenous veins** begin at the lateral aspect of the dorsal venous arches of the foot. They pass posterior to the lateral malleolus of the fibula and ascend deep to the skin along the posterior aspect of the leg. They empty into the popliteal veins in the popliteal fossa, posterior to the knee. Along their length, the small saphenous veins have from 9 to 12 valves. The small saphenous veins drain the foot and posterior aspect of the leg. They may communicate with the great saphenous veins in the proximal thigh.
Deep veins	
Posterior tibial veins (TIB-ē-al)	The **plantar digital veins** on the plantar surfaces of the toes unite to form the **plantar metatarsal veins,** which parallel the metatarsals. They unite to form the **deep plantar venous arches.** From each arch emerges the **medial** and **lateral plantar veins.** The medial and lateral plantar veins, posterior to the medial malleolus of the tibia, form the paired **posterior tibial veins,** which sometimes merge into a single vessel. They accompany the posterior tibial artery through the leg. They ascend deep to the muscles in the posterior aspect of the leg and drain the foot and posterior compartment muscles. About two-thirds of the way up the leg, the posterior tibial veins drain blood from the **fibular (peroneal) veins,** which drain the lateral and posterior leg muscles. The posterior tibial veins unite with the anterior tibial veins just inferior to the popliteal fossa to form the popliteal veins.
Anterior tibial veins	The paired **anterior tibial veins** arise in the dorsal venous arch and accompany the anterior tibial artery. They ascend in the interosseous membrane between the tibia and fibula and unite with the posterior tibial veins to form the popliteal vein. The anterior tibial veins drain the ankle joint, knee joint, tibiofibular joint, and anterior portion of the leg.
Popliteal veins (pop′-li-TĒ-al = pertaining to the hollow behind knee)	The anterior and posterior tibial veins unite to form the **popliteal veins.** The popliteal veins also receive blood from the small saphenous veins and tributaries that correspond to branches of the popliteal artery. The popliteal veins drain the knee joint and the skin, muscles, and bones of portions of the calf and thigh around the knee joint.
Femoral veins (FEM-o-ral)	The **femoral veins** accompany the femoral arteries and are the continuations of the popliteal veins just superior to the knee. The femoral veins extend up the posterior surface of the thighs and drain the muscles of the thighs, femurs, external genitals, and superficial lymph nodes. The largest tributaries of the femoral veins are the **deep femoral veins.** Just before penetrating the abdominal wall, the femoral veins receive the deep femoral veins and the great saphenous veins. The veins formed from this union penetrate the body wall and enter the pelvic cavity. Here they are known as the **external iliac veins.**

SCHEME OF DRAINAGE

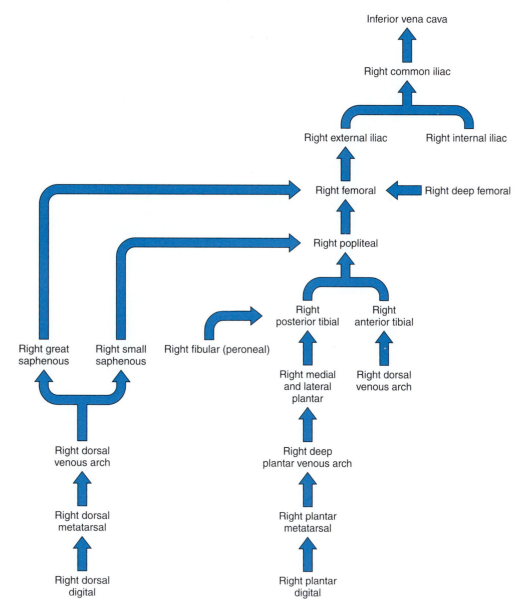

(continues)

Exhibit 21.12 Veins of the Lower Limbs (Figure 21.28) *(continued)*

Figure 21.28 Principal veins of the pelvis and lower limbs.

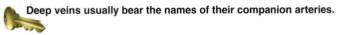 Deep veins usually bear the names of their companion arteries.

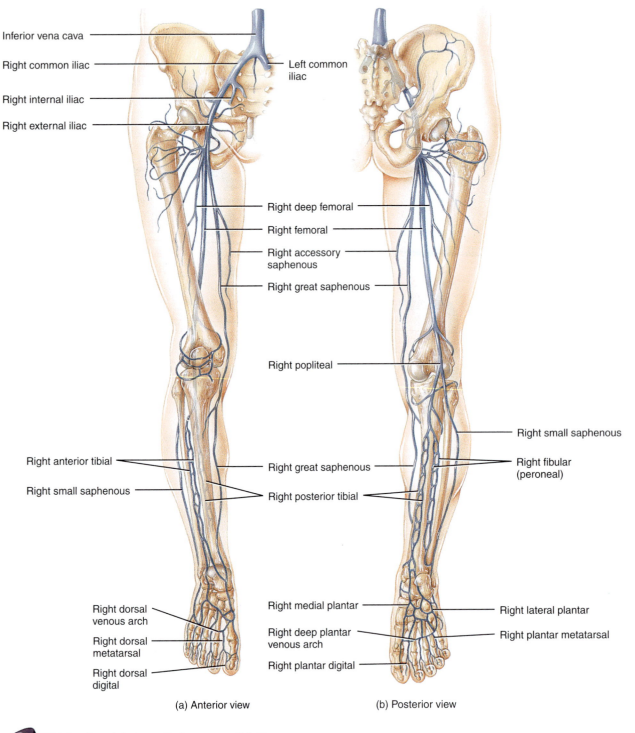

(a) Anterior view   (b) Posterior view

Which veins of the lower limb are superficial?

The Hepatic Portal Circulation

The **hepatic portal circulation** carries venous blood from gastrointestinal organs and spleen to the liver. A vein that carries blood from one capillary network to another is called a **portal vein.** The **hepatic portal vein** (*hepat-* = liver) receives blood from capillaries of gastrointestinal organs and the spleen and delivers it to the sinusoids of the liver (Figure 21.29). After a meal, hepatic portal blood is rich in nutrients absorbed from the gastrointestinal tract. The liver stores some of them and modifies others before they pass into the general circulation. For example, the liver converts glucose into glycogen for storage, thereby reducing blood glucose level shortly after a meal. The liver also detoxifies harmful substances, such as alcohol, that have been absorbed from the gastrointestinal tract and destroys bacteria by phagocytosis. While the liver is receiving nutrient-rich but deoxygenated blood via the hepatic portal vein, it also is receiving oxygenated blood via the hepatic artery, a branch of the celiac trunk. The oxygenated blood mixes with the deoxygenated blood in sinusoids. Eventually, blood leaves the

Figure 21.29 Hepatic portal circulation.   A schematic diagram of blood flow through the liver, including arterial circulation, is shown in (b). As usual, deoxygenated blood is indicated in blue, oxygenated blood in red.

🔑 The hepatic portal circulation delivers venous blood from the organs of the gastrointestinal tract and spleen to the liver.

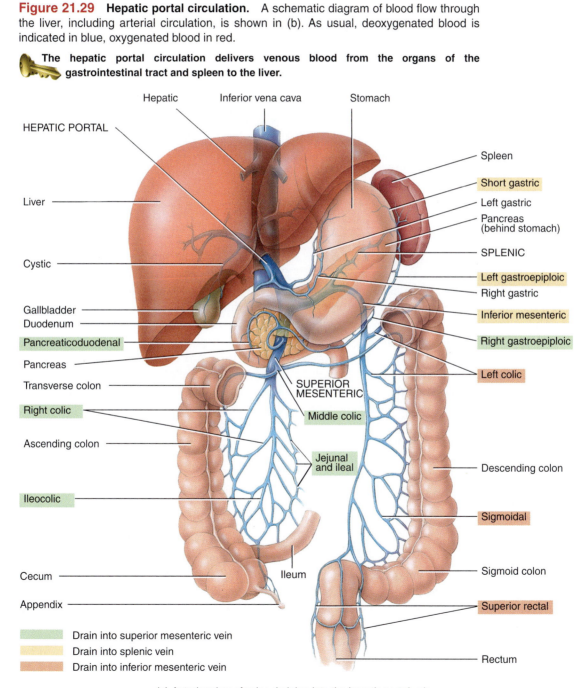

Drain into superior mesenteric vein
Drain into splenic vein
Drain into inferior mesenteric vein

(a) Anterior view of veins draining into the hepatic portal vein

(continues)

Figure 21.29 *(continued)*

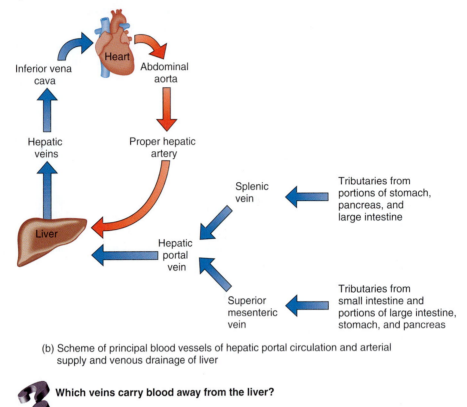

(b) Scheme of principal blood vessels of hepatic portal circulation and arterial supply and venous drainage of liver

? Which veins carry blood away from the liver?

sinusoids of the liver through the **hepatic veins,** which drain into the inferior vena cava.

The superior mesenteric and splenic veins unite to form the hepatic portal vein. The **superior mesenteric vein** drains blood from the small intestine and portions of the large intestine, stomach, and pancreas through the *jejunal, ileal, ileocolic, right colic, middle colic, pancreaticoduodenal,* and *right gastroepiploic veins.* The **splenic vein** drains blood from the stomach, pancreas, and portions of the large intestine through the *short gastric, left gastroepiploic, pancreatic,* and *inferior mesenteric veins.* The inferior mesenteric vein, which passes into the splenic vein, drains portions of the large intestine through the *superior rectal, sigmoidal,* and *left colic veins.* The *right* and *left gastric veins,* which open directly into the hepatic portal vein, drain the stomach. The *cystic vein,* which also opens into the hepatic portal vein, drains the gallbladder.

The Pulmonary Circulation

The **pulmonary circulation** (*pulmo-* = lung) carries deoxygenated blood from the right ventricle to the air sacs (alveoli) within the lungs and returns oxygenated blood from the air sacs to the left atrium (Figure 21.30). The **pulmonary trunk** emerges from the right ventricle and passes superiorly, posteriorly, and to the left. It then divides into two branches: the **right pulmonary artery** to the right lung and the **left pulmonary artery** to the left lung. After birth, the pulmonary arteries are the only arteries that carry deoxygenated blood. On entering the lungs, the

branches divide and subdivide until finally they form capillaries around the air sacs (alveoli) within the lungs. CO_2 passes from the blood into the air sacs and is exhaled. Inhaled O_2 passes from the air within the lungs into the blood. The pulmonary capillaries unite to form venules and eventually **pulmonary veins,** which exit the lungs and carry the oxygenated blood to the left atrium. Two left and two right pulmonary veins enter the left atrium. After birth, the pulmonary veins are the only veins that carry oxygenated blood. Contractions of the left ventricle then eject the oxygenated blood into the systemic circulation.

The Fetal Circulation

The circulatory system of a fetus, called the **fetal circulation,** exists only in the fetus and contains special structures that allow the developing fetus to exchange materials with its mother (Figure 21.31 on pages 754–755). It differs from the postnatal (after birth) circulation because the lungs, kidneys, and gastrointestinal organs do not begin to function until birth. The fetus obtains O_2 and nutrients from and eliminates CO_2 and other wastes into the maternal blood.

The exchange of materials between fetal and maternal circulations occurs through the **placenta** (pla-SEN-ta), which forms inside the mother's uterus and attaches to the umbilicus (navel) of the fetus by the **umbilical cord** (um-BIL-i-kal). The placenta communicates with the mother's cardiovascular system through many small blood vessels that emerge from the uterine wall. The umbilical cord contains blood vessels that branch into capillaries

Figure 21.30 Pulmonary circulation.

The pulmonary circulation brings deoxygenated blood from the right ventricle to the lungs and returns oxygenated blood from the lungs to the left atrium.

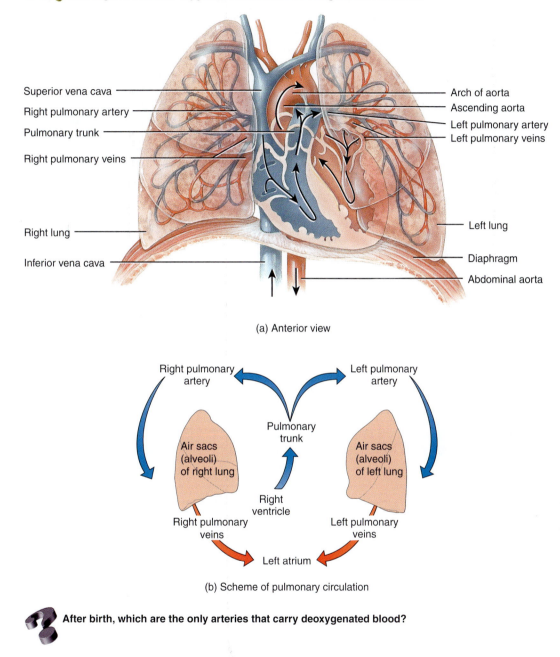

(a) Anterior view

(b) Scheme of pulmonary circulation

After birth, which are the only arteries that carry deoxygenated blood?

in the placenta. Wastes from the fetal blood diffuse out of the capillaries, into spaces containing maternal blood (intervillous spaces) in the placenta, and finally into the mother's uterine veins. Nutrients travel the opposite route—from the maternal blood vessels to the intervillous spaces to the fetal capillaries. Normally, there is no direct mixing of maternal and fetal blood because all exchanges occur by diffusion through capillary walls.

Blood passes from the fetus to the placenta via two **umbilical arteries** (Figure 21.31a, c). These branches of the internal iliac (hypogastric) arteries are within the umbilical cord. At the placenta, fetal blood picks up O_2 and nutrients and eliminates CO_2 and wastes. The oxygenated blood returns from the placenta via a single **umbilical vein.** This vein ascends to the liver of the fetus, where it divides into two branches. Whereas some blood flows through the branch that joins the hepatic portal vein and enters the liver, most of the blood flows into the second branch, the **ductus venosus** (DUK-tus ve-NŌ-sus), which drains into the inferior vena cava.

Deoxygenated blood returning from lower body regions of the fetus mingles with oxygenated blood from the ductus venosus in the inferior vena cava. This mixed blood then enters the right atrium. Deoxygenated blood returning from upper body regions of the fetus enters the superior vena cava and passes into the right atrium.

Figure 21.31 Fetal circulation and changes at birth. The boxes between parts (a) and (b) describe the fate of certain fetal structures once postnatal circulation is established.

🔑 **The lungs and gastrointestinal organs do not begin to function until birth.**

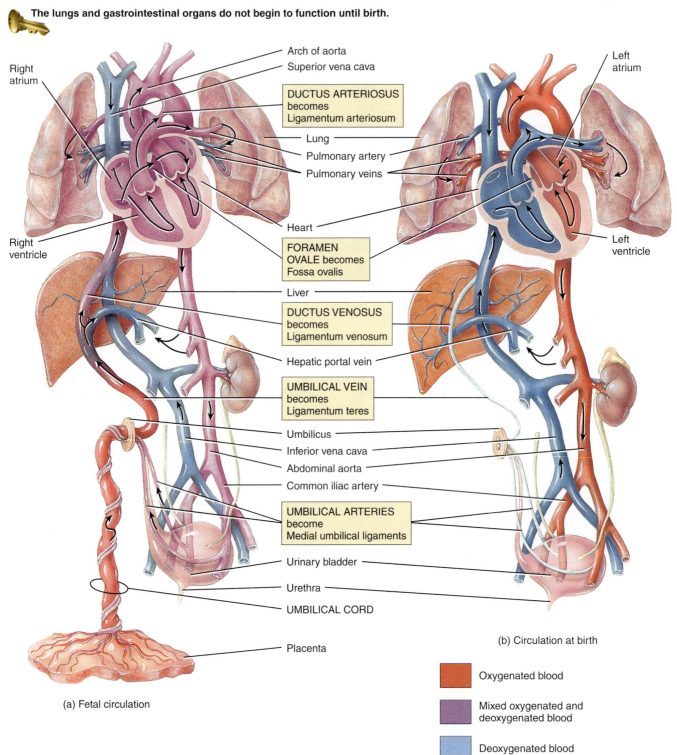

(a) Fetal circulation

(b) Circulation at birth

■ Oxygenated blood

■ Mixed oxygenated and deoxygenated blood

■ Deoxygenated blood

Most of the fetal blood does not pass from the right ventricle to the lungs, as it does in postnatal circulation, because an opening called the **foramen ovale** (fō-RĀ-men ō-VAL-ē) exists in the septum between the right and left atria. About one-third of the blood that enters the right atrium passes through the foramen ovale into the left atrium and joins the systemic circulation. The blood that does pass into the right ventricle is pumped into the pulmonary trunk, but little of this blood reaches the nonfunction-

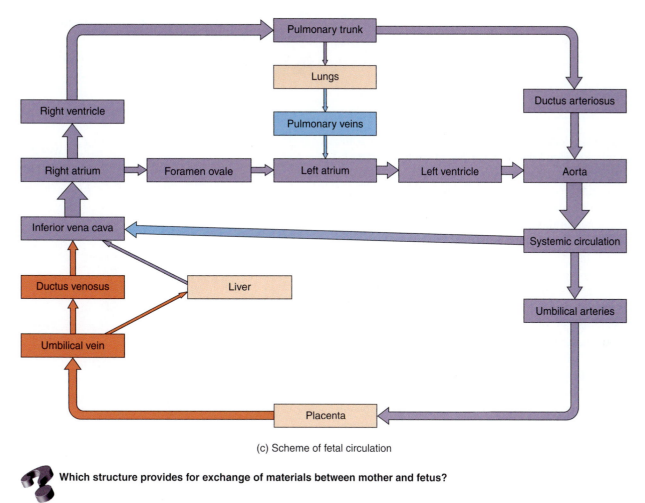

(c) Scheme of fetal circulation

Which structure provides for exchange of materials between mother and fetus?

ing fetal lungs. Instead, most is sent through the **ductus arterio- sus** (ar-tē-rē-Ō-sus), a vessel that connects the pulmonary trunk with the aorta, so that most blood bypasses the fetal lungs. The blood in the aorta is carried to all fetal tissues through the systemic circulation. When the common iliac arteries branch into the external and internal iliacs, part of the blood flows into the internal iliacs, into the umbilical arteries, and back to the placenta for another exchange of materials.

After birth, when pulmonary (lung), renal, and digestive functions begin, the following vascular changes occur (Figure 21.31b):

1. When the umbilical cord is tied off, blood no longer flows through the umbilical arteries, they fill with connective tissue, and the distal portions of the umbilical arteries become fibrous cords called the **medial umbilical ligaments.** Although the arteries are closed functionally only a few minutes after birth, complete obliteration of the lumens may take 2 to 3 months.

2. The umbilical vein collapses but remains as the **ligamentum teres (round ligament),** a structure that attaches the umbilicus to the liver.

3. The ductus venosus collapses but remains as the **ligamentum venosum,** a fibrous cord on the inferior surface of the liver.

4. The placenta is expelled as the **"afterbirth."**

5. The foramen ovale normally closes shortly after birth to become the **fossa ovalis,** a depression in the interatrial septum. When an infant takes its first breath, the lungs expand and blood flow to the lungs increases. Blood returning from the lungs to the heart increases pressure in the left atrium. This closes the foramen ovale by pushing the valve that guards it against the interatrial septum. Permanent closure occurs in about a year.

6. The ductus arteriosus closes by vasoconstriction almost immediately after birth and becomes the **ligamentum arteriosum.** Complete anatomical obliteration of the lumen takes 1 to 3 months.

▶ **CHECKPOINT**

22. Prepare a diagram to show the hepatic portal circulation. Why is this route important?

23. Prepare a diagram to show the route of the pulmonary circulation.

24. Discuss the anatomy and physiology of the fetal circulation. Indicate the function of the umbilical arteries, umbilical vein, ductus venosus, foramen ovale, and ductus arteriosus.

DEVELOPMENT OF BLOOD VESSELS AND BLOOD

▶ OBJECTIVE

• Describe the development of blood vessels and blood.

The human yolk sac and ovum have little yolk to nourish the developing embryo. Blood and blood vessel formation starts as early as 15–16 days in the **mesoderm** of the yolk sac, chorion, and connecting stalk.

Blood vessels develop from mesenchyme that differentiates into cells called **angioblasts.** The cells aggregate to form isolated masses of cells called **blood islands** (Figure 21.32). Spaces soon appear in the islands and become the lumens of the blood vessels. Some of the angioblasts immediately around the spaces give rise to the *endothelial lining of the blood vessels.* Angioblasts around the endothelium form the *tunics* (interna, media, and externa) of the larger blood vessels. Growth and fusion of blood islands form an extensive network of blood vessels throughout the embryo.

Endothelial cells outside the embryo produce *blood plasma* and *blood cells,* which appear in the blood vessels of the yolk sac, chorion, and allantois at about three weeks. Blood formation in the embryo itself begins at about the fifth week in the liver and the twelfth week in the spleen, red bone marrow, and thymus.

▶ CHECKPOINT

25. What are the sites of blood cell production outside the embryo and within the embryo?

AGING AND THE CARDIOVASCULAR SYSTEM

▶ OBJECTIVE

• Explain the effects of aging on the cardiovascular system.

General changes in the cardiovascular system associated with aging include decreased compliance of the aorta and large arteries, reduction in cardiac muscle fiber size, progressive loss of cardiac muscular strength, reduced cardiac output, a decline in maximum heart rate, and an increase in systolic blood pressure. There is an increase in the incidence of coronary artery disease (CAD), the major cause of heart disease and death in older Americans. Congestive heart failure, a set of symptoms associated with impaired pumping of the heart, is also prevalent in older individuals. Atherosclerosis of blood vessels that serve brain tissue reduces brain blood flow and thus decreases nourish-

Figure 21.32 Development of blood vessels and blood cells from blood islands.

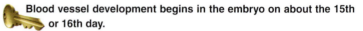 Blood vessel development begins in the embryo on about the 15th or 16th day.

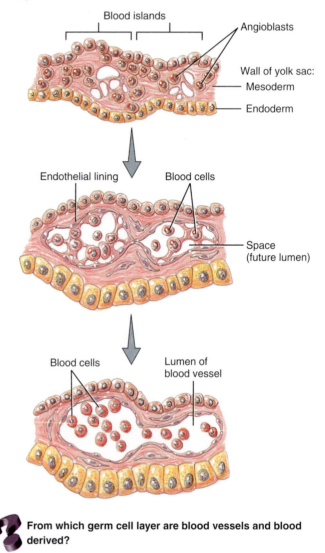

From which germ cell layer are blood vessels and blood derived?

ment to the brain. As a result, some brain cells malfunction or die. By age 80, cerebral blood flow is 20% less and renal blood flow is 50% less than in the same person at age 30.

• • •

To appreciate the many ways the blood, heart, and blood vessels contribute to homeostasis of other body systems, examine *Focus on Homeostasis: The Cardiovascular System.*

Body System	Contribution of the Cardiovascular System
For all body systems	The heart pumps blood through blood vessels to body tissues, delivering oxygen and nutrients and removing wastes by means of capillary exchange. Circulating blood keeps body tissues at a proper temperature.
Integumantary system	Blood delivers clotting factors and white blood cells that aid in hemostasis when skin is damaged and contribute to repair of injured skin. Changes in skin blood flow contribute to body temperature regulation by adjusting the amount of heat loss via the skin. Blood flowing in skin may give skin a pink hue.
Skeletal system	Blood delivers calcium and phosphate ions that are needed for building bone matrix, hormones that govern building and breakdown of bone matrix, and erythropoietin that stimulates production of red blood cells by red bone marrow.
Muscular system	Blood circulating through exercising muscle removes heat and lactic acid.
Nervous system	Endothelial cells lining choroid plexuses in brain ventricles help produce cerebrospinal fluid (CSF) and contribute to the blood–brain barrier.
Endocrine system	Circulating blood delivers most hormones to their target tissues. Atrial cells secrete atrial natriuretic peptide.
Lymphatic and immune system	Circulating blood distributes lymphocytes, antibodies, and macrophages that carry out immune functions. Lymph forms from excess interstitial fluid, which filters from blood plasma due to blood pressure generated by heart.
Respiratory system	Circulating blood transports oxygen from the lungs to body tissues and carbon dioxide to the lungs for exhalation.
Digestive system	Blood carries newly absorbed nutrients and water to liver. Blood distributes hormones that aid digestion.
Urinary system	Heart and blood vessels deliver 20% of the resting cardiac output to the kidneys, where blood is filtered, needed substances are reabsorbed, and unneeded substances remain as part of urine, which is excreted.
Reproductive systems	Vasodilation of arterioles in penis and clitoris cause erection during sexual intercourse. Blood distributes hormones that regulate reproductive functions.

DISORDERS: HOMEOSTATIC IMBALANCES

Hypertension

Hypertension, or persistently high blood pressure, is defined as systolic blood pressure of 140 mmHg or greater and diastolic blood pressure of 90 mmHg or greater. Recall that a blood pressure of 120/80 is normal and desirable in a healthy adult. In industrialized societies, hypertension is the most common disorder affecting the heart and blood vessels; it is a major cause of heart failure, kidney disease, and stroke. Blood pressure values for adults are classified as follows:

Optimal	Systolic less than 120 mmHg; diastolic less than 80 mmHg
Normal	Systolic less than 130 mmHg; diastolic less than 85 mmHg
High-normal	Systolic 130–139 mmHg; diastolic 85–89 mmHg
Hypertension	Systolic 140 mmHg or greater; diastolic 90 mmHg or greater
Stage 1	Systolic 140–159 mmHg; diastolic 90–99 mmHg
Stage 2	Systolic 160–179 mmHg; diastolic 100–109 mmHg
Stage 3	Systolic 180 mmHg or greater; diastolic 110 mmHg or greater

Types and Causes of Hypertension

Between 90% and 95% of all cases of hypertension are **primary hypertension,** which is a persistently elevated blood pressure that cannot be attributed to any identifiable cause. The remaining 5–10% of cases are **secondary hypertension,** which has an identifiable underlying cause. Several disorders cause secondary hypertension:

* *Obstruction of renal blood flow* or disorders that damage renal tissue may cause the kidneys to release excessive amounts of renin into the blood. The resulting high level of angiotensin II causes vasoconstriction, thus increasing systemic vascular resistance.

* *Hypersecretion of aldosterone*—resulting, for instance, from a tumor of the adrenal cortex—stimulates excess reabsorption of salt and water by the kidneys, which increases the volume of body fluids.

* *Hypersecretion of epinephrine and norepinephrine* by a **pheochromocytoma** (fē-ō-krō′-mō-sī-TŌ-ma), a tumor of the adrenal medulla. Epinephrine and norepinephrine increase heart rate and contractility and increase systemic vascular resistance.

Damaging Effects of Untreated Hypertension

High blood pressure is known as the "silent killer" because it can cause considerable damage to the blood vessels, heart, brain, and kidneys before it causes pain or other noticeable symptoms. It is a major risk factor for the number-one (heart disease) and number-three (stroke) causes of death in the United States. In blood vessels, hypertension causes thickening of the tunica media, accelerates development of atherosclerosis and coronary artery disease, and increases systemic vascular resistance. In the heart, hypertension increases the afterload, which forces the ventricles to work harder to eject blood.

The normal response to an increased workload due to vigorous and regular exercise is hypertrophy of the myocardium, especially in the wall of the left ventricle. An increased afterload, however, leads to myocardial hypertrophy that is accompanied by muscle damage and fibrosis (a buildup of collagen fibers between the muscle fibers). As a result, the left ventricle enlarges, weakens, and dilates. Because arteries in the brain are usually less protected by surrounding tissues than are the major arteries in other parts of the body, prolonged hypertension can eventually cause them to rupture, and a stroke occurs due to a brain hemorrhage. Hypertension also damages kidney arterioles, causing them to thicken, which narrows the lumen; because the blood supply to the kidneys is thereby reduced, the kidneys secrete more renin, which elevates the blood pressure even more.

Lifestyle Changes to Reduce Hypertension

Although several categories of drugs (described next) can reduce elevated blood pressure, the following lifestyle changes are also effective in managing hypertension:

* *Lose weight.* This is the best treatment for high blood pressure short of using drugs. Loss of even a few pounds helps reduce blood pressure in overweight hypertensive individuals.

* *Limit alcohol intake.* Drinking in moderation may lower the risk of coronary heart disease, mainly among males over 45 and females over 55. Moderation is defined as no more than one 12-oz beer per day for females and no more than two 12-oz beers per day for males.

* *Exercise.* Becoming more physically fit by engaging in moderate activity (such as brisk walking) several times a week for 30 to 45 minutes can lower systolic blood pressure by about 10 mmHg.

* *Reduce intake of sodium (salt).* Roughly half the people with hypertension are "salt sensitive." For them, a high-salt diet appears to promote hypertension, and a low-salt diet can lower their blood pressure.

* *Maintain recommended dietary intake of potassium, calcium, and magnesium.* Higher levels of potassium, calcium, and magnesium in the diet are associated with a lower risk of hypertension.

* *Don't smoke.* Smoking has devastating effects on the heart and can augment the damaging effects of high blood pressure by promoting vasoconstriction.

* *Manage stress.* Various meditation and biofeedback techniques help some people reduce high blood pressure. These methods may work by decreasing the daily release of epinephrine and norepinephrine by the adrenal medulla.

Drug Treatment of Hypertension

Drugs having several different mechanisms of action are effective in lowering blood pressure. Many people are successfully treated with *diuretics,* agents that decrease blood pressure by decreasing blood volume because they increase elimination of water and salt in the urine. *ACE (angiotensin converting enzyme) inhibitors* block formation of angiotensin II and thereby promote vasodilation and decrease the liberation of aldosterone. *Beta blockers* reduce blood pressure by inhibiting the secretion of renin and by decreasing heart rate and contractility. *Vasodilators* relax the smooth muscle in arterial walls, causing vasodilation and lowering blood pressure by lowering systemic vascular resistance. An important category of vasodilators are the *calcium channel blockers,* which slow the inflow of Ca^{2+} into vascular smooth muscle cells. They reduce the heart's workload by slowing Ca^{2+} entry into pacemaker cells and regular myocardial fibers, thereby decreasing heart rate and the force of myocardial contraction.

MEDICAL TERMINOLOGY

Aneurysm (AN-ū-rizm) A thin, weakened section of the wall of an artery or a vein that bulges outward, forming a balloonlike sac. Common causes are atherosclerosis, syphilis, congenital blood vessel defects, and trauma. If untreated, the aneurysm enlarges and the blood vessel wall becomes so thin that it bursts. The result is massive hemorrhage with shock, severe pain, stroke, or death.

Angiogenesis (an′-jē-ō-JEN-e-sis) Formation of new blood vessels.

Aortography (a′-or-TOG-ra-fē) X-ray examination of the aorta and its main branches after injection of a radiopaque dye.

Arteritis (ar′-te-RĪ-tis; -itis = inflammation of) Inflammation of an artery, probably due to an autoimmune response.

Carotid endarterectomy (ka-ROT-id end′-ar-ter-EK-tō-mē) The removal of atherosclerotic plaque from the carotid artery to restore greater blood flow to the brain.

Claudication (klaw′-di-KĀ-shun) Pain and lameness or limping caused by defective circulation of the blood in the vessels of the limbs.

Deep venous thrombosis The presence of a thrombus (blood clot) in a deep vein of the lower limbs. It may lead to (1) pulmonary embolism, if the thrombus dislodges and then lodges within the pulmonary arterial blood flow, and (2) postphlebitic syndrome, which consists of edema, pain, and skin changes due to destruction of venous valves.

Hypotension (hī-pō-TEN-shun) Low blood pressure; most commonly used to describe an acute drop in blood pressure, as occurs during excessive blood loss.

Normotensive (nor′-mō-TEN-siv) Characterized by normal blood pressure.

Occlusion (ō-KLOO-shun) The closure or obstruction of the lumen of a structure such as a blood vessel. An example is an atherosclerotic plaque in an artery.

Orthostatic hypotension (or′-thō-STAT-ik; ortho- = straight; -static = causing to stand) An excessive lowering of systemic blood pressure when a person assumes an erect or semi-erect posture; it is usually a sign of a disease. May be caused by excessive fluid loss, certain drugs, and cardiovascular or neurogenic factors. Also called **postural hypotension.**

Phlebitis (fle-BĪ-tis; phleb- = vein) Inflammation of a vein, often in a leg.

Thrombectomy (throm-BEK-tō-mē; thrombo- = clot) An operation to remove a blood clot from a blood vessel.

Thrombophlebitis (throm′-bō-fle-BĪ-tis) Inflammation of a vein involving clot formation. Superficial thrombophlebitis occurs in veins under the skin, especially in the calf.

White coat (office) hypertension A stress-induced syndrome found in patients who have elevated blood pressure when being examined by health-care personnel, but otherwise have normal blood pressure.

STUDY OUTLINE

STRUCTURE AND FUNCTION OF BLOOD VESSELS (p. 697)

1. Arteries carry blood away from the heart. The wall of an artery consists of a tunica interna, a tunica media (which maintains elasticity and contractility), and a tunica externa.
2. Large arteries are termed elastic (conducting) arteries, and medium-sized arteries are called muscular (distributing) arteries.
3. Many arteries anastomose: The distal ends of two or more vessels unite. An alternate blood route from an anastomosis is called collateral circulation. Arteries that do not anastomose are called end arteries.
4. Arterioles are small arteries that deliver blood to capillaries.
5. Through constriction and dilation, arterioles assume a key role in regulating blood flow from arteries into capillaries and in altering arterial blood pressure.
6. Capillaries are microscopic blood vessels through which materials are exchanged between blood and tissue cells; some capillaries are continuous, whereas others are fenestrated.
7. Capillaries branch to form an extensive network throughout a tissue. This network increases surface area, allowing a rapid exchange of large quantities of materials.
8. Precapillary sphincters regulate blood flow through capillaries.
9. Microscopic blood vessels in the liver are called sinusoids.
10. Venules are small vessels that continue from capillaries and merge to form veins.
11. Veins consist of the same three tunics as arteries but have a thinner tunica interna and media. The lumen of a vein is also larger than that of a comparable artery.
12. Veins contain valves to prevent backflow of blood.
13. Weak valves can lead to varicose veins.
14. Vascular (venous) sinuses are veins with very thin walls.
15. Systemic veins are collectively called blood reservoirs because they hold a large volume of blood. If the need arises, this blood can be shifted into other blood vessels through vasoconstriction of veins.
16. The principal blood reservoirs are the veins of the abdominal organs (liver and spleen) and skin.

CAPILLARY EXCHANGE (p. 703)

1. Substances enter and leave capillaries by diffusion, transcytosis, or bulk flow.
2. The movement of water and solutes (except proteins) through capillary walls depends on hydrostatic and osmotic pressures.
3. The near equilibrium between filtration and reabsorption in capillaries is called Starling's law of the capillaries.
4. Edema is an abnormal increase in interstitial fluid.

HEMODYNAMICS: FACTORS AFFECTING BLOOD FLOW (p. 705)

1. The velocity of blood flow is inversely related to the cross-sectional area of blood vessels; blood flows slowest where cross-sectional area is greatest.
2. The velocity of blood flow decreases from the aorta to arteries to capillaries and increases in venules and veins.
3. Blood pressure and resistance determine blood flow.
4. Blood flows from regions of higher to lower pressure. The higher the resistance, however, the lower the blood flow.
5. Cardiac output equals the mean arterial blood pressure divided by total resistance (CO = MABP ÷ R).
6. Blood pressure is the pressure exerted on the walls of a blood vessel.
7. Factors that affect blood pressure are cardiac output, blood volume, viscosity, resistance, and the elasticity of arteries.
8. As blood leaves the aorta and flows through the systemic circulation, its pressure progressively falls to 0 mmHg by the time it reaches the right ventricle.
9. Resistance depends on blood vessel diameter, blood viscosity, and total blood vessel length.
10. Venous return depends on pressure differences between the venules and the right ventricle.
11. Blood return to the heart is maintained by several factors, including skeletal muscular contractions, valves in veins (especially in the limbs), and pressure changes associated with breathing.

CONTROL OF BLOOD PRESSURE AND BLOOD FLOW (p. 709)

1. The cardiovascular (CV) center is a group of neurons in the medulla oblongata that regulates heart rate, contractility, and blood vessel diameter.
2. The cardiovascular center receives input from higher brain regions and sensory receptors (baroreceptors and chemoreceptors).
3. Output from the cardiovascular center flows along sympathetic and parasympathetic axons. Sympathetic impulses propagated along cardioaccelerator nerves increase heart rate and contractility, whereas parasympathetic impulses propagated along vagus nerves decrease heart rate.
4. Baroreceptors monitor blood pressure, and chemoreceptors monitor blood levels of O_2, CO_2, and hydrogen ions. The carotid sinus reflex helps regulate blood pressure in the brain. The aortic reflex regulates general systemic blood pressure.
5. Hormones that help regulate blood pressure are epinephrine, norepinephrine, ADH (vasopressin), angiotensin II, and ANP.
6. Autoregulation refers to local, automatic adjustments of blood flow in a given region to meet a particular tissue's need.
7. O_2 level is the principal stimulus for autoregulation.

CHECKING CIRCULATION (p. 713)

1. Pulse is the alternate expansion and elastic recoil of an artery wall with each heartbeat. It may be felt in any artery that lies near the surface or over a hard tissue.
2. A normal resting pulse (heart) rate is 70–80 beats/min.
3. Blood pressure is the pressure exerted by blood on the wall of an artery when the left ventricle undergoes systole and then diastole. It is measured by the use of a sphygmomanometer.

4. Systolic blood pressure (SBP) is the arterial blood pressure during ventricular contraction. Diastolic blood pressure (DBP) is the arterial blood pressure during ventricular relaxation. Normal blood pressure is 120/80 mmHg.
5. Pulse pressure is the difference between systolic and diastolic blood pressure. It normally is about 40 mmHg.

SHOCK AND HOMEOSTASIS (p. 715)

1. Shock is a failure of the cardiovascular system to deliver enough O_2 and nutrients to meet the metabolic needs of cells.
2. Types of shock include hypovolemic, cardiogenic, vascular, and obstructive.
3. Signs and symptoms of shock include systolic blood pressure less than 90 mmHg; rapid resting heart rate; weak, rapid pulse; clammy, cool, pale skin; sweating; hypotension; altered mental state; decreased urinary output; thirst; and acidosis.

CIRCULATORY ROUTES (p. 717)

1. The two main circulatory routes are the systemic and pulmonary circulations.
2. Among the subdivisions of the systemic circulation are the coronary (cardiac) circulation and the hepatic portal circulation.
3. The systemic circulation carries oxygenated blood from the left ventricle through the aorta to all parts of the body, including some lung tissue, but *not* the air sacs (alveoli) of the lungs, and returns the deoxygenated blood to the right atrium.
4. The aorta is divided into the ascending aorta, the arch of the aorta, and the descending aorta. Each section gives off arteries that branch to supply the whole body.
5. Blood returns to the heart through the systemic veins. All veins of the systemic circulation drain into the superior or inferior venae cavae or the coronary sinus, which, in turn, empty into the right atrium.
6. The major blood vessels of the systemic circulation may be reviewed in Exhibits 21.1–21.12.
7. The hepatic portal circulation directs venous blood from the gastrointestinal organs and spleen into the hepatic portal vein of the liver before it returns to the heart. It enables the liver to utilize nutrients and detoxify harmful substances in the blood.
8. The pulmonary circulation takes deoxygenated blood from the right ventricle to the alveoli within the lungs and returns oxygenated blood from the alveoli to the left atrium.
9. Fetal circulation exists only in the fetus. It involves the exchange of materials between fetus and mother via the placenta.
10. The fetus derives O_2 and nutrients from and eliminates CO_2 and wastes into maternal blood. At birth, when pulmonary (lung), digestive, and liver functions begin, the special structures of fetal circulation are no longer needed.

DEVELOPMENT OF BLOOD VESSELS AND BLOOD (p. 756)

1. Blood vessels develop from angioblasts, which form isolated masses called blood islands.
2. Blood is produced outside the embryo by the endothelium of blood vessels of the yolk sac, chorion, and connecting stalk at about three weeks. Within the embryo, blood is produced by the liver at about the fifth week and in the spleen, red bone marrow, and thymus at about the twelfth week.

AGING AND THE CARDIOVASCULAR SYSTEM (p. 756)

1. General changes associated with aging include reduced compliance of blood vessels, reduction in cardiac muscle size, reduced cardiac output, and increased systolic blood pressure.

2. The incidence of coronary artery disease (CAD), congestive heart failure (CHF), and atherosclerosis increases with age.

SELF-QUIZ QUESTIONS

Fill in the blanks in the following statements.

1. The _____ reflex helps maintain normal blood pressure in the brain; the _____ reflex governs general systemic blood pressure.

2. In addition to the pressure created by contraction of the left ventricle, venous return is aided by the _____ and the _____, both of which depend on the presence of valves in the veins.

Indicate whether the following statements are true or false.

3. The most important method of capillary exchange is simple diffusion.

4. The overall function of the cardiovascular system is to ensure adequate circulation of blood to all body tissues to enable capillary exchange between blood plasma, interstitial fluid, and tissue cells.

Choose the one best answer to the following questions.

5. Which of the following pairs are mismatched? (a) muscular arteries: conducting arteries, (b) arterioles: resistance vessels, (c) elastic arteries: pressure reservoirs, (d) capillaries: exchange vessels, (e) veins: blood reservoirs.

6. The main vessels that permit the exchange of nutrients and wastes between the blood and tissue cells through interstitial fluid are (a) arteries, (b) arterioles, (c) capillaries, (d) venules, (e) veins.

7. Which of the following are causes of edema? (1) increased blood colloid osmotic pressure, (2) increased blood hydrostatic pressure in capillaries, (3) decreased plasma protein concentration, (4) increased capillary permeability, (5) blockage of lymph vessels. (a) 1, 2, and 3, (b) 2, 3, and 4, (c) 3, 4, and 5, (d) 1, 3, 4, and 5, (e) 2, 3, 4, and 5.

8. Systemic vascular resistance depends on which of the following factors? (1) blood viscosity, (2) total blood vessel length, (3) size of lumen, (4) type of blood vessel, (5) oxygen concentration of the blood. (a) 1, 2, and 3, (b) 2, 3, and 4, (c) 3, 4, and 5, (d) 1, 3, and 5, (e) 2, 4, and 5.

9. Which of the following help regulate blood pressure and help control regional blood flow? (1) baroreceptor and chemoreceptor reflexes, (2) hormones, (3) autoregulation, (4) H^+ concentration of blood, (5) oxygen concentration of the blood. (a) 1, 2, and 4, (b) 2, 4, and 5, (c) 1, 4, and 5, (d) 1, 2, 3, 4, and 5, (e) 3, 4, and 5.

10. Which of the following could cause vasoconstriction? (1) atrial natriuretic peptide, (2) angiotensin II, (3) lactic acid, (4) increased sympathetic stimulation to smooth muscle, (5) increase in body temperature. (a) 1, 2, 3, 4, and 5, (b) 1, 2, 3, and 4, (c) 2, 4, and 5, (d) 1, 3, and 4, (e) 2 and 4.

11. Substances enter and leave capillaries by (1) diffusion, (2) bulk flow, (3) vasomotion, (4) transcytosis, (5) filtration. (a) 1, 2, 3, 4, and 5, (b) 1, 4, and 5, (c) 1, 2, 4, and 5, (d) 2, 3, and 4, (e) 2, 4, and 5.

12. Match the following:
_____ (a) pressure generated by the pumping of the heart; pushes fluid out of capillaries
_____ (b) pressure created by proteins present in the interstitial fluid; pulls fluid out of capillaries
_____ (c) balance of pressure; determines whether blood volume and interstitial fluid remain steady or changes
_____ (d) force due to presence of plasma proteins; pulls fluid into capillaries from interstitial spaces
_____ (e) pressure due to fluid in interstitial spaces; pushes fluid back into capillaries

(1) net filtration pressure
(2) blood hydrostatic pressure
(3) interstitial fluid hydrostatic pressure
(4) blood colloid osmotic pressure
(5) interstitial fluid osmotic pressure

13. Match the following:
_____ (a) supplies blood to the kidney
_____ (b) drains blood from the small intestine, portions of the large intestine, stomach, and pancreas
_____ (c) supply and drain blood from the heart muscle
_____ (d) supply blood to the lower limbs
_____ (e) drain oxygenated blood from the lungs and carry it to the left atrium
_____ (f) supplies blood to the stomach, liver, and pancreas
_____ (g) supply blood to the brain
_____ (h) supplies blood to the large intestine
_____ (i) drain blood from the head
_____ (j) detours venous blood from the gastrointestinal organs and spleen through the liver before it returns to the heart
_____ (k) drain most of the thorax and abdominal wall; can serve as a bypass for the inferior vena cava
_____ (l) a part of the venous circulation of the leg; a vessel used in heart bypass surgery

(1) superior mesenteric vein
(2) inferior mesenteric artery
(3) pulmonary veins
(4) coronary vessels
(5) hepatic portal circulation
(6) carotid arteries
(7) jugular veins
(8) celiac trunk
(9) common iliac arteries
(10) azygous veins
(11) renal artery
(12) saphenous vein

14. Match the following:

(1) shock (2) pulse (3) tachycardia (4) bradycardia
(5) systolic blood pressure (6) diastolic blood pressure

____ (a) a traveling pressure wave created by the alternate expansion and recoil of elastic arteries after each systole of the left ventricle

____ (b) the lowest blood pressure in arteries during ventricular relaxation

____ (c) a slow resting heart rate or pulse rate

____ (d) an inadequate cardiac output that results in a failure of the cardiovascular system to deliver enough oxygen and nutrients to meet the metabolic needs of body cells

____ (e) a rapid resting heart rate or pulse rate

____ (f) the highest force with which blood pushes against arterial walls as a result of ventricular contraction

15. Match the following:

(1) ductus venosus (2) ductus arteriosus (3) foramen ovale
(4) umbilical arteries (5) umbilical vein

____ (a) returns oxygenated blood from the placenta

____ (b) an opening in the septum between the right and left atria

____ (c) becomes the ligamentum venosum after birth

____ (d) passes blood from the fetus to the placenta

____ (e) bypasses the nonfunctioning lungs; becomes the ligamentum arteriosum at birth

____ (f) become the medial umbilical ligaments at birth

____ (g) transports oxygenated blood into the inferior vena cava

____ (h) becomes the ligamentum teres at birth

CRITICAL THINKING QUESTIONS

1. Which structures that are present in the fetal circulation are absent in the adult circulatory system? Why do these changes occur?

HINT *A mother may want to do everything for her baby after he's born, but she can't breathe for him.*

2. Josef is thinking about the route that his blood follows on its way from his heart to his hands. He realizes that the route is a bit different for blood flowing to the left as compared to the right hand. Trace the pathway of blood from the heart to the right and left hands.

HINT *The lack of symmetry is closer to the heart than to the hand.*

3. Chris spent the week before the exam lying on a beach studying the back of his eyelids. "What was that question about sinuses doing on the cardiovascular system test?" he complained. How would you enlighten Chris?

HINT *In bone, sinuses are filled with air; these aren't.*

ANSWERS TO FIGURE QUESTIONS

21.1 The femoral artery has the thicker wall; the femoral vein has the wider lumen.

21.2 Due to atherosclerosis, less energy is stored in the less-compliant elastic arteries during systole; thus, the heart must pump harder to maintain the same rate of blood flow.

21.3 Metabolically active tissues use O_2 and produce wastes more rapidly than inactive tissues.

21.4 Materials cross capillary walls through intercellular clefts and fenestrations, via transcytosis in pinocytic vesicles, and through the plasma membranes of endothelial cells.

21.5 When you are standing, gravity causes pooling of blood in the veins of the limbs. The valves prevent backflow as the blood proceeds toward the right atrium after each heartbeat. When you are erect, gravity aids the flow of blood in neck veins back toward the heart.

21.6 Blood volume in veins is about 60% of 5 liters, or 3 liters; blood volume in capillaries is about 5% of 5 liters, or 250 mL.

21.7 Blood colloid osmotic pressure is lower than normal in a person with a low level of plasma proteins, and therefore capillary reabsorption is low. The result is edema (see page 705).

21.8 Mean blood pressure in the aorta is closer to diastolic than to systolic pressure; here it is about 93 mmHg.

21.9 The skeletal muscle and respiratory pumps aid venous return.

21.10 Vasodilation and vasoconstriction of arterioles is the main regulator of systemic vascular resistance.

21.11 Velocity of blood flow is fastest in the aorta and arteries.

21.12 The effector tissues regulated by the cardiovascular center are cardiac muscle in the heart and smooth muscle in blood vessel walls.

21.13 Impulses to the cardiovascular center pass from baroreceptors in the carotid sinuses via the glossopharyngeal nerves (cranial nerve IX) and from baroreceptors in the arch of the aorta via the vagus nerves (cranial nerve X).

21.14 It represents a change that occurs when you stand up because gravity causes pooling of blood in leg veins once you are upright, decreasing the blood pressure in your upper body.

21.15 Diastolic blood pressure = 95 mmHg; systolic blood pressure = 142 mmHg; pulse pressure = 47 mmHg. This person has stage I hypertension because the systolic blood pressure is greater than 140 mmHg and the diastolic blood pressure is greater than 90 mmHg.

21.16 Not necessarily; if systemic vascular resistance has increased greatly, tissue perfusion may be inadequate.

21.17 The two main circulatory routes are the systemic and pulmonary circulations.

21.18 The subdivisions of the aorta are the ascending aorta, arch of the aorta, thoracic aorta, and abdominal aorta.

21.19 The coronary arteries arise from the ascending aorta.

21.20 Branches of the arch of aorta are the brachiocephalic trunk, left common carotid artery, and left subclavian artery.

21.21 The thoracic aorta begins at the level of the intervertebral disc between T4 and T5.

21.22 The abdominal aorta begins at the aortic hiatus in the diaphragm.

21.23 The abdominal aorta divides into the common iliac arteries at about the level of L4.

21.24 The superior vena cava drains regions above the diaphragm, and the inferior vena cava drains regions below the diaphragm.

21.25 All venous blood in the brain drains into the internal jugular veins.

21.26 The median cubital vein is often used for withdrawing blood.

21.27 The inferior vena cava returns blood from abdominopelvic viscera to the heart.

21.28 Superficial veins of the lower limbs are the dorsal venous arch and the great saphenous and small saphenous veins.

21.29 The hepatic veins carry blood away from the liver.

21.30 The pulmonary arteries carry deoxygenated blood.

21.31 Exchange of materials between mother and fetus occurs across the placenta.

21.32 Blood vessels and blood are derived from mesoderm.

The Lymphatic and Immune

System and Resistance

to Disease

A Practitioner's View

It might seem like a stretch to say that a detailed knowledge of the lymphatic system and the body's immune response are essential to the practice of pharmacy, but as both a clinical pharmacist and a retail pharmacist, I use the fundamentals presented in this chapter on a daily basis.

In my retail pharmacy practice, I am called on daily to triage patients who may or may not need to see a physician. The degree to which non-specific indicators of infection are present, or absent, is always the first place to start. Pain, swelling, redness, vomiting, diarrhea, fever, or a change in mucosal secretions are good indicators of whether or not to encourage someone to see a doctor. Likewise, when a pharmacist can advise a patient against pursuing an antibiotic prescription for a viral infection, we cut down on opportunities for developing antibiotic resistance. Understanding the relationships between antigens and antibodies as presented in this chapter allows me as a pharmacist to be able to communicate to my patients how a medication will or won't work for them. If I can increase a patient's understanding of their disease and how their medications work to treat it, they will be more compliant in completing their course of therapy leading to more successful outcomes.

Antoinette E. Sheridan, R.Ph., Pharm.D.
Pharmacist
St. Vincent Williamsport Hospital

THE LYMPHATIC SYSTEM, DISEASE RESISTANCE, AND HOMEOSTASIS

The lymphatic system contributes to homeostasis by draining interstitial fluid as well providing the mechanisms for defense against disease.

ENERGIZE YOUR STUDY
FOUNDATIONS CD
Anatomy Overview
- The Lymphatic System and Disease Resistance
- Connective Tissue: Blood

Animation
- Systems Contributions to Homeostasis

Concepts and Connections and Exercises reinforce your understanding

www.wiley.com/college/apcentral
INSIGHTS AND EXPLORATIONS

Did you know that by some estimates about two-thirds of the food products in grocery stores contain genetically engineered crops? Perhaps you recall news reports in the fall of 2000 about how some genetically engineered corn approved for use in cattle feed had found its way into corn used for human food products such as Taco Bell taco shells. This incident could cost the company that developed the recombinant corn over $100 million due to recall of contaminated stocks and suspension of sales. The government was concerned that the recombinant corn might contain proteins that could cause an allergic reaction in humans. We will explore in this web-based activity what allergic reactions are and how it is that recombinant food might pose a health threat.

Maintaining homeostasis in the body requires ways to continually combat harmful agents in our environment. Despite constant exposure to a variety of **pathogens** (PATH-ō-jens), which are disease-producing microbes such as bacteria and viruses, most people remain healthy. The body surface also endures cuts and bumps, exposure to ultraviolet rays in sunlight, chemical toxins, and minor burns with an array of defensive ploys. **Resistance** is the ability to ward off damage or disease through our defenses. Vulnerability or lack of resistance is termed **susceptibility.**

The two general types of resistance are (1) nonspecific resistance or innate defenses and (2) specific resistance or immunity. **Nonspecific resistance** or **innate defenses** are present at birth and include defense mechanisms that provide *immediate* but *general* protection against invasion by a wide range of pathogens. Mechanical and chemical barriers of the skin and mucous membranes provide the first line of defense in nonspecific resistance. The acidity of gastric juice in the stomach, for example, kills many bacteria in food. **Specific resistance** or **immunity** develops in response to contact with a *particular* invader. It occurs more *slowly* than nonspecific resistance mechanisms and involves activation of specific lymphocytes that can combat a specific invader.

The body system responsible for specific resistance (and some aspects of nonspecific resistance) is the lymphatic and immune system. This system is closely allied with the cardiovascular system, and it also functions with the digestive system in the absorption of fatty foods. In this chapter, we will explore the mechanisms that provide defenses against intruders and promote the repair of damaged body tissues.

LYMPHATIC AND IMMUNE SYSTEM

▶ **OBJECTIVES**

- Describe the components and major functions of the lymphatic and immune system.
- Describe the organization of lymphatic vessels.
- Describe the formation and flow of lymph.
- Compare the structure and functions of the primary and secondary lymphatic organs and tissues.

The **lymphatic and immune system** (lim-FAT-ik) consists of a fluid called lymph, vessels called lymphatic vessels to transport the fluid, a number of structures and organs containing lymphatic tissue, and red bone marrow, where stem cells develop into various types of blood cells including lymphocytes (Figure 22.1). It assists in circulating body fluids and helps defend the body against disease-causing agents. As you will see shortly, most components of blood plasma filter through blood capillary walls to form interstitial fluid. After interstitial fluid passes into lymphatic vessels, it is called **lymph** (LIMF = clear fluid). Therefore, interstitial fluid and lymph are very similar; the major difference between the two is location. Whereas interstitial fluid is found between cells, lymph is located within lymphatic vessels and lymphatic tissue.

Lymphatic tissue is a specialized form of reticular connective tissue (see Table 4.3C on page 123) that contains large numbers of lymphocytes. Recall from Chapter 19 that lymphocytes are agranular white blood cells. Two types of lymphocytes participate in immune responses: *B cells* and *T cells*.

Functions of the Lymphatic and Immune System

The lymphatic and immune system has three primary functions:

1. ***Draining excess interstitial fluid.*** Lymphatic vessels drain excess interstitial fluid from tissue spaces and return it to the blood.

2. ***Transporting dietary lipids.*** Lymphatic vessels transport the lipids and lipid-soluble vitamins (A, D, E, and K) absorbed by the gastrointestinal tract to the blood.

3. ***Carrying out immune responses.*** Lymphatic tissue initiates highly specific responses directed against particular microbes or abnormal cells. Lymphocytes, aided by macrophages, recognize foreign cells, microbes, toxins, and cancer cells and respond to them in two basic ways: (1) In cell-mediated immune responses, T cells destroy the intruders by causing them to rupture or by releasing cytotoxic (cell-killing) substances. (2) In antibody-mediated immune responses, B cells differentiate into plasma cells that protect us against disease by producing antibodies, proteins that combine with and cause destruction of specific foreign substances.

Lymphatic Vessels and Lymph Circulation

Lymphatic vessels begin as lymphatic capillaries. These tiny vessels are closed at one end and located in the spaces between cells (Figure 22.2 on page 767). Just as blood capillaries converge to form venules and then veins, lymphatic capillaries unite to form larger lymphatic vessels (see Figure 22.1), which resemble veins in structure but have thinner walls and more valves. At intervals along the lymphatic vessels, lymph flows through lymph nodes, encapsulated masses of B cells and T cells. In the skin, lymphatic vessels lie in the subcutaneous tissue and generally follow veins; lymphatic vessels of the viscera generally follow arteries, forming plexuses (networks) around them. Tissues that lack lymphatic capillaries include avascular tissues (such as cartilage, the epidermis, and the cornea of the eye), the central nervous system, portions of the spleen, and bone marrow.

Figure 22.1 Components of the lymphatic and immune system.

The lymphatic and immune system consists of lymph, lymphatic vessels, lymphatic tissues, and red bone marrow.

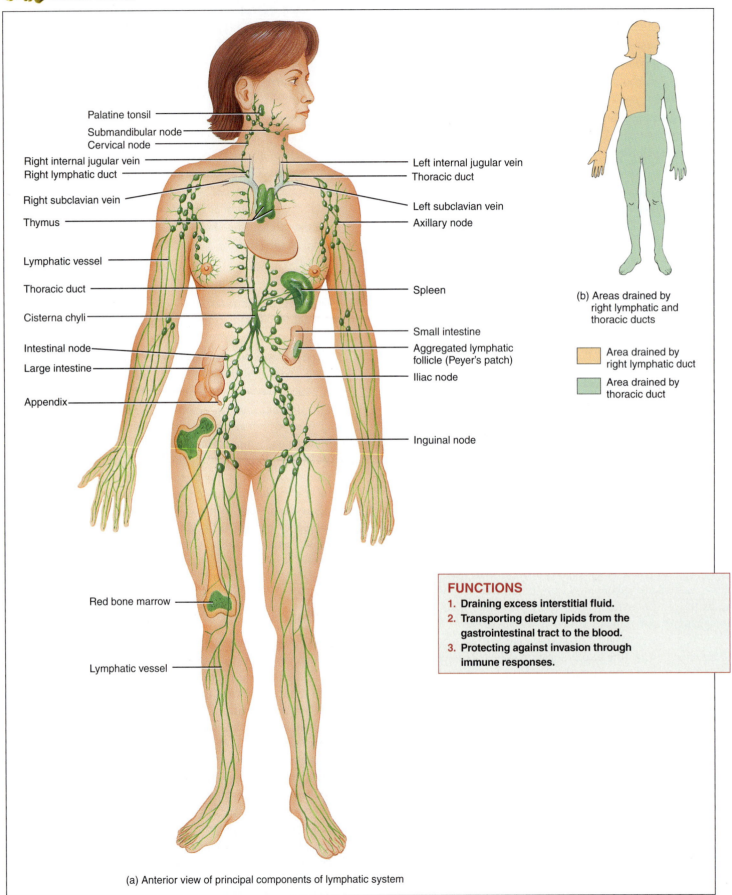

Palatine tonsil
Submandibular node
Cervical node
Right internal jugular vein
Right lymphatic duct
Right subclavian vein
Thymus
Lymphatic vessel
Thoracic duct
Cisterna chyli
Intestinal node
Large intestine
Appendix
Red bone marrow
Lymphatic vessel

Left internal jugular vein
Thoracic duct
Left subclavian vein
Axillary node
Spleen
Small intestine
Aggregated lymphatic follicle (Peyer's patch)
Iliac node
Inguinal node

(b) Areas drained by right lymphatic and thoracic ducts

Area drained by right lymphatic duct

Area drained by thoracic duct

FUNCTIONS
1. **Draining excess interstitial fluid.**
2. **Transporting dietary lipids from the gastrointestinal tract to the blood.**
3. **Protecting against invasion through immune responses.**

(a) Anterior view of principal components of lymphatic system

What tissue contains stem cells that develop into lymphocytes?

Figure 22.2 Lymphatic capillaries.

🔑 Lymphatic capillaries are found throughout the body except in avascular tissues, the central nervous system, portions of the spleen, and bone marrow.

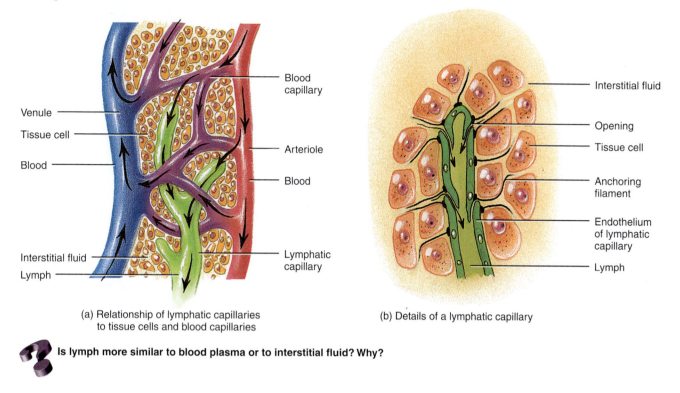

(a) Relationship of lymphatic capillaries to tissue cells and blood capillaries

(b) Details of a lymphatic capillary

❓ **Is lymph more similar to blood plasma or to interstitial fluid? Why?**

Lymphatic Capillaries

Lymphatic capillaries are slightly larger in diameter than blood capillaries and have a unique structure that permits interstitial fluid to flow into them but not out. The ends of endothelial cells that make up the wall of a lymphatic capillary overlap (Figure 22.2b). When pressure is greater in the interstitial fluid than in lymph, the cells separate slightly, like the opening of a one-way swinging door, and interstitial fluid enters the lymphatic capillary. When pressure is greater inside the lymphatic capillary, the cells adhere more closely, and lymph cannot escape back into interstitial fluid. Attached to the lymphatic capillaries are *anchoring filaments,* which contain elastic fibers. They extend out from the lymphatic capillary, attaching lymphatic endothelial cells to surrounding tissues. When excess interstitial fluid accumulates and causes tissue swelling, the anchoring filaments are pulled, making the openings between cells even larger so that more fluid can flow into the lymphatic capillary.

In the small intestine, specialized lymphatic capillaries called **lacteals** (LAK-tē-als; *lact-* = milky) carry dietary lipids into lymphatic vessels and ultimately into the blood. The presence of these lipids causes the lymph draining the small intestine to appear creamy white; such lymph is referred to as **chyle** (KĪL = juice). Elsewhere, lymph is a clear, pale-yellow fluid.

Lymph Trunks and Ducts

Lymph passes from lymphatic capillaries into lymphatic vessels and then through lymph nodes. Lymphatic vessels exiting lymph nodes pass lymph either toward another node within the same group or on to another group of nodes. From the most proximal group of each chain of nodes, the exiting vessels unite to form **lymph trunks.** The principal trunks are the **lumbar, intestinal, bronchomediastinal, subclavian,** and **jugular trunks** (Figure 22.3). Lymph passes from lymph trunks into two main channels, the thoracic duct and the right lymphatic duct, and then drains into venous blood.

The **thoracic (left lymphatic) duct** is about 38–45 cm (15–18 in.) long and begins as a dilation called the **cisterna chyli** (sis-TER-na KĪ-lē; *cisterna* = cavity or reservoir) anterior to the second lumbar vertebra. The thoracic duct is the main duct for return of lymph to blood. It receives lymph from the left side of the head, neck, and chest, the left upper limb, and the entire body inferior to the ribs (see Figure 22.16). The thoracic duct drains lymph into venous blood via the **left subclavian vein.**

The cisterna chyli receives lymph from the right and left lumbar trunks and from the intestinal trunk. The lumbar trunks drain lymph from the lower limbs, the wall and viscera of the pelvis, the kidneys, the adrenal glands, and the deep lymphatic vessels that drain lymph from most of the abdominal wall. The intestinal trunk drains lymph from the stomach, intestines, pancreas, spleen, and part of the liver.

In the neck, the thoracic duct also receives lymph from the left jugular, left subclavian, and left bronchomediastinal trunks. The left jugular trunk drains lymph from the left side of the head and neck, and the left subclavian trunk drains lymph from the

Figure 22.3   Routes for drainage of lymph from lymph trunks into the thoracic and right lymphatic ducts.

All lymph returns to the bloodstream through the thoracic (left) lymphatic duct and right lymphatic duct.

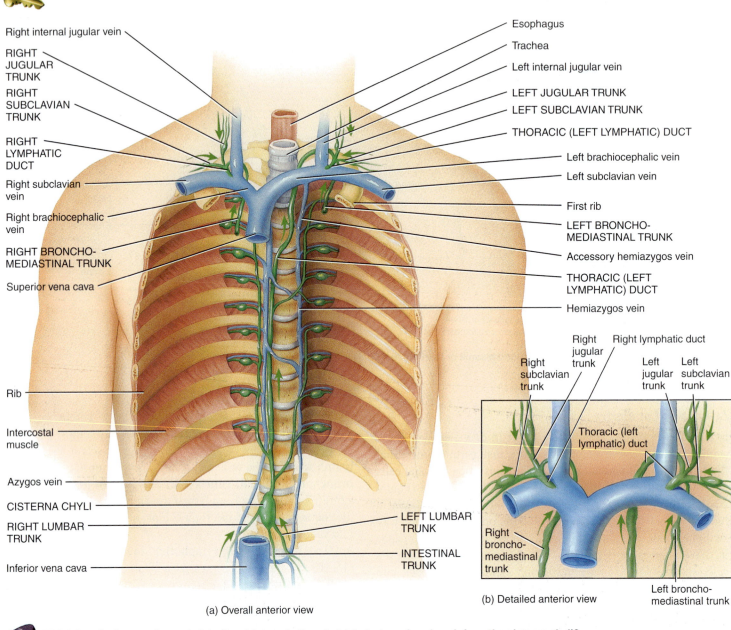

Right internal jugular vein

RIGHT JUGULAR TRUNK

RIGHT SUBCLAVIAN TRUNK

RIGHT LYMPHATIC DUCT

Right subclavian vein

Right brachiocephalic vein

RIGHT BRONCHO-MEDIASTINAL TRUNK

Superior vena cava

Rib

Intercostal muscle

Azygos vein

CISTERNA CHYLI

RIGHT LUMBAR TRUNK

Inferior vena cava

Esophagus

Trachea

Left internal jugular vein

LEFT JUGULAR TRUNK

LEFT SUBCLAVIAN TRUNK

THORACIC (LEFT LYMPHATIC) DUCT

Left brachiocephalic vein

Left subclavian vein

First rib

LEFT BRONCHO-MEDIASTINAL TRUNK

Accessory hemiazygos vein

THORACIC (LEFT LYMPHATIC) DUCT

Hemiazygos vein

LEFT LUMBAR TRUNK

INTESTINAL TRUNK

(a) Overall anterior view

Right jugular trunk   Right lymphatic duct

Right subclavian trunk

Left jugular trunk   Left subclavian trunk

Thoracic (left lymphatic) duct

Right broncho-mediastinal trunk

Left broncho-mediastinal trunk

(b) Detailed anterior view

Which lymphatic vessels empty into the cisterna chyli, and which duct receives lymph from the cisterna chyli?

left upper limb. The left bronchomediastinal trunk drains lymph from the left side of the deeper parts of the anterior thoracic wall, the superior part of the anterior abdominal wall, the anterior part of the diaphragm, the left lung, and the left side of the heart.

The **right lymphatic duct** (Figure 22.3) is about 1.2 cm (0.5 in.) long and drains lymph from the upper-right side of the body into venous blood via the **right subclavian vein** (see Figure 22.1b). Three lymphatic trunks drain into the right lymphatic duct: (1) The right jugular trunk drains the right side of the head and neck. (2) The right subclavian trunk drains the right upper

limb. (3) The right bronchomediastinal trunk drains the right side of the thorax, the right lung, the right side of the heart, and part of the liver.

Formation and Flow of Lymph

Most components of blood plasma freely filter through the capillary walls to form interstitial fluid. More fluid filters out of blood capillaries, however, than returns to them by reabsorption (see Figure 21.7 on page 704). The excess filtered fluid—about 3 liters per day—drains into lymphatic vessels and becomes lymph. Because most plasma proteins are too large to leave blood vessels, interstitial fluid contains only a small amount of protein. Proteins that do leave blood plasma, however, cannot return to the blood directly by diffusion because the concentration gradient (high level of proteins inside blood capillaries, low level outside) opposes such movement. The proteins, however, can move readily through the more permeable lymphatic capillaries into lymph. Thus, an important function of lymphatic vessels is to return lost plasma proteins to the bloodstream.

Like veins, lymphatic vessels contain valves, which ensure the one-way movement of lymph. Ultimately, lymph drains into venous blood through the right lymphatic duct and the thoracic duct at the junction of the internal jugular and subclavian veins (Figure 22.3). Thus, the sequence of fluid flow is blood capillaries (blood) → interstitial spaces (interstitial fluid) → lymphatic capillaries (lymph) → lymphatic vessels (lymph) → lymphatic ducts (lymph) → subclavian veins (blood). Figure 22.4 illustrates this sequence, as well as the relationship of the lymphatic and cardiovascular systems.

The same two "pumps" that aid return of venous blood to the heart maintain the flow of lymph.

1. **Skeletal muscle pump.** The "milking action" of skeletal muscle contractions (see Figure 21.9 on page 707) compresses lymphatic vessels (as well as veins) and forces lymph toward the subclavian veins.

2. **Respiratory pump.** Lymph flow is also maintained by pressure changes that occur during inhalation (breathing in). Lymph flows from the abdominal region, where the pressure is higher, toward the thoracic region, where it is lower. When the pressures reverse during exhalation (breathing out), the valves prevent backflow of lymph. In addition, when a lymphatic vessel distends, the smooth muscle in its wall contracts, which helps move lymph from one segment of the vessel to the next.

Edema and Lymph Flow

Edema, an excessive accumulation of interstitial fluid in tissue spaces, may be caused by an obstruction to lymph flow, such as an infected lymph node or a blocked lymphatic vessel. Edema may also result from increased capillary blood pressure, which causes excess interstitial fluid to form faster than it can pass into lymphatic vessels or be reabsorbed back into the capillaries. ■

Figure 22.4 **Schematic diagram showing the relationship of the lymphatic and immune system to the cardiovascular system.**

 The sequence of fluid flow is blood capillaries (blood) → interstitial spaces (interstitial fluid) → lymphatic capillaries (lymph) → lymphatic vessels (lymph) → lymphatic ducts (lymph) → subclavian veins (blood).

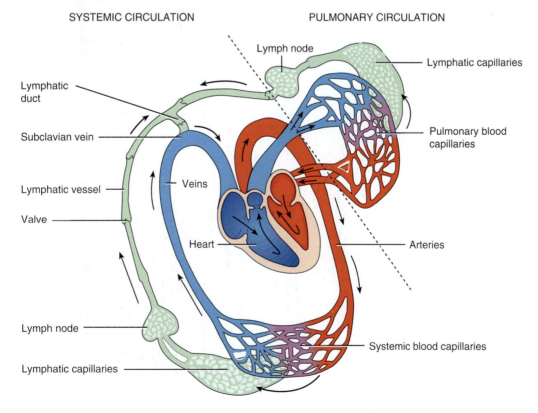

Arrows show direction of flow of lymph and blood

 Does inhalation promote or hinder the flow of lymph?

Lymphatic Organs and Tissues

Lymphatic organs and tissues, which are widely distributed throughout the body, are classified into two groups based on their functions. **Primary lymphatic organs** are the sites where stem cells divide and become **immunocompetent,** that is, capable of mounting an immune response. The primary lymphatic organs are the **red bone marrow** (in flat bones and the epiphyses of long bones of adults) and the **thymus.** Pluripotent stem cells in red bone marrow give rise to mature, immunocompetent B cells and to pre-T cells, which migrate to and become immunocompetent T cells in the thymus. The **secondary lymphatic organs** and **tissues** are the sites where most immune responses occur. They include **lymph nodes,** the **spleen,** and **lymphatic nodules (follicles).** The thymus, lymph nodes, and spleen are considered organs because each is surrounded by a connective tissue capsule; lymphatic nodules, in contrast, are not organs because they lack a capsule.

Thymus

The **thymus** is a bilobed organ located in the mediastinum between the sternum and the aorta (Figure 22.5a). An enveloping layer of connective tissue holds the two lobes closely together, but a connective tissue **capsule** encloses each lobe separately. Extensions of the capsule, called **trabeculae** (tra-BEK-ū-lē = little beams) penetrate inward and divide the lobes into **lobules** (Figure 22.5b).

Each thymic lobule consists of a deeply staining outer cortex and a lighter-staining central medulla (Figure 22.5b). The **cortex** is composed of large numbers of T cells and scattered dendritic cells, epithelial cells, and macrophages. Immature T cells (pre-T cells) migrate from red bone marrow to the cortex of the thymus, where they proliferate and begin to mature. **Dendritic cells** (*dendr-* =a tree), so-named because they have long, branched projections that resemble the dendrites of a neuron, assist the maturation process. As you will see shortly, dendritic

Figure 22.5   Thymus.   (See Tortora, *A Photographic Atlas of the Human Body,* Figure 7.2.)

🔑 **The bilobed thymus is largest at puberty and then atrophies with age.**

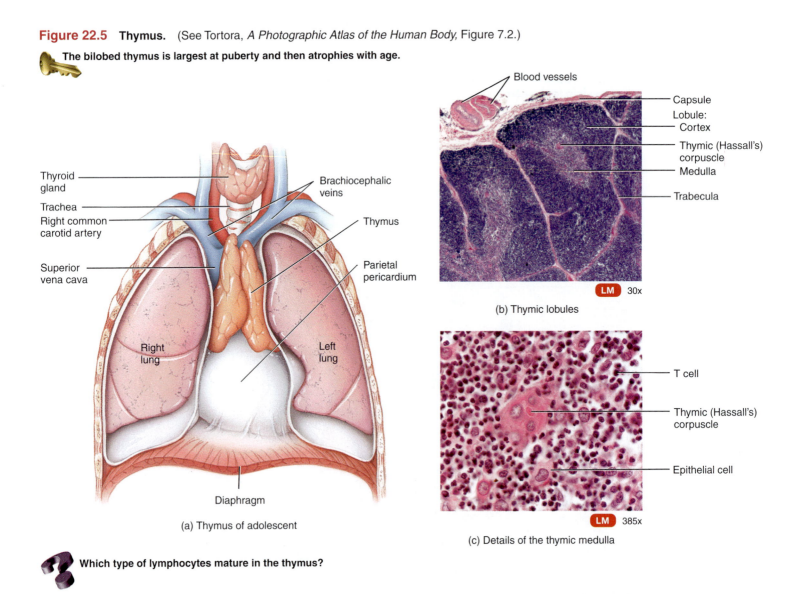

(a) Thymus of adolescent

(b) Thymic lobules

(c) Details of the thymic medulla

❓ **Which type of lymphocytes mature in the thymus?**

cells in other parts of the body, such as lymph nodes, play another key role in immune responses. Each of the specialized **epithelial cells** in the cortex has several long processes that surround and serve as a framework for as many as 50 T cells. These epithelial cells help "educate" the pre-T cells in a process known as positive selection (see Figure 22.20). Additionally, they produce thymic hormones that are thought to aid in the maturation of T cells. Only about 2% of developing T cells survive in the cortex. The remaining cells die via apoptosis (programmed cell death). Thymic macrophages help clear out the debris of dead and dying cells. The surviving T cells enter the medulla.

The **medulla** consists of widely scattered, more mature T cells, epithelial cells, dendritic cells, and macrophages (Figure 22.5c). Some of the epithelial cells become arranged into concentric layers of flat cells that degenerate and become filled with keratohyalin granules and keratin. These clusters are called **thymic (Hassall's) corpuscles.** Although their role is uncertain, they may serve as sites of T cell death in the medulla. T cells that leave the thymus via the blood are carried to lymph nodes, the spleen, and other lymphatic tissues where they colonize parts of these organs and tissues.

In infants, the thymus is large, having a mass of about 70 g (2.3 oz). After puberty, adipose and areolar connective tissue begin to replace the thymic tissue. By the time a person reaches maturity, the gland has atrophied considerably, and in old age it may weigh only 3 g (0.1 oz). Before the thymus atrophies, it populates the secondary lymphatic organs and tissues with T cells. However, some T cells continue to proliferate in the thymus throughout an individual's lifetime.

Lymph Nodes

Located along lymphatic vessels are about 600 bean-shaped **lymph nodes.** They are scattered throughout the body, both superficially and deep, and usually occur in groups (see Figure 22.1). Large groups of lymph nodes are present near the mammary glands and in the axillae and groin. Lymph nodes are 1–25 mm (0.04–1 in.) long and are covered by a **capsule** of dense connective tissue that extends into the node (Figure 22.6). The capsular extensions, called **trabeculae,** divide the node into compartments, provide support, and provide a route for blood vessels into the interior of a node. Internal to the capsule is a supporting network of reticular fibers and fibroblasts. The capsule, trabeculae, reticular fibers, and fibroblasts constitute the *stroma,* or framework, of a lymph node.

The *parenchyma* of a lymph node is divided into a superficial cortex and a deep medulla. Within the **outer cortex** are egg-shaped aggregates of B cells called **lymphatic nodules (follicles).** A lymphatic nodule consisting chiefly of B cells is called a *primary lymphatic nodule.* Most lymphatic nodules in the outer cortex are *secondary lymphatic nodules* (Figure 22.6), which form in response to an antigenic challenge and are sites of plasma cell and memory B cell formation. After B cells in a primary lymphatic nodule recognize an antigen, the primary lymphatic nodule develops into a secondary lymphatic nodule. The center of a secondary lymphatic nodule contains a region of light-staining cells called a *germinal center.* In the germinal center are B cells, follicular dendritic cells (a special type of dendritic cell), and macrophages. When follicular dendritic cells "present" an antigen, B cells proliferate and develop into antibody-producing plasma cells or develop into memory B cells. Memory B cells persist after an immune response and "remember" having encountered a specific antigen. B cells that do not develop properly undergo **apoptosis** (programmed cell death) and are destroyed by macrophages. The region of a secondary lymphatic nodule surrounding the germinal center is composed of dense accumulations of B cells that have migrated away from their site of origin within the nodule.

The **inner cortex,** also called the *paracortex,* does not contain lymphatic nodules. It consists mainly of T cells and dendritic cells that enter a lymph node from other tissues. The dendritic cells present antigens to T cells, causing their proliferation. The newly formed T cells then migrate from the lymph node to areas of the body where there is antigenic activity.

The **medulla** of a lymph node contains B cells, antibody-producing plasma cells that have migrated out of the cortex into the medulla, and macrophages. The various cells are embedded in a network of reticular fibers and reticular cells.

Lymph flows through a node in one direction only (Figure 22.6a). It enters through **afferent lymphatic vessels** (*afferent* = to carry toward), which penetrate the convex surface of the node at several points. The afferent vessels contain valves that open toward the center of the node, such that the lymph is directed *inward.* Within the node, lymph enters **sinuses,** which are a series of irregular channels that contain branching reticular fibers, lymphocytes, and macrophages. From the afferent lymphatic vessels, lymph flows into the **subcapsular sinus,** immediately beneath the capsule. Then lymph flows through **trabecular sinuses,** which extend through the cortex parallel to the trabeculae, and into **medullary sinuses,** which extend through the medulla. The medullary sinuses drain into one or two **efferent lymphatic vessels** (*efferent* = to carry away), which are wider than afferent vessels and fewer in number. They contain valves that open away from the center of the node to convey lymph, antibodies secreted by plasma cells, and activated T cells *out* of the node. Efferent lymphatic vessels emerge from one side of the lymph node at a slight depression called a **hilus** (HĪ-lus). Blood vessels also enter and leave the node at the hilus.

Lymph nodes function to filter lymph. As lymph enters one end of a node, foreign substances are trapped by the reticular fibers within the sinuses of the node. Then macrophages destroy some foreign substances by phagocytosis while lymphocytes destroy others by immune responses. Filtered lymph then leaves the other end of the node.

Figure 22.6 **Structure of a lymph node.** Arrows indicate the direction of lymph flow through a lymph node. (See Tortora, *A Photographic Atlas of the Human Body,* Figure 7.4.)

🔑 **Lymph nodes are present throughout the body, usually clustered in groups.**

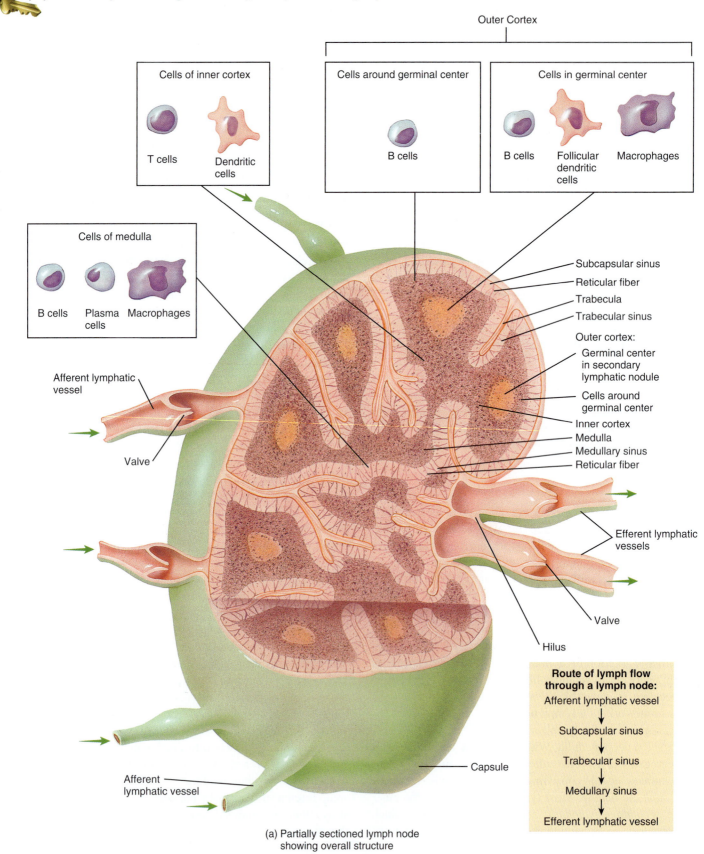

Outer Cortex

Cells of inner cortex

T cells   Dendritic cells

Cells around germinal center

B cells

Cells in germinal center

B cells   Follicular dendritic cells   Macrophages

Cells of medulla

B cells   Plasma cells   Macrophages

Afferent lymphatic vessel

Valve

Subcapsular sinus
Reticular fiber
Trabecula
Trabecular sinus
Outer cortex:
Germinal center in secondary lymphatic nodule
Cells around germinal center
Inner cortex
Medulla
Medullary sinus
Reticular fiber

Efferent lymphatic vessels

Valve

Hilus

Capsule

Afferent lymphatic vessel

Route of lymph flow through a lymph node:
Afferent lymphatic vessel
↓
Subcapsular sinus
↓
Trabecular sinus
↓
Medullary sinus
↓
Efferent lymphatic vessel

(a) Partially sectioned lymph node showing overall structure

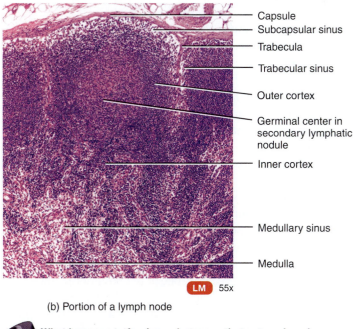

Capsule
Subcapsular sinus
Trabecula
Trabecular sinus
Outer cortex
Germinal center in secondary lymphatic nodule
Inner cortex
Medullary sinus
Medulla

LM 55x

(b) Portion of a lymph node

What happens to foreign substances that enter a lymph node in lymph?

Metastasis Through Lymphatic Vessels

Metastasis (me-TAS-ta-sis; *meta-* = beyond; *stasis* = to stand), the spread of a disease from one part of the body to another, can occur via lymphatic vessels. All malignant tumors eventually exhibit metastasis. Cancer cells may travel in the blood or lymph and establish new tumors where they lodge. When metastasis occurs via lymphatic vessels, secondary tumor sites can be predicted according to the direction of lymph flow from the primary tumor site. Cancerous lymph nodes feel enlarged, firm, nontender, and fixed to underlying structures. By contrast, most lymph nodes that are enlarged due to an infection are not firm, are moveable, and are very tender. ■

Spleen

The oval **spleen** is the largest single mass of lymphatic tissue in the body, measuring about 12 cm (5 in.) in length (Figure 22.7a). It is located in the left hypochondriac region between the stomach and diaphragm. The superior surface of the spleen is smooth and convex and conforms to the concave surface of the

Figure 22.7   Structure of the spleen.   (See Tortora, *A Photographic Atlas of the Human Body,* Figure 7.3.)

The spleen is the largest single mass of lymphatic tissue in the body.

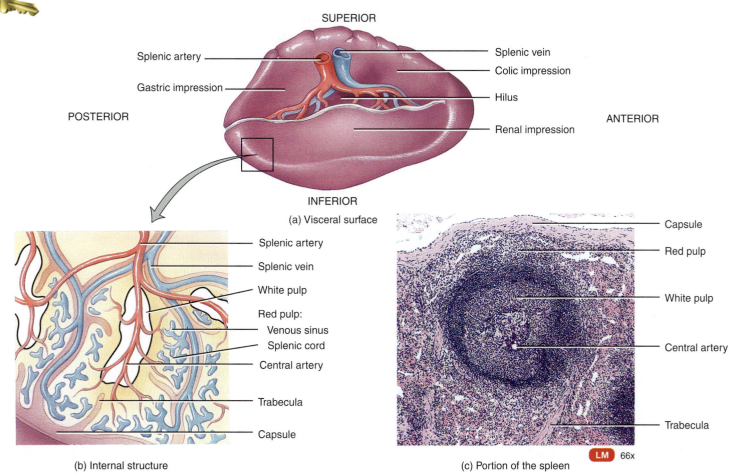

SUPERIOR

Splenic artery
Gastric impression
POSTERIOR
Splenic vein
Colic impression
Hilus
ANTERIOR
Renal impression

INFERIOR

(a) Visceral surface

Splenic artery
Splenic vein
White pulp
Red pulp:
  Venous sinus
  Splenic cord
Central artery
Trabecula
Capsule

(b) Internal structure

Capsule
Red pulp
White pulp
Central artery
Trabecula

LM 66x

(c) Portion of the spleen

After birth, what are the main functions of the spleen?

diaphragm. Neighboring organs make indentations in the visceral surface of the spleen—the gastric impression (stomach), the renal impression (left kidney), and the colic impression (left flexure of colon). Like lymph nodes, the spleen has a hilus. Through it pass the splenic artery, splenic vein, and efferent lymphatic vessels.

A capsule of dense connective tissue surrounds the spleen and is covered, in turn, by a serous membrane, the visceral peritoneum. Trabeculae extend inward from the capsule. The capsule plus trabeculae, reticular fibers, and fibroblasts constitute the stroma of the spleen; the parenchyma of the spleen consists of two different kinds of tissue called white pulp and red pulp (Figure 22.7b, c). **White pulp** is lymphatic tissue, consisting mostly of lymphocytes and macrophages arranged around branches of the splenic artery called **central arteries.** The **red pulp** consists of blood-filled **venous sinuses** and cords of splenic tissue called **splenic (Billroth's) cords.** Splenic cords consist of red blood cells, macrophages, lymphocytes, plasma cells, and granulocytes. Veins are closely associated with the red pulp.

Blood flowing into the spleen through the splenic artery enters the central arteries of the white pulp. Within the white pulp, B cells and T cells carry out immune functions, similar to lymph nodes, while spleen macrophages destroy blood-borne pathogens by phagocytosis. Within the red pulp, the spleen performs three functions related to blood cells: (1) removal by macrophages of ruptured, worn out, or defective blood cells and platelets; (2) storage of platelets, up to one-third of the body's supply; and (3) production of blood cells (hemopoiesis) during fetal life.

Ruptured Spleen

The spleen is the organ most often damaged in cases of abdominal trauma. Severe blows over the inferior left chest or superior abdomen can fracture the protecting ribs. Such crushing injury may rupture the spleen, which causes severe intraperitoneal hemorrhage and shock. Prompt removal of the spleen, called a **splenectomy,** is needed to prevent death due to bleeding. Other structures, particularly red bone marrow and the liver, can take over some functions normally carried out by the spleen. Immune functions, however, decrease in the absence of a spleen. The spleen's absence also places the patient at higher risk for **sepsis** (a blood infection) due to loss of the filtering and phagocytic functions of the spleen. To reduce the risk of sepsis, patients with a history of a splenectomy take prophylactic antibiotics before any invasive procedures. ■

Lymphatic Nodules

Lymphatic nodules are egg-shaped masses of lymphatic tissue that are not surrounded by a capsule. Because they are scattered throughout the lamina propria (connective tissue) of mucous membranes lining the gastrointestinal, urinary, and reproductive tracts and the respiratory airways, lymphatic nodules in these areas are also referred to as **mucosa-associated lymphatic tissue (MALT).**

Although many lymphatic nodules are small and solitary, some occur in multiple large aggregations in specific parts of the body. Among these are the tonsils in the pharyngeal region and the aggregated lymphatic follicles (Peyer's patches) in the ileum of the small intestine. Aggregations of lymphatic nodules also occur in the appendix. Usually there are five **tonsils,** which form a ring at the junction of the oral cavity and oropharynx and at the junction of the nasal cavity and nasopharynx (see Figure 23.2b on page 808). The tonsils are strategically positioned to participate in immune responses against inhaled or ingested foreign substances. The single **pharyngeal tonsil** (fa-RIN-jē-al) or **adenoid** is embedded in the posterior wall of the nasopharynx. The two **palatine tonsils** (PAL-a-tīn) lie at the posterior region of the oral cavity, one on either side; these are the tonsils commonly removed in a tonsillectomy. The paired **lingual tonsils** (LIN-gwal), located at the base of the tongue, may also require removal during a tonsillectomy.

▶ **CHECKPOINT**

1. How are interstitial fluid and lymph similar, and how do they differ?

2. How do lymphatic vessels differ in structure from veins?

3. Construct a diagram that shows the route of lymph circulation.

4. What is the role of the thymus in immunity?

5. What functions do lymph nodes serve?

6. Describe the functions of the spleen and tonsils.

DEVELOPMENT OF LYMPHATIC TISSUES

▶ **OBJECTIVE**

• **Describe the development of lymphatic tissues.**

Lymphatic tissues begin to develop by the end of the fifth week of embryonic life. *Lymphatic vessels* develop from **lymph sacs** that arise from developing veins, which are derived from **mesoderm.**

The first lymph sacs to appear are the paired **jugular lymph sacs** at the junction of the internal jugular and subclavian veins (Figure 22.8). From the jugular lymph sacs, lymphatic capillary plexuses spread to the thorax, upper limbs, neck, and head. Some of the plexuses enlarge and form lymphatic vessels in their respective regions. Each jugular lymph sac retains at least one connection with its jugular vein, the left one developing into the superior portion of the thoracic duct (left lymphatic duct).

The next lymph sac to appear is the unpaired **retroperitoneal lymph sac** at the root of the mesentery of the intestine. It

Figure 22.8 Development of lymphatic tissues.

Lymphatic tissues are derived from mesoderm.

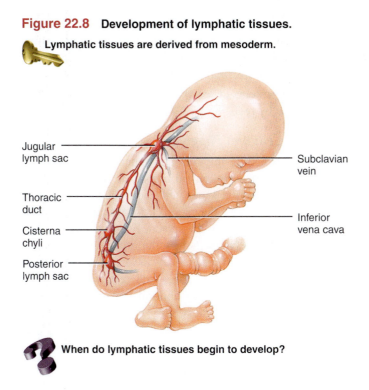

Jugular lymph sac

Thoracic duct

Cisterna chyli

Posterior lymph sac

Subclavian vein

Inferior vena cava

When do lymphatic tissues begin to develop?

develops from the primitive vena cava and mesonephric (primitive kidney) veins. Capillary plexuses and lymphatic vessels spread from the retroperitoneal lymph sac to the abdominal viscera and diaphragm. The sac establishes connections with the cisterna chyli but loses its connections with neighboring veins.

At about the time the retroperitoneal lymph sac is developing, another lymph sac, the **cisterna chyli,** develops inferior to the diaphragm on the posterior abdominal wall. It gives rise to the inferior portion of the *thoracic duct* and the *cisterna chyli* of the thoracic duct. Like the retroperitoneal lymph sac, the cisterna chyli also loses its connections with surrounding veins.

The last of the lymph sacs, the paired **posterior lymph sacs,** develop from the iliac veins. The posterior lymph sacs produce capillary plexuses and lymphatic vessels of the abdominal wall, pelvic region, and lower limbs. The posterior lymph sacs join the cisterna chyli and lose their connections with adjacent veins.

With the exception of the anterior part of the sac from which the cisterna chyli develops, all lymph sacs become invaded by **mesenchymal cells** and are converted into groups of *lymph nodes.*

The *spleen* develops from **mesenchymal cells** between layers of the dorsal mesentery of the stomach. The *thymus* arises as an outgrowth of the **third pharyngeal pouch** (see Figure 18.21a on page 623).

▶ CHECKPOINT

7. What are the names of the four lymph sacs from which lymphatic vessels develop?

NONSPECIFIC RESISTANCE: INNATE DEFENSES

▶ OBJECTIVE

• Describe the mechanisms of nonspecific resistance to disease.

Although several mechanisms contribute to innate defenses or nonspecific resistance to disease, they all have two things in common. They are present at birth, and they offer immediate protection against a wide variety of pathogens and foreign substances. Nonspecific resistance, as its name suggests, lacks specific responses to specific invaders; instead, its protective mechanisms function the same way, regardless of the type of invader. Innate defense mechanisms include the external physical and chemical barriers provided by the skin and mucous membranes. They also include various internal nonspecific defenses, such as antimicrobial proteins, natural killer cells and phagocytes, inflammation, and fever.

First Line of Defense: Skin and Mucous Membranes

The skin and mucous membranes of the body are the first line of defense against pathogens. Both physical and chemical barriers discourage pathogens and foreign substances from penetrating the body and causing disease.

With its many layers of closely packed, keratinized cells, the outer epithelial layer of the skin—the **epidermis**—provides a formidable physical barrier to the entrance of microbes (see Figure 5.1 on page 141). In addition, periodic shedding of epidermal cells helps remove microbes at the skin surface. Bacteria rarely penetrate the intact surface of healthy epidermis. If this surface is broken by cuts, burns, or punctures, however, pathogens can penetrate the epidermis and invade adjacent tissues or circulate in the blood to other parts of the body.

The epithelial layer of **mucous membranes,** which line body cavities, secretes a fluid called **mucus** that lubricates and moistens the cavity surface. Because mucus is slightly viscous, it traps many microbes and foreign substances. The mucous membrane of the nose has mucus-coated **hairs** that trap and filter microbes, dust, and pollutants from inhaled air. The mucous membrane of the upper respiratory tract contains **cilia,** microscopic hairlike projections on the surface of the epithelial cells. The waving action of cilia propels inhaled dust and microbes that have become trapped in mucus toward the throat. Coughing and sneezing accelerate movement of mucus and its entrapped pathogens out of the body.

Other fluids produced by various organs also help protect epithelial surfaces of the skin and mucous membranes. The **lacrimal apparatus** (LAK-ri-mal) of the eyes (see Figure 16.5 on page 533) manufactures and drains away tears in response to irritants. Blinking spreads tears over the surface of the eyeball, and the continual washing action of tears helps to dilute microbes and keep them from settling on the surface of the eye. **Saliva,** produced by the salivary glands, washes microbes from the surfaces of the teeth and from the mucous membrane of the mouth, much as tears wash the eyes. The flow of saliva reduces colonization of the mouth by microbes.

The cleansing of the urethra by the **flow of urine** retards microbial colonization of the urinary system. **Vaginal secretions** likewise move microbes out of the body in females. **Defecation** and **vomiting** also expel microbes. For example, in response to some microbial toxins, the smooth muscle of the lower gastrointestinal tract contracts vigorously; the resulting diarrhea rapidly expels many of the microbes.

Certain chemicals also contribute to the high degree of resistance of the skin and mucous membranes to microbial invasion. Sebaceous (oil) glands of the skin secrete an oily substance called **sebum** that forms a protective film over the surface of the skin. The unsaturated fatty acids in sebum inhibit the growth of certain pathogenic bacteria and fungi.

The acidity of the skin (pH 3–5) is caused in part by the secretion of fatty acids and lactic acid. **Perspiration** helps flush microbes from the surface of the skin, and it also contains **lysozyme,** an enzyme capable of breaking down the cell walls of certain bacteria. Lysozyme is also present in tears, saliva, nasal secretions, and tissue fluids, where it also exhibits antimicrobial activity. **Gastric juice,** produced by the glands of the stomach, is a mixture of hydrochloric acid, enzymes, and mucus. The strong acidity of gastric juice (pH 1.2–3.0) destroys many bacteria and most bacterial toxins. Vaginal secretions also are slightly acidic, which discourages bacterial growth.

Second Line of Defense: Internal Defenses

When pathogens penetrate the mechanical and chemical barriers of the skin and mucous membranes, they encounter a second line of defense: internal antimicrobial proteins, phagocytes, natural killer cells, inflammation, and fever.

Antimicrobial Proteins

Blood and interstitial fluids contain three main types of **antimicrobial proteins** that discourage microbial growth: interferons, complement, and transferrins.

1. Lymphocytes, macrophages, and fibroblasts infected with viruses produce proteins called **interferons** (in′-ter-FĒR-ons), or **IFNs.** Once released by virus-infected cells, IFNs diffuse to uninfected neighboring cells, where they induce synthesis of antiviral proteins that interfere with viral replication. Although IFNs do not prevent viruses from attaching to and penetrating host cells, they do stop replication. Viruses can cause disease only if they can replicate within body cells. IFNs are an important defense against infection by many different viruses. The three types of interferon are alpha-, beta-, and gamma-IFN.

2. A group of normally inactive proteins in blood plasma and on plasma membranes makes up the **complement system.** When activated, these proteins "complement" or enhance certain immune, allergic, and inflammatory reactions.

3. Iron-binding proteins called **transferrins** inhibit the growth of certain bacteria by reducing the amount of available iron.

Natural Killer Cells and Phagocytes

When microbes penetrate the skin and mucous membranes or bypass the antimicrobial proteins in blood, the next nonspecific defense consists of natural killer cells and phagocytes. About 5–10% of lymphocytes in the blood are **natural killer (NK) cells.** They also are present in the spleen, lymph nodes, and red bone marrow. Although NK cells lack the membrane molecules that identify B cells and T cells, they have the ability to kill a wide variety of infectious microbes and certain tumor cells. NK cells attack cells that display abnormal plasma membrane proteins called major histocompatibility complex (MHC) antigens, which are described on page 782. NK cells cause cellular destruction in at least two ways:

1. NK cells release **perforins,** chemicals that when inserted into the plasma membrane of a microbe make the membrane so leaky that cytolysis occurs.

2. Upon encountering a virus-infected or tumor cell, NK cells release molecules that enter the target cells and cause apoptosis.

Phagocytes (*phago-* = eat; *-cytes* = cells) are specialized cells that perform **phagocytosis** (*-osis* = process), the ingestion of microbes or other particles such as cellular debris (see Figure 3.14 on page 73). The two major types of phagocytes are **neutrophils** and **macrophages** (MAK-rō-fā-jez). When an infection occurs, neutrophils and monocytes migrate to the infected area. During this migration, the monocytes enlarge and develop into actively phagocytic macrophages. **Wandering macrophages** leave the blood and migrate to infected areas. Other macrophages, called **fixed macrophages,** stand guard in specific tissues. Among the fixed macrophages are histiocytes in the skin and subcutaneous layer, stellate reticuloendothelial cells (Kupffer cells) in the liver, alveolar macrophages in the lungs, microglia in the nervous system, and tissue macrophages in the spleen, lymph nodes, and red bone marrow. In addition to being an innate defense mechanism, phagocytosis plays a vital role in immunity, as discussed later in the chapter.

Phagocytosis occurs in five phases: chemotaxis, adherence, ingestion, digestion, and killing (Figure 22.9):

1 Phagocytosis begins with **chemotaxis,** a chemically stimulated movement of phagocytes to a site of damage. Chemicals that attract phagocytes might come from the microbes, white blood cells, damaged tissue cells, or activated complement proteins.

2 Attachment of the phagocyte to the microbe or other foreign material is termed **adherence.** The binding of complement to the invading pathogen enhances adherence.

3 Following adherence, the plasma membrane of the phagocyte extends projections, called **pseudopods,** that engulf the microbe in a process called **ingestion.** When the pseudopods meet, they fuse, surrounding the microorganism with a sac called a **phagosome.**

4 The phagosome enters the cytoplasm and merges with lysosomes to form a single, larger structure called a **phagolysosome.** The lysosome contributes lysozyme, which breaks down microbial cell walls, and other digestive enzymes that degrade carbohydrates, proteins, lipids, and nucleic acids. The phagocyte also forms lethal oxidants, such as superoxide anion (O_2^-), hypochlorite anion (OCl^-), and hydrogen peroxide (H_2O_2), in a process called an **oxidative burst.**

5 The chemical onslaught provided by lysozyme, digestive enzymes, and oxidants within a phagolysosome quickly kills many types of microbes. Any materials that cannot be degraded further remain in structures called **residual bodies.**

Microbial Evasion of Phagocytosis

Some microbes, such as the bacteria that cause pneumonia, have extracellular structures called capsules that prevent adherence. This makes it physically difficult for phagocytes to engulf the microbes. Other microbes, such as the toxin-producing bacteria that cause one kind of food poisoning, may be ingested but not killed; instead, the toxins they produce (leukocidins) may kill the phagocytes by causing the release the phagocyte's own lysosomal enzymes into its cytoplasm. Still other microbes—such as the bacteria that cause tuberculosis—inhibit fusion of phagosomes and lysosomes and thus prevent exposure of the microbes to lysosomal enzymes. These bacteria can also apparently use chemicals in their cell walls to counter the effects of lethal oxidants produced by phagocytes. Subsequent multiplication of the microbes within phagosomes may eventually destroy the phagocyte. ■

Inflammation

Inflammation is a nonspecific, defensive response of the body to tissue damage. Among the conditions that may produce inflammation are pathogens, abrasions, chemical irritations, distortion or disturbances of cells, and extreme temperatures. The four characteristic signs and symptoms of inflammation are **redness, pain, heat,** and **swelling.** Inflammation can also cause the **loss of function** in the injured area, depending on the site and extent of the injury. Inflammation is an attempt to dispose of microbes, toxins, or foreign material at the site of injury, to prevent their spread to other tissues, and to prepare the site for tissue repair. Thus, it helps restore tissue homeostasis.

Because inflammation is one of the body's innate defenses, the response of a tissue to, say, a cut is similar to the response to

Figure 22.9   Phagocytosis of a microbe.

The major types of phagocytes are neutrophils and macrophages.

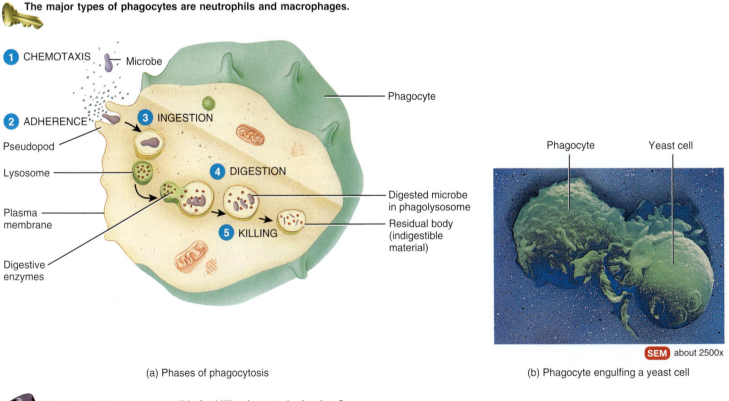

(a) Phases of phagocytosis

(b) Phagocyte engulfing a yeast cell

What chemicals are responsible for killing ingested microbes?

damage caused by burns, radiation, or bacterial or viral invasion. In each case, the inflammatory response has three basic stages: (1) vasodilation and increased permeability of blood vessels, (2) emigration (movement) of phagocytes from the blood into interstitial fluid, and, ultimately, (3) tissue repair.

VASODILATION AND INCREASED PERMEABILITY OF BLOOD VESSELS   Two immediate changes occur in the blood vessels in a region of tissue injury: **vasodilation** (increase in the diameter) of arterioles and **increased permeability** of capillaries (Figure 22.10). Increased permeability means that substances normally retained in blood are permitted to pass from the blood vessels. Vasodilation allows more blood to flow through the damaged area, and increased permeability permits defensive proteins such as antibodies and clotting factors to enter the injured area from the blood. The increased blood flow also helps remove microbial toxins and dead cells.

Among the substances that contribute to vasodilation, increased permeability, and other aspects of the inflammatory response are the following:

- *Histamine.* In response to injury, mast cells in connective tissue and basophils and platelets in blood release histamine. Neutrophils and macrophages attracted to the site of injury also stimulate the release of histamine, which causes vasodilation and increased permeability of blood vessels.
- *Kinins.* These polypeptides, formed in blood from inactive precursors called kininogens, induce vasodilation and increased permeability and serve as chemotactic agents for phagocytes.
- *Prostaglandins (PGs).* These lipids, especially those of the E series, are released by damaged cells and intensify the effects of histamine and kinins. PGs also may stimulate the emigration of phagocytes through capillary walls.
- *Leukotrienes (LTs).* Produced by basophils and mast cells by breakdown of membrane phospholipids, LTs cause increased permeability; they also function in adherence of phagocytes to pathogens and as chemotactic agents that attract phagocytes.
- *Complement.* Different components of the complement system stimulate histamine release, attract neutrophils by chemotaxis, and promote phagocytosis; some components can also destroy bacteria.

Dilation of arterioles and increased permeability of capillaries produce three of the symptoms of inflammation, heat, redness (erythema), and swelling (edema). Heat and redness result from the large amount of blood that accumulates in the damaged area. As the local temperature rises slightly, metabolic reactions proceed more rapidly and release additional heat. Edema results from increased permeability of blood vessels, which permits more fluid to move from blood plasma into tissue spaces.

Pain is a prime symptom of inflammation. It results from injury to neurons and from toxic chemicals released by microbes. Kinins affect some nerve endings, causing much of the pain associated with inflammation. Prostaglandins intensify and prolong the pain associated with inflammation. Pain may also be due to increased pressure from edema.

The increased permeability of capillaries allows leakage of blood-clotting factors into tissues. The clotting cascade is set into motion, and fibrinogen is ultimately converted to an insoluble, thick mesh of fibrin threads that localizes and traps invading microbes and blocks their spread.

EMIGRATION OF PHAGOCYTES   Generally, within an hour after the inflammatory process starts, phagocytes appear on the scene. As large amounts of blood accumulate, neutrophils begin to stick to the inner surface of the endothelium (lining) of blood vessels (Figure 22.10). Then the neutrophils begin to squeeze through the wall of the blood vessel to reach the damaged area. This process, called **emigration,** depends on chemotaxis. Neutrophils attempt to destroy the invading microbes by phagocytosis. A steady stream of neutrophils is ensured by the production and release of additional cells from bone marrow.

Figure 22.10   Inflammation.

The three stages of inflammation are as follows: (1) vasodilation and increased permeability of blood vessels, (2) phagocyte emigration, and (3) tissue repair.

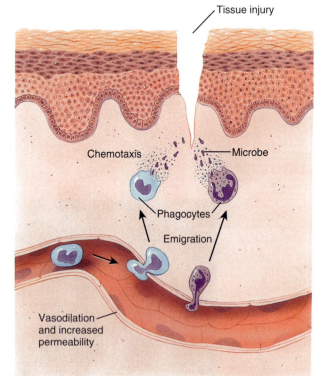

Phagocytes migrate from blood to site of tissue injury

What causes each of the following signs and symptoms of inflammation: redness, pain, heat, and swelling?

Such an increase in white blood cells in the blood is termed **leukocytosis.**

Although neutrophils predominate in the early stages of infection, they die off rapidly. As the inflammatory response continues, monocytes follow the neutrophils into the infected area. Once in the tissue, monocytes transform into wandering macrophages that add to the phagocytic activity of the fixed macrophages already present. True to their name, macrophages are much more potent phagocytes than are neutrophils. They are large enough to engulf damaged tissue, worn-out neutrophils, and invading microbes.

Eventually, macrophages also die. Within a few days, a pocket of dead phagocytes and damaged tissue forms; this collection of dead cells and fluid is called **pus.** Pus formation occurs in most inflammatory responses and usually continues until the infection subsides. At times, pus reaches the surface of the body or drains into an internal cavity and is dispersed; on other occasions the pus remains even after the infection is terminated. In this case, the pus is gradually destroyed over a period of days and is absorbed.

Abscesses and Ulcers

If pus cannot drain out of an inflamed region, the result is an **abscess**—an excessive accumulation of pus in a confined space. Common examples are pimples and boils. When superficial inflamed tissue sloughs off the surface of an organ or tissue, the resulting open sore is called an **ulcer.** People with poor circulation—for instance, diabetics with advanced atherosclerosis—are susceptible to ulcers in the tissues of their legs. These ulcers, which are called stasis ulcers, develop because of poor oxygen and nutrient supply to tissues that then become very susceptible to even a very mild injury or an infection. ■

Fever

Fever is an abnormally high body temperature that occurs because the hypothalamic thermostat is reset. It commonly occurs during infection and inflammation. Many bacterial toxins elevate body temperature, sometimes by triggering release of fever-causing cytokines such as interleukin-1. Elevated body temperature intensifies the effects of interferons, inhibits the growth of some microbes, and speeds up body reactions that aid repair.

Table 22.1 summarizes the components of nonspecific resistance.

▶ **CHECKPOINT**

8. What physical and chemical factors provide protection from disease in the skin and mucous membranes?

9. What internal defenses provide protection against microbes that penetrate the skin and mucous membranes?

10. How are the activities of natural killer cells and phagocytes similar and different?

11. What are the main signs, symptoms, and stages of inflammation?

Table 22.1	Summary of Nonspecific Resistance (Innate Defenses)
Component	**Functions**
First Line of Defense: Skin and Mucous Membranes	
Physical Factors	
Epidermis of skin	Forms a physical barrier to the entrance of microbes.
Mucous membranes	Inhibit the entrance of many microbes, but not as effective as intact skin.
Mucus	Traps microbes in respiratory and gastrointestinal tracts.
Hairs	Filter out microbes and dust in nose.
Cilia	Together with mucus, trap and remove microbes and dust from upper respiratory tract.
Lacrimal apparatus	Tears dilute and wash away irritating substances and microbes.
Saliva	Washes microbes from surfaces of teeth and mucous membranes of mouth.
Urine	Washes microbes from urethra.
Defecation and vomiting	Expel microbes from body.
Chemical Factors	
Sebum	Forms a protective acidic film over the skin surface that inhibits growth of many microbes.
Lysozyme	Antimicrobial substance in perspiration, tears, saliva, nasal secretions, and tissue fluids.
Gastric juice	Destroys bacteria and most toxins in stomach.
Vaginal secretions	Slight acidity discourages bacterial growth; flush microbes out of vagina.
Second Line of Defense: Internal Defenses	
Antimicrobial Proteins	
Interferons (IFNs)	Protect uninfected host cells from viral infection.
Complement system	Causes cytolysis of microbes, promotes phagocytosis, and contributes to inflammation.
Natural killer (NK) cells	Kill a wide variety of microbes and certain tumor cells.
Phagocytes	Ingest foreign particulate matter.
Inflammation	Confines and destroys microbes and initiates tissue repair.
Fever	Intensifies the effects of interferons, inhibits growth of some microbes, and speeds up body reactions that aid repair.

SPECIFIC RESISTANCE: IMMUNITY

► O B J E C T I V E S

- **Define immunity, and describe how T cells and B cells arise.**
- **Explain the relationship between an antigen and an antibody.**
- **Compare the functions of cell-mediated immunity and antibody-mediated immunity.**

The ability of the body to defend itself against specific invading agents such as bacteria, toxins, viruses, and foreign tissues is called **specific resistance** or **immunity.** Substances that are recognized as foreign and provoke immune responses are called **antigens (Ags).** Two properties distinguish immunity from nonspecific defenses: (1) *specificity* for particular foreign molecules (antigens), which also involves distinguishing self from nonself molecules, and (2) *memory* for most previously encountered antigens such that a second encounter prompts an even more rapid and vigorous response. The branch of science that deals with the responses of the body when challenged by antigens is called **immunology** (im'-ū-NOL-ō-jē ; *immuno-* = free from service or exempt; *-logy* = study of). The **immune system** includes the cells and tissues that carry out immune responses.

Maturation of T Cells and B Cells

The cells that develop **immunocompetence,** the ability to carry out immune responses if properly stimulated, are lymphocytes called B cells and T cells. Both develop in primary lymphatic organs (red bone marrow and the thymus) from pluripotent stem cells that originate in red bone marrow (see Figure 19.3 on page 638). B cells complete their development in bone marrow, a process that continues throughout life. T cells develop from pre-T cells that migrate from bone marrow into the thymus, where they mature (Figure 22.11). Most T cells arise before puberty, but they continue to mature and leave the thymus throughout life.

Before T cells leave the thymus or B cells leave red bone marrow, they begin to make several distinctive proteins that are inserted into the plasma membrane. Some of these proteins function as **antigen receptors**—molecules capable of recognizing specific antigens (Figure 22.11). In addition, T cells exit the thymus as either CD4$^+$ or CD8$^+$ cells, which means their plasma membrane includes either a protein called CD4 or one called CD8. As we will see later in this chapter, these two types of T cells have very different functions.

Types of Immune Responses

Immunity consists of two kinds of closely allied responses, both triggered by antigens. In **cell-mediated immune responses,** CD8$^+$ T cells proliferate into cytotoxic T cells that directly attack the invading antigen. In **antibody-mediated immune responses,** B cells transform into plasma cells, which synthesize and secrete specific proteins called **antibodies (Abs)** or **immunoglobulins** (im'-ū-nō-GLOB-ū-lins). A given antibody can bind to and inactivate a specific antigen. Most CD4$^+$ T cells become helper T cells that aid both cell-mediated and antibody-mediated immune responses.

To some extent, each type of immune response specializes in dealing with certain types of invaders. Cell-mediated immunity is particularly effective against (1) intracellular pathogens that reside within host cells (primarily fungi, parasites, and viruses); (2) some cancer cells; and (3) foreign tissue transplants. Thus, cell-mediated immunity always involves cells attacking cells. Antibody-mediated immunity works mainly against (1) antigens present in body fluids and (2) extracellular pathogens that multiply in body fluids but rarely enter body cells (primarily bacteria). However, a given pathogen often provokes both types of immune responses.

Antigens and Antigen Receptors

Antigens have two important characteristics: immunogenicity and reactivity. **Immunogenicity** (im'-ū -nō-je-NIS-i-tē; *-genic* = producing) is the ability to provoke an immune response by stimulating the production of specific antibodies, the proliferation of specific T cells, or both. The term *antigen* derives from its function as an *anti*body *gen*erator. **Reactivity** is the ability of the antigen to react specifically with the antibodies or cells it provoked. Strictly speaking, immunologists define antigens as substances that have reactivity; substances with both immunogenicity and reactivity are considered **complete antigens.** Commonly, however, the term *antigen* implies both immunogenicity and reactivity, and we use the word in this way.

Entire microbes or parts of microbes may act as antigens. Chemical components of bacterial structures such as flagella, capsules, and cell walls are antigenic, as are bacterial toxins. Nonmicrobial examples of antigens include chemical components of pollen, egg white, incompatible blood cells, and transplanted tissues and organs. The huge variety of antigens in the environment provides myriad opportunities for provoking immune responses. Typically, just certain parts of a large antigen molecule act as the triggers for immune responses. The small parts of antigen molecules that initiate immune responses are called **epitopes** (EP-i-tōps), or *antigenic determinants* (Figure 22.12). Most antigens have many epitopes, each of which induces production of a specific antibody or activates a specific T cell.

Antigens that get past the nonspecific defenses generally follow one of three routes into lymphatic tissue: (1) Most antigens that enter the bloodstream (for example, through an injured blood vessel) are trapped as they flow through the spleen. (2) Antigens that penetrate the skin enter lymphatic vessels and lodge in lymph nodes. (3) Antigens that penetrate mucous membranes are entrapped by mucosa-associated lymphatic tissue (MALT).

Chemical Nature of Antigens

Antigens are large, complex molecules. Most often, they are proteins. However, nucleic acids, lipoproteins, glycoproteins, and

Figure 22.11 **B cells and pre-T cells arise from pluripotent stem cells in red bone marrow.** B cells and T cells develop in primary lymphatic tissues (red bone marrow and the thymus) and are activated in secondary lymphatic tissues (lymph nodes, spleen, and lymphatic nodules).

🔑 The two types of immune responses are cell-mediated immune responses and antibody-mediated immune responses.

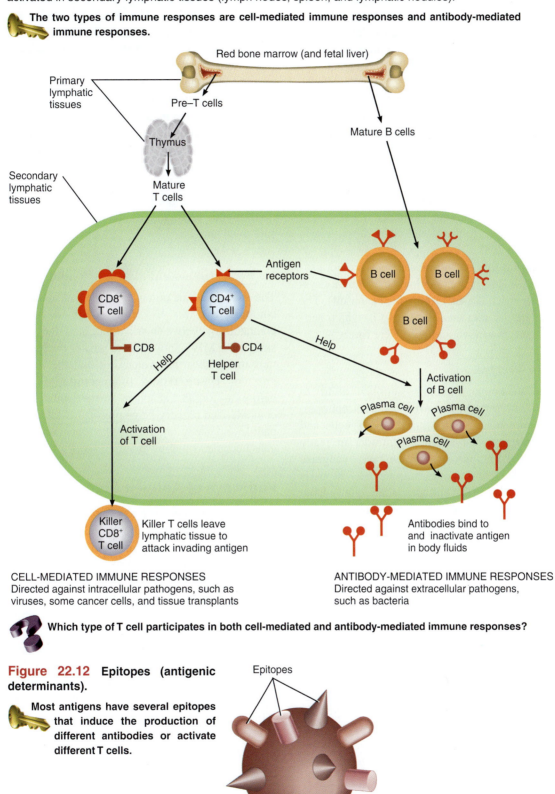

CELL-MEDIATED IMMUNE RESPONSES
Directed against intracellular pathogens, such as viruses, some cancer cells, and tissue transplants

ANTIBODY-MEDIATED IMMUNE RESPONSES
Directed against extracellular pathogens, such as bacteria

❓ Which type of T cell participates in both cell-mediated and antibody-mediated immune responses?

Figure 22.12 **Epitopes (antigenic determinants).**

🔑 Most antigens have several epitopes that induce the production of different antibodies or activate different T cells.

Epitopes

Antigen

❓ What is the difference between an epitope and a hapten?

certain large polysaccharides may also act as antigens. T cells respond only to antigens that include protein; B cells respond to antigens made of proteins, certain lipids, carbohydrates, and nucleic acids. Complete antigens usually have large molecular weights of 10,000 daltons or more, but large molecules that have simple, repeating subunits—for example, cellulose and most plastics—are not usually antigenic. This is why plastic materials can be used in artificial heart valves or joints.

A smaller substance that has reactivity but lacks immunogenicity is called a **hapten** (HAP-ten = to grasp). A hapten can stimulate an immune response only if it is attached to a larger carrier molecule. An example is the small lipid toxin in poison ivy, which triggers an immune response after combining with a body protein. Likewise, some drugs, such as penicillin, may combine with proteins in the body to form immunogenic complexes. Such hapten-stimulated immune responses are responsible for some allergic reactions to drugs and other substances in the environment (see page 798).

As a rule, antigens are foreign substances; they are not usually part of body tissues. However, sometimes the immune system fails to distinguish "friend" (self) from "foe" (nonself). The result is an autoimmune disorder (see page 799) in which self-molecules or cells are attacked as though they were foreign.

Diversity of Antigen Receptors

An amazing feature of the human immune system is its ability to recognize and bind to at least a billion (10^9) different epitopes. Before a particular antigen ever enters the body, T cells and B cells that can recognize and respond to that intruder are ready and waiting. Cells of the immune system can even recognize artificially made molecules that do not exist in nature. The basis for the ability to recognize so many epitopes is an equally large diversity of antigen receptors. Given that human cells contain only about 35,000 genes, how could a billion or more different antigen receptors possibly be generated?

The answer to this puzzle turned out to be simple in concept. The diversity of antigen receptors in both B cells and T cells is the result of shuffling and rearranging a few hundred versions of several small gene segments. This process is called **genetic recombination.** The gene segments are put together in different combinations as the lymphocytes are developing from stem cells in red bone marrow and the thymus. The situation is similar to shuffling a deck of 52 cards and then dealing out three cards. If you did this over and over, you could generate many more than 52 different sets of three cards. Because of genetic recombination, each B cell or T cell has a unique set of gene segments that codes for its unique antigen receptor. After transcription and translation, the receptor molecules are inserted into the plasma membrane.

Major Histocompatibility Complex Antigens

Located in the plasma membrane of body cells are "self-antigens," the **major histocompatibility complex (MHC)** anti-

gens. These transmembrane glycoproteins are also called *human leukocyte antigens (HLA)* because they were first identified on white blood cells. Unless you have an identical twin, your MHC antigens are unique. Thousands to several hundred thousand MHC molecules mark the surface of each of your body cells except red blood cells. Although MHC antigens are the reason that tissues may be rejected when they are transplanted from one person to another, their normal function is to help T cells recognize that an antigen is foreign, not self. Such recognition is an important first step in any immune response.

The two types of major histocompatibility complex antigens are class I and class II. Class I MHC (MHC-I) molecules are built into the plasma membranes of all body cells except red blood cells. Class II MHC (MHC-II) molecules appear only on the surface of antigen-presenting cells (described in the next section), thymic cells, and T cells that have been activated by exposure to an antigen.

Histocompatibility Testing

The success of an organ or tissue transplant depends on **histocompatibility** (his'-tō-kom-pat-i-BIL-i-tē)—that is, the tissue compatibility between the donor and the recipient. The more similar the MHC antigens, the greater the histocompatibility, and thus the greater the probability that the transplant will not be rejected. **Tissue typing (histocompatibility testing)** is done before any organ transplant. In the United States, a nationwide computerized registry helps physicians select the most histocompatible and needy organ transplant recipients whenever donor organs become available. ■

Pathways of Antigen Processing

For an immune response to occur, B cells and T cells must recognize that a foreign antigen is present. Whereas B cells can recognize and bind to antigens in lymph, interstitial fluid, or blood plasma, T cells only recognize fragments of antigenic proteins that are "presented" together with major histocompatibility complex self-antigens. Presentation occurs in the following way. As proteins inside body cells are being broken down, some peptide fragments associate with a peptide-binding groove of newly synthesized MHC molecules. This association stabilizes the MHC molecule and aids its proper folding, so that the MHC molecule can be inserted into the plasma membrane. When a peptide fragment comes from a *self-protein,* T cells ignore the peptide–MHC complex. When the peptide fragment is from a *foreign protein,* however, a few T cells recognize it as an intruder, and an immune response ensues. Preparation of a foreign antigen for display at the cell surface is termed *processing and presenting* of antigen. It occurs in two ways, depending on whether the antigen is located inside or outside body cells.

Processing of Exogenous Antigens

Foreign antigens that are present in fluids outside body cells are termed *exogenous antigens.* They include intruders such as bacteria and bacterial toxins, worm parasites, inhaled pollen and dust, and viruses that have not yet infected a body cell. A special class of cells called **antigen-presenting cells (APCs)** process and present exogenous antigens. APCs include macrophages, B cells, and dendritic cells. They are strategically located in places where antigens are likely to penetrate nonspecific defenses and enter the body, such as the epidermis and dermis of the skin (Langerhans cells are a type of dendritic cell); mucous membranes that line the respiratory, gastrointestinal, urinary, and reproductive tracts; and lymph nodes. After processing an antigen, APCs migrate from tissues via lymphatic vessels to lymph nodes.

The steps in the processing and presenting of an exogenous antigen by an antigen-presenting cell occur as follows (Figure 22.13):

1. ***Ingestion of the antigen.*** Antigen-presenting cells ingest antigens by phagocytosis or endocytosis. Ingestion could occur almost anywhere in the body that invaders, such as microbes, have penetrated the nonspecific defenses.

2. ***Digestion of antigen into peptide fragments.*** Within the phagosome or endosome, digestive enzymes split large antigens into short peptide fragments. At the same time, the antigen-presenting cell is synthesizing MHC-II molecules and packaging them into vesicles. The MCH-II molecules are part of the inner membrane surface of the vesicles.

3. ***Fusion of vesicles.*** The vesicles containing antigen peptide fragments and MHC-II molecules merge and fuse.

4. ***Binding of peptide fragments to MHC-II molecules.*** After fusion of the two types of vesicles, antigen peptide fragments bind to MHC-II molecules.

5. ***Insertion of antigen–MHC-II complex into the plasma membrane.*** The combined vesicle that contains antigen–MHC-II complexes undergoes exocytosis. As a result, the antigen–MHC-II complexes are inserted into the plasma membrane.

After processing an antigen, the antigen-presenting cell migrates to lymphatic tissue to present the antigen to T cells. Within lymphatic tissue, a small number of T cells that have compatibly shaped receptors recognize and bind to the antigen fragment–MHC-II complex, triggering either a cell-mediated or an antibody-mediated immune response. The presentation of exogenous antigen together with MHC-II molecules by antigen-presenting cells informs T cells that intruders are present in the body and that combative action should begin.

Processing of Endogenous Antigens

Foreign antigens that are synthesized within a body cell are termed *endogenous antigens.* Such antigens may be viral proteins produced after a virus infects the cell and takes over the cell's metabolic machinery, or they may be abnormal proteins synthesized by a cancerous cell. Fragments of endogenous antigens associate with major histocompatibility complex-I molecules inside infected cells. The resulting endogenous antigen

Figure 22.13 **Processing and presenting of exogenous antigen by an antigen-presenting cell (APC).**

Except for identical twins, major histocompatibility complex (MHC) molecules are unique in each person. They help T cells recognize foreign invaders.

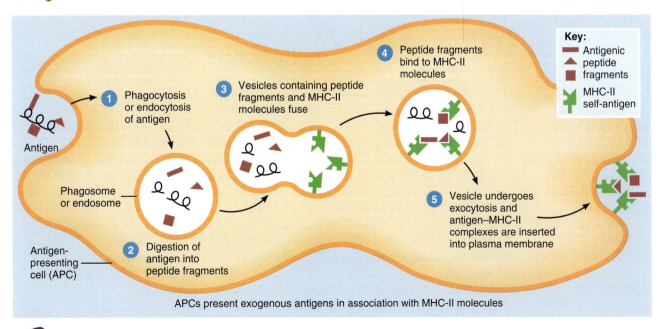

APCs present exogenous antigens in association with MHC-II molecules

What types of cells are APCs, and where in the body are they found?

fragment–MHC-I complex then moves to the plasma membrane, where it is displayed at the surface of the cell. Most cells of the body can process and present endogenous antigens. The display of an endogenous antigen bound to an MHC-I molecule signals that a cell has been infected and needs help.

Cytokines

Cytokines are small protein hormones that stimulate or inhibit many normal cell functions, such as cell growth and differentiation. Lymphocytes and antigen-presenting cells secrete cytokines, as do fibroblasts, endothelial cells, monocytes, hepatocytes, and kidney cells. Some cytokines stimulate proliferation of progenitor blood cells in red bone marrow. Others regulate activities of cells involved in nonspecific defenses or immune responses, as described in Table 22.2.

Cytokine Therapy

Cytokine therapy is the use of cytokines to treat medical conditions. Interferons were the first cytokines shown to be effective against a human cancer. Alpha-interferon (Intron A) is approved in the United States for treating Kaposi (KAP-ō-sē)

sarcoma, a cancer that often occurs in patients infected with HIV, the virus that causes AIDS. Other approved uses for alpha-interferon include treating genital herpes caused by herpes virus; treating hepatitis B and C, caused by the hepatitis B and C viruses; and treating hairy cell leukemia. A form of beta-interferon (Betaseron) slows the progression of multiple sclerosis (MS) and lessens the frequency and severity of MS attacks. Of the interleukins, the one most widely used to fight cancer is interleukin-2. Although this treatment is effective in causing tumor regression in some patients, it also can be very toxic. Among the adverse effects are high fever, severe weakness, difficulty breathing due to pulmonary edema, and hypotension leading to shock. ∎

▶ **CHECKPOINT**

12. What is immunocompetence, and which body cells display it?
13. How do the major histocompatibility complex class I and class II self-antigens function?
14. How do antigens arrive at lymphatic tissues?
15. How do antigen-presenting cells process exogenous and endogenous antigens?
16. What are cytokines, what is their origin, and how do they function?

Table 22.2	Summary of Cytokines Participating in Immune Responses
Cytokine	**Origins and Functions**
Interleukin-1 (IL-1)	Produced by monocytes and macrophages; promotes proliferation of helper T cells; acts on hypothalamus to cause fever.
Interleukin-2 (IL-2) (T cell growth factor)	Secreted by helper T cells; costimulates the proliferation of helper T cells, cytotoxic T cells, and B cells; activates NK cells.
Interleukin-4 (IL-4) (B cell stimulating factor)	Produced by activated helper T cells; costimulator for B cells; causes plasma cells to secrete IgE antibodies (see Table 22.3); promotes growth of T cells.
Interleukin-5 (IL-5)	Produced by certain activated CD4$^+$ T cells and activated mast cells; costimulator for B cells; causes plasma cells to secrete IgA antibodies.
Tumor necrosis factor (TNF)	Produced mainly by macrophages; stimulates accumulation of neutrophils and macrophages at sites of inflammation and stimulates their killing of microbes; stimulates macrophages to produce IL-1; induces synthesis of colony-stimulating factors by endothelial cells and fibroblasts; exerts an interferon-like protective effect against viruses; and functions as an endogenous pyrogen to induce fever.
Transforming growth factor beta (TGF-β)	Secreted by T cells and macrophages; has some positive effects but is thought to be important for turning off immune responses; inhibits proliferation of T cells and activation of macrophages.
Gamma-interferon (γ-IFN)	Secreted by helper and cytotoxic T cells and NK cells; strongly stimulates phagocytosis by neutrophils and macrophages; activates NK cells; enhances both cell-mediated and antibody-mediated immune responses.
Alpha- and beta-interferons (α-IFN and β-IFN)	Produced by virus-infected cells to inhibit viral replication in uninfected cells; produced by antigen-stimulated macrophages to stimulate T cell growth; activate NK cells, inhibit cell growth, and suppress formation of some tumors.
Lymphotoxin (LT)	Secreted by cytotoxic T cells; kills cells by causing fragmentation of DNA.
Perforin	Secreted by cytotoxic T cells and perhaps by NK cells; perforates cell membranes of target cells, which causes cytolysis.
Macrophage migration inhibiting factor	Produced by T cells; prevents macrophages from leaving site of infection.

CELL-MEDIATED IMMUNITY

• **Describe the steps of a cell-mediated immune response.**

A cell-mediated immune response begins with *activation* of a small number of T cells by a specific antigen. Once a T cell has been activated, it undergoes *proliferation* and *differentiation* into a clone of **effector cells,** a population of identical cells that can recognize the same antigen and carry out some aspect of the immune attack. Finally, the immune response results in *elimination* of the intruder.

Activation, Proliferation, and Differentiation of T Cells

Antigen receptors on the surface of T cells, called **T-cell receptors (TCRs),** recognize and bind to specific foreign antigen fragments that are presented in antigen–MHC complexes. There are millions of different T cells; each has its own unique TCRs that can recognize a specific antigen–MHC complex. At any given time, most T cells are inactive. When an antigen enters the body, only a few T cells have TCRs that can recognize and bind to the antigen. Antigen recognition by a TCR is the *first signal* in activation of a T cell.

A T cell becomes activated only if it binds to the foreign antigen and at the same time receives a *second signal,* a process known as **costimulation.** Of the more than 20 known costimulators, some are cytokines, such as **interleukin-2.** Other costimulators include pairs of plasma membrane molecules, one on the surface of the T cell and a second on the surface of an antigen-presenting cell, that enable the two cells to adhere to one another for a period of time.

The need for two signals is a little like starting and driving a car: When you insert the correct key (antigen) in the ignition (TCR) and turn it, the car starts (recognition of specific antigen), but it cannot move forward until you move the gear shift into drive (costimulation). The need for costimulation may prevent immune responses from occurring accidentally. Different co-stimulators affect the activated T cell in different ways, just as shifting a car into reverse has a different effect than shifting it into drive. Moreover, recognition (antigen binding to a receptor) without costimulation leads to a prolonged *state of inactivity* called **anergy** in both T cells and B cells. Anergy is rather like a car in neutral with its engine running until it's out of gas!

Once a T cell has received two signals (antigen recognition and costimulation), it is **activated.** An activated T cell enlarges and begins to **proliferate** (divide several times) and to **differentiate** (form more highly specialized cells). The result is a population of identical cells, termed a **clone,** that can recognize the same specific antigen. Before the first exposure to a given antigen, only a few T cells might be able to recognize it, but once an immune response has begun, there are thousands. Activation,

proliferation, and differentiation of T cells occur in the secondary lymphatic organs and tissues. If you have ever noticed swollen tonsils or lymph nodes in your neck, the proliferation of lymphocytes participating in an immune response was likely the cause.

Types of T Cells

The three main types of differentiated T cells are helper T cells, cytotoxic T cells, and memory T cells.

Helper T Cells

Most T cells that display CD4 develop into **helper T (T$_H$) cells,** also known as **CD4$^+$ T cells.** Resting (inactive) T$_H$ cells recognize antigen fragments associated with major histocompatibility complex class II (MHC-II) molecules at the surface of an antigen-presenting cell (Figure 22.14a). Several different pairs of molecules, one attached to the membrane of the APC and the other attached to the membrane of the helper T cell, interact with each other to hold the two cells tightly together and achieve costimulation.

Within hours after costimulation, helper T cells start secreting a variety of cytokines (see Table 22.2). Different subsets of T$_H$ cells specialize in the production of particular cytokines. One very important cytokine produced by T$_H$ cells is interleukin-2 (IL-2), which is needed for virtually all immune responses and is the prime trigger of T cell proliferation. IL-2 can act as a costimulator for resting T$_H$ or cytotoxic T cells, and it enhances activation and proliferation of T cells, B cells, and natural killer cells.

Some actions of interleukin-2 provide a good example of a beneficial positive feedback system. As noted earlier, activation of a helper T cell stimulates it to start secreting IL-2, which then acts in an autocrine manner by binding to IL-2 receptors on the plasma membrane of the cell that secreted it. One effect is stimulation of cell division. As the T$_H$ cells proliferate, a positive feedback effect occurs because they secrete more IL-2, which causes further cell division. IL-2 may also act in a paracrine manner by binding to IL-2 receptors on neighboring T$_H$ cells, cytotoxic T cells, or B cells. If any of these cells have already bound an antigen, IL-2 serves as a costimulator to activate them.

Cytotoxic T Cells

T cells that display CD8 develop into **cytotoxic T (T$_C$) cells,** also termed **CD8$^+$ cells.** T$_C$ cells recognize foreign antigens combined with major histocompatibility complex class I (MHC-I) molecules on the surfaces of (1) body cells infected by viruses, (2) some tumor cells, and (3) cells of a tissue transplant (Figure 22.14b). However, to become cytolytic (able to lyse cells) T$_C$ cells need costimulation by interleukin-2 or other cytokines produced by helper T cells. (Recall that T$_H$ cells are activated by antigen associated with MHC-II molecules.) Thus, maximal activation of T$_C$ cells requires presentation of antigen associated with both MHC-I and MHC-II molecules.

Figure 22.14   Activation, proliferation, and differentiation of T cells.

The binding of CD4 to MHC-II and CD8 to MHC-I helps anchor the T-cell receptor (TCR)–antigen interaction so that antigen recognition can occur.

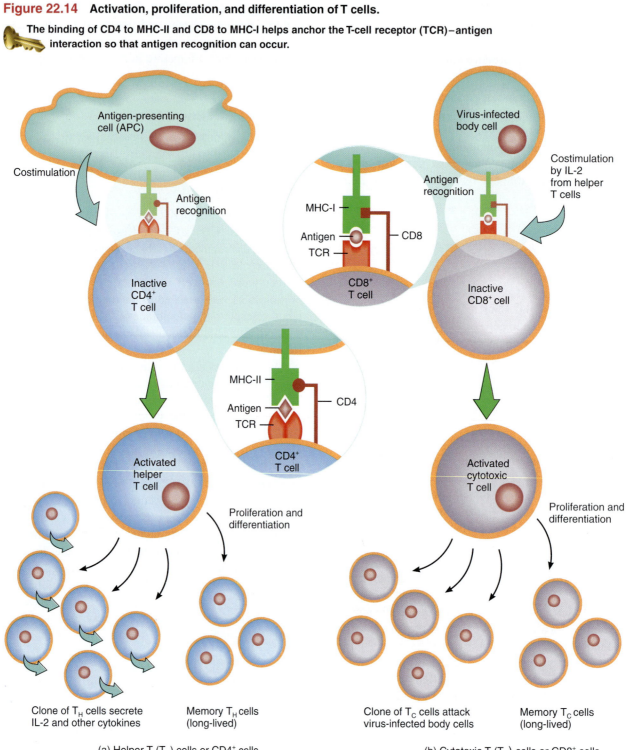

Antigen-presenting cell (APC)

Costimulation

Antigen recognition

Inactive CD4⁺ T cell

MHC-I

Antigen

TCR

CD8

CD8⁺ T cell

Virus-infected body cell

Antigen recognition

Costimulation by IL-2 from helper T cells

Inactive CD8⁺ cell

MHC-II

Antigen

TCR

CD4

CD4⁺ T cell

Activated helper T cell

Proliferation and differentiation

Activated cytotoxic T cell

Proliferation and differentiation

Clone of T_H cells secrete IL-2 and other cytokines

Memory T_H cells (long-lived)

Clone of T_C cells attack virus-infected body cells

Memory T_C cells (long-lived)

(a) Helper T (T_H) cells or CD4⁺ cells

(b) Cytotoxic T (T_C) cells or CD8⁺ cells

What are the first and second signals in activation of a T cell?

Memory T Cells

T cells that remain from a proliferated clone after a cell-mediated immune response are termed **memory T cells.** Should a pathogen bearing the same foreign antigen invade the body later, thousands of memory cells are available to initiate a far swifter reaction than occurred during the first invasion. The second response usually is so fast and so vigorous that the pathogens are destroyed before any signs or symptoms of disease can occur.

Elimination of Invaders

Cytotoxic T cells are the soldiers that march forth to do battle with foreign invaders in cell-mediated immune responses. They leave secondary lymphatic organs and tissues and migrate to the site of invasion, infection, or tumor formation. They recognize and attach to target cells bearing the antigen that stimulated activation and proliferation of their progenitor cells. Then, the cytotoxic T cells deliver a "lethal hit" that kills the target cells without damaging themselves (Figure 22.15). After detaching from a target cell, a cytotoxic T cell can seek out and destroy another invader that displays the same antigen.

Cytotoxic T cells use two killing mechanisms. In the first, granules containing the protein perforin undergo exocytosis from the cytotoxic T cell. **Perforin** forms holes in the plasma membrane of the target cell, which allow extracellular fluid to flow in, and the cell bursts—a process known as **cytolysis.** In the second mechanism, the cytotoxic T cell secretes a toxic molecule known as **lymphotoxin** that activates enzymes within the target cell. These enzymes cause the target cell's DNA to fragment, and the cell dies. In these ways, cytotoxic T cells destroy infected cells. Also, cytotoxic T cells secrete gamma-interferon, which activates phagocytic cells at the scene of the battle. Cytotoxic T cells are especially effective against slowly developing bacterial diseases (such as tuberculosis and brucellosis), some viruses, fungi, cancer cells associated with viral infection, and transplanted cells.

Immunological Surveillance

When a normal cell transforms into a cancerous cell, it often displays novel cell surface components called **tumor antigens.** These molecules are rarely, if ever, displayed on the surface of normal cells. If the immune system recognizes a tumor antigen as nonself, it can destroy the cancer cells carrying them. Such immune responses, called **immunological surveillance,** are carried out by cytotoxic T cells, macrophages, and natural killer cells. Immunological surveillance is most effective in eliminating tumor cells due to cancer-causing viruses. For this reason, transplant recipients who are taking immunosuppressive drugs to prevent transplant rejection have an increased incidence of virus-associated cancers. Their risk for other types of cancer is not increased.

Figure 22.15 Activity of cytotoxic T cells. After delivering a "lethal hit," a cytotoxic T cell can detach and attack another target cell displaying the same antigen.

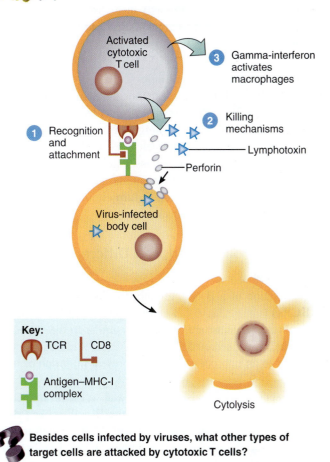

Cytotoxic T cells kill microbes directly by secreting perforin and lymphotoxin.

Besides cells infected by viruses, what other types of target cells are attacked by cytotoxic T cells?

Graft Rejection

Organ transplantation involves the replacement of an injured or diseased organ, such as the heart, liver, kidney, lungs, or pancreas, with an organ donated by another individual. Usually, the immune system recognizes the proteins in the transplanted organ as foreign and mounts both cell-mediated and antibody-mediated immune responses against them. This phenomenon is known as **graft rejection.** The more closely matched are the major histocompatibility complex proteins of the donor and recipient, the weaker is the graft rejection response. To reduce the risk of graft rejection, organ transplant recipients receive immunosuppressive drugs. One such drug is *cyclosporine,* derived from a fungus, which inhibits secretion of interleukin-2 by helper T cells but has only a minimal effect on B cells. Thus, the risk of rejection is diminished while maintaining resistance to some diseases. ■

▶ **CHECKPOINT**

17. What are the functions of helper, cytotoxic, and memory T cells?

18. How do cytotoxic T cells kill their targets?

19. How is immunological surveillance useful?

ANTIBODY-MEDIATED IMMUNITY

▶ **O B J E C T I V E S**

- **Describe the steps in an antibody-mediated immune response.**
- **Describe the chemical characteristics and actions of antibodies.**

The body contains not only millions of different T cells but also millions of different B cells, each capable of responding to a specific antigen. Whereas cytotoxic T cells leave lymphatic tissues to seek out and destroy a foreign antigen, B cells stay put. In the presence of a foreign antigen, specific B cells in lymph nodes, the spleen, or mucosa-associated lymphatic tissue become activated. They then differentiate into plasma cells that secrete specific antibodies, which in turn circulate in the lymph and blood to reach the site of invasion.

Activation, Proliferation, and Differentiation of B Cells

During activation of a B cell, an antigen binds to **B-cell receptors (BCRs)** (Figure 22.16). B-cell receptors, which are integral transmembrane proteins, are chemically similar to the antibodies that eventually are secreted by plasma cells. Although B cells can respond to an unprocessed antigen present in lymph or interstitial fluid, their response is much more intense when nearby antigen-presenting cells also process and present antigen to them. Some antigen is then taken into the B cell, broken down into peptide fragments and combined with MHC-II self-antigen, and moved to the B cell plasma membrane. Helper T cells recognize the antigen–MHC-II complex and deliver the costimulation needed for B cell proliferation and differentiation. The helper T cell produces interleukin-2 and other cytokines that function as costimulators to activate B cells. Interleukin-4 and interleukin-6, also produced by helper T cells, enhance B cell proliferation, B cell differentiation into plasma cells, and secretion of antibodies by plasma cells.

Some of the activated B cells enlarge, divide, and differentiate into a clone of antibody-secreting **plasma cells.** A few days after exposure to an antigen, a plasma cell is secreting hundreds of millions of antibodies daily and secretion occurs for about 4 or 5 days, until the plasma cell dies. Most antibodies travel in lymph and blood to the invasion site. Some activated B cells do not differentiate into plasma cells but rather remain as **memory B cells** that are ready to respond more rapidly and forcefully should the same antigen reappear at a future time.

Different antigens stimulate different B cells to develop into plasma cells and their accompanying memory B cells. The B cells of a particular clone are capable of secreting only one kind of antibody, which is identical in specificity to the antigen receptor displayed by the B cell that first responded to the antigen. Each specific antigen activates only those B cells that are predestined (by the combination of gene segments they carry) to secrete antibody specific to that antigen. Antibodies produced by a clone of plasma cells enter the circulation and form

Figure 22.16 Activation, proliferation, and differentiation of B cells into plasma cells and memory cells. Plasma cells are actually much larger than B cells.

🔑 **Plasma cells secrete antibodies.**

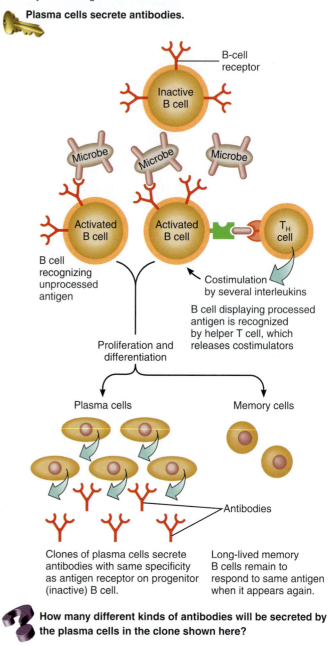

How many different kinds of antibodies will be secreted by the plasma cells in the clone shown here?

antigen–antibody complexes with the antigen that initiated their production.

Antibodies

An **antibody (Ab)** can combine specifically with the epitope on the antigen that triggered its production. The antibody's structure matches its antigen much as a lock fits a specific key. In theory, plasma cells could secrete as many different antibodies as there are different B-cell receptors because the same recombined gene segments code for both the BCR and the antibodies eventually secreted by plasma cells.

Antibody Structure

Antibodies belong to a group of glycoproteins called globulins, and for this reason they are also known as **immunoglobulins (Igs).** Most antibodies contain four polypeptide chains (Figure 22.17). Two of the chains are identical to each other and are called **heavy (H) chains;** each consists of about 450 amino acids. Short carbohydrate chains are attached to each heavy polypeptide chain. The two other polypeptide chains, also identical to each other, are called **light (L) chains,** and each consists of about 220 amino acids. A disulfide bond (S–S) holds each light chain to a heavy chain. Two disulfide bonds also link the midregion of the two heavy chains; this part of the antibody displays considerable flexibility and is called the **hinge region.** Because the antibody "arms" can move somewhat as the hinge region bends, an antibody can assume either a T shape (Figure 22.17a) or a Y shape (Figure 22.17b). Beyond the hinge region, parts of the two heavy chains form the **stem region.**

Within each H and L chain are two distinct regions. The tips of the H and L chains, called the **variable (V) regions,** constitute the **antigen-binding site.** The variable region, which is different for each kind of antibody, is the part of the antibody that recognizes and attaches specifically to a particular antigen. Because most antibodies have two antigen binding sites, they are said to be bivalent. Flexibility at the hinge allows the antibody to simultaneously bind to two epitopes that are some distance apart—for example, on the surface of a microbe.

The remainder of each H and L chain, called the **constant (C) region,** is nearly the same in all antibodies of the same class and is responsible for the type of antigen–antibody reaction that occurs. However, the constant region of the H chain differs from one class of antibody to another, and its structure serves as a basis for distinguishing five different classes, designated IgG, IgA, IgM, IgD, and IgE. Each class has a distinct chemical structure and a specific biological role. Because they appear first and are relatively short-lived, IgM antibodies indicate a recent invasion. In a sick patient, the responsible pathogen may be suggested by the presence of high levels of IgM specific to a particular organism. Resistance of the fetus and newborn baby to infection stems mainly from maternal IgG antibodies that cross the placenta before birth and IgA antibodies in breast milk after birth. Table 22.3 summarizes the structures and functions of the five classes of antibodies.

Antibody Actions

Even though the five classes of immunoglobulins have actions that differ somewhat, all act in some way to disable antigens. Actions of antibodies include:

- **Neutralizing antigen.** The reaction of antibody with antigen blocks or neutralizes some bacterial toxins and prevents attachment of some viruses to body cells.

- **Immobilizing bacteria.** If antibodies form against antigens on the cilia or flagella of motile bacteria, the antigen–antibody reaction may cause the bacteria to lose their motility, which limits their spread into nearby tissues.

- **Agglutinating and precipitating antigen.** Because antibodies have two or more sites for binding to antigen, the antigen–antibody reaction may cross-link pathogens to one another, causing agglutination (clumping together). Phagocytic cells ingest agglutinated microbes more readily.

Figure 22.17 Chemical structure of the immunoglobulin G (IgG) class of antibody. Each molecule is composed of four polypeptide chains (two heavy and two light) plus a short carbohydrate chain attached to each heavy chain. In (a), each circle represents one amino acid. In (b), V_L = variable regions of light chain, C_L = constant region of light chain, V_H = variable region of heavy chain, and C_H = constant region of heavy chain.

An antibody combines only with the epitope on the antigen that triggered its production.

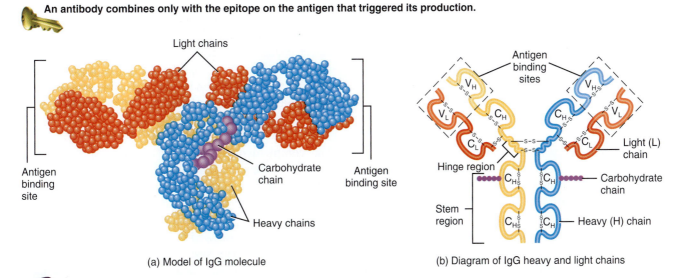

(a) Model of IgG molecule

(b) Diagram of IgG heavy and light chains

What is the function of the variable regions in an antibody molecule?

Table 22.3	Classes of Immunoglobulins (Igs)
Name and Structure	**Characteristics and Functions**
IgG	Most abundant, about 80% of all antibodies in the blood; found in blood, lymph, and the intestines; monomer (one-unit) structure. Protects against bacteria and viruses by enhancing phagocytosis, neutralizing toxins, and triggering the complement system. It is the only class of antibody to cross the placenta from mother to fetus, thereby conferring considerable immune protection in newborns.
IgA	Found mainly in sweat, tears, saliva, mucus, breast milk and gastrointestinal secretions. Smaller quantities are present in blood and lymph. Makes up 10–15% of all antibodies in the blood; occurs as monomers and dimers (two units). Levels decrease during stress, lowering resistance to infection. Provides localized protection on mucous membranes against bacteria and viruses.
IgM	About 5–10% of all antibodies in the blood; also found in lymph. Occurs as pentamers (five units); first antibody class to be secreted by plasma cells after an initial exposure to any antigen. Activates complement and causes agglutination and lysis of microbes. Also present as monomers on the surfaces of B cells, where they serve as antigen receptors. In blood plasma, the anti-A and anti-B antibodies of the ABO blood group, which bind to A and B antigens during incompatible blood transfusions, are also IgM antibodies (see Figure 19.12 on page 652).
IgD	Mainly found on the surfaces of B cells as antigen receptors, where it occurs as monomers; involved in activation of B cells. About 0.2% of all antibodies in the blood.
IgE	Less than 0.1% of all antibodies in the blood; occurs as monomers; located on mast cells and basophils. Involved in allergic and hypersensitivity reactions; provides protection against parasitic worms.

Likewise, soluble antigens may come out of solution and form a more-easily phagocytized precipitate when cross-linked by antibodies.

- *Activating complement.* Antigen–antibody complexes initiate the classical pathway of the complement system (discussed shortly).

- *Enhancing phagocytosis.* The stem region of an antibody acts as a "flag" that attracts phagocytes once antigens have bound to the antibody's variable region. Antibodies enhance the activity of phagocytes by causing agglutination and precipitation, by activating complement, and by coating microbes so that they are more susceptible to phagocytosis.

Monoclonal Antibodies

The antibodies produced against a given antigen by plasma cells can be harvested from an individual's blood. However, because an antigen typically has many epitopes, several different clones of plasma cells produce different antibodies against the antigen. If a single plasma cell could be isolated and induced to proliferate into a clone of identical plasma cells, then a large quantity of identical antibodies could be produced. Unfortunately, lymphocytes and plasma cells are difficult to grow in culture, so scien-

tists sidestepped this difficulty by fusing B cells with tumor cells that grow easily and proliferate endlessly. The resulting hybrid cell is called a **hybridoma** (hī-bri-DŌ-ma). Hybridomas are long-term sources of large quantities of pure, identical antibodies, called **monoclonal antibodies (MAbs)** because they come from a single clone of identical cells. One clinical use of monoclonal antibodies is for measuring levels of a drug in a patient's blood. Other uses include the diagnosis of strep throat, pregnancy, allergies, and diseases such as hepatitis, rabies, and some sexually transmitted diseases. MAbs have also been used to detect cancer at an early stage and to ascertain the extent of metastasis. They may also be useful in preparing vaccines to counteract the rejection associated with transplants, to treat autoimmune diseases, and perhaps to treat AIDS. ■

Role of the Complement System in Immunity

The **complement system** is a defensive system consisting of plasma proteins that attack and destroy microbes. The system can be activated via two pathways (classical and alternative), both of which are an ordered sequence or cascade of reactions. Both pathways lead to the same events: inflammation, enhancement of phagocytosis, and bursting of microbes.

The complement system consists of more than 30 different proteins in blood plasma and in plasma membranes. Included are proteins called C1 through C9 (the C stands for complement) and proteins called factors B, D, and P (properdin). The **classical pathway,** which is so-named because it was discovered first, begins by the binding of antibody to antigen (Figure 22.18). The antigen could be a bacterium or other foreign cell. The antigen–antibody attachment activates complement protein C1 through binding to the stem region of the antibody. In the ensuing cascade of enzymatic reactions, inactive complement proteins are split into active fragments termed C2a, C2b, and so forth. The **alternative pathway** does not involve antibodies. It begins by the interaction between polysaccharides on the surface of a microbe and factors B, D, and P (Figure 22.18). This interaction activates complement protein C3 and the cascade begins.

The consequences of activation of the classical and alternative pathways are as follows:

❶ Activation of inflammation. Some complement proteins (C3a, C4a, and C5a) contribute to the development of inflammation: They dilate arterioles, which increases blood flow to the area, and cause the release of histamine from mast cells, basophils, and platelets. Because histamine increases the permeability of blood capillaries, white blood cells can more easily move into tissues to combat infection or allergy. Other complement proteins serve as chemotactic agents that attract phagocytes to the site of microbial invasion.

❷ Opsonization. Complement fragment C3b binds to the surface of a microbe and then interacts with receptors on phagocytes, which promotes phagocytosis. Making microbes more susceptible to phagocytosis is known as **opsonization** (op-sō-ni-ZĀ-shun).

❸ Cytolysis. Several complement proteins (C5b, C6, C7, C8, and C9) come together to form a **membrane attack complex (MAC)** that inserts into the plasma membrane of the microbe, forming large holes. The ensuing leakiness allows fluid to flow into the cell's interior, and the microbe swells and bursts (cytolysis).

Figure 22.18   The classical and alternative pathways of the complement system.

When activated, complement proteins enhance certain immune, allergic, and inflammatory reactions.

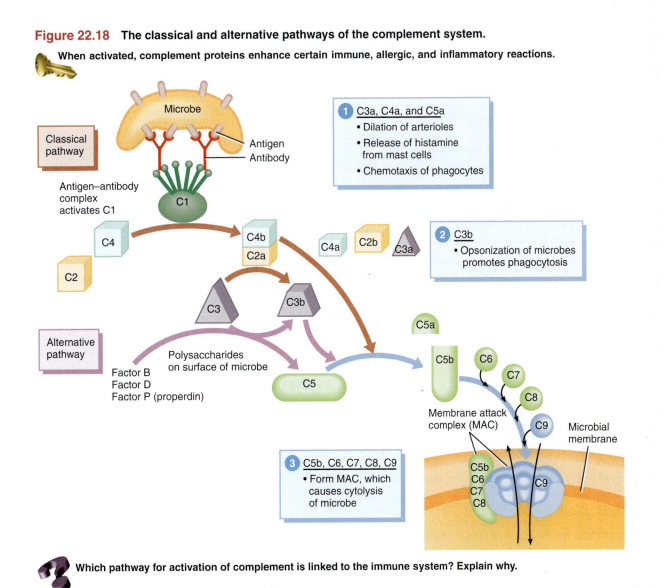

Which pathway for activation of complement is linked to the immune system? Explain why.

Immunological Memory

A hallmark of immune responses is memory for specific antigens that have triggered immune responses in the past. Immunological memory is due to the presence of long-lasting antibodies and very long-lived lymphocytes that arise during proliferation and differentiation of antigen-stimulated B cells and T cells.

Immune responses, whether cell-mediated or antibody-mediated, are much quicker and more intense after a second or subsequent exposure to an antigen than after the first exposure. Initially, only a few cells have the correct specificity to respond, and the immune response may take several days to build to maximum intensity. Because thousands of memory cells exist after an initial encounter with an antigen, the next time the same antigen appears they can proliferate and differentiate into plasma cells or cytotoxic T cells within hours.

One measure of immunological memory is the amount of antibody in serum, called the *antibody titer* (TĪ-ter). After an initial contact with an antigen, no antibodies are present for a period of several days. Then, a slow rise in the antibody titer occurs, first IgM and then IgG, followed by a gradual decline in antibody titer (Figure 22.19). This is the **primary response.**

Memory cells may remain for decades. Every new encounter with the same antigen results in a rapid proliferation of memory cells. The antibody titer after subsequent encounters is far greater than during a primary response and consists mainly of IgG antibodies. This accelerated, more intense response is called the **secondary response.** Antibodies produced during a secondary response have an even higher affinity for the antigen than those produced during a primary response, and thus they are more successful in disposing of it.

Primary and secondary responses occur during microbial infection. When you recover from an infection without taking antimicrobial drugs, it is usually because of the primary response. If the same microbe infects you later, the secondary response could be so swift that the microbes are destroyed before you exhibit any signs or symptoms of infection.

Immunological memory provides the basis for immunization by vaccination against certain diseases (for example, polio). When you receive the vaccine, which may contain weakened or killed whole microbes or portions of microbes, your B cells and T cells are activated. Should you subsequently encounter the living pathogen as an infecting microbe, your body initiates a secondary response. Table 22.4 summarizes the various types of antigen encounters that provide naturally and artificially acquired immunity.

▶ **C H E C K P O I N T**

20. How do the five classes of antibodies differ in structure and function?

21. How are cell-mediated and antibody-mediated immune responses similar and different?

22. In what ways does the complement system augment antibody-mediated immune responses?

23. How is the secondary response to an antigen different from the primary response?

Figure 22.19 Production of antibodies in the primary (after first exposure) and secondary (after second exposure) responses to a given antigen.

🔑 Immunological memory is the basis for successful immunization by vaccination.

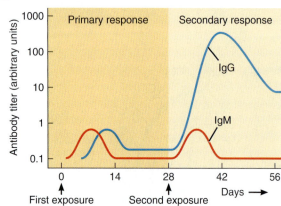

According to this graph, how much more IgG is circulating in the blood in the secondary response than in the primary response? (*Hint:* Notice that each mark on the antibody titer axis represents a 10-fold increase.)

Table 22.4	Types of Immunity
Type of Immunity	**How Acquired**
Naturally acquired active immunity	Antigen recognition by B cells and T cells and costimulation lead to antibody-secreting plasma cells, cytotoxic T cells, and B and T memory cells.
Naturally acquired passive immunity	Transfer of IgG antibodies from mother to fetus across placenta, or of IgA antibodies from mother to baby in milk during breast-feeding.
Artificially acquired active immunity	Antigens introduced during a vaccination stimulate cell-mediated and antibody-mediated immune responses, leading to production of memory cells. The antigens are pretreated to be immunogenic but not pathogenic; that is, they will trigger an immune response but not cause significant illness.
Artificially acquired passive immunity	Intravenous injection of immunoglobulins (antibodies).

SELF-RECOGNITION AND SELF-TOLERANCE

▶ O B J E C T I V E

• Describe how self-recognition and self-tolerance develop.

To function properly, your T cells must have two traits: (1) They must be able to *recognize* your own major histocompatibility complex (MHC) proteins, a process known as **self-recognition,** and (2) they must *lack reactivity* to peptide fragments from your own proteins, a condition known as **self-tolerance** (Figure 22.20). B cells also display self-tolerance. Loss of self-tolerance leads to the development of autoimmune diseases (see page 799).

Pre-T cells in the thymus develop the capability for self-recognition via **positive selection** (Figure 22.20a). In this process, some pre-T cells express T-cell receptors (TCRs) that interact with self-MHC proteins on epithelial cells in the thymic cortex. Because of this interaction, the T cells can recognize the MHC part of an antigen–MHC complex. These T cells survive. Other immature T cells that fail to interact with thymic epithelial cells are not able to recognize self-MHC proteins. These cells undergo apoptosis.

The development of self-tolerance, in contrast, occurs by a weeding-out process called **negative selection** in which the T cells interact with dendritic cells located at the junction of the cortex and medulla in the thymus. In this process, T cells with receptors that recognize self-peptide fragments or other self-

Figure 22.20 Development of self-recognition and self-tolerance. MHC = major histocompatibility complex. TCR = T-cell receptor.

Positive selection allows recognition of self-MHC proteins; negative selection provides self-tolerance of your own peptides and other self-antigens.

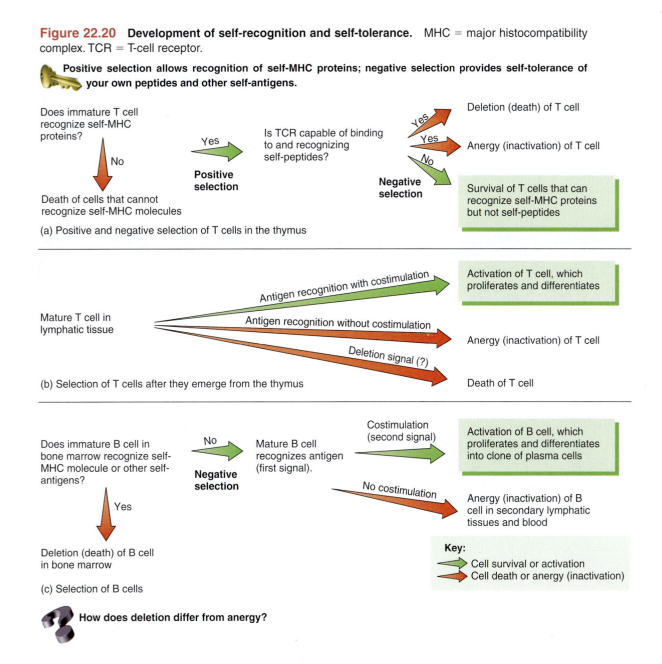

(a) Positive and negative selection of T cells in the thymus

(b) Selection of T cells after they emerge from the thymus

(c) Selection of B cells

How does deletion differ from anergy?

antigens are eliminated or inactivated (Figure 22.20a). The T cells selected to survive do not respond to self-antigens, the fragments of molecules that are normally present in the body. Negative selection occurs via both deletion and anergy. In **deletion,** self-reactive T cells undergo apoptosis and die, whereas in **anergy** they remain alive but are unresponsive to antigenic stimulation. Only 1–5% of the immature T cells in the thymus receive the proper signals to survive apoptosis during both positive and negative selection and emerge as mature, immunocompetent T cells.

Once T cells have emerged from the thymus, they may still encounter an unfamiliar self-protein; in such cases they may also become anergic if there is no costimulator (Figure 22.20b). Deletion of self-reactive T cells may also occur after they leave the thymus.

B cells also develop tolerance through deletion and anergy (Figure 22.20c). While B cells are developing in bone marrow, those cells exhibiting antigen receptors that recognize common self-antigens (such as MHC proteins or blood group antigens) are deleted. Once B cells are released into the blood, however, anergy appears to be the main mechanism for preventing responses to self-proteins. When B cells encounter an antigen not associated with an antigen-presenting cell, the necessary costimulation signal often is missing. In this case, the B cell is likely to become anergic (inactivated) rather than activated.

Table 22.5 summarizes the activities of cells involved in immune responses.

Cancer Immunology

Although the immune system responds to cancerous cells, for example, in tumor surveillance, often immunity provides inadequate protection, as evidenced by the number of people dying each year from cancer. Over the past 25 years, considerable research has focused on *cancer immunology,* the study of ways to use immune responses for detecting, monitoring, and treating cancer. For example, some tumors of the colon release *carcinoembryonic antigen (CEA)* into the blood and prostate cancer cells release *prostate-specific antigen (PSA).* Detecting these antigens in blood does not provide definitive diagnosis of cancer, however, because both antigens are also released in certain noncancerous conditions. Still, high levels of cancer-related antigens in the blood often do indicate the presence of a malignant tumor.

Finding ways to induce our immune system to mount vigorous attacks against cancerous cells has been an elusive goal. Many different techniques have been tried, with only modest success. In one method, inactive lymphocytes are removed in a blood sample and cultured with interleukin-2. The resulting *lymphokine-activated killer (LAK) cells* are then transfused back into the patient's blood. Although LAK cells have produced dramatic improvement in a few cases, severe complications affect most patients. In another method, lymphocytes procured from a small biopsy sample of a tumor are cultured with interleukin-2. After their proliferation in culture, such *tumor-infiltrating lymphocytes (TILs)* are reinjected. About a quarter of patients with

Table 22.5	Summary of Functions of Cells Participating in Immune Responses
Cell	**Functions**
Antigen-Presenting Cells (APCs)	
Macrophage	Phagocytosis; processing and presentation of foreign antigens to T cells; secretion of interleukin-1, which stimulates secretion of interleukin-2 by helper T cells and induces proliferation of B cells; secretion of interferons that stimulate T cell growth.
Dendritic cell	Processes and presents antigen to T cells and B cells; found in mucous membranes, skin, and lymph nodes.
B cell	Processes and presents antigen to helper T cells.
Lymphocytes	
Cytotoxic T (T_C) cell or CD8$^+$ T cell	Causes lysis and death of foreign cells by releasing perforin and lymphotoxin; releases other cytokines that attract macrophages and increase their phagocytic activity (gamma-IFN) and prevent macrophage migration from site of action (macrophage migration inhibition factor).
Helper T (T_H) cell or CD4$^+$ T cell	Cooperates with B cells to amplify antibody production by plasma cells and secretes interleukin-2, which stimulates proliferation of T cells and B cells. May secrete gamma-IFN and tumor necrosis factor (TNF), which stimulate inflammatory response.
Memory T cell	Remains in lymphatic tissue and recognizes original invading antigens, even years after the first encounter.
B cell	Differentiates into antibody-producing plasma cell.
Plasma cell	Descendant of B cell that produces and secretes antibodies.
Memory B cell	Descendent of B cell that remains after an immune response and is ready to respond rapidly and forcefully should the same antigen enter the body in the future.

malignant melanoma and renal-cell carcinoma who received TIL therapy showed significant improvement. The many studies currently underway provide reason to hope that immune-based methods will eventually lead to cures for cancer. ■

STRESS AND IMMUNITY

▶ **OBJECTIVE**

• **Describe the effects of stress on immunity.**

The field of **psychoneuroimmunology (PNI)** deals with communication pathways that link the nervous, endocrine, and immune systems. PNI research appears to justify what people have long observed: Your thoughts, feelings, moods, and beliefs influence your level of health and the course of disease. For example, cortisol, a hormone secreted by the adrenal cortex in association with the stress response, inhibits immune system activity.

If you want to observe the relationship between lifestyle and immune function, visit a college campus. As the semester progresses and the workload accumulates, an increasing number of students can be found in the waiting rooms of student health services. When work and stress pile up, health habits can change. Many people smoke or consume more alcohol when stressed, two habits detrimental to optimal immune function. Under stress, people are less likely to eat well or exercise regularly, two habits that enhance immunity.

People resistant to the negative health effects of stress are more likely to experience a sense of control over the future, a commitment to their work, expectations of generally positive outcomes for themselves, and feelings of social support. To increase your stress resistance, cultivate an optimistic outlook, get involved in your work, and build good relationships with others.

Adequate sleep and relaxation are especially important for a healthy immune system. But when there aren't enough hours in the day, you may be tempted to steal some from the night. While skipping sleep may give you a few more hours of productive time in the short run, in the long run you end up even farther behind, especially if getting sick keeps you out of commission for several days, blurs your concentration, and blocks your creativity.

Even if you make time to get eight hours of sleep, stress can cause insomnia. If you find yourself tossing and turning at night, it's time to improve your stress management and relaxation skills! Be sure to unwind from the day before going to bed.

▶ **CHECKPOINT**

24. What do positive selection, negative selection, and anergy accomplish?

25. Have you ever observed a connection between stress and illness in your own life?

AGING AND THE IMMUNE SYSTEM

▶ **OBJECTIVE**

• **Describe the effects of aging on the immune system.**

With advancing age, most people become more susceptible to all types of infections and malignancies. Their response to vaccines is decreased, and they tend to produce more autoantibodies (antibodies against their body's own molecules). In addition, the immune system exhibits lowered levels of function. For example, T cells become less responsive to antigens, and fewer T cells respond to infections. This may result from age-related atrophy of the thymus or decreased production of thymic hormones. Because the T cell population decreases with age, B cells are also less responsive. Consequently, antibody levels do not increase as rapidly in response to a challenge by an antigen, resulting in increased susceptibility to various infections. It is for this key reason that elderly individuals are encouraged to get influenza (flu) vaccinations each year.

• • •

To appreciate the many ways that the lymphatic and immune system contributes to homeostasis of other body systems, examine *Focus on Homeostasis: The Lymphatic and Immune System* on page 796.

Next, in Chapter 23, we will explore the structure and function of the respiratory system and see how its operation is regulated by the nervous system. Most importantly, the respiratory system provides for gas exchange—taking in oxygen and blowing off carbon dioxide. The cardiovascular system aids gas exchange by transporting blood containing the gases between the lungs and tissue cells.

Focus on Homeostasis:

The Lymphatic and Immune System

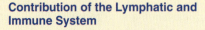

Body System	Contribution of the Lymphatic and Immune System
For all body systems 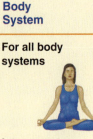	B cells, T cells, and antibodies protect all body systems from attack by harmful foreign invaders (pathogens), foreign cells, and cancer cells.
Integumentary system	Lymphatic vessels drain excess interstitial fluid and leaked plasma proteins from dermis of skin. Immune system cells (Langerhans cells) in skin help protect skin. Lymphatic tissue provides IgA antibodies in sweat.
Skeletal system	Lymphatic vessels drain excess interstitial fluid and leaked plasma proteins from connective tissue around bones.
Muscular system	Lymphatic vessels drain excess interstitial fluid and leaked plasma proteins from muscles.
Endocrine system	Flow of lymph helps distribute some hormones and cytokines. Lymphatic vessels drain excess interstitial fluid and leaked plasma proteins from endocrine glands.
Cardiovascular system	Lymph returns excess fluid filtered from blood capillaries and leaked plasma proteins to venous blood. Macrophages in spleen destroy aged red blood cells and remove debris in blood.
Respiratory system	Tonsils, alveolar macrophages, and MALT (mucosa-associated lymphatic tissue) help protect lungs from pathogens. Lymphatic vessels drain excess interstitial fluid from lungs.
Digestive system	Tonsils and MALT help defend against toxins and pathogens that penetrate the body from the gastrointestinal tract. Digestive system provides IgA antibodies in saliva and gastrointestinal secretions. Lymphatic vessels pick up absorbed dietary lipids and fat-soluble vitamins from the small intestine and transport them to the blood. Lymphatic vessels drain excess interstitial fluid and leaked plasma proteins from organs of the digestive system.
Urinary system	Lymphatic vessels drain excess interstitial fluid and leaked plasma proteins from organs of the urinary system. MALT helps defend against toxins and pathogens that penetrate the body via the urethra.
Reproductive systems	Lymphatic vessels drain excess interstitial fluid and leaked plasma proteins from organs of the reproductive system. MALT helps defend against toxins and pathogens that penetrate the body via the vagina and penis. In females, sperm deposited in the vagina are not attacked as foreign invaders due to inhibition of immune responses. IgG antibodies can cross the placenta to provide protection to a developing fetus. Lymphatic tissue provides IgA antibodies in milk of nursing mother.

DISORDERS: HOMEOSTATIC IMBALANCES

AIDS: Acquired Immunodeficiency Syndrome

Acquired immunodeficiency syndrome is a condition in which a person experiences a telltale assortment of infections due to the progressive destruction of immune system cells by the **human immunodeficiency virus (HIV).** As we will see, AIDS represents the end stage of infection by HIV. A person who is infected with HIV may be symptom free for many years, even while the virus is actively attacking the immune system. HIV infection is serious and usually fatal because it takes control of and destroys the very cells that the body deploys to attack the virus. In the two decades after the first five cases were reported in 1981, 22 million people died of AIDS. Worldwide, 35 to 40 million people are currently infected with HIV.

Epidemiology

Because HIV is present in the blood and some body fluids, it is most effectively transmitted by actions or practices that involve the exchange of blood or body fluids between people. Within most populations, HIV is transmitted in semen or vaginal fluid during unprotected (without a condom) anal, vaginal, or oral sex. The likelihood of transmission of HIV from an infected person to an uninfected person varies significantly depending on the type of exposure or contact involved. The risk of becoming infected with HIV through unprotected oral sex is lower than that of unprotected anal or vaginal sex. HIV also is transmitted by direct blood-to-blood contact, such as occurs among intravenous drug users who share hypodermic needles. People at high risk are the sexual partners of HIV-infected individuals; those at lesser risk include healthcare professionals who may be accidentally stuck by HIV-contaminated hypodermic needles. In addition, HIV may be transmitted from a mother to her baby during a vaginal delivery or during breast-feeding.

In the United States and Europe before 1985, HIV was unknowingly spread by the transfusion of blood and blood products containing the virus. Effective HIV screening of blood instituted after 1985 has largely eliminated this mode of HIV transmission in the United States and other developed nations. By contrast, in sub-Saharan Africa, where two-thirds of those infected with HIV live, 25% of transfused blood is not screened for HIV.

Most people in economically advantaged, industrial nations who are diagnosed as having AIDS are either homosexual men who have engaged in unprotected anal intercourse or intravenous drug users. The rate of new HIV infections in the United States paints a different and disturbing picture: The greatest increases in new HIV infections are among African-Americans, Hispanics, women, and teenagers. In developing nations, HIV is largely transmitted during unprotected heterosexual intercourse, but the problem is compounded by a contaminated blood supply and unsanitary health care practices.

HIV is a very fragile virus; it cannot survive for long outside the human body. The virus is not transmitted by insect bites. One cannot become infected by casual physical contact with an HIV-infected person, such as by hugging or sharing household items. The virus can be eliminated from personal care items and medical equipment by exposing them to heat (135°F for 10 minutes) or by cleaning them with common disinfectants such as hydrogen peroxide, rubbing alcohol, Lysol® household bleach, or germicidal cleansers such as Betadine or Hibiclens. Standard dishwashing and clothes washing also kills HIV.

The chance of transmitting or of being infected by HIV during vaginal or anal intercourse can be greatly reduced—although not entirely eliminated—by the use of latex condoms. Public health programs aimed at encouraging drug users not to share needles have proved effective at checking the increase in new HIV infections in this population. Also, the prophylactic administration of certain drugs such as AZT (discussed shortly) to pregnant HIV-infected women has proved remarkably effective in reducing the transmission of the virus to their babies.

Pathogenesis of HIV Infection

HIV consists of two identical strands of genetic material surrounded by a protective protein coat. To replicate, a virus must enter a host cell, where it uses the cell's enzymes, ribosomes, and nutrients to reproduce copies of its genetic information. Unlike most viruses, however, HIV is a form of **retrovirus,** a virus whose genetic information is carried in RNA instead of DNA. Once inside the host cell, a viral enzyme called **reverse transcriptase** reads the viral RNA strand and makes a DNA copy.

The coat that surrounds HIV's RNA and reverse transcriptase is composed of many molecules of a protein termed P24. In addition, the coat is wrapped by an envelope composed of a lipid bilayer penetrated by a distinctive assortment of glycoproteins (Figure 22.21). One of these glycoproteins, GP120, functions as the docking protein that binds to CD4 molecules on T cells, macrophages, and dendritic cells. In addition, HIV must simultaneously attach to a coreceptor in the host cell's plasma membrane—a molecule designated CCR5 in antigen-presenting cells

Figure 22.21 Human immunodeficiency virus (HIV), the causative agent of AIDS. The core contains RNA and reverse transcriptase, plus several other enzymes. The protein coat (capsid) around the core consists of multiple copies of a protein called P24. The envelope consists of a lipid bilayer studded with glycoproteins (GP120 and GP41) that assist HIV's binding to and entering target cells.

🔑 **HIV is most effectively transmitted by practices that involve the exchange of body fluids.**

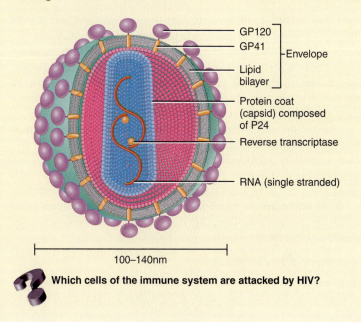

GP120
GP41 ⎫
Lipid bilayer ⎬ Envelope
Protein coat (capsid) composed of P24
Reverse transcriptase
RNA (single stranded)

100–140nm

Which cells of the immune system are attacked by HIV?

and CXCR4 in T cells—before it can gain entry into the cell. Another glycoprotein, GP41, helps the viral lipid bilayer fuse with the host cell's lipid bilayer.

The binding of HIV's docking proteins with the receptors and coreceptors in the host cell's plasma membrane causes receptor-mediated endocytosis, which brings the virus into the cell's cytoplasm. Once inside the cell, HIV sheds its protein coat. The reverse transcriptase enzyme makes a DNA copy of the viral RNA, and the viral DNA copy becomes integrated into the cell's DNA. Thus, the viral DNA is duplicated along with the host cell's DNA during normal cell division. In addition, the viral DNA can cause the infected cell to begin producing millions of copies of viral RNA and to assemble new protein coats for each copy. The new HIV copies bud off from the cell's plasma membrane and circulate in the blood to infect other cells.

HIV mainly damages helper (CD4$^+$) T cells, and it does so in various ways. Over 10 billion viral copies may be produced each day. The viruses bud so rapidly from an infected cell's plasma membrane that cell lysis eventually occurs. In addition, the body's defenses attack the infected cells, killing them but not all the viruses they harbor. In most HIV-infected individuals, CD4$^+$ T cells are initially replaced as fast as they are destroyed. After several years, however, the body's ability to replace CD4$^+$ T cells is slowly exhausted, and the number of CD4$^+$ T cells in circulation progressively declines.

Signs, Symptoms, and Diagnosis of HIV Infection

Immediately after being infected with HIV, most people experience a brief flu-like illness. Common signs and symptoms are fever, fatigue, rash, headache, joint pain, sore throat, and swollen lymph nodes. In addition, about 50% of infected people have night sweats. After three to four weeks, plasma cells begin secreting antibodies to components of the HIV protein coat. These antibodies are detectable in blood plasma and form the basis for some of the screening tests for HIV. When people test "HIV-positive," this usually means they have antibodies to HIV epitopes in their bloodstream. If an acute HIV infection is suspected but the antibody test is negative, laboratory tests based on detection of HIV's RNA or P24 coat protein in blood plasma can confirm the presence of HIV.

Progression to AIDS

After a period of 2 to 10 years, the virus destroys enough CD4$^+$ T cells that most infected people begin to experience symptoms of immunodeficiency. HIV-infected people commonly have enlarged lymph nodes and experience persistent fatigue, involuntary weight loss, night sweats, skin rashes, diarrhea, and various lesions of the mouth and gums. In addition, the virus may begin to infect neurons in the brain, affecting the person's memory and producing visual disturbances.

As the immune system slowly collapses, an HIV-infected person becomes susceptible to a host of *opportunistic infections*. These are diseases caused by microorganisms that are normally held in check but now proliferate because of the defective immune system. AIDS is diagnosed when the CD4$^+$ T cell count drops below 200 cells per microliter (= cubic millimeter) of blood or when opportunistic infections arise, whichever occurs first. Typically, opportunistic infections eventually are the cause of death.

Treatment of HIV Infection

At present, infection with HIV cannot be cured, and despite intensive research, no effective vaccine is yet available to provide immunity against HIV. However, two categories of drugs have proved successful in extending the life of many HIV-infected individuals. The first category, *reverse transcriptase inhibitors,* interferes with the action of the reverse transcriptase enzyme that the virus uses to convert its RNA into a DNA copy. Among the drugs in this category are zidovudine (also known as AZT), didanosine (ddI), dideoxycytidine (ddC), and stavudine (d4T). Trizivir, approved in 2000 for treatment of HIV infection, combines three reverse transcriptase inhibitors (abacavir, lamivudine, and zidovudine) in one pill. The second category of drugs is the *protease inhibitors.* These drugs interfere with the action of protease, a viral enzyme that cuts proteins into pieces to assemble the coat of newly produced HIV particles. Drugs in this category include nelfinavir, saquinavir, ritonavir, and indinavir.

In 1996, physicians treating HIV-infected patients widely adopted *highly active antiretroviral therapy (HAART),* also known as *triple therapy*—a combination of two differently acting reverse transcriptase inhibitors and one protease inhibitor. This regimen has been used extensively for treating patients at virtually all stages of HIV infection, regardless of the CD4$^+$ T cell count and viral load (the number of copies of HIV RNA in a milliliter of plasma). Most HIV-infected individuals receiving HAART experience a drastic reduction in viral load and an increase in the number of CD4$^+$ T cells in their blood. Not only does HAART delay the progression of HIV infection to AIDS, but many individuals with AIDS have seen the remission or disappearance of opportunistic infections and an apparent return to health. Unfortunately, triple therapy is very costly (exceeding $10,000 per year), the dosing schedule is grueling, and not all people can tolerate the toxic side effects of these drugs. Data published in 2001 suggests that for some patients HAART could be postponed until the CD4$^+$ T cell count drops to 200 cells per microliter.

Allergic Reactions

A person who is overly reactive to a substance that is tolerated by most other people is said to be **hypersensitive (allergic).** Whenever an allergic reaction takes place, some tissue injury occurs. The antigens that induce an allergic reaction are called **allergens.** Common allergens include certain foods (milk, peanuts, shellfish, eggs), antibiotics (penicillin, tetracycline), vaccines (pertussis, typhoid), venoms (honeybee, wasp, snake), cosmetics, chemicals in plants such as poison ivy, pollens, dust, molds, iodine-containing dyes used in certain x-ray procedures, and even microbes.

There are four basic types of hypersensitivity reactions: type I (anaphylactic), type II (cytotoxic), type III (immune-complex), and type IV (cell-mediated). The first three are antibody-mediated immune responses; the last is a cell-mediated immune response.

Type I (anaphylactic) reactions are the most common and occur within a few minutes after a person sensitized to an allergen is reexposed to it. In response to certain allergens, some people produce IgE antibodies that bind to the surface of mast cells and basophils. The next time the same allergen enters the body, it attaches to the IgE antibodies already present. In response, the mast cells and basophils release histamine, prostaglandins, leukotrienes, and kinins. Collectively, these mediators cause vasodilation, increased blood capillary permeability, increased smooth muscle contraction in the airways of the lungs, and increased mucus secretion. As a result, a person may experience inflammatory responses, difficulty in breathing through the constricted airways, and a runny nose from excess mucus secretion. In **anaphylactic**

shock, which may occur in a susceptible individual who has just received a triggering drug or been stung by a wasp, wheezing and shortness of breath as airways constrict are usually accompanied by shock due to vasodilation and fluid loss from blood. This life-threatening emergency is usually treated by injecting epinephrine to dilate the airways and strengthen the heartbeat.

Type II (cytotoxic) reactions are caused by antibodies (IgG or IgM) directed against antigens on a person's blood cells (red blood cells, lymphocytes, or platelets) or tissue cells. The reaction of antibodies and antigens usually leads to activation of complement. Type II reactions, which may occur in incompatible blood transfusion reactions, damage cells by causing lysis.

Type III (immune-complex) reactions involve antigens, antibodies (IgA or IgM), and complement. When certain ratios of antigen to antibody occur, the immune complexes are small enough to escape phagocytosis, but they become trapped in the basement membrane under the endothelium of blood vessels, where they activate complement and cause inflammation. Glomerulonephritis and rheumatoid arthritis (RA) arise in this way.

Type IV (cell-mediated) reactions or **delayed hypersensitivity reactions** usually appear 12–72 hours after exposure to an allergen. Type IV reactions occur when allergens are taken up by antigen-presenting cells (such as Langerhans cells in the skin) that migrate to lymph nodes and present the allergen to T cells, which then proliferate. Some of the new T cells return to the site of allergen entry into the body, where they produce gamma-interferon, which activates macrophages, and tumor necrosis factor, which stimulates an inflammatory response. Intracellular bacteria such as *Mycobacterium tuberculosis* trigger this type of cell-mediated immune response, as do certain haptens, such as poison ivy toxin. The skin test for tuberculosis also is a delayed hypersensitivity reaction.

Autoimmune Diseases

In an **autoimmune disease** (aw-tō-i-MŪN) or **autoimmunity,** the immune system fails to display self-tolerance and attacks the person's own tissues. Autoimmune diseases usually arise in early adulthood and are common, afflicting an estimated 5% of adults in North America and Europe. Females suffer autoimmune diseases twice as often as males. Recall that self-reactive B cells and T cells normally are deleted or undergo anergy during negative selection (see Figure 22.20). Apparently, this process is not 100% effective. Under the influence of unknown environmental triggers and certain genes that make some people more susceptible, self-tolerance breaks down, leading to activation of self-reactive clones of T cells and B cells. These cells then generate cell-mediated or antibody-mediated immune responses against self-antigens.

A variety of mechanisms produce different autoimmune diseases. Some involve production of **autoantibodies,** antibodies that bind to and stimulate or block self-antigens. For example, autoantibodies that mimic TSH (thyroid-stimulating hormone) are present in Graves disease and stimulate secretion of thyroid hormones (thus producing hyperthyroidism), whereas autoantibodies that bind to and block acetylcholine receptors cause muscle weakness in myasthenia gravis. Other autoimmune diseases involve activation of cytotoxic T cells that destroy certain body cells. Examples include type 1 diabetes mellitus, in which T cells attack the insulin-producing pancreatic beta cells, and multiple sclerosis (MS), in which T cells attack myelin sheaths around

axons of neurons. Inappropriate activation of helper T cells or excessive production of gamma-interferon also occur in certain autoimmune diseases. Other autoimmune disorders include rheumatoid arthritis (RA), systemic lupus erythematosus (SLE), rheumatic fever, hemolytic and pernicious anemias, Addison's disease, Hashimoto's thyroiditis, and ulcerative colitis.

Therapies for various autoimmune diseases include removal of the thymus gland (thymectomy), injections of interferon beta, immunosuppressive drugs, and plasmapheresis, in which the person's blood plasma is filtered to remove antibodies and antigen–antibody complexes.

Severe Combined Immunodeficiency Disease

Severe combined immunodeficiency disease (SCID) is a rare inherited disorder in which both B cells and T cells are missing or inactive in providing immunity. Scientists have now identified mutations in several genes that are responsible for some types of SCID. One gene codes for an enzyme known as adenosine deaminase (ADA). Deficiency of ADA leads to accumulations of adenosine, ATP (adenosine triphosphate), and deoxy-ATP, which inhibit activity of an enzyme needed for synthesis of DNA. Thus, lymphocytes that lack ADA cannot proliferate to carry out their immune functions. Other mutations affect the enzymes needed to shuffle and rearrange gene segments to generate diverse B-cell and T-cell receptors. Another gene, on the X chromosome, codes for one of the three polypeptide chains (called the gamma chain) that together form the interleukin-2 receptor. This chain also is part of the receptors for Il-4, Il-7, Il-11, and Il-15. In this type of SCID, which afflicts boys exclusively, T cells fail to mature because interleukin signals cannot be received.

The most famous patient with SCID was David, the "bubble boy." He was placed in a sterile see-through chamber shortly after birth in the 1970s to protect him from microbes that his body could not fight. At age 12, David underwent a bone marrow transplant in an effort to correct his disorder. Eighty days after the transplant and still in a germ-free environment, David developed some of the symptoms of infectious mononucleosis, a condition caused by the Epstein–Barr virus. He was brought out of isolation for easier treatment, with the hope that the transplant would provide the same protection as his sterile plastic chamber. It probably was successful but, unfortunately, David died 4 months later from cancer associated with Epstein–Barr virus. David's legacy has continued, however, because some of his cells were used in the experiments that localized the genetic defect to the gene that codes for the gamma chain of the interleukin-2 receptor.

In some cases, an infusion of red bone marrow cells from a sibling having very similar MHC (HLA) antigens can provide normal stem cells that give rise to normal B and T cells. The result can be a complete cure. Less than 30% of afflicted patients, however, have a compatible sibling who could be the donor. Another approach for treatment of SCID is to provide good copies of the defective gene through gene therapy. In 1990 and 1991, two girls with ADA deficiency were the first persons in the United States to receive gene therapy for a life-threatening disorder, and they have continued to improve while receiving infusions of gene-corrected T cells. Because the procedure involves infusion of mature T cells, however, patients must be retreated every couple of months for life.

Infectious Mononucleosis

Infectious mononucleosis (IM) is a contagious disease caused by the *Epstein–Barr virus (EBV).* "Mono" occurs mainly in children and young adults, and more often in females than in males by a 3:1 ratio. The virus most commonly enters the body through intimate oral contact such as kissing. It then multiplies in lymphatic tissues and spreads into the blood, where it infects and multiplies in B cells. Because of this infection, the B cells become enlarged and abnormal in appearance such that they resemble monocytes, the primary reason for the term *mononucleosis*. Signs and symptoms include an elevated white blood cell count with an abnormally high percentage of lymphocytes, fatigue, headache, dizziness, sore throat, enlarged and tender lymph nodes, and fever. There is no cure for infectious mononucleosis, but the disease usually runs its course in a few weeks.

Lymphomas

Lymphomas (lim-FŌ-mas; *lymph-* = clear water; *-oma* = tumor) are cancers of the lymphatic organs, especially the lymph nodes. Most have no known cause. The two main types of lymphomas are Hodgkin disease and non-Hodgkin lymphoma.

Hodgkin disease (HD) is characterized by a painless, nontender enlargement of one or more lymph nodes, most commonly in the neck, chest, and axilla. If the disease has metastasized from these sites, fever, night sweats, weight loss, and bone pain also occur. HD primarily affects individuals between ages 15 and 35 and those over 60, and it is more common in males. If diagnosed early, HD has a 90–95% cure rate.

Non-Hodgkin lymphoma (NHL), which is more common than HD, occurs in all age groups, the incidence increasing with age to a maximum between ages 45 and 70. NHL may start the same way as HD but may also include an enlarged spleen, anemia, and general malaise. Up to half of all individuals with NHL are cured or survive for a lengthy period. Treatment options for both HD and NHL include radiation therapy, chemotherapy, and bone marrow transplantation.

Systemic Lupus Erythematosus

Systemic lupus erythematosus (er-e′-thēm-a-TŌ-sus), **SLE,** or **lupus** (*lupus* = wolf) is a chronic autoimmune, inflammatory disease that affects multiple body systems. Most cases of SLE occur in women between the ages of 15 and 25, more often in blacks than in whites. Although the cause of SLE is not known, both a genetic predisposition to the disease and environmental factors contribute. Sex hormones appear to influence the development of SLE. Females are nine times more likely than males to suffer from SLE. The disorder often occurs in females who exhibit extremely low levels of androgens.

Signs and symptoms of SLE include joint pain, slight fever, fatigue, oral ulcers, weight loss, enlarged lymph nodes and spleen, photosensitivity, rapid loss of large amounts of scalp hair, and sometimes an eruption across the bridge of the nose and cheeks called a "butterfly rash." The erosive nature of some of the SLE skin lesions was thought to resemble the damage inflicted by the bite of a wolf—thus, the term lupus.

Two immunological features of SLE are excessive activation of B cells and inappropriate production of autoantibodies against DNA (anti-DNA antibodies) and other components of cellular nuclei such as histone proteins. Triggers of B cell activation are thought to include various chemicals and drugs, viral and bacterial antigens, and exposure to sunlight. Circulating complexes of abnormal autoantibodies and their "antigens" cause damage in tissues throughout the body. Kidney damage occurs as the complexes become trapped in the basement membrane of kidney capillaries thereby obstructing blood filtering. Renal failure is the most common cause of death.

MEDICAL TERMINOLOGY

Adenitis (ad′-e-NĪ-tis; *aden-* = gland; *-itis* = inflammation of) Enlarged, tender, and inflamed lymph nodes resulting from an infection.

Allograft (AL-ō-graft; *allo-* = other) A transplant between genetically distinct individuals of the same species. Skin transplants from other people and blood transfusions are allografts.

Autograft (AW-tō-graft; *auto-* = self) A transplant in which one's own tissue is grafted to another part of the body (such as skin grafts for burn treatment or plastic surgery).

Chronic fatigue syndrome (CFS) A disorder, usually occurring in young adults and primarily in females, characterized by (1) extreme fatigue that impairs normal activities for at least 6 months and (2) the absence of other known diseases (cancer, infections, drug abuse, toxicity, or psychiatric disorders) that might produce similar symptoms.

Congenital thymic hypoplasia (DiGeorge syndrome) A disorder in which the thymus is absent or much smaller than normal; caused by faulty development of the third and fourth pharyngeal pouches (see Figure 18.21 on page 623). Thymic deficiency decreases cell-mediated immune responses, but antibody-mediated immune responses are more or less normal. Transplantation of a fetal thymus provides excellent treatment of babies that completely lack a thymus.

Gamma globulin (GLOB-ū-lin) Suspension of immunoglobulins from blood consisting of antibodies that react with a specific pathogen. It is prepared by injecting the pathogen into animals, removing blood from the animals after antibodies have been produced, isolating the antibodies, and injecting them into a human to provide short-term immunity.

Hypersplenism (hī-per-SPLĒN-izm; *hyper-* = over) Abnormal splenic activity due to splenic enlargement and associated with an increased rate of destruction of normal blood cells and platelets.

Lymphadenopathy (lim-fad′-e-NOP-a-thē; *lymph-* = clear fluid; *-pathy* = disease) Enlarged, sometimes tender lymph glands.

Lymphedema (lim′-fe-DĒ-ma; *edema* = swelling) Accumulation of lymph producing subcutaneous tissue swelling.

Splenomegaly (splē′-nō-MEG-a-lē; *mega-* = large) Enlarged spleen.

Tonsillectomy (ton′-si-LEK-tō-mē; *-ectomy* = excision) Removal of a tonsil.

Xenograft (ZEN-ō-graft; *xeno-* = strange or foreign) A transplant between animals of different species. Xenografts from porcine (pig) or bovine (cow) tissue may be used in a human as a physiological dressing for severe burns.

STUDY OUTLINE

INTRODUCTION (p. 765)

1. The ability to ward off disease is called resistance. Lack of resistance is called susceptibility.
2. Nonspecific resistance refers to a wide variety of body responses against a wide range of pathogens; specific resistance or immunity involves activation of specific lymphocytes to combat a particular foreign substance.

THE LYMPHATIC AND IMMUNE SYSTEM (p. 765)

1. The lymphatic and immune system carries out immune responses and consists of lymph, lymphatic vessels, and structures and organs that contain lymphatic tissue (specialized reticular tissue containing many lymphocytes).
2. It drains interstitial fluid, transports dietary lipids, and protects against invasion through immune responses.
3. Lymphatic vessels begin as closed-ended lymphatic capillaries in tissue spaces between cells.
4. Interstitial fluid drains into lymphatic capillaries, thus forming lymph.
5. Lymph capillaries merge to form larger vessels, called lymphatic vessels, which convey lymph into and out of lymph nodes.
6. The route of lymph flow is from lymph capillaries to lymphatic vessels to lymph trunks to the thoracic duct (or right lymphatic duct) to the subclavian veins.
7. Lymph flows because of skeletal muscle contractions and respiratory movements. Valves in lymphatic vessels also aid flow of lymph.
8. The primary lymphatic organs are red bone marrow and the thymus. Secondary lymphatic organs are lymph nodes, spleen, and lymphatic nodules.
9. The thymus lies between the sternum and the large blood vessels above the heart. It is the site of T cell maturation.
10. Lymph nodes are encapsulated, egg-shaped structures located along lymphatic vessels.
11. Lymph enters nodes through afferent lymphatic vessels, is filtered, and exits through efferent lymphatic vessels.
12. Lymph nodes are the site of proliferation of plasma cells and T cells.
13. The spleen is the largest single mass of lymphatic tissue in the body. It is a site of B cell proliferation into plasma cells and phagocytosis of bacteria and worn-out red blood cells.
14. Lymphatic nodules are scattered throughout the mucosa of the gastrointestinal, respiratory, urinary, and reproductive tracts. This lymphatic tissue is termed mucosa-associated lymphatic tissue (MALT).

DEVELOPMENT OF LYMPHATIC TISSUES (p. 774)

1. Lymphatic vessels develop from lymph sacs, which arise from developing veins. Thus, they are derived from mesoderm.
2. Lymph nodes develop from lymph sacs that become invaded by mesenchymal cells.

NONSPECIFIC RESISTANCE: INNATE DEFENSES (p. 775)

1. Mechanisms of nonspecific resistance include mechanical factors, chemical factors, antimicrobial proteins, natural killer cells, phagocytes, inflammation, and fever.
2. The skin and mucous membranes are the first line of defense against entry of pathogens.
3. Antimicrobial proteins include interferons, the complement system, and transferrins.
4. Natural killer cells and phagocytes attack and kill pathogens and defective cells in the body.
5. Inflammation aids disposal of microbes, toxins, or foreign material at the site of an injury, and prepares the site for tissue repair.
6. Fever intensifies the antiviral effects of interferons, inhibits growth of some microbes, and speeds up body reactions that aid repair.
7. Table 22.1 on page 779 summarizes the components of nonspecific resistance.

SPECIFIC RESISTANCE: IMMUNITY (p. 780)

1. Specific resistance to disease involves the production of a specific lymphocyte or antibody against a specific antigen and is called immunity.
2. B cells and T cells derive from stem cells in red bone marrow.
3. T cells complete their maturation and develop immunocompetence in the thymus.
4. In cell-mediated immune responses, cytotoxic T cells directly attack the invading antigen, whereas in antibody-mediated immune responses, plasma cells secrete antibodies.
5. Antigens (Ags) are chemical substances that are recognized as foreign by the immune system.
6. Antigen receptors exhibit great diversity due to genetic recombination.
7. "Self-antigens" termed major histocompatibility complex (MHC) antigens are unique to each person's body cells. All cells except red blood cells display MHC-I molecules; some cells also display MHC-II molecules.
8. Cells called antigen-presenting cells (APCs), which include macrophages, B cells, and dendritic cells, process antigens.
9. Exogenous antigens (formed outside the body) are presented together with MHC-II molecules to T cells, whereas endogenous (formed inside a body cell) antigens are presented together with MHC-I molecules.
10. Cytokines are small protein hormones needed for many normal cell functions. Some of them regulate immune responses (see Table 22.2 on page 784).

CELL-MEDIATED IMMUNITY (p. 785)

1. In a cell-mediated immune response, an antigen is recognized, specific T cells proliferate and differentiate into effector cells, and the antigen is eliminated.
2. T-cell receptors (TCRs) recognize antigen fragments associated with MHC molecules on the surface of a body cell.
3. Proliferation of T cells requires costimulation, either by cytokines such as interleukin-2 or by pairs of plasma membrane molecules.
4. T cells consist of several subpopulations. Helper T cells display CD4 protein, recognize antigen fragments associated with MHC-II molecules, and secrete several cytokines, most importantly interleukin-2, which acts as a costimulator for other helper T cells, cytotoxic T cells, and B cells. Cytotoxic T cells display CD8 protein and recognize antigen fragments associated with MHC-I molecules. Memory T cells remain after a cell-

mediated immune response and initiate a faster response when a pathogen bearing the same foreign antigen invades the body again.

5. Cytotoxic T cells eliminate invaders by secreting perforin, which causes cytolysis, and lymphotoxin, which causes fragmentation of the DNA of a target cell.

6. Cytotoxic T cells, macrophages, and natural killer cells carry out immunological surveillance, recognizing and destroying cancerous cells that display tumor antigens.

ANTIBODY-MEDIATED IMMUNITY (p. 788)

1. B cells can respond to unprocessed antigens, but their response is more intense when dendritic cells present antigen to them. Interleukin-2 and other cytokines secreted by helper T cells provide costimulation for the proliferation of B cells.

2. An activated B cell develops into a clone of antibody-producing plasma cells.

3. An antibody (Ab) is a protein that combines specifically with the antigen that triggered its production.

4. Antibodies consist of heavy and light chains and variable and constant regions.

5. Based on chemistry and structure, antibodies are grouped into five principal classes, (IgG, IgA, IgM, IgD, and IgE), each with specific biological roles.

6. Actions of antibodies include neutralization of antigen, immobilization of bacteria, agglutination and precipitation of antigen, activation of complement, and enhancement of phagocytosis.

7. Complement is a group of proteins whose functions complement immune responses and help clear antigens from the body.

8. Immunization against certain microbes is possible because memory B cells and memory T cells remain after a primary response to an antigen. The secondary response provides protection should the same microbe enter the body again.

SELF-RECOGNITION AND SELF-TOLERANCE (p. 793)

1. T cells undergo positive selection to ensure that they can recognize self-MHC proteins (self-recognition), and negative selection to ensure that they do not react to other self-proteins (self-tolerance). Negative selection involves both deletion and anergy.

2. B cells develop tolerance through deletion and anergy.

STRESS AND IMMUNITY (p. 795)

1. Psychoneuroimmunology (PNI) deals with communication pathways that link the nervous, endocrine, and immune systems. Thoughts, feelings, moods, and beliefs influence health and the course of disease.

2. Under stress, people are less likely to eat well or exercise regularly, two habits that enhance immunity.

AGING AND THE IMMUNE SYSTEM (p. 795)

1. With advancing age, individuals become more susceptible to infections and malignancies, respond less well to vaccines, and produce more autoantibodies.

2. Immune responses also diminish with age.

SELF-QUIZ QUESTIONS

Fill in the blanks in the following statements.

1. The first line of nonspecific defense against pathogens are the _____ and _____; the second line of nonspecific defense are the _____, _____, and _____.

2. The two distinguishing features of immunity are _____ and _____.

Indicate whether the following statements are true or false.

3. The sequence of fluid flow from blood vessel to blood vessel by way of the lymphatic system is arteries (blood) to blood capillaries (blood) to interstitial spaces (interstitial fluid) to lymphatic capillaries (lymph) to lymphatic vessels (lymph) to lymphatic ducts (lymph) to subclavian veins (blood).

4. A person's T cells must be able to recognize the person's own MHC molecules, a process known as self-recognition, and lack reactivity to peptide fragments from the person's own proteins, a condition known as self-tolerance.

Choose the one best answer to the following questions.

5. Which of the following are functions of the lymphatic system? (1) draining interstitial fluid, (2) draining intracellular fluid, (3) transporting dietary lipids, (4) transporting nucleic acids, (5) protecting against invasion. (a) 1, 2, and 3, (b) 2, 3 and 4, (c) 3, 4, and 5, (d) 1, 2, and 4, (e) 1, 3 and 5.

6. Which of the following statements are *correct*? (1) Lymphatic vessels are found throughout the body, execept in avascular tissues, the CNS, portions of the spleen, and bone marrow. (2) Lymphatic capillaries allow interstitial fluid to flow into them but not out of them. (3) Anchoring filaments attach lymphatic endothelial cells to surrounding tissues. (4) Lymphatic vessels freely receive all the components of blood, including the formed elements. (5) Lymph vessels directly connect to blood vessels by way of the subclavian veins. (a) 1, 3, 4 and 5, (b) 2, 3, 4 and 5, (c) 1, 2, 3 and 4, (d) 1, 2, 4 and 5, (e) 1, 2, 3, and 5.

7. Which of the following are mechanical factors that help fight pathogens and disease? (1) tight junctions of epidermal cells, (2) mucus of mucous membranes, (3) saliva, (4) interferons, (5) complement. (a) 1, 3, and 4, (b) 2, 4, and 5, (c) 1, 4, and 5, (d) 1, 2, and 3, (e) 1, 2, and 4.

8. Which of the following are processes involved in inflammation? (1) vasodilation and increased permeability of blood vessels, (2) phagocyte emigration, (3) tissue repair, (4) opsonization, (5) adherence. (a) 1, 2, and 3, (b) 2, 3, and 4, (c) 3, 4, and 5, (d) 1, 3, and 5, (e) 2, 4, and 5.

9. Antibody-mediated immunity works mainly aganist (a) foregin tissue transplants, (b) intracellular pathogens, (c) extracellular pathogens, (d) cancer cells, (e) viruses.

10. Which of the following are functions of antibodies? (1) neutralization of antigens, (2) immobilization of bacteria, (3) agglutination and precipitation of antigens, (4) activation of complement, (5) enhancement of phagocytosis. (a) 1, 3 and 4, (b) 2, 4, and 5, (c) 1, 2, 3 and 4, (d) 1, 2, 3 and 5, (e) 1, 2, 3, 4, and 5.

11. Which of the following are *true*? (1) Lymphatic vessels resemble arteries. (2) Lymph is very similar to interstitial fluid. (3) Lacteals are specialized lymphatic capillaries responsible for transporting dietary lipids. (4) Lymph is normally a cloudy, pale yellow fluid. (5) The thoracic duct drains lymph from the upper right side of the body. (6) Lymph flow is maintained by skeletal muscle contractions, one-way valves, and breathing movements. (a) 1, 2, 5, and 6, (b) 2, 3, and 6, (c) 2, 3, 4, and 6, (d) 2, 4, and 6, (e) 3, 5, and 6.

12. Place the stages of phagocytosis in the correct order of occurrence. (1) formation of phagolysosome, (2) adherence to microbe, (3) destruction of microbe, (4) ingestion to form a phagosome, (5) chemotaxic attraction of phagocyte. (a) 2, 5, 4, 1, 3, (b) 4, 5, 2, 1, 3, (c) 5, 2, 4, 1, 3, (d) 5, 4, 2, 3, 1, (e) 2, 5, 1, 4, 3.

13. Match the following:
    (1) red bone marrow   (2) thymus   (3) lymph nodes
    (4) spleen   (5) lymphatic nodules
    ____ (a) bean-shaped structures located along the length of lymphatic vessels; contain T cells, macrophages, and follicular dendritic cells; filter lymph
    ____ (b) produces pre-T cells and B cells; found in flat bones and epiphyses of long bones
    ____ (c) the single largest mass of lymphatic tissue in the body
    ____ (d) responsible for the maturation of T cells
    ____ (e) nonencapsulated clusters of lymphocytes located in all mucous membranes

14. Match the following:
    ____ (a) recognize foreign antigens combined with MHC-1 molecules on the surface of body cells infected by viruses, some turmor cells, and cells of a tissue transplant
    ____ (b) are programmed to recognize the reappearance of a previously encountered antigen
    ____ (c) differentiate into plasma cells that secrete specific antibodies
    ____ (d) process and present exogenous antigens; include macrophages, B cells, and dendritic cells
    ____ (e) secrete cytokines as costimulators
    ____ (f) ingest microbes or any foreign particulate matter; include neutrophils and macrophages
    ____ (g) lymphocytes that have the ability to kill a wide variety of infectious microbes plus certain spontaneously arising tumor cells; lack antigen receptors

    (1) helper T cells
    (2) cytotoxic T cells
    (3) memory T cells
    (4) B cells
    (5) NK cells
    (6) phagocytes
    (7) antigen-presenting cells

15. Match the following (answers may be used more than once):
    ____ (a) participate in inflammation, opsonization, and cytolysis
    ____ (b) stimulate histamine release, attract neutrophills by chemotaxis, promote phagocytosis, and destroy bacteria
    ____ (c) glycoproteins that mark the surface of all body cells except for RBCs; distinguish self from nonself
    ____ (d) foreign antigens present in fluids outside body cells
    ____ (e) foreign antigens synthesized within body cells
    ____ (f) small protein hormones that stimulate or inhibit many normal cell functions; serve as costimulators for B cell and T cell activity
    ____ (g) a substance that has reactivity but lacks immunogenicity
    ____ (h) causes vasodilation and increased permeability of blood vessels; is found in mast cells in connective tissue and in basophills and platelets in blood
    ____ (i) polypeptides formed in blood; induce vasodilation and increased permeability of blood vessels; serve as chemotaxic agents for phagocytes
    ____ (j) small parts of antigens that initiate immune responses
    ____ (k) produced by virus-infected cells; they interfere with viral replication in host cells
    ____ (l) chemicals released by NK and cytotoxic T cells that can cause cytolysis in microbes
    ____ (m) glycoproteins that contain four polypeptide chains, two of which are identical to each other and two of which are variable and contain the anitgen-binding site

    (1) exogenous antigens
    (2) endogenous antigens
    (3) interferons
    (4) hapten
    (5) cytokines
    (6) perforins
    (7) histamine
    (8) major histocompatibility complex (MHC) antigens
    (9) kinins
    (10) antibodies
    (11) complement proteins
    (12) epitopes

CRITICAL THINKING QUESTIONS

1. Three-year-old Tariq was running barefoot through the grass when he felt a sharp pain and ran crying to his mother about a stepping on a "buzzy bug." Mom pulled an insect stinger out of his foot. Thirty minutes later, the sole of Tariq's foot was swollen around the stinger hole and itchy. What type of immune response is Tariq exhibiting?
   HINT *Ice and topical cortisone ointment relieved the symptoms.*

2. On the second day of Cara's wilderness backpacking trip, she got a splinter stuck in her right thumb. It didn't bother her too much (her feet hurt more than her thumb), so she put a bandage on it and ignored it. At the end of the trip, Cara had red streaks running along her right arm and tender swollen bumps in her right armpit. What happened to her arm?
   HINT *She should have paid more attention to that splinter.*

3. Esperanza watched as her mother got her "flu shot." "Why do you need a shot if you're not sick?" she asked. "So I won't get sick," answered her mom. Explain how the influenza vaccination prevents illness.
   HINT *A flu shot won't protect against measles, chicken pox, or pneumonia; it only works against the flu.*

ANSWERS TO FIGURE QUESTIONS

22.1 Red bone marrow contains stem cells that develop into lymphocytes.

22.2 Lymph is more similar to interstitial fluid than to blood plasma because the protein content of lymph is low.

22.3 The left and right lumbar trunks and the intestinal trunk empty into the cisterna chyli, which then drains into the thoracic duct.

22.4 Inhalation promotes the movement of lymph from abdominal lymphatic vessels toward the thoracic region.

22.5 T cells mature in the thymus.

22.6 Foreign substances that enter a lymph node in lymph may be phagocytized by macrophages or attacked by lymphocytes that mount immune responses.

22.7 White pulp of the spleen functions in immunity; red pulp of the spleen performs functions related to blood cells.

22.8 Lymphatic tissues begin to develop by the end of the fifth week of gestation.

22.9 Lysozyme, digestive enzymes, and oxidants can kill microbes.

22.10 *Redness* results from increased blood flow due to vasodilation; *pain,* from injury of nerve fibers, irritation by microbial toxins, kinins, and prostaglandins, and pressure due to edema; *heat,* from increased blood flow and heat released by locally increased metabolic reactions; *swelling,* from leakage of fluid from capillaries due to increased permeability.

22.11 Helper T cells participate in both cell-mediated and antibody-mediated immune responses.

22.12 Epitopes are small immunogenic parts of a larger antigen; haptens are small molecules that become immunogenic only when they attach to a body protein.

22.13 APCs include macrophages in tissues throughout the body, B cells in blood and lymphatic tissue, and dendritic cells in mucous membranes and the skin.

22.14 The first signal in T cell activation is antigen binding to a TCR; the second signal is a costimulator, such as a cytokine or another pair of plasma membrane molecules.

22.15 Cytotoxic T cells attack some tumor cells and transplanted tissue cells, as well as cells infected by viruses.

22.16 A clone of plasma cells secretes just one kind of antibody.

22.17 The variable regions recognize and bind to a specific antigen.

22.18 The classical pathway for the activation of complement is linked to antibody-mediated immunity because Ag–Ab complexes activate C1.

22.19 At peak secretion, approximately 1000 times more IgG is produced in the secondary response than in the primary response.

22.20 In deletion, self-reactive T cells or B cells die; in anergy, T cells or B cells are alive but are unresponsive to antigenic stimulation.

22.21 HIV attacks helper T cells.

The Respiratory System

THE RESPIRATORY SYSTEM AND HOMEOSTASIS

The respiratory system contributes to homeostasis by providing for the exchange of gases — oxygen and carbon dioxide — between the atmospheric air, blood, and tissues cells. It also helps adjust the pH of body fluids.

ENERGIZE YOUR STUDY

FOUNDATIONS CD

Anatomy Overview
  • The Respiratory System

Animations
  • Systems Contributions to Homeostasis
  • The Case of the Worried Smoker
  • The Case of the Coughing Boy

Concepts and Connections and Exercises reinforce your understanding

www.wiley.com/college/apcentral

INSIGHTS AND EXPLORATIONS

The American Lung Association reports that nearly 335,000 Americans die of lung disease every year. One out of every seven deaths is due to lung disease. Perhaps most disturbing is the claim that lung disease and other breathing problems are the number one killer of infants less than one-year old. Come explore in this web-based exercise the many respiratory disorders that affect people of all ages and genders.

Cells continually use oxygen (O_2) for the metabolic reactions that release energy from nutrient molecules and produce ATP. At the same time, these reactions release carbon dioxide (CO_2). Because an excessive amount of CO_2 produces acidity that can be toxic to cells, excess CO_2 must be eliminated quickly and efficiently. The two systems that cooperate to supply O_2 and eliminate CO_2 are the cardiovascular and respiratory systems. The respiratory system provides for gas exchange—intake of O_2 and elimination of CO_2—whereas the cardiovascular system transports blood containing the gases between the lungs and body cells. Failure of either system disrupts homeostasis by causing rapid death of cells from oxygen starvation and buildup of waste products. In addition to functioning in gas exchange, the respiratory system also participates in regulating blood pH, contains receptors for the sense of smell, filters inspired air, produces sounds, and rids the body of some water and heat in exhaled air.

RESPIRATORY SYSTEM ANATOMY

▶ OBJECTIVES

- **Describe the anatomy and histology of the nose, pharynx, larynx, trachea, bronchi, and lungs.**
- **Identify the functions of each respiratory system structure.**

The **respiratory system** consists of the nose, pharynx (throat), larynx (voice box), trachea (windpipe), bronchi, and lungs (Figure 23.1). Structurally, the respiratory system consists of two parts: (1) The **upper respiratory system** includes the nose, pharynx, and associated structures. (2) The **lower respiratory system** includes the larynx, trachea, bronchi, and lungs. Functionally, the respiratory system also consists of two parts: (1) The **conducting portion** consists of a series of interconnecting cavities and tubes both outside and within the lungs—the nose, pharynx, larynx, trachea, bronchi, bronchioles, and terminal bronchioles—that filter, warm, and moisten air and conduct it into the lungs. (2) The **respiratory portion** consists of tissues within the lungs where gas exchange occurs—the respiratory bronchioles, alveolar ducts, alveolar sacs, and alveoli, the main sites of gas exchange between air and blood. The volume of the conducting portion in an adult is about 150 mL; that of the respiratory portion is 5 to 6 liters.

The branch of medicine that deals with the diagnosis and treatment of diseases of the ears, nose, and throat is called **otorhinolaryngology** (ō′-tō-rī′-nō-lar′-in-GOL-ō-jē; *oto-* = ear; *rhino-* = nose; *laryngo-* = voice box; *-logy* = study of). A **pulmonologist** is a specialist in the diagnosis and treatment of diseases of the lungs.

Nose

The **nose** can be divided into external and internal portions. The external nose consists of a supporting framework of bone and hyaline cartilage covered with muscle and skin and lined by a mucous membrane. The frontal bone, nasal bones, and maxillae form the bony framework of the external nose (Figure 23.2a on page 808). The cartilaginous framework of the external nose consists of the **septal cartilage,** which forms the anterior portion of the nasal septum; the **lateral nasal cartilages** inferior to the nasal bones; and the **alar cartilages,** which form a portion of the walls of the nostrils. Because it consists of pliable hyaline cartilage, the cartilaginous framework of the external nose is somewhat flexible. On the undersurface of the external nose are two openings called the **external nares** (NĀ-rez; singular is **naris**) or **nostrils.** Figure 23.3 on page 809 shows the surface anatomy of the nose. The interior structures of the external nose have three functions: (1) warming, moistening, and filtering incoming air; (2) detecting olfactory stimuli; and (3) modifying speech vibrations as they pass through the large, hollow resonating chambers. *Resonance* refers to prolonging, amplifying, or modifying a sound by vibration.

The internal nose is a large cavity in the anterior aspect of the skull that lies inferior to the nasal bone and superior to the mouth; it also includes muscle and mucous membrane. Anteriorly, the internal nose merges with the external nose, and posteriorly it communicates with the pharynx through two openings called the **internal nares** or **choanae** (kō-Ā-nē) (see Figure 23.2b). Ducts from the paranasal sinuses (frontal, sphenoidal, maxillary, and ethmoidal paranasal sinuses) and the nasolacrimal ducts also open into the internal nose. The lateral walls of the internal nose are formed by the ethmoid, maxillae, lacrimal, palatine, and inferior nasal conchae bones (see Figure 7.9 on page 197); the ethmoid also forms the roof. The palatine bones and palatine processes of the maxillae, which together constitute the hard palate, form the floor of the internal nose.

The space within the internal nose is called the **nasal cavity.** The anterior portion of the nasal cavity, just inside the nostrils, is called the **vestibule** and is surrounded by cartilage; the superior part of the nasal cavity is surrounded by bone. A vertical partition, the **nasal septum,** divides the nasal cavity into right and left sides. The anterior portion of the septum consists primarily of hyaline cartilage; the remainder is formed by the vomer, perpendicular plate of the ethmoid, maxillae, and palatine bones (see Figure 7.14 on page 201).

When air enters the nostrils, it passes first through the vestibule, which is lined by skin containing coarse hairs that filter out large dust particles. Three shelves formed by projections of the superior, middle, and inferior nasal conchae extend out of each lateral wall of the cavity. The conchae, almost reaching the septum, subdivide each side of the nasal cavity into a series of groovelike passageways—the **superior, middle,** and **inferior meatuses** (mē-

Figure 23.1   **Structures of the respiratory system.**

The upper respiratory system includes the nose, pharynx, and associated structures; the lower respiratory system includes the larynx, trachea, bronchi, and lungs.

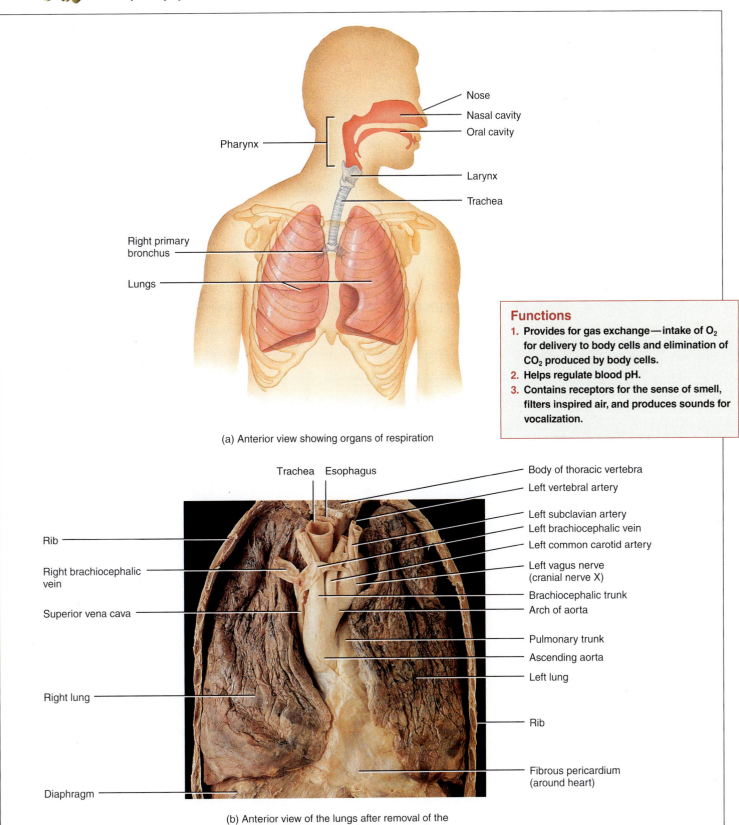

(a) Anterior view showing organs of respiration

Functions

1. **Provides for gas exchange—intake of O_2 for delivery to body cells and elimination of CO_2 produced by body cells.**
2. **Helps regulate blood pH.**
3. **Contains receptors for the sense of smell, filters inspired air, and produces sounds for vocalization.**

(b) Anterior view of the lungs after removal of the anterolateral thoracic wall and parietal pleura

Which structures are part of the conducting portion of the respiratory system?

Figure 23.2 **Respiratory structures in the head and neck.** (See Tortora, *A Photographic Atlas of the Human Body,* Figures 11.2 and 11.3.)

As air passes through the nose, it is warmed, filtered, and moistened.

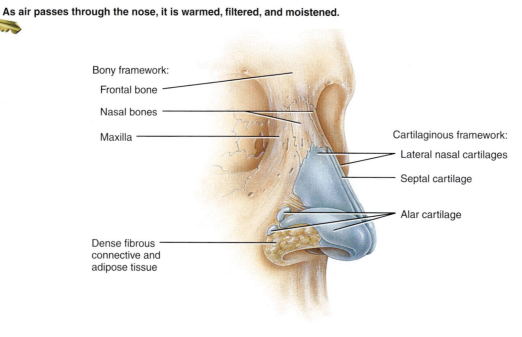

Bony framework:
Frontal bone
Nasal bones
Maxilla

Cartilaginous framework:
Lateral nasal cartilages
Septal cartilage
Alar cartilage

Dense fibrous connective and adipose tissue

(a) Anterolateral view of external portion of nose showing cartilaginous and bony framework

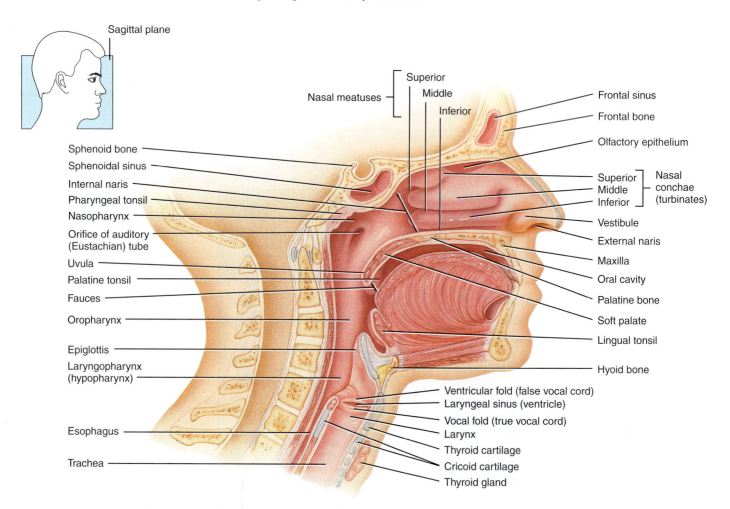

Sagittal plane

Nasal meatuses
Superior
Middle
Inferior

Frontal sinus
Frontal bone
Olfactory epithelium

Superior
Middle
Inferior
Nasal conchae (turbinates)

Vestibule
External naris
Maxilla
Oral cavity
Palatine bone
Soft palate
Lingual tonsil

Hyoid bone

Ventricular fold (false vocal cord)
Laryngeal sinus (ventricle)
Vocal fold (true vocal cord)
Larynx
Thyroid cartilage
Cricoid cartilage
Thyroid gland

Sphenoid bone
Sphenoidal sinus
Internal naris
Pharyngeal tonsil
Nasopharynx
Orifice of auditory (Eustachian) tube
Uvula
Palatine tonsil
Fauces
Oropharynx
Epiglottis
Laryngopharynx (hypopharynx)
Esophagus
Trachea

(b) Sagittal section of the left side of the head and neck showing the location of respiratory structures

What is the path taken by air molecules into and through the nose?

Figure 23.3 Surface anatomy of the nose.

The external nose has a cartilaginous and a bony framework.

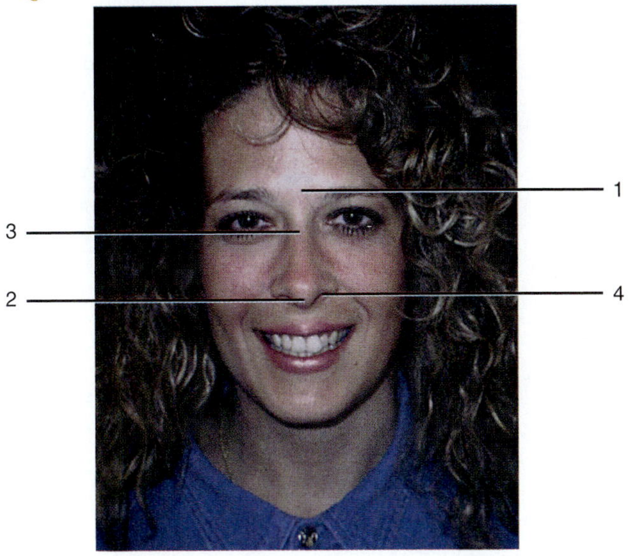

Anterior view

1. **Root**: Superior attachment of the nose to the frontal bone
2. **Apex**: Tip of nose
3. **Bridge**: Bony framework of nose formed by nasal bones
4. **External naris**: Nostril; external opening into nasal cavity

What part of the nose is attached to the frontal bone?

Ā-tus-ez = openings or passages). Mucous membrane lines the cavity and its shelves. The arrangement of conchae and meatuses increases surface area in the internal nose and prevents dehydration by acting as a baffle that traps water droplets during exhalation.

The olfactory receptors lie in the membrane lining the superior nasal conchae and adjacent septum. This region is called the **olfactory epithelium.** Inferior to the olfactory epithelium, the mucous membrane contains capillaries and pseudostratified ciliated columnar epithelium with many goblet cells. As inspired air whirls around the conchae and meatuses, it is warmed by blood in the capillaries. Mucus secreted by the goblet cells moistens the air and traps dust particles. Drainage from the nasolacrimal ducts and perhaps secretions from the paranasal sinuses also help moisten the air. The cilia move the mucus and trapped dust particles toward the pharynx, at which point they can be swallowed or spit out, thus removing particles from the respiratory tract.

Rhinoplasty

Rhinoplasty (RĪ-nō-plas′-tē; -*plasty* = to mold or to shape), commonly called a "nose job," is a surgical procedure in which the structure of the external nose is altered. Although rhinoplasty is often done for cosmetic reasons, it is sometimes performed to repair a fractured nose or a deviated nasal septum. In the proce-

dure, both a local and general anesthetic are given, and with instruments inserted through the nostrils, the nasal cartilage is reshaped, and the nasal bones are fractured and repositioned, to achieve the desired shape. An internal packing and splint are inserted to keep the nose in the desired position while it heals. ■

Pharynx

The **pharynx** (FAIR-inks), or throat, is a funnel-shaped tube about 13 cm (5 in.) long that starts at the internal nares and extends to the level of the cricoid cartilage, the most inferior cartilage of the larynx (voice box) (Figure 23.4). The pharynx lies just posterior to the nasal and oral cavities, superior to the larynx, and just anterior to the cervical vertebrae. Its wall is composed of skeletal muscles and is lined with a mucous membrane. The pharynx functions as a passageway for air and food, provides a resonating chamber for speech sounds, and houses the tonsils, which participate in immunological reactions against foreign invaders.

The pharynx can be divided into three anatomical regions: (1) nasopharynx, (2) oropharynx, and (3) laryngopharynx. (See the lower orientation diagram in Figure 23.4.) The muscles of the entire pharynx are arranged in two layers, an outer circular layer and an inner longitudinal layer.

The superior portion of the pharynx, called the **nasopharynx,** lies posterior to the nasal cavity and extends to the plane of the soft palate. There are five openings in its wall: two internal nares, two openings that lead into the auditory (pharyngotympanic) tubes (commonly known as the Eustachian tubes), and the opening into the oropharynx. The posterior wall also contains the **pharyngeal tonsil.** Through the internal nares, the nasopharynx receives air from the nasal cavity and receives packages of dust-laden mucus. The nasopharynx is lined with pseudostratified ciliated columnar epithelium, and the cilia move the mucus down toward the most inferior part of the pharynx. The nasopharynx also exchanges small amounts of air with the auditory (Eustachian) tubes to equalize air pressure between the pharynx and the middle ear.

The intermediate portion of the pharynx, the **oropharynx,** lies posterior to the oral cavity and extends from the soft palate inferiorly to the level of the hyoid bone. It has only one opening, the **fauces** (FAW-sēz = throat), the opening from the mouth. This portion of the pharynx has both respiratory and digestive functions because it is a common passageway for air, food, and drink. Because the oropharynx is subject to abrasion by food particles, it is lined with nonkeratinized stratified squamous epithelium. Two pairs of tonsils, the **palatine** and **lingual tonsils,** are found in the oropharynx.

The inferior portion of the pharynx, the **laryngopharynx** (la-rin′-gō-FAIR-inks), or **hypopharynx,** begins at the level of the hyoid bone. It opens into the esophagus (food tube) posteriorly and the larynx (voice box) anteriorly. Like the oropharynx, the laryngopharynx is both a respiratory and a digestive pathway and is lined by nonkeratinized stratified squamous epithelium.

Figure 23.4   Pharynx.   (See Tortora, *A Photographic Atlas of the Human Body,* Figure 11.4.)

🔑 The three subdivisions of the pharynx are the (1) nasopharynx, (2) oropharynx, and (3) laryngopharynx.

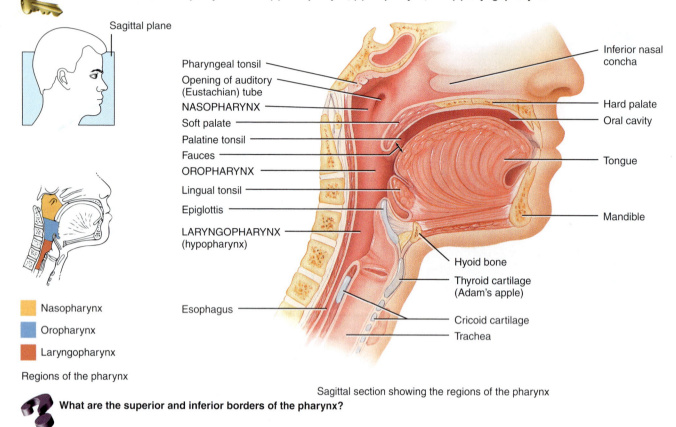

Sagittal plane

Pharyngeal tonsil
Opening of auditory (Eustachian) tube
NASOPHARYNX
Soft palate
Palatine tonsil
Fauces
OROPHARYNX
Lingual tonsil
Epiglottis
LARYNGOPHARYNX (hypopharynx)

Esophagus

Inferior nasal concha
Hard palate
Oral cavity
Tongue
Mandible
Hyoid bone
Thyroid cartilage (Adam's apple)
Cricoid cartilage
Trachea

■ Nasopharynx
■ Oropharynx
■ Laryngopharynx

Regions of the pharynx

Sagittal section showing the regions of the pharynx

❓ **What are the superior and inferior borders of the pharynx?**

Larynx

The **larynx** (LAIR-inks), or voice box, is a short passageway that connects the laryngopharynx with the trachea. It lies in the midline of the neck anterior to the fourth through sixth cervical vertebrae (C4–C6).

The wall of the larynx is composed of nine pieces of cartilage (Figure 23.5). Three occur singly (thyroid cartilage, epiglottis, and cricoid cartilage), and three occur in pairs (arytenoid, cuneiform, and corniculate cartilages). Of the paired cartilages, the arytenoid cartilages are the most important because they influence the positions and tensions of the vocal folds (true vocal cords). Whereas the extrinsic muscles of the larynx connect the cartilages to other structures in the throat, the intrinsic muscles connect the cartilages to each other.

The **thyroid cartilage (Adam's apple)** consists of two fused plates of hyaline cartilage that form the anterior wall of the larynx and give it a triangular shape. It is usually larger in males than in females due to the influence of male sex hormones on its growth during puberty. The ligament that connects the thyroid cartilage to the hyoid bone is called the **thyrohyoid membrane.**

The **epiglottis** (*epi-* = over; *glottis* = tongue) is a large, leaf-shaped piece of elastic cartilage that is covered with epithe-lium (see also Figure 23.4). The "stem" of the epiglottis is the tapered inferior portion that is attached to the anterior rim of the thyroid cartilage and hyoid bone. The broad superior "leaf" portion of the epiglottis is unattached and is free to move up and down like a trap door. During swallowing, the pharynx and larynx rise. Elevation of the pharynx widens it to receive food or drink; elevation of the larynx causes the epiglottis to move down and form a lid over the glottis, closing it off. The **glottis** consists of a pair of folds of mucous membrane, the vocal folds (true vocal cords) in the larynx, and the space between them called the **rima glottidis** (RĪ-ma GLOT-ti-dis). The closing of the larynx in this way during swallowing routes liquids and foods into the esophagus and keeps them out of the larynx and airways inferior to it. When small particles of dust, smoke, food, or liquids pass into the larynx, a cough reflex occurs, usually expelling the material.

The **cricoid cartilage** (KRĪ-koyd = ringlike) is a ring of hyaline cartilage that forms the inferior wall of the larynx. It is attached to the first ring of cartilage of the trachea by the **crico-tracheal ligament.** The thyroid cartilage is connected to the cricoid cartilage by the **cricothyroid ligament.** The cricoid cartilage is the landmark for making an emergency airway (a tra-cheostomy; see page 813).

Figure 23.5 Larynx. (See Tortora, *A Photographic Atlas of the Human Body,* Figures 11.5 and 11.6.)

The larynx is composed of nine pieces of cartilage.

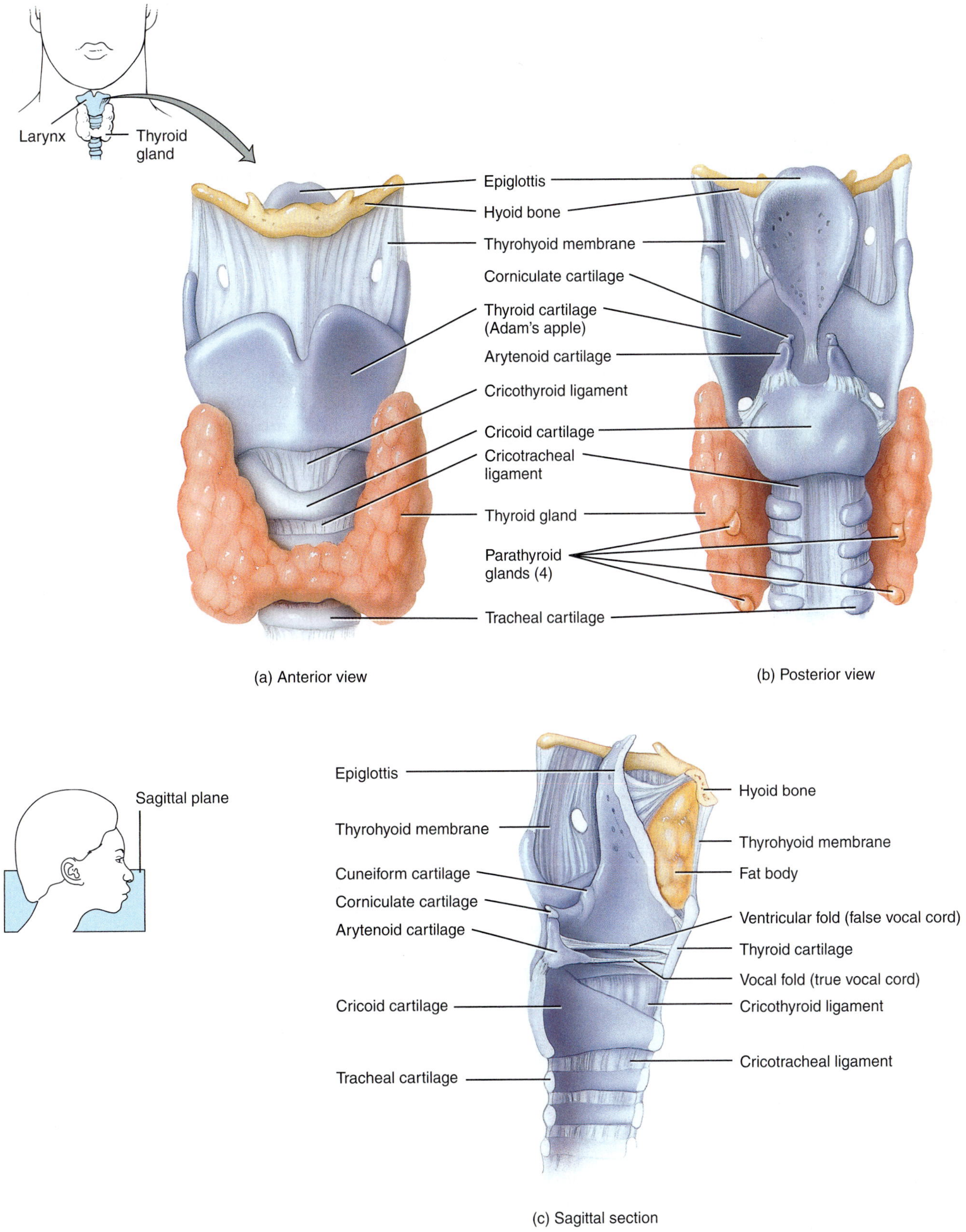

Larynx

Thyroid gland

Epiglottis
Hyoid bone
Thyrohyoid membrane
Corniculate cartilage
Thyroid cartilage (Adam's apple)
Arytenoid cartilage
Cricothyroid ligament
Cricoid cartilage
Cricotracheal ligament
Thyroid gland
Parathyroid glands (4)
Tracheal cartilage

(a) Anterior view

(b) Posterior view

Sagittal plane

Epiglottis
Thyrohyoid membrane
Cuneiform cartilage
Corniculate cartilage
Arytenoid cartilage
Cricoid cartilage
Tracheal cartilage

Hyoid bone
Thyrohyoid membrane
Fat body
Ventricular fold (false vocal cord)
Thyroid cartilage
Vocal fold (true vocal cord)
Cricothyroid ligament
Cricotracheal ligament

(c) Sagittal section

How does the epiglottis prevent aspiration of foods and liquids?

The paired **arytenoid cartilages** (ar'-i-TĒ-noyd = ladle-like) are triangular pieces of mostly hyaline cartilage located at the posterior, superior border of the cricoid cartilage. They attach to the vocal folds and intrinsic pharyngeal muscles. Supported by the arytenoid cartilages, the intrinsic pharyngeal muscles contract and thus move the vocal folds.

The paired **corniculate cartilages** (kor-NIK-ū-lāt = shaped like a small horn), which are horn-shaped pieces of elastic cartilage, are located at the apex of each arytenoid cartilage. The paired **cuneiform cartilages** (KŪ-nē-i-form = wedge-shaped), which are club-shaped elastic cartilages anterior to the corniculate cartilages, support the vocal folds and lateral aspects of the epiglottis.

The lining of the larynx superior to the vocal folds is nonkeratinized stratified squamous epithelium. The lining of the larynx inferior to the vocal folds is pseudostratified ciliated columnar epithelium consisting of ciliated columnar cells, goblet cells, and basal cells. Its mucus helps trap dust not removed in the upper passages. Whereas the cilia in the upper respiratory tract move mucus and trapped particles *down* toward the pharynx, the cilia in the lower respiratory tract move them *up* toward the pharynx.

The Structures of Voice Production

The mucous membrane of the larynx forms two pairs of folds (Figure 23.5c): a superior pair called the **ventricular folds (false vocal cords)** and an inferior pair called simply the **vocal folds (true vocal cords)**. The space between the ventricular folds is known as the **rima vestibuli**. The **laryngeal sinus (ventricle)** is a lateral expansion of the middle portion of the laryngeal cavity between the ventricular folds above and the vocal folds below (see Figure 23.2b).

When the ventricular folds are brought together, they function in holding the breath against pressure in the thoracic cavity, such as might occur when a person strains to lift a heavy object. Deep to the mucous membrane of the vocal folds, which is lined by nonkeratinized stratified squamous epithelium, are bands of elastic ligaments stretched between pieces of rigid cartilage like the strings on a guitar. Intrinsic laryngeal muscles attach to both the rigid cartilage and the vocal folds. When the muscles contract, they pull the elastic ligaments tight and stretch the vocal folds out into the airways so that the rima glottidis is narrowed. If air is directed against the vocal folds, they vibrate and produce sounds (phonation) and set up sound waves in the column of air in the pharynx, nose, and mouth. The greater the pressure of air, the louder the sound.

When the intrinsic muscles of the larynx contract, they pull on the arytenoid cartilages, which causes them to pivot. Contraction of the posterior cricoarytenoid muscles, for example, moves the vocal folds apart (abduction), thereby opening the rima glottidis (Figure 23.6a). By contrast, contraction of the lateral cricoarytenoid muscles moves the vocal folds together (adduction), thereby closing the rima glottidis (Figure 23.6b). Other

intrinsic muscles can elongate (and place tension on) or shorten (and relax) the vocal folds.

Pitch is controlled by the tension on the vocal folds. If they are pulled taut by the muscles, they vibrate more rapidly, and a higher pitch results. Decreasing the muscular tension on the vocal folds produces lower-pitch sounds. Due to the influence of androgens (male sex hormones), vocal folds are usually thicker and longer in males than in females, and therefore they vibrate more slowly. Thus, men's voices generally have a lower range of pitch than women's.

Sound originates from the vibration of the vocal folds, but other structures are necessary for converting the sound into recognizable speech. The pharynx, mouth, nasal cavity, and paranasal sinuses all act as resonating chambers that give the voice its human and individual quality. We produce the vowel sounds by constricting and relaxing the muscles in the wall of the pharynx. Muscles of the face, tongue, and lips help us enunciate words.

Whispering is accomplished by closing all but the posterior portion of the rima glottidis. Because the vocal folds do not vibrate during whispering, there is no pitch to this form of speech. However, we can still produce intelligible speech while whispering by changing the shape of the oral cavity as we enunciate. As the size of the oral cavity changes, its resonance qualities change, which imparts a vowel-like pitch to the air as it rushes toward the lips.

Laryngitis and Cancer of the Larynx

Laryngitis is an inflammation of the larynx that is most often caused by a respiratory infection or irritants such as cigarette smoke. Inflammation of the vocal folds causes hoarseness or loss of voice by interfering with the contraction of the folds or by causing them to swell to the point where they cannot vibrate freely. Many long-term smokers acquire a permanent hoarseness from the damage done by chronic inflammation. **Cancer of the larynx** is found almost exclusively in individuals who smoke. The condition is characterized by hoarseness, pain on swallowing, or pain radiating to an ear. Treatment consists of radiation therapy and/or surgery. ■

Trachea

The **trachea** (TRĀ-kē-a = sturdy), or windpipe, is a tubular passageway for air that is about 12 cm (5 in.) long and 2.5 cm (1 in.) in diameter. It is located anterior to the esophagus (Figure 23.7 on page 814) and extends from the larynx to the superior border of the fifth thoracic vertebra (T5), where it divides into right and left primary bronchi (see Figure 23.8).

The layers of the tracheal wall, from deep to superficial, are (1) mucosa, (2) submucosa, (3) hyaline cartilage, and (4) adventitia, which is composed of areolar connective tissue. The mucosa of the trachea consists of an epithelial layer of pseudostratified ciliated columnar epithelium and an underlying layer

Figure 23.6 **Movement of the vocal folds.** (See Tortora, *A Photographic Atlas of the Human Body*, Figure 11.7.)

🔑 The glottis consists of a pair of folds of mucous membrane in the larynx (the vocal folds) and the space between them (the rima glottidis).

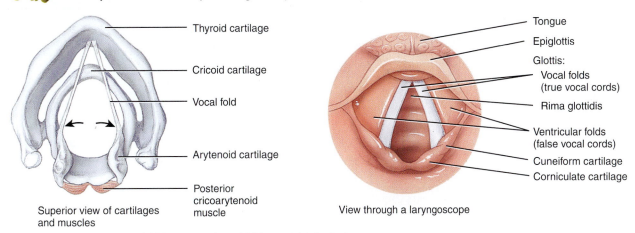

Thyroid cartilage

Cricoid cartilage

Vocal fold

Arytenoid cartilage

Posterior cricoarytenoid muscle

Superior view of cartilages and muscles

Tongue

Epiglottis

Glottis:
Vocal folds (true vocal cords)

Rima glottidis

Ventricular folds (false vocal cords)

Cuneiform cartilage

Corniculate cartilage

View through a laryngoscope

(a) Movement of vocal folds apart (abduction)

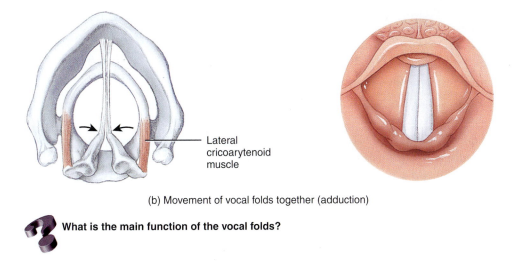

Lateral cricoarytenoid muscle

(b) Movement of vocal folds together (adduction)

❓ **What is the main function of the vocal folds?**

of lamina propria that contains elastic and reticular fibers. The epithelium consists of ciliated columnar cells and goblet cells that reach the luminal surface, plus basal cells that do not (see Table 4.1I on page 113). The epithelium provides the same protection against dust as the membrane lining the nasal cavity and larynx. The submucosa consists of areolar connective tissue that contains seromucous glands and their ducts.

The 16–20 incomplete, horizontal rings of hyaline cartilage resemble the letter **C** and are stacked one on top of another. They may be felt through the skin inferior to the larynx. The open part of each **C**-shaped cartilage ring faces the esophagus (Figure 23.7), an arrangement that accommodates slight expansion of the esophagus into the trachea during swallowing. Transverse smooth muscle fibers, called the **trachealis muscle,** and elastic connective tissue stabilize the open ends of the cartilage rings. The solid **C**-shaped cartilage rings provide a semirigid support so that the tracheal wall does not collapse inward (espe-

cially during inhalation) and obstruct the air passageway. The adventitia of the trachea consists of areolar connective tissue that joins the trachea to surrounding tissues.

🩺 **Tracheostomy and Intubation**

Several conditions may block airflow by obstructing the trachea. For example, the rings of cartilage that support the trachea may collapse due to a crushing injury to the chest, inflammation of the mucous membrane may cause it to swell so much that the airway closes, or vomit or a foreign object may be aspirated into it. Two methods are used to reestablish airflow past a tracheal obstruction. If the obstruction is superior to the level of the larynx, a **tracheostomy** (trā-kē-OS-tō-mē), an operation to make an opening into the trachea, may be performed. In this procedure, a skin incision is followed by a short longitudinal incision into the trachea inferior to the cricoid cartilage. The patient can then breathe through

Figure 23.7 Location of the trachea in relation to the esophagus.

The trachea is anterior to the esophagus and extends from the larynx to the superior border of the fifth thoracic vertebra.

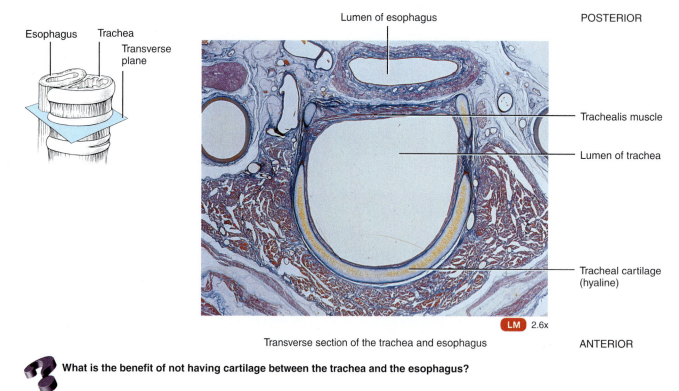

Transverse section of the trachea and esophagus

What is the benefit of not having cartilage between the trachea and the esophagus?

a metal or plastic tracheal tube inserted through the incision. The second method is **intubation,** in which a tube is inserted into the mouth or nose and passed inferiorly through the larynx and trachea. The firm wall of the tube pushes aside any flexible obstruction, and the lumen of the tube provides a passageway for air; any mucus clogging the trachea can be suctioned out through the tube. ■

Bronchi

At the superior border of the fifth thoracic vertebra, the trachea divides into a **right primary bronchus** (BRON-kus = windpipe), which goes into the right lung, and a **left primary bronchus,** which goes into the left lung (Figure 23.8). The right primary bronchus is more vertical, shorter, and wider than the left. As a result, an aspirated object is more likely to enter and lodge in the right primary bronchus than the left. Like the trachea, the primary bronchi (BRON-kē) contain incomplete rings of cartilage and are lined by pseudostratified ciliated columnar epithelium.

At the point where the trachea divides into right and left primary bronchi is an internal ridge called the **carina** (ka-RĪ-na = keel of a boat). It is formed by a posterior and somewhat inferior projection of the last tracheal cartilage. The mucous membrane of the carina is one of the most sensitive areas of the entire larynx and trachea for triggering a cough reflex. Widening and distortion of the carina is a serious sign because it usually

indicates a carcinoma of the lymph nodes around the region where the trachea divides.

On entering the lungs, the primary bronchi divide to form smaller bronchi—the **secondary (lobar) bronchi,** one for each lobe of the lung. (The right lung has three lobes; the left lung has two.) The secondary bronchi continue to branch, forming still smaller bronchi, called **tertiary (segmental) bronchi,** that divide into **bronchioles.** Bronchioles, in turn, branch repeatedly, and the smallest ones branch into even smaller tubes called **terminal bronchioles.** This extensive branching from the trachea resembles an inverted tree and is commonly referred to as the **bronchial tree.**

As the branching becomes more extensive in the bronchial tree, several structural changes may be noted. First, the mucous membrane in the bronchial tree changes from pseudostratified ciliated columnar epithelium in the primary bronchi, secondary bronchi, and tertiary bronchi to ciliated simple columnar epithelium with some goblet cells in larger bronchioles, to mostly ciliated simple cuboidal epithelium with no goblet cells in smaller bronchioles, to mostly nonciliated simple cuboidal epithelium in terminal bronchioles. (In regions where simple nonciliated cuboidal epithelium is present, inhaled particles are removed by macrophages.) Second, plates of cartilage gradually replace the incomplete rings of cartilage in primary bronchi and finally disappear in the distal bronchioles. Third, as the amount of cartilage decreases, the amount of smooth muscle increases. Smooth

Figure 23.8 Branching of airways from the trachea: the bronchial tree.
(See Tortora, *A Photographic Atlas of the Human Body,* Figure 11.8.)

The bronchial tree begins at the trachea and ends at the terminal bronchioles.

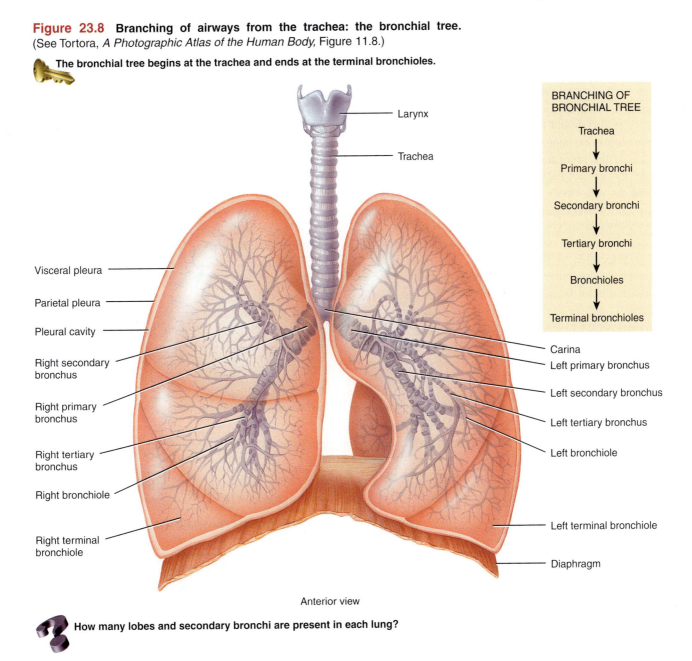

BRANCHING OF
BRONCHIAL TREE

Trachea
↓
Primary bronchi
↓
Secondary bronchi
↓
Tertiary bronchi
↓
Bronchioles
↓
Terminal bronchioles

Larynx

Trachea

Visceral pleura

Parietal pleura

Pleural cavity

Right secondary bronchus

Right primary bronchus

Right tertiary bronchus

Right bronchiole

Right terminal bronchiole

Carina
Left primary bronchus

Left secondary bronchus

Left tertiary bronchus

Left bronchiole

Left terminal bronchiole

Diaphragm

Anterior view

How many lobes and secondary bronchi are present in each lung?

muscle encircles the lumen in spiral bands. Because there is no supporting cartilage, however, muscle spasms can close off the airways. This is what happens during an asthma attack, and it can be a life-threatening situation. During exercise, activity in the sympathetic division of the autonomic nervous system (ANS) increases and the adrenal medulla releases the hormones epinephrine and norepinephrine, both of which cause relaxation of smooth muscle in the bronchioles, which dilates the airways. The result is improved lung ventilation because air reaches the alveoli more quickly. The parasympathetic division of the ANS and mediators of allergic reactions such as histamine cause contraction of bronchiolar smooth muscle and result in constriction of distal bronchioles.

▶ **CHECKPOINT**

1. What functions do the respiratory and cardiovascular systems have in common?

2. What structural and functional features are different in the upper and lower respiratory systems?

3. How do the structure and functions of the external nose and the internal nose compare?

4. What are the three anatomical regions of the pharynx? List the roles of each in respiration.

5. How does the larynx function in respiration and voice production?

6. Describe the location, structure, and function of the trachea.

7. What is the bronchial tree? Describe its structure.

Lungs

The **lungs** (= lightweights, because they float) are paired cone-shaped organs in the thoracic cavity. They are separated from each other by the heart and other structures in the mediastinum, which separates the thoracic cavity into two anatomically distinct chambers. As a result, should trauma cause one lung to collapse, the other may remain expanded. Two layers of serous membrane, collectively called the **pleural membrane** (PLOOR-al; *pleur-* = side), enclose and protect each lung. The superficial layer lines the wall of the thoracic cavity and is called the **parietal pleura;** the deep layer, the **visceral pleura,** covers the lungs themselves (Figure 23.9). Between the visceral and parietal pleurae is a small space, the **pleural cavity,** which contains a small amount of lubricating fluid secreted by the membranes. This fluid reduces friction between the membranes, allowing them to slide easily over one another during breathing. Pleural fluid also causes the two membranes to adhere to one another just as a film of water causes two glass slides to stick together, a phenomenon called surface tension. Separate pleural cavities surround the left and right lungs. Inflammation of the pleural membrane, called **pleurisy** or **pleuritis,** may in its early stages cause pain due to friction between the parietal and visceral layers of the pleura. If the inflammation persists, excess fluid accumulates in the pleural space, a condition known as **pleural effusion.**

The lungs extend from the diaphragm to just slightly superior to the clavicles and lie against the ribs anteriorly and posteriorly (Figure 23.10a). The broad inferior portion of the lung, the **base,** is concave and fits over the convex area of the diaphragm. The narrow superior portion of the lung is the **apex.** The surface of the lung lying against the ribs, the **costal surface,** matches the rounded curvature of the ribs. The **mediastinal (medial) surface** of each lung contains a region, the **hilus,** through which bronchi, pulmonary blood vessels, lymphatic vessels, and nerves enter and exit (Figure 23.10e). These structures are held together by the pleura and connective tissue and constitute the **root** of the lung. Medially, the left lung also contains a concavity, the **cardiac notch,** in which the heart lies. Due to the space occupied by the heart, the left lung is about 10% smaller than the right lung. Although the right lung is thicker and broader, it is

Figure 23.9   Relationship of the pleural membranes to the lungs.   The arrow in the inset indicates the direction from which the lungs are viewed (superior).

🔑   The parietal pleura lines the thoracic cavity, whereas the visceral pleura covers the lungs.

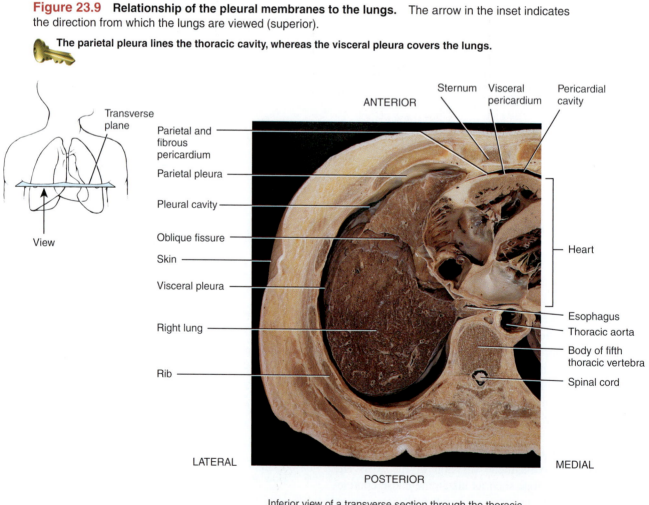

Inferior view of a transverse section through the thoracic cavity showing the pleural cavity and pleural membranes

What type of membrane is the pleural membrane?

Figure 23.10   Surface anatomy of the lungs.   (See Tortora, *A Photographic Atlas of the Human Body,* Figures 11.12 and 11.14.)

The oblique fissure divides the left lung into two lobes. The oblique and horizontal fissures divide the right lung into three lobes.

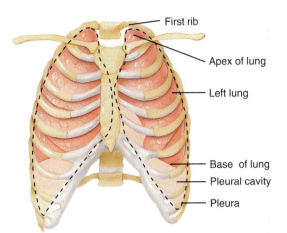

First rib
Apex of lung
Left lung
Base of lung
Pleural cavity
Pleura

(a) Anterior view of lungs and pleurae in thorax

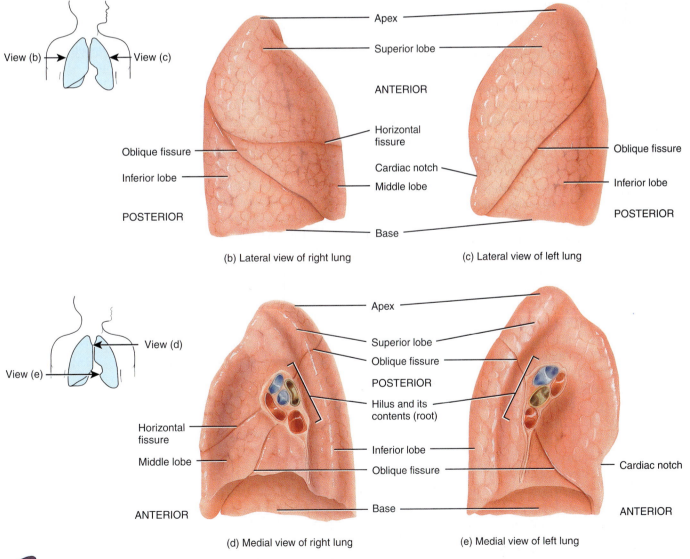

View (b)   View (c)

Apex
Superior lobe
ANTERIOR
Oblique fissure
Horizontal fissure
Inferior lobe
Cardiac notch
Middle lobe
Oblique fissure
Inferior lobe
POSTERIOR
POSTERIOR
Base

(b) Lateral view of right lung   (c) Lateral view of left lung

View (d)
View (e)

Apex
Superior lobe
Oblique fissure
POSTERIOR
Hilus and its contents (root)
Horizontal fissure
Middle lobe
Inferior lobe
Oblique fissure
Cardiac notch
ANTERIOR
Base
ANTERIOR

(d) Medial view of right lung   (e) Medial view of left lung

Why are the right and left lungs slightly different in size and shape?

also somewhat shorter than the left lung because the diaphragm is higher on the right side, accommodating the liver that lies inferior to it.

The lungs almost fill the thorax (see Figure 23.10a). The apex of the lungs lies superior to the medial third of the clavicles and is the only area that can be palpated. The anterior, lateral, and posterior surfaces of the lungs lie against the ribs. The base of the lungs extends from the sixth costal cartilage anteriorly to the spinous process of the tenth thoracic vertebra posteriorly. The pleura extends about 5 cm below the base from the sixth costal cartilage anteriorly to the twelfth rib posteriorly. Thus, the lungs do not completely fill the pleural cavity in this area. Removal of excessive fluid in the pleural cavity can be accomplished without injuring lung tissue by inserting the needle posteriorly through the seventh intercostal space, a procedure termed **thoracentesis** (thor′-a-sen-TĒ-sis; -*centesis* = puncture).

Lobes, Fissures, and Lobules

One or two fissures divide each lung into lobes (Figure 23.10b–e). Both lungs have an **oblique fissure,** which extends inferiorly and anteriorly; the right lung also has a **horizontal fissure.** The oblique fissure in the left lung separates the **superior lobe** from the **inferior lobe.** In the right lung, the superior part of the oblique fissure separates the superior lobe from the inferior lobe, whereas the inferior part of the oblique fissure separates the inferior lobe from the **middle lobe.** The horizontal fissure of the right lung subdivides the superior lobe, thus forming a middle lobe.

Each lobe receives its own secondary (lobar) bronchus. Thus, the right primary bronchus gives rise to three secondary (lobar) bronchi called the **superior, middle,** and **inferior secondary (lobar) bronchi,** whereas the left primary bronchus gives rise to **superior** and **inferior secondary (lobar) bronchi.** Within the substance of the lung, the secondary bronchi give rise to the **tertiary (segmental) bronchi,** which are constant in both origin and distribution—there are ten tertiary bronchi in each lung. The segment of lung tissue that each tertiary bronchus supplies is called a **bronchopulmonary segment.** Bronchial and pulmonary disorders (such as tumors or abscesses) that are localized in a bronchopulmonary segment may be surgically removed without seriously disrupting the surrounding lung tissue.

Each bronchopulmonary segment of the lungs has many small compartments called **lobules,** each of which is wrapped in elastic connective tissue and contains a lymphatic vessel, an arteriole, a venule, and a branch from a terminal bronchiole (Figure 23.11a). Terminal bronchioles subdivide into microscopic branches called **respiratory bronchioles** (Figure 23.11b). As the respiratory bronchioles penetrate more deeply into the lungs, the epithelial lining changes from simple cuboidal to simple squamous. Respiratory bronchioles, in turn, subdivide into several (2–11) **alveolar ducts.** The respiratory passages from the trachea to the alveolar ducts contain about 25 orders of branching; that is, branching—from the trachea into primary bronchi (first-order branching) into secondary bronchi (second-order branching) and so on down to the alveolar ducts—occurs about 25 times.

Alveoli

Around the circumference of the alveolar ducts are numerous alveoli and alveolar sacs. An **alveolus** (al-VĒ-ō-lus) is a cup-shaped outpouching lined by simple squamous epithelium and supported by a thin elastic basement membrane; an **alveolar sac** consists of two or more alveoli that share a common opening (Figure 23.11a, b). The walls of alveoli consist of two types of alveolar epithelial cells (Figure 23.12 on page 820). **Type I alveolar cells,** the predominant cells, are simple squamous epithelial cells that form a nearly continuous lining of the alveolar wall. **Type II alveolar cells,** also called **septal cells,** are fewer in number and are found between type I alveolar cells. The thin type I alveolar cells are the main sites of gas exchange. Type II alveolar cells, which are rounded or cuboidal epithelial cells whose free surfaces contain microvilli, secrete alveolar fluid, which keeps the surface between the cells and the air moist. Included in the alveolar fluid is **surfactant** (sur-FAK-tant), a complex mixture of phospholipids and lipoproteins. Surfactant lowers the surface tension of alveolar fluid, which reduces the tendency of alveoli to collapse. Associated with the alveolar wall are **alveolar macrophages (dust cells),** wandering phagocytes that remove fine dust particles and other debris in the alveolar spaces. Also present are fibroblasts that produce reticular and elastic fibers. Underlying the layer of type I alveolar cells is an elastic basement membrane. On the outer surface of the alveoli, the lobule's arteriole and venule disperse into a network of blood capillaries (see Figure 23.11a) that consist of a single layer of endothelial cells and basement membrane.

The exchange of O_2 and CO_2 between the air spaces in the lungs and the blood takes place by diffusion across the alveolar and capillary walls, which together form the **respiratory membrane.** Extending from the alveolar air space to blood plasma, the respiratory membrane consists of four layers (see Figure 23.12b):

1. A layer of type I and type II alveolar cells and associated alveolar macrophages that constitutes the **alveolar wall.**

2. An **epithelial basement membrane** underlying the alveolar wall.

3. A **capillary basement membrane** that is often fused to the epithelial basement membrane.

4. The **endothelial cells** of the capillary.

Despite having several layers, the respiratory membrane is very thin—only $0.5\mu m$ thick, about one-sixteenth the diameter of a red blood cell. This thinness allows rapid diffusion of gases. Moreover, it has been estimated that the lungs contain 300 million alveoli, providing an immense surface area of 70 m^2 (750 ft^2)—about the size of a handball court—for the exchange of gases.

Nebulization

Many respiratory disorders are treated by means of **nebulization** (neb-ū-li-ZA-shun). This procedure consists of administering medication in the form of droplets that are suspended in air into the respiratory tract. The patient inhales the medication as a fine mist. Nebulization therapy can be used with many different types of drugs, such as chemicals that relax the smooth muscle of the airways, chemicals that reduce the thickness of mucus, and antibiotics. ■

Blood Supply to the Lungs

The lungs receive blood via two sets of arteries: pulmonary arteries and bronchial arteries. Deoxygenated blood passes through the pulmonary trunk, which divides into a left pulmonary artery that enters the left lung and a right pulmonary artery that enters the right lung. Return of the oxygenated blood to the heart occurs by way of the four pulmonary veins, which drain into the left atrium (see Figure 21.30 on page 753). A unique feature of pulmonary blood vessels is their constriction in response to localized hypoxia (low O_2 level). In all other body tissues, hypoxia causes dilation of blood vessels, which serves to increase blood flow to a tissue that is not receiving adequate O_2. In the lungs, however, vasoconstriction in response to hypoxia diverts pulmonary blood from poorly ventilated areas to well-ventilated regions of the lungs. This phenomenon is known as **ventilation-perfusion coupling** because the perfusion (blood flow) to each area of the lungs matches the extent of ventilation (airflow) to alveoli in that area.

Figure 23.11 **Microscopic anatomy of a lobule of the lungs.**

🔑 Alveolar sacs consist of two or more alveoli that share a common opening.

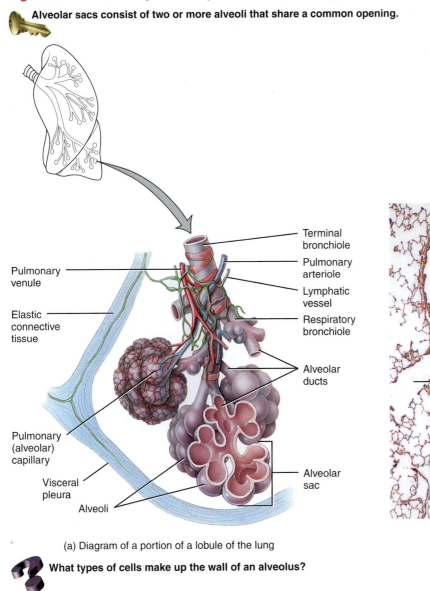

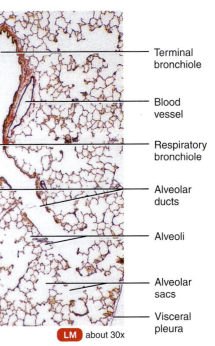

(a) Diagram of a portion of a lobule of the lung

(b) Lung lobule

LM about 30x

What types of cells make up the wall of an alveolus?

Figure 23.12 Structural components of an alveolus. The respiratory membrane consists of a layer of type I and type II alveolar cells, an epithelial basement membrane, a capillary basement membrane, and the capillary endothelium.

🔑 The exchange of respiratory gases occurs by diffusion across the respiratory membrane.

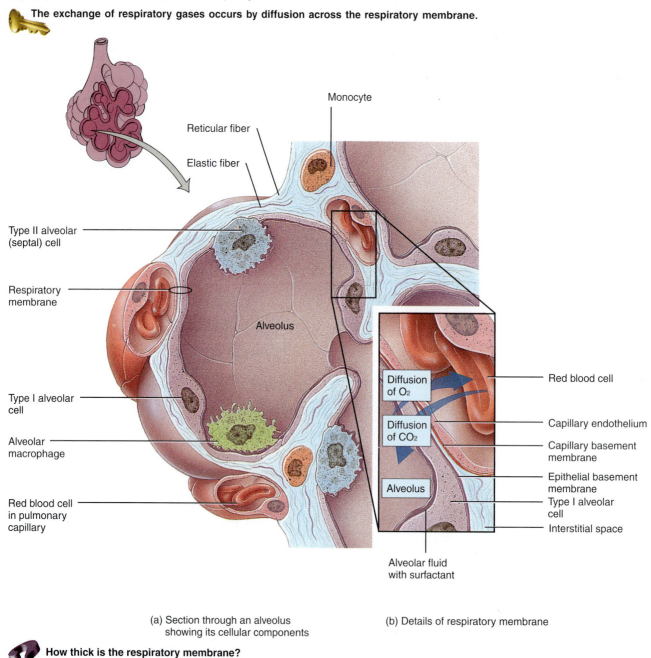

(a) Section through an alveolus showing its cellular components

(b) Details of respiratory membrane

❓ How thick is the respiratory membrane?

Bronchial arteries, which branch from the aorta, deliver oxygenated blood to the lungs. This blood mainly perfuses the walls of the bronchi and bronchioles. Connections exist between branches of the bronchial arteries and branches of the pulmonary arteries, however, and most blood returns to the heart via pulmonary veins. Some blood, however, drains into bronchial veins, branches of the azygos system, and returns to the heart via the superior vena cava.

▶ CHECKPOINT

8. Where are the lungs located? Distinguish the parietal pleura from the visceral pleura.

9. Define each of the following parts of a lung: base, apex, costal surface, medial surface, hilus, root, cardiac notch, lobe, and lobule.

10. What is a bronchopulmonary segment?

11. Describe the histology and function of the respiratory membrane.

PULMONARY VENTILATION

OBJECTIVE

- **Describe the events that cause inhalation and exhalation.**

The process of gas exchange in the body, called **respiration,** occurs in three basic steps:

1. **Pulmonary ventilation** (*pulmon-* = lung), or breathing, is the mechanical flow of air into (inhalation) and out of (exhalation) the lungs.

2. **External respiration** is the exchange of gases between the air spaces of the lungs and the blood in pulmonary capillaries. In this process, pulmonary capillary blood gains O_2 and loses CO_2.

3. **Internal respiration** is the exchange of gases between blood in systemic capillaries and tissue cells. The blood loses O_2 and gains CO_2. Within cells, the metabolic reactions that consume O_2 and give off CO_2 during the production of ATP are termed *cellular respiration* (discussed in Chapter 25).

Pulmonary ventilation (breathing) is the process by which gases flow between the atmosphere and lung alveoli. Air flows between the atmosphere and the lungs because of alternating pressure differences created by contraction and relaxation of respiratory muscles. The rate of airflow and the amount of effort needed for breathing is also influenced by alveolar surface tension, compliance of the lungs, and airway resistance.

Pressure Changes During Pulmonary Ventilation

Air moves into the lungs when the air pressure inside the lungs is less than the air pressure in the atmosphere and out of the lungs when the pressure inside the lungs is greater than the pressure in the atmosphere.

Inhalation

Breathing in is called **inhalation (inspiration).** Just before each inhalation, the air pressure inside the lungs is equal to the pressure of the atmosphere, which at sea level is about 760 millimeters of mercury (mmHg), or 1 atmosphere (atm). For air to flow into the lungs, the pressure inside the alveoli must become lower than the atmospheric pressure. This condition is achieved by increasing the volume of the lungs.

The pressure of a gas in a closed container is inversely proportional to the volume of the container. This means that if the size of a closed container is increased, the pressure of the gas inside the container decreases, and that if the size of the container is decreased, then the pressure inside it increases. This inverse relationship between volume and pressure, called **Boyle's law,** may be demonstrated as follows (Figure 23.13): Suppose we place a gas in a cylinder that has a movable piston and a pressure gauge, and that the initial pressure created by the

Figure 23.13 Boyle's law.

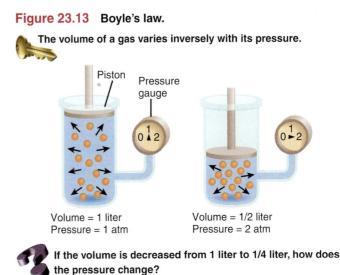

The volume of a gas varies inversely with its pressure.

Volume = 1 liter
Pressure = 1 atm

Volume = 1/2 liter
Pressure = 2 atm

If the volume is decreased from 1 liter to 1/4 liter, how does the pressure change?

gas molecules striking the wall of the container is 1 atm. If the piston is pushed down, the gas is compressed into a smaller volume, so that the same number of gas molecules strike less wall area. The gauge shows that the pressure doubles as the gas is compressed to half its original volume. In other words, the same number of molecules in half the volume produces twice the pressure. Conversely, if the piston is raised to increase the volume, the pressure decreases. Thus, the pressure of a gas varies inversely with volume. Boyle's law also applies to everyday activities such as the operation of a bicycle pump or the blowing up of a balloon.

Differences in pressure caused by changes in lung volume force air into our lungs when we inhale and out when we exhale. For inhalation to occur, the lungs must expand, which increases lung volume and thus decreases the pressure in the lungs to below atmospheric pressure. The first step in expanding the lungs during normal quiet inhalation involves contraction of the main muscles of inhalation, the diaphragm and external intercostals (Figure 23.14).

The most important muscle of inhalation is the diaphragm, the dome-shaped skeletal muscle that forms the floor of the thoracic cavity. It is innervated by fibers of the phrenic nerves, which emerge from the spinal cord at cervical levels 3, 4, and 5. Contraction of the diaphragm causes it to flatten and increases the vertical dimension of the thoracic cavity. During normal quiet inhalation, the diaphragm descends about 1 cm (0.4 in.), producing a pressure difference of 1–3 mmHg and the inhalation of about 500 mL of air. In strenuous breathing, the diaphragm may descend 10 cm (4 in.), which produces a pressure difference of 100 mmHg and the inhalation of 2–3 liters of air. Contraction of the diaphragm is responsible for about 75% of the air that enters the lungs during quiet breathing.

The next most important muscles of inhalation are the external intercostals. When these muscles contract, they elevate the

Figure 23.14 Muscles of inhalation and exhalation and their actions. The pectoralis minor muscle (not shown here) is illustrated in Figure 11.14a on page 343.

During deep, labored breathing, accessory muscles of inhalation (sternocleidomastoids, scalenes, and pectoralis minors) participate.

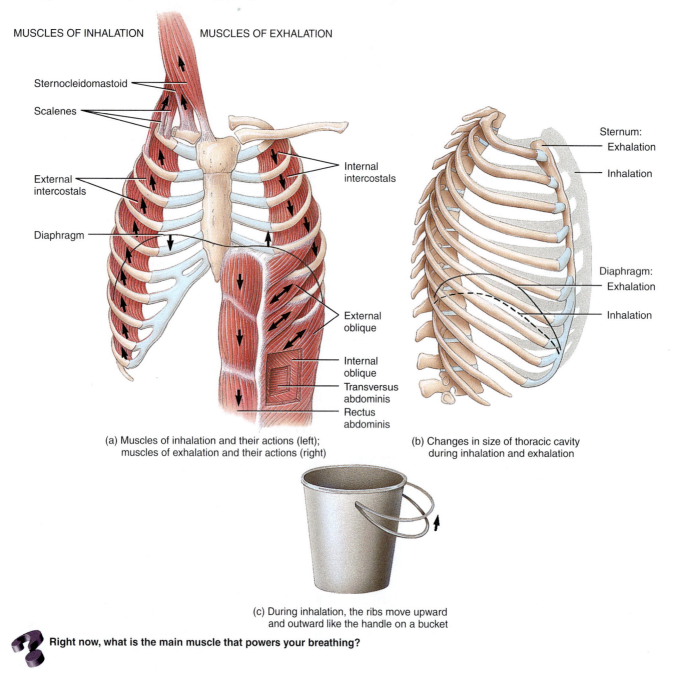

MUSCLES OF INHALATION   MUSCLES OF EXHALATION

Sternocleidomastoid

Scalenes

External intercostals

Diaphragm

Internal intercostals

External oblique

Internal oblique

Transversus abdominis

Rectus abdominis

Sternum:
Exhalation
Inhalation

Diaphragm:
Exhalation
Inhalation

(a) Muscles of inhalation and their actions (left); muscles of exhalation and their actions (right)

(b) Changes in size of thoracic cavity during inhalation and exhalation

(c) During inhalation, the ribs move upward and outward like the handle on a bucket

Right now, what is the main muscle that powers your breathing?

ribs. As a result, there is an increase in the anteroposterior and lateral diameters of the chest cavity. Contraction of the external intercostals is responsible for about 25% of the air that enters the lungs during normal quiet breathing.

During quiet inhalations, the pressure between the two pleural layers, called **intrapleural pressure,** is always subatmospheric (lower than atmospheric pressure). Just before in-

halation, it is about 4 mmHg less than the atmospheric pressure, or about 756 mmHg if the atmospheric pressure is 760 mmHg (Figure 23.15). As the diaphragm and external intercostals contract and the overall size of the thoracic cavity increases, the volume of the pleural cavity also increases, which causes intrapleural pressure to decrease to about 754 mmHg. During expansion of the thorax, the parietal and visceral pleurae

Figure 23.15 **Pressure changes in pulmonary ventilation.** During inhalation, the diaphragm contracts, the chest expands, the lungs are pulled outward, and alveolar pressure decreases. During exhalation, the diaphragm relaxes, the lungs recoil inward, and alveolar pressure increases, forcing air out of the lungs.

🔑 **Air moves into the lungs when alveolar pressure is less than atmospheric pressure, and out of the lungs when alveolar pressure is greater than atmospheric pressure.**

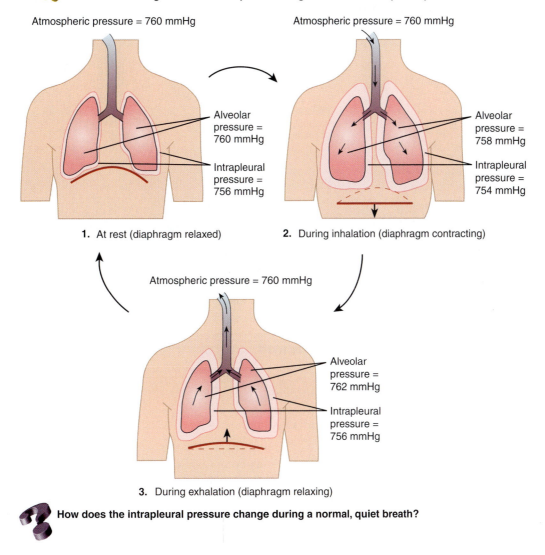

1. At rest (diaphragm relaxed)

2. During inhalation (diaphragm contracting)

3. During exhalation (diaphragm relaxing)

❓ **How does the intrapleural pressure change during a normal, quiet breath?**

normally adhere tightly because of the subatmospheric pressure between them and because of the surface tension created by their moist adjoining surfaces. As the thoracic cavity expands, the parietal pleura lining the cavity is pulled outward in all directions, and the visceral pleura and lungs are pulled along with it.

As the volume of the lungs increases in this way, the pressure inside the lungs, called the **alveolar (intrapulmonic) pressure,** drops from 760 to 758 mmHg. A pressure difference is thus established between the atmosphere and the alveoli. Because air always flows from a region of higher pressure to a region of lower pressure, inhalation takes place. Air continues to flow into the lungs as long as a pressure difference exists.

During deep, forceful inhalations, accessory muscles of inspiration also participate in increasing the size of the thoracic cavity (see Figure 23.14a). The muscles are so-named because they make little, if any, contribution during normal quiet inhalation, but during exercise or forced ventilation they may contract vigorously. The accessory muscles of inhalation include the sternocleidomastoid muscles, which elevate the sternum; the scalene muscles, which elevate the first two ribs; and the pectoralis minor muscles, which elevate the third through fifth ribs. Because both normal quiet inhalation and inhalation during exercise or forced ventilation involve muscular contraction, the process of inhalation is said to be *active.*

Figure 23.16a summarizes the events of inhalation.

Figure 23.16 **Summary of events of inhalation and exhalation.**

Inhalation and exhalation are caused by changes in alveolar pressure.

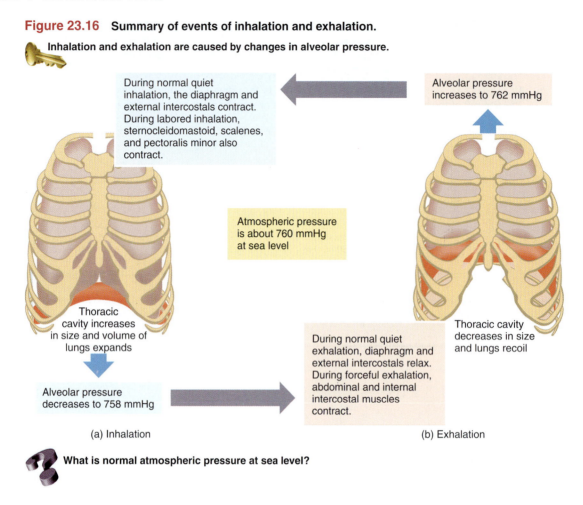

During normal quiet inhalation, the diaphragm and external intercostals contract. During labored inhalation, sternocleidomastoid, scalenes, and pectoralis minor also contract.

Alveolar pressure increases to 762 mmHg

Atmospheric pressure is about 760 mmHg at sea level

Thoracic cavity increases in size and volume of lungs expands

Alveolar pressure decreases to 758 mmHg

During normal quiet exhalation, diaphragm and external intercostals relax. During forceful exhalation, abdominal and internal intercostal muscles contract.

Thoracic cavity decreases in size and lungs recoil

(a) Inhalation

(b) Exhalation

What is normal atmospheric pressure at sea level?

Exhalation

Breathing out, called **exhalation (expiration),** is also due to a pressure gradient, but in this case the gradient is in the opposite direction: The pressure in the lungs is greater than the pressure of the atmosphere. Normal exhalation during quiet breathing, unlike inhalation, is a *passive process* because no muscular contractions are involved. Instead, exhalation results from **elastic recoil** of the chest wall and lungs, both of which have a natural tendency to spring back after they have been stretched. Two inwardly directed forces contribute to elastic recoil: (1) the recoil of elastic fibers that were stretched during inhalation and (2) the inward pull of surface tension due to the film of alveolar fluid.

Exhalation starts when the inspiratory muscles relax. As the diaphragm relaxes, its dome moves superiorly owing to its elasticity. As the external intercostals relax, the ribs are depressed. These movements decrease the vertical, lateral, and anteroposterior dimensions of the thoracic cavity, which decreases lung volume. In turn, the alveolar pressure increases, to about 762 mmHg. Air then flows from the area of higher pressure in the alveoli to the area of lower pressure in the atmosphere (see Figure 23.15).

Exhalation becomes active only during forceful breathing, as occurs while playing a wind instrument or during exercise. During these times, muscles of exhalation—the abdominals and internal intercostals (see Figure 23.14a)—contract, which increases pressure in the abdominal region and thorax. Contraction of the abdominal muscles moves the inferior ribs downward and compresses the abdominal viscera, thereby forcing the diaphragm superiorly. Contraction of the internal intercostals, which extend inferiorly and posteriorly between adjacent ribs, pulls the ribs inferiorly. Although intrapleural pressure is always less than alveolar pressure, it may briefly exceed atmospheric pressure during a forceful exhalation, such as during a cough.

Figure 23.16b summarizes the events of exhalation.

Other Factors Affecting Pulmonary Ventilation

Whereas air pressure differences drive airflow during inhalation and exhalation, three other factors affect the rate of airflow and the ease of pulmonary ventilation: surface tension of the alveolar fluid, compliance of the lungs, and airway resistance.

Surface Tension of Alveolar Fluid

As noted earlier, a thin layer of alveolar fluid coats the luminal surface of alveoli and exerts a force known as **surface tension.** Surface tension arises at all air–water interfaces because the polar water molecules are more strongly attracted to each other than they are to gas molecules in the air. When liquid surrounds a sphere of air, as in an alveolus or a soap bubble, surface tension produces an inwardly directed force. Soap bubbles "burst" because they collapse inward due to surface tension. In the lungs, surface tension causes the alveoli to assume the smallest possible diameter. During breathing, surface tension must be overcome to expand the lungs during each inhalation. Surface tension also accounts for two-thirds of lung elastic recoil, which decreases the size of alveoli during exhalation.

The surfactant present in alveolar fluid reduces its surface tension below the surface tension of pure water. A deficiency of surfactant in premature infants causes *respiratory distress syndrome,* in which the surface tension of alveolar fluid is greatly increased, so that many alveoli collapse at the end of each exhalation. Great effort is then needed at the next inhalation to reopen the collapsed alveoli.

Pneumothorax

The pleural cavities are sealed off from the outside environment, which prevents them from equalizing their pressure with that of the atmosphere. Injuries of the chest wall that allow air to enter the intrapleural space either from the outside or from the alveoli cause **pneumothorax** (*pneumo-* = air), filling of the pleural cavity with air. Because the intrapleural pressure is then equal to atmospheric pressure rather than subatmospheric, surface tension and recoil of elastic fibers cause the lung to collapse. ■

Compliance of the Lungs

Compliance refers to how much effort is required to stretch the lungs and chest wall. High compliance means that the lungs and chest wall expand easily; low compliance means that they resist expansion. By analogy, a thin balloon that is easy to inflate has high compliance, whereas a heavy and stiff balloon that takes a lot of effort to inflate has low compliance. In the lungs, compliance is related to two principal factors: elasticity and surface tension. The lungs normally have high compliance and expand easily because elastic fibers in lung tissue are easily stretched and surfactant in alveolar fluid reduces surface tension. Decreased compliance is a common feature in pulmonary conditions that (1) scar lung tissue (for example, tuberculosis), (2) cause lung tissue to become filled with fluid (pulmonary edema), (3) produce a deficiency in surfactant, or (4) impede lung expansion in any way (for example, paralysis of the intercostal muscles). An increase in lung compliance occurs in emphysema due to destruction of elastic fibers in alveolar walls.

Airway Resistance

As is true for blood flow through blood vessels, the rate of airflow through the airways depends on both the pressure difference and the resistance: Airflow equals the pressure difference between the alveoli and the atmosphere divided by the resistance. The walls of the airways, especially the bronchioles, offer some resistance to the normal flow of air into and out of the lungs. As the lungs expand during inhalation, the bronchioles enlarge because their walls are pulled outward in all directions. Larger-diameter airways have decreased resistance. Airway resistance then increases during exhalation as the diameter of bronchioles decreases. Also, the degree of contraction or relaxation of smooth muscle in the walls of the airways regulates airway diameter and thus resistance. Increased signals from the sympathetic division of the autonomic nervous system cause relaxation of this smooth muscle, which results in bronchodilation and decreased resistance.

Any condition that narrows or obstructs the airways increases resistance, and more pressure is required to maintain the same airflow. The hallmark of asthma or chronic obstructive pulmonary disease (COPD)—emphysema or chronic bronchitis—is increased airway resistance due to obstruction or collapse of airways.

Breathing Patterns and Modified Respiratory Movements

The term for the normal pattern of quiet breathing is **eupnea** (ūp-NĒ-a; *eu-* = good, easy, or normal; *-pnea* = breath). Eupnea can consist of shallow, deep, or combined shallow and deep breathing. A pattern of shallow (chest) breathing, called **costal breathing,** consists of an upward and outward movement of the chest due to contraction of the external intercostal muscles. A pattern of deep (abdominal) breathing, called **diaphragmatic breathing,** consists of the outward movement of the abdomen due to the contraction and descent of the diaphragm.

Respirations also provide humans with methods for expressing emotions such as laughing, sighing, and sobbing. Moreover, respiratory air can be used to expel foreign matter from the lower air passages through actions such as sneezing and coughing. Respiratory movements are also modified and controlled during talking and singing. Some of the modified respiratory movements that express emotion or clear the airways are listed in Table 23.1. All these movements are reflexes, but some of them also can be initiated voluntarily.

▶ **CHECKPOINT**

12. What are the basic differences among pulmonary ventilation, external respiration, and internal respiration?

13. Compare what happens during quiet versus forceful pulmonary ventilation.

14. Describe how alveolar surface tension, compliance, and airway resistance affect pulmonary ventilation.

15. Define the various kinds of modified respiratory movements.

Table 23.1	Modified Respiratory Movements
Movement	**Description**
Coughing	A long-drawn and deep inhalation followed by a complete closure of the rima glottidis, which results in a strong exhalation that suddenly pushes the rima glottidis open and sends a blast of air through the upper respiratory passages. Stimulus for this reflex act may be a foreign body lodged in the larynx, trachea, or epiglottis.
Sneezing	Spasmodic contraction of muscles of exhalation that forcefully expels air through the nose and mouth. Stimulus may be an irritation of the nasal mucosa.
Sighing	A long-drawn and deep inhalation immediately followed by a shorter but forceful exhalation.
Yawning	A deep inhalation through the widely opened mouth producing an exaggerated depression of the mandible. It may be stimulated by drowsiness, fatigue, or someone else's yawning, but precise cause is unknown.
Sobbing	A series of convulsive inhalations followed by a single prolonged exhalation. The rima glottidis closes earlier than normal after each inhalation so only a little air enters the lungs with each inhalation.
Crying	An inhalation followed by many short convulsive exhalations, during which the rima glottidis, remains open and the vocal folds vibrate; accompanied by characteristic facial expressions and tears.
Laughing	The same basic movements as crying, but the rhythm of the movements and the facial expressions usually differ from those of crying. Laughing and crying are sometimes indistinguishable.
Hiccupping	Spasmodic contraction of the diaphragm followed by a spasmodic closure of the rima glottidis, which produces a sharp sound on inhalation. Stimulus is usually irritation of the sensory nerve endings of the gastrointestinal tract.
Valsalva maneuver	Forced exhalation against a closed rima glottidis as may occur during periods of straining while defecating.

LUNG VOLUMES AND CAPACITIES

► **O B J E C T I V E**

• **Define the various lung volumes and capacities.**

While at rest, a healthy adult averages 12 breaths a minute, with each inhalation and exhalation moving about 500 mL of air into and out of the lungs. The volume of one breath is called the **tidal volume** (V_T). Thus, the **minute ventilation (MV)**—the total volume of air inhaled and exhaled each minute—is respiratory rate multiplied by tidal volume:

$$MV = 12 \text{ breaths/min} \times 500 \text{ mL/breath}$$
$$= 6 \text{ liters/min}$$

A lower-than-normal minute ventilation usually is a sign of pulmonary malfunction. The apparatus commonly used to measure the volume of air exchanged during breathing and the respiratory rate is a **spirometer** (*spiro-* = breathe; *meter* = measuring device) or **respirometer.** The record is called a **spirogram.** Inhalation is recorded as an upward deflection, and exhalation is recorded as a downward deflection (Figure 23.17).

Tidal volume varies considerably from one person to another and in the same person at different times. In a typical adult, about 70% of the tidal volume (350 mL) actually reaches the respiratory portion of the respiratory system—the respiratory bronchioles, alveolar ducts, alveolar sacs, and alveoli—and participates in external respiration. The other 30% (150 mL) remains in the conducting airways of the nose, pharynx, larynx, trachea, bronchi, bronchioles, and terminal bronchioles. Collectively, these conducting airways are known as the **anatomic dead space.** (An easy rule of thumb for determining the volume of your anatomic dead space in milliliters is that it is about the same as your ideal weight in pounds.) Not all of the minute ventilation can be used in gas exchange because some of it remains in the anatomic dead space. The **alveolar ventilation rate** is the volume of air per minute that reaches the alveoli and other respiratory portions. In the example just given, alveolar ventilation rate would be 350 mL/breath × 12 breaths/min = 4200 mL/min.

Several other lung volumes are defined relative to forceful breathing. In general, these volumes are larger in males, taller individuals, and younger adults, and smaller in females, shorter individuals, and the elderly. Various disorders also may be diagnosed by comparison of actual and predicted normal values for one's gender, height, and age. The values given here are averages for young adults.

By taking a very deep breath, you can inhale a good deal more than 500 mL. This additional inhaled air, called the **inspiratory reserve volume,** is about 3100 mL (Figure 23.17). Even more air can be inhaled if inhalation follows forced exhalation. If you inhale normally and then exhale as forcibly as possible, you should be able to push out 1200 mL of air in addition to the 500 mL of tidal volume. The extra 1200 mL is called the **expiratory reserve volume.** The **$FEV_{1.0}$** is the **forced expiratory volume in 1 second,** which is the volume of air that can be exhaled from the lungs in 1 second with maximal effort following a maximal inhalation. Typically, chronic obstructive pulmonary disease (COPD) greatly reduces $FEV_{1.0}$. That is because COPD increases airway resistance.

Even after the expiratory reserve volume is exhaled, considerable air remains in the lungs because the subatmospheric intrapleural pressure keeps the alveoli slightly inflated, and some air also remains in the noncollapsible airways. This volume,

Figure 23.17 Spirogram of lung volumes and capacities (average values for a healthy adult).
Note that the spirogram is read from right (start of record) to left (end of record).

🔑 **Lung capacities are combinations of various lung volumes.**

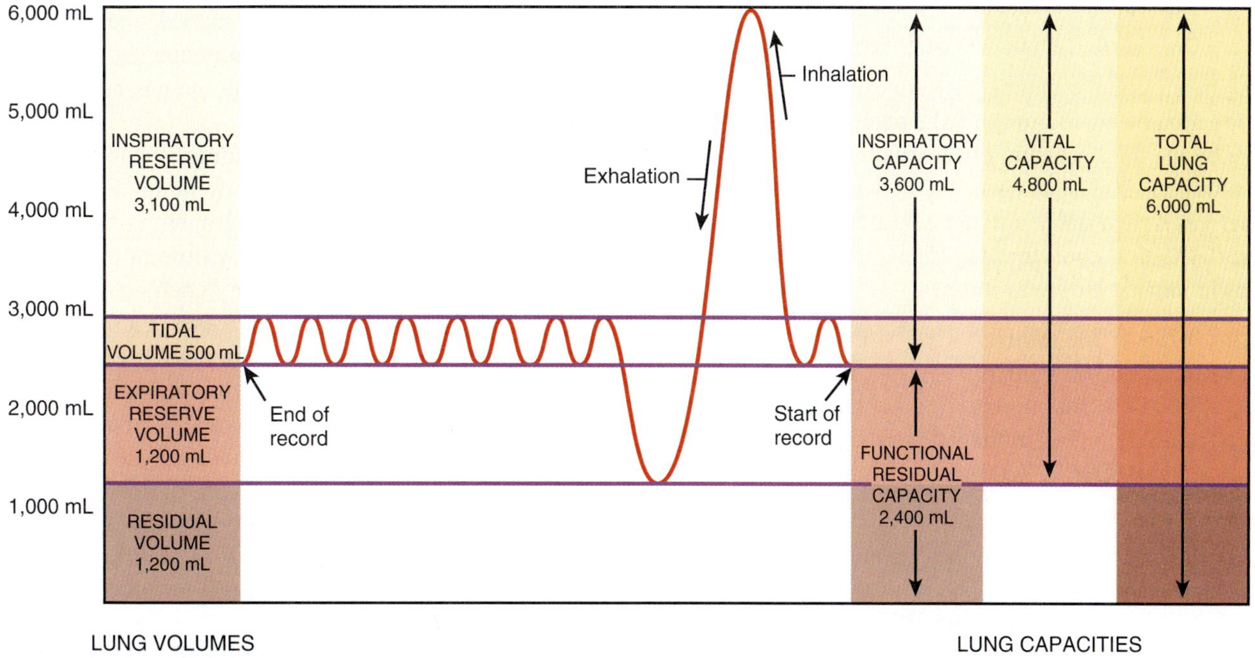

LUNG VOLUMES   LUNG CAPACITIES

❓ **If you breathe in as deeply as possible and then exhale as much air as you can, which lung capacity have you demonstrated?**

---

which cannot be measured by spirometry, is called the **residual volume** and amounts to about 1200 mL.

If the thoracic cavity is opened, the intrapleural pressure rises to equal the atmospheric pressure and forces out some of the residual volume. The air remaining is called the **minimal volume.** Minimal volume provides a medical and legal tool for determining whether a baby was born dead or died after birth. The presence of minimal volume can be demonstrated by placing a piece of lung in water and observing if it floats. Fetal lungs contain no air, and so the lung of a stillborn baby will not float in water.

Lung capacities are combinations of specific lung volumes (see Figure 23.17). **Inspiratory capacity** is the sum of tidal volume and inspiratory reserve volume (500 mL + 3100 mL = 3600 mL). **Functional residual capacity** is the sum of residual volume and expiratory reserve volume (1200 mL + 1200 mL = 2400 mL). **Vital capacity** is the sum of inspiratory reserve volume, tidal volume, and expiratory reserve volume (4800 mL). Finally, **total lung capacity** is the sum of vital capacity and residual volume (4800 mL + 1200 mL = 6000 mL).

▶ **CHECKPOINT**

16. What is a spirometer?
17. What is the difference between a lung volume and a lung capacity?
18. How is minute ventilation calculated?
19. Define alveolar ventilation rate and $FEV_{1.0}$.

EXCHANGE OF OXYGEN AND CARBON DIOXIDE

▶ **OBJECTIVES**

- **Explain Dalton's law and Henry's law.**
- **Describe the exchange of oxygen and carbon dioxide in external and internal respiration.**

The exchange of oxygen and carbon dioxide between alveolar air and pulmonary blood occurs via passive diffusion, which is governed by the behavior of gases as described by Dalton's law and Henry's law. Whereas Dalton's law is important for understanding how gases move down their pressure differences by diffusion, Henry's law helps explain how the solubility of a gas relates to its diffusion.

Gas Laws: Dalton's Law and Henry's Law

According to **Dalton's law,** each gas in a mixture of gases exerts its own pressure as if no other gases were present. The pressure of a specific gas in a mixture is called its *partial pressure* and is denoted as P_x, where the subscript is the chemical formula of the gas. The total pressure of the mixture is calculated simply by adding all the partial pressures. Atmospheric air is a mixture of gases—nitrogen (N_2), oxygen (O_2), water vapor (H_2O), and carbon dioxide, plus other gases present in small quantities. Atmospheric pressure is the sum of the pressures of all these gases:

$$\text{Atmospheric pressure (760 mmHg)} =$$
$$P_{N_2} + P_{O_2} + P_{H_2O} + P_{CO_2} + P_{\text{other gases}}$$

We can determine the partial pressure exerted by each component in the mixture by multiplying the percentage of the gas in the mixture by the total pressure of the mixture. Atmospheric air is 78.6% nitrogen, 20.9% oxygen, 0.04% carbon dioxide, and 0.06% other gases; a variable amount of water vapor is also present, about 0.4% on a cool, dry day. Thus, the partial pressures of the gases in inhaled air are as follows:

$$P_{N_2} = 0.786 \times 760 \text{ mmHg} = 597.4 \text{ mmHg}$$
$$P_{O_2} = 0.209 \times 760 \text{ mmHg} = 158.8 \text{ mmHg}$$
$$P_{H_2O} = 0.004 \times 760 \text{ mmHg} = 3.0 \text{ mmHg}$$
$$P_{CO_2} = 0.0004 \times 760 \text{ mmHg} = 0.3 \text{ mmHg}$$
$$P_{\text{other gases}} = 0.0006 \times 760 \text{ mmHg} = 0.5 \text{ mmHg}$$
$$\text{Total} = 760 \text{ mmHg}$$

These partial pressures govern the movement of O_2 and CO_2 between the atmosphere and lungs, between the lungs and blood, and between the blood and body cells. When a mixture of gases diffuses across a permeable membrane, each gas diffuses from the area where its partial pressure is greater to the area where its partial pressure is less. The greater the difference in partial pressure, the faster the rate of diffusion. Each gas behaves as if the other gases in the mixture were not present and diffuses at a rate determined by its own partial pressure.

Compared with inhaled air, alveolar air has less O_2 (20.9% versus 13.6%) and more CO_2 (0.04% versus 5.2%) for two reasons. First, gas exchange in the alveoli increases the CO_2 content and decreases the O_2 content of alveolar air. Second, when air is inhaled it becomes humidified as it passes along the moist mucosal linings. As water vapor content of the air increases, the relative percentage that is O_2 decreases. In contrast, exhaled air contains more O_2 than alveolar air (16% versus 13.6%) and less CO_2 (4.5% versus 5.2%) because some of the exhaled air was in the anatomic dead space and did not participate in gas exchange. Exhaled air is a mixture of alveolar air and inhaled air that was in the anatomic dead space.

Henry's law states that the quantity of a gas that will dissolve in a liquid is proportional to the partial pressure of the gas and its solubility coefficient. In body fluids, the ability of a gas to stay in solution is greater when its partial pressure is higher and when it has a high solubility coefficient in water. The higher the partial pressure of a gas over a liquid and the higher the solubility coefficient, the more gas will stay in solution. In comparison to oxygen, much more CO_2 is dissolved in blood plasma because the solubility coefficient of CO_2 is 24 times greater than that of O_2. Even though the air we breathe contains almost 79% N_2, this gas has no known effect on bodily functions, and at sea level pressure very little of it dissolves in blood plasma because its solubility coefficient is very low.

An everyday experience gives a demonstration of Henry's law. You have probably noticed that a soft drink makes a hissing sound when the top of the container is removed, and bubbles rise to the surface for some time afterward. The gas dissolved in carbonated beverages is CO_2. Because the soft drink is bottled or canned under high pressure and capped, the CO_2 remains dissolved so long as the container is unopened. Once you remove the cap, the pressure decreases and the gas begins to bubble out of solution.

Henry's law explains two conditions resulting from changes in the solubility of nitrogen in body fluids. Even though the air we breathe contains about 79% nitrogen, this gas has no known effect on bodily functions, and very little of it dissolves in blood plasma because of its low solubility coefficient at sea level pressure. As the total air pressure increases, the partial pressures of all its gases increase. When a scuba diver breathes air under high pressure, the nitrogen in the mixture can have serious negative effects. Because the partial pressure of nitrogen is higher in a mixture of compressed air than in air at sea level pressure, a considerable amount of nitrogen dissolves in plasma and interstitial fluid. Excessive amounts of dissolved nitrogen may produce giddiness and other symptoms similar to alcohol intoxication. The condition is called **nitrogen narcosis** or "rapture of the depths."

If a diver comes to the surface slowly, the dissolved nitrogen can be eliminated by exhaling it. However, if the ascent is too rapid, nitrogen comes out of solution too quickly and forms gas bubbles in the tissues, resulting in **decompression sickness** (the **bends**). The effects of decompression sickness typically result from bubbles in nervous tissue and can be mild or severe, depending on the number of bubbles formed. Symptoms include joint pain, especially in the arms and legs, dizziness, shortness of breath, extreme fatigue, paralysis, and unconsciousness.

Hyperbaric Oxygenation

A major clinical application of Henry's law is **hyperbaric oxygenation** (*hyper* = over; *baros* = pressure). Using pressure to cause more O_2 to dissolve in the blood is an effective technique in treating patients infected by anaerobic bacteria, such as those that cause tetanus and gangrene. (Anaerobic bacteria cannot live in the presence of free O_2.) A person undergoing hyperbaric oxygenation is placed in a hyperbaric chamber, which contains O_2 at a pressure of 3 to 4 atmospheres (2280–3040 mmHg). As body tissues pick up the O_2, the bacteria are killed. Hyperbaric chambers may also be used for treating certain heart disorders, carbon monoxide poisoning, gas embolisms, crush injuries, cerebral edema, certain hard-to-treat bone infections caused by anaerobic bacteria, smoke inhalation, near-drowning, asphyxia, vascular insufficiencies, and burns. ■

External and Internal Respiration

External respiration or **pulmonary gas exchange** is the diffusion of O_2 from air in the alveoli of the lungs to blood in pulmonary capillaries and the diffusion of CO_2 in the opposite direction (Figure 23.18a). External respiration in the lungs con-

Figure 23.18 Changes in partial pressures of oxygen and carbon dioxide (in mmHg) during external and internal respiration.

Gases diffuse from areas of higher partial pressure to areas of lower partial pressure.

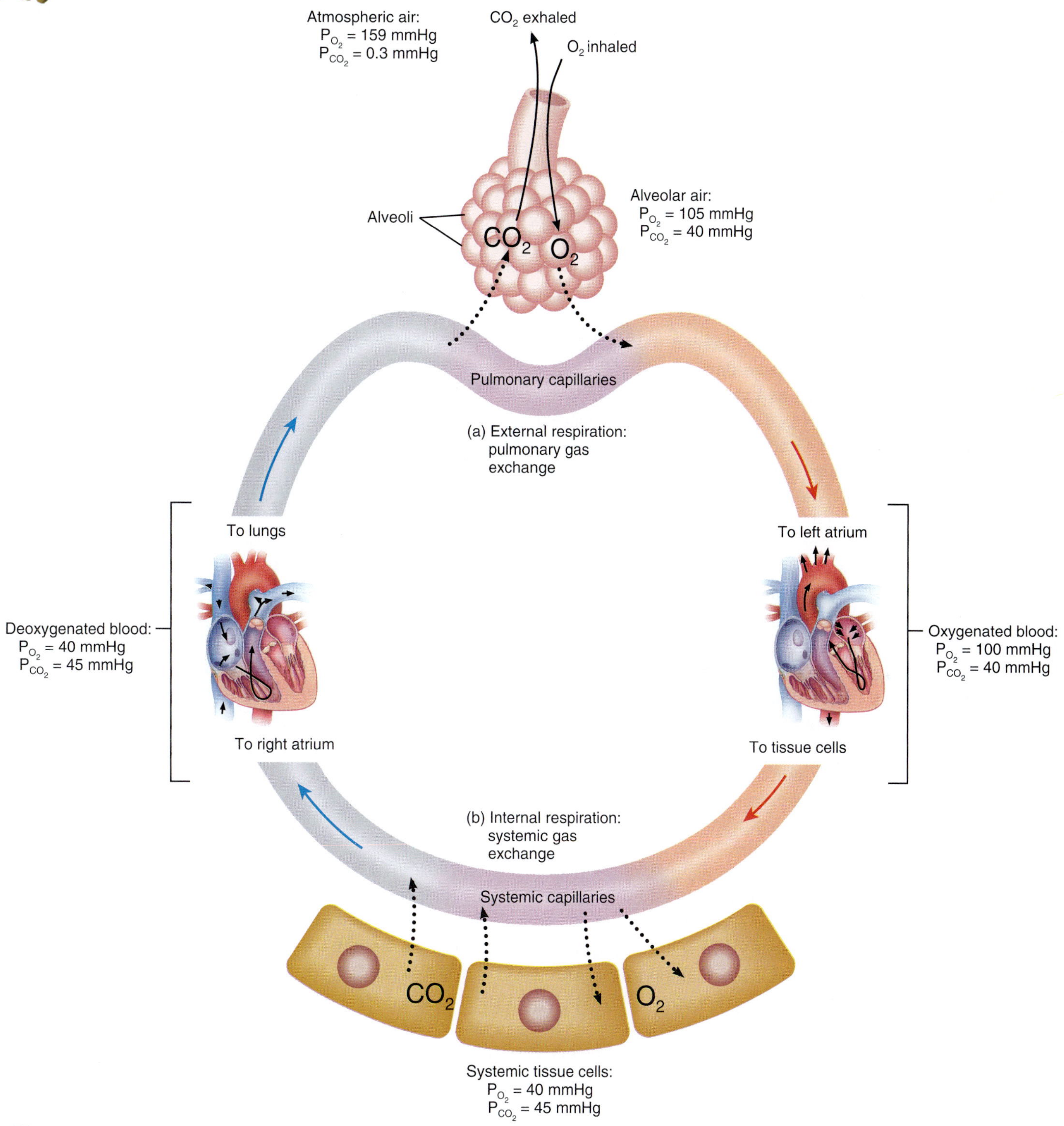

Atmospheric air:
P_{O_2} = 159 mmHg
P_{CO_2} = 0.3 mmHg

CO$_2$ exhaled

O$_2$ inhaled

Alveoli

Alveolar air:
P_{O_2} = 105 mmHg
P_{CO_2} = 40 mmHg

CO_2 O_2

Pulmonary capillaries

(a) External respiration: pulmonary gas exchange

To lungs

To left atrium

Deoxygenated blood:
P_{O_2} = 40 mmHg
P_{CO_2} = 45 mmHg

Oxygenated blood:
P_{O_2} = 100 mmHg
P_{CO_2} = 40 mmHg

To right atrium

To tissue cells

(b) Internal respiration: systemic gas exchange

Systemic capillaries

CO_2 O_2

Systemic tissue cells:
P_{O_2} = 40 mmHg
P_{CO_2} = 45 mmHg

What causes oxygen to enter pulmonary capillaries from alveoli and to enter tissue cells from systemic capillaries?

verts **deoxygenated blood** (depleted of some O$_2$) coming from the right side of the heart into **oxygenated blood** (saturated with O$_2$) that returns to the left side of the heart (see Figure 21.30 on page 753). As blood flows through the pulmonary capillaries, it picks up O$_2$ from alveolar air and unloads CO$_2$ into alveolar air. Although this process is commonly called an "exchange" of gases, each gas diffuses independently from the area where its partial pressure is higher to the area where its partial pressure is lower.

As Figure 23.18a shows, O_2 diffuses from alveolar air, where its partial pressure is 105 mmHg, into the blood in pulmonary capillaries, where P_{O_2} is only 40 mmHg in a resting person. If you have been exercising, the P_{O_2} will be even lower because contracting muscle fibers are using more O_2. Diffusion continues until the P_{O_2} of pulmonary capillary blood increases to match the P_{O_2} of alveolar air, 105 mmHg. Because blood leaving pulmonary capillaries near alveolar air spaces mixes with a small volume of blood that has flowed through conducting portions of the respiratory system, where gas exchange does not occur, the P_{O_2} of blood in the pulmonary veins is slightly less than the P_{O_2} in pulmonary capillaries, about 100 mmHg.

While O_2 is diffusing from alveolar air into deoxygenated blood, CO_2 is diffusing in the opposite direction. The P_{CO_2} of deoxygenated blood is 45 mmHg in a resting person, whereas P_{CO_2} of alveolar air is 40 mmHg. Because of this difference in P_{CO_2}, carbon dioxide diffuses from deoxygenated blood into the alveoli until the P_{CO_2} of the blood decreases to 40 mmHg. Exhalation keeps alveolar P_{CO_2} at 40 mmHg. Oxygenated blood returning to the left side of the heart in the pulmonary veins thus has a P_{CO_2} of 40 mmHg.

The number of capillaries near alveoli in the lungs is very large, and blood flows slowly enough through these capillaries that it picks up a maximal amount of O_2. During vigorous exercise, when cardiac output is increased, blood flows more rapidly through both the systemic and pulmonary circulations. As a result, blood's transit time in the pulmonary capillaries is shorter. Still, the P_{O_2} of blood in the pulmonary veins normally reaches 100 mmHg. In diseases that decrease the rate of gas diffusion, however, the blood may not come into full equilibrium with alveolar air, especially during exercise. When this happens, the P_{O_2} declines and P_{CO_2} rises in systemic arterial blood.

The left ventricle pumps oxygenated blood into the aorta and through the systemic arteries to systemic capillaries. The exchange of O_2 and CO_2 between systemic capillaries and tissue cells is called **internal respiration** or **systemic gas exchange** (Figure 23.18b). As O_2 leaves the bloodstream, oxygenated blood is converted into deoxygenated blood. Unlike external respiration, which occurs only in the lungs, internal respiration occurs in tissues throughout the body.

The P_{O_2} of blood pumped into systemic capillaries is higher (105 mmHg) than the P_{O_2} in tissue cells (40 mmHg at rest) because the cells constantly use O_2 to produce ATP. Due to this pressure difference, oxygen diffuses out of the capillaries into tissue cells and blood P_{O_2} drops to 40 mmHg by the time the blood exits systemic capillaries.

While O_2 diffuses from the systemic capillaries into tissue cells, CO_2 diffuses in the opposite direction. Because tissue cells are constantly producing CO_2, the P_{CO_2} of cells (45 mmHg at rest) is higher than that of systemic capillary blood (40 mmHg). As a result, CO_2 diffuses from tissue cells through interstitial fluid into systemic capillaries until the P_{CO_2} in the blood increases to 45 mmHg. The deoxygenated blood then returns to the heart and is pumped to the lungs for another cycle of external respiration.

In a person at rest, tissue cells, on average, need only 25% of the available O_2 in oxygenated blood and deoxygenated blood retains 75% of its O_2 content. During exercise, more O_2 diffuses from the blood into metabolically active cells, such as contracting skeletal muscle fibers. Active cells use more O_2 for ATP production, and the O_2 content of deoxygenated blood drops below 75%.

The *rate* of pulmonary and systemic gas exchange depends on several factors.

- *Partial pressure difference of the gases.* Alveolar P_{O_2} must be higher than blood P_{O_2} for oxygen to diffuse from alveolar air into the blood. The rate of diffusion is faster when the difference between P_{O_2} in alveolar air and pulmonary capillary blood is larger; diffusion is slower when the difference is smaller. The differences between P_{O_2} and P_{CO_2} in alveolar air versus pulmonary blood increase during exercise. The larger partial pressure differences accelerate the rates of gas diffusion. The partial pressures of O_2 and CO_2 in alveolar air also depend on the rate of airflow into and out of the lungs. Certain drugs (such as morphine) slow ventilation, thereby decreasing the amount of O_2 and CO_2 that can exchanged between alveolar air and blood. With increasing altitude, the total atmospheric pressure decreases, as does the partial pressure of O_2—from 159 mmHg at sea level, to 110 mmHg at 10,000 ft, to 73 mmHg at 20,000 ft. Although O_2 still is 20.9% of the total, the P_{O_2} of inhaled air decreases with increasing altitude. Alveolar P_{O_2} decreases correspondingly, and O_2 diffuses into the blood more slowly. The common signs and symptoms of **high altitude sickness**—shortness of breath, headache, fatigue, insomnia, nausea, and dizziness—are due to a lower level of oxygen in the blood.

- *Surface area available for gas exchange.* The surface area of the alveoli is huge (about 70 m^2 or 750 ft^2). In addition, many capillaries surround each alveolus, so many that as much as 900 mL of blood is able to participate in gas exchange at any instant. Any pulmonary disorder that decreases the functional surface area of the respiratory membranes decreases the rate of external respiration. In emphysema, for example, alveolar walls disintegrate, so surface area is smaller than normal and pulmonary gas exchange is slowed.

- *Diffusion distance.* The respiratory membrane is very thin, so diffusion occurs quickly. Also, the capillaries are so narrow that the red blood cells must pass through them in single file, which minimizes the diffusion distance from an alveolar air space to hemoglobin inside red blood cells. Buildup of interstitial fluid between alveoli, as occurs in pulmonary edema, slows the rate of gas exchange because it increases diffusion distance.

- *Molecular weight and solubility of the gases.* Because O_2 has a lower molecular weight than CO_2, it could be expected to diffuse across the respiratory membrane about 1.2 times faster. However, the solubility of CO_2 in the fluid portions of the respiratory membrane is about 24 times greater than that of O_2. Taking both of these factors into account, net outward CO_2 diffusion occurs 20 times more rapidly than net inward O_2 diffusion. Consequently, when diffusion is slower than normal, for example, in emphysema or pulmonary edema, O_2 insufficiency (hypoxia) typically occurs before there is significant retention of CO_2 (hypercapnia).

▶ **CHECKPOINT**

20. How does the partial pressure of oxygen change as altitude changes?

21. What are the diffusion paths of oxygen and carbon dioxide during external and internal respiration?

TRANSPORT OF OXYGEN AND CARBON DIOXIDE

▶ **OBJECTIVE**

- **Describe how the blood transports oxygen and carbon dioxide.**

The blood transports gases between the lungs and body tissues. When O_2 and CO_2 enter the blood, certain chemical reactions occur that aid in gas transport and gas exchange.

Oxygen Transport

Oxygen does not dissolve easily in water, and therefore only about 1.5% of the O_2 is dissolved in blood plasma, which is mostly water. About 98.5% of blood O_2 is bound to hemoglobin in red blood cells (Figure 23.19). Each 100 mL of oxygenated blood contains the equivalent of 20 mL of gaseous O_2. The amount dissolved in the plasma is 0.3 mL and the amount bound to hemoglobin is 19.7 mL.

The heme portion of hemoglobin contains four atoms of iron, each capable of binding to a molecule of O_2 (see Figure 19.4b, c on page 640). Oxygen and hemoglobin bind in an easily reversible reaction to form **oxyhemoglobin:**

$$\text{Hb} \quad + \quad O_2 \underset{\substack{\text{Dissociation} \\ \text{of } O_2}}{\overset{\substack{\text{Binding of } O_2}}{\rightleftharpoons}} \text{Hb}-O_2$$

| Reduced hemoglobin (deoxyhemoglobin) | Oxygen | Dissociation of O_2 | Oxyhemoglobin |

Because 98.5% of the O_2 is bound to hemoglobin and therefore trapped inside RBCs, only the dissolved O_2 (1.5%) can diffuse out of tissue capillaries into tissue cells. Thus, it is important to understand the factors that promote O_2 binding to and dissociation (separation) from hemoglobin.

The Relation Between Hemoglobin and Oxygen Partial Pressure

The most important factor that determines how much O_2 binds to hemoglobin is the P_{O_2}; the higher the P_{O_2}, the more O_2 combines with Hb. When reduced hemoglobin (Hb) is completely converted to oxyhemoglobin (Hb$-O_2$), the hemoglobin is said to be **fully saturated;** when hemoglobin consists of a mixture of Hb and Hb$-O_2$, it is **partially saturated.** The **percent saturation of hemoglobin** expresses the average saturation of hemoglobin with oxygen. For instance, if each hemoglobin molecule has bound two O_2 molecules, then the hemoglobin is 50% saturated because each Hb can bind a maximum of four O_2.

The relation between the percent saturation of hemoglobin and P_{O_2} is illustrated in the oxygen–hemoglobin dissociation curve in Figure 23.20 on page 833. Note that when the P_{O_2} is high, hemoglobin binds with large amounts of O_2 and is almost 100% saturated. When P_{O_2} is low, hemoglobin is only partially saturated. In other words, the greater the P_{O_2}, the more O_2 will bind to hemoglobin, until all the available hemoglobin molecules are saturated. Therefore, in pulmonary capillaries, where P_{O_2} is high, a lot of O_2 binds to hemoglobin. In tissue capillaries, where the P_{O_2} is lower, hemoglobin does not hold as much O_2, and the O_2 is unloaded via diffusion into tissue cells. Note that hemoglobin is still 75% saturated with O_2 at a P_{O_2} of 40 mmHg, the average P_{O_2} of tissue cells in a person at rest. This is the basis for the earlier statement that only 25% of the available O_2 unloads from hemoglobin and is used by tissue cells under resting conditions.

When the P_{O_2} is between 60 and 100 mmHg, hemoglobin is 90% or more saturated with O_2 (Figure 23.20). Thus, blood picks up a nearly full load of O_2 from the lungs even when the P_{O_2} of alveolar air is as low as 60 mmHg. The Hb$-P_{O_2}$ curve explains why people can still perform well at high altitudes or when they have certain cardiac and pulmonary diseases, even though P_{O_2} may drop as low as 60 mmHg. Note also in the curve that at a considerably lower P_{O_2} of 40 mmHg, hemoglobin is still 75% saturated with O_2. However, oxygen saturation of Hb drops to 35% at 20 mmHg. Between 40 and 20 mmHg, large amounts of O_2 are released from hemoglobin in response to only small decreases in P_{O_2}. In active tissues such as contracting muscles, P_{O_2} may drop well below 40 mmHg. Then, a large percentage of the O_2 is released from hemoglobin, providing more O_2 to metabolically active tissues.

Other Factors Affecting Hemoglobin's Affinity for Oxygen

Although P_{O_2} is the most important factor that determines the percent O_2 saturation of hemoglobin, several other factors influence the tightness or **affinity** with which hemoglobin binds O_2. In effect, these factors shift the entire curve either to the left (higher affinity) or to the right (lower affinity). The changing affinity of hemoglobin for O_2 is another example of how homeostatic mechanisms adjust body activities to cellular needs. Each one makes sense if you keep in mind that metabolically active

Figure 23.19 Transport of oxygen (O_2) and carbon dioxide (CO_2) in the blood.

Most O_2 is transported by hemoglobin as oxyhemoglobin (Hb–O_2) within red blood cells; most CO_2 is transported in blood plasma as bicarbonate ions (HCO_3^-).

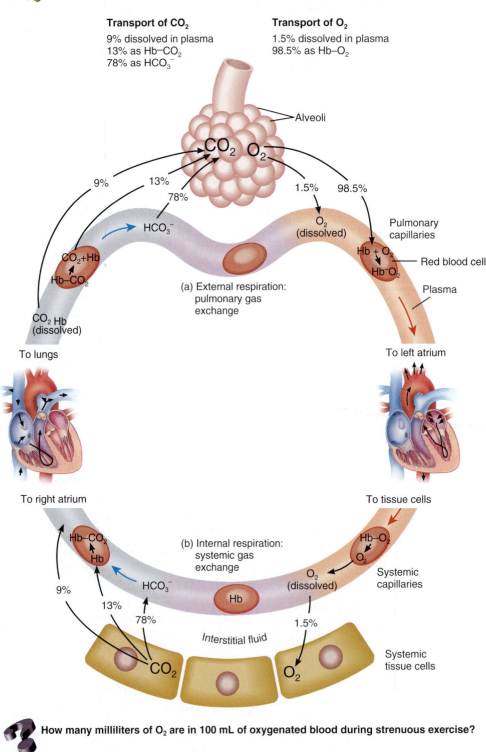

Transport of CO_2
9% dissolved in plasma
13% as Hb–CO_2
78% as HCO_3^-

Transport of O_2
1.5% dissolved in plasma
98.5% as Hb–O_2

Alveoli

CO_2 O_2

9%   13%   78%   1.5%   98.5%

HCO_3^- O_2 (dissolved)   Pulmonary capillaries

CO_2+Hb Hb + O_2 Red blood cell
Hb–CO_2 Hb–O_2

(a) External respiration: pulmonary gas exchange   Plasma

CO_2 Hb (dissolved)

To lungs   To left atrium

To right atrium   To tissue cells

Hb–CO_2 Hb–O_2
Hb O_2

(b) Internal respiration: systemic gas exchange

9%   13%   78% HCO_3^- O_2 (dissolved)   Systemic capillaries

Hb

1.5%

Interstitial fluid

CO_2 O_2 Systemic tissue cells

How many milliliters of O_2 are in 100 mL of oxygenated blood during strenuous exercise?

tissue cells need O_2 and produce acids, CO_2, and heat as wastes. The following four factors affect hemoglobin's affinity for O_2:

1. Acidity (pH). As acidity increases (pH decreases), the affinity of hemoglobin for O_2 decreases, and O_2 dissociates more

readily from hemoglobin (Figure 23.21a). In other words, increasing acidity enhances the unloading of oxygen from hemoglobin. The main acids produced by metabolically active tissues are lactic acid and carbonic acid. When pH decreases, the entire oxygen–hemoglobin dissociation curve shifts to the right; at any

Figure 23.20 Oxygen–hemoglobin dissociation curve showing the relationship between hemoglobin saturation and P_{O_2} at normal body temperature.

🔑 As P_{O_2} increases, more O_2 combines with hemoglobin.

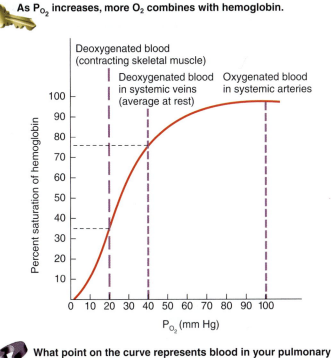

What point on the curve represents blood in your pulmonary veins right now? In your pulmonary veins if you were jogging?

given P_{O_2}, Hb is less saturated with O_2, a change termed the **Bohr effect.** The Bohr effect works both ways: An increase in H^+ in blood causes O_2 to unload from hemoglobin, and the binding of O_2 to hemoglobin causes unloading of H^+ from hemoglobin. The explanation for the Bohr effect is that hemoglobin can act as a buffer for hydrogen ions (H^+). But when H^+ bind to amino acids in hemoglobin, they slightly alter its structure and thereby decrease its oxygen-carrying capacity. Thus, lowered pH drives O_2 off hemoglobin, making more O_2 available for tissue cells. By contrast, elevated pH increases the affinity of hemoglobin for O_2 and shifts the oxygen–hemoglobin dissociation curve to the left.

2. Partial pressure of carbon dioxide. CO_2 also can bind to hemoglobin, and the effect is similar to that of H^+ (shifting the curve to the right). As P_{CO_2} rises, hemoglobin releases O_2 more readily (Figure 23.21b). P_{CO_2} and pH are related factors because low blood pH (acidity) results from high P_{CO_2}. As CO_2 enters the blood, much of it is temporarily converted to carbonic acid (H_2CO_3), a reaction catalyzed by an enzyme in red blood cells called *carbonic anhydrase (CA):*

$$CO_2 + H_2O \overset{CA}{\rightleftharpoons} H_2CO_3 \rightleftharpoons H^+ + HCO_3^-$$

Carbon dioxide Water Carbonic acid Hydrogen ion Bicarbonate ion

The carbonic acid thus formed in red blood cells dissociates into hydrogen ions and bicarbonate ions. As the H^+ concentra-

Figure 23.21 Oxygen–hemoglobin dissociation curves showing the relationship of (a) pH and (b) P_{CO_2} to hemoglobin saturation at normal body temperature. As pH increases or P_{CO_2} decreases, O_2 combines more tightly with hemoglobin, so that less is available to tissues. The broken lines emphasize these relationships.

🔑 As pH decreases or P_{CO_2} increases, the affinity of hemoglobin for O_2 declines, so less O_2 combines with hemoglobin and more is available to tissues.

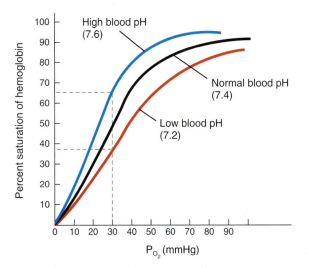

(a) Effect of pH on affinity of hemoglobin for oxygen

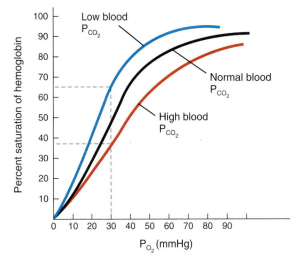

(b) Effect of P_{CO_2} on affinity of hemoglobin for oxygen

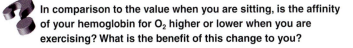

In comparison to the value when you are sitting, is the affinity of your hemoglobin for O_2 higher or lower when you are exercising? What is the benefit of this change to you?

tion increases, pH decreases. Thus, an increased P_{CO_2} produces a more acidic environment, which helps release O_2 from hemoglobin. During exercise, lactic acid—a byproduct of anaerobic metabolism within muscles—also decreases blood pH. Decreased P_{CO_2} (and elevated pH) shifts the saturation curve to the left.

3.   *Temperature.* Within limits, as temperature increases, so does the amount of O_2 released from hemoglobin (Figure 23.22). Heat is a byproduct of the metabolic reactions of all cells, and the heat released by contracting muscle fibers tends to raise body temperature. Metabolically active cells require more O_2 and liberate more acids and heat. The acids and heat, in turn, promote release of O_2 from oxyhemoglobin. Fever produces a similar result. In contrast, during hypothermia (lowered body temperature) cellular metabolism slows, the need for O_2 is reduced, and more O_2 remains bound to hemoglobin (a shift to the left in the saturation curve).

4.   *BPG.* A substance found in red blood cells called **2,3-bis-phosphoglycerate (BPG),** previously called diphosphoglycerate (DPG), decreases the affinity of hemoglobin for O_2 and thus helps unload O_2 from hemoglobin. BPG is formed in red blood cells when they break down glucose to produce ATP in a process called glycolysis (described on page 911). When BPG combines with hemoglobin, by binding to the terminal amino groups of the two beta globin chains, the hemoglobin binds O_2 less tightly at the heme group sites. The greater the level of BPG, the more O_2 is unloaded from hemoglobin. Certain hormones, such as thyroxine, human growth hormone, epinephrine, norepinephrine, and testosterone, increase the formation of BPG. The level of BPG also is higher in people living at higher altitudes.

Oxygen Affinity of Fetal and Adult Hemoglobin

Fetal hemoglobin (Hb-F) differs from **adult hemoglobin (Hb-A)** in structure and in its affinity for O_2. Hb-F has a higher affinity for O_2 because it binds BPG less strongly. Thus, when P_{O_2} is low, Hb-F can carry up to 30% more O_2 than maternal Hb-A (Figure 23.23). As the maternal blood enters the placenta, O_2 is readily transferred to fetal blood. This is very important because the O_2 saturation in maternal blood in the placenta is quite low, and the fetus might suffer hypoxia were it not for the greater affinity of fetal hemoglobin for O_2.

Carbon Monoxide Poisoning

Carbon monoxide (CO) is a colorless and odorless gas found in exhaust fumes from automobiles, gas furnaces, and space heaters and in tobacco smoke. It is a byproduct of the combustion of carbon-containing materials such as coal, gas, and wood. CO binds to the heme group of hemoglobin, just as O_2 does, except that the binding of carbon monoxide to hemoglobin is over 200 times as strong as the binding of O_2 to hemoglobin. Thus, at a concentration as small as 0.1% ($P_{CO} = 0.5$ mmHg), CO will combine with half the available hemoglobin molecules and reduce the oxygen-carrying capacity of the blood by 50%. Elevated blood levels of CO cause **carbon monoxide poisoning,** one sign of which is a bright, cherry-red color of the lips and oral mucosa. Administering pure oxygen, which speeds up the separation of carbon monoxide from hemoglobin, may rescue the person. ■

Carbon Dioxide Transport

Under normal resting conditions, each 100 mL of deoxygenated blood contains the equivalent of 53 mL of gaseous CO_2, which is transported in the blood in three main forms (see Figure 23.19):

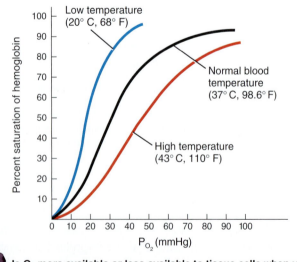

Figure 23.22 Oxygen–hemoglobin dissociation curves showing the effect of temperature changes.

 As temperature increases, the affinity of hemoglobin for O_2 decreases.

Is O_2 more available or less available to tissue cells when you have a fever? Why?

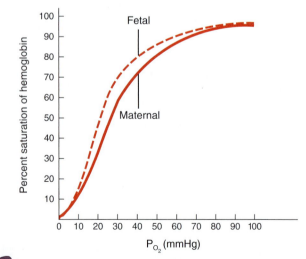

Figure 23.23 Oxygen–hemoglobin dissociation curves comparing fetal and maternal hemoglobin.

Fetal hemoglobin has a higher affinity for O_2 than does adult hemoglobin.

 The P_{O_2} of placental blood is about 40 mmHg. What are the O_2 saturations of maternal and fetal hemoglobin at this P_{O_2}?

1. **Dissolved CO_2.** The smallest percentage—about 9%—is dissolved in blood plasma. Upon reaching the lungs, it diffuses into alveolar air and is exhaled.

2. **Carbamino compounds.** A somewhat higher percentage, about 13%, combines with the amino groups of amino acids and proteins in blood to form **carbamino compounds.** Because the most prevalent protein in blood is hemoglobin (inside red blood cells), most of the CO_2 transported in this manner is bound to hemoglobin. The main CO_2 binding sites are the terminal amino acids in the two alpha and two beta globin chains. Previous higher estimates of the amount of carbamino compounds formed in blood ignored the competing effect of bis-phosphoglycerate (BPG), which also binds to the terminal amino acids of the two beta chains. Hemoglobin that has bound CO_2 is termed **carbaminohemoglobin (Hb–CO_2):**

$$\underset{\text{Hemoglobin}}{\text{Hb}} + \underset{\text{Carbon dioxide}}{CO_2} \rightleftharpoons \underset{\text{Carbaminohemoglobin}}{\text{Hb–}CO_2}$$

The formation of carbaminohemoglobin is greatly influenced by P_{CO_2}. For example, in tissue capillaries P_{CO_2} is relatively high, which promotes formation of carbaminohemoglobin. But in pulmonary capillaries, P_{CO_2} is relatively low, and the CO_2 readily splits apart from globin and enters the alveoli by diffusion.

3. **Bicarbonate ions.** The greatest percentage of CO_2—about 78%—is transported in blood plasma as **bicarbonate ions (HCO_3^-).** As CO_2 diffuses into systemic capillaries and enters red blood cells, it reacts with water in the presence of the enzyme carbonic anhydrase (CA) to form carbonic acid, which dissociates into H^+ and HCO_3^-:

$$\underset{\substack{\text{Carbon} \\ \text{dioxide}}}{CO_2} + \underset{\text{Water}}{H_2O} \overset{CA}{\rightleftharpoons} \underset{\substack{\text{Carbonic} \\ \text{acid}}}{H_2CO_3} \rightleftharpoons \underset{\substack{\text{Hydrogen} \\ \text{ion}}}{H^+} + \underset{\substack{\text{Bicarbonate} \\ \text{ion}}}{HCO_3^-}$$

Thus, as blood picks up CO_2, HCO_3^- accumulates inside RBCs. Some HCO_3^- moves out into the blood plasma, down its concentration gradient. In exchange, chloride ions (Cl^-) move from plasma into the RBCs. This exchange of negative ions, which maintains the electrical balance between blood plasma and RBC cytosol, is known as the **chloride shift** (see Figure 23.24b). The net effect of these reactions is that CO_2 is removed from tissue cells and transported in blood plasma as HCO_3^-. As blood passes through pulmonary capillaries in the lungs, all these reactions reverse as CO_2 is exhaled.

The amount of CO_2 that can be transported in the blood is influenced by the percent saturation of hemoglobin with oxygen. The lower the amount of oxyhemoglobin (Hb–O_2), the higher the CO_2 carrying capacity of the blood, a relationship known as the **Haldane effect.** Two characteristics of deoxyhemoglobin give rise to the Haldane effect: (1) Deoxyhemoglobin binds to and thus transports more CO_2 than does Hb–O_2. (2) Deoxyhemoglobin also buffers more H^+ than does Hb–O_2, thereby removing H^+ from solution and promoting conversion of CO_2 to HCO_3^- via the reaction catalyzed by carbonic anhydrase.

Summary of Gas Exchange and Transport in Lungs and Tissues

Deoxygenated blood returning to the pulmonary capillaries in the lungs (Figure 23.24a) contains CO_2 dissolved in blood plasma, CO_2 combined with globin as carbaminohemoglobin (Hb–CO_2), and CO_2 incorporated into HCO_3^- within RBCs. The RBCs have also picked up H^+, some of which binds to and therefore is buffered by hemoglobin (Hb–H). As blood passes through the pulmonary capillaries, molecules of CO_2 dissolved in blood plasma and CO_2 that dissociates from the globin portion of hemoglobin diffuse into alveolar air and are exhaled. At the same time, inhaled O_2 is diffusing from alveolar air into RBCs and is binding to hemoglobin to form oxyhemoglobin (Hb–O_2). Carbon dioxide also is released from HCO_3^- when H^+ combines with HCO_3^- inside RBCs. The H_2CO_3 formed from this reaction then splits into CO_2, which is exhaled, and H_2O. As the concentration of HCO_3^- declines inside RBCs in pulmonary capillaries, HCO_3^- diffuses in from the blood plasma, in exchange for Cl^-. In sum, oxygenated blood leaving the lungs has increased O_2 content and decreased amounts of CO_2 and H^+. In systemic capillaries, as cells use O_2 and produce CO_2, the chemical reactions reverse (Figure 23.24b).

▶ **CHECKPOINT**

22. In a resting person, how many O_2 molecules are attached to each hemoglobin molecule, on average, in blood in the pulmonary arteries? In blood in the pulmonary veins?

23. What is the relationship between hemoglobin and P_{O_2}? How do temperature, H^+, P_{CO_2}, and BPG influence the affinity of Hb for O_2?

24. Why can hemoglobin unload more oxygen as blood flows through capillaries of metabolically active tissues, such as skeletal muscle during exercise, than is unloaded at rest?

CONTROL OF RESPIRATION

▶ **OBJECTIVE**

• **Explain how the nervous system controls breathing and list the factors that can alter the rate and depth of breathing.**

At rest, about 200 mL of O_2 are used each minute by body cells. During strenuous exercise, however, O_2 use typically increases 15- to 20-fold in normal healthy adults, and as much as 30-fold in elite endurance-trained athletes. Several mechanisms help match respiratory effort to metabolic demand.

Respiratory Center

The basic rhythm of respiration is controlled by groups of neurons in the brain stem. The area from which nerve impulses are sent to respiratory muscles is called the **respiratory center** and consists of groups of neurons functionally divided into three

Figure 23.24   Summary of chemical reactions that occur during gas exchange.   (a) As carbon dioxide (CO_2) is exhaled, hemoglobin (Hb) inside red blood cells in pulmonary capillaries unloads CO_2 and picks up O_2 from alveolar air. Binding of O_2 to Hb–H releases hydrogen ions (H^+). Bicarbonate ions (HCO_3^-) pass into the RBC and bind to released H^+, forming carbonic acid (H_2CO_3). The H_2CO_3 dissociates into water (H_2O) and CO_2, and the CO_2 diffuses from blood into alveolar air. To maintain electrical balance, a chloride ion (Cl^-) exits the RBC for each HCO_3^- that enters (reverse chloride shift). (b) CO_2 diffuses out of tissue cells that produce it and enters red blood cells, where some of it binds to hemoglobin, forming carbaminohemoglobin (Hb–CO_2). This reaction causes O_2 to dissociate from oxyhemoglobin (Hb–O_2). Other molecules of CO_2 combine with water to produce bicarbonate ions (HCO_3^-) and hydrogen ions (H^+). As Hb buffers H^+, the Hb releases O_2 (Bohr effect). To maintain electrical balance, a chloride ion (Cl^-) enters the RBC for each HCO_3^- that exits (chloride shift).

🔑 **Hemoglobin inside red blood cells transports O_2, CO_2, and H^+.**

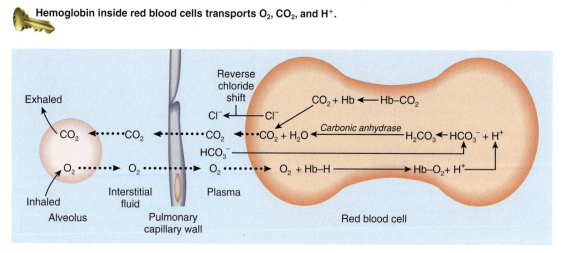

(a) Exchange of O_2 and CO_2 in pulmonary capillaries (external respiration)

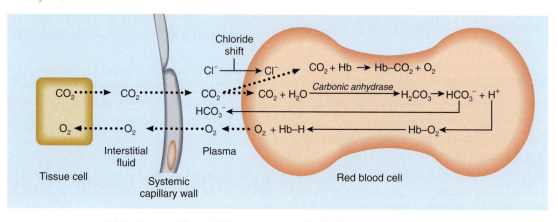

(b) Exchange of O_2 and CO_2 in systemic capillaries (internal respiration)

Would you expect the concentration of HCO_3^- to be higher in blood plasma taken from a systemic artery or a systemic vein?

areas: the medullary rhythmicity area, the pneumotaxic area, and the apneustic area (Figure 23.25).

Medullary Rhythmicity Area

The **medullary rhythmicity area** (rith-MIS-i-tē) in the medulla oblongata controls the basic rhythm of respiration. Within the medullary rhythmicity area are both inspiratory and expiratory areas. Figure 23.26 shows the relationships of the inspiratory and expiratory areas during normal quiet breathing and forceful breathing.

During quiet breathing, inhalation lasts for about 2 seconds and exhalation lasts for about 3 seconds. Nerve impulses generated in the **inspiratory area** establish the basic rhythm of breathing. While the inspiratory area is active, it generates nerve

Figure 23.25 Locations of areas of the respiratory center.

The respiratory center is composed of neurons in the medullary rhythmicity area in the medulla oblongata plus the pneumotaxic and apneustic areas in the pons.

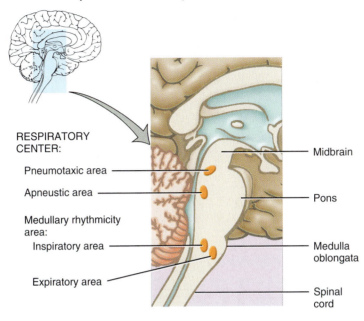

RESPIRATORY CENTER:

Pneumotaxic area

Apneustic area

Medullary rhythmicity area:

Inspiratory area

Expiratory area

Midbrain

Pons

Medulla oblongata

Spinal cord

Sagittal section of brain stem

Which area contains neurons that are active and then inactive in a repeating cycle?

impulses for about 2 seconds (Figure 23.26a). The impulses propagate to the external intercostal muscles via intercostal nerves and to the diaphragm via the phrenic nerves. When the nerve impulses reach the diaphragm and external intercostal muscles, the muscles contract and inhalation occurs. Even when all incoming nerve connections to the inspiratory area are cut or blocked, neurons in this area still rhythmically discharge impulses that cause inhalation. At the end of 2 seconds, the inspiratory area becomes inactive and nerve impulses cease. With no impulses arriving, the diaphragm and external intercostal muscles relax for about 3 seconds, allowing passive elastic recoil of the lungs and thoracic wall. Then, the cycle repeats.

The neurons of the expiratory area remain inactive during quiet breathing. However, during forceful breathing nerve impulses from the inspiratory area activate the expiratory area (Figure 23.26b). Impulses from the expiratory area cause contraction of the internal intercostal and abdominal muscles, which decreases the size of the thoracic cavity and causes forceful exhalation.

Pneumotaxic Area

Although the medullary rhythmicity area controls the basic rhythm of respiration, other sites in the brain stem help coordinate the transition between inhalation and exhalation. One of these sites is the **pneumotaxic area** (noo-mō-TAK-sik; *pneumo-* = air or breath; *-taxic* = arrangement) in the upper pons (see Figure 23.25), which transmits inhibitory impulses to the inspiratory area. The major effect of these nerve impulses is to help turn off the inspiratory area before the lungs become too

Figure 23.26 Roles of the medullary rhythmicity area in controlling (a) the basic rhythm of respiration and (b) forceful breathing.

During normal, quiet breathing, the expiratory area is inactive; during forceful breathing, the inspiratory area activates the expiratory area.

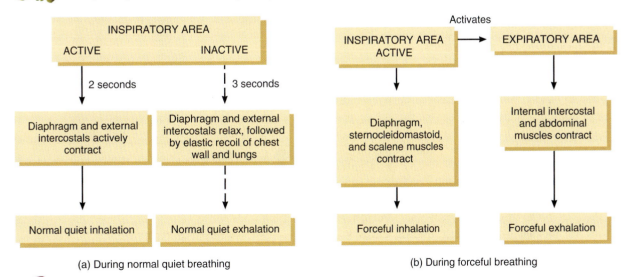

(a) During normal quiet breathing

(b) During forceful breathing

Which nerves convey impulses from the respiratory center to the diaphragm?

full of air. In other words, the impulses shorten the duration of inhalation. When the pneumotaxic area is more active, breathing rate is more rapid.

Apneustic Area

Another part of the brain stem that coordinates the transition between inhalation and exhalation is the **apneustic area** (ap-NOO-stik) in the lower pons (see Figure 23.25). This area sends stimulatory impulses to the inspiratory area that activate it and prolong inhalation. The result is a long, deep inhalation. When the pneumotaxic area is active, it overrides the apneustic area.

Regulation of the Respiratory Center

Although the basic rhythm of respiration is set and coordinated by the inspiratory area, the rhythm can be modified in response to inputs from other brain regions, receptors in the peripheral nervous system, and other factors.

Cortical Influences on Respiration

Because the cerebral cortex has connections with the respiratory center, we can voluntarily alter our pattern of breathing. We can even refuse to breathe at all for a short time. Voluntary control is protective because it enables us to prevent water or irritating gases from entering the lungs. The ability to not breathe, however, is limited by the buildup of CO_2 and H^+ in the body. When P_{CO_2} and H^+ concentrations increase to a certain level, the inspiratory area is strongly stimulated, nerve impulses are sent along the phrenic and intercostal nerves to inspiratory muscles, and breathing resumes, whether the person wants it or not. It is impossible for people to kill themselves by voluntarily holding their breath. Even if breath is held long enough to cause fainting, breathing resumes when consciousness is lost. Nerve impulses from the hypothalamus and limbic system also stimulate the respiratory center, allowing emotional stimuli to alter respirations as, for example, in laughing and crying.

Chemoreceptor Regulation of Respiration

Certain chemical stimuli modulate how quickly and how deeply we breathe. The respiratory system functions to maintain proper levels of CO_2 and O_2 and is very responsive to changes in the levels of either in body fluids. Sensory neurons that are responsive to chemicals are termed **chemoreceptors.** Chemoreceptors in two locations monitor levels of CO_2, H^+, and O_2 and provide input to the respiratory center (Figure 23.27). **Central chemoreceptors** are located in the medulla oblongata in the *central* nervous system. They respond to changes in H^+ concentration or P_{CO_2}, or both, in cerebrospinal fluid. **Peripheral chemoreceptors** are located in the **aortic bodies,** clusters of chemoreceptors located in the wall of the arch of the aorta, and in the **carotid bodies,** which are oval nodules in the wall of the left and right common carotid arteries where they divide into the internal and external carotid arteries. (The chemoreceptors

Figure 23.27   Locations of peripheral chemoreceptors.

Chemoreceptors are sensory neurons that respond to changes in the levels of certain chemicals in the body.

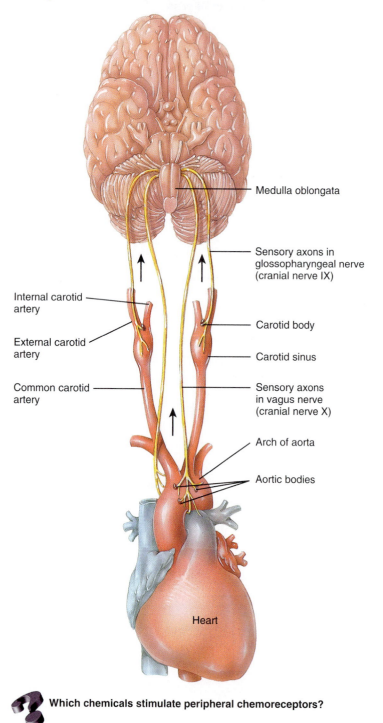

Which chemicals stimulate peripheral chemoreceptors?

of the aortic bodies are located close to the aortic baroreceptors, and the carotid bodies are located close to the carotid sinus baroreceptors. Recall that baroreceptors monitor blood pressure.) These chemoreceptors are part of the *peripheral* nervous system and are sensitive to changes in P_{O_2}, H^+, and P_{CO_2} in the blood. Axons of sensory neurons from the aortic bodies are part of the vagus nerve (cranial nerve X), whereas those from the carotid bodies are part of the right and left glossopharyngeal nerves (cranial nerve IX).

Because CO_2 is lipid soluble, it easily diffuses into cells, where in the presence of carbonic anhydrase it combines with water (H_2O) to form carbonic acid (H_2CO_3). Carbonic acid quickly breaks down into H^+ and HCO_3^-. Thus, an increase in CO_2 in the blood causes an increase in H^+ inside cells, and a decrease in CO_2 causes a decrease in H^+.

Normally, the P_{CO_2} in arterial blood is 40 mmHg. If even a slight increase in P_{CO_2} occurs—a condition called **hypercapnia** or **hypercarbia**—the central chemoreceptors are stimulated and respond vigorously to the resulting increase in H^+ level. The peripheral chemoreceptors also are stimulated by both the high P_{CO_2} and the rise in H^+. In addition, the peripheral chemoreceptors (but not the central chemoreceptors) respond to a deficiency of O_2. When P_{O_2} in arterial blood falls from a normal level of 105 mmHg but is still above 50 mmHg, the peripheral chemoreceptors are stimulated. Severe deficiency of O_2 depresses activity of the central chemoreceptors and inspiratory area, which then do not respond well to any inputs and send fewer impulses to the muscles of inhalation. As the breathing rate decreases or breathing ceases altogether, P_{O_2} falls lower and lower, thereby establishing a positive feedback cycle with a possibly fatal result.

The chemoreceptors participate in a negative feedback system that regulates the levels of CO_2, O_2, and H^+ in the blood (Figure 23.28). As a result of increased P_{CO_2}, decreased pH (increased H^+), or decreased P_{O_2}, input from the central and peripheral chemoreceptors causes the inspiratory area to become highly active, and the rate and depth of breathing increase. Rapid and deep breathing, called **hyperventilation,** allows the inhalation of more O_2 and exhalation of more CO_2 until P_{CO_2} and H^+ are lowered to normal.

If arterial P_{CO_2} is lower than 40 mmHg—a condition called **hypocapnia** or **hypocarbia**—the central and peripheral chemoreceptors are not stimulated, and stimulatory impulses are not sent to the inspiratory area. As a result, the area sets its own moderate pace until CO_2 accumulates and the P_{CO_2} rises to 40 mmHg. The inspiratory center is more strongly stimulated when P_{CO_2} is rising above normal than when P_{O_2} is falling below normal. As a result, people who hyperventilate voluntarily and cause hypocapnia can hold their breath for an unusually long period. Swimmers were once encouraged to hyperventilate just before diving in to compete. However, this practice is risky because the O_2 level may fall dangerously low and cause fainting before the P_{CO_2} rises high enough to stimulate inhalation. A person who faints on land may suffer bumps and bruises, but one who faints in the water may drown.

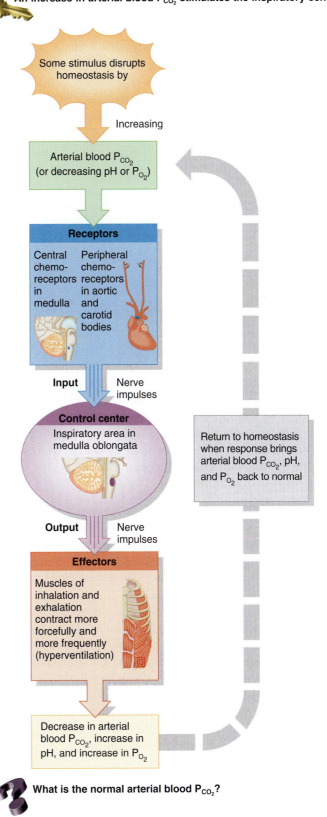

Figure 23.28 Regulation of breathing in response to changes in blood P_{CO_2}, P_{O_2}, and pH (H^+ concentration) via negative feedback control.

An increase in arterial blood P_{CO_2} stimulates the inspiratory center.

Some stimulus disrupts homeostasis by

Increasing

Arterial blood P_{CO_2} (or decreasing pH or P_{O_2})

Receptors
Central chemoreceptors in medulla
Peripheral chemoreceptors in aortic and carotid bodies

Input Nerve impulses

Control center
Inspiratory area in medulla oblongata

Output Nerve impulses

Effectors
Muscles of inhalation and exhalation contract more forcefully and more frequently (hyperventilation)

Decrease in arterial blood P_{CO_2}, increase in pH, and increase in P_{O_2}

Return to homeostasis when response brings arterial blood P_{CO_2}, pH, and P_{O_2} back to normal

What is the normal arterial blood P_{CO_2}?

Hypoxia

Hypoxia (hī-POK-sē-a; *hypo-* = under) is a deficiency of O_2 at the tissue level. Based on the cause, we can classify hypoxia into four types, as follows:

1.   Hypoxic hypoxia is caused by a low P_{O_2} in arterial blood as a result of high altitude, airway obstruction, or fluid in the lungs.

2.   In anemic hypoxia, too little functioning hemoglobin is present in the blood, which reduces O_2 transport to tissue cells. Among the causes are hemorrhage, anemia, and failure of hemoglobin to carry its normal complement of O_2, as in carbon monoxide poisoning.

3.   In ischemic hypoxia, blood flow to a tissue is so reduced that too little O_2 is delivered to it, even though P_{O_2} and oxyhemoglobin level are normal.

4.   In histotoxic hypoxia, the blood delivers adequate O_2 to tissues, but the tissues are unable to use it properly because of the action of some toxic agent. One cause is cyanide poisoning, in which cyanide blocks an enzyme needed for O_2 utilization during ATP synthesis. ■

Proprioceptor Stimulation of Respiration

As soon as you start exercising, your rate and depth of breathing increase, even before changes in P_{O_2}, P_{CO_2}, or H^+ level occur. The main stimulus for these quick changes in respiratory effort is input from proprioceptors, which monitor movement of joints and muscles. Nerve impulses from the proprioceptors stimulate the inspiratory area of the medulla oblongata. At the same time, axon collaterals (branches) of upper motor neurons that originate in the primary motor cortex (precentral gyrus) also feed excitatory impulses into the inspiratory area.

The Inflation Reflex

Located in the walls of bronchi and bronchioles are stretch-sensitive receptors called **baroreceptors** or **stretch receptors.** When these receptors become stretched during overinflation of the lungs, nerve impulses are sent along the cranial nerve X (vagus) to the inspiratory and apneustic areas. In response, the inspiratory area is inhibited directly, and the apneustic area is inhibited from activating the inspiratory area. As a result, exhalation begins. As air leaves the lungs during exhalation, the lungs deflate and the stretch receptors are no longer stimulated. Thus, the inspiratory and apneustic areas are no longer inhibited, and a new inhalation begins. This reflex, referred to as the **inflation (Hering–Breuer) reflex,** is mainly a protective mechanism for preventing excessive inflation of the lungs rather than a key component in the normal regulation of respiration.

Other Influences on Respiration

Other factors that contribute to regulation of respiration include the following:

- **Limbic system stimulation.** Anticipation of activity or emotional anxiety may stimulate the limbic system, which then sends excitatory input to the inspiratory area, increasing the rate and depth of ventilation.

- **Temperature.** An increase in body temperature, as occurs during a fever or vigorous muscular exercise, increases the rate of respiration. A decrease in body temperature decreases respiratory rate. A sudden cold stimulus (such as plunging into cold water) causes temporary **apnea** (AP-nē-a; *a-* = without; *-pnea* = breath), an absence of breathing.

- **Pain.** A sudden, severe pain brings about brief apnea, but a prolonged somatic pain increases respiratory rate. Visceral pain may slow respiratory rate.

- **Stretching the anal sphincter muscle.** This action increases the respiratory rate and is sometimes used to stimulate respiration in a newborn baby or a person who has stopped breathing.

- **Irritation of airways.** Physical or chemical irritation of the pharynx or larynx brings about an immediate cessation of breathing followed by coughing or sneezing.

- **Blood pressure.** The carotid and aortic baroreceptors that detect changes in blood pressure have a small effect on respiration. A sudden rise in blood pressure decreases the rate of respiration, and a drop in blood pressure increases the respiratory rate.

Table 23.2 summarizes the stimuli that affect the rate and depth of ventilation.

▶ **CHECKPOINT**

25. How does the medullary rhythmicity area function in regulating respiration?

26. How are the apneustic and pneumotaxic areas related to the control of respiration?

27. How do the cerebral cortex, levels of CO_2 and O_2, proprioceptors, inflation reflex, temperature changes, pain, and irritation of the airways modify respiration?

EXERCISE AND THE RESPIRATORY SYSTEM

▶ **OBJECTIVE**

- **Describe the effects of exercise on the respiratory system.**

During exercise, the respiratory and cardiovascular systems make adjustments in response to both the intensity and duration of the exercise. The effects of exercise on the heart are discussed in Chapter 20. Here we focus on how exercise affects the respiratory system.

Recall that the heart pumps the same amount of blood to the lungs as to all the rest of the body. Thus, as cardiac output rises,

Table 23.2	Summary of Stimuli that Affect Ventilation Rate and Depth
Stimuli that Increase Ventilation Rate and Depth	**Stimuli that Decrease Ventilation Rate and Depth**
Voluntary hyperventilation controlled by cerebral cortex and anticipation of activity by stimulation of the limbic system.	Voluntary hypoventilation controlled by cerebral cortex.
Increase in arterial blood P_{CO_2} above 40 mmHg (causes an increase in H^+) detected by peripheral and central chemoreceptors.	Decrease in arterial blood P_{CO_2} below 40 mmHg (causes a decrease in H^+) detected by peripheral and central chemoreceptors.
Decrease in arterial blood P_{O_2} from 105 mmHg to 50 mmHg.	Decrease in arterial blood P_{O_2} below 50 mmHg.
Increased activity of proprioceptors.	Decreased activity of proprioceptors.
Increase in body temperature.	Decrease in body temperature decreases rate of respiration, and sudden cold stimulus causes apnea.
Prolonged pain.	Severe pain causes apnea.
Decrease in blood pressure.	Increase in blood pressure.
Stretching anal sphincter.	Irritation of pharynx or larynx by touch or chemicals causes brief apnea followed by coughing or sneezing.

the blood flow to the lungs, termed **pulmonary perfusion,** increases as well. In addition, the **O_2 diffusing capacity,** a measure of the rate at which O_2 can diffuse from alveolar air into the blood, may increase threefold during maximal exercise because more pulmonary capillaries become maximally perfused. As a result, there is a greater surface area available for diffusion of O_2 into pulmonary blood capillaries.

When muscles contract during exercise, they consume large amounts of O_2 and produce large amounts of CO_2. During vigorous exercise, O_2 consumption and pulmonary ventilation both increase dramatically. At the onset of exercise, an abrupt increase in pulmonary ventilation is followed by a more gradual increase. With moderate exercise, the increase is due mostly to an increase in the depth of ventilation rather than to increased breathing rate. When exercise is more strenuous, the frequency of breathing also increases.

The abrupt increase in ventilation at the start of exercise is due to *neural* changes that send excitatory impulses to the inspiratory area in the medulla oblongata. These changes include (1) anticipation of the activity, which stimulates the limbic system; (2) sensory impulses from proprioceptors in muscles, tendons, and joints; and (3) motor impulses from the primary motor cortex (precentral gyrus). The more gradual increase in ventila-

tion during moderate exercise is due to *chemical* and *physical* changes in the bloodstream, including (1) slightly decreased P_{O_2}, due to increased O_2 consumption; (2) slightly increased P_{CO_2}, due to increased CO_2 production by contracting muscle fibers; and (3) increased temperature, due to liberation of more heat as more O_2 is utilized. Moreover, during strenuous exercise, HCO_3^- buffers H^+ released by lactic acid in a reaction that liberates CO_2, which further increases P_{CO_2}.

At the end of an exercise session, an abrupt decrease in pulmonary ventilation is followed by a more gradual decline to the resting level. The initial decrease is due mainly to changes in neural factors when movement stops or slows, whereas the more gradual phase reflects the slower return of blood chemistry levels and temperature to the resting state.

Smoking Lowers Respiratory Efficiency

Smoking may cause a person to become easily "winded" with even moderate exercise because several factors decrease respiratory efficiency in smokers: (1) Nicotine constricts terminal bronchioles, which decreases airflow into and out of the lungs. (2) Carbon monoxide in smoke binds to hemoglobin and reduces its oxygen-carrying capability. (3) Irritants in smoke cause increased mucus secretion by the mucosa of the bronchial tree and swelling of the mucosal lining, both of which impede airflow into and out of the lungs. (4) Irritants in smoke also inhibit the movement of cilia and destroy cilia in the lining of the respiratory system. Thus, excess mucus and foreign debris are not easily removed, which further adds to the difficulty in breathing. (5) With time, smoking leads to destruction of elastic fibers in the lungs and is the prime cause of emphysema (described on page 844). These changes cause collapse of small bronchioles and trapping of air in alveoli at the end of exhalation. The result is less efficient gas exchange. ■

DEVELOPMENT OF THE RESPIRATORY SYSTEM

▶ OBJECTIVE

• **Describe the development of the respiratory system.**

The development of the mouth and pharynx are discussed in Chapter 24; here we consider the development of the remainder of the respiratory system. At about four weeks of development, the respiratory system begins as an outgrowth of the **endoderm** of the foregut (precursor of some digestive organs) just posterior to the pharynx. This outgrowth is called the **laryngotracheal diverticulum** (see Figure 18.21a on page 623). As the diverticulum grows, it elongates and differentiates into the future epithelial lining of the *larynx* and other structures. Its proximal end maintains a slitlike opening into the pharynx called the *rima* *glottis.* The middle portion of the bud gives rise to the epithelial lining of the *trachea.* The distal portion divides into two **lung buds,** which grow into the epithelial lining of the *bronchi* and *lungs* (Figure 23.29).

As the lung buds develop, they branch repeatedly and give rise to all the *bronchial tubes.* After the sixth month, the closed terminal portions of the tubes dilate and become the *alveoli* of the lungs. **Mesenchymal (mesodermal) cells** form the smooth muscle, cartilage, and connective tissues of the bronchial tubes and the pleural sacs of the lungs.

AGING AND THE RESPIRATORY SYSTEM

▶ **OBJECTIVE**

• **Describe the effects of aging on the respiratory system.**

With advancing age, the airways and tissues of the respiratory tract, including the alveoli, become less elastic and more rigid; the chest wall becomes more rigid as well. The result is a decrease in lung capacity. In fact, vital capacity (the maximum amount of air that can be expired after maximal inhalation) can decrease as much as 35% by age 70. Moreover, a decrease in blood level of O_2, decreased activity of alveolar macrophages, and diminished ciliary action of the epithelium lining the respiratory tract occur. Owing to all these age-related factors, elderly people are more susceptible to pneumonia, bronchitis, emphysema, and other pulmonary disorders.

▶ **CHECKPOINT**

28. How does exercise affect the inspiratory area?

29. What structures develop from the laryngotracheal bud?

30. What accounts for the decrease in lung capacity with aging?

• • •

To appreciate the many ways that the respiratory system contributes to homeostasis of other body systems, examine *Focus on Homeostasis: The Respiratory System.* Next, in Chapter 24, we will see how the digestive system makes nutrients available to body cells so that oxygen provided by the respiratory system can be used for ATP production.

Figure 23.29 **Development of the bronchial tubes and lungs.**

The respiratory system develops from endoderm and mesoderm.

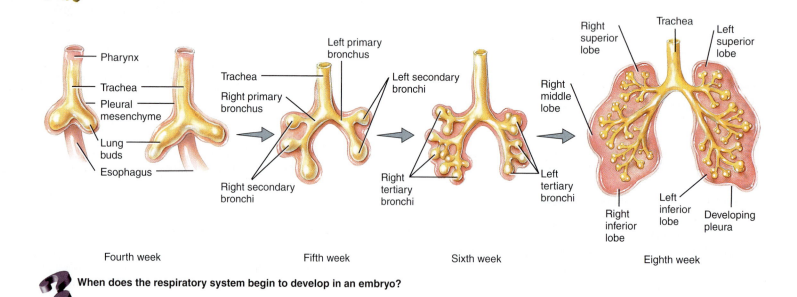

Fourth week   Fifth week   Sixth week   Eighth week

When does the respiratory system begin to develop in an embryo?

Focus on Homeostasis:

The Respiratory System

Body System	Contribution of the Respiratory System
For all body systems	Provides oxygen and removes carbon dioxide. Helps adjust pH of body fluids through exhalation of carbon dioxide.
Muscular system	Increased rate and depth of breathing support increased activity of skeletal muscles during exercise.
Nervous system	Nose contains receptors for sense of smell (olfaction). Vibrations of air flowing across vocal cords produce sounds for speech.
Endocrine system	Angiotensin converting enzyme (ACE) in lungs catalyzes formation of the hormone angiotensin II from angiotensin I.
Cardiovascular system	During inhalations, respiratory pump aids return of venous blood to the heart.
Lymphatic and immune system	Hairs in nose, cilia and mucus in trachea, bronchi, and smaller airways, and alveolar macrophages contribute to nonspecific resistance to disease. Pharynx (throat) contains lymphatic tissue (tonsils). Respiratory pump (during inhalation) promotes flow of lymph.
Digestive system	Forceful contraction of respiratory muscles can assist in defecation.
Urinary system	Together, respiratory and urinary systems regulate pH of body fluids.
Reproductive systems	Increased rate and depth of breathing support activity during sexual intercourse. Internal respiration provides oxygen to developing fetus.

DISORDERS: HOMEOSTATIC IMBALANCES

Asthma

Asthma (AZ-ma = panting) is a disorder characterized by chronic airway inflammation, airway hypersensitivity to a variety of stimuli, and airway obstruction that is at least partially reversible, either spontaneously or with treatment. It affects 3–5% of the U.S. population and is more common in children than in adults. Airway obstruction may be due to smooth muscle spasms in the walls of smaller bronchi and bronchioles, edema of the mucosa of the airways, increased mucus secretion, and damage to the epithelium of the airway.

Individuals with asthma typically react to low concentrations of agents that do not normally cause symptoms in people without asthma. Sometimes the trigger is an allergen such as pollen, house dust mites, molds, or a particular food. Other common triggers of asthma attacks are emotional upset, aspirin, sulfiting agents (used in wine and beer and to keep greens fresh in salad bars), exercise, and breathing cold air or cigarette smoke. In the early phase (acute) response, smooth muscle spasm is accompanied by excessive secretion of mucus that may clog the bronchi and bronchioles and worsen the attack. The late phase (chronic) response is characterized by inflammation, fibrosis, edema, and necrosis (death) of bronchial epithelial cells. A host of mediator chemicals, including leukotrienes, prostaglandins, thromboxane, platelet-activating factor, and histamine, take part.

Symptoms include difficult breathing, coughing, wheezing, chest tightness, tachycardia, fatigue, moist skin, and anxiety. An acute attack is treated by giving an inhaled beta$_2$-adrenergic agonist (albuterol) to help relax smooth muscle in the bronchioles and open up the airways. However, long-term therapy of asthma strives to suppress the underlying inflammation. The anti-inflammatory drugs that are used most often are inhaled corticosteroids (glucocorticoids), cromolyn sodium (Intal), and leukotriene blockers (Accolate).

Chronic Obstructive Pulmonary Disease

Chronic obstructive pulmonary disease (COPD) is a type of respiratory disorder characterized by chronic and recurrent obstruction of airflow, which increases airway resistance. COPD affects about 30 million Americans and is the fourth leading cause of death behind heart disease, cancer, and cerebrovascular disease. The principal types of COPD are emphysema and chronic bronchitis. In most cases, COPD is preventable because its most common cause is cigarette smoking or breathing secondhand smoke. Other causes include air pollution, pulmonary infection, occupational exposure to dusts and gases, and genetic factors. Because men, on average, have more years of exposure to cigarette smoke than women, men are twice as likely as women to suffer from COPD; still, the incidence of COPD in women has risen sixfold in the past 50 years, a reflection of increased smoking among women.

Emphysema

Emphysema (em′-fi-SĒ-ma′ = blown up or full of air) is a disorder characterized by destruction of the walls of the alveoli, producing abnormally large air spaces that remain filled with air during exhalation. With less surface area for gas exchange, O$_2$ diffusion across the damaged respiratory membrane is reduced. Blood O$_2$ level is somewhat lowered, and any mild exercise that raises the O$_2$ requirements of the cells leaves the patient breathless. As increasing numbers of alveolar walls are damaged, lung elastic recoil decreases due to loss of elastic fibers, and an increasing amount of air becomes trapped in the lungs at the end of exhalation. Over several years, added inspiratory exertion increases the size of the chest cage, resulting in a "barrel chest."

Emphysema is generally caused by a long-term irritation; cigarette smoke, air pollution, and occupational exposure to industrial dust are the most common irritants. Some destruction of alveolar sacs may be caused by an imbalance between enzymes called proteases (for example, elastase) and *alpha-1-antitrypsin*, a molecule that inhibits proteases. When decreased production by the liver of the plasma protein alpha-1-antitrypsin occurs, elastase is not inhibited and is free to attack the connective tissue in the walls of alveolar sacs. Cigarette smoke not only deactivates a protein that is apparently crucial in preventing emphysema; it also prevents the repair of affected lung tissue.

Treatment consists of cessation of smoking and removing other environmental irritants, exercise training under careful medical supervision, breathing exercises, use of bronchodilators, and oxygen therapy.

Chronic Bronchitis

Chronic bronchitis is a disorder characterized by excessive secretion of bronchial mucus and accompanied by a productive cough (sputum is raised) that lasts for at least three months of the year for two successive years. Cigarette smoking is the leading cause of chronic bronchitis. Inhaled irritants lead to chronic inflammation with an increase in the size and number of mucous glands and goblet cells in the airway epithelium. The thickened and excessive mucus produced narrows the airway and impairs ciliary function. Thus, inhaled pathogens become embedded in airway secretions and multiply rapidly. Besides a productive cough, symptoms of chronic bronchitis are shortness of breath, wheezing, cyanosis, and pulmonary hypertension. Treatment for chronic bronchitis is similar to that for emphysema.

Lung Cancer

In the United States **lung cancer** is the leading cause of cancer death in both males and females, accounting for 160,000 deaths annually. At the time of diagnosis, lung cancer is usually well advanced, with distant metastases present in about 55% of patients, and regional lymph node involvement in an additional 25%. Most people with lung cancer die within a year of the initial diagnosis; the overall survival rate is only 10–15%. Cigarette smoke is the most common cause of lung cancer. Roughly 85% of lung cancer cases are related to smoking, and the disease is 10 to 30 times more common in smokers than nonsmokers. Exposure to secondhand smoke is also associated with lung cancer and heart disease. In the United States, secondhand smoke causes an estimated 4000 deaths a year from lung cancer, and nearly 40,000 deaths a year from heart disease. Other causes of lung cancer are ionizing radiation and inhaled irritants, such as asbestos and radon gas. Emphysema is a common precursor to the development of lung cancer.

The most common type of lung cancer, **bronchogenic carcinoma,** starts in the epithelium of the bronchial tubes. Bronchogenic tumors are named based on where they arise. For example, *adenocarcinomas* develop in peripheral areas of the lungs from bronchial glands and alveolar cells, *squamous cell carcinomas* develop from the epithelium of larger bronchial tubes, and *small (oat) cell carcinomas* develop from epithelial cells in primary bronchi near the hilus of the lungs and tend to involve the mediastinum early on. Depending on the type of bronchogenic tumors, they may be aggressive, locally invasive, and undergo widespread metastasis. The tumors begin as epithelial lesions that grow to form masses that obstruct the bronchial tubes or invade adjacent lung tissue. Bronchogenic carcinomas metastasize to lymph nodes, the brain, bones, liver, and other organs.

Symptoms of lung cancer are related to the location of the tumor. These may include a chronic cough, spitting blood from the respiratory tract, wheezing, shortness of breath, chest pain, hoarseness, difficulty swallowing, weight loss, anorexia, fatigue, bone pain, confusion, problems with balance, headache, anemia, thrombocytopenia, and jaundice.

Treatment consists of partial or complete surgical removal of a diseased lung (pulmonectomy), radiation therapy, and chemotherapy.

Pneumonia

Pneumonia is an acute infection or inflammation of the alveoli. It is the most common infectious cause of death in the United States, where an estimated 4 million cases occur annually. When certain microbes enter the lungs of susceptible individuals, they release damaging toxins, stimulating inflammation and immune responses that have damaging side effects. The toxins and immune response damage alveoli and bronchial mucous membranes; inflammation and edema cause the alveoli to fill with debris and exudate, interfering with ventilation and gas exchange.

The most common cause of pneumonia is the pneumococcal bacterium *Streptococcus pneumoniae,* but other microbes may also cause pneumonia. Those who are most susceptible to pneumonia are the elderly, infants, immunocompromised individuals (those having AIDS or a malignancy, or those taking immunosuppressive drugs), cigarette smokers, and individuals with an obstructive lung disease. Most cases of pneumonia are preceded by an upper respiratory infection that often is viral. Individuals then develop fever, chills, productive or dry cough, malaise, chest pain, and sometimes dyspnea (difficult breathing) and hemoptysis (spitting blood).

Treatment may involve antibiotics, bronchodilators, oxygen therapy, increased fluid intake, and chest physiotherapy (percussion, vibration, and postural drainage).

Tuberculosis

The bacterium *Mycobacterium tuberculosis* produces an infectious, communicable disease called **tuberculosis (TB)** that most often affects the lungs and the pleurae but may involve other parts of the body. Once the bacteria are inside the lungs, they multiply and cause inflammation, which stimulates neutrophils and macrophages to migrate to the area and engulf the bacteria to prevent their spread. If the immune system is not impaired, the bacteria remain dormant for life, but impaired immunity may enable the bacteria to escape into blood and lymph to infect other organs. In many people, symptoms—fatigue, weight loss, lethargy, anorexia, a low-grade fever, night sweats, cough, dyspnea, chest pain, and hemoptysis—do not develop until the disease is advanced.

During the past several years, the incidence of TB in the United States has risen dramatically. Perhaps the single most important factor related to this increase is the presence of the human immunodeficiency virus (HIV). People infected with HIV are much more likely to develop tuberculosis because their immune systems are impaired. Among the other factors that have contributed to the increased number of cases are homelessness, increased drug abuse, increased immigration from countries with a high prevalence of tuberculosis, increased crowding in housing among the poor, and airborne transmission of tuberculosis in prisons and shelters. In addition, recent outbreaks of tuberculosis involving multi-drug-resistant strains of *Mycobacterium tuberculosis* have occurred because patients fail to complete their antibiotic and other treatment regimens.

Coryza and Influenza

Hundreds of viruses can cause **coryza** (ko-RĪ-za) or the **common cold.** But a group of viruses called *rhinoviruses* is responsible for about 40% of all colds in adults. Typical symptoms include sneezing, excessive nasal secretion, dry cough, and congestion. The uncomplicated common cold is not usually accompanied by a fever. Complications include sinusitis, asthma, bronchitis, ear infections, and laryngitis. Recent investigations suggest an association between emotional stress and the common cold. The higher the stress level, the greater the frequency and duration of colds.

Influenza (flu) is also caused by a virus. Its symptoms include chills, fever (usually higher than $101°F = 39°C$), headache, and muscular aches. Coldlike symptoms appear as the fever subsides.

Pulmonary Edema

Pulmonary edema is an abnormal accumulation of fluid in the interstitial spaces and alveoli of the lungs. The edema may arise from increased pulmonary capillary permeability (pulmonary origin) or increased pulmonary capillary pressure (cardiac origin); the latter cause may coincide with congestive heart failure. The most common symptom is dyspnea. Others include wheezing, tachypnea (rapid breathing rate), restlessness, a feeling of suffocation, cyanosis, pallor (paleness), and diaphoresis (excessive perspiration). Treatment consists of administering oxygen, drugs that dilate the bronchioles and lower blood pressure, diuretics to rid the body of excess fluid, and drugs that correct acid–base imbalance; suctioning of airways; and mechanical ventilation.

Cystic Fibrosis

Cystic fibrosis (CF) is an inherited disease of secretory epithelia that affects the airways, liver, pancreas, small intestine, and sweat glands. It is the most common lethal genetic disease in whites: 5% of the population are thought to be genetic carriers. The cause of cystic fibrosis is a genetic mutation affecting a transporter protein that carries chloride ions across the plasma membranes of many epithelial cells. Because dysfunction of sweat glands causes perspiration to contain excessive sodium chloride (salt), measurement of the excess chloride is one index for diagnosing CF. The mutation also disrupts the normal functioning of several organs by causing ducts within them to become obstructed by thick mucus secretions that do not drain easily from the passageways. Buildup of these secretions leads to inflammation and replacement of injured cells with connective tissue that further blocks the ducts. Clogging and infection of the airways leads to difficulty in breathing and eventual destruction of lung tissue. Lung disease accounts for most deaths from CF. Obstruction of small bile ducts in the liver interferes with digestion and disrupts liver function; clogging of pancreatic ducts prevents digestive enzymes from reaching the small intestine. Because pancreatic juice contains the main fat-digesting enzyme, the person fails to absorb fats or fat-soluble vitamins and thus suffers from vitamin A, D, and K deficiency diseases. With respect to the reproductive systems, blockage of the ductus (vas) deferens leads to infertility in males; the formation of dense mucus plugs in the vagina restricts the entry of sperm into the uterus and can lead to infertility in females.

A child suffering from cystic fibrosis is given pancreatic extract and large doses of vitamins A, D, and K. The recommended diet is high in calories, fats, and proteins, with vitamin supplementation and liberal use of salt.

Acute Respiratory Distress Syndrome

Acute respiratory distress syndrome (ARDS) is a form of respiratory failure characterized by excessive leakiness of the respiratory membranes and severe hypoxia. Situations that can cause ARDS include near drowning, aspiration of acidic gastric juice, drug reactions, inhalation of an irritating gas such as ammonia, allergic reactions, various lung infections such as pneumonia or TB, and pulmonary hypertension. ARDS strikes about 250,000 people a year in the United States, and about half of them die despite intensive medical care.

MEDICAL TERMINOLOGY

Abdominal thrust (Heimlich) maneuver (HĪM-lik ma-NOO-ver) First-aid procedure designed to clear the airways of obstructing objects. It is performed by applying a quick upward thrust between the navel and costal margin that causes sudden elevation of the diaphragm and forceful, rapid expulsion of air in the lungs; this action forces air out the trachea to eject the obstructing object. The Heimlich maneuver is also used to expel water from the lungs of near-drowning victims before resuscitation is begun.

Asphyxia (as-FIK-sē-a; *sphyxia* = pulse) Oxygen starvation due to low atmospheric oxygen or interference with ventilation, external respiration, or internal respiration.

Aspiration (as′-pi-RĀ-shun) Inhalation of a foreign substance such as water, food, or a foreign body into the bronchial tree; also, the drawing of a substance in or out by suction.

Atelectasis (at′-ē-LEK-ta-sis; *atel-* = incomplete; *ectasis* = expansion) Incomplete expansion of a lung or a portion of a lung caused by airway obstruction, lung compression, or inadequate pulmonary surfactant.

Bronchiectasis (bron′-kē-EK-ta-sis) A chronic dilation of the bronchi or bronchioles.

Bronchography (bron-KOG-ra-fē) An imaging technique used to visualize the bronchial tree using x rays. After an opaque contrast medium is inhaled through an intratracheal catheter, radiographs of the chest in various positions are taken, and the developed film, a **bronchogram** (BRON-kō-gram), provides a picture of the bronchial tree.

Bronchoscopy (bron-KOS-kō-pē) Visual examination of the bronchi through a **bronchoscope,** an illuminated, flexible tubular instrument that is passed through the mouth (or nose), larynx, and trachea into the bronchi. The examiner can view the interior of the trachea and bronchi to biopsy a tumor, clear an obstructing object or secretions from an airway, take cultures or smears for microscopic examination, stop bleeding, or deliver drugs.

Cheyne–Stokes respiration (CHĀN STŌKS res′-pi-RĀ-shun) A repeated cycle of irregular breathing that begins with shallow breaths that increase in depth and rapidity and then decrease and cease altogether for 15 to 20 seconds. Cheyne–Stokes is normal in infants; it is also often seen just before death from pulmonary, cerebral, cardiac, and kidney disease.

Dyspnea (DISP-nē-a; *dys-* = painful, difficult; *-pnea* = breath) Painful or labored breathing.

Epistaxis (ep′-i-STAK-sis) Loss of blood from the nose due to trauma, infection, allergy, malignant growths, or bleeding disorders. It can be arrested by cautery with silver nitrate, electrocautery, or firm packing. Also called **nosebleed.**

Hemoptysis (hē-MOP-ti-sis; *hemo-* = blood; *-ptysis* = spit) Spitting of blood from the respiratory tract.

Hypoventilation (*hypo-* = below) Slow and shallow breathing.

Rales (RĀLS) Sounds sometimes heard in the lungs that resemble bubbling or rattling. Rales are to the lungs what murmurs are to the heart. Different types are due to the presence of an abnormal type or amount of fluid or mucus within the bronchi or alveoli, or to bronchoconstriction that causes turbulent airflow.

Respirator (RES-pi-rā′-tor) An apparatus fitted to a mask over the nose and mouth, or hooked directly to an endotracheal or tracheotomy tube, that is used to assist or support ventilation or to provide nebulized medication to the air passages.

Respiratory failure A condition in which the respiratory system either cannot supply sufficient O_2 to maintain metabolism or cannot eliminate enough CO_2 to prevent respiratory acidosis (a lower-than-normal pH in extracellular fluid).

Rhinitis (rī-NĪ-tis; *rhin-* = nose) Chronic or acute inflammation of the mucous membrane of the nose.

Sudden infant death syndrome (SIDS) Death of infants between the ages of 1 week and 12 months thought to be due to hypoxia while sleeping in a prone position (on the stomach) and the rebreathing of exhaled air trapped in a depression of the mattress. It is now recommended that normal newborns be placed on their backs for sleeping. Remember: "back to sleep."

Tachypnea (tak′-ip-NĒ-a; *tachy-* = rapid; *-pnea* = breath) Rapid breathing rate.

STUDY OUTLINE

RESPIRATORY SYSTEM ANATOMY (p. 806)

1. The respiratory system consists of the nose, pharynx, larynx, trachea, bronchi, and lungs. They act with the cardiovascular system to supply oxygen (O_2) and remove carbon dioxide (CO_2) from the blood.

2. The external portion of the nose is made of cartilage and skin and is lined with a mucous membrane. Openings to the exterior are the external nares.

3. The internal portion of the nose communicates with the paranasal sinuses and nasopharynx through the internal nares.

4. The nasal cavity is divided by a septum. The anterior portion of the cavity is called the vestibule. The nose warms, moistens, and filters air and functions in olfaction and speech.

5. The pharynx (throat) is a muscular tube lined by a mucous membrane. The anatomic regions are the nasopharynx, oropharynx, and laryngopharynx.

6. The nasopharynx functions in respiration. The oropharynx and laryngopharynx function both in digestion and in respiration.

7. The larynx (voice box) is a passageway that connects the pharynx with the trachea. It contains the thyroid cartilage (Adam's apple); the epiglottis, which prevents food from entering the larynx; the cricoid cartilage, which connects the larynx and trachea; and the paired arytenoid, corniculate, and cuneiform cartilages.

8. The larynx contains vocal folds, which produce sound as they vibrate. Taut folds produce high pitches, and relaxed ones produce low pitches.

9. The trachea (windpipe) extends from the larynx to the primary bronchi. It is composed of C-shaped rings of cartilage and smooth muscle and is lined with pseudostratified ciliated columnar epithelium.

10. The bronchial tree consists of the trachea, primary bronchi, secondary bronchi, tertiary bronchi, bronchioles, and terminal bronchioles. Walls of bronchi contain rings of cartilage; walls of bronchioles contain increasingly smaller plates of cartilage and increasing amounts of smooth muscle.

11. Lungs are paired organs in the thoracic cavity enclosed by the pleural membrane. The parietal pleura is the superficial layer that lines the thoracic cavity; the visceral pleura is the deep layer that covers the lungs.

12. The right lung has three lobes separated by two fissures; the left lung has two lobes separated by one fissure and a depression, the cardiac notch.

13. Secondary bronchi give rise to branches called segmental bronchi, which supply segments of lung tissue called bronchopulmonary segments.

14. Each bronchopulmonary segment consists of lobules, which contain lymphatics, arterioles, venules, terminal bronchioles, respiratory bronchioles, alveolar ducts, alveolar sacs, and alveoli.

15. Alveolar walls consist of type I alveolar cells, type II alveolar cells, and associated alveolar macrophages.

16. Gas exchange occurs across the respiratory membranes.

PULMONARY VENTILATION (p. 821)

1. Pulmonary ventilation, or breathing, consists of inhalation and exhalation.

2. The movement of air into and out of the lungs depends on pressure changes governed in part by Boyle's law, which states that the volume of a gas varies inversely with pressure, assuming that temperature remains constant.

3. Inhalation occurs when alveolar pressure falls below atmospheric pressure. Contraction of the diaphragm and external intercostals increases the size of the thorax, thereby decreasing the intrapleural pressure so that the lungs expand. Expansion of the lungs decreases alveolar pressure so that air moves down a pressure gradient from the atmosphere into the lungs.

4. During forceful inhalation, accessory muscles of inhalation (sternocleidomastoids, scalenes, and pectoralis minors) are also used.

5. Exhalation occurs when alveolar pressure is higher than atmospheric pressure. Relaxation of the diaphragm and external intercostals results in elastic recoil of the chest wall and lungs, which increases intrapleural pressure; lung volume decreases and alveolar pressure increases, so air moves from the lungs to the atmosphere.

6. Forceful exhalation involves contraction of the internal intercostal and abdominal muscles.

7. The surface tension exerted by alveolar fluid is decreased by the presence of surfactant.

8. Compliance is the ease with which the lungs and thoracic wall can expand.

9. The walls of the airways offer some resistance to breathing.

10. Normal quiet breathing is termed eupnea; other patterns are costal breathing and diaphragmatic breathing. Modified respiratory movements, such as coughing, sneezing, sighing, yawning, sobbing, crying, laughing, and hiccupping, are used to express emotions and to clear the airways. See Table 23.1 on page 826.

LUNG VOLUMES AND CAPACITIES (p. 826)

1. Lung volumes exchanged during breathing and the rate of respiration are measured with a spirometer.

2. Lung volumes measured by spirometry include tidal volume, minute ventilation, alveolar ventilation rate, inspiratory reserve volume, expiratory reserve volume, and $FEV_{1.0}$. Other lung volumes are anatomic dead space, residual volume, and minimal volume.

3. Lung capacities, the sum of two or more volumes, include inspiratory, functional residual, vital, and total lung capacities.

EXCHANGE OF OXYGEN AND CARBON DIOXIDE (p. 827)

1. The partial pressure of a gas is the pressure exerted by that gas in a mixture of gases. It is symbolized by P_x, where the subscript is the chemical formula of the gas.

2. According to Dalton's law, each gas in a mixture of gases exerts its own pressure as if all the other gases were not present.

3. Henry's law states that the quantity of a gas that will dissolve in a liquid is proportional to the partial pressure of the gas and its solubility coefficient (given that the temperature remains constant).

4. In internal and external respiration, O_2 and CO_2 diffuse from areas of higher partial pressures to areas of lower partial pressures.

5. External respiration or pulmonary gas exchange is the exchange of gases between alveoli and pulmonary blood capillaries. It depends

on partial pressure differences, a large surface area for gas exchange, a small diffusion distance across the respiratory membrane, and the rate of airflow into and out of the lungs.

6. Internal respiration or systemic gas exchange is the exchange of gases between systemic blood capillaries and tissue cells.

TRANSPORT OF OXYGEN AND CARBON DIOXIDE (p. 831)

1. In each 100 mL of oxygenated blood, 1.5% of the O_2 is dissolved in blood plasma and 98.5% is bound to hemoglobin as oxyhemoglobin ($Hb-O_2$).
2. The binding of O_2 to hemoglobin is affected by P_{O_2}, acidity (pH), P_{CO_2}, temperature, and BPG.
3. Fetal hemoglobin differs from adult hemoglobin in structure and has a higher affinity for O_2.
4. In each 100 mL of deoxygenated blood, 9% of CO_2 is dissolved in blood plasma, 13% combines with hemoglobin as carbaminohemoglobin ($Hb-CO_2$), and 78% is converted to bicarbonate ions (HCO_3^-).
5. In an acidic environment, hemoglobin's affinity for O_2 is lower, and O_2 dissociates more readily from it (Bohr effect).
6. In the presence of O_2, less CO_2 binds to hemoglobin (Haldane effect).

CONTROL OF RESPIRATION (p. 835)

1. The respiratory center consists of a medullary rhythmicity area in the medulla oblongata and a pneumotaxic area and an apneustic area in the pons.
2. The inspiratory area sets the basic rhythm of respiration.
3. The pneumotaxic and apneustic areas coordinate the transition between inhalation and exhalation.

4. Respirations may be modified by a number of factors, including cortical influences; the inflation reflex; chemical stimuli, such as O_2 and CO_2 and H^+ levels; proprioceptor input; blood pressure changes; limbic system stimulation; temperature; pain; and irritation to the airways. (See Table 23.2 on page 841.)

EXERCISE AND THE RESPIRATORY SYSTEM (p. 840)

1. The rate and depth of ventilation change in response to both the intensity and duration of exercise.
2. An increase in pulmonary perfusion and O_2-diffusing capacity occurs during exercise.
3. The abrupt increase in ventilation at the start of exercise is due to neural changes that send excitatory impulses to the inspiratory area in the medulla oblongata. The more gradual increase in ventilation during moderate exercise is due to chemical and physical changes in the bloodstream.

DEVELOPMENT OF THE RESPIRATORY SYSTEM (p. 841)

1. The respiratory system begins as an outgrowth of endoderm called the laryngotracheal bud.
2. Smooth muscle, cartilage, and connective tissue of the bronchial tubes and pleural sacs develop from mesoderm.

AGING AND THE RESPIRATORY SYSTEM (p. 843)

1. Aging results in decreased vital capacity, decreased blood level of O_2, and diminished alveolar macrophage activity.
2. Elderly people are more susceptible to pneumonia, emphysema, bronchitis, and other pulmonary disorders.

SELF-QUIZ QUESTIONS

Fill in the blanks in the following statements.

1. The three basic steps of respiration are _____, _____, and _____.

2. For inhalation to occur, air pressure in the alveoli must be _____ than atmospheric pressure; for exhalation to occur, air pressure in the alveoli must be _____ than atmospheric pressure.

3. Oxygen in blood is carried primarily in the form of _____; carbon dioxide is carried as _____, _____, and _____.

4. Write the equation for the chemical reaction that occurs for transport of carbon dioxide as bicarbonate ions in blood: _____.

Indicate whether the following statements are true or false.

5. The carina is the landmark used for an emergency tracheostomy.

6. The most important muscle of inhalation is the diaphragm.

Choose the one best answer to the following questions.

7. Which of the following statements are *correct*? (1) Normal exhalation during quiet breathing is an active process involving intensive muscle contraction. (2) Passive exhalation results from elastic recoil of the chest wall and lungs. (3) Air flow during breathing is due to a pressure gradient between the lungs and the atmospheric air. (4) During normal breathing, the pressure between the two pleural layers (intrapleural pressure) is always subatmospheric. (5) Surface tension of alveolar fluid facilitates inhalation. (a) 1, 2, and 3, (b) 2, 3, and 4, (c) 3, 4, and 5, (d) 1, 3, and 5 (e) 2, 3, and 5.

8. Which of the following factors affect the rate of external respiration? (1) partial pressure differences of the gases, (2) surface area for gas exchange, (3) diffusion distance, (4) solubility and molecular weight of the gases, (5) presence of bis-phosphoglycerate (BPG). (a) 1, 2, and 3, (b) 2, 4, and 5, (c) 1, 2, 4, and 5, (d) 1, 2, 3, and 4 (e) 2, 3, 4, and 5.

9. The most important factor in determining the percent oxygen saturation of hemoglobin is (a) the partial pressure of oxygen, (b) acidity, (c) the partial pressure of carbon dioxide, (d) temperature, (e) BPG.

10. Which gas law that states that each gas in a mixture exerts its own pressure as if all the other gases were not present? (a) Boyle's law, (b) Henry's law, (c) Dalton's law, (d) Haldane's law, (e) Bohr's law.

11. Which of the following statements are *true*? (1) It is impossible for people to kill themselves by holding their breath. (2) The cerebral cortex provides voluntary control over breathing. (3) Emotional stimuli can alter respiration. (4) Certain chemical stimuli can alter the rate and depth of breathing. (5) At the onset of exercise, an abrupt increase in breathing is followed by a more gradual increase in breathing. (a) 1, 2, 4, and 5, (b) 2, 3, and 5, (c) 1, 3, and 5, (d) 2, 3, 4, and 5, (e) 1, 2, ,3, 4, and 5.

12. Match the following:

_____ (a) functions as a passageway for air and food, provides a resonating chamber for speech sounds, and houses the tonsils

_____ (b) site of external respiration

_____ (c) connects the laryngopharynx with the trachea; houses the vocal cords

_____ (d) serous membranes that surround the lungs

_____ (e) functions in warming, moistening, and filtering air; receives olfactory stimuli; is a resonating chamber for sound

_____ (f) form a continuous lining of the alveolar wall; sites of gas exchange

_____ (g) a tubular passageway for air connecting the larynx to the bronchi

_____ (h) secrete alveolar fluid and surfactant

_____ (i) prevents food or fluid from entering the airways

_____ (j) air passageways entering the lungs

(1) nose
(2) pharynx
(3) larynx
(4) epiglottis
(5) trachea
(6) bronchi
(7) pleurae
(8) alveoli
(9) type I alveolar cells
(10) type II alveolar cells

13. Match the following:

_____ (a) a deficiency of oxygen at the tissue level

_____ (b) above-normal partial pressure of carbon dioxide

_____ (c) normal quiet breathing

_____ (d) deep, abdominal breathing

_____ (e) the ease with which the lungs and thoracic wall can be expanded

_____ (f) absence of breathing

_____ (g) rapid and deep breathing

_____ (h) shallow, chest breathing

(1) eupnea
(2) apnea
(3) hyperventilation
(4) costal breathing
(5) diaphragmatic breathing
(6) compliance
(7) hypoxia
(8) hypercapnia

14. Match the following:

_____ (a) total volume of air inhaled and exhaled each minute

_____ (b) tidal volume + inspiratory reserve volume + expiratory reserve volume

_____ (c) additional amount of air inhaled beyond tidal volume when taking a very deep breath

_____ (d) amount of air remaining in lungs after expiratory reserve volume is expelled

_____ (e) tidal volume + inspiratory reserve volume

_____ (f) volume of air in one breath

_____ (g) amount of air exhaled in forced exhalation

_____ (h) provides a medical and legal tool for determining if a baby was born dead or died after birth

(1) tidal volume
(2) residual volume
(3) minute ventilation
(4) expiratory reserve volume
(5) inspiratory reserve volume
(6) minimal volume
(7) inspiratory capacity
(8) vital capacity

15. Match the following:

_____ (a) prevents excessive inflation of the lungs

_____ (b) controls the basic rhythm of respiration

_____ (c) sends stimulatory impulses to the inspiratory area that activate it and prolong inhalation

_____ (d) as acidity increases, the affinity of hemoglobin for oxygen decreases and oxygen dissociates more readily from hemoglobin

_____ (e) transmits inhibitory impulses to turn off the inspiratory area before the lungs become too full of air

_____ (f) the quantity of gas that dissolves in a liquid is proportional to the partial pressure of the gas and its solubility coefficient

_____ (g) relates to the partial pressure of a gas in a mixture of gases

(1) Bohr effect
(2) Dalton's law
(3) medullary rhythmicity area
(4) Henry's law
(5) apneustic area
(6) pneumotaxic area
(7) Hering-Breuer reflex

CRITICAL THINKING QUESTIONS

1. Leung, a member of the college swim team, was measuring his lung volumes during his A&P lab. Although his tidal volume was average for a male of his age and size, his inspiratory reserve volume was the highest in the class. List the normal male values for lung volumes that can be measured with spirometer. Would you expect the average female volumes to be the same? Why or why not?
   HINT *Total lung capacity includes a volume that cannot be measured by spirometry.*

2. Samir stepped up to the plate and lifted his bat. The pitch was wild, and the ball hit him right in the nose! The x ray revealed a break in both medial bones of the external nose. Describe the structure of the external nose, and identify the location of the break.
   HINT *Samir may need a rhinoplasty to repair his nose.*

3. Her mother calls Kirsten a "challenging personality," but you have some other adjectives in mind. She's threatening to hold her breath until she, as she puts it, "turns blue, falls down, and dies—and won't you be sorry!" Do you need to worry about the "dies" part?
   HINT *Only some aspects of respiration are partly under voluntary control.*

ANSWERS TO FIGURE QUESTIONS

23.1 The conducting portion of the respiratory system includes the nose, pharynx, larynx, trachea, bronchi, and bronchioles (except the respiratory bronchioles).

23.2 Air's path is external nares → vestibule → nasal cavity → internal nares.

23.3 The root of the nose attaches it to the frontal bone.

23.4 The superior border of the pharynx is the internal nares; the inferior border of the pharynx is the cricoid cartilage.

23.5 During swallowing, the epiglottis closes over the rima glottidis, the entrance to the trachea.

23.6 The main function of the vocal folds is voice production.

23.7 Because the tissues between the esophagus and trachea are soft, the esophagus can bulge and press against the trachea during swallowing.

23.8 The left lung has two lobes and secondary bronchi; the right lung has three of each.

23.9 The pleural membrane is a serous membrane.

23.10 Because two-thirds of the heart lies to the left of the midline, the left lung contains a cardiac notch to accommodate the presence of the heart.

23.11 The wall of an alveolus is made up of type I alveolar cells, type II alveolar cells, and associated alveolar macrophages.

23.12 The respiratory membrane averages 0.5 μm in thickness.

23.13 The pressure would increase to 4 atm.

23.14 If you are at rest while reading, your diaphragm is responsible for about 75% of each inhalation.

23.15 At the start of inhalation, intrapleural pressure is about 756 mmHg. With contraction of the diaphragm, it decreases to about 754 mmHg as the volume of the space between the two pleural layers expands. With relaxation of the diaphragm, it increases back to 756 mmHg.

23.16 Normal atmospheric pressure at sea level is 760 mmHg.

23.17 Breathing in and then exhaling as much air as possible demonstrates vital capacity.

23.18 In both places, a difference in P_{O_2} promotes oxygen diffusion.

23.19 During strenuous exercise, oxygenated blood contains 20 mL of O_2 per 100 mL of blood (the same as at rest).

23.20 In both cases, hemoglobin in your pulmonary veins would be fully saturated with O_2, a point which is at the upper right of the curve.

23.21 Because lactic acid (lactate) and CO_2 are produced by active skeletal muscles, blood pH decreases slightly and P_{CO_2} increases when you are actively exercising. The result is lowered affinity of hemoglobin for O_2, so more O_2 is available to the working muscles.

23.22 O_2 is more available when you have a fever because the affinity of hemoglobin for O_2 decreases with increasing temperature.

23.23 Fetal Hb is 80% saturated with O_2, whereas maternal Hb is about 75% saturated at a P_{O_2} of 40 mmHg.

23.24 Blood in a systemic vein would have a higher concentration of HCO_3^-.

23.25 The medullary inspiratory area contains autorhythmic neurons.

23.26 The phrenic nerves innervate the diaphragm.

23.27 Peripheral chemoreceptors are responsive to changes in levels of oxygen partial pressure, carbon dioxide partial pressure, and H^+.

23.28 Normal arterial P_{CO_2} is 40 mmHg.

23.29 The respiratory system begins to develop about 4 weeks after fertilization.

The Digestive System

A Student's View

Realizing that a particular body system does not work alone—that all systems are connected and that structure relates to function—are perhaps the hardest concepts for A&P students to grasp. I am currently enrolled in the Cardiopulmonary Technologies Program, finishing my second semester of A&P. As a Cardiovascular Technologist I will not only have to be able to recognize and diagnose various disease processes, but understand normal processes as well. In addition, knowing how the cardiovascular system contributes to homeostasis of other body systems—how all body systems are integrated—is key.

This chapter does an excellent job of reinforcing this subject. It introduces the organs of the gastrointestinal tract, and in doing so they cover the muscles and bones—the structures/tissues—that comprise these organs, again building on previously introduced material and showing how the systems intertwine. The tables provided throughout the chapter also follow this theme. Take Table 24.4 Major Hormones that Control Digestion as an example. It illustrates how the endocrine system works together with the digestive system to achieve balance and proper functioning. The tables in general are a great feature because they organize/summarize the information in an easy to understand manner that is very helpful to review. Another great feature is the fact that clear definitions are provided whenever new terms are introduced.

The Focus on Homeostasis section is extremely valuable because, again, while reviewing information introduced in the chapter, it also illustrates the many ways that one system, in this case the digestive system, contributes to homeostasis of other body systems.

Sabrina von Brueckwitz
Edison Community College

THE DIGESTIVE SYSTEM AND HOMEOSTASIS

The digestive system contributes to homeostasis by breaking down food into forms that can be absorbed and used by body cells. It absorbs water, vitamins, and minerals and eliminates wastes from the body.

ENERGIZE YOUR STUDY
FOUNDATIONS CD

Anatomy Overview
- The Digestive System

Animations
- Systems Contributions to Homeostasis
- The Case of the Coughing Boy
- The Case of the Girl with the Fruity Breath

Concepts and Connections and Exercises reinforce your understanding

www.wiley.com/college/apcentral

INSIGHTS AND EXPLORATIONS

Most of us are aware that bacteria can cause diarrheal disease. Did you also know that there is a bacterium that can cause peptic ulcers? We used to think that peptic ulcers were the result of too much worry or an inability to handle a stressful life style. Scientists have now shown that peptic ulcers can be caused by the bacterium Helicobacter pylori. In this web-based activity we will explore more about this bacterium, how it can survive the acidity of the stomach, and how a breathalyzer test can be used to diagnose this disease.

The food we eat contains a variety of nutrients, which are used for building new body tissues and repairing damaged tissues. Food is also vital for life because it is our only source of chemical energy. As consumed, however, most food cannot be used either to build tissues or to power the energy-requiring chemical reactions of body cells. Rather, foods must be broken down into molecules that are small enough to enter body cells, a process known as **digestion.** The passage of these smaller molecules through the plasma membranes of cells lining the stomach and intestines and then into the blood and lymph is termed **absorption.** The organs that collectively perform these functions—the **digestive system**—are the focus of this chapter.

The medical specialty that deals with the structure, function, diagnosis, and treatment of diseases of the stomach and intestines is called **gastroenterology** (gas'-trō-en'-ter-OL-ō-jē; *gastro-* = stomach; *entero-* = intestines; *-logy* = study of). The medical specialty that deals with the diagnosis and treatment of disorders of the rectum and anus is called **proctology** (prok-TOL-ō-jē; *proct-* = rectum).

OVERVIEW OF THE DIGESTIVE SYSTEM

▶ **OBJECTIVES**

- **Identify the organs of the digestive system.**
- **Describe the basic processes performed by the digestive system.**

Two groups of organs compose the digestive system (Figure 24.1): the gastrointestinal (GI) tract and the accessory digestive organs. The **gastrointestinal (GI) tract,** or **alimentary canal** (*alimentary* = nourishment), is a continuous tube that extends from the mouth to the anus through the ventral body cavity. Organs of the gastrointestinal tract include the mouth, most of the pharynx, esophagus, stomach, small intestine, and large intestine. The length of the GI tract taken from a cadaver is about 9 m (30 ft). In a living person, it is much shorter because the muscles along the walls of GI tract organs are in a state of tonus (sustained contraction). The **accessory digestive organs** are the teeth, tongue, salivary glands, liver, gallbladder, and pancreas. Teeth aid in the physical breakdown of food, and the tongue assists in chewing and swallowing. The other accessory digestive organs, however, never come into direct contact with food. They produce or store secretions that flow into the GI tract through ducts; the secretions aid in the chemical breakdown of food.

The GI tract contains and processes food from the time it is eaten until it is digested and absorbed or eliminated. In some parts of the GI tract, muscular contractions in the wall physically break down the food by churning it. The contractions also help to dissolve foods by mixing them with fluids secreted into the tract. Enzymes secreted by accessory structures and cells that line the tract break down the food chemically. Wavelike contractions of the smooth muscle in the wall of the GI tract propel the food along the tract, from the esophagus to the anus.

Overall, the digestive system performs six basic processes:

1. ***Ingestion.*** This process involves taking foods and liquids into the mouth (eating).

2. ***Secretion.*** Each day, cells within the walls of the GI tract and accessory digestive organs secrete a total of about 7 liters of water, acid, buffers, and enzymes into the lumen of the tract.

3. ***Mixing and propulsion.*** Alternating contraction and relaxation of smooth muscle in the walls of the GI tract mix food and secretions and propel them toward the anus. This capability of the GI tract to mix and move material along its length is termed **motility.**

4. ***Digestion.*** Mechanical and chemical processes break down ingested food into small molecules. In **mechanical digestion** the teeth cut and grind food before it is swallowed, and then smooth muscles of the stomach and small intestine churn the food. As a result, food molecules become dissolved and thoroughly mixed with digestive enzymes. In **chemical digestion** the large carbohydrate, lipid, protein, and nucleic acid molecules in food are split into smaller molecules by hydrolysis (see Figure 2.15 on page 44). Digestive enzymes produced by the salivary glands, tongue, stomach, pancreas, and small intestine catalyze these catabolic reactions. A few substances in food can be absorbed without chemical digestion. These include amino acids, cholesterol, glucose, vitamins, minerals, and water.

5. ***Absorption.*** The entrance of ingested and secreted fluids, ions, and the small molecules that are products of digestion into the epithelial cells lining the lumen of the GI tract is called **absorption.** The absorbed substances pass into blood or lymph and circulate to cells throughout the body.

6. ***Defecation.*** Wastes, indigestible substances, bacteria, cells sloughed from the lining of the GI tract, and digested materials that were not absorbed leave the body through the anus in a process called **defecation.** The eliminated material is termed **feces.**

▶ **CHECKPOINT**

1. Which components of the digestive system are GI tract organs and which are accessory digestive organs?

2. Which organs of the digestive system come in contact with food, and what are some of their digestive functions?

3. Which kinds of food molecules undergo chemical digestion, and which do not?

Figure 24.1   Organs of the digestive system.

Organs of the gastrointestinal (GI) tract are the mouth, pharynx, esophagus, stomach, small intestine, and large intestine. Accessory digestive organs are the teeth, tongue, salivary glands, liver, gallbladder, and pancreas.

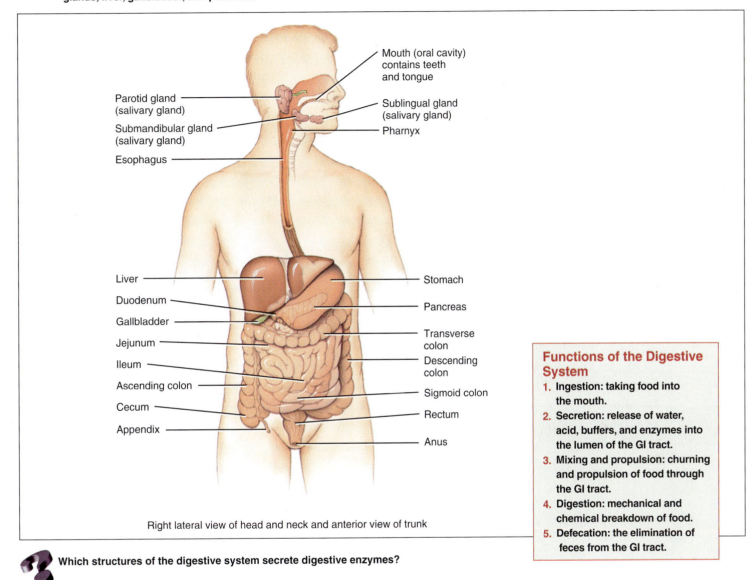

Right lateral view of head and neck and anterior view of trunk

Functions of the Digestive System

1. **Ingestion:** taking food into the mouth.
2. **Secretion:** release of water, acid, buffers, and enzymes into the lumen of the GI tract.
3. **Mixing and propulsion:** churning and propulsion of food through the GI tract.
4. **Digestion:** mechanical and chemical breakdown of food.
5. **Defecation:** the elimination of feces from the GI tract.

Which structures of the digestive system secrete digestive enzymes?

LAYERS OF THE GI TRACT

▶ **OBJECTIVE**

- **Describe the layers that form the wall of the gastrointestinal tract.**

The wall of the GI tract from the lower esophagus to the anal canal has the same basic, four-layered arrangement of tissues. The four layers of the tract, from deep to superficial, are the mucosa, submucosa, muscularis, and serosa (Figure 24.2).

Mucosa

The **mucosa,** or inner lining of the GI tract, is a mucous membrane. It is composed of a layer of epithelium in direct contact with the contents of the GI tract, areolar connective tissue, and a thin layer of smooth muscle (muscularis mucosae).

1. The **epithelium** in the mouth, pharynx, esophagus, and anal canal is mainly nonkeratinized stratified squamous epithelium that serves a protective function. Simple columnar epithelium, which functions in secretion and absorption, lines the stomach

Figure 24.2 Layers of the gastrointestinal tract. Variations in this basic plan may be seen in the esophagus (Figure 24.9), stomach (Figure 24.12), small intestine (Figure 24.23), and large intestine (Figure 24.28).

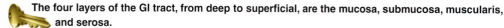

The four layers of the GI tract, from deep to superficial, are the mucosa, submucosa, muscularis, and serosa.

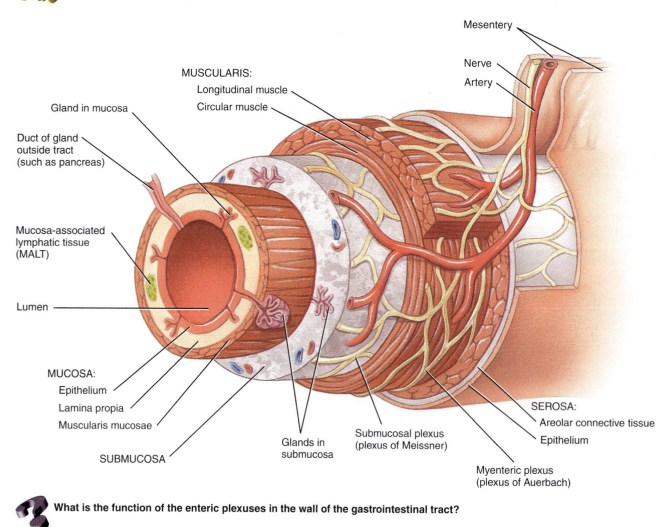

What is the function of the enteric plexuses in the wall of the gastrointestinal tract?

and intestines. Neighboring simple columnar epithelial cells are firmly sealed to each other by tight junctions that restrict leakage between the cells. The rate of renewal of GI tract epithelial cells is rapid: Every 5 to 7 days they slough off and are replaced by new cells. Located among the absorptive epithelial cells are exocrine cells that secrete mucus and fluid into the lumen of the tract, and several types of endocrine cells, collectively called **enteroendocrine cells,** that secrete hormones into the bloodstream.

2. The **lamina propria** (*lamina* = thin, flat plate; *propria* = one's own) is areolar connective tissue containing many blood and lymphatic vessels, which are the routes by which nutrients absorbed into the GI tract reach the other tissues of the body. This layer supports the epithelium and binds it to the muscularis mucosae (discussed next). The lamina propria also contains the

majority of the cells of the **mucosa-associated lymphatic tissue (MALT).** These prominent lymphatic nodules contain immune system cells that protect against disease. MALT is present all along the GI tract, especially in the tonsils, small intestine, appendix, and large intestine, and it contains about as many immune cells as are present in all the rest of the body. The lymphocytes and macrophages in MALT mount immune responses against microbes, such as bacteria, that may penetrate the epithelium.

3. A thin layer of smooth muscle fibers called the **muscularis mucosae** throws the mucous membrane of the stomach and small intestine into many small folds, which increase the surface area for digestion and absorption. Movements of the muscularis mucosae ensure that all absorptive cells are fully exposed to the contents of the GI tract.

Submucosa

The **submucosa** consists of areolar connective tissue that binds the mucosa to the muscularis. It contains many blood and lymphatic vessels that receive absorbed food molecules. Also located in the submucosa is the **submucosal plexus** or *plexus of Meissner,* an extensive network of neurons. The neurons are part of the **enteric nervous system (ENS),** the "brain of the gut." The ENS consists of about 100 million neurons in two plexuses that extend the entire length of the GI tract. The submucosal plexus contains sensory and motor enteric neurons, plus parasympathetic and sympathetic postganglionic neurons that innervate the mucosa and submucosa. Enteric nerves in the submucosa regulate movements of the mucosa and vasoconstriction of blood vessels. Because its neurons also innervate secretory cells of mucosal and submucosal glands, the ENS is important in controlling secretions by the GI tract. The submucosa may also contain glands and lymphatic tissue.

Muscularis

The **muscularis** of the mouth, pharynx, and superior and middle parts of the esophagus contains *skeletal muscle* that produces voluntary swallowing. Skeletal muscle also forms the external anal sphincter, which permits voluntary control of defecation. Throughout the rest of the tract, the muscularis consists of *smooth muscle* that is generally found in two sheets: an inner sheet of circular fibers and an outer sheet of longitudinal fibers. Involuntary contractions of the smooth muscle help break down food physically, mix it with digestive secretions, and propel it along the tract. Between the layers of the muscularis is a second plexus of the enteric nervous system—the **myenteric plexus** (*my-* = muscle) or *plexus of Auerbach.* The myenteric plexus contains enteric neurons, parasympathetic ganglia, parasympathetic postganglionic neurons, and sympathetic postganglionic neurons that innervate the muscularis. This plexus mostly controls GI tract motility, in particular the frequency and strength of contraction of the muscularis.

Serosa

The **serosa** is the superficial layer of those portions of the GI tract that are suspended in the abdominopelvic cavity. It is a serous membrane composed of areolar connective tissue and simple squamous epithelium. As we will see shortly, the esophagus, which passes through the mediastinum, has a superficial layer called the *adventitia* composed of areolar connective tissue. Inferior to the diaphragm, the serosa is also called the **visceral peritoneum;** it forms a portion of the peritoneum, which we examine in detail next.

▶ C H E C K P O I N T

4. Where along the GI tract is the muscularis composed of skeletal muscle? Is control of this skeletal muscle voluntary or involuntary?

5. What two plexuses form the enteric nervous system, and where are they located?

PERITONEUM

▶ O B J E C T I V E

• Describe the peritoneum and its folds.

The **peritoneum** (per′-i-tō-NĒ-um; *peri-* = around) is the largest serous membrane of the body; it consists of a layer of simple squamous epithelium (mesothelium) with an underlying supporting layer of connective tissue. The peritoneum is divided into the **parietal peritoneum,** which lines the wall of the abdominopelvic cavity, and the **visceral peritoneum,** which covers some of the organs in the cavity and is their serosa (Figure 24.3a). The slim space between the parietal and visceral portions of the peritoneum is called the **peritoneal cavity,** which contains serous fluid. In certain diseases, the peritoneal cavity may become distended by the accumulation of several liters of fluid, a condition called **ascites** (a-SĪ-tēz).

As we will see, some organs lie on the posterior abdominal wall and are covered by peritoneum only on their anterior surfaces. Such organs, including the kidneys and pancreas, are said to be **retroperitoneal** (*retro-* = behind).

Unlike the pericardium and pleurae, which smoothly cover the heart and lungs, the peritoneum contains large folds that weave between the viscera. The folds bind the organs to each other and to the walls of the abdominal cavity. They also contain blood vessels, lymphatic vessels, and nerves that supply the abdominal organs.

The **greater omentum** (ō-MEN-tum = fat skin), the largest peritoneal fold, drapes over the transverse colon and coils of the small intestine like a "fatty apron" (Figure 24.3a, d). Because the greater omentum is a double sheet that folds back upon itself, it is a four-layered structure. From attachments along the stomach and duodenum, the greater omentum extends downward anterior to the small intestine, then turns and extends upward and attaches to the transverse colon. The greater omentum normally contains a considerable amount of adipose tissue. Its adipose tissue content can greatly expand with weight gain, giving rise to the characteristic "beer belly" seen in some overweight individuals. The many lymph nodes of the greater omentum contribute macrophages and antibody-producing plasma cells that help combat and contain infections of the GI tract.

The **falciform ligament** (FAL-si-form; *falc-* = sickle-shaped) attaches the liver to the anterior abdominal wall and diaphragm (Figure 24.3b). The liver is the only digestive organ that is attached to the anterior abdominal wall.

The **lesser omentum** arises as two folds in the serosa of the stomach and duodenum, and it suspends the stomach and duodenum from the liver (Figure 24.3a, c). It contains some lymph nodes.

Another fold of the peritoneum, called the **mesentery** (MEZ-en-ter′-ē; *mes-* = middle), is fan-shaped and binds the small intestine to the posterior abdominal wall (Figure 24.3a, d). It extends from the posterior abdominal wall to wrap around the small intestine and then returns to its origin, forming a double-

Salivation

Secretion of saliva, or **salivation** (sal-i-VĀ-shun), is controlled by the autonomic nervous system. Amounts of saliva secreted daily vary considerably but average 1000–1500 mL (1–1.6 qt). Normally, parasympathetic stimulation promotes continuous secretion of a moderate amount of saliva, which keeps the mucous membranes moist and lubricates the movements of the tongue and lips during speech. The saliva is then swallowed and helps moisten the esophagus. Eventually, most components of saliva are reabsorbed, which prevents fluid loss. Sympathetic stimulation dominates during stress, resulting in dryness of the mouth. During dehydration, the salivary glands stop secreting saliva to conserve water; the resulting dryness in the mouth contributes to the sensation of thirst. Drinking will then not only restore the homeostasis of body water but also moisten the mouth.

The feel and taste of food also are potent stimulators of salivary gland secretions. Chemicals in the food stimulate receptors in taste buds on the tongue, and impulses propagate from the taste buds to the superior and inferior salivatory nuclei in the brain stem. Impulses in parasympathetic neurons of the facial nerve (cranial nerve VII) and glossopharyngeal nerve (cranial nerve IX) then stimulate the secretion of saliva. Saliva continues to be heavily secreted for some time after food is swallowed; this flow of saliva washes out the mouth and dilutes and buffers the remnants of irritating chemicals.

The smell, sight, sound, or thought of food may also stimulate secretion of saliva. These stimuli constitute psychological activation and involve learned behavior. When memories that associate the stimuli with food occur, nerve impulses from the cerebral cortex reach the salivatory nuclei in the brain stem, and the salivary glands are activated. Psychological activation of the glands has some benefit to the body because it allows chemical digestion to start in the mouth as soon as the food is ingested. Salivation also occurs in response to swallowing irritating foods or during nausea due to reflexes originating in the stomach and upper small intestine. This mechanism presumably helps dilute or neutralize the irritating substance.

Mumps

Although any of the salivary glands may be the target of a nasopharyngeal infection, the mumps virus (a myxovirus) typically attacks the parotid glands. **Mumps** is an inflammation and enlargement of the parotid glands accompanied by moderate fever, malaise (general discomfort), and extreme pain in the throat, especially when swallowing sour foods or acidic juices. Swelling occurs on one or both sides of the face, just anterior to the ramus of the mandible. In about 30% of males past puberty, the testes may also become inflamed; sterility rarely occurs because testicular involvement is usually unilateral (one testis only). Since a vaccine became available for mumps in 1967, the incidence of the disease has declined. ■

Tongue

The **tongue** is an accessory digestive organ composed of skeletal muscle covered with mucous membrane. Together with its associated muscles, it forms the floor of the oral cavity. The tongue is divided into symmetrical lateral halves by a median septum that extends its entire length, and it is attached inferiorly to the hyoid bone, styloid process of the temporal bone, and mandible. Each half of the tongue consists of an identical complement of extrinsic and intrinsic muscles.

The **extrinsic muscles** of the tongue, which originate outside the tongue (attach to bones in the area) and insert into connective tissues in the tongue, include the hyoglossus, genioglossus, and styloglossus muscles (see Figure 11.7 on page 327). The extrinsic muscles move the tongue from side to side and in and out to maneuver food for chewing, shape the food into a rounded mass, and force the food to the back of the mouth for swallowing. They also form the floor of the mouth and hold the tongue in position. The **intrinsic muscles** originate in and insert into connective tissue within the tongue. They alter the shape and size of the tongue for speech and swallowing. The intrinsic muscles include the longitudinalis superior, longitudinalis inferior, transversus linguae, and verticalis linguae muscles. The **lingual frenulum** (*lingua* = the tongue), a fold of mucous membrane in the midline of the undersurface of the tongue, is attached to the floor of the mouth and aids in limiting the movement of the tongue posteriorly (see Figures 24.4 and 24.5). If a person's lingual frenulum is abnormally short or rigid—a condition called **ankyloglossia** (ang′-kē-lō-GLOSS-ēa)—then eating and speaking are impaired such that the person is said to be "tongue-tied."

The dorsum (upper surface) and lateral surfaces of the tongue are covered with **papillae** (pa-PIL-ē = nipple shaped projections), projections of the lamina propria covered with keratinized epithelium (see Figure 16.2a on page 530). Many papillae contain taste buds, the receptors for gustation (taste). **Fungiform papillae** (FUN-ji-form = mushroomlike) are mushroomlike elevations distributed among the filiform papillae that are more numerous near the tip of the tongue. They appear as red dots on the surface of the tongue, and most of them contain taste buds. **Vallate (circumvallate) papillae** (VAL-āt = wall-like) are arranged in an inverted V shape on the posterior surface of the tongue; all of them contain taste buds. **Foliate papillae** (FŌ-l-ē-āt = leaflike) are located in small trenches on the lateral margins of the tongue but most of their taste buds degenerate in early childhood. **Filiform papillae** (FIL-i-form = threadlike) are pointed, threadlike projections distributed in parallel rows over the anterior two-thirds of the tongue. Although filiform papillae lack taste buds, they contain receptors for touch and increase friction between the tongue and food, making it easier for the tongue to move food in the oral cavity. **Lingual glands** in the lamina propria secrete both mucus and a watery serous fluid that contains the enzyme **lingual lipase,** which acts on triglycerides.

Teeth

The **teeth,** or **dentes** (Figure 24.6), are accessory digestive organs located in sockets of the alveolar processes of the mandible and maxillae. The alveolar processes are covered by the **gingivae** (JIN-ji-vē), or gums, which extend slightly into each socket to form the gingival sulcus. The sockets are lined by the **periodontal ligament** or **membrane** (*odont-* = tooth), which consists of dense fibrous connective tissue and is attached to the socket walls and the cemental surface of the roots. Thus, it anchors the teeth in position and acts as a shock absorber during chewing.

A typical tooth has three major regions. The **crown** is the visible portion above the level of the gums. Embedded in the socket are one to three **roots.** The **neck** is the constricted junction of the crown and root near the gum line.

Teeth are composed primarily of **dentin,** a calcified connective tissue that gives the tooth its basic shape and rigidity. It is harder than bone because of its higher content of calcium salts (70% of dry weight). The dentin encloses a cavity. The enlarged part of the cavity, the **pulp cavity,** lies within the crown and is filled with **pulp,** a connective tissue containing blood vessels, nerves, and lymphatic vessels. Narrow extensions of the pulp cavity, called **root canals,** run through the root of the tooth. Each root canal has an opening at its base, the **apical foramen,** through which blood vessels, lymphatic vessels, and nerves extend.

The dentin of the crown is covered by **enamel** that consists primarily of calcium phosphate and calcium carbonate. Enamel, the hardest substance in the body and the richest in calcium salts (about 95% of dry weight), protects the tooth from the wear and tear of chewing. It also protects against acids that can easily dissolve dentin. The dentin of the root is covered by **cementum,** another bonelike substance, which attaches the root to the periodontal ligament.

The branch of dentistry that is concerned with the prevention, diagnosis, and treatment of diseases that affect the pulp, root, periodontal ligament, and alveolar bone is known as **endodontics** (en'-dō-DON-tiks; *endo-* = within). **Orthodontics** (or'-thō-DON-tiks; *ortho-* = straight) is a branch of dentistry that is concerned with the prevention and correction of abnormally aligned teeth, whereas **periodontics** (per'-ē-ō-DON-tiks) is a branch of dentistry concerned with the treatment of abnormal conditions of the tissues immediately surrounding the teeth.

Humans have two **dentitions,** or sets of teeth: deciduous and permanent. The first of these—the **deciduous teeth** (*decidu-* = falling out), also called **primary teeth, milk teeth,** or **baby teeth**—begin to erupt at about 6 months of age, and one pair of teeth appears at about each month thereafter, until all 20 are present (Figure 24.7a). The incisors, which are closest to the midline, are chisel-shaped and adapted for cutting into food. They are referred to as either **central** or **lateral incisors** based on their position. Next to the incisors, moving posteriorly, are the **cuspids (canines),** which have a pointed surface called a cusp. Cuspids are used to tear and shred food. Incisors and cuspids have only one root apiece. Posterior to them lie the **first and second molars,** which have four cusps. Maxillary (upper) molars have three roots; mandibular (lower) molars have two roots. The molars crush and grind food.

All the deciduous teeth are lost—generally between ages 6 and 12 years—and are replaced by the **permanent (secondary) teeth** (Figure 24.7b). The permanent dentition contains 32 teeth that erupt between age 6 and adulthood. The pattern resembles the deciduous dentition, with the following exceptions. The deciduous molars are replaced by the **first** and **second premolars (bicuspids),** which have two cusps and one root (upper first bicuspids have two roots) and are used for crushing and grinding. The permanent molars, which erupt into the mouth posterior to the bicuspids, do not replace any deciduous teeth and erupt as the jaw grows to accommodate them—the **first molars** at age 6, the **second molars** at age 12, the **third molars (wisdom teeth)** after age 17.

Often the human jaw does not have enough room posterior to the second molars to accommodate the eruption of the third molars. In this case, the third molars remain embedded in the

Figure 24.6   A typical tooth and surrounding structures.

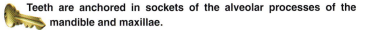 Teeth are anchored in sockets of the alveolar processes of the mandible and maxillae.

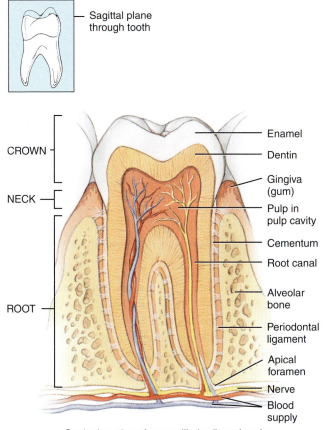

Sagittal section of a mandibular (lower) molar

What type of tissue is the main component of teeth?

Figure 24.7   Dentitions and times of eruptions (indicated in parentheses).   A designated letter (deciduous teeth) or number (permanent teeth) uniquely identifies each tooth. Deciduous teeth begin to erupt at 6 months of age, and one pair of teeth appears about each month thereafter, until all 20 are present. (See Tortora, *A Photographic Atlas of the Human Body,* Figure 12.7.)

🔑 **There are 20 teeth in a complete deciduous set and 32 teeth in a complete permanent set.**

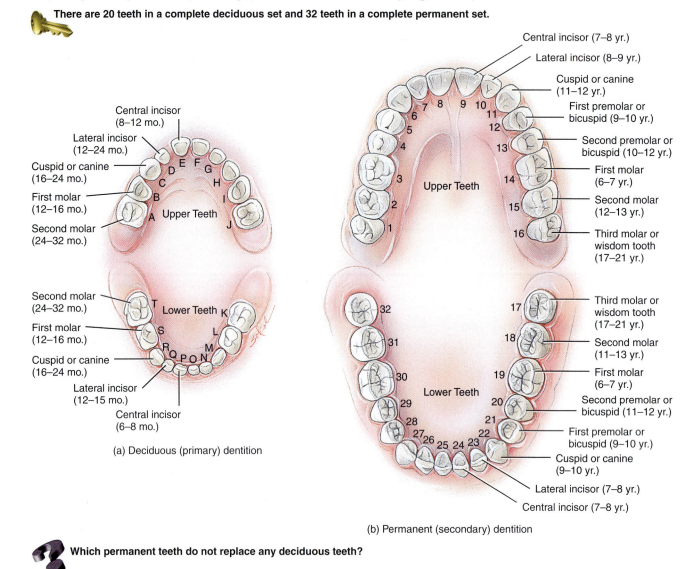

(a) Deciduous (primary) dentition

(b) Permanent (secondary) dentition

❓ **Which permanent teeth do not replace any deciduous teeth?**

alveolar bone and are said to be "impacted." They often cause pressure and pain and must be removed surgically. In some people, third molars may be dwarfed in size or may not develop at all.

🩺 **Root Canal Therapy**

Root canal therapy is a procedure, accomplished in several phases, in which all traces of pulp tissue are removed from the pulp cavity and root canals of a badly diseased tooth. After a hole is made in the tooth, the root canals are filed out and irrigated to remove bacteria. Then, the canals are treated with medication and sealed tightly. The damaged crown is then repaired. ■

Mechanical and Chemical Digestion in the Mouth

Mechanical digestion in the mouth results from chewing, or **mastication** (mas′-ti-KĀ-shun = to chew), in which food is manipulated by the tongue, ground by the teeth, and mixed with saliva. As a result, the food is reduced to a soft, flexible, easily swallowed mass called a **bolus** (= lump). Food molecules begin to dissolve in the water in saliva, an important activity because enzymes can react with food molecules in a liquid medium only.

Two enzymes, salivary amylase and lingual lipase, contribute to chemical digestion in the mouth. **Salivary amylase** initiates the breakdown of starch. Dietary carbohydrates are either monosaccharide and disaccharide sugars or complex polysaccharides such as starches. Most of the carbohydrates we eat are starches, but only monosaccharides can be absorbed into the bloodstream. Thus, ingested disaccharides and starches must be broken down into monosaccharides. The function of salivary amylase is to break certain chemical bonds between glucose units in the starches, which reduces the long-chain polysaccharides to the disaccharide maltose, the trisaccharide maltotriose, and short-chain glucose polymers called α-dextrins. Even though food is usually swallowed too quickly for all the starches to be reduced to disaccharides in the mouth, salivary amylase in the swallowed food continues to act on the starches for about another hour, at which time stomach acids inactivate it. Saliva also contains **lingual lipase,** which is secreted by glands in the tongue. This enzyme becomes activated in the acidic environment of the stomach and thus starts to work after food is swallowed. It breaks down dietary triglycerides into fatty acids and diglycerides.

Table 24.1 summarizes the digestive activities in the mouth.

▶ **C H E C K P O I N T**

8. What structures form the mouth (oral cavity)?

9. How are the major salivary glands distinguished on the basis of location?

10. How is saliva secretion regulated?

11. What functions do incisors, cuspids, premolars, and molars perform?

12. What is a bolus? How is it formed?

Table 24.1	Summary of Digestive Activities in the Mouth	
Structure	**Activity**	**Result**
Cheeks and Lips	Keep food between teeth.	Foods uniformly chewed during mastication.
Salivary glands	Secrete saliva.	Lining of mouth and pharynx moistened and lubricated. Saliva softens, moistens, and dissolves food and cleanses mouth and teeth. Salivary amylase splits starch into smaller fragments.
Tongue		
Extrinsic tongue muscles	Move tongue from side to side and in and out.	Food maneuvered for mastication, shaped into bolus, and maneuvered for swallowing.
Intrinsic tongue muscles	Alter shape of tongue.	Swallowing and speech.
Taste buds	Serve as receptors for gustation (taste) and presence of food in mouth.	Secretion of saliva stimulated by nerve impulses from taste buds to salivary nuclei in brain stem to salivary glands.
Lingual glands	Secrete lingual lipase.	Triglycerides broken down.
Teeth	Cut, tear, and pulverize food.	Solid foods reduced to smaller particles for swallowing.

PHARYNX

▶ **O B J E C T I V E**

- **Describe the location and function of the pharynx.**

When food is first swallowed, it passes from the mouth into the **pharynx** (= throat), a funnel-shaped tube that extends from the internal nares to the esophagus posteriorly and to the larynx anteriorly (see Figure 23.4 on page 810). The pharynx is composed of skeletal muscle and lined by mucous membrane. Whereas the nasopharynx functions only in respiration, both the oropharynx and laryngopharynx have digestive as well as respiratory functions. Swallowed food passes from the mouth into the oropharynx and laryngopharynx, the muscular contractions of which help propel food into the esophagus and then into the stomach.

The movement of food from the mouth into the stomach is achieved by the act of swallowing, or **deglutition** (dē-gloo-TISH-un) (Figure 24.8). Deglutition is facilitated by saliva and mucus and involves the mouth, pharynx, and esophagus. Swallowing occurs in three stages: (1) the voluntary stage, in which the bolus is passed into the oropharynx; (2) the pharyngeal stage, which is the involuntary passage of the bolus through the

pharynx into the esophagus; and (3) the esophageal stage (described in the section on the esophagus), which is the involuntary passage of the bolus through the esophagus into the stomach.

Swallowing starts when the bolus is forced to the back of the oral cavity and into the oropharynx by the movement of the tongue upward and backward against the palate; these actions constitute the **voluntary stage** of swallowing. With the passage of the bolus into the oropharynx, the involuntary **pharyngeal stage** of swallowing begins (Figure 24.8b). The respiratory passageways close, and breathing is temporarily interrupted. The bolus stimulates receptors in the oropharynx, which send impulses to the **deglutition center** in the medulla oblongata and lower pons of the brain stem. The returning impulses cause the soft palate and uvula to move upward to close off the nasopharynx, and the larynx is pulled forward and upward under the tongue. As the larynx rises, the epiglottis moves backward and downward and seals off the rima glottidis. The movement of the larynx also pulls the vocal cords together, further sealing off the respiratory tract, and widens the opening between the laryngopharynx and esophagus. The bolus passes through the laryngopharynx and enters the esophagus in 1 to 2 seconds. The respiratory passageways then reopen, and breathing resumes.

Figure 24.8 Deglutition (swallowing). During the pharyngeal stage of deglutition (b), the tongue rises against the palate, the nasopharynx is closed off, the larynx rises, the epiglottis seals off the larynx, and the bolus is passed into the esophagus.

🔑 **Deglutition is a mechanism that moves food from the mouth into the stomach.**

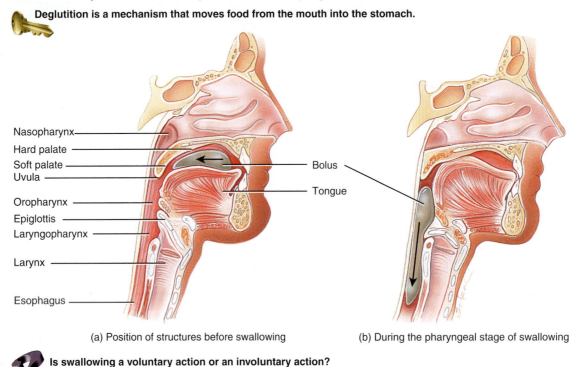

(a) Position of structures before swallowing

(b) During the pharyngeal stage of swallowing

❓ **Is swallowing a voluntary action or an involuntary action?**

▶ **CHECKPOINT**

13. What does deglutition mean?

14. How does a bolus pass from the mouth into the stomach?

15. What occurs during the voluntary and pharyngeal stages of swallowing?

ESOPHAGUS

▶ **OBJECTIVE**

- **Describe the location, anatomy, histology, and functions of the esophagus.**

The **esophagus** (e-SOF-a-gus = eating gullet) is a collapsible muscular tube that lies posterior to the trachea. It is about 25 cm (10 in.) long. The esophagus begins at the inferior end of the laryngopharynx and passes through the mediastinum anterior to the vertebral column. Then, it pierces the diaphragm through an opening called the **esophageal hiatus,** and ends in the superior portion of the stomach (see Figure 24.1). Sometimes, part of the stomach protrudes above the diaphragm through the esophageal hiatus. This condition is termed **hiatal hernia** (HER-nē-a).

Histology of the Esophagus

The **mucosa** of the esophagus consists of nonkeratinized stratified squamous epithelium, lamina propria (areolar connective tissue), and a muscularis muscosae (smooth muscle) (Figure 24.9). Near the stomach, the mucosa of the esophagus also contains mucous glands. The stratified squamous epithelium associated with the lips, mouth, tongue, oropharynx, laryngopharynx, and esophagus affords considerable protection against abrasion and wear-and-tear from food particles that are chewed, mixed with secretions, and swallowed. The **submucosa** contains areolar connective tissue, blood vessels, and mucous glands. The **muscularis** of the superior third of the esophagus is skeletal muscle, the intermediate third is skeletal and smooth muscle, and the inferior third is smooth muscle. The superficial layer is known as the **adventitia** (ad-ven-TISH-a), rather than the serosa, because the areolar connective tissue of this layer is not covered by mesothelium and because the connective tissue merges with the connective tissue of surrounding structures of the mediastinum, through which is passes. The adventitia attaches the esophagus to surrounding structures.

Physiology of the Esophagus

The esophagus secretes mucus and transports food into the stomach. It does not produce digestive enzymes, and it does not carry on absorption. The passage of food from the laryngopharynx into the esophagus is regulated at the entrance to the esophagus by a sphincter (a circular band or ring of muscle that is normally contracted) called the **upper esophageal sphincter** (e-sof′-a-JĒ-al). It consists of the cricopharyngeus muscle

Figure 24.9 Histology of the esophagus. A higher magnification view of nonkeratinized, stratified squamous epithelium is shown in Table 4.1E on page 111. (See Tortora, *A Photographic Atlas of the Human Body,* Figure 12.8.)

🔑 The esophagus secretes mucus and transports food to the stomach.

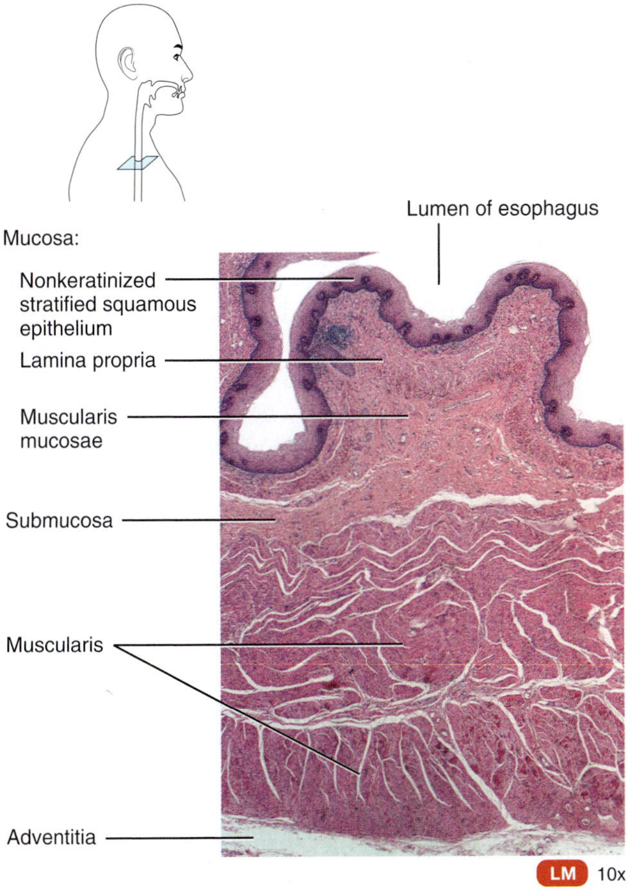

Mucosa:

Nonkeratinized stratified squamous epithelium

Lamina propria

Muscularis mucosae

Submucosa

Muscularis

Adventitia

Lumen of esophagus

LM 10x

Wall of the esophagus

❓ In which layers of the esophagus are the glands that secrete lubricating mucus located?

Figure 24.10 Peristalsis during the esophageal stage of deglutition (swallowing).

🔑 Peristalsis consists of progressive, wavelike contractions of the muscularis.

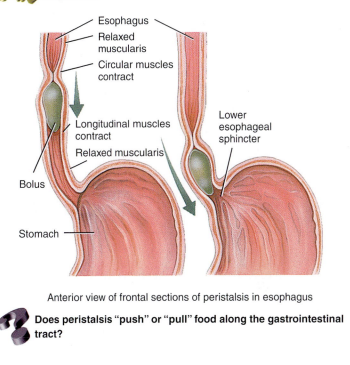

Esophagus

Relaxed muscularis

Circular muscles contract

Longitudinal muscles contract

Relaxed muscularis

Bolus

Stomach

Lower esophageal sphincter

Anterior view of frontal sections of peristalsis in esophagus

❓ Does peristalsis "push" or "pull" food along the gastrointestinal tract?

attached to the cricoid cartilage. The elevation of the larynx during the pharyngeal stage of swallowing causes the sphincter to relax, and the bolus enters the esophagus. This sphincter also relaxes during exhalation.

During the **esophageal stage** of swallowing, **peristalsis** (per′-is-STAL-sis; *stalsis* = constriction), a progression of coordinated contractions and relaxations of the circular and longitudinal layers of the muscularis, pushes the food bolus onward (Figure 24.10). (Peristalsis occurs in other tubular structures, including other parts of the GI tract and the ureters, bile ducts, and uterine tubes; in the esophagus it is controlled by the medulla oblongata.) In the section of the esophagus just superior to the bolus, the circular muscle fibers contract, constricting the esophageal wall and squeezing the bolus toward the stomach. Meanwhile, longitudinal fibers inferior to the bolus also con-

tract, which shortens this inferior section and pushes its walls outward so it can receive the bolus. The contractions are repeated in a wave that pushes the food toward the stomach. Mucus secreted by esophageal glands lubricates the bolus and reduces friction. The passage of solid or semisolid food from the mouth to the stomach takes 4 to 8 seconds; very soft foods and liquids pass through in about 1 second.

Just superior to the diaphragm, the esophagus narrows slightly due to a maintained contraction of the muscularis at the lowest part of the esophagus. This physiological sphincter, known as the **lower esophageal sphincter,** relaxes during swallowing and thus allows the bolus to pass from the esophagus into the stomach.

Table 24.2 summarizes the digestive activities of the pharynx and esophagus.

🩺 **Gastroesophageal Reflux Disease**

If the lower esophageal sphincter fails to close adequately after food has entered the stomach, the stomach contents can reflux, or back up, into the inferior portion of the esophagus. This condition is known as **gastroesophageal reflux disease (GERD).** Hydrochloric acid (HCl) from the stomach contents can irritate the esophageal wall, resulting in a burning sensation called **heartburn** because it is experienced in a region very near the heart, even though it is unrelated to any cardiac problem. Drinking alcohol and smoking can cause the sphincter to relax,

Table 24.2	Summary of Digestive Activities in the Pharynx and Esophagus	
Structure	**Activity**	**Result**
Pharynx	Pharyngeal stage of deglutition.	Moves bolus from oropharynx to laryngopharynx and into esophagus; closes air passageways.
Esophagus	Relaxation of upper esophageal sphincter.	Permits entry of bolus from laryngopharynx into esophagus.
	Esophageal stage of deglutition (peristalsis).	Pushes bolus down esophagus.
	Relaxation of lower esophageal sphincter.	Permits entry of bolus into stomach.
	Secretion of mucus.	Lubricates esophagus for smooth passage of bolus.

worsening the problem. The symptoms of GERD often can be controlled by avoiding foods that strongly stimulate stomach acid secretion (coffee, chocolate, tomatoes, fatty foods, orange juice, peppermint, spearmint, and onions). Other acid-reducing strategies include taking over-the-counter histamine-2 (H_2) blockers such as Tagamet HB or Pepcid AC 30 to 60 minutes before eating, and neutralizing already secreted acid with antacids such as Tums or Maalox. Symptoms are less likely to occur if food is eaten in smaller amounts and if the person does not lie down immediately after a meal. GERD may be associated with cancer of the esophagus. ■

▶ CHECKPOINT

16. Describe the location and histology of the esophagus. What is its role in digestion?

17. Explain the operation of the upper and lower esophageal sphincters.

STOMACH

▶ OBJECTIVE

• Describe the location, anatomy, histology, and functions of the stomach.

The **stomach** is typically a J-shaped enlargement of the GI tract directly inferior to the diaphragm in the epigastric, umbilical, and left hypochondriac regions of the abdomen (see Figure 1.13a on page 20). The stomach connects the esophagus to the duodenum, the first part of the small intestine (Figure 24.11). Because a meal can be eaten much more quickly than the intestines can digest and absorb it, one of the functions of the stomach is to serve as a mixing chamber and holding reservoir. At appropriate intervals after food is ingested, the stomach forces a small quantity of material into the first portion of the small intestine. The position and size of the stomach vary continually; the diaphragm pushes it inferiorly with each inspiration and pulls it superiorly with each expiration. Empty, it is about the size of a large sausage, but it is the most distensible part of the GI tract and can accommodate a large quantity of food. In the stomach, digestion of starch continues, digestion of proteins and triglycerides begins, the semisolid bolus is converted to a liquid, and certain substances are absorbed.

Anatomy of the Stomach

The stomach has four main regions: the cardia, fundus, body, and pylorus (Figure 24.11). The **cardia** (CAR-dē-a) surrounds the superior opening of the stomach. The rounded portion superior to and to the left of the cardia is the **fundus** (FUN-dus). Inferior to the fundus is the large central portion of the stomach, called the **body.** The region of the stomach that connects to the duodenum is the **pylorus** (pī-LOR-us; *pyl-* = gate; *-orus* = guard); it has two parts, the **pyloric antrum** (AN-trum = cave), which connects to the body of the stomach, and the **pyloric canal,** which leads into the duodenum. When the stomach is empty, the mucosa lies in large folds, called **rugae** (ROO-gē = wrinkles), that can be seen with the unaided eye. The pylorus communicates with the duodenum of the small intestine via a sphincter called the **pyloric sphincter.** The concave medial border of the stomach is called the **lesser curvature,** and the convex lateral border is called the **greater curvature.**

Pylorospasm and Pyloric Stenosis

Two abnormalities of the pyloric sphincter can occur in infants. In **pylorospasm** the muscle fibers of the sphincter fail to relax normally, so food does not pass easily from the stomach to the small intestine, the stomach becomes overly full, and the infant vomits often to relieve the pressure. Pylorospasm is treated by drugs that relax the muscle fibers of the sphincter. **Pyloric stenosis** is a narrowing of the pyloric sphincter, and it must be corrected surgically. The hallmark symptom is projectile vomiting—the spraying of liquid vomitus some distance from the infant. ■

Histology of the Stomach

The stomach wall is composed of the same four basic layers as the rest of the GI tract, with certain modifications. The surface of the **mucosa** is a layer of simple columnar epithelial cells called **surface mucous cells** (Figure 24.12a on page 868). The mucosa contains a **lamina propria** (areolar connective tissue) and a **muscularis mucosae** (smooth muscle). Epithelial cells extend down into the lamina propria, where they form columns of secretory cells called **gastric glands** that line many narrow channels called **gastric pits.** Secretions from several gastric glands flow into each gastric pit and then into the lumen of the stomach.

Figure 24.11 **External and internal anatomy of the stomach.** (See Tortora, *A Photographic Atlas of the Human Body,* Figure 12.9.)

🔑 The four regions of the stomach are the cardia, fundus, body, and pylorus.

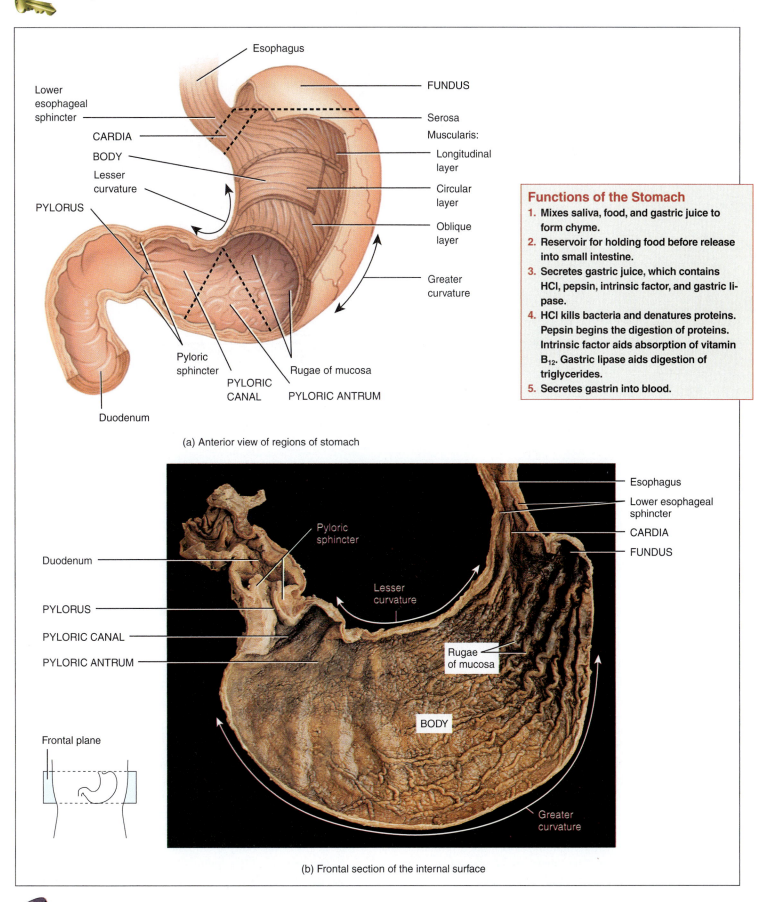

Functions of the Stomach

1. Mixes saliva, food, and gastric juice to form chyme.
2. Reservoir for holding food before release into small intestine.
3. Secretes gastric juice, which contains HCl, pepsin, intrinsic factor, and gastric lipase.
4. HCl kills bacteria and denatures proteins. Pepsin begins the digestion of proteins. Intrinsic factor aids absorption of vitamin B_{12}. Gastric lipase aids digestion of triglycerides.
5. Secretes gastrin into blood.

(a) Anterior view of regions of stomach

(b) Frontal section of the internal surface

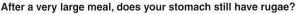 **After a very large meal, does your stomach still have rugae?**

Figure 24.12   Histology of the stomach.

Gastric juice is the combined secretions of mucous cells, parietal cells, and chief cells.

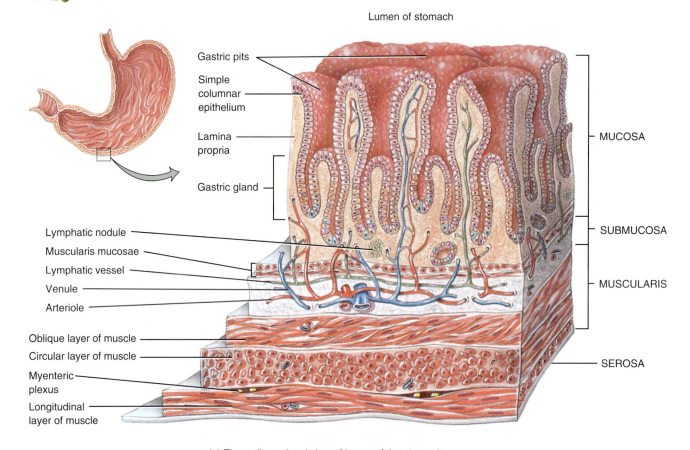

(a) Three dimensional view of layers of the stomach

The gastric glands contain three types of *exocrine gland cells* that secrete their products into the stomach lumen: mucous neck cells, chief cells, and parietal cells. Both surface mucous cells and **mucous neck cells** secrete mucus (Figure 24.12b). **Parietal cells** produce intrinsic factor, which is needed for absorption of vitamin B$_{12}$, and hydrochloric acid. The **chief cells** secrete pepsinogen and gastric lipase. The secretions of the mucous, parietal, and chief cells form **gastric juice,** which totals 2000–3000 mL (roughly 2–3 qt.) per day. In addition, gastric glands include a type of enteroendocrine cell, the **G cell,** which is located mainly in the pyloric antrum and secretes the hormone gastrin into the bloodstream. As we will see shortly, this hormone stimulates several aspects of gastric activity.

Three additional layers lie deep to the mucosa. The **submucosa** of the stomach is composed of areolar connective tissue. The **muscularis** has three (rather than two) layers of smooth muscle: an outer longitudinal layer, a middle circular layer, and an inner oblique layer. The oblique layer is limited primarily to the body of the stomach. The **serosa** is composed of simple squamous epithelium (mesothelium) and areolar connective tis-

sue, and the portion of the serosa covering the stomach is part of the visceral peritoneum. At the lesser curvature of the stomach, the visceral peritoneum extends upward to the liver as the lesser omentum. At the greater curvature of the stomach, the visceral peritoneum continues downward as the greater omentum and drapes over the intestines.

Mechanical and Chemical Digestion in the Stomach

Several minutes after food enters the stomach, gentle, rippling, peristaltic movements called **mixing waves** pass over the stomach every 15 to 25 seconds. These waves macerate food, mix it with secretions of the gastric glands, and reduce it to a soupy liquid called **chyme** (KĪM = juice). Few mixing waves are observed in the fundus, which primarily has a storage function. As digestion proceeds in the stomach, more vigorous mixing waves begin at the body of the stomach and intensify as they reach the pylorus. The pyloric sphincter normally remains almost, but not completely, closed. As food reaches the pylorus,

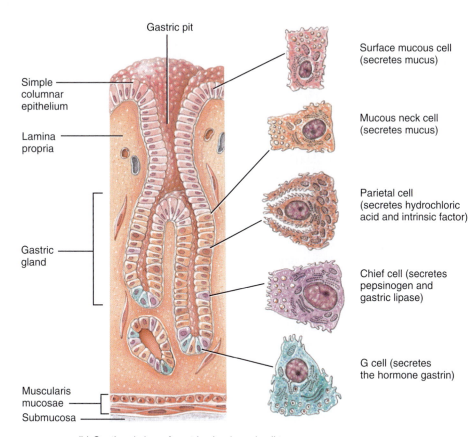

Gastric pit

Simple columnar epithelium

Lamina propria

Gastric gland

Muscularis mucosae

Submucosa

Surface mucous cell (secretes mucus)

Mucous neck cell (secretes mucus)

Parietal cell (secretes hydrochloric acid and intrinsic factor)

Chief cell (secretes pepsinogen and gastric lipase)

G cell (secretes the hormone gastrin)

(b) Sectional view of gastric glands and cell types

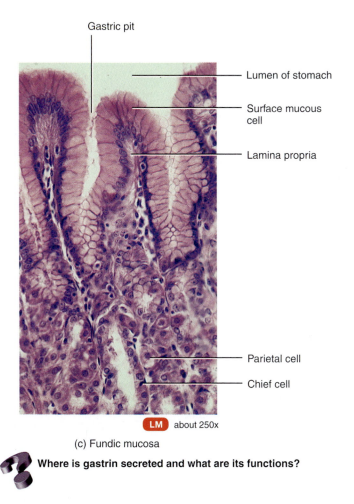

Gastric pit

Lumen of stomach

Surface mucous cell

Lamina propria

Parietal cell

Chief cell

LM about 250x

(c) Fundic mucosa

Where is gastrin secreted and what are its functions?

each mixing wave forces several milliliters of chyme into the duodenum through the pyloric sphincter. Most of the chyme is forced back into the body of the stomach, where mixing continues. The next wave pushes the chyme forward again and forces a little more into the duodenum. These forward and backward movements of the gastric contents are responsible for most mixing in the stomach.

Foods may remain in the fundus for about an hour without becoming mixed with gastric juice. During this time, digestion by salivary amylase continues. Soon, however, the churning action mixes chyme with acidic gastric juice, inactivating salivary amylase and activating lingual lipase, which starts to digest triglycerides into fatty acids and diglycerides.

Even though parietal cells secrete hydrogen ions (H^+) and chloride ions (Cl^-) separately into the stomach lumen, the net effect is secretion of hydrochloric acid (HCl). **Proton pumps** powered by H^+/K^+ ATPases actively transport H^+ into the lumen while bringing potassium ions (K^+) into the cell (Figure 24.13). At the same time Cl^- and K^+ diffuse out through Cl^- and K^+ channels in the apical membrane (next to the lumen). The enzyme *carbonic anhydrase*, which is especially plentiful in parietal cells, catalyzes the formation of carbonic acid (H_2CO_3) from water (H_2O) and carbon dioxide (CO_2). As carbonic acid dissociates, it provides a ready source of H^+ for the proton pumps but also generates bicarbonate ions (HCO_3^-). As HCO_3^- builds up in the cytosol, it exits the parietal cell in exchange for Cl^- via Cl^-/HCO_3^- antiporters in the basolateral membrane (next to the lamina propria). HCO_3^- diffuses into nearby blood

Figure 24.13 Secretion of HCl (hydrochloric acid) by parietal cells in the stomach.

Proton pumps, powered by ATP, secrete the H^+; Cl^- diffuses into the stomach lumen through Cl^- channels.

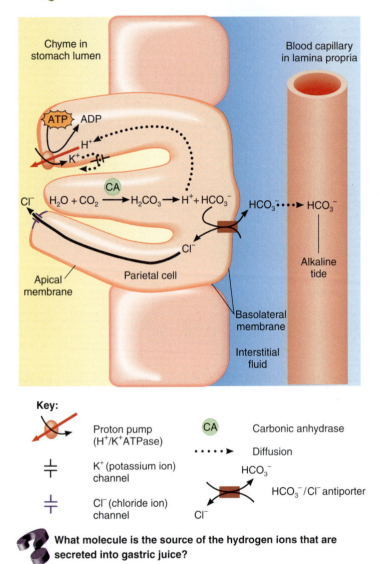

Key:

Proton pump (H^+/K^+ATPase)

K^+ (potassium ion) channel

Cl^- (chloride ion) channel

CA Carbonic anhydrase

Diffusion

HCO_3^-/Cl^- antiporter

What molecule is the source of the hydrogen ions that are secreted into gastric juice?

capillaries. This "alkaline tide" of bicarbonate ions entering the bloodstream after a meal may be large enough to slightly elevate blood pH and make urine more alkaline.

The strongly acidic fluid of the stomach kills many microbes in food, and HCl partially denatures (unfolds) proteins in food and stimulates the secretion of hormones that promote the flow of bile and pancreatic juice. Enzymatic digestion of proteins also begins in the stomach. The only proteolytic (protein-digesting) enzyme in the stomach is **pepsin,** which is secreted by chief cells. Because pepsin breaks certain peptide bonds between the amino acids making up proteins, a protein chain of many amino acids is broken down into smaller peptide fragments. Pepsin is most effective in the very acidic environment of the stomach (pH 2); it becomes inactive at higher pH.

What keeps pepsin from digesting the protein in stomach cells along with the food? First, pepsin is secreted in an inactive

form called *pepsinogen;* in this form, it cannot digest the proteins in the chief cells that produce it. Pepsinogen is not converted into active pepsin until it comes in contact with active pepsin molecules or hydrochloric acid secreted by parietal cells. Second, the stomach epithelial cells are protected from gastric juices by a 1–3 mm thick layer of alkaline mucus secreted by surface mucous cells and mucous neck cells.

Another enzyme of the stomach is **gastric lipase,** which splits the short-chain triglycerides in fat molecules found in milk into fatty acids and monoglycerides. This enzyme, which has a limited role in the adult stomach, operates best at a pH of 5–6. More important than either lingual lipase or gastric lipase is pancreatic lipase, an enzyme secreted by the pancreas into the small intestine.

Only a small amount of absorption occurs in the stomach because its epithelial cells are impermeable to most materials. However, mucous cells of the stomach absorb some water, ions, and short-chain fatty acids, as well as certain drugs (especially aspirin) and alcohol.

Table 24.3 summarizes the digestive activities of the stomach.

Regulation of Gastric Secretion and Motility

Both neural and hormonal mechanisms control the secretion of gastric juice and the contraction of smooth muscle in the stomach wall. Events in gastric digestion occur in three overlapping phases: the cephalic, gastric, and intestinal phases (Figure 24.14).

Cephalic Phase

The **cephalic phase** of gastric digestion consists of reflexes initiated by sensory receptors in the head. Even before food enters the stomach, the sight, smell, taste, or thought of food initiates this reflex. The cerebral cortex and the feeding center in the hypothalamus send nerve impulses to the medulla oblongata. The medulla then transmits impulses to parasympathetic preganglionic neurons in the vagus nerves (cranial nerve X), which stimulate parasympathetic postganglionic neurons in the submucosal plexus. In turn, impulses from parasympathetic postganglionic neurons stimulate the gastric glands to secrete pepsinogen, hydrochloric acid, and mucus into stomach chyme, and gastrin into the blood. Impulses from parasympathetic neurons also increase stomach motility. Emotions such as anger, fear, and anxiety may slow digestion in the stomach because they stimulate the sympathetic nervous system, which inhibits gastric activity.

Gastric Phase

Once food reaches the stomach, sensory receptors in the stomach initiate both neural and hormonal mechanisms to ensure that gastric secretion and motility continue. This is the **gastric phase** of gastric digestion (Figure 24.14). Food of any kind distends (stretches) the stomach and stimulates stretch receptors in its

Table 24.3	Summary of Digestive Activities in the Stomach	
Structure	**Activity**	**Result**
Mucosa		
Chief cells	Secrete pepsinogen.	Pepsin, the activated form, breaks certain peptide bonds in proteins.
	Secrete gastric lipase.	Splits short-chain triglycerides into fatty acids and monoglycerides.
Parietal cells	Secrete hydrochloric acid.	Kills microbes in food; denatures proteins; converts pepsinogen into pepsin.
	Secrete intrinsic factor.	Needed for absorption of vitamin B_{12}, which is used in red blood cell formation (erythropoiesis).
Surface mucous cells and mucous neck cells	Secrete mucus.	Forms a protective barrier that prevents digestion of stomach wall.
	Absorption.	Small quantity of water, ions, short-chain fatty acids, and some drugs enter the bloodstream.
G cells	Secrete gastrin.	Stimulates parietal cells to secrete HCl and chief cells to secrete pepsinogen; contracts lower esophageal sphincter, increases motility of the stomach, and relaxes pyloric sphincter.
Muscularis	Mixing waves.	Macerate food and mix it with gastric juice, forming chyme.
	Peristalsis.	Forces chyme through pyloric sphincter.
Pyloric sphincter	Opens to permit passage of chyme into duodenum.	Regulates passage of chyme from stomach to duodenum; prevents backflow of chyme from duodenum to stomach.

Figure 24.14 The cephalic, gastric, and intestinal phases of gastric digestion. Reflexes initiated by sensory receptors in the head, stomach, and small intestine constitute the cephalic, gastric, and intestinal phases of gastric digestion, respectively.

During the cephalic and gastric phases, digestion in the stomach is stimulated whereas during the intestinal phase, gastric juice secretion and gastric peristalsis are inhibited.

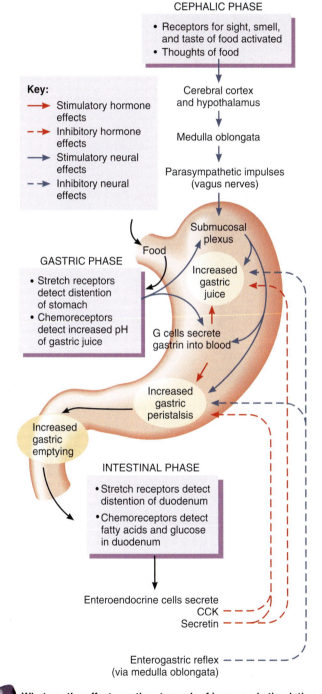

CEPHALIC PHASE
• Receptors for sight, smell, and taste of food activated
• Thoughts of food

Cerebral cortex and hypothalamus

Medulla oblongata

Parasympathetic impulses (vagus nerves)

Key:
→ Stimulatory hormone effects
⇢ Inhibitory hormone effects
→ Stimulatory neural effects
⇢ Inhibitory neural effects

Submucosal plexus

Food

Increased gastric juice

GASTRIC PHASE
• Stretch receptors detect distention of stomach
• Chemoreceptors detect increased pH of gastric juice

G cells secrete gastrin into blood

Increased gastric peristalsis

Increased gastric emptying

INTESTINAL PHASE
• Stretch receptors detect distention of duodenum
• Chemoreceptors detect fatty acids and glucose in duodenum

Enteroendocrine cells secrete
CCK
Secretin

Enterogastric reflex (via medulla oblongata)

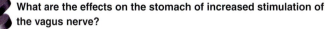
What are the effects on the stomach of increased stimulation of the vagus nerve?

walls. Chemoreceptors in the stomach monitor the pH of the stomach chyme. When the stomach walls are distended or pH increases because proteins have entered the stomach and buffered some of the stomach acid, the stretch receptors and chemoreceptors are activated, and a neural negative feedback loop is set in motion (Figure 24.15). From the stretch receptors and chemoreceptors, nerve impulses propagate to the submucosal plexus, where they activate parasympathetic and enteric neurons. The resulting nerve impulses cause waves of peristalsis and continue to stimulate the flow of gastric juice from parietal cells, chief cells, and mucous cells.

The peristaltic waves mix the food with gastric juice, and when the waves become strong enough, a small quantity of chyme—about 10–15 mL (2–3 teaspoons)—squirts past the pyloric sphincter into the duodenum. As the pH of the stomach chyme becomes more acidic again and the stomach walls are less

Figure 24.15 Neural negative feedback regulation of the pH of gastric juice and gastric motility during the gastric phase of gastric digestion.

🔑 Food entering the stomach stimulates secretion of gastric juice and causes vigorous waves of peristalsis.

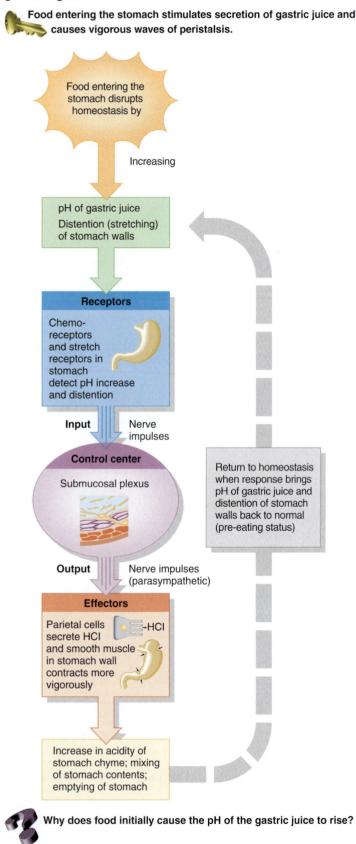

Food entering the stomach disrupts homeostasis by

Increasing

pH of gastric juice
Distention (stretching) of stomach walls

Receptors

Chemo-receptors and stretch receptors in stomach detect pH increase and distention

Input   Nerve impulses

Control center

Submucosal plexus

Return to homeostasis when response brings pH of gastric juice and distention of stomach walls back to normal (pre-eating status)

Output   Nerve impulses (parasympathetic)

Effectors

Parietal cells secrete HCl and smooth muscle in stomach wall contracts more vigorously

Increase in acidity of stomach chyme; mixing of stomach contents; emptying of stomach

❓ Why does food initially cause the pH of the gastric juice to rise?

distended because chyme has passed into the small intestine, this negative feedback cycle suppresses secretion of gastric juice.

Hormonal negative feedback also regulates gastric secretion during the gastric phase (see Figure 24.14). Partially digested proteins buffer H^+, thus increasing pH, and ingested food distends the stomach. Chemoreceptors and stretch receptors monitoring these changes stimulate parasympathetic neurons to release acetylcholine. In turn, acetylcholine stimulates secretion of the hormone **gastrin** by G cells, enteroendocrine cells in the mucosa of the pyloric antrum. (A small amount of gastrin is also secreted by enteroendocrine cells in the small intestine; moreover, some chemicals in food—for example, caffeine—directly stimulate gastrin release.) Gastrin enters the bloodstream and finally reaches its target cells, the gastric glands.

Gastrin stimulates growth of the gastric glands and secretion of large amounts of gastric juice. It also strengthens contraction of the lower esophageal sphincter, increases motility of the stomach, and relaxes the pyloric and ileocecal sphincters (described later). Gastrin secretion is inhibited when the pH of gastric juice drops below 2.0 and is stimulated when the pH rises. This negative feedback mechanism helps provide an optimal low pH for the functioning of pepsin, the killing of microbes, and the denaturing of proteins in the stomach.

Acetylcholine (ACh) released by parasympathetic neurons and gastrin secreted by G cells stimulate parietal cells to secrete more HCl in the presence of histamine. In other words, histamine, a paracrine substance that is released by mast cells in the lamina propria and acts on nearby parietal cells, acts synergistically with acetylcholine and gastrin to enhance their effects. Receptors for all three substances are present in the plasma membrane of parietal cells. The histamine receptors on parietal cells are called H_2 receptors; they mediate different responses than do the H_1 receptors involved in allergic responses.

Intestinal Phase

The **intestinal phase** of gastric digestion is due to activation of receptors in the small intestine. Whereas reflexes initiated during the cephalic and gastric phases stimulate stomach secretory activity and motility, those occurring during the intestinal phase have inhibitory effects (see Figure 24.14) that slow the exit of chyme from the stomach and prevent overloading of the duodenum with more chyme than it can handle. In addition, responses occurring during the intestinal phase promote the continued digestion of foods that have reached the small intestine. When chyme containing fatty acids and glucose leaves the stomach and enters the small intestine, it triggers enteroendocrine cells in the small intestinal mucosa to release into the blood two hormones that affect the stomach—**secretin** (se-KRĒ-tin) and **cholecystokinin** (kō′-lē-sis′-tō-KĪN-in) or **CCK.** With respect to the stomach, secretin mainly decreases gastric secretions, whereas CCK mainly inhibits stomach emptying. Both hormones have other important effects on the pancreas, liver, and gallbladder (explained shortly) that contribute to the regulation of digestive processes.

Regulation of Gastric Emptying

Gastric emptying, the periodic release of chyme from the stomach into the duodenum, is regulated by both neural and hormonal reflexes, as follows (Figure 24.16a):

1 Stimuli such as distention of the stomach and the presence of partially digested proteins, alcohol, and caffeine initiate gastric emptying.

2 These stimuli increase the secretion of gastrin and generate parasympathetic impulses in the vagus nerves.

3 Gastrin and nerve impulses stimulate contraction of the lower esophageal sphincter, increase motility of the stomach, and relax the pyloric sphincter.

4 The net effect of these actions is gastric emptying.

Neural and hormonal reflexes also help ensure that the stomach does not release more chyme than the small intestine can process. The neural reflex called the **enterogastric reflex** and the hormone cholecystokinin inhibit gastric emptying, as follows (Figure 24.16b):

1 Stimuli such as distention of the duodenum and the presence of fatty acids, glucose, and partially digested proteins in the duodenal chyme inhibit gastric emptying.

2 These stimuli then initiate the enterogastric reflex. Nerve impulses propagate from the duodenum to the medulla oblongata, where they inhibit parasympathetic stimulation and stimulate sympathetic activity in the stomach. The same stimuli also increase secretion of cholecystokinin.

3 Increased sympathetic impulses and cholecystokinin both decrease gastric motility.

4 The net effect of these actions is inhibition of gastric emptying.

Within 2 to 4 hours after eating a meal, the stomach has emptied its contents into the duodenum. Foods rich in carbohydrate spend the least time in the stomach; high-protein foods remain somewhat longer, and emptying is slowest after a fat-laden meal containing large amounts of triglycerides. The reason for slower emptying after eating triglycerides is that fatty acids in chyme stimulate release of cholecystokinin, which slows stomach emptying.

Vomiting

Vomiting or *emesis* is the forcible expulsion of the contents of the upper GI tract (stomach and sometimes duodenum) through the mouth. The strongest stimuli for vomiting are irritation and distension of the stomach; other stimuli include unpleasant sights, general anesthesia, dizziness, and certain drugs such as morphine and derivatives of digitalis. Nerve impulses are transmitted to the vomiting center in the medulla oblongata, and returning impulses propagate to the upper GI tract organs, diaphragm, and abdominal muscles. Vomiting involves squeezing the stomach between the diaphragm and abdominal muscles

Figure 24.16 Neural and hormonal regulation of gastric emptying.

Foods rich in carbohydrates are the first to leave the stomach.

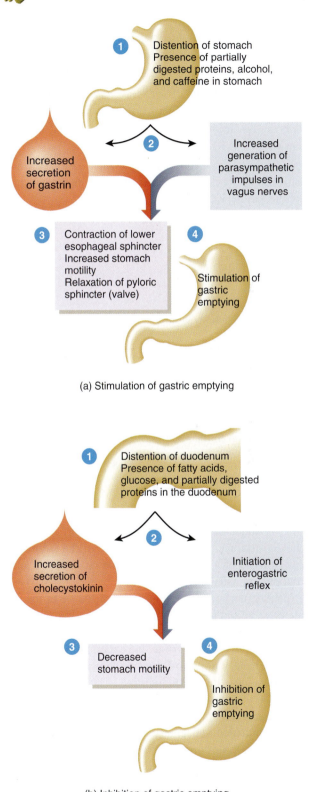

(a) Stimulation of gastric emptying

(b) Inhibition of gastric emptying

 What effect would bilateral vagotomy (cutting of axons in both vagus nerves) have on gastric emptying?

and expelling the contents through open esophageal sphincters. Prolonged vomiting, especially in infants and elderly people, can be serious because the loss of acidic gastric juice can lead to alkalosis (higher than normal blood pH). ■

▶ **CHECKPOINT**

18. Compare the epithelium of the esophagus with that of the stomach. How is each adapted to the function of the organ?

19. What is the importance of rugae, surface mucous cells, mucous neck cells, chief cells, parietal cells, and G cells in the stomach?

20. What is the role of pepsin? Why is it secreted in an inactive form?

21. What are the functions of gastric lipase and lingual lipase in the stomach?

22. Describe the factors that stimulate and inhibit gastric secretion and motility. Why are the three phases of gastric digestion named the cephalic, gastric, and intestinal phases?

PANCREAS

▶ **OBJECTIVE**

- **Describe the location, anatomy, histology, and function of the pancreas.**

From the stomach, chyme passes into the small intestine. Because chemical digestion in the small intestine depends on activities of the pancreas, liver, and gallbladder, we first consider the activities of these accessory digestive organs and their contributions to digestion in the small intestine.

Anatomy of the Pancreas

The **pancreas** (*pan-* = all; *-creas* = flesh), a retroperitoneal gland that is about 12–15 cm (5–6 in.) long and 2.5 cm (1 in.) thick, lies posterior to the greater curvature of the stomach. The pancreas consists of a head, a body, and a tail and is connected to the duodenum usually by two ducts (Figure 24.17). The **head** is the expanded portion of the organ near the curve of the duodenum; superior to and to the left of the head are the central **body** and the tapering **tail.**

Pancreatic juices pass from the exocrine cells into small ducts that ultimately unite to form two larger ducts that convey the secretions into the small intestine. The larger of the two ducts is called the **pancreatic duct (duct of Wirsung).** In most people, the pancreatic duct joins the common bile duct from the liver and gallbladder and enters the duodenum as a common duct called the **hepatopancreatic ampulla (ampulla of Vater).** The ampulla opens on an elevation of the duodenal mucosa known as the **major duodenal papilla,** which lies about 10 cm (4 in.) inferior to the pyloric sphincter of the stomach. The smaller of the two ducts, the **accessory duct (duct of Santorini),** leads from the pancreas and empties into the duodenum about 2.5 cm (1 in.) superior to the hepatopancreatic ampulla.

Histology of the Pancreas

The pancreas is made up of small clusters of glandular epithelial cells, about 99% of which are arranged in clusters called **acini** (AS-i-nē) and constitute the *exocrine* portion of the organ (see Figure 18.18b, c on page 615). The cells within acini secrete a mixture of fluid and digestive enzymes called **pancreatic juice.** The remaining 1% of the cells are organized into clusters called **pancreatic islets (islets of Langerhans),** the *endocrine* portion of the pancreas. These cells secrete the hormones glucagon, insulin, somatostatin, and pancreatic polypeptide. The functions of these hormones are discussed in Chapter 18.

Composition and Functions of Pancreatic Juice

Each day the pancreas produces 1200–1500 mL (about 1.2–1.5 qt) of pancreatic juice, which is a clear, colorless liquid consisting mostly of water, some salts, sodium bicarbonate, and several enzymes. The sodium bicarbonate gives pancreatic juice a slightly alkaline pH (7.1–8.2) that buffers acidic gastric juice in chyme, stops the action of pepsin from the stomach, and creates the proper pH for the action of digestive enzymes in the small intestine. The enzymes in pancreatic juice include a carbohydrate-digesting enzyme called **pancreatic amylase;** several protein-digesting enzymes called **trypsin** (TRIP-sin), **chymotrypsin** (kī′-mō-TRIP-sin), **carboxypeptidase** (kar-bok′-sē-PEP-ti-dās), and **elastase** (ē-LAS-tās); the principal triglyceride-digesting enzyme in adults, called **pancreatic lipase;** and nucleic acid–digesting enzymes called **ribonuclease** and **deoxyribonuclease.**

Just as pepsin is produced in the stomach in an inactive form (pepsinogen), so too are the protein-digesting enzymes of the pancreas. Because they are inactive, the enzymes do not digest cells of the pancreas itself. Trypsin is secreted in an inactive form called **trypsinogen** (trip-SIN-ō-jen). Pancreatic acinar cells also secrete a protein called **trypsin inhibitor,** that combines with any trypsin formed accidentally in the pancreas or in pancreatic juice and blocks its enzymatic activity. When trypsinogen reaches the lumen of the small intestine, it encounters an activating brush-border enzyme called **enterokinase** (en′-ter-ō-KĪ-nās), which splits off part of the trypsinogen molecule to form trypsin. In turn, trypsin acts on the inactive precursors (called **chymotrypsinogen, procarboxypeptidase,** and **proelastase**) to produce chymotrypsin, carboxypeptidase, and elastase, respectively.

 **Pancreatitis**

Inflammation of the pancreas, as may occur in association with alcohol abuse or chronic gallstones, is called **pancreatitis** (pan′-kre-a-TĪ-tis). In a more severe condition known as **acute pancreatitis,** which is associated with heavy alcohol intake or biliary tract obstruction, the pancreatic cells may release either trypsin instead of trypsinogen or insufficient amounts of trypsin

Figure 24.17 **Relation of the pancreas to the liver, gallbladder, and duodenum.** The inset shows details of the common bile duct and pancreatic duct forming the hepatopancreatic ampulla (ampulla of Vater) and emptying into the duodenum. (See Tortora, *A Photographic Atlas of the Human Body,* Figures 12.10, 12.11.)

🔑 **Pancreatic enzymes digest starches (polysaccharides), proteins, triglycerides, and nucleic acids.**

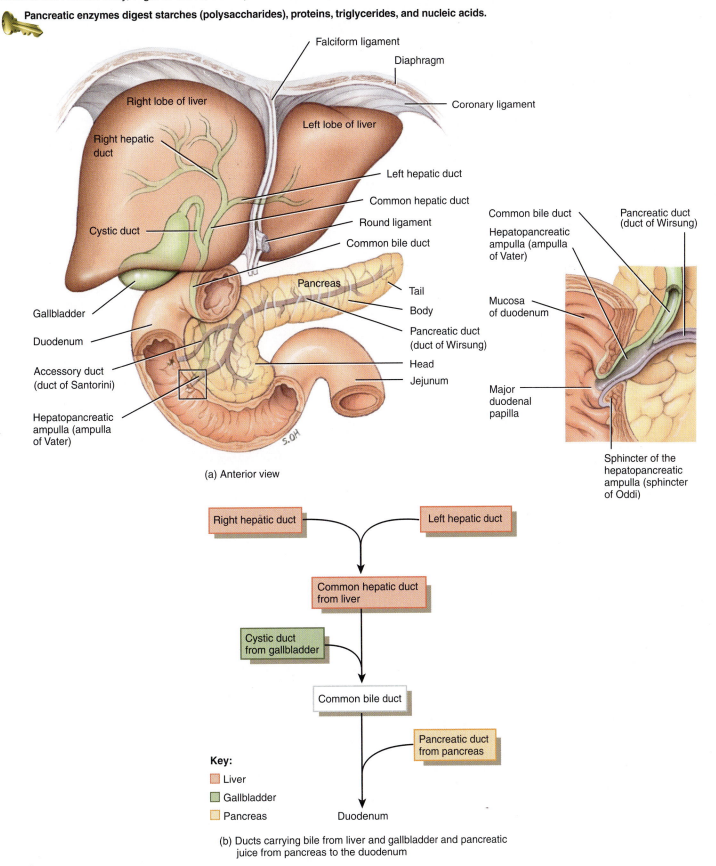

(a) Anterior view

(b) Ducts carrying bile from liver and gallbladder and pancreatic juice from pancreas to the duodenum

Key:
- 🟥 Liver
- 🟩 Gallbladder
- 🟧 Pancreas

❓ **What type of fluid is found in the pancreatic duct? The common bile duct? The hepatopancreatic ampulla?**

inhibitor, and the trypsin begins to digest the pancreatic cells. Patients with acute pancreatitis usually respond to treatment, but recurrent attacks are the rule. ■

Regulation of Pancreatic Secretions

Pancreatic secretion, like gastric secretion, is regulated by both neural and hormonal mechanisms as follows (Figure 24.18).

1. During the cephalic and gastric phases of gastric digestion, parasympathetic impulses are transmitted along the vagus nerves (cranial nerve X) to the pancreas.

2. These parasympathetic nerve impulses stimulate increased secretion of pancreatic enzymes.

3. Acidic chyme containing partially digested fats and proteins enters the small intestine.

4. In response to fatty acids and amino acids, some enteroendocrine cells in the small intestine secrete cholecystokinin (CCK) into the blood. In response to acidic chyme, other enteroendocrine cells in the small intestinal mucosa liberate secretin into the blood.

5. Secretin stimulates the flow of pancreatic juice that is rich in bicarbonate ions.

6. Cholecystokinin (CCK) stimulates a pancreatic secretion that is rich in digestive enzymes.

▶ CHECKPOINT

23. Describe the duct system connecting the pancreas to the duodenum.

24. What are pancreatic acini? How do their functions differ from those of the pancreatic islets (islets of Langerhans)?

25. What digestive functions do the components of pancreatic juice have?

26. How is pancreatic juice secretion regulated?

LIVER AND GALLBLADDER

▶ OBJECTIVE

- Describe the location, anatomy, histology, and functions of the liver and gallbladder.

The **liver** is the heaviest gland of the body, weighing about 1.4 kg (about 3 lb) in an average adult, and after the skin it is the second-largest organ of the body. It is inferior to the diaphragm and occupies most of the right hypochondriac and part of the epigastric regions of the abdominopelvic cavity (see Figure 1.13a on page 20).

The **gallbladder** (*gall-* = bile) is a pear-shaped sac that is located in a depression of the posterior surface of the liver. It is 7–10 cm (3–4 in.) long and typically hangs from the anterior inferior margin of the liver (see Figure 24.17).

Figure 24.18 Neural and hormonal stimulation of pancreatic juice secretion.

Parasympathetic stimulation (via the vagus nerves) and acidic chyme in the small intestine stimulate secretin release into the blood. Vagal stimulation and fatty acids and amino acids in chyme within the small intestine stimulate cholecystokinin (CCK) release into the blood.

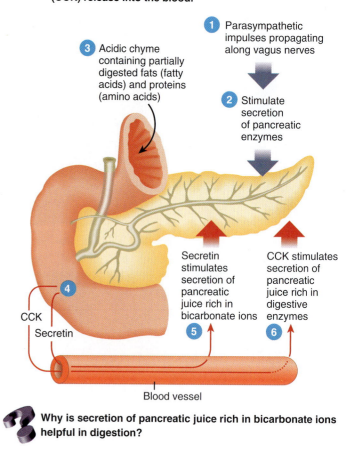

1 Parasympathetic impulses propagating along vagus nerves

2 Stimulate secretion of pancreatic enzymes

3 Acidic chyme containing partially digested fats (fatty acids) and proteins (amino acids)

4 CCK / Secretin

5 Secretin stimulates secretion of pancreatic juice rich in bicarbonate ions

6 CCK stimulates secretion of pancreatic juice rich in digestive enzymes

Blood vessel

Why is secretion of pancreatic juice rich in bicarbonate ions helpful in digestion?

Anatomy of the Liver and Gallbladder

The liver is almost completely covered by visceral peritoneum and is completely covered by a dense irregular connective tissue layer that lies deep to the peritoneum. The liver is divided into two principal lobes—a large **right lobe** and a smaller **left lobe**—by the **falciform ligament** (see Figure 24.17). Even though the right lobe is considered by many anatomists to include an inferior **quadrate lobe** and a posterior **caudate lobe,** based on internal morphology (primarily the distribution of blood vessels), the quadrate and caudate lobes more appropriately belong to the left lobe. The falciform ligament, a fold of the parietal peritoneum, extends from the undersurface of the diaphragm between the two principal lobes of the liver to the superior surface of the liver, helping to suspend the liver. In the free border of the falciform ligament is the **ligamentum teres (round ligament),** a fibrous cord that is a remnant of the umbilical vein of the fetus (see Figure 21.31a, b on page 754); it extends from the liver to the umbilicus. The right and left **coro-**

nary ligaments are narrow reflections of the parietal peritoneum that suspend the liver from the diaphragm.

The parts of the gallbladder are the broad *fundus,* which projects downward beyond the inferior border of the liver; the *body,* the central portion; and the *neck,* the tapered portion. The body and neck project upward.

Histology of the Liver and Gallbladder

The lobes of the liver are made up of many functional units called **lobules** (Figure 24.19). A lobule is typically a six-sided structure (hexagon) that consists of specialized epithelial cells, called **hepatocytes** (*hepat-* = liver; *-cytes* = cells), arranged in irregular, branching, interconnected plates around a **central vein.** Instead of capillaries, the liver has larger, endothelium-lined spaces called **sinusoids,** through which blood passes. Also present in the sinusoids are fixed phagocytes called **stellate reticuloendothelial (Kupffer) cells,** which destroy worn-out white blood cells and red blood cells, bacteria, and other foreign matter in the venous blood draining from the gastrointestinal tract.

Bile, which is secreted by hepatocytes, enters **bile canaliculi** (kan′-a-LIK-ū-lī = small canals), which are narrow intercellular canals that empty into small *bile ductules* (Figure 24.19a). The ductules pass bile into *bile ducts* at the periphery of the lobules. The bile ducts merge and eventually form the larger **right** and **left hepatic ducts,** which unite and exit the liver as the **common hepatic duct** (see Figure 24.17). Farther on, the common hepatic duct joins the **cystic duct** (*cystic* = bladder) from the gallbladder to form the **common bile duct.** Bile enters the cystic duct and is temporarily stored in the gallbladder.

The mucosa of the gallbladder consists of simple columnar epithelium arranged in rugae resembling those of the stomach. The wall of the gallbladder lacks a submucosa. The middle, muscular coat of the wall consists of smooth muscle fibers. Contraction of the smooth muscle fibers ejects the contents of the gallbladder into the **cystic duct.** The gallbladder's outer coat is the visceral peritoneum. The functions of the gallbladder are to store and concentrate bile (up to tenfold) until it is needed in the small intestine. In the concentration process, water and ions are absorbed by the gallbladder mucosa.

Jaundice

Jaundice (JAWN-dis = yellowed) is a yellowish coloration of the sclerae (whites of the eyes), skin, and mucous membranes due to a buildup of a yellow compound called bilirubin. After bilirubin is formed from the breakdown of the heme pigment in aged red blood cells, it is transported to the liver, where it is processed and eventually excreted into bile. The three main categories of jaundice are (1) *prehepatic jaundice,* due to excess production of bilirubin; (2) *hepatic jaundice,* due to congenital liver disease, cirrhosis of the liver, or hepatitis; and (3) *extrahepatic jaundice,* due to blockage of bile drainage by gallstones or cancer of the bowel or the pancreas.

Because the liver of a newborn functions poorly for the first week or so, many babies experience a mild form of jaundice called *neonatal (physiological) jaundice* that disappears as the liver matures. Usually, it is treated by exposing the infant to blue light, which converts bilirubin into substances the kidneys can excrete. ■

Blood Supply of the Liver

The liver receives blood from two sources (Figure 24.20 on page 879). From the hepatic artery it obtains oxygenated blood, and from the hepatic portal vein it receives deoxygenated blood containing newly absorbed nutrients, drugs, and possibly microbes and toxins from the gastrointestinal tract (see Figure 21.29 on page 751). Branches of both the hepatic artery and the hepatic portal vein carry blood into liver sinusoids, where oxygen, most of the nutrients, and certain toxic substances are taken up by the hepatocytes. Products manufactured by the hepatocytes and nutrients needed by other cells are secreted back into the blood, which then drains into the central vein and eventually passes into a hepatic vein. Because blood from the gastrointestinal tract passes through the liver as part of the hepatic portal circulation, the liver is often a site for metastasis of cancer that originates in the GI tract. Branches of the hepatic portal vein, hepatic artery, and bile duct typically accompany each other in their distribution through the liver. Collectively, these three structures are called a **portal triad** (see Figure 24.19). Portal triads are located at the corners of the liver lobules.

Role and Composition of Bile

Each day, hepatocytes secrete 800–1000 mL (about 1 qt) of **bile,** a yellow, brownish, or olive-green liquid. It has a pH of 7.6–8.6 and consists mostly of water and bile acids, bile salts, cholesterol, a phospholipid called lecithin, bile pigments, and several ions.

Bile is partially an excretory product and partially a digestive secretion. Bile salts, which are sodium salts and potassium salts of bile acids (mostly cholic acid and chenodeoxycholic acid), play a role in **emulsification,** the breakdown of large lipid globules into a suspension of droplets about 1 μm in diameter, and in the absorption of lipids following their digestion. The tiny lipid droplets present a very large surface area that allows pancreatic lipase to more rapidly accomplish digestion of triglycerides. Cholesterol is made soluble in bile by bile salts and lecithin.

The principal bile pigment is **conjugated bilirubin.** The phagocytosis of aged red blood cells liberates iron, globin, and bilirubin (derived from heme). The iron and globin are recycled, and some of the bilirubin is converted to **conjugated bilirubin,** in which the bilirubin is attached to glucuronic acid molecules. Conjugated bilirubin is then secreted into the bile and is eventually broken down in the intestine. One of its breakdown products—**stercobilin**—gives feces their normal brown color.

Figure 24.19 Histology of a lobule, the functional unit of the liver.

A lobule consists of hepatocytes arranged around a central vein.

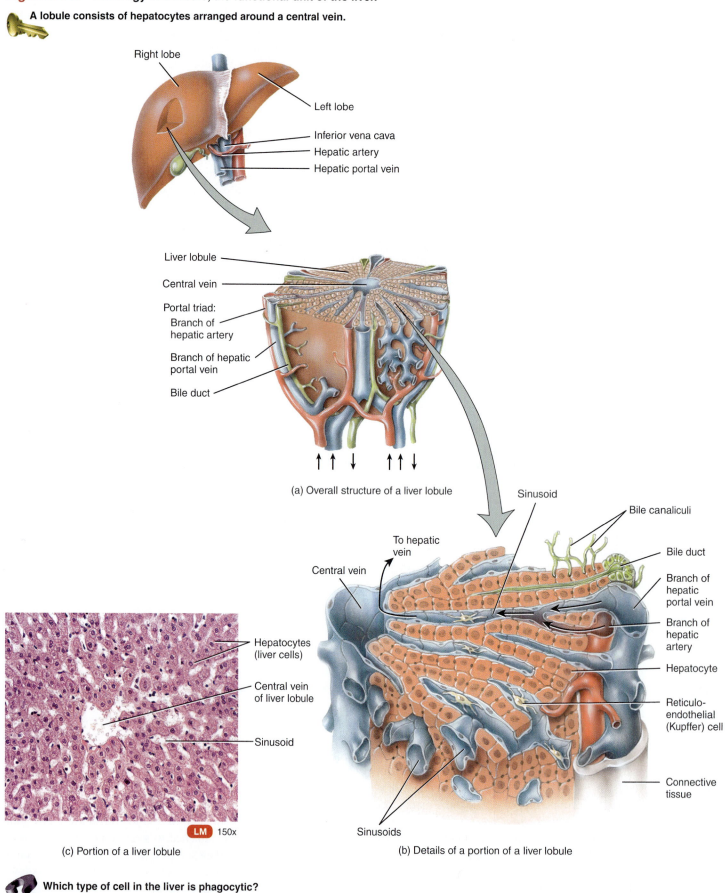

(a) Overall structure of a liver lobule

(c) Portion of a liver lobule

LM 150x

(b) Details of a portion of a liver lobule

Which type of cell in the liver is phagocytic?

Figure 24.20 Hepatic blood flow: sources, path through the liver, and return to the heart.

The liver receives oxygenated blood via the hepatic artery and nutrient-rich deoxygenated blood via the hepatic portal vein.

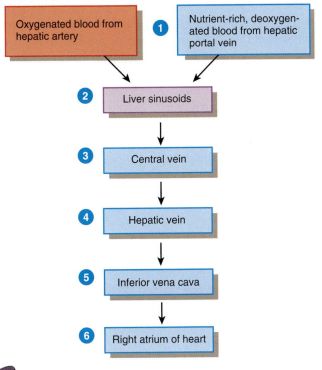

During the first few hours after a meal, how does the chemical composition of blood change as it flows through the liver sinusoids?

Regulation of Bile Secretion

After they have served their function as emulsifying agents, most bile salts are reabsorbed by active transport in the final segment of the small intestine (ileum) and enter portal blood flowing toward the liver. Although hepatocytes continually release bile, they increase production and secretion when the portal blood contains more bile acids; thus, as digestion and absorption continue in the small intestine, bile release increases. Between meals, after most absorption has occurred, bile flows into the gallbladder for storage because the **sphincter of the hepatopancreatic ampulla** (*sphincter of Oddi;* see Figure 24.17) closes off the entrance to the duodenum. After a meal, several neural and hormonal stimuli promote production and release of bile (Figure 24.21):

1 Parasympathetic impulses propagate along axons of the vagus nerve (cranial nerve X) and stimulate the liver to increase bile production.

2 Fatty acids and amino acids in chyme entering the duodenum stimulate some duodenal enteroendocrine cells to secrete the hormone cholecystokinin (CCK) into the blood. Acidic chyme entering the duodenum stimulates other enteroendocrine cells to secrete the hormone secretin into the blood.

3 CCK causes contraction of the wall of the gallbladder, which squeezes stored bile out of the gallbladder into the cystic duct and through the common bile duct. CCK also causes relaxation of the sphincter of the hepatopancreatic ampulla, which allows bile to flow into the duodenum.

Figure 24.21 Neural and hormonal stimuli that promote production and release of bile.

Bile salts in bile promote absorption of dietary lipids by emulsifying them.

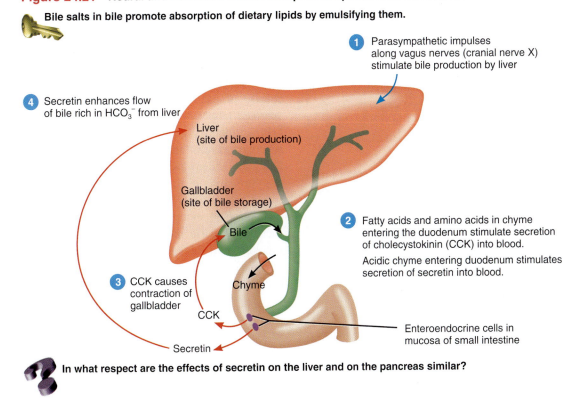

In what respect are the effects of secretin on the liver and on the pancreas similar?

4 Secretin, which stimulates secretion of pancreatic juice that is rich in HCO_3^-, also stimulates the secretion of HCO_3^- by hepatocytes into bile.

Functions of the Liver

Besides secreting bile, which is needed for absorption of dietary fats, the liver performs many other vital functions:

- **Carbohydrate metabolism.** The liver is especially important in maintaining a normal blood glucose level. When blood glucose is low, the liver can break down glycogen to glucose and release glucose into the bloodstream. The liver can also convert certain amino acids and lactic acid to glucose, and it can convert other sugars, such as fructose and galactose, into glucose. When blood glucose is high, as occurs just after eating a meal, the liver converts glucose to glycogen and triglycerides for storage.

- **Lipid metabolism.** Hepatocytes store some triglycerides; break down fatty acids to generate ATP; synthesize lipoproteins, which transport fatty acids, triglycerides, and cholesterol to and from body cells; synthesize cholesterol; and use cholesterol to make bile salts.

- **Protein metabolism.** Hepatocytes deaminate (remove the amino group, NH_2, from) amino acids so that the amino acids can be used for ATP production or converted to carbohydrates or fats. The resulting toxic ammonia (NH_3) is then converted into the much less toxic urea, which is excreted in urine. Hepatocytes also synthesize most plasma proteins, such as alpha and beta globulins, albumin, prothrombin, and fibrinogen.

- **Processing of drugs and hormones.** The liver can detoxify substances such as alcohol or excrete drugs such as penicillin, erythromycin, and sulfonamides into bile. It can also chemically alter or excrete thyroid hormones and steroid hormones such as estrogens and aldosterone.

- **Excretion of bilirubin.** As previously noted, bilirubin, derived from the heme of aged red blood cells, is absorbed by the liver from the blood and secreted into bile. Most of the bilirubin in bile is metabolized in the small intestine by bacteria and eliminated in feces.

- **Synthesis of bile salts.** Bile salts are used in the small intestine for the emulsification and absorption of lipids, cholesterol, phospholipids, and lipoproteins.

- **Storage.** In addition to glycogen, the liver is a prime storage site for certain vitamins (A, B_{12}, D, E, and K) and minerals (iron and copper), which are released from the liver when needed elsewhere in the body.

- **Phagocytosis.** The stellate reticuloendothelial (Kupffer) cells of the liver phagocytize aged red blood cells and white blood cells and some bacteria.

- **Activation of vitamin D.** The skin, liver, and kidneys participate in synthesizing the active form of vitamin D.

The liver functions related to metabolism are discussed more fully in Chapter 25.

Gallstones

If bile contains either insufficient bile salts or lecithin or excessive cholesterol, the cholesterol may crystallize to form **gallstones.** As they grow in size and number, gallstones may cause minimal, intermittent, or complete obstruction to the flow of bile from the gallbladder into the duodenum. Treatment consists of using gallstone-dissolving drugs, lithotripsy (shock-wave therapy), or surgery. For people with recurrent gallstones or for whom drugs or lithotripsy is not indicated, *cholecystectomy*—the removal of the gallbladder and its contents—is necessary. More than half a million cholecystectomies are performed each year in the United States. ■

▶ CHECKPOINT

27. Draw and label a diagram of a liver lobule.
28. Describe the pathways of blood flow into, through, and out of the liver.
29. How are the liver and gallbladder connected to the duodenum?
30. Once bile has been formed by the liver, how is it collected and transported to the gallbladder for storage?
31. What is the function of bile?
32. How is bile secretion regulated?

SUMMARY: DIGESTIVE HORMONES

▶ OBJECTIVE

- **Describe the site of secretion and the actions of gastrin, secretin, and cholecystokinin.**

The effects of the three major digestive hormones—gastrin, secretin, and cholecystokinin—and the stimuli that promote release of each are summarized in Table 24.4. All three are secreted into the blood by enteroendocrine cells located in the GI tract mucosa. Gastrin exerts its major effects on the stomach, whereas secretin and cholecystokinin affect the pancreas, liver, and gallbladder most strongly.

Stretching of the stomach as it receives food and the buffering of gastric acids by proteins in food trigger the release of gastrin. Gastrin, in turn, promotes secretion of gastric juice and increases gastric motility so that ingested food becomes well mixed into a thick, soupy chyme. Reflux of the acidic chyme into the esophagus is prevented by tonic contraction of the lower esophageal sphincter, which is enhanced by gastrin.

The major stimulus for secretion of secretin is acidic chyme (high concentration of H^+) entering the small intestine. In turn, secretin promotes secretion of bicarbonate ions (HCO_3^-) into pancreatic juice and bile (see Figures 24.18 and 24.21). HCO_3^- acts to buffer ("soak up") excess H^+. Besides these major

Table 24.4	Major Hormones that Control Digestion	
Hormone	**Stimulus and Site of Secretion**	**Actions**
Gastrin	Distension of stomach, partially digested proteins and caffeine in stomach, and high pH of stomach chyme stimulate gastrin secretion by enteroendocrine G cells, located mainly in the mucosa of pyloric antrum of stomach.	*Major effects:* Promotes secretion of gastric juice, increases gastric motility, and promotes growth of gastric mucosa. *Minor effects:* Constricts lower esophageal sphincter; relaxes pyloric sphincter and ileocecal sphincter.
Secretin	Acidic (high H^+ level) chyme that enters the small intestine stimulates secretion of secretin by enteroendocrine S cells in the mucosa of the duodenum.	*Major effects:* Stimulates secretion of pancreatic juice and bile that are rich in HCO_3^- (bicarbonate ions). *Minor effects:* Inhibits secretion of gastric juice, promotes normal growth and maintenance of the pancreas, and enhances effects of CCK.
Cholecystokinin (CCK)	Partially digested proteins (amino acids), triglycerides and fatty acids that enter the small intestine stimulate secretion of CCK by enteroendocrine CCK cells in the mucosa of the small intestine; CCK is also released in the brain.	*Major effects:* Stimulates secretion of pancreatic juice rich in digestive enzymes, causes ejection of bile from the gallbladder and opening of the sphincter of the hepatopancreatic ampulla (sphincter of Oddi), and induces satiety (feeling full to satisfaction). *Minor effects:* Inhibits gastric emptying, promotes normal growth and maintenance of the pancreas, and enhances effects of secretin.

effects, secretin inhibits secretion of gastric juice, promotes normal growth and maintenance of the pancreas, and enhances the effects of CCK. Overall, secretin causes buffering of acid in chyme that reaches the duodenum and slows production of acid in the stomach.

Amino acids from partially digested proteins and fatty acids from partially digested triglycerides stimulate secretion of cholecystokinin by enteroendocrine cells in the mucosa of the small intestine. CCK stimulates secretion of pancreatic juice that is rich in digestive enzymes (see Figure 24.18) and ejection of bile into the duodenum (see Figure 24.21), slows gastric emptying by promoting contraction of the pyloric sphincter, and produces satiety (feeling full to satisfaction) by acting on the hypothalamus in the brain. Like secretin, CCK promotes normal growth and maintenance of the pancreas; it also enhances the effects of secretin.

Besides these three hormones, at least ten other so-called "gut hormones" are secreted by and have effects on the GI tract. They include *motilin, substance P,* and *bombesin,* which stimulate motility of the intestines; *vasoactive intestinal polypeptide (VIP),* which stimulates secretion of ions and water by the intestines and inhibits gastric acid secretion; *gastrin-releasing peptide,* which stimulates release of gastrin; and *somatostatin,* which inhibits gastrin release. Some of these hormones are thought to act as local hormones (paracrines), whereas others are secreted into the blood or even into the lumen of the GI tract. The physiological roles of these and other gut hormones are still under investigation.

▶ **CHECKPOINT**

33. What are the stimuli that trigger secretion of gastrin, secretin, and cholecystokinin?

SMALL INTESTINE

▶ **OBJECTIVE**

- **Describe the location, anatomy, histology, and functions of the small intestine.**

Now that we have examined the sources of the main digestive enzymes and hormones, we can proceed to the next part of the GI tract and pick up the story of digestion and absorption. The major events of digestion and absorption occur in a long tube called the **small intestine.** Because most digestion and absorption of nutrients occur in the small intestine, its structure is specially adapted for these functions. Its length alone provides a large surface area for digestion and absorption, and that area is further increased by circular folds, villi, and microvilli. The small intestine begins at the pyloric sphincter of the stomach, coils through the central and inferior part of the abdominal cavity, and eventually opens into the large intestine. It averages 2.5 cm (1 in.) in diameter; its length is about 3 m (10 ft) in a living person and about 6.5 m (21 ft) in a cadaver due to the loss of smooth muscle tone after death.

Anatomy of the Small Intestine

The small intestine is divided into three regions (Figure 24.22). The **duodenum** (doo-ō-DĒ-num), the shortest region, is retroperitoneal. It starts at the pyloric sphincter of the stomach and extends about 25 cm (10 in.) until it merges with the jejunum. *Duodenum* means "12"; it is so-named because it is about as long as the width of 12 fingers. The **jejunum** (je-JOO-num) is about 1 m (3 ft) long and extends to the ileum. *Jejunum* means "empty," which is how it is found at death. The final and

Figure 24.22　Anatomy of the small intestine. (a) Regions of the small intestine are the duodenum, jejunum, and ileum. (See Tortora, *A Photographic Atlas of the Human Body,* Figure 12.12.) (b) Circular folds increase the surface area for digestion and absorption in the small intestine.

 Most digestion and absorption occur in the small intestine.

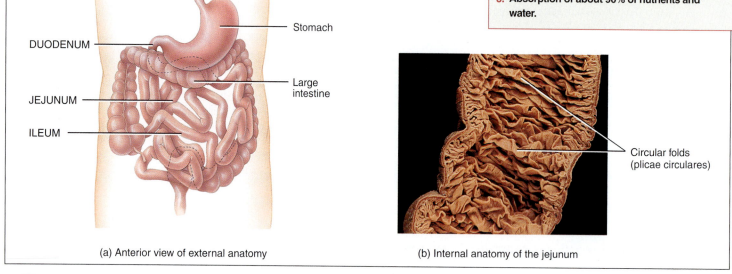

Functions of the Small Intestine 1. Segmentations mix chyme with digestive juices and bring food into contact with the mucosa for absorption; peristalsis propels chyme through the small intestine. 2. Completes the digestion of carbohydrates, proteins, and lipids; begins and completes the digestion of nucleic acids. 3. Absorption of about 90% of nutrients and water.

DUODENUM

JEJUNUM

ILEUM

Stomach

Large intestine

Circular folds (plicae circulares)

(a) Anterior view of external anatomy

(b) Internal anatomy of the jejunum

Which portion of the small intestine is the longest?

longest region of the small intestine, the **ileum** (IL-ē-um = twisted), measures about 2 m (6 ft) and joins the large intestine at the **ileocecal sphincter** (il′-ē-ō-SĒ-kal).

Projections called **circular folds** are 10 mm (0.4 in.) permanent ridges in the mucosa (Figure 24.22b). The circular folds begin near the proximal portion of the duodenum and end at about the midportion of the ileum; some extend all the way around the circumference of the intestine, and others extend only part of the way around. They enhance absorption by increasing surface area and causing the chyme to spiral, rather than move in a straight line, as it passes through the small intestine.

Histology of the Small Intestine

Even though the wall of the small intestine is composed of the same four coats that make up most of the GI tract, special features of both the mucosa and the submucosa facilitate the process of digestion and absorption. The mucosa forms a series of fingerlike **villi** (= tufts of hair), projections that are 0.5–1 mm long (Figure 24.23a). The large number of villi (20–40 per square millimeter) vastly increases the surface area of the epithelium available for absorption and digestion and gives the intestinal mucosa a velvety appearance. Each villus (singular form) has a core of lamina propria (areolar connective tissue); embedded in this connective tissue are an arteriole, a venule, a blood capillary network, and a **lacteal** (LAK-tē-al = milky), which is a lymphatic capillary. Nutrients absorbed by

the epithelial cells covering the villus pass through the wall of a capillary or a lacteal to enter blood or lymph, respectively.

The epithelium of the mucosa consists of simple columnar epithelium that contains absorptive cells, goblet cells, enteroendocrine cells, and Paneth cells (Figure 24.23b). The apical (free) membrane of absorptive cells features **microvilli** (mī′-krō-VIL-ī; *micro-* = small); each microvillus is a 1 μm-long cylindrical, membrane-covered projection that contains a bundle of 20–30 actin filaments. In a photomicrograph taken through a light microscope, the microvilli are too small to be seen individually; instead they form a fuzzy line, called the **brush border,** extending into the lumen of the small intestine (see Figure 24.24d). There are an estimated 200 million microvilli per square millimeter of small intestine. Because the microvilli greatly increase the surface area of the plasma membrane, larger amounts of digested nutrients can diffuse into absorptive cells in a given period. The brush border also contains several brush-border enzymes that have digestive functions (discussed shortly). Goblet cells secrete mucus.

The mucosa contains many deep crevices lined with glandular epithelium. Cells lining the crevices form the **intestinal glands (crypts of Lieberkühn)** and secrete intestinal juice. Many of the epithelial cells in the mucosa are goblet cells, which secrete mucus. **Paneth cells,** found in the deepest parts of the intestinal glands, secrete lysozyme, a bactericidal enzyme, and are capable of phagocytosis. They may have a role in regulating the microbial population in the intestines. Three types of

Figure 24.23   **Histology of the small intestine.**

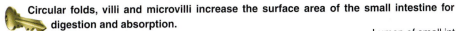 Circular folds, villi and microvilli increase the surface area of the small intestine for digestion and absorption.

Lumen of small intestine

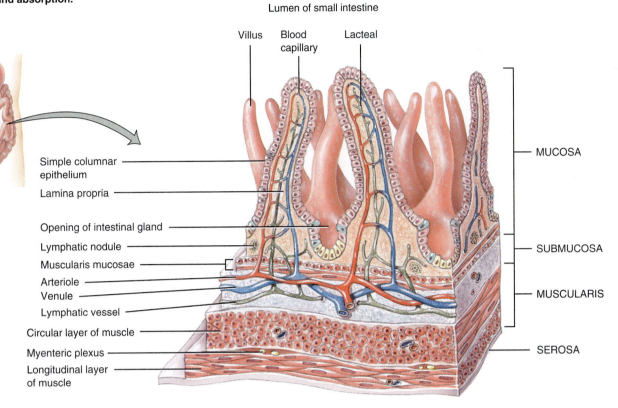

Villus   Blood capillary   Lacteal

Simple columnar epithelium

Lamina propria

Opening of intestinal gland

Lymphatic nodule

Muscularis mucosae

Arteriole

Venule

Lymphatic vessel

Circular layer of muscle

Myenteric plexus

Longitudinal layer of muscle

MUCOSA

SUBMUCOSA

MUSCULARIS

SEROSA

(a) Three-dimensional view of layers of the small intestine showing villi

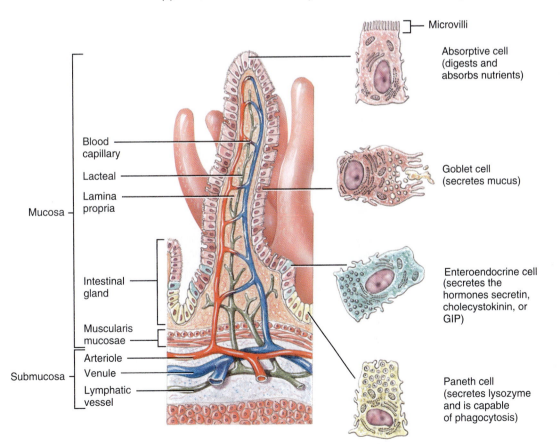

Microvilli

Absorptive cell (digests and absorbs nutrients)

Blood capillary

Lacteal

Lamina propria

Mucosa

Intestinal gland

Muscularis mucosae

Arteriole

Venule

Submucosa

Lymphatic vessel

Goblet cell (secretes mucus)

Enteroendocrine cell (secretes the hormones secretin, cholecystokinin, or GIP)

Paneth cell (secretes lysozyme and is capable of phagocytosis)

(b) Enlarged villus showing lacteal, capillaries, intestinal glands, and cell types

What is the functional significance of the blood capillary network and lacteal in the center of each villus?

enteroendocrine cells, also in the deepest part of the intestinal glands, secrete hormones: secretin (by S cells), cholecystokinin (by CCK cells), and glucose-dependent insulinotropic peptide (by K cells). The lamina propria of the small intestine has an abundance of mucosa-associated lymphoid tissue (MALT). **Solitary lymphatic nodules** are most numerous in the distal part of the ileum (see Figure 24.24c). Groups of lymphatic nodules referred to as **aggregated lymphatic follicles (Peyer's patches)** are also present in the ileum. The muscularis mucosae consists of smooth muscle. The submucosa of the duodenum contains **duodenal (Brunner's) glands** (Figure 24.24a), which secrete an alkaline mucus that helps neutralize gastric acid in the chyme. Sometimes the lymphatic tissue of the lamina propria extends through the muscularis mucosae into the submucosa.

The **muscularis** of the small intestine consists of two layers of smooth muscle. The outer, thinner layer contains longitudinal fibers; the inner, thicker layer contains circular fibers. Except for a major portion of the duodenum, the serosa (or visceral peritoneum) completely surrounds the small intestine.

Role of Intestinal Juice and Brush-Border Enzymes

Intestinal juice is a clear yellow fluid secreted in amounts of 1–2 liters (about 1–2 qt) a day. It contains water and mucus and is slightly alkaline (pH 7.6). Together, pancreatic and intestinal juices provide a liquid medium that aids the absorption of substances from chyme as they come in contact with the microvilli. The absorptive epithelial cells synthesize several digestive enzymes, called **brush-border enzymes,** and insert them in the plasma membrane of the microvilli. Thus, some enzymatic digestion occurs at the surface of the epithelial cells that line the villi, rather than in the lumen exclusively, as occurs in other parts of the GI tract. Among the brush-border enzymes are four carbohydrate-digesting enzymes called α-dextrinase, maltase, sucrase, and lactase; protein-digesting enzymes called peptidases (aminopeptidase and dipeptidase); and two types of nucleotide-digesting enzymes, nucleosidases and phosphatases. Also, as cells slough off into the lumen of the small intestine, they break apart and release enzymes that help digest nutrients in the chyme.

Mechanical Digestion in the Small Intestine

The two types of movements of the small intestine—segmentations and a type of peristalsis called migrating motility complexes—are governed mainly by the myenteric plexus. **Segmentations** are localized, mixing contractions that occur in portions of intestine distended by a large volume of chyme. Segmentations mix chyme with the digestive juices and bring the particles of food into contact with the mucosa for absorption; they do not push the intestinal contents along the tract. A seg-

Figure 24.24 Histology of the duodenum and ileum.

🔑 Microvilli in the small intestine contain several brush-border enzymes that help digest nutrients.

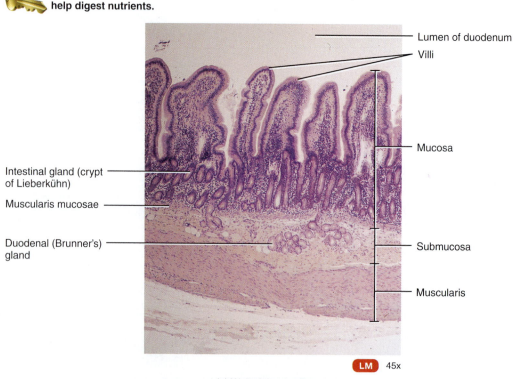

Lumen of duodenum
Villi
Mucosa
Intestinal gland (crypt of Lieberkühn)
Muscularis mucosae
Submucosa
Duodenal (Brunner's) gland
Muscularis

LM 45x

(a) Wall of the duodenum

mentation starts with the contractions of circular muscle fibers in a portion of the small intestine, an action that constricts the intestine into segments. Next, muscle fibers that encircle the middle of each segment also contract, dividing each segment again. Finally, the fibers that first contracted relax, and each small segment unites with an adjoining small segment so that large segments are formed again. As this sequence of events repeats, the chyme sloshes back and forth. Segmentations occur most rapidly in the duodenum, about 12 times per minute, and progressively slow to about 8 times per minute in the ileum. This movement is similar to alternately squeezing the middle and then the ends of a capped tube of toothpaste.

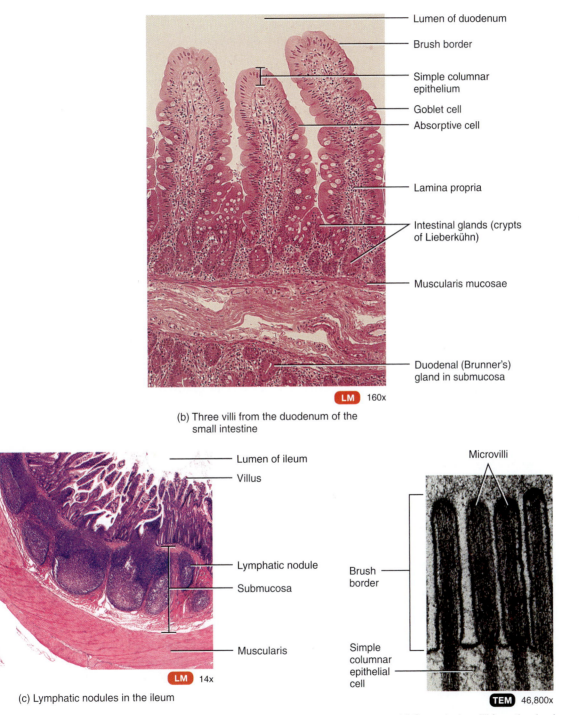

(b) Three villi from the duodenum of the small intestine

(c) Lymphatic nodules in the ileum

(d) Several microvilli from the duodenum

 What is the function of the fluid secreted by duodenal (Brunner's) glands?

After most of a meal has been absorbed, which lessens distention of the wall of the small intestine, segmentation stops and peristalsis begins. The type of peristalsis that occurs in the small intestine, termed a **migrating motility complex (MMC),** begins in the lower portion of the stomach and pushes chyme forward along a short stretch of small intestine before dying out. The MMC slowly migrates down the small intestine, reaching the end of the ileum in 90–120 minutes. Then another MMC begins in the stomach. Altogether, chyme remains in the small intestine for 3–5 hours.

Chemical Digestion in the Small Intestine

In the mouth, salivary amylase converts starch (a polysaccharide) to maltose (a disaccharide), maltotriose (a trisaccharide), and α-dextrins (short-chain, branched fragments of starch with five to ten glucose units). In the stomach, pepsin converts proteins to peptides (small fragments of proteins), and lingual and gastric lipases convert some triglycerides into fatty acids, diglycerides, and monoglycerides. Thus, chyme entering the small intestine contains partially digested carbohydrates, proteins, and lipids. The completion of the digestion of carbohydrates, proteins, and lipids is a collective effort of pancreatic juice, bile, and intestinal juice in the small intestine.

Digestion of Carbohydrates

Even though the action of **salivary amylase** may continue in the stomach for a while, the acidic pH of the stomach destroys salivary amylase and ends its activity. Thus, only a few starches are reduced to maltose by the time chyme leaves the stomach. Those starches not already broken down into maltose, maltotriose, and α-dextrins are cleaved by **pancreatic amylase,** an enzyme in pancreatic juice that acts in the small intestine. Although amylase acts on both glycogen and starches, it does not act on another polysaccharide called cellulose, which is an indigestible plant fiber. After amylase (either salivary or pancreatic) has split starch into smaller fragments, a brush-border enzyme called **α-dextrinase** acts on the resulting α-dextrins, clipping off one glucose unit at a time.

Ingested molecules of sucrose, lactose, and maltose—three disaccharides—are not acted on until they reach the small intestine. Three brush-border enzymes digest the disaccharides into monosaccharides. **Sucrase** breaks sucrose into a molecule of glucose and a molecule of fructose; **lactase** digests lactose into a molecule of glucose and a molecule of galactose; and **maltase** splits maltose and maltotriose into two or three molecules of glucose, respectively. Digestion of carbohydrates ends with the production of monosaccharides, as mechanisms exist for their absorption.

Lactose Intolerance

In some people the mucosal cells of the small intestine fail to produce enough lactase, which is essential for the digestion of lactose. This results in a condition called **lactose intolerance,** in which undigested lactose in chyme retains fluid in the feces, and bacterial fermentation of lactose results in the production of gases. Symptoms of lactose intolerance include diarrhea, gas, bloating, and abdominal cramps after consumption of milk and other dairy products. The severity of symptoms varies from relatively minor to sufficiently serious to require medical attention. Persons with lactose intolerance can take dietary supplements to aid in the digestion of lactose. ■

Digestion of Proteins

Protein digestion starts in the stomach, where proteins are fragmented into peptides by the action of **pepsin.** Enzymes in pancreatic juice—**trypsin, chymotrypsin, carboxypeptidase,** and **elastase**—continue to break down proteins into peptides. Although all these enzymes convert whole proteins into peptides, their actions differ somewhat because each splits peptide bonds between different amino acids. Trypsin, chymotrypsin, and elastase all cleave the peptide bond between a specific amino acid and its neighbor; carboxypeptidase breaks the peptide bond that attaches the terminal amino acid to the carboxyl (acid) end of the peptide. Protein digestion is completed by two **peptidases** in the brush border: aminopeptidase and dipeptidase. **Aminopeptidase** acts on peptides by breaking the peptide bond that attaches the terminal amino acid to the amino end of the peptide. **Dipeptidase** splits dipeptides (two amino acids joined by a peptide bond) into single amino acids.

Digestion of Lipids

The most abundant lipids in the diet are triglycerides, which consist of a molecule of glycerol bonded to three fatty acid molecules (see Figure 2.17 on page 46). Enzymes that split triglycerides and phospholipids are called **lipases.** In adults, most lipid digestion occurs in the small intestine, although some occurs in the stomach through the action of **lingual** and **gastric lipases.** When chyme enters the small intestine, bile salts emulsify the globules of triglycerides into droplets about 1 μm in diameter, which increases the surface area exposed to **pancreatic lipase,** another enzyme in pancreatic juice. This enzyme hydrolyzes triglycerides into fatty acids and monoglycerides, the main end products of triglyceride digestion. Pancreatic and gastric lipases remove two of the three fatty acids from glycerol; the third remains attached to the glycerol, forming a monoglyceride.

Digestion of Nucleic Acids

Pancreatic juice contains two nucleases: **ribonuclease,** which digests RNA, and **deoxyribonuclease,** which digests DNA. The nucleotides that result from the action of the two nucleases are further digested by brush-border enzymes called **nucleosidases** and **phosphatases** into pentoses, phosphates, and nitrogenous bases. These products are absorbed via active transport. Table 24.5 summarizes the sources, substrates, and products of the digestive enzymes.

Table 24.5 Summary of Digestive Enzymes

Enzyme	Source	Substrates	Products
Saliva			
Salivary amylase	Salivary glands.	Starches (polysaccharides).	Maltose (disaccharide), maltotriose (trisaccharide), and α-dextrins.
Lingual lipase	Glands in the tongue.	Triglycerides (fats and oils) and other lipids.	Fatty acids and diglycerides.
Gastric Juice			
Pepsin (activated from pepsinogen by pepsin and hydrochloric acid)	Stomach chief (zymogenic) cells.	Proteins.	Peptides.
Gastric lipase	Stomach chief (zymogenic) cells.	Short-chain triglycerides (fats and oils) in fat molecules in milk.	Fatty acids and monoglycerides.
Pancreatic Juice			
Pancreatic amylase	Pancreatic acinar cells.	Starches (polysaccharides).	Maltose (disaccharide), maltotriose (trisaccharide), and α-dextrins.
Trypsin (activated from trypsinogen by enterokinase)	Pancreatic acinar cells.	Proteins.	Peptides.
Chymotrypsin (activated from chymotrypsinogen by trypsin)	Pancreatic acinar cells.	Proteins.	Peptides.
Elastase (activated from proelastase by trypsin)	Pancreatic acinar cells.	Proteins.	Peptides.
Carboxypeptidase (activated from procarboxypeptidase by trypsin)	Pancreatic acinar cells.	Terminal amino acid at carboxyl (acid) end of peptides.	Peptides and amino acids.
Pancreatic lipase	Pancreatic acinar cells.	Triglycerides (fats and oils) that have been emulsified by bile salts.	Fatty acids and monoglycerides.
Nucleases			
Ribonuclease	Pancreatic acinar cells.	Ribonucleic acid.	Nucleotides.
Deoxyribonuclease	Pancreatic acinar cells.	Deoxyribonucleic acid.	Nucleotides.
Brush Border			
α-Dextrinase	Small intestine.	α-Dextrins.	Glucose.
Maltase	Small intestine.	Maltose.	Glucose.
Sucrase	Small intestine.	Sucrose.	Glucose and fructose.
Lactase	Small intestine.	Lactose.	Glucose and galactose.
Enterokinase	Small intestine.	Trypsinogen.	Trypsin.
Peptidases			
Aminopeptidase	Small intestine.	Terminal amino acid at amino end of peptides.	Peptides and amino acids.
Dipeptidase	Small intestine.	Dipeptides.	Amino acids.
Nucleosidases and phosphatases	Small intestine.	Nucleotides.	Nitrogenous bases, pentoses, and phosphates.

Regulation of Intestinal Secretion and Motility

The most important mechanisms that regulate small intestinal secretion and motility are enteric reflexes that respond to the presence of chyme; vasoactive intestinal polypeptide (VIP) also stimulates the production of intestinal juice. Segmentation move- ments depend mainly on intestinal distention, which initiates nerve impulses to the enteric plexuses and the central nervous system. Enteric reflexes and returning parasympathetic impulses from the CNS increase motility; sympathetic impulses decrease intestinal motility. Migrating motility complexes strengthen when most nutrients and water have been absorbed—that is, when the walls of the small intestine are less distended. With more vigorous peristalsis, the chyme moves along toward the large intestine as fast as 10 cm/sec. The first remnants of a meal reach the beginning of the large intestine in about 4 hours.

Absorption in the Small Intestine

All the chemical and mechanical phases of digestion from the mouth through the small intestine are directed toward changing food into forms that can pass through the epithelial cells lining the mucosa and into the underlying blood and lymphatic vessels. These forms are monosaccharides (glucose, fructose, and galactose) from carbohydrates; single amino acids, dipeptides, and tripeptides from proteins; and fatty acids, glycerol, and monoglycerides from triglycerides. Passage of these digested nutrients from the gastrointestinal tract into the blood or lymph is called **absorption.**

Absorption of materials occurs via diffusion, facilitated diffusion, osmosis, and active transport. About 90% of all absorption of nutrients occurs in the small intestine; the other 10% occurs in the stomach and large intestine. Any undigested or unabsorbed material left in the small intestine passes on to the large intestine.

Absorption of Monosaccharides

All carbohydrates are absorbed as monosaccharides. The capacity of the small intestine to absorb monosaccharides is huge—an estimated 120 grams per hour. As a result, all dietary carbohydrates that are digested normally are absorbed, leaving only indigestible cellulose and fibers in the feces. Monosaccharides pass from the lumen through the apical membrane via *facilitated diffusion* or *active transport.* Fructose, a monosaccharide found in fruits, is transported via *facilitated diffusion;* glucose and galactose are transported into epithelial cells of the villi via *secondary active transport* that is coupled to the active transport of Na^+ (Figure 24.25a). The transporter has binding sites for one glucose molecule and two sodium ions; unless all three sites are filled, neither substance is transported. Galactose competes with glucose to ride the same transporter. (Because both Na^+ and glucose or galactose move in the same direction, this is a *symporter.* The same type of Na^+–glucose symporter reabsorbs filtered blood glucose in the tubules of the kidneys; see Figure 26.12 on

Figure 24.25 Absorption of digested nutrients in the small intestine. For simplicity, all digested foods are shown in the lumen of the small intestine, even though some nutrients are digested by brush-border enzymes.

Long-chain fatty acids and monoglycerides are absorbed into lacteals; other products of digestion enter blood capillaries.

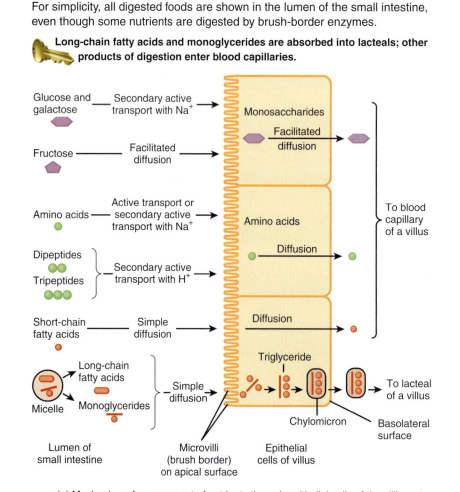

(a) Mechanisms for movement of nutrients through epithelial cells of the villi

page 967.) Monosaccharides then move out of the epithelial cells through their basolateral surfaces via *facilitated diffusion* and enter the capillaries of the villi (see Figure 24.24a, b).

Absorption of Amino Acids, Dipeptides, and Tripeptides

Most proteins are absorbed as amino acids via *active transport* processes that occur mainly in the duodenum and jejunum. About half of the absorbed amino acids are present in food, whereas the other half come from proteins in digestive juices and dead cells that slough off the mucosal surface! Normally, 95–98% of the protein present in the small intestine is digested and absorbed. Several transporters carry different types of amino acids. Some amino acids enter epithelial cells of the villi via Na^+-dependent secondary active transport processes that are similar to the glucose transporter; other amino acids are actively transported by themselves. At least one symporter brings in dipeptides and tripeptides together with H^+; the peptides then are hydrolyzed to single amino acids inside the epithelial cells. Amino acids move out of the epithelial cells via diffusion and enter capillaries of the villus (Figure 24.25a, b). Both monosaccharides and amino acids are transported in the blood to the liver by way of the hepatic portal system. If not removed by hepatocytes, they enter the general circulation.

Absorption of Lipids

All dietary lipids are absorbed via *simple diffusion*. Adults absorb about 95% of the lipids present in the small intestine; due to their lower production of bile, newborn infants absorb only about 85% of lipids. As a result of their emulsification and digestion, triglycerides are broken down into monoglycerides and fatty acids. Recall that lingual and pancreatic lipases remove two of the three fatty acids from glycerol during digestion of a triglyceride; the other fatty acid remains attached to glycerol, thus forming a monoglyceride. The small amount of short-chain fatty acids (having fewer than 10–12 carbon atoms) in the diet passes into the epithelial cells via simple diffusion and follows the same route taken by monosaccharides and amino acids into a blood capillary of a villus (Figure 24.25a, b).

Most dietary fatty acids, however, are long-chain fatty acids. They and monoglycerides reach the bloodstream by a different route and require bile for adequate absorption. Bile salts are amphipathic; they have both polar (hydrophilic) and nonpolar (hydrophobic) portions. Thus, they can form tiny spheres called **micelles** (mī-SELZ = small morsels), which are 2–10 nm in diameter and include 20–50 bile salt molecules. Because they are small and have the polar portions of bile salt molecules at their surface, micelles can dissolve in the water of intestinal fluid. In contrast, partially digested dietary lipids can dissolve in the nonpolar central core of micelles. It is in this form that fatty acids and monoglycerides reach the epithelial cells of the villi.

At the apical surface of the epithelial cells, fatty acids and monoglycerides diffuse into the cells, leaving the micelles behind in chyme. The micelles continually repeat this ferrying function. When chyme reaches the ileum, 90–95% of the bile salts are reabsorbed and returned by the blood to the liver

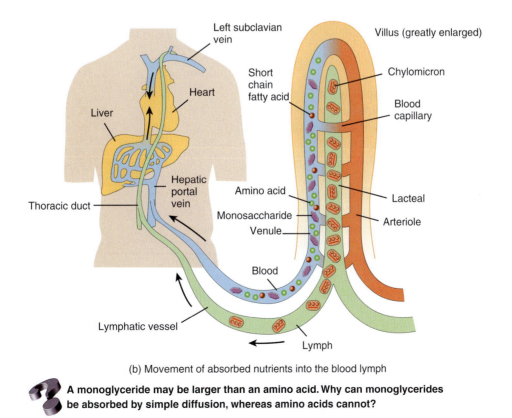

(b) Movement of absorbed nutrients into the blood lymph

A monoglyceride may be larger than an amino acid. Why can monoglycerides be absorbed by simple diffusion, whereas amino acids cannot?

through the hepatic portal system for recycling. This cycle of bile salt secretion by hepatocytes into bile, reabsorption by the ileum, and resecretion into bile is called the **enterohepatic circulation.** Insufficient bile salts, due either to obstruction of the bile ducts or removal of the gallbladder, can result in the loss of up to 40% of dietary lipids in feces due to diminished lipid absorption. Moreover, when lipids are not absorbed properly, the fat-soluble vitamins—A, D, E, and K—are not adequately absorbed.

Within the epithelial cells, many monoglycerides are further digested by lipase to glycerol and fatty acids. The fatty acids and glycerol are then recombined to form triglycerides, which aggregate into globules along with phospholipids and cholesterol and become coated with proteins. These large (about 80 nm in diameter) spherical masses are called **chylomicrons.** The hydrophilic protein coat keeps the chylomicrons suspended and prevents them from sticking to each other. Chylomicrons leave the epithelial cell via exocytosis. Because they are so large and bulky, chylomicrons cannot enter blood capillaries in the small intestine; instead, they enter the much leakier lacteals. From there they are transported by way of lymphatic vessels to the thoracic duct and enter the blood at the left subclavian vein (see Figure 24.25b).

Within 10 minutes after their absorption, about half of the chylomicrons have already been removed from the blood as they pass through blood capillaries in the liver and adipose tissue. This removal is accomplished by an enzyme in capillary endothelial cells, called **lipoprotein lipase,** that breaks down triglycerides in chylomicrons and other lipoproteins into fatty acids and glycerol. The fatty acids diffuse into hepatocytes and adipose cells and combine with glycerol during resynthesis of triglycerides. Two or three hours after a meal, few chylomicrons remain in the blood.

Absorption of Electrolytes

Many of the electrolytes absorbed by the small intestine come from gastrointestinal secretions, and some are part of ingested foods and liquids. Sodium ions are actively transported out of intestinal epithelial cells by sodium-potassium pumps (Na^+/K^+ ATPase) after they have moved into epithelial cells via diffusion and secondary active transport. Thus, most of the sodium ions in gastrointestinal secretions are reclaimed and not lost in the feces. Negatively charged bicarbonate, chloride, iodide, and nitrate ions can passively follow Na^+ or be actively transported. Calcium ions also are absorbed actively in a process stimulated by calcitriol. Other electrolytes such as iron, potassium, magnesium, and phosphate ions also are absorbed via active transport mechanisms.

Absorption of Vitamins

The fat-soluble vitamins A, D, E, and K are included with ingested dietary lipids in micelles and are absorbed via simple diffusion. Most water-soluble vitamins, such as most B vitamins and vitamin C, also are absorbed via simple diffusion. Vitamin B_{12}, however, combines with intrinsic factor produced by the

stomach, and the combination is absorbed in the ileum via an active transport mechanism.

Absorption of Water

The total volume of fluid that enters the small intestine each day—about 9.3 liters (9.8 qt)—comes from ingestion of liquids (about 2.3 liters) and from various gastrointestinal secretions (about 7.0 liters). Figure 24.26 depicts the amounts of fluid ingested, secreted, absorbed, and excreted by the GI tract. The small intestine absorbs about 8.3 liters of the fluid; the remainder passes into the large intestine, where most of the rest of it—about 0.9 liter—is also absorbed. Only 0.1 liter (100 mL) of water is excreted in the feces each day.

All water absorption in the GI tract occurs via *osmosis* from the lumen of the intestines through epithelial cells and into blood capillaries. Because water can move across the intestinal mucosa in both directions, the absorption of water from the small intes-

Figure 24.26 Daily volumes of fluid ingested, secreted, absorbed, and excreted from the GI tract.

All water absorption in the GI tract occurs via osmosis.

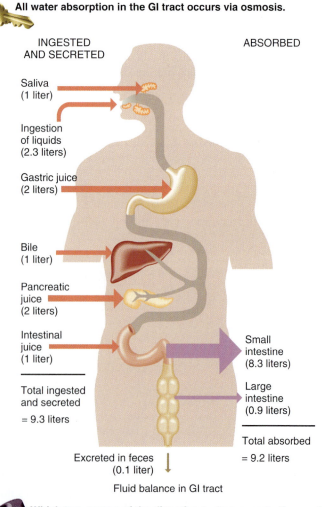

INGESTED AND SECRETED

ABSORBED

Saliva (1 liter)

Ingestion of liquids (2.3 liters)

Gastric juice (2 liters)

Bile (1 liter)

Pancreatic juice (2 liters)

Intestinal juice (1 liter)

Small intestine (8.3 liters)

Large intestine (0.9 liters)

Total ingested and secreted = 9.3 liters

Total absorbed = 9.2 liters

Excreted in feces (0.1 liter)

Fluid balance in GI tract

Which two organs of the digestive system secrete the most fluid?

tine depends on the absorption of electrolytes and nutrients to maintain an osmotic balance with the blood. The absorbed electrolytes, monosaccharides, and amino acids establish a concentration gradient for water that promotes water absorption via osmosis.

Table 24.6 summarizes the digestive activities of the pancreas, liver, gallbladder, and small intestine.

Table 24.6	Summary of Digestive Activities in the Pancreas, Liver, Gallbladder, and Small Intestine
Structure	**Activity**
Pancreas	Delivers pancreatic juice into the duodenum via the pancreatic duct (see Table 24.5 for pancreatic enzymes and their functions).
Liver	Produces bile (bile salts) necessary for emulsification and absorption of lipids.
Gallbladder	Stores, concentrates, and delivers bile into the duodenum via the common bile duct.
Small Intestine	Major site of digestion and absorption of nutrients and water in the gastrointestinal tract.
Mucosa/ submucosa	
Intestinal glands	Secrete intestinal juice.
Duodenal (Brunner's) glands	Secrete alkaline fluid to buffer stomach acids, and mucus for protection and lubrication.
Microvilli	Microscopic, membrane-covered projections of epithelial cells that contain brush-border enzymes (listed in Table 24.5) and that increase the surface area for digestion and absorption.
Villi	Fingerlike projections of mucosa that are the sites of absorption of digested food and that increase the surface area for digestion and absorption.
Circular folds	Folds of mucosa and submucosa that increase the surface area for digestion and absorption.
Muscularis	
Segmentation	Consists of alternating contractions of circular smooth muscle fibers that produce segmentation and resegmentation of sections of the small intestine; mixes chyme with digestive juices and brings food into contact with the mucosa for absorption.
Migrating motility complex (MMC)	A type of peristalsis consisting of waves of contraction and relaxation of circular and longitudinal smooth muscle fibers passing the length of the small intestine; moves chyme toward ileocecal sphincter.

Absorption of Alcohol

The intoxicating and incapacitating effects of alcohol depend on the blood alcohol level. Because it is lipid soluble, alcohol begins to be absorbed in the stomach. However, the surface area available for absorption is much greater in the small intestine than in the stomach, so when alcohol passes into the duodenum, it is absorbed more rapidly. Thus, the longer the alcohol remains in the stomach, the more slowly blood alcohol level rises. Because fatty acids in chyme slow gastric emptying, blood alcohol level will rise more slowly when fat-rich foods, such as pizza, hamburgers, or nachos, are consumed with alcoholic beverages. Also, the enzyme alcohol dehydrogenase, which is present in gastric mucosa cells, breaks down some of the alcohol to acetaldehyde, which is not intoxicating. When the rate of gastric emptying is slower, proportionally more alcohol will be absorbed and converted to acetaldehyde in the stomach, and thus less alcohol will reach the bloodstream. Given identical consumption of alcohol, females often develop higher blood alcohol levels (and therefore experience greater intoxication) than males of comparable size because the activity of gastric alcohol dehydrogenase is up to 60% lower in females than in males. Asian males may also have lower levels of this gastric enzyme. ■

► **CHECKPOINT**

34. What are the regions of the small intestine?

35. In what ways are the mucosa and submucosa of the small intestine adapted for digestion and absorption?

36. Describe the types of movement that occur in the small intestine.

37. Explain the function of each digestive enzyme.

38. How is small intestinal secretion regulated?

39. Define absorption. How are the end products of carbohydrate and protein digestion absorbed? How are the end products of lipid digestion absorbed?

40. By what routes do absorbed nutrients reach the liver?

41. Describe the absorption of electrolytes, vitamins, and water by the small intestine.

LARGE INTESTINE

► OBJECTIVE

• **Describe the anatomy, histology, and functions of the large intestine.**

The large intestine is the terminal portion of the GI tract and is divided into four principal regions. The overall functions of the large intestine are the completion of absorption, the production of certain vitamins, the formation of feces, and the expulsion of feces from the body.

Anatomy of the Large Intestine

The **large intestine,** which is about 1.5 m (5 ft) long and 6.5 cm (2.5 in.) in diameter, extends from the ileum to the anus. It is attached to the posterior abdominal wall by its **mesocolon,** which is a double layer of peritoneum. Structurally, the four major regions of the large intestine are the cecum, colon, rectum, and anal canal (Figure 24.27a).

The opening from the ileum into the large intestine is guarded by a fold of mucous membrane called the **ileocecal sphincter (valve)**, which allows materials from the small intestine to pass into the large intestine. Hanging inferior to the ileocecal valve is the **cecum,** a blind pouch about 6 cm (2.4 in.) long. Attached to the cecum is a twisted, coiled tube, measuring about 8 cm (3 in.) in length, called the **appendix** or **vermiform appendix** (*vermiform* = worm-shaped; *appendix* = appendage). The mesentery of the appendix, called the **mesoappendix,** attaches the appendix to the inferior part of the mesentery of the ileum.

The open end of the cecum merges with a long tube called the **colon** (= food passage), which is divided into ascending, transverse, descending, and sigmoid portions. Both the ascending and descending colon are retroperitoneal, whereas the trans-verse and sigmoid colon are not. The **ascending colon** ascends on the right side of the abdomen, reaches the inferior surface of the liver, and turns abruptly to the left to form the **right colic (hepatic) flexure.** The colon continues across the abdomen to the left side as the **transverse colon.** It curves beneath the inferior end of the spleen on the left side as the **left colic (splenic) flexure** and passes inferiorly to the level of the iliac crest as the **descending colon.** The **sigmoid colon** (*sigm-* = S-shaped) begins near the left iliac crest, projects medially to the midline, and terminates as the rectum at about the level of the third sacral vertebra.

The **rectum,** the last 20 cm (8 in.) of the GI tract, lies anterior to the sacrum and coccyx. The terminal 2–3 cm (1 in.) of the rectum is called the **anal canal** (Figure 24.27b). The mucous membrane of the anal canal is arranged in longitudinal folds called **anal columns** that contain a network of arteries and veins. The opening of the anal canal to the exterior, called the **anus,** is guarded by an **internal anal sphincter** of smooth muscle (involuntary) and an

Figure 24.27 Anatomy of the large intestine. (See Tortora, *A Photographic Atlas of the Human Body*, Figure 12.13.)

The regions of the large intestine are the cecum, colon, rectum, and anal canal.

Functions of the Large Intestine

1. **Haustral churning, peristalsis, and mass peristalsis drive the contents of the colon into the rectum.**
2. **Bacteria in the large intestine produce some B vitamins and vitamin K.**
3. **Absorption of some water, ions, and vitamins.**
4. **Defecation (emptying of the rectum).**

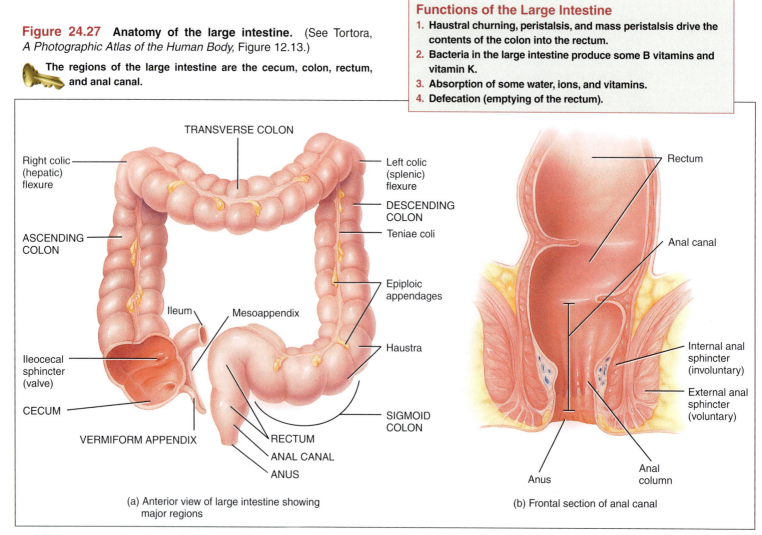

(a) Anterior view of large intestine showing major regions

(b) Frontal section of anal canal

Which portions of the colon are retroperitoneal?

external anal sphincter of skeletal muscle (voluntary). Normally these sphincters keep the anus closed except during the elimination of feces.

Appendicitis

Inflammation of the appendix, termed **appendicitis,** is preceded by obstruction of the lumen of the appendix by chyme, inflammation, a foreign body, a carcinoma of the cecum, stenosis, or kinking of the organ. It is characterized by high fever, elevated white blood cell count, and a neutrophil count higher than 75%. The infection that follows may result in edema and ischemia and may progress to gangrene and perforation within 24 hours. Typically, appendicitis begins with referred pain in the umbilical region of the abdomen, followed by anorexia (loss of appetite), nausea, and vomiting. After several hours the pain localizes in the right lower quadrant (RLQ) and is continuous, dull or severe, and intensified by coughing, sneezing, or body movements. Early appendectomy (removal of the appendix) is recommended because it is safer to operate than to risk rupture, peritonitis, and gangrene. ■

Histology of the Large Intestine

The wall of the large intestine differs from that of the small intestine in several respects. No villi or permanent circular folds are found in the **mucosa,** which consists of simple columnar epithelium, lamina propria (areolar connective tissue), and muscularis mucosae (smooth muscle) (Figure 24.28a). The epithelium contains mostly absorptive and goblet cells (Figure 24.28b and c). The absorptive cells function primarily in water absorption, whereas the goblet cells secrete mucus that lubricates the passage of the colonic contents. Both absorptive and goblet cells are located in long, straight, tubular intestinal glands that extend the full thickness of the mucosa. Solitary lymphatic nodules are also found in the lamina propria of the mucosa and may extend through the muscularis mucosae into the submucosa. The **submucosa** of the large intestine is similar to that found in the rest of the GI tract. The **muscularis** consists of an external layer of longitudinal smooth muscle and an internal layer of circular smooth muscle. Unlike other parts of the GI tract, portions of the longitudinal muscles are thickened, forming three conspicuous longitudinal bands called the **teniae coli** (TĒ-nē-ē KŌ-lī; *teniae* = flat bands), that run most of the length of the large intestine (see Figure 24.27a). The teniae coli are separated by portions of the wall with less or no longitudinal muscle. Tonic contractions of the bands gather the colon into a series of pouches called **haustra** (HAWS-tra = shaped like pouches; singular is **haustrum**), which give the colon a puckered appearance. A single layer of circular smooth muscle lies between teniae coli. The **serosa** of the large intestine is part of the visceral peritoneum. Small pouches of visceral peritoneum filled with fat are attached to teniae coli and are called **epiploic appendages.**

Figure 24.28 Histology of the large intestine.

Intestinal glands formed by simple columnar epithelial cells and goblet cells extend the full thickness of the mucosa.

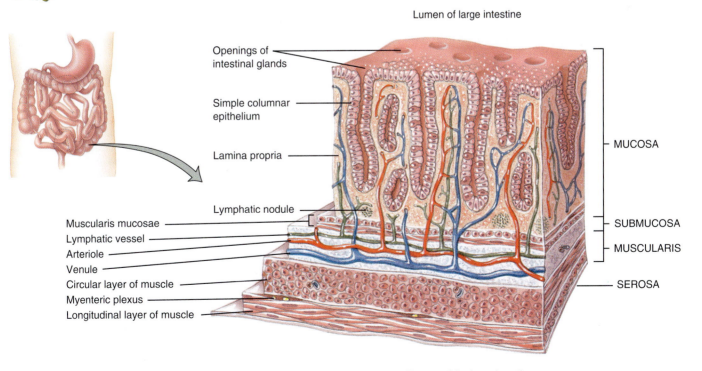

Lumen of large intestine

Openings of intestinal glands

Simple columnar epithelium

Lamina propria

Lymphatic nodule

Muscularis mucosae

Lymphatic vessel

Arteriole

Venule

Circular layer of muscle

Myenteric plexus

Longitudinal layer of muscle

MUCOSA

SUBMUCOSA

MUSCULARIS

SEROSA

(a) Three-dimensional view of layers of the large intestine

(continues)

Figure 24.28 *(continued)*

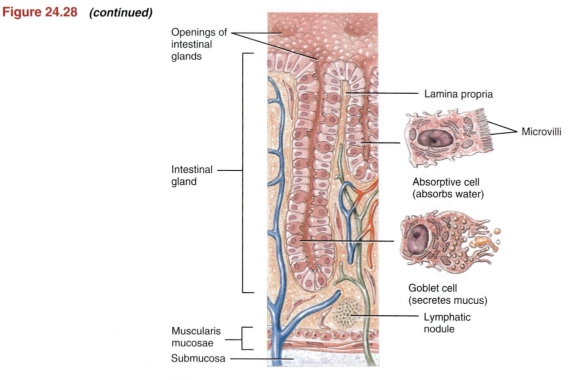

Openings of intestinal glands

Intestinal gland

Muscularis mucosae

Submucosa

Lamina propria

Microvilli

Absorptive cell (absorbs water)

Goblet cell (secretes mucus)

Lymphatic nodule

(b) Sectional view of intestinal glands and cell types

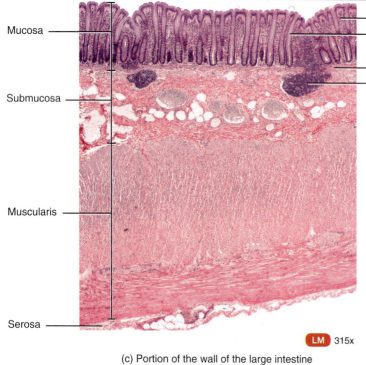

Mucosa

Submucosa

Muscularis

Serosa

Lumen of large intestine

Lamina propria

Intestinal gland

Muscularis mucosae

Lymphatic nodule

LM 315x

(c) Portion of the wall of the large intestine

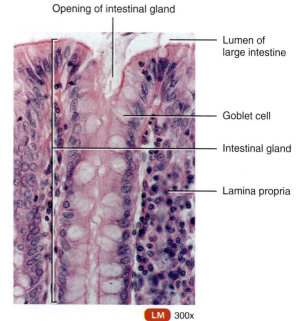

Opening of intestinal gland

Lumen of large intestine

Goblet cell

Intestinal gland

Lamina propria

LM 300x

(d) Details of mucosa of large intestine

What is the function of the goblet cells in the large intestine?

Mechanical Digestion in the Large Intestine

The passage of chyme from the ileum into the cecum is regulated by the action of the ileocecal sphincter. Normally, the valve remains partially closed so that the passage of chyme into the cecum usually occurs slowly. Immediately after a meal, a **gastroileal reflex** intensifies ileal peristalsis and forces any chyme in the ileum into the cecum. The hormone gastrin also relaxes the sphincter. Whenever the cecum is distended, the degree of contraction of the ileocecal sphincter intensifies.

Movements of the colon begin when substances pass the ileocecal sphincter. Because chyme moves through the small intestine at a fairly constant rate, the time required for a meal to pass into the colon is determined by gastric emptying time. As food passes through the ileocecal sphincter, it fills the cecum and accumulates in the ascending colon.

One movement characteristic of the large intestine is **haustral churning.** In this process, the haustra remain relaxed and become distended while they fill up. When the distension reaches a certain point, the walls contract and squeeze the contents into the next haustrum. **Peristalsis** also occurs, although at a slower rate (3–12 contractions per minute) than in more proximal portions of the tract. A final type of movement is **mass peristalsis,** a strong peristaltic wave that begins at about the middle of the transverse colon and quickly drives the contents of the colon into the rectum. Because food in the stomach initiates this **gastrocolic reflex** in the colon, mass peristalsis usually takes place three or four times a day, during or immediately after a meal.

Chemical Digestion in the Large Intestine

The final stage of digestion occurs in the colon through the activity of bacteria that inhabit the lumen. Mucus is secreted by the glands of the large intestine, but no enzymes are secreted. Chyme is prepared for elimination by the action of bacteria, which ferment any remaining carbohydrates and release hydrogen, carbon dioxide, and methane gases. These gases contribute to flatus (gas) in the colon, termed *flatulence* when it is excessive. Bacteria also convert any remaining proteins to amino acids and break down the amino acids into simpler substances: indole, skatole, hydrogen sulfide, and fatty acids. Some of the indole and skatole is eliminated in the feces and contributes to their odor; the rest is absorbed and transported to the liver, where these compounds are converted to less toxic compounds and excreted in the urine. Bacteria also decompose bilirubin to simpler pigments, including stercobilin, which give feces their brown color. Several vitamins needed for normal metabolism, including some B vitamins and vitamin K, are bacterial products that are absorbed in the colon.

Absorption and Feces Formation in the Large Intestine

By the time chyme has remained in the large intestine 3–10 hours, it has become solid or semisolid because of water absorption and is now called **feces.** Chemically, feces consist of water, inorganic salts, sloughed-off epithelial cells from the mucosa of the gastrointestinal tract, bacteria, products of bacterial decomposition, unabsorbed digested materials, and indigestible parts of food.

Although 90% of all water absorption occurs in the small intestine, the large intestine absorbs enough to make it an important organ in maintaining the body's water balance. Of the 0.5–1.0 liter of water that enters the large intestine, all but about 100–200 mL is absorbed via osmosis. The large intestine also absorbs ions, including sodium and chloride, and some vitamins.

Occult Blood

The term **occult blood** refers to blood that is hidden; it is not detectable by the human eye. The main diagnostic value of occult blood testing is to screen for colorectal cancer. Two substances often examined for occult blood are feces and urine. Several types of products are available for at-home testing for hidden blood in feces. The tests are based on color changes when reagents are added to feces. The presence of occult blood in urine may be detected at home by using dip-and-read reagent strips. ∎

The Defecation Reflex

Mass peristaltic movements push fecal material from the sigmoid colon into the rectum. The resulting distention of the rectal wall stimulates stretch receptors, which initiates a **defecation reflex** that empties the rectum. The defecation reflex occurs as follows: In response to distention of the rectal wall, the receptors send sensory nerve impulses to the sacral spinal cord. Motor impulses from the cord travel along parasympathetic nerves back to the descending colon, sigmoid colon, rectum, and anus. The resulting contraction of the longitudinal rectal muscles shortens the rectum, thereby increasing the pressure within it. This pressure, along with voluntary contractions of the diaphragm and abdominal muscles, plus parasympathetic stimulation, open the internal anal sphincter.

The external anal sphincter is voluntarily controlled. If it is voluntarily relaxed, defecation occurs and the feces are expelled through the anus; if it is voluntarily constricted, defecation can be postponed. Voluntary contractions of the diaphragm and abdominal muscles aid defecation by increasing the pressure within the abdomen, which pushes the walls of the sigmoid colon and rectum inward. If defecation does not occur, the feces back up into the sigmoid colon until the next wave of mass peristalsis again stimulates the stretch receptors, further creating the urge to defecate. In infants, the defecation reflex causes automatic emptying of the rectum because voluntary control of the external anal sphincter has not yet developed.

Diarrhea (dī-a-RĒ-a; *dia-* = through; *rrhea* = flow) is an increase in the frequency, volume, and fluid content of the feces caused by increased motility of and decreased absorption by the intestines. When chyme passes too quickly through the small intestine and feces pass too quickly through the large intestine, there is not enough time for absorption. Frequent diarrhea can result in dehydration and electrolyte imbalances. Excessive motility may be caused by lactose intolerance, stress, and microbes that irritate the gastrointestinal mucosa.

Constipation (kon-sti-PĀ-shun; *con-* = together; *stip-* = to press) refers to infrequent or difficult defecation caused by decreased motility of the intestines. Because the feces remain in the colon for prolonged periods, excessive water absorption occurs, and the feces become dry and hard. Constipation may be caused by poor habits (delaying defecation), spasms of the colon, insufficient fiber in the diet, inadequate fluid intake, lack of exercise, emotional stress, and certain drugs. A common treatment is a mild laxative, such as milk of magnesia, which induces defecation. However, many physicians maintain that laxatives are habit-forming, and that adding fiber to the diet, increasing the amount of exercise, and increasing fluid intake are safer ways of controlling this common problem.

Table 24.7 summarizes the digestive activities in the large intestine, and Table 24.8 summarizes the functions of all digestive system organs.

Dietary Fiber

Dietary fiber consists of indigestible plant carbohydrates—such as cellulose, lignin, and pectin—found in fruits, vegetables,

Table 24.7	Summary of Digestive Activities in the Large Intestine	
Structure	**Activity**	**Function(s)**
Lumen	Bacterial activity.	Breaks down undigested carbohydrates, proteins, and amino acids into products that can be expelled in feces or absorbed and detoxified by liver; synthesizes certain B vitamins and vitamin K.
Mucosa	Secretes mucus.	Lubricates colon and protects mucosa.
	Absorption.	Water absorption solidifies feces and contributes to the body's water balance; solutes absorbed include ions and some vitamins.
Muscularis	Haustral churning.	Moves contents from haustrum to haustrum by muscular contractions.
	Peristalsis.	Moves contents along length of colon by contractions of circular and longitudinal muscles.
	Mass peristalsis.	Forces contents into sigmoid colon and rectum.
	Defecation reflex.	Eliminates feces by contractions in sigmoid colon and rectum.

Table 24.8	Summary of Organs of the Digestive System and Their Functions
Organ	**Functions**
Mouth	See the functions of the tongue, salivary glands, and teeth, all of which are in the mouth. Additionally, the lips and cheeks keep food between the teeth during mastication, and buccal glands lining the mouth produce saliva.
Tongue	Maneuvers food for mastication, shapes food into a bolus, maneuvers food for deglutition, detects taste and touch sensations, and initiates digestion of triglycerides.
Salivary glands	Saliva softens, moistens, and dissolves foods; cleanses mouth and teeth; and initiates the digestion of starch.
Teeth	Cut, tear, and pulverize food to reduce solids to smaller particles for swallowing.
Pharynx	Receives a bolus from the oral cavity and passes it into the esophagus.
Esophagus	Receives a bolus from the pharynx and moves it into the stomach. This requires relaxation of the upper esophageal sphincter and secretion of mucus.
Stomach	Mixing waves macerate food, mix it with secretions of gastric glands (gastric juice), and reduce food to chyme. Gastric juice activates pepsin and kills many microbes in food. Intrinsic factor aids absorption of vitamin B_{12}. The stomach serves as a reservoir for food before releasing it into the small intestine.
Pancreas	Pancreatic juice buffers acidic gastric juice in chyme (creating the proper pH for digestion in the small intestine), stops the action of pepsin from the stomach, and contains enzymes that digest carbohydrates, proteins, triglycerides, and nucleic acids.
Liver	Produces bile, which is needed for the emulsification and absorption of lipids in the small intestine.
Gallbladder	Stores and concentrates bile and releases it into the small intestine.
Small intestine	Segmentations mix chyme with digestive juices; migrating motility complexes propel chyme toward the ileocecal sphincter; digestive secretions from the small intestine, pancreas, and liver complete the digestion of carbohydrates, proteins, lipids, and nucleic acids; circular folds, villi, and microvilli increase surface area for absorption; site where about 90% of nutrients and water are absorbed.
Large intestine	Haustral churning, peristalsis, and mass peristalsis drive the contents of the colon into the rectum; bacteria produce some B vitamins and vitamin K; absorption of some water, ions, and vitamins, defecation.

grains, and beans. **Insoluble fiber,** which does not dissolve in water, includes the woody or structural parts of plants such as the skins of fruits and vegetables and the bran coating around wheat and corn kernels. Insoluble fiber passes through the GI tract largely unchanged but speeds up the passage of material through the tract. **Soluble fiber,** which does dissolve in water, forms a gel that slows the passage of material through the tract. It is found in abundance in beans, oats, barley, broccoli, prunes, apples, and citrus fruits.

People who choose a fiber-rich diet may reduce their risk of developing obesity, diabetes, atherosclerosis, gallstones, hemorrhoids, diverticular disease, appendicitis, and colorectal cancer. Soluble fiber also may help lower blood cholesterol because the fiber binds bile salts and retards their reabsorption. As a result, more cholesterol is used to replace the bile salts lost in the feces. ■

► CHECKPOINT

42. What are the major regions of the large intestine?

43. How does the muscularis of the large intestine differ from that of the rest of the gastrointestinal tract? What are haustra?

44. Describe the mechanical movements that occur in the large intestine.

45. What is defecation and how does it occur?

46. What activities occur in the large intestine to change its contents into feces?

DEVELOPMENT OF THE DIGESTIVE SYSTEM

► OBJECTIVE

• Describe the development of the digestive system.

During the fourth week of development, the cells of the **endoderm** form a cavity called the **primitive gut,** the forerunner of the gastrointestinal tract. See Figure 29.12b on page 1078. Soon afterwards the mesoderm forms and splits into two layers (somatic and splanchnic), as shown in Figure 29.9d on page 1074. The splanchnic mesoderm associates with the endoderm of the primitive gut; as a result, the primitive gut has a double-layered wall. The **endodermal layer** gives rise to the *epithelial lining* and *glands* of most of the gastrointestinal tract; the **mesodermal layer** produces the *smooth muscle* and *connective tissue* of the tract.

The primitive gut elongates and differentiates into an anterior **foregut,** an intermediate **midgut,** and a posterior **hindgut** (see Figure 29.12c on page 1078). Until the fifth week of development, the midgut opens into the yolk sac; after that time, the yolk sac constricts and detaches from the midgut, and the midgut seals. In the region of the foregut, a depression consisting of ectoderm, the **stomodeum** (stō-mō-DĒ-um), appears (see Figure 29.12d on page 1078). This develops into the *oral cavity.* The **oropharyngeal membrane** is a depression of fused ectoderm and endoderm on the surface of the embryo that separates the foregut from the stomodeum. The membrane ruptures during the fourth week of development, so that the foregut is continuous with the outside of the embryo through the oral cavity. Another depression consisting of ectoderm, the **proctodeum** (prok-tō-DĒ-um), forms in the hindgut and goes on to develop into the *anus.* See Figure 29.12d on page 1078. The **cloacal membrane** (klō-Ā-kul) is a fused membrane of ectoderm and endoderm that separates the hindgut from the proctodeum. After it ruptures during the seventh week, the hindgut is continuous with the outside of the embryo through the anus. Thus, the gastrointestinal tract forms a continuous tube from mouth to anus.

The foregut develops into the *pharynx, esophagus, stomach,* and *part of the duodenum.* The midgut is transformed into the *remainder of the duodenum,* the *jejunum,* the *ileum,* and *portions of the large intestine* (cecum, appendix, ascending colon, and most of the transverse colon). The hindgut develops into the *remainder of the large intestine,* except for a portion of the anal canal that is derived from the proctodeum.

As development progresses, the endoderm at various places along the foregut develops into hollow buds that grow into the mesoderm. These buds will develop into the *salivary glands, liver, gallbladder,* and *pancreas.* Each of these organs retains a connection with the gastrointestinal tract through ducts.

► CHECKPOINT

47. What structures develop from the foregut, midgut, and hindgut?

AGING AND THE DIGESTIVE SYSTEM

► OBJECTIVE

• Describe the effects of aging on the digestive system.

Overall changes of the digestive system associated with aging include decreased secretory mechanisms, decreased motility of the digestive organs, loss of strength and tone of the muscular tissue and its supporting structures, changes in neurosensory feedback regarding enzyme and hormone release, and diminished response to pain and internal sensations. In the upper portion of the GI tract, common changes include reduced sensitivity to mouth irritations and sores, loss of taste, periodontal disease, difficulty in swallowing, hiatal hernia, gastritis, and peptic ulcer disease. Changes that may appear in the small intestine include duodenal ulcers, malabsorption, and maldigestion. Other pathologies that increase in incidence with age are appendicitis, gallbladder problems, jaundice, cirrhosis, and acute pancreatitis. Large intestinal changes such as constipation, hemorrhoids, and diverticular disease may also occur. Cancer of the colon or rectum is quite common.   • • •

Now that our exploration of the digestive system is completed, you can appreciate the many ways that this system contributes to homeostasis of other body systems by examining *Focus on Homeostasis: The Digestive System* on page 898. Next, in Chapter 25, you will discover how the nutrients absorbed by the GI tract enter into metabolic reactions in the body tissues.

Focus on Homeostasis:
The Digestive System

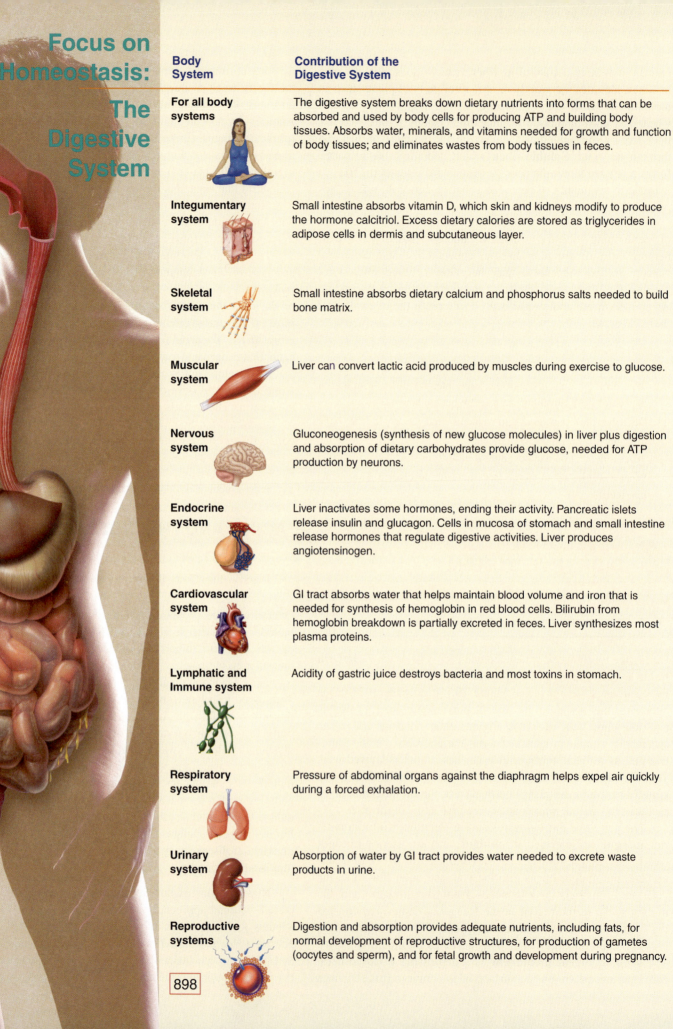

Body System	Contribution of the Digestive System
For all body systems	The digestive system breaks down dietary nutrients into forms that can be absorbed and used by body cells for producing ATP and building body tissues. Absorbs water, minerals, and vitamins needed for growth and function of body tissues; and eliminates wastes from body tissues in feces.
Integumentary system	Small intestine absorbs vitamin D, which skin and kidneys modify to produce the hormone calcitriol. Excess dietary calories are stored as triglycerides in adipose cells in dermis and subcutaneous layer.
Skeletal system	Small intestine absorbs dietary calcium and phosphorus salts needed to build bone matrix.
Muscular system	Liver can convert lactic acid produced by muscles during exercise to glucose.
Nervous system	Gluconeogenesis (synthesis of new glucose molecules) in liver plus digestion and absorption of dietary carbohydrates provide glucose, needed for ATP production by neurons.
Endocrine system	Liver inactivates some hormones, ending their activity. Pancreatic islets release insulin and glucagon. Cells in mucosa of stomach and small intestine release hormones that regulate digestive activities. Liver produces angiotensinogen.
Cardiovascular system	GI tract absorbs water that helps maintain blood volume and iron that is needed for synthesis of hemoglobin in red blood cells. Bilirubin from hemoglobin breakdown is partially excreted in feces. Liver synthesizes most plasma proteins.
Lymphatic and Immune system	Acidity of gastric juice destroys bacteria and most toxins in stomach.
Respiratory system	Pressure of abdominal organs against the diaphragm helps expel air quickly during a forced exhalation.
Urinary system	Absorption of water by GI tract provides water needed to excrete waste products in urine.
Reproductive systems	Digestion and absorption provides adequate nutrients, including fats, for normal development of reproductive structures, for production of gametes (oocytes and sperm), and for fetal growth and development during pregnancy.

898

DISORDERS: HOMEOSTATIC IMBALANCES

Dental Caries

Dental caries, or tooth decay, involve a gradual demineralization (softening) of the enamel and dentin. If untreated, microorganisms may invade the pulp, causing inflammation and infection, with subsequent death of the pulp and abscess of the alveolar bone surrounding the root's apex. Such teeth are treated by root canal therapy.

Dental caries begin when bacteria, acting on sugars, produce acids that demineralize the enamel. **Dextran,** a sticky polysaccharide produced from sucrose, causes the bacteria to stick to the teeth. Masses of bacterial cells, dextran, and other debris adhering to teeth constitute **dental plaque.** Saliva cannot reach the tooth surface to buffer the acid because the plaque covers the teeth. Brushing the teeth after eating removes the plaque from flat surfaces before the bacteria can produce acids. Dentists also recommend that the plaque between the teeth be removed every 24 hours with dental floss.

Periodontal Disease

Periodontal disease is a collective term for a variety of conditions characterized by inflammation and degeneration of the gingivae, alveolar bone, periodontal ligament, and cementum. One such condition is called **pyorrhea,** the initial symptoms of which are enlargement and inflammation of the soft tissue and bleeding of the gums. Without treatment, the soft tissue may deteriorate and the alveolar bone may be resorbed, causing loosening of the teeth and recession of the gums. Periodontal diseases are often caused by poor oral hygiene; by local irritants, such as bacteria, impacted food, and cigarette smoke; or by a poor "bite."

Peptic Ulcer Disease

In the United States, 5–10% of the population develops **peptic ulcer disease (PUD).** An **ulcer** is a craterlike lesion in a membrane; ulcers that develop in areas of the GI tract exposed to acidic gastric juice are called **peptic ulcers.** The most common complication of peptic ulcers is bleeding, which can lead to anemia if enough blood is lost. In acute cases, peptic ulcers can lead to shock and death. Three distinct causes of PUD are recognized: (1) the bacterium *Helicobacter pylori;* (2) non-steroidal anti-inflammatory drugs (NSAIDs) such as aspirin; and (3) hypersecretion of HCl, as occurs in Zollinger–Ellison syndrome, a gastrin-producing tumor usually of the pancreas.

Helicobacter pylori (previously named *Campylobacter pylori*) is the most frequent cause of PUD. The bacterium produces an enzyme called urease, which splits urea into ammonia and carbon dioxide. While shielding the bacterium from the acidity of the stomach, the ammonia also damages the protective mucous layer of the stomach and the underlying gastric cells. *H. pylori* also produces catalase, an enzyme that may protect the microbe from phagocytosis by neutrophils, plus several adhesion proteins that allow the bacterium to attach itself to gastric cells.

Several therapeutic approaches are helpful in the treatment of PUD. Cigarette smoke, alcohol, caffeine, and NSAIDs should be avoided because they can impair mucosal defensive mechanisms, which increases mucosal susceptibility to the damaging effects of HCl. In cases associated with *H. pylori,* treatment with an antibiotic drug often resolves the problem. Oral antacids such as Tums or Maalox can help temporarily by buffering gastric acid. When hypersecretion of HCl is the cause of PUD, H_2 blockers (such as Tagamet) or proton pump inhibitors such as omeprazole (Prilosec), which block secretion of H^+ from parietal cells, may be used.

Diverticular Disease

In **diverticular disease,** saclike outpouchings of the wall of the colon, termed **diverticula,** occur in places where the muscularis has weakened and may become inflamed. Development of diverticula is known as **diverticulosis.** Many people who develop diverticulosis have no symptoms and experience no complications. Of those people known to have diverticulosis, 10–25% eventually develop an inflammation known as **diverticulitis.** This condition may be characterized by pain, either constipation or increased frequency of defecation, nausea, vomiting, and low-grade fever. Because diets low in fiber contribute to development of diverticulitis, patients who change to high-fiber diets show marked relief of symptoms. In severe cases, affected portions of the colon may require surgical removal. If diverticula rupture, the release of bacteria into the abdominal cavity can cause peritonitis.

Colorectal Cancer

Colorectal cancer is among the deadliest of malignancies, ranking second to lung cancer in males and third after lung cancer and breast cancer in females. Genetics plays a very important role in that an inherited predisposition contributes to more than half of all cases of colorectal cancer. Intake of alcohol and diets high in animal fat and protein are associated with increased risk of colorectal cancer, whereas dietary fiber, retinoids, calcium, and selenium may be protective. Signs and symptoms of colorectal cancer include diarrhea, constipation, cramping, abdominal pain, and rectal bleeding, either visible or occult. Precancerous growths on the mucosal surface, called **polyps,** also increase the risk of developing colorectal cancer. Screening for colorectal cancer includes testing for blood in the feces, digital rectal examination, sigmoidoscopy, colonoscopy, and barium enema. Tumors may be removed endoscopically or surgically.

Hepatitis

Hepatitis is an inflammation of the liver that can be caused by viruses, drugs, and chemicals, including alcohol. Clinically, several types of viral hepatitis are recognized. **Hepatitis A (infectious hepatitis)** is caused by hepatitis A virus and is spread via fecal contamination of objects such as food, clothing, toys, and eating utensils (fecal–oral route). It is generally a mild disease of children and young adults characterized by loss of appetite, malaise, nausea, diarrhea, fever, and chills. Eventually, jaundice appears. This type of hepatitis does not cause lasting liver damage. Most people recover in 4 to 6 weeks.

Hepatitis B is caused by hepatitis B virus and is spread primarily by sexual contact and contaminated syringes and transfusion equipment. It can also be spread via saliva and tears. Hepatitis B virus can be present for years or even a lifetime, and it can produce cirrhosis and possibly cancer of the liver. Individuals who harbor the active hepatitis B virus also become carriers. Vaccines produced through recombinant DNA technology (for example, Recombivax HB) are available to prevent hepatitis B infection.

Hepatitis C, caused by hepatitis C virus, is clinically similar to hepatitis B. Hepatitis C can cause cirrhosis and possibly liver cancer. In developed nations, donated blood is screened for the presence of hepatitis B and C.

Hepatitis D is caused by hepatitis D virus. It is transmitted like hepatitis B and, in fact, a person must have been co-infected with hepatitis B before contracting hepatitis D. Hepatitis D results in severe liver damage and has a higher fatality rate than infection with hepatitis B virus alone.

Hepatitis E is caused by hepatitis E virus and is spread like hepatitis A. Although it does not cause chronic liver disease, hepatitis E virus is responsible for a very high mortality rate in pregnant women.

Anorexia Nervosa

Anorexia nervosa is a chronic disorder characterized by self-induced weight loss, negative perception of body image, and physiological changes that result from nutritional depletion. Patients with anorexia nervosa have a fixation on weight control and often insist on having a bowel movement every day despite inadequate food intake. They often abuse laxatives, which worsens the fluid and electrolyte imbalances and nutrient deficiencies. The disorder is found predominantly in young, single females, and it may be inherited. Abnormal patterns of menstruation, amenorrhea (absence of menstruation), and a lowered basal metabolic rate reflect the depressant effects of starvation. Individuals may become emaciated and may ultimately die of starvation or one of its complications. Also associated with the disorder are osteoporosis, depression, and brain abnormalities coupled with impaired mental performance. Treatment consists of psychotherapy and dietary regulation.

MEDICAL TERMINOLOGY

Achalasia (ak′-a-LĀ-zē-a; *a-* = without; *chalasis* = relaxation) A condition, caused by malfunction of the myenteric plexus, in which the lower esophageal sphincter fails to relax normally as food approaches. A whole meal may become lodged in the esophagus and enter the stomach very slowly. Distension of the esophagus results in chest pain that is often confused with pain originating from the heart.

Borborygmus (bor′-bō-RIG-mus) A rumbling noise caused by the propulsion of gas through the intestines.

Bulimia (boo-LIM-ē-a; *bu-* = ox; *limia* = hunger) or **binge–purge syndrome** A disorder that typically affects young, single, middle-class, white females, characterized by overeating at least twice a week followed by purging by self-induced vomiting, strict dieting or fasting, vigorous exercise, or use of laxatives or diuretics. It occurs in response to fears of being overweight, or to stress, depression, and physiological disorders such as hypothalamic tumors.

Canker sore (KANG-ker) Painful ulcer on the mucous membrane of the mouth that affects females more often than males, usually between ages 10 and 40. It may be an autoimmune reaction or a food allergy.

Cholecystitis (kō′-lē-sis-TĪ-tis; *chole-* = bile; *cyst-* = bladder; *-itis* = inflammation of) In some cases, an autoimmune inflammation of the gallbladder; other cases are caused by obstruction of the cystic duct by bile stones.

Cirrhosis (sir-RŌ-sis; *cirrh-* = orange-colored) Distorted or scarred liver as a result of chronic inflammation due to hepatitis, chemicals that destroy hepatocytes, parasites that infect the liver, or alcoholism; the hepatocytes are replaced by fibrous or adipose connective tissue. Symptoms include jaundice, edema in the legs, uncontrolled bleeding, and increased sensitivity to drugs.

Colitis (ko-LĪ-tis) Inflammation of the mucosa of the colon and rectum in which absorption of water and salts is reduced, producing watery, bloody feces and, in severe cases, dehydration and salt depletion. Spasms of the irritated muscularis produce cramps. It is thought to be an autoimmune disorder.

Colostomy (kō-LOS-tō-mē; *-stomy* = provide an opening) The diversion of the fecal stream through an opening in the colon, creating a surgical "stoma" (artificial opening) that is affixed to the exterior of the abdominal wall. This opening is a substitute anus through which feces are eliminated into a bag worn on the abdomen.

Dysphagia (dis-FĀ-jē-a; *dys-* = abnormal; *phagia* = to eat) Difficulty in swallowing that may be caused by inflammation, paralysis, obstruction, or trauma.

Enteritis (en′-ter-Ī-tis; *enter-* = intestine) An inflammation of the intestine, particularly the small intestine.

Flatus (FLĀ-tus) Air (gas) in the stomach or intestine, usually expelled through the anus. If the gas is expelled through the mouth, it is called **eructation** or **belching** (burping). Flatus may result from gas released during the breakdown of foods in the stomach or from swallowing air or gas-containing substances such as carbonated drinks.

Gastrectomy (gas-TREK-tō-mē; *gastr-* = stomach; *-ectomy* = to cut out) Removal of all or a portion of the stomach.

Gastroscopy (gas-TROS-kō-pē; *-scopy* = to view with a lighted instrument) Endoscopic examination of the stomach in which the examiner can view the interior of the stomach directly to evaluate an ulcer, tumor, inflammation, or source of bleeding.

Hernia (HER-nē-a) Protrusion of all or part of an organ through a membrane or cavity wall, usually the abdominal cavity. *Diaphragmatic (hiatal) hernia* is the protrusion of the lower esophagus, stomach, or intestine into the thoracic cavity through the esophageal hiatus. *Inguinal hernia* is the protrusion of the hernial sac into the inguinal opening; it may contain a portion of the bowel in an advanced stage and may extend into the scrotal compartment in males, causing strangulation of the herniated part.

Hemorrhoids (HEM-ō-royds = veins likely to bleed) Varicose (enlarged and inflamed) veins in the rectum. Hemorrhoids develop when the veins are put under pressure and become engorged with blood. If the pressure continues, the wall of the vein stretches. Such a distended vessel oozes blood, and bleeding or itching are usually the first signs that a hemorrhoid has developed. Stretching of a vein also favors clot formation, further aggravating swelling and pain. Hemorrhoids may be caused by constipation, which may be brought on by low-fiber diets. Also, repeated straining during defecation forces blood down into the rectal veins, increasing pressure in these veins and possibly causing hemorrhoids. Also called **piles.**

Inflammatory bowel disease (in-FLAM-a-tō′-rē BOW-el) Inflammation of the gastrointestinal tract that exists in two forms. (1) Crohn's disease is an inflammation of any part of the gastrointestinal tract in which the inflammation extends from the mucosa through the submucosa, muscularis, and serosa. (2) Ulcerative colitis is an inflammation of the mucosa of the colon and rectum, usually accompanied by rectal bleeding. Curiously, cigarette smoking increases one's risk for Crohn's disease but decreases the risk of ulcerative colitis.

Irritable bowel syndrome (IBS) Disease of the entire gastrointestinal tract in which a person reacts to stress by developing symptoms (such as cramping and abdominal pain) associated with alternating patterns of diarrhea and constipation. Excessive amounts of mucus may appear in feces, and other symptoms include flatulence, nausea, and loss of appetite. The condition is also known as **irritable colon** or **spastic colitis.**

Malocclusion (mal′-ō-KLOO-zhun; *mal-* = bad; *occlusion* = to fit together) Condition in which the surfaces of the maxillary (upper) and mandibular (lower) teeth fit together poorly.

Traveler's diarrhea Infectious disease of the gastrointestinal tract that results in loose, urgent bowel movements, cramping, abdominal pain, malaise, nausea, and occasionally fever and dehydration. It is acquired through ingestion of food or water contaminated with fecal material that contains bacteria (especially *Escherichia coli*); viruses or protozoan parasites in feces are less common causes.

STUDY OUTLINE

INTRODUCTION (p. 852)

1. The breaking down of larger food molecules into smaller molecules is called digestion; the passage of these smaller molecules into blood and lymph is termed absorption.
2. The organs that collectively perform digestion and absorption constitute the digestive system and are usually composed of two main groups: the gastrointestinal (GI) tract and accessory digestive organs.
3. The GI tract is a continuous tube extending from the mouth to the anus.
4. The accessory digestive organs include the teeth, tongue, salivary glands, liver, gallbladder, and pancreas.

OVERVIEW OF THE DIGESTIVE SYSTEM (p. 852)

1. Digestion includes six basic processes: ingestion, secretion, mixing and propulsion, mechanical and chemical digestion, absorption, and defecation.
2. Mechanical digestion consists of mastication and movements of the gastrointestinal tract that aid chemical digestion.
3. Chemical digestion is a series of hydrolysis reactions that break down large carbohydrates, lipids, proteins, and nucleic acids in foods into smaller molecules that are usable by body cells.

LAYERS OF THE GI TRACT (p. 853)

1. The basic arrangement of layers in most of the gastrointestinal tract, from deep to superficial, is the mucosa, submucosa, muscularis, and serosa.
2. Associated with the lamina propria of the mucosa are extensive patches of lymphatic tissue called mucosa-associated lymphoid tissue (MALT).

PERITONEUM (p. 855)

1. The peritoneum is the largest serous membrane of the body; it lines the wall of the abdominal cavity and covers some abdominal organs.
2. Folds of the peritoneum include the mesentery, mesocolon, falciform ligament, lesser omentum, and greater omentum.

MOUTH (p. 857)

1. The mouth is formed by the cheeks, hard and soft plates, lips, and tongue.
2. The vestibule is the space bounded externally by the cheeks and lips and internally by the teeth and gums.
3. The oral cavity proper extends from the vestibule to the fauces.
4. The tongue, together with its associated muscles, forms the floor of the oral cavity. It is composed of skeletal muscle covered with mucous membrane.
5. The upper surface and sides of the tongue are covered with papillae, some of which contain taste buds.
6. The major portion of saliva is secreted by the major salivary glands, which lie outside the mouth and pour their contents into ducts that empty into the oral cavity.
7. There are three pairs of major salivary glands: parotid, submandibular, and sublingual glands.
8. Saliva lubricates food and starts the chemical digestion of carbohydrates.
9. Salivation is controlled by the nervous system.
10. The teeth (dentes) project into the mouth and are adapted for mechanical digestion.
11. A typical tooth consists of three principal regions: crown, root, and neck.
12. Teeth are composed primarily of dentin and are covered by enamel, the hardest substance in the body.
13. There are two dentitions: deciduous and permanent.
14. Through mastication, food is mixed with saliva and shaped into a soft, flexible mass called a bolus.
15. Salivary amylase begins the digestion of starches, and lingual lipase acts on triglycerides.

1 *Glycolysis* is a set of reactions in which one glucose molecule is oxidized and two molecules of pyruvic acid are produced. The reactions also produce two molecules of ATP and two energy-containing $NADH + H^+$. Because glycolysis does not require oxygen, it is a way to produce ATP anaerobically (without oxygen) and is known as **anaerobic cellular respiration.**

2 *Formation of acetyl coenzyme A* is a transition step that prepares pyruvic acid for entrance into the Krebs cycle. This step also produces energy-containing $NADH + H^+$ plus carbon dioxide (CO_2).

3 *Krebs cycle reactions* oxidize acetyl coenzyme A and produce CO_2, ATP, energy-containing $NADH + H^+$, and $FADH_2$.

4 *Electron transport chain reactions* oxidize $NADH + H^+$ and $FADH_2$ and transfer their electrons through a series of electron carriers. The Krebs cycle and the electron transport chain together require oxygen to produce ATP and are known as **aerobic cellular respiration.**

Glycolysis

During **glycolysis** (glī-KOL-i-sis; *-lysis* = breakdown), chemical reactions split a six-carbon molecule of glucose into two three-carbon molecules of pyruvic acid (Figure 25.3). Even though glycolysis consumes two ATP molecules, it produces four ATP molecules, for a net gain of two ATP molecules for each glucose molecule that is oxidized.

Figure 25.4 shows the 10 reactions that comprise glycolysis. In the first half of the sequence (reactions **1** through **5**), energy in the form of ATP is "invested" and the 6-carbon glucose is split into two 3-carbon molecules of glyceraldehyde 3-phosphate. *Phosphofructokinase,* the enzyme that catalyzes step **3**, is the key regulator of the rate of glycolysis. The activity of this enzyme is high when ADP concentration is high, in which case ATP is produced rapidly. When the activity of phosphofructokinase is low, most glucose does not enter the reactions of glycolysis but instead undergoes conversion to glycogen for storage. In the second half (reactions **6** through **10**), the two glyceraldehyde 3-phosphate molecules are converted to two pyruvic acid molecules and ATP is generated.

Figure 25.3 Cellular respiration begins with glycolysis.

During glycolysis, each molecule of glucose is converted to two molecules of pyruvic acid.

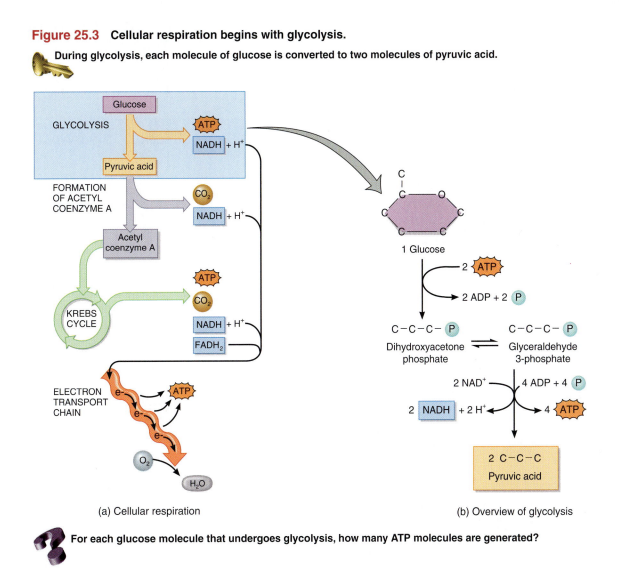

(a) Cellular respiration

(b) Overview of glycolysis

For each glucose molecule that undergoes glycolysis, how many ATP molecules are generated?

Figure 25.4 **The ten reactions of glycolysis.** **①** The first reaction phosphorylates glucose, using a phosphate group from an ATP molecule to form glucose 6-phosphate. **②** The second reaction converts glucose 6-phosphate to fructose 6-phosphate. **③** The third reaction uses a second ATP to add a second phosphate group to form fructose 1, 6-bisphosphate. **④** and **⑤** Fructose splits into two three-carbon molecules, glyceraldehyde 3-phosphate (G 3-P) and dihydroxyacetone phosphate, each having one phosphate group. **⑥** Oxidation occurs as two molecules of NAD⁺ accept two pairs of electrons and hydrogen ions from two molecules of G 3-P, forming two molecules each of NADH. Many body cells use the two NADH produced in this step to generate four ATPs in the electron transport chain. Hepatocytes, kidney cells, and cardiac muscle fibers, can generate six ATPs from the two NADH. A second phosphate group attaches to G 3-P, forming 1,3-bisphosphoglyceric acid (BPG). **⑦** through **⑩** These reactions generate four molecules of ATP and produce two molecules of pyruvic acid (pyruvate*).

🔑 **Glycolysis results in a net gain of two ATP, two NADH, and two H⁺.**

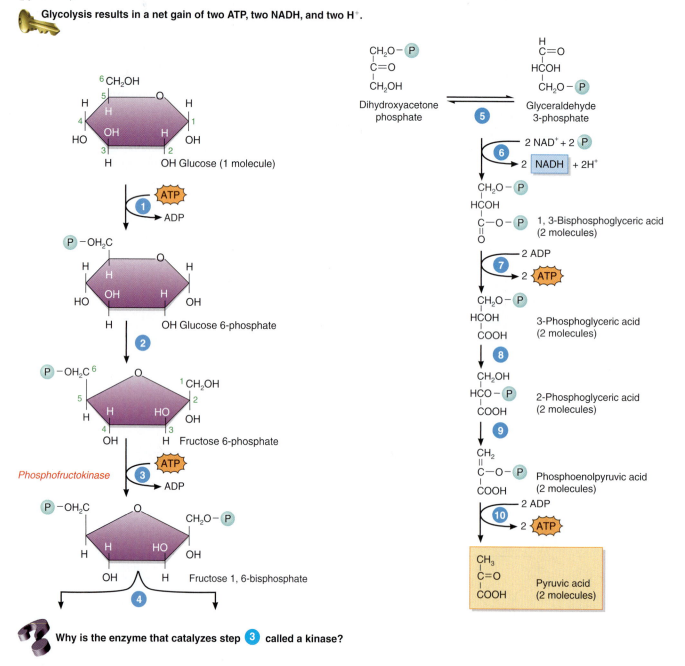

Why is the enzyme that catalyzes step ③ called a kinase?

---

*The carboxyl groups (—COOH) of intermediates in glycolysis and in the citric acid cycle are mostly ionized at the pH of body fluids to —COO⁻. The suffix "-ic acid" indicates the non-ionized form, whereas the ending "-ate" indicates the ionized form. Although the "-ate" names are more correct, we will use the "acid" names because these terms are more familiar.

The Fate of Pyruvic Acid

The fate of pyruvic acid produced during glycolysis depends on the availability of oxygen (Figure 25.5). If oxygen is scarce (anaerobic conditions)—for example, in skeletal muscle fibers during strenuous exercise—then pyruvic acid is reduced via an anaerobic pathway by the addition of two hydrogen atoms to form lactic acid (lactate):

$$2 \text{ Pyruvic acid} + 2 \text{ NADH} + 2 \text{ H}^+ \longrightarrow 2 \text{ Lactic acid} + 2 \text{ NAD}^+$$

Oxidized   Reduced

This reaction regenerates the NAD^+ that was used in the oxidation of glyceraldehyde 3-phosphate (see step **6** in Figure 25.4) and thus allows glycolysis to continue. As lactic acid is produced, it rapidly diffuses out of the cell and enters the blood. Hepatocytes remove lactic acid from the blood and convert it back to pyruvic acid.

When oxygen is plentiful (aerobic conditions), most cells convert pyruvic acid to acetyl coenzyme A. This molecule links glycolysis, which occurs in the cytosol, with the Krebs cycle, which occurs in the matrix of mitochondria. Pyruvic acid enters the mitochondrial matrix with the help of a special transporter protein. Because they lack mitochondria, red blood cells can produce ATP through glycolysis only.

Formation of Acetyl Coenzyme A

Each step in the oxidation of glucose requires a different enzyme, and often a coenzyme as well. The coenzyme used at this point in cellular respiration is **coenzyme A (CoA),** which is derived from pantothenic acid, a B vitamin. During the transitional step between glycolysis and the Krebs cycle, pyruvic acid is prepared for entrance into the cycle. The enzyme *pyruvate dehydrogenase,* which is located exclusively in the mitochondrial matrix, converts pyruvate to a two-carbon fragment by removing a molecule of carbon dioxide (Figure 25.5). The loss of a molecule of CO_2 by a substance is called **decarboxylation** (dē-kar-bok′-si-LĀ-shun). This is the first reaction in cellular respiration that releases CO_2. During this reaction, pyruvic acid is also oxidized. Each pyruvic acid loses two hydrogen atoms in the form of one hydride ion (H^-) plus one hydrogen ion (H^+). The coenzyme NAD^+ is reduced as it picks up the H^- from pyruvic acid; the H^+ is released into the mitochondrial matrix. The reduction of NAD^+ to $NADH + H^+$ is indicated in Figure 25.5 by the curved arrow entering and then leaving the reaction. Recall that the oxidation of one glucose molecule produces two molecules of pyruvic acid, so for each molecule of glucose, two molecules of carbon dioxide are lost and two $NADH + H^+$ are produced. The two-carbon fragment, called an **acetyl group,** attaches to coenzyme A, and the whole complex is called **acetyl coenzyme A (acetyl CoA).**

The Krebs Cycle

Once the pyruvic acid has undergone decarboxylation and the remaining acetyl group has attached to CoA, the resulting compound (acetyl CoA) is ready to enter the Krebs cycle (Figure 25.6). The **Krebs cycle**—named for the biochemist Hans Krebs, who described these reactions in the 1930s—is also known as the **citric acid cycle,** for the first molecule formed when an

Figure 25.5 Fate of pyruvic acid.

🔑 **When oxygen is plentiful, pyruvic acid enters mitochondria, is converted to acetyl coenzyme A, and enters the Krebs cycle (aerobic pathway). When oxygen is scarce, most pyruvic acid is converted to lactic acid via an anaerobic pathway.**

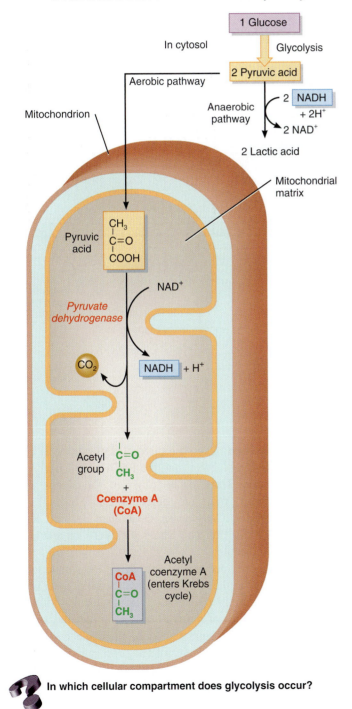

In which cellular compartment does glycolysis occur?

Figure 25.6 **After formation of acetyl coenzyme A, the next stage of cellular respiration is the Krebs cycle.**

Reactions of the Krebs cycle occur in the matrix of mitochondria.

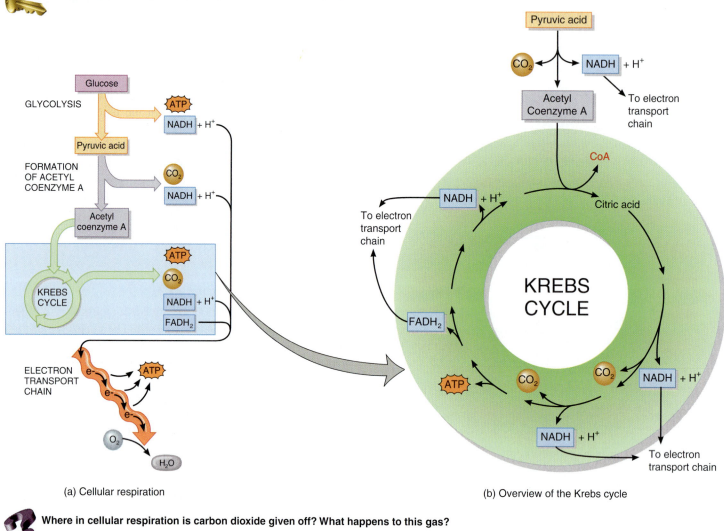

(a) Cellular respiration

(b) Overview of the Krebs cycle

Where in cellular respiration is carbon dioxide given off? What happens to this gas?

acetyl group joins the cycle. The reactions occur in the matrix of mitochondria and consist of a series of oxidation–reduction reactions and decarboxylation reactions that release CO_2. In the Krebs cycle, the oxidation–reduction reactions transfer chemical energy, in the form of electrons, to two coenzymes—NAD^+ and FAD. The pyruvic acid derivatives are oxidized, whereas the coenzymes are reduced. In addition, one step generates ATP. Figure 25.7 shows the reactions of the Krebs cycle in more detail.

The reduced coenzymes ($NADH$ and $FADH_2$) are the most important outcome of the Krebs cycle because they contain the energy originally stored in glucose and then in pyruvic acid. Overall, for every acetyl CoA that enters the Krebs cycle, three $NADH$, three H^+, and one $FADH_2$ are produced by oxidation–reduction reactions, and one molecule of ATP is generated by substrate-level phosphorylation (see Figure 25.6).

In the electron transport chain, the three $NADH$ + three H^+ will later yield nine ATP molecules, and the $FADH_2$, will later yield two ATP molecules. Thus, each "turn" of the Krebs cycle eventually generates 12 molecules of ATP. Because each glucose provides two acetyl CoA, glucose catabolism via the Krebs cycle and the electron transport chain yields 24 molecules of ATP per glucose molecule.

Liberation of CO_2 occurs as pyruvic acid is converted to acetyl CoA and during the two decarboxylation reactions of the Krebs cycle (see Figure 25.6). Because each molecule of glucose generates two molecules of pyruvic acid, six molecules of CO_2 are liberated from each original glucose molecule catabolized along this pathway. The molecules of CO_2 diffuse out of the mitochondria, through the cytosol and plasma membrane, and then into the blood. Blood transports the CO_2 to the lungs, where it eventually is exhaled.

Figure 25.7 **The eight reactions of the Krebs cycle.** ① *Entry of the acetyl group.* The chemical bond that attaches the acetyl group to coenzyme A (CoA) breaks, and the two-carbon acetyl group attaches to a four-carbon molecule of oxaloacetic acid to form a six-carbon molecule called citric acid. CoA is free to combine with another acetyl group from pyruvic acid and repeat the process. ② *Isomerization.* Citric acid undergoes isomerization to isocitric acid, which has the same molecular formula as citrate. Notice, however, that the hydroxyl group (–OH) is attached to a different carbon. ③ *Oxidative decarboxylation.* Isocitric acid is oxidized and loses a molecule of CO_2, forming alpha-ketoglutaric acid. The H^- from the oxidation is passed on to NAD^+, which is reduced to $NADH + H^+$. ④ *Oxidative decarboxylation.* Alpha-ketoglutaric acid is oxidized, loses a molecule of CO_2, and picks up CoA to form succinyl CoA. ⑤ *Substrate-level phosphorylation.* CoA is displaced by a phosphate group, which is then transferred to guanosine diphosphate (GDP) to form guanosine triphosphate (GTP). GTP can donate a phosphate group to ADP to form ATP. ⑥ *Dehydrogenation.* Succinic acid is oxidized to fumaric acid as two of its hydrogen atoms are transferred to the coenzyme flavin adenine nucleotide (FAD), which is reduced to $FADH_2$. ⑦ *Hydration.* Fumaric acid is converted to malic acid by the addition of a molecule of water. ⑧ *Dehydrogenation.* In the final step in the cycle, malic acid is oxidized to re-form oxaloacetic acid. Two hydrogen atoms are removed and one is transferred to NAD^+, which is reduced to $NADH + H^+$. The regenerated oxaloacetic acid can combine with another molecule of acetyl CoA, beginning a new cycle.

🔑 The three main results of the Krebs cycle are the production of reduced coenzymes ($NADH + H^+$ and $FADH_2$), which contain stored energy; the generation of GTP, a high-energy compound that is used to produce ATP; and the formation of CO_2, which is transported to the lungs and exhaled.

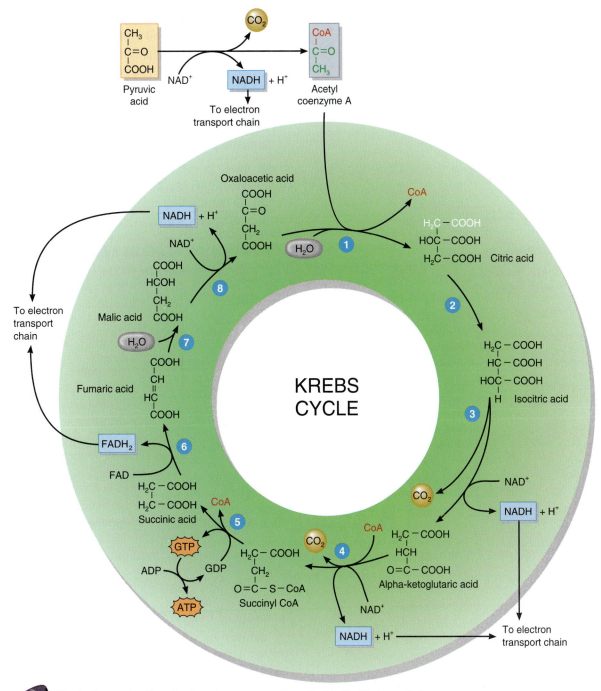

❓ Why is the production of reduced coenzymes important in the Krebs cycle?

The Electron Transport Chain

The **electron transport chain** is a series of **electron carriers** that are integral membrane proteins in the inner mitochondrial membrane. This membrane is folded into cristae that increase its surface area, thus accommodating thousands of copies of the transport chain in each mitochondrion. Each carrier in the chain is reduced as it picks up electrons and is oxidized as it gives up electrons. As electrons pass through the chain, exergonic reactions release small amounts of energy in a series of steps, and this energy is used to form ATP. In aerobic cellular respiration, the final electron acceptor of the chain is oxygen. Because this mechanism of ATP generation links chemical reactions (the passage of electrons along the transport chain) with the pumping of hydrogen ions, it is called **chemiosmosis** (kem′-ē-oz-MŌ-sis; *chemi-* = chemical; *-osmosis* = pushing). Briefly, chemiosmosis works as follows (Figure 25.8):

1 Energy from $NADH + H^+$ passes along the electron transport chain and is used to pump H^+ (protons) from the matrix of the mitochondrion into the space between the inner and outer mitochondrial membranes.

2 A high concentration of H^+ accumulates between the inner and outer mitochondrial membranes.

3 ATP synthesis then occurs as hydrogen ions flow back into

Figure 25.8   Chemiosmosis.

🔑 In chemiosmosis, ATP is produced when hydrogen ions diffuse back into the mitochondrial matrix.

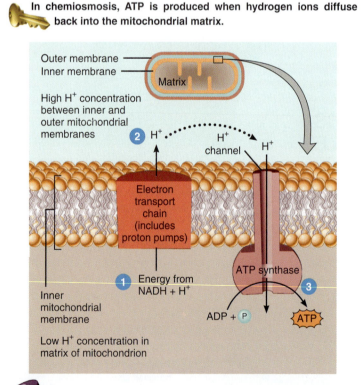

❓ What is the energy source that powers the proton pumps?

the mitochondrial matrix through a special type of H^+ channel in the inner membrane.

ELECTRON CARRIERS   Several types of molecules and atoms serve as electron carriers:

• **Flavin mononucleotide (FMN),** like FAD (flavin adenine dinucleotide), is a flavoprotein derived from riboflavin (vitamin B_2).

• **Cytochromes** (SĪ-tō-krōmz) are proteins with an iron-containing group (heme) capable of existing alternately in a reduced form (Fe^{2+}) and an oxidized form (Fe^{3+}). The several cytochromes involved in the electron transport chain are cytochromes *b* (cyt *b*), cytochromes c_1 (cyt c_1), cytochrome *c* (cyt *c*), cytochrome *a* (cyt *a*), and cytochrome a_3 (cyt a_3).

• **Iron-sulfur (Fe-S) centers** contain either two or four iron atoms bound to sulfur atoms that form an electron transfer center within a protein.

• **Copper (Cu) atoms** bound to two proteins in the chain also participate in electron transfer.

• **Coenzyme Q,** symbolized **Q,** is a nonprotein, low-molecular-weight carrier that is mobile in the lipid bilayer of the inner membrane.

STEPS IN ELECTRON TRANSPORT AND CHEMIOSMOTIC ATP GENERATION   Within the inner mitochondrial membrane, the carriers of the electron transport chain are clustered into three complexes, each of which acts as a **proton pump** that expels H^+ from the mitochondrial matrix and helps create an electrochemical gradient of H^+. Each of the three proton pumps is a complex that transports electrons and pumps H^+, as shown in Figure 25.9. Notice that oxygen is used to help form water in step **3**. This is the only point in aerobic cellular respiration where O_2 is consumed. Cyanide is a deadly poison because it binds to the cytochrome oxidase complex and blocks this last step in electron transport.

The pumping of H^+ produces both a concentration gradient of protons and an electrical gradient. The buildup of H^+ on one side of the inner mitochondrial membrane makes that side positively charged compared with the other side. The resulting electrochemical gradient has potential energy, called the *proton motive force.* Included in the inner mitochondrial membrane are proton channels that allow H^+ to flow back across the membrane, driven by the proton motive force. As H^+ flow back, they generate ATP because the H^+ channels also include an enzyme called **ATP synthase.** The enzyme uses the proton motive force to synthesize ATP from ADP and Ⓟ. This process of chemiosmosis is responsible for most of the ATP produced during cellular respiration.

Summary of Cellular Respiration

The various electron transfers in the electron transport chain generate either 32 or 34 ATP molecules from each molecule of

Figure 25.9 The actions of the three proton pumps and ATP synthase in the inner membrane of mitochondria. Each pump is a complex of three or more electron carriers. **①** The first proton pump is the *NADH dehydrogenase complex,* which contains flavin mononucleotide (FMN) and five or more Fe-S centers. $NADH + H^+$ is oxidized to NAD^+, and FMN is reduced to $FMNH_2$, which in turn is oxidized as it passes electrons to the iron-sulfur centers. Q, which is mobile in the membrane, shuttles electrons to the second pump complex. **②** The second proton pump is the *cytochrome b-c_1 complex,* which contains cytochromes and an iron-sulfur center.

Electrons are passed successively from Q to cyt *b*, to Fe-S, to cyt c_1. The mobile shuttle that passes electrons from the second pump complex to the third is cytochrome *c* (cyt *c*). **③** The third proton pump is the *cytochrome oxidase complex,* which contains cytochromes *a* and a_3 and two copper atoms. Electrons pass from cyt *c*, to Cu, to cyt *a*, and finally to cyt a_3. Cyt a_3 passes its electrons to one-half of a molecule of oxygen (O_2), which becomes negatively charged and then picks up two H^+ from the surrounding medium to form H_2O.

As the three proton pumps pass electrons from one carrier to the next, they also move protons (H^+) from the matrix into the space between the inner and outer mitochondrial membranes. As protons flow back into the mitochondrial matrix through the H^+ channel in ATP synthase, ATP is synthesized.

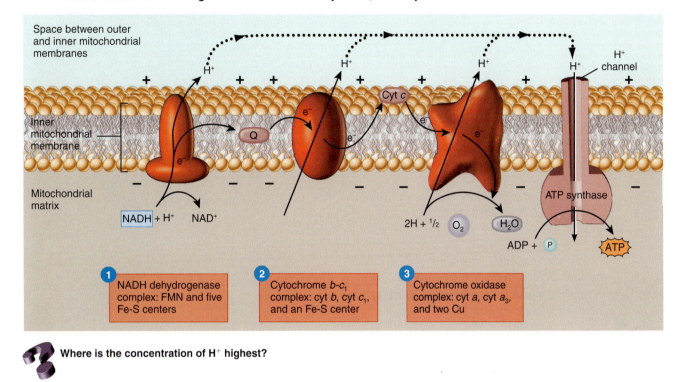

Where is the concentration of H^+ highest?

glucose that is oxidized: either 28 or 30* from the ten molecules of $NADH + H^+$ and two from each of the two molecules of $FADH_2$ (four total). Thus, during cellular respiration, 36 or 38 ATPs can be generated from one molecule of glucose. Note that two of those ATPs come from substrate-level phosphorylation in glycolysis, and two come from substrate-level phosphorylation in the Krebs cycle. The overall reaction is:

$$C_6H_{12}O_6 + 6\ O_2 + 36\ \text{or}\ 38\ \text{ADPs} + 36\ \text{or}\ 38\ \text{\textcircled{P}} \longrightarrow$$
Glucose   Oxygen

$$6\ CO_2 + 6\ H_2O + 36\ \text{or}\ 38\ \text{ATPs}$$
Carbon   Water
dioxide

Table 25.1 summarizes the ATP yield during cellular respiration. A schematic depiction of the principal reactions of cellular respiration is presented in Figure 25.10. The actual ATP yield may be lower than 36 or 38 ATPs per glucose. One uncertainty is the exact number of H^+ that must be pumped out to generate one ATP during chemiosmosis. Moreover, the ATP generated in mitochondria must be transported out of these organelles into the cytosol for use elsewhere in a cell. Exporting ATP in exchange for the inward movement of ADP that is formed from metabolic reactions in the cytosol uses up part of the proton motive force.

---

* The two NADH produced in the cytosol during glycolysis cannot enter mitochondria. Instead, they donate their electrons to one of two transfer molecules, known as the malate shuttle and the glycerol phosphate shuttle. In cells of the liver, kidneys, and heart, use of the malate shuttle results in three ATPs synthesized for each NADH. In other body cells, such as skeletal muscle fibers and neurons, use of the glycerol phosphate shuttle results in two ATPs synthesized for each NADH.

Table 25.1	Summary of ATP Produced in Cellular Respiration
Source	**ATP Yield Per Glucose Molecule (Process)**
Glycolysis	
Oxidation of one glucose molecule to two pyruvic acid molecules	2 ATPs (substrate-level phosphorylation)
Production of 2 NADH + H$^+$	4 or 6 ATPs (oxidative phosphorylation in electron transport chain)
Formation of Two Molecules of Acetyl Coenzyme A	
2 NADH + 2 H$^+$	6 ATPs (oxidative phosphorylation in electron transport chain)
Krebs Cycle and Electron Transport Chain	
Oxidation of succinyl CoA to succinic acid	2 GTPs that are converted to 2 ATPs (substrate-level phosphorylation)
Production of 6 NADH + 6 H$^+$	18 ATPs (oxidative phosphorylation in electron transport chain)
Production of 2 FADH$_2$	4 ATPs (oxidative phosphorylation in electron transport chain)
Total:	36 or 38 ATPs per glucose molecule (theoretical maximum)

Glycolysis, the Krebs cycle, and especially the electron transport chain provide all the ATP for cellular activities. Because the Krebs cycle and electron transport chain are aerobic processes, cells cannot carry on their activities for long if oxygen is lacking.

Glucose Anabolism

Even though most of the glucose in the body is catabolized to generate ATP, glucose may take part in or be formed via several anabolic reactions. One is the synthesis of glycogen; another is the synthesis of new glucose molecules from some of the products of protein and lipid breakdown.

Glucose Storage: Glycogenesis

If glucose is not needed immediately for ATP production, it combines with many other molecules of glucose to form **glycogen,** a polysaccharide that is the only stored form of carbohydrate in our bodies. The hormone insulin, from pancreatic beta cells, stimulates hepatocytes and skeletal muscle cells to carry out **glycogenesis,** the synthesis of glycogen (Figure 25.11). The body can store about 500 g (about 1.1 lb) of glycogen, roughly 75% in skeletal muscle fibers and the rest in liver cells. During glycogenesis, glucose is first phosphorylated to glucose 6-phosphate by hexokinase. Glucose 6-phosphate is converted to glucose 1-phosphate, then to uridine diphosphate glucose, and finally to glycogen.

Figure 25.10 **Summary of the principal reactions of cellular respiration.** ETC = electron transport chain and chemiosmosis.

Except for glycolysis, which occurs in the cytosol, all other reactions of cellular respiration occur within mitochondria.

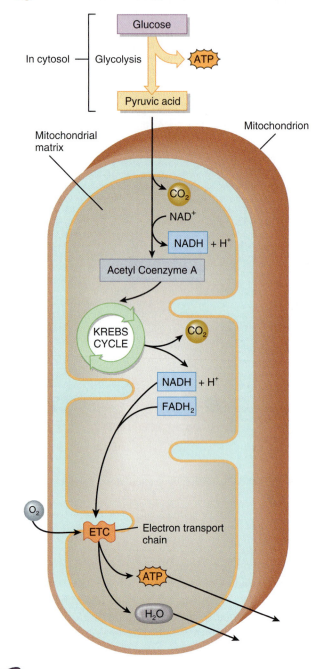

 How many molecules of O$_2$ are used and how many molecules of CO$_2$ are produced during the complete oxidation of one glucose molecule?

Figure 25.11 Glycogenesis and glycogenolysis.

The glycogenesis pathway converts glucose into glycogen, whereas the glycogenolysis pathway breaks down glycogen into glucose.

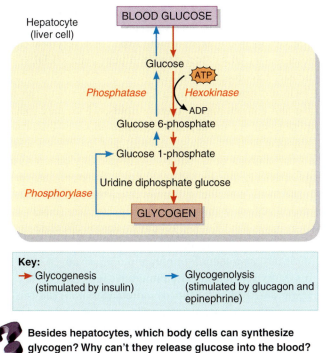

Key:
→ Glycogenesis (stimulated by insulin)
→ Glycogenolysis (stimulated by glucagon and epinephrine)

Besides hepatocytes, which body cells can synthesize glycogen? Why can't they release glucose into the blood?

Glucose Release: Glycogenolysis

When body activities require ATP, glycogen stored in hepatocytes is broken down into glucose and released into the blood to be transported to cells, where it will be catabolized. The process of splitting glycogen into its glucose subunits is called **glycogenolysis** (glī′-kō-je-NOL-e-sis). (Note: Do not confuse *glycogenolysis,* which is the breakdown of glycogen to glucose, with *glycolysis,* the ten reactions that convert glucose to pyruvic acid.)

Glycogenolysis is not a simple reversal of the steps of glycogenesis (Figure 25.11). It begins by splitting glucose molecules off the branched glycogen molecule via phosphorylation to form glucose 1-phosphate. Phosphorylase, the enzyme that catalyzes this reaction, is activated by glucagon from pancreatic alpha cells and epinephrine from the adrenal medulla. Glucose 1-phosphate is then converted to glucose 6-phosphate and finally to glucose, which leaves hepatocytes via glucose transporters (GluT) in the plasma membrane. Phosphorylated glucose molecules cannot ride aboard the GluT transporters, however, and **phosphatase,** the enzyme that converts glucose 6-phosphate into glucose, is absent in skeletal muscle cells. Thus, hepatocytes, which have phosphatase, can release glucose derived from glycogen to the bloodstream, whereas skeletal muscle cells cannot. In skeletal muscle cells, glycogen is broken down into glucose 1-phosphate, which is then catabolized for ATP production via glycolysis and the Krebs cycle. However, the lactic acid pro-duced by glycolysis in muscle cells can be converted to glucose in the liver. In this way, muscle glycogen can be an indirect source of blood glucose.

Carbohydrate Loading

The amount of glycogen stored in the liver and skeletal muscles varies and can be completely exhausted during long-term athletic endeavors. Thus, many marathon runners and other endurance athletes follow a precise exercise and dietary regimen that includes eating large amounts of complex carbohydrates, such as pasta and potatoes, in the three days before an event. This practice, called **carbohydrate loading,** helps maximize the amount of glycogen available for ATP production in muscles. For athletic events lasting more than an hour, carbohydrate loading has been shown to increase an athlete's endurance. ■

Formation of Glucose from Proteins and Fats: Gluconeogenesis

When your liver runs low on glycogen, it is time to eat. If you don't, your body starts catabolizing triglycerides (fats) and proteins. Actually, the body normally catabolizes some of its triglycerides and proteins, but large-scale triglyceride and protein catabolism does not happen unless you are starving, eating very few carbohydrates, or suffering from an endocrine disorder.

The glycerol part of triglycerides, lactic acid, and certain amino acids can be converted in the liver to glucose (Figure 25.12). The process by which glucose is formed from these non-carbohydrate sources is called **gluconeogenesis** (gloo′-kō-nē′-ō-JEN-e-sis; *neo* = new). An easy way to distinguish this term from glycogenesis or glycogenolysis is to remember that in this

Figure 25.12 Gluconeogenesis, the conversion of noncarbohydrate molecules (amino acids, lactic acid, and glycerol) into glucose.

About 60% of the amino acids in the body can be used for gluconeogenesis.

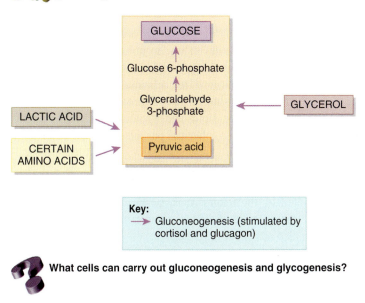

Key:
→ Gluconeogenesis (stimulated by cortisol and glucagon)

What cells can carry out gluconeogenesis and glycogenesis?

case glucose is not converted back from glycogen, but is instead *newly formed.* About 60% of the amino acids in the body can be used for gluconeogenesis. Amino acids such as alanine, cysteine, glycine, serine, and threonine, and lactic acid are converted to pyruvic acid, which then may be synthesized into glucose or enter the Krebs cycle. Glycerol may be converted into glyceraldehyde 3-phosphate, which may form pyruvic acid or be used to synthesize glucose.

Gluconeogenesis is stimulated by cortisol, the main glucocorticoid hormone of the adrenal cortex, and by glucagon from the pancreas. In addition, cortisol stimulates the breakdown of proteins into amino acids, thus expanding the pool of amino acids available for gluconeogenesis. Thyroid hormones (thyroxine and triiodothyronine) also mobilize proteins and may mobilize triglycerides from adipose tissue, thereby making glycerol available for gluconeogenesis.

▶ CHECKPOINT

5. How can glucose move into or out of body cells?

6. What happens during glycolysis?

7. How is acetyl coenzyme A formed?

8. Outline the principal events and outcomes of the Krebs cycle.

9. What happens in the electron transport chain and why is this process called chemiosomosis?

10. Which reactions produce ATP during the complete oxidation of a molecule of glucose?

11. Under what circumstances do glycogenesis and glycogenolysis occur?

12. What is gluconeogenesis, and why is it important?

LIPID METABOLISM

▶ OBJECTIVES

• **Describe the lipoproteins that transport lipids in the blood.**
• **Describe the fate, metabolism, and functions of lipids.**

Transport of Lipids by Lipoproteins

Most lipids, such as triglycerides, are nonpolar and therefore very hydrophobic molecules. They do not dissolve in water. To be transported in watery blood, such molecules first must be made more water soluble. This chemical transformation is accomplished by combining them with proteins produced by the liver and intestine. The lipid and protein combinations thus formed are **lipoproteins,** spherical particles with an outer shell of proteins, phospholipids, and cholesterol molecules surrounding an inner core of triglycerides and other lipids (Figure 25.13). The proteins in the outer shell are called **apoproteins (apo)** and are designated by the letters A, B, C, D, and E plus a number. Besides helping to solubilize the lipoprotein in body fluids, each apoprotein also has specific functions.

There are several types of lipoproteins, each having different functions, but all essentially are transport vehicles. They provide

delivery and pickup services so that lipids can be available when cells need them or be removed from circulation when they are not needed. Lipoproteins are categorized and named mainly according to their density, which varies with the ratio of lipids (which have a low density) to proteins (which have a high density). From largest and lightest to smallest and heaviest, the four major classes of lipoproteins are chylomicrons, very low-density lipoproteins (VLDLs), low-density lipoproteins (LDLs), and high-density lipoproteins (HDLs).

Chylomicrons form in mucosal epithelial cells of the small intestine and contain mainly dietary lipids, which they transport to adipose tissue for storage. They contain about 1–2% proteins, 85% triglycerides, 7% phospholipids, and 6–7% cholesterol, plus a small amount of fat-soluble vitamins. Chylomicrons enter lacteals of intestinal villi and are carried by lymph into venous blood and then into the systemic circulation. Their presence gives blood plasma a milky appearance, but they remain in the blood for only a few minutes. As chylomicrons circulate through the capillaries of adipose tissue, one of their apoproteins, **apo C-2,** activates *endothelial lipoprotein lipase,* an enzyme that removes fatty acids from chylomicron triglycerides. The free fatty acids are then taken up by adipocytes for synthesis and storage of triglycerides and by muscle cells for ATP production. Hepatocytes remove chylomicron remnants from the blood via receptor-mediated endocytosis, in which another chylomicron apoprotein, **apo E,** is the docking protein.

Very low-density lipoproteins (VLDLs) form in hepatocytes and contain mainly *endogenous* (made in the body) lipids.

Figure 25.13   A lipoprotein.   Shown here is a VLDL.

🔑 **A single layer of amphipathic phospholipids, cholesterol, and proteins surrounds a core of nonpolar lipids.**

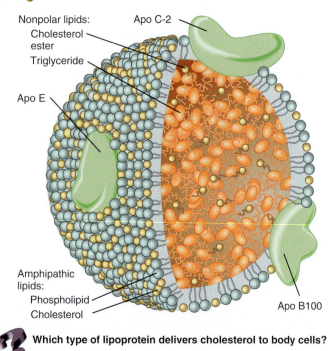

Nonpolar lipids:
Cholesterol ester
Triglyceride

Apo E

Amphipathic lipids:
Phospholipid
Cholesterol

Apo C-2

Apo B100

Which type of lipoprotein delivers cholesterol to body cells?

VLDLs contain about 10% proteins, 50% triglycerides, 20% phospholipids, and 20% cholesterol. VLDLs transport triglycerides synthesized in hepatocytes to adipocytes for storage. Like chylomicrons, they lose triglycerides as their apo C-2 activates endothelial lipoprotein lipase, and the resulting fatty acids are taken up by adipocytes for storage and by muscle cells for ATP production. As they deposit some of their triglycerides in adipose cells, VLDLs are converted to LDLs.

Low-density lipoproteins (LDLs) contain 25% proteins, 5% triglycerides, 20% phospholipids, and 50% cholesterol. They carry about 75% of the total cholesterol in blood and deliver it to cells throughout the body for use in repair of cell membranes and synthesis of steroid hormones and bile salts. The only apoprotein LDLs contain is **apo B100,** which is the docking protein that binds to LDL receptors for receptor-mediated endocytosis of the LDL into a body cell. Within the cell, the LDL is broken down, and the cholesterol is released to serve the cell's needs. Once a cell has sufficient cholesterol for its activities, a negative feedback system inhibits the cell's synthesis of new LDL receptors.

When present in excessive numbers, LDLs also deposit cholesterol in and around smooth muscle fibers in arteries, forming fatty plaques that increase the risk of coronary artery disease (see page 447). For this reason, the cholesterol in LDLs, called LDL-cholesterol, is known as "bad" cholesterol. Because some people have too few LDL receptors, their body cells remove LDL from the blood less efficiently; as a result, their plasma LDL level is abnormally high, and they are more likely to develop fatty plaques. Furthermore, eating a high-fat diet increases the production of VLDLs, which elevates the LDL level and increases the formation of fatty plaques.

High-density lipoproteins (HDLs), which contain 40–45% proteins, 5–10% triglycerides, 30% phospholipids, and 20% cholesterol, remove excess cholesterol from body cells and the blood and transport it to the liver for elimination. Because HDLs prevent accumulation of cholesterol in the blood, a high HDL level is associated with decreased risk of coronary artery disease. For this reason, HDL-cholesterol is known as "good" cholesterol.

Sources and Significance of Blood Cholesterol

There are two sources of cholesterol in the body. Some is present in foods (eggs, dairy products, organ meats, beef, pork, and processed luncheon meats), but most is synthesized by hepatocytes. Fatty foods that don't contain any cholesterol at all can still dramatically increase blood cholesterol level in two ways. First, a high intake of dietary fats stimulates reabsorption of cholesterol-containing bile back into the blood, so less cholesterol is lost in the feces. Second, when saturated fats are broken down in the body, hepatocytes use some of the breakdown products to make cholesterol.

A lipid profile test usually measures total cholesterol (TC), HDL-cholesterol, and triglycerides (VLDLs). LDL-cholesterol

then is calculated by using the following formula: LDL-cholesterol = TC − HDL-cholesterol − (triglycerides/5). In the United States, blood cholesterol is usually measured in milligrams per deciliter (mg/dL); a deciliter is 0.1 liter or 100mL. For adults, desirable levels of blood cholesterol are total cholesterol under 200mg/dL, LDL-cholesterol under 130mg/dL, and HDL-cholesterol over 40 mg/dL. Normally, triglycerides are in the range of 10–190mg/dL.

As total cholesterol level increases, the risk of coronary artery disease begins to rise. When total cholesterol is above 200 mg/dL (5.2 mmol/liter), the risk of a heart attack doubles with every 50 mg/dL (1.3mmol/liter) increase in total cholesterol. Total cholesterol of 200–23 mg/dL and LDL of 130–159 mg/dL are borderline-high, whereas total cholesterol above 239 mg/dL and LDL above 159 mg/dL are classified as high blood cholesterol. The ratio of total cholesterol to HDL-cholesterol predicts the risk of developing coronary artery disease. For example, a person with a total cholesterol of 180 mg/dL and HDL of 60 mg/dL has a risk ratio of 3. Ratios above 4 are considered undesirable; the higher the ratio, the greater the risk of developing coronary artery disease.

Among the therapies used to reduce blood cholesterol level are exercise, diet, and drugs. Regular physical activity at aerobic and nearly aerobic levels raises HDL level. Dietary changes are aimed at reducing the intake of total fat, saturated fats, and cholesterol. Drugs used to treat high blood cholesterol levels include cholestyramine (Questran) and colestipol (Colestid), which promote excretion of bile in the feces; nicotinic acid (Liponicin); and the "statin" drugs—atorvastatin (Lipitor), lovastatin (Mevacor), and simvastatin (Zocor), which block the key enzyme (HMG-CoA reductase) needed for cholesterol synthesis.

The Fate of Lipids

Lipids, like carbohydrates, may be oxidized to produce ATP. If the body has no immediate need to use lipids in this way, they are stored in adipose tissue (fat depots) throughout the body and in the liver. A few lipids are used as structural molecules or to synthesize other essential substances. Some examples include phospholipids, which are constituents of plasma membranes; lipoproteins, which are used to transport cholesterol throughout the body; thromboplastin, which is needed for blood clotting; and myelin sheaths, which speed up nerve impulse conduction. Two **essential fatty acids** that the body cannot synthesize are linoleic acid and linolenic acid. Dietary sources include vegetable oils and leafy vegetables. The various functions of lipids in the body may be reviewed in Table 2.7 on page 45.

Triglyceride Storage

A major function of adipose tissue is to remove triglycerides from chylomicrons and VLDLs and store them until they are needed for ATP production in other parts of the body. Triglycerides stored in adipose tissue constitute 98% of all body energy reserves. They are stored more readily than is glycogen, in part because triglycerides are hydrophobic and do not exert osmotic pressure on cell membranes. Adipose tissue also insulates and protects various parts of the body. Adipocytes in the subcutaneous layer contain about 50% of the stored triglycerides. Other adipose tissues account for the other half: about 12% around the kidneys, 10–15% in the omenta, 15% in genital areas, 5–8% between muscles, and 5% behind the eyes, in the sulci of the heart, and attached to the outside of the large intestine. Triglycerides in adipose tissue are continually broken down and resynthesized. Thus, the triglycerides stored in adipose tissue today are not the same molecules that were present last month because they are continually released from storage, transported in the blood, and redeposited in other adipose tissue cells.

Lipid Catabolism: Lipolysis

Muscle, liver, and adipose tissue routinely oxidize fatty acids derived from triglycerides to produce ATP. First, the triglycerides must be split into glycerol and fatty acids, a process called **lipolysis** (li-POL-i-sis) that is catalyzed by enzymes called **lipases.** Two hormones that enhance triglyceride breakdown into fatty acids and glycerol are epinephrine and norepinephrine, which are released when sympathetic tone increases, as occurs, for example, during exercise. Other lipolytic hormones are cortisol, thyroid hormones, and insulinlike growth factors. By contrast, insulin itself inhibits lipolysis.

The glycerol and fatty acids that result from lipolysis are catabolized via different pathways (Figure 25.14). Glycerol is converted by many cells of the body to glyceraldehyde 3-phosphate, one of the compounds also formed during the catabolism of glucose. If ATP supply in a cell is high, glyceraldehyde 3-phosphate is converted into glucose, an example of gluconeogenesis. If ATP supply in a cell is low, glyceraldehyde 3-phosphate enters the catabolic pathway to pyruvic acid.

Fatty acids are catabolized differently than is glycerol and yield more ATP. The first stage in fatty acid catabolism is a series of reactions, collectively called **beta oxidation,** that occurs in the matrix of mitochondria. Enzymes remove two carbon atoms at a time from the long chain of carbon atoms composing a fatty acid and attach the resulting two-carbon fragment to coenzyme A, forming acetyl CoA. Then, acetyl CoA enters the Krebs cycle (Figure 25.14). A 16-carbon fatty acid such as palmitic acid can yield as many as 129 ATPs upon its complete

Figure 25.14 Pathways of lipid metabolism.   Glycerol may be converted to glyceraldehyde 3-phosphate, which can then be converted to glucose or enter the Krebs cycle for oxidation. Fatty acids undergo beta oxidation and enter the Krebs cycle via acetyl coenzyme A. The synthesis of lipids from glucose or amino acids is called lipogenesis.

Glycerol and fatty acids are catabolized in separate pathways.

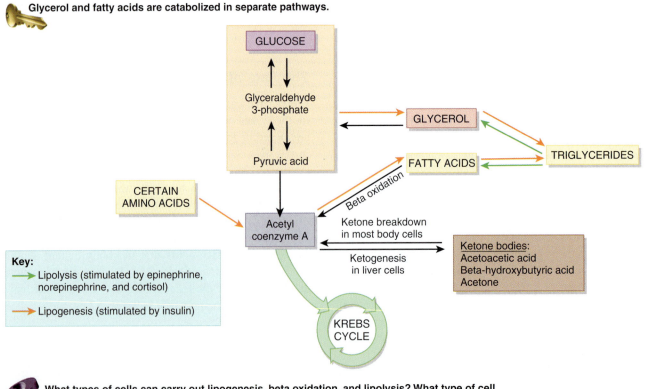

What types of cells can carry out lipogenesis, beta oxidation, and lipolysis? What type of cell can carry out ketogenesis?

oxidation via beta oxidation, the Krebs cycle, and the electron transport chain.

As part of normal fatty acid catabolism, hepatocytes can take two acetyl CoA molecules at a time and condense them to form **acetoacetic acid.** This reaction liberates the bulky CoA portion, which cannot diffuse out of cells. Some acetoacetic acid is converted into **beta-hydroxybutyric acid** and **acetone.** The formation of these three substances, collectively known as **ketone bodies,** is called **ketogenesis** (Figure 25.14). Because ketone bodies freely diffuse through plasma membranes, they leave hepatocytes and enter the bloodstream.

Other cells take up acetoacetic acid and attach its four carbons to two coenzyme A molecules to form two acetyl CoA molecules, which can then enter the Krebs cycle for oxidation. Heart muscle and the cortex (outer part) of the kidneys use acetoacetic acid in preference to glucose for generating ATP. Hepatocytes, which make acetoacetic acid, cannot use it for ATP production because they lack the enzyme that transfers acetoacetic acid back to coenzyme A.

Lipid Anabolism: Lipogenesis

Liver cells and adipose cells can synthesize lipids from glucose or amino acids through **lipogenesis** (Figure 25.14), which is stimulated by insulin. Lipogenesis occurs when individuals consume more calories than are needed to satisfy their ATP needs. Excess dietary carbohydrates, proteins, and fats all have the same fate — they are converted into triglycerides. Certain amino acids can undergo the following reactions: amino acids → acetyl CoA → fatty acids → triglycerides. The use of glucose to form lipids takes place via two pathways: (1) glucose → glyceraldehyde 3-phosphate → glycerol and (2) glucose → glyceraldehyde 3-phosphate → acetyl CoA → fatty acids. The resulting glycerol and fatty acids can undergo anabolic reactions to become stored triglycerides, or they can go through a series of anabolic reactions to produce other lipids such as lipoproteins, phospholipids, and cholesterol.

Ketosis

The level of ketone bodies in the blood normally is very low because other tissues use them for ATP production as fast as they are generated from the breakdown of fatty acids in the liver. During periods of excessive beta oxidation, however, the production of ketone bodies exceeds their uptake and use by body cells. This might occur after a meal rich in triglycerides, or during fasting or starvation, because few carbohydrates are available for catabolism. Excessive beta oxidation may also occur in poorly controlled or untreated diabetes mellitus for two reasons: (1) Because adequate glucose cannot get into cells, triglycerides are used for ATP production, and (2) because insulin normally inhibits lipolysis, a lack of insulin accelerates the pace of lipolysis. When the concentration of ketone bodies in the blood rises

above normal — a condition called **ketosis** — the ketone bodies, most of which are acids, must be buffered. If too many accumulate, they decrease the concentration of buffers such as bicarbonate ions, and blood pH falls. Thus, extreme or prolonged ketosis can lead to **acidosis (ketoacidosis),** an abnormally low blood pH. When a diabetic becomes seriously insulin deficient, one of the telltale signs is the sweet smell on the breath from the ketone body acetone. ■

▶ **CHECKPOINT**

13. What are the functions of the apoproteins in lipoproteins?

14. Which lipoprotein particles contain "good" and "bad" cholesterol, and why are these terms used?

15. Where are triglycerides stored in the body?

16. Explain the principal events of the catabolism of glycerol and fatty acids.

17. What are ketone bodies? What is ketosis?

18. Define lipogenesis and explain its importance.

PROTEIN METABOLISM

▶ **OBJECTIVE**

• **Describe the fate, metabolism, and functions of proteins.**

During digestion, proteins are broken down into amino acids. Unlike carbohydrates and triglycerides, which are stored, proteins are not warehoused for future use. Instead, amino acids are either oxidized to produce ATP or used to synthesize new proteins for body growth and repair. Excess dietary amino acids are not excreted in the urine or feces but instead are converted into glucose (gluconeogenesis) or triglycerides (lipogenesis).

The Fate of Proteins

The active transport of amino acids into body cells is stimulated by insulinlike growth factors (IGFs) and insulin. Almost immediately after digestion, amino acids are reassembled into proteins. Many proteins function as enzymes; others are involved in transportation (hemoglobin) or serve as antibodies, clotting chemicals (fibrinogen), hormones (insulin), or contractile elements in muscle fibers (actin and myosin). Several proteins serve as structural components of the body (collagen, elastin, and keratin). The various functions of proteins in the body may be reviewed in Table 2.8 on page 48.

Protein Catabolism

A certain amount of protein catabolism occurs in the body each day, stimulated mainly by cortisol from the adrenal cortex. Proteins from worn-out cells (such as red blood cells) are broken down into amino acids. Some amino acids are converted into other amino acids, peptide bonds are re-formed, and new proteins are synthesized as part of the recycling process. Hepatocytes convert some amino acids to fatty acids, ketone bodies, or glucose. Also, cells throughout the body oxidize a small amount of amino acids to generate ATP via the Krebs cycle and the electron transport chain. However, before amino acids can be oxidized, they must first be converted to molecules that are part of the Krebs cycle or can enter the Krebs cycle, such as acetyl CoA

(Figure 25.15). Before amino acids can enter the Krebs cycle, their amino group (NH_2) must first be removed—a process called **deamination** (dē-am′-i-NĀ-shun). Deamination occurs in hepatocytes and produces ammonia (NH_3). The liver cells then convert the highly toxic ammonia to urea, a relatively harmless substance that is excreted in the urine. The conversion of amino acids into glucose (gluconeogenesis) may be reviewed in Figure 25.12; the conversion of amino acids into fatty acids (lipogenesis) or ketone bodies (ketogenesis) is shown in Figure 25.14.

Figure 25.15 Various points at which amino acids (shown in yellow boxes) enter the Krebs cycle for oxidation.

Before amino acids can be catabolized, they must first be converted to various substances that can enter the Krebs cycle.

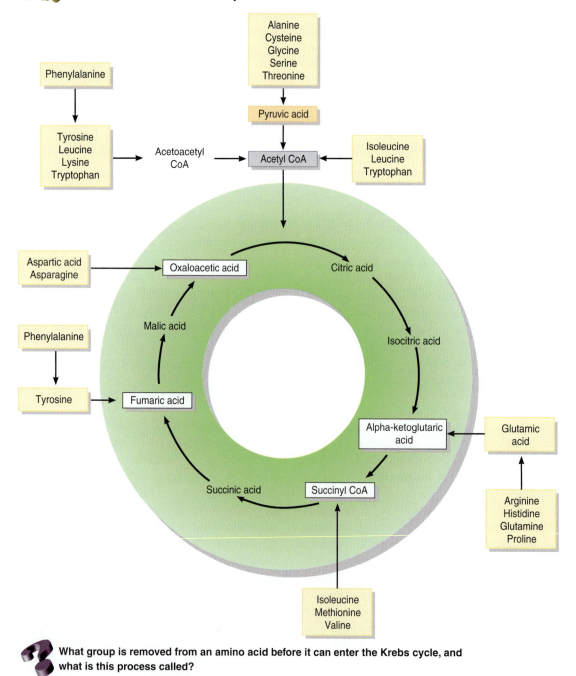

What group is removed from an amino acid before it can enter the Krebs cycle, and what is this process called?

Protein Anabolism

Protein anabolism, the formation of peptide bonds between amino acids to produce new proteins, is carried out on the ribosomes of almost every cell in the body, directed by the cells' DNA and RNA (see Figure 3.30 on page 89). Insulinlike growth factors, thyroid hormones (T3 and T4), insulin, estrogen, and testosterone stimulate protein synthesis. Because proteins are a main component of most cell structures, adequate dietary protein is especially essential during the growth years, during pregnancy, and when tissue has been damaged by disease or injury. Once dietary intake of protein is adequate, eating more protein will not increase bone or muscle mass; only a regular program of forceful, weight-bearing muscular activity accomplishes that goal.

Of the 20 amino acids in the human body, ten are **essential amino acids:** They must be present in the diet because they cannot be synthesized in the body in adequate amounts. Humans are unable to synthesize seven amino acids (isoleucine, leucine, methionine, phenylalanine, threonine, tryptophan, and valine) and synthesize two others (arginine and histidine) in inadequate amounts, especially in childhood. Because plants synthesize all amino acids, humans can obtain the essential ones by eating plants or animals that eat plants. **Nonessential amino acids** can be synthesized by body cells. They are formed by **transamination,** the transfer of an amino group from an amino acid to pyruvic acid or to an acid in the Krebs cycle. Once the appropriate essential and nonessential amino acids are present in cells, protein synthesis occurs rapidly.

Phenylketonuria

Phenylketonuria (fen′-il-kē′-tō-NOO-rē-a) or **PKU** is a genetic error of protein metabolism characterized by elevated blood levels of the amino acid phenylalanine. Most children with phenylketonuria have a mutation in the gene that codes for the enzyme phenylalanine hydroxylase, the enzyme needed to convert phenylalanine into the amino acid tyrosine, which can enter the Krebs cycle (see Figure 25.15). Because the enzyme is deficient, phenylalanine cannot be metabolized, and what is not used in protein synthesis builds up in the blood. If untreated, the disorder causes vomiting, rashes, seizures, growth deficiency, and severe mental retardation. Newborns are screened for PKU, and mental retardation can be prevented by restricting the child to a diet that supplies only the amount of phenylalanine needed for growth, although learning disabilities may still ensue. Because the artificial sweetener aspartame (NutraSweet) contains phenylalanine, its consumption must be restricted in children with PKU. ■

▶ **CHECKPOINT**

19. What is deamination and why does it occur?

20. What are the possible fates of the amino acids from protein catabolism?

21. How are essential and nonessential amino acids different?

KEY MOLECULES AT METABOLIC CROSSROADS

▶ **OBJECTIVE**

• **Identify the key molecules in metabolism, and describe the reactions and the products they may form.**

Although there are thousands of different chemicals in cells, three molecules—glucose 6-phosphate, pyruvic acid, and acetyl coenzyme A—play pivotal roles in metabolism (Figure 25.16). These molecules stand at "metabolic crossroads;" the reactions that occur depend on the nutritional or activity status of the individual. Reactions ❶ through ❼ in Figure 25.16 occur in the cytosol, reactions ❽ and ❾ occur inside mitochondria, and reactions indicated by ❿ occur on smooth endoplasmic reticulum.

The Role of Glucose 6-Phosphate

Shortly after glucose enters a body cell, a kinase converts it to **glucose 6-phosphate.** Four possible fates await glucose 6-phosphate (see Figure 25.14):

❶ *Synthesis of glycogen.* When glucose is abundant in the bloodstream, such as just after a meal, considerable glucose 6-phosphate is used to synthesize glycogen, the storage form of carbohydrate in animals. Subsequent breakdown of glycogen into glucose 6-phosphate occurs through a slightly different series of reactions. Synthesis and breakdown of glycogen occur mainly in skeletal muscle fibers and hepatocytes.

❷ *Release of glucose into the bloodstream.* If the enzyme glucose 6-phosphatase is present and active, glucose 6-phosphate can be dephosphorylated to glucose. Once glucose is released from the phosphate group, it can leave the cell and enter the bloodstream. Hepatocytes are the main cells that can provide glucose to the bloodstream in this way.

❸ *Synthesis of nucleic acids.* Glucose 6-phosphate is the precursor used by cells throughout the body to make ribose 5-phosphate, a five-carbon sugar that is needed for synthesis of RNA (ribonucleic acid) and DNA (deoxyribonucleic acid). The same sequence of reactions also produces NADPH. This molecule is a hydrogen and electron donor in certain reduction reactions, such as synthesis of fatty acids and steroid hormones.

❹ *Glycolysis.* Some ATP is produced anaerobically via glycolysis, in which glucose 6-phosphate is converted to pyruvic acid, another key molecule in metabolism. Most body cells carry out glycolysis.

The Role of Pyruvic Acid

Each 6-carbon molecule of glucose that undergoes glycolysis yields two 3-carbon molecules of **pyruvic acid.** This molecule

Figure 25.16 Summary of the roles of the key molecules in metabolic pathways. Double-headed arrows indicate that reactions between two molecules may proceed in either direction, if the appropriate enzymes are present and the conditions are favorable; single-headed arrows signify the presence of an irreversible step.

🔑 **Three molecules—glucose 6-phosphate, pyruvic acid, and acetyl coenzyme A—stand at "metabolic crossroads." They can undergo different reactions depending on one's nutritional or activity status.**

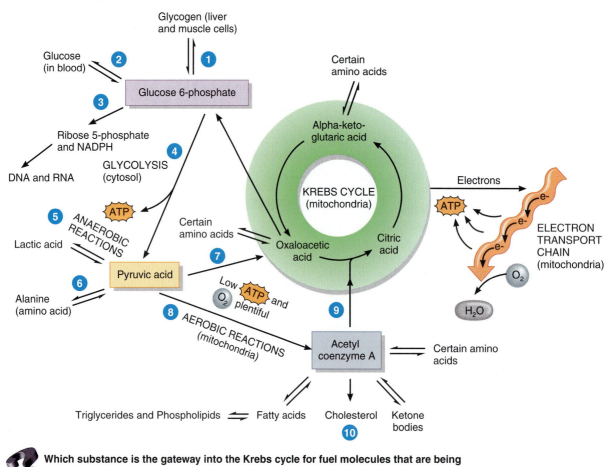

Which substance is the gateway into the Krebs cycle for fuel molecules that are being oxidized to generate ATP?

stands at a metabolic crossroads: Given enough oxygen, the aerobic (oxygen-consuming) reactions of cellular respiration can proceed; if oxygen is in short supply, anaerobic reactions can occur (Figure 25.16):

5 *Production of lactic acid.* When oxygen is in short supply in a tissue, as in actively contracting skeletal or cardiac muscle, some pyruvic acid is changed to lactic acid. The lactic acid then diffuses into the bloodstream and is taken up by hepatocytes, which eventually convert it back to pyruvic acid.

6 *Production of alanine.* One link between carbohydrate and protein metabolism occurs at pyruvic acid. Through transamination, an amino group ($-NH_3$) can either be added to pyruvic acid (a carbohydrate) to produce the amino acid alanine, or be removed from alanine to generate pyruvic acid.

7 *Gluconeogenesis.* Pyruvic acid and certain amino acids also can be converted to oxaloacetic acid, one of the Krebs cycle intermediates, which, in turn, can be used to form glucose 6-phosphate. This sequence of gluconeogenesis reactions bypasses certain one-way reactions of glycolysis.

The Role of Acetyl Coenzyme A

8 When the ATP level in a cell is low but oxygen is plentiful, most pyruvic acid streams toward ATP-producing reactions—the Krebs cycle and electron transport chain—via conversion to **acetyl coenzyme A.**

9 *Entry into the Krebs cycle.* Acetyl CoA is the vehicle for two-carbon acetyl groups to enter the Krebs cycle. Oxidative

Krebs cycle reactions convert acetyl CoA to CO_2 and produce reduced coenzymes (NADH and $FADH_2$) that transfer electrons into the electron transport chain. Oxidative reactions in the electron transport chain, in turn, generate ATP. Most fuel molecules that will be oxidized to generate ATP—glucose, fatty acids, and ketone bodies—are first converted to acetyl CoA.

10 *Synthesis of lipids.* Acetyl CoA also can be used for synthesis of certain lipids, including fatty acids, ketone bodies, and cholesterol. Because pyruvic acid can be converted to acetyl CoA, carbohydrates can be turned into triglycerides; this is a metabolic pathway for storage of some excess carbohydrate calories as fat. Mammals, including humans, cannot reconvert acetyl CoA to pyruvic acid, however, so fatty acids cannot be used to generate glucose or other carbohydrate molecules.

Table 25.2 summarizes carbohydrate, lipid, and protein metabolism.

► **CHECKPOINT**

22. What are the possible fates of glucose 6-phosphate, pyruvic acid, and acetyl coenzyme A in a cell?

METABOLIC ADAPTATIONS

► **OBJECTIVE**

• **Compare metabolism during the absorptive and postabsorptive states.**

Regulation of metabolic reactions depends both on the chemical environment within body cells, such as the levels of ATP and oxygen, and on signals from the nervous and endocrine systems. Some aspects of metabolism depend on how much time has passed since the last meal. During the **absorptive state,** ingested nutrients are entering the bloodstream, and glucose is readily available for ATP production. During the **postabsorptive state,** absorption of nutrients from the GI tract is complete, and energy needs must be met by fuels already in the body. A typical meal requires about 4 hours for complete absorption, and given three meals a day, the absorptive state exists for about 12 hours each day. Assuming no between-meal snacks, the other 12 hours— typically late morning, late afternoon, and most of the night— are spent in the postabsorptive state.

Because the nervous system and red blood cells continue to depend on glucose for ATP production during the postabsorptive

Table 25.2	Summary of Metabolism
Process	**Comments**
Carbohydrates	
Glucose catabolism	Complete oxidation of glucose (cellular respiration) is the chief source of ATP in cells and consists of glycolysis, Krebs cycle, and electron transport chain. Complete oxidation of 1 molecule of glucose yields a maximum of 36 or 38 molecules of ATP.
Glycolysis	Conversion of glucose into pyruvic acid results in the production of some ATP. Reactions do not require oxygen (anaerobic cellular respiration).
Krebs cycle	Cycle includes series of oxidation—reduction reactions in which coenzymes (AND$^+$ and FAD) pick up hydrogen ions and hydride ions from oxidized organic acids, and some ATP is produced. CO_2 and H_2O are byproducts. Reactions are aerobic.
Electron transport chain	Third set of reactions in glucose catabolism is another series of oxidation—reduction reactions, in which electrons are passed from one carrier to the next, and most of the ATP is produced. Reactions require oxygen (aerobic cellular respiration).
Glucose anabolism	Some glucose is converted into glycogen (glycogenesis) for storage if not needed immediately for ATP production. Glycogen can be reconverted to glucose (glycogenolysis). The conversion of amino acids, glycerol, and lactic acid into glucose is called gluconeogenesis.
Lipids	
Triglyceride catabolism	Triglycerides are broken down into glycerol and fatty acids. Glycerol may be converted into glucose (gluconeogenesis) or catabolized via glycolysis. Fatty acids are catabolized via beta oxidation into acetyl coenzyme A that can enter the Krebs cycle for ATP production or be converted into ketone bodies (ketogenesis).
Triglyceride anabolism	The synthesis of triglycerides from glucose and fatty acids is called lipogenesis. Triglycerides are stored in adipose tissue.
Proteins	
Protein catabolism	Amino acids are oxidized via the Krebs cycle after deamination. Ammonia resulting from deamination is converted into urea in the liver, passed into blood, and excreted in urine. Amino acids may be converted into glucose (gluconeogenesis), fatty acids, or ketone bodies.
Protein anabolism	Protein synthesis is directed by DNA and utilizes cells' RNA and ribosomes.

state, maintaining a steady blood glucose level is critical during this period. Hormones are the major regulators of metabolism in each state. The effects of insulin dominate in the absorptive state, whereas several other hormones regulate metabolism in the postabsorptive state. During fasting and starvation, many body cells increasingly use ketone bodies for ATP production.

Metabolism During the Absorptive State

Soon after a meal, nutrients start to enter the blood. Recall that ingested food reaches the bloodstream mainly as glucose, amino acids, and triglycerides (in chylomicrons). Two metabolic hall-

marks of the absorptive state are the oxidation of glucose for ATP production, which occurs in most body cells, and the storage of excess fuel molecules for future between-meal use, which occurs mainly in hepatocytes, adipocytes, and skeletal muscle fibers.

Absorptive State Reactions

The following reactions dominate during the absorptive state (Figure 25.17):

1. About 50% of the glucose absorbed from a typical meal is oxidized by cells throughout the body to produce ATP via glycolysis, the Krebs cycle, and the electron transport chain.

Figure 25.17   Principal metabolic pathways during the absorptive state.

During the absorptive state, most body cells produce ATP by oxidizing glucose to CO_2 and H_2O.

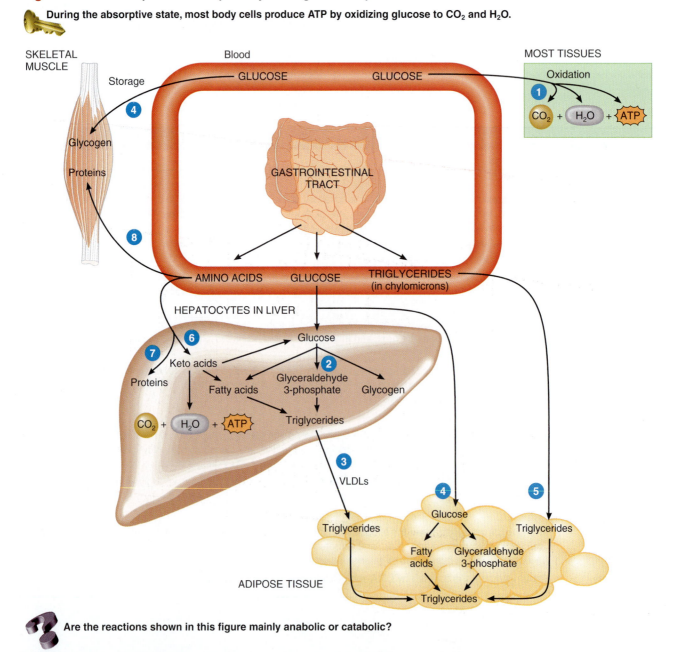

Are the reactions shown in this figure mainly anabolic or catabolic?

2 Most glucose that enters hepatocytes is converted to glycogen. Small amounts may be used for synthesis of fatty acids and glyceraldehyde 3-phosphate.

3 Some fatty acids and triglycerides synthesized in the liver remain there, but hepatocytes package most into VLDLs, which carry lipids to adipose tissue for storage.

4 Adipocytes also take up glucose not picked up by the liver and convert it into triglycerides for storage. Overall, about 40% of the glucose absorbed from a meal is converted to triglycerides, and about 10% is stored as glycogen in skeletal muscles and hepatocytes.

5 Most dietary lipids (mainly triglycerides and fatty acids) are stored in adipose tissue; only a small portion is used for synthesis reactions. Adipocytes obtain the lipids from chylomicrons, from VLDLs, and from their own synthesis reactions.

6 Many absorbed amino acids that enter hepatocytes are deaminated to keto acids, which can either enter the Krebs cycle for ATP production or be used to synthesize glucose or fatty acids.

7 Some amino acids that enter hepatocytes are used to synthesize proteins (for example, plasma proteins).

8 Amino acids not taken up by hepatocytes are used in other body cells (such as muscle cells) for synthesis of proteins or regulatory chemicals such as hormones or enzymes.

Regulation of Metabolism During the Absorptive State

Soon after a meal, glucose-dependent insulinotropic peptide (GIP), plus the rising blood levels of glucose and certain amino acids, stimulate pancreatic beta cells to release insulin. In general, insulin increases the activity of enzymes needed for anabolism and the synthesis of storage molecules; at the same time it decreases the activity of enzymes needed for catabolic or breakdown reactions. Insulin promotes the entry of glucose and amino acids into cells of many tissues, and it stimulates the phosphorylation of glucose in hepatocytes and the conversion of glucose 6-phosphate to glycogen in both liver and muscle cells. In liver and adipose tissue, insulin enhances the synthesis of triglycerides, and in cells throughout the body insulin stimulates protein synthesis. (See page 614 to review the effects of insulin.) Insulinlike growth factors and the thyroid hormones (T3 and T4) also stimulate protein synthesis. Table 25.3 summarizes the hormonal regulation of metabolism in the absorptive state.

Metabolism During the Postabsorptive State

About 4 hours after the last meal, absorption of nutrients from the small intestine is nearly complete, and blood glucose level starts to fall because glucose continues to leave the bloodstream and enter body cells while none is being absorbed from the GI tract. Thus, the main metabolic challenge during the postabsorptive state is to maintain the normal blood glucose level of 70–110 mg/100 mL (3.9–6.1 mmol/liter). Homeostasis of blood

Table 25.3	Hormonal Regulation of Metabolism in the Absorptive State	
Process	**Location(s)**	**Main Stimulating Hormone(s)**
Facilitated diffusion of glucose into cells	Most cells.	Insulin.*
Active transport of amino acids into cells	Most cells.	Insulin.
Glycogenesis (glycogen synthesis)	Hepatocytes and muscle fibers.	Insulin.
Protein synthesis	All body cells.	Insulin, thyroid hormones, and insulinlike growth factors.
Lipogenesis (triglyceride synthesis)	Adipose cells and hepatocytes.	Insulin.

*Facilitated diffusion of glucose into hepatocytes (liver cells) and neurons is always "turned on" and does not require insulin.

glucose concentration is especially important for the nervous system and for red blood cells for the following reasons:

- The dominant fuel molecule for ATP production in the nervous system is glucose, because fatty acids are unable to pass the blood–brain barrier.

- Red blood cells derive all their ATP from glycolysis of glucose because they have no mitochondria, and thus the Krebs cycle and the electron transport chain are not available to them.

Postabsorptive State Reactions

During the postabsorptive state, both *glucose production* and *glucose conservation* help maintain blood glucose level: Hepatocytes produce glucose molecules and export them into the blood, and other body cells switch from glucose to alternative fuels for ATP production to conserve scarce glucose. The major reactions of the postabsorptive state that produce glucose are the following (Figure 25.18):

1 *Breakdown of liver glycogen.* During fasting, a major source of blood glucose is liver glycogen, which can provide about a 4-hour supply of glucose. Liver glycogen is continually being formed and broken down as needed.

2 *Lipolysis.* Glycerol, produced by breakdown of triglycerides in adipose tissue, is also used to form glucose.

3 *Gluconeogenesis using lactic acid.* During exercise, skeletal muscle tissue breaks down stored glycogen (see step **9**) and produces some ATP anaerobically via glycolysis. Some of the pyruvic acid that results is converted to acetyl CoA, and some is converted to lactic acid, which diffuses into the blood. In the liver, lactic acid can be used for gluconeogenesis, and the resulting glucose is released into the blood.

4 *Gluconeogenesis using amino acids.* Modest breakdown of proteins in skeletal muscle and other tissues releases large amounts of amino acids, which then can be converted to glucose by gluconeogenesis in the liver.

Despite these ways to produce glucose, blood glucose level cannot be maintained for very long without further metabolic changes. Thus, a major adjustment must be made during the postabsorptive state to produce ATP while conserving glucose. The following reactions produce ATP without using glucose:

5 *Oxidation of fatty acids.* The fatty acids released by lipolysis of triglycerides cannot be used for glucose production because acetyl CoA cannot readily be converted to pyruvic acid. But most cells can oxidize the fatty acids directly, feed them into the Krebs cycle as acetyl CoA, and produce ATP through the electron transport chain.

6 *Oxidation of lactic acid.* Cardiac muscle can produce ATP aerobically from lactic acid.

7 *Oxidation of amino acids.* In hepatocytes, amino acids may be oxidized directly to produce ATP.

8 *Oxidation of ketone bodies.* Hepatocytes also convert fatty acids to ketone bodies, which can be used by the heart, kidneys, and other tissues for ATP production.

9 *Breakdown of muscle glycogen.* Skeletal muscle cells break down glycogen to glucose 6-phosphate, which undergoes glycolysis and provides ATP for muscle contraction.

Regulation of Metabolism During the Postabsorptive State

Both hormones and the sympathetic division of the autonomic nervous system (ANS) regulate metabolism during the postabsorptive state. The hormones that regulate postabsorptive state metabolism sometimes are called anti-insulin hormones because they counter the effects of insulin that dominate during the absorptive state. As blood glucose level declines, the secretion of insulin falls and the release of anti-insulin hormones rises.

Figure 25.18 **Principal metabolic pathways during the postabsorptive state.**

The principal function of postabsorptive state reactions is to maintain a normal blood glucose level.

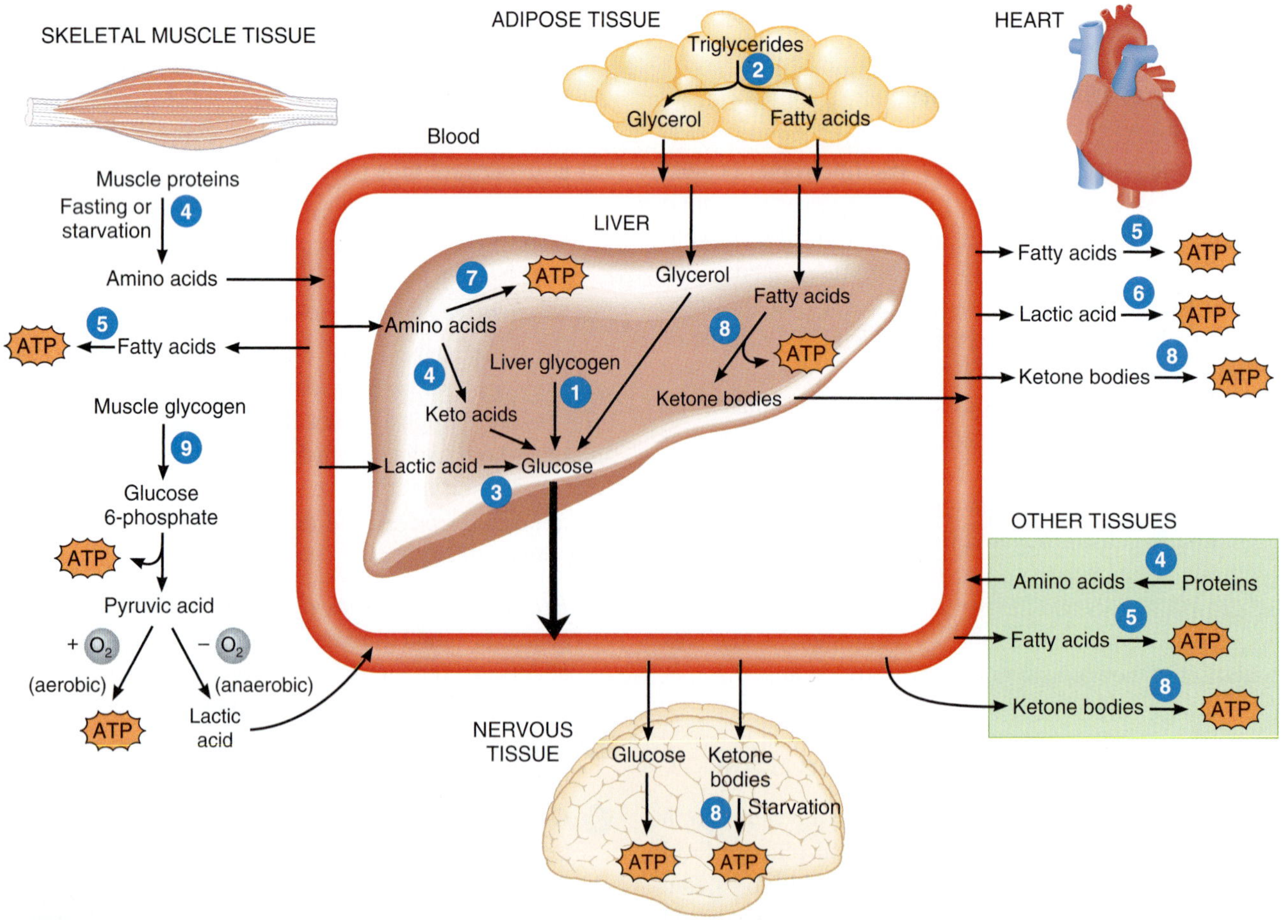

What processes directly elevate blood glucose during this state, and where does each occur?

When blood glucose concentration starts to drop, the pancreatic alpha cells release glucagon at a faster rate, and the beta cells secrete insulin more slowly. The primary target tissue of glucagon is the liver; the major effect is increased release of glucose into the bloodstream due to gluconeogenesis and glycogenolysis.

Low blood glucose also activates the sympathetic branch of the ANS. Glucose-sensitive neurons in the hypothalamus detect low blood glucose and increase sympathetic output. As a result, sympathetic nerve endings release the neurotransmitter norepinephrine, and the adrenal medulla releases two catecholamine hormones—epinephrine and norepinephrine—into the bloodstream. Like glucagon, epinephrine stimulates glycogen breakdown. Epinephrine and norepinephrine are both potent stimulators of lipolysis. These actions of the catecholamines help to increase glucose and free fatty acid levels in the blood. As a result, muscle uses more fatty acids for ATP production, and more glucose is available to the nervous system. Table 25.4 summarizes the hormonal regulation of metabolism in the postabsorptive state.

Metabolism During Fasting and Starvation

The term **fasting** means going without food for many hours or a few days, whereas **starvation** implies weeks or months of food deprivation or inadequate food intake. People can survive without food for two months or more if they drink enough water to prevent dehydration. Although glycogen stores are depleted within a few hours of beginning a fast, catabolism of stored triglycerides and structural proteins can provide energy for several weeks. The amount of adipose tissue the body contains determines the lifespan possible without food.

During fasting and starvation, nervous tissue and RBCs continue to use glucose for ATP production. There is a ready supply of amino acids for gluconeogenesis because lowered insulin and increased cortisol levels slow the pace of protein synthesis and promote protein catabolism. Most cells in the body, especially skeletal muscle cells because they contain so much protein to begin with, can spare a fair amount of protein before their performance is adversely affected. During the first few days of fasting, protein catabolism outpaces protein synthesis by about 75 grams daily as some of the "old" amino acids are being deaminated and used for gluconeogenesis and "new" (that is, dietary) amino acids are lacking.

By the second day of a fast, blood glucose level has stabilized at about 65 mg/100 mL (3.6 mmol/liter), whereas the level of fatty acids in plasma has risen fourfold. Lipolysis of triglycerides in adipose tissue releases glycerol, which is used for gluconeogenesis, and fatty acids. The fatty acids diffuse into muscle fibers and other body cells, where they are used to produce acetyl-CoA, which enters the Krebs cycle. ATP then is synthesized as oxidation proceeds via the Krebs cycle and the electron transport chain.

The most dramatic metabolic change that occurs with fasting and starvation is the increase in the formation of ketone bodies by hepatocytes. During fasting, only small amounts of glucose undergo glycolysis to pyruvic acid, which in turn can be converted to oxaloacetic acid. Acetyl-CoA enters the Krebs cycle by combining with oxaloacetic acid (see Figure 25.16); when oxaloacetic acid is scarce due to fasting, only some of the available acetyl-CoA can enter the Krebs cycle. Surplus acetyl-CoA is used for ketogenesis, mainly in hepatocytes. Ketone body production thus increases as catabolism of fatty acids rises. Lipid-soluble ketone bodies can diffuse through plasma membranes and across the blood–brain barrier and be used as an alternate fuel for ATP production, especially by cardiac and skeletal muscle fibers and neurons. Normally, only a trace of ketone bodies (0.01 mmol/liter) are present in the blood, and thus they are a negligible fuel source. After two days of fasting, however, the level of ketones is 100–300 times higher and supplies roughly a third of the brain's fuel for ATP production. By 40 days of starvation, ketones provide up to two-thirds of the brain's energy needs. The presence of ketones actually reduces the use of glucose for ATP production, which in turn decreases the demand for gluconeogenesis and slows the catabolism of muscle proteins later in starvation to about 20 grams daily.

Table 25.4	Hormonal Regulation of Metabolism in the Postabsorptive State	
Process	**Location(s)**	**Main Stimulating Hormone(s)**
Glycogenolysis (glycogen breakdown)	Hepatocytes and skeletal muscle fibers.	Glucagon and epinephrine.
Lipolysis (triglyceride breakdown)	Adipocytes.	Epinephrine, norepinephrine, cortisol, insulinlike growth factors, thyroid hormones, and others.
Protein breakdown	Most body cells, but especially skeletal muscle fibers.	Cortisol.
Gluconeogenesis (synthesis of glucose from noncarbohydrates)	Hepatocytes and kidney cortex cells.	Glucagon and cortisol.

► **CHECKPOINT**

23. What are the roles of insulin, glucagon, epinephrine, insulinlike growth factors, thyroxine, cortisol, estrogen, and testosterone in regulation of metabolism?

24. Why is ketogenesis more significant during fasting or starvation than during normal absorptive and postabsorptive states?

HEAT AND ENERGY BALANCE

▶ **O B J E C T I V E S**

- **Define basal metabolic rate (BMR), and explain several factors that affect it.**
- **Describe the factors that influence body heat production.**
- **Explain how normal body temperature is maintained by negative feedback loops involving the hypothalamic thermostat.**

The body produces more or less heat depending on the rates of metabolic reactions. Homeostasis of body temperature can be maintained only if the rate of heat loss from the body equals the rate of heat production by metabolism. Thus, it is important to understand the ways in which heat can be lost, gained, or conserved. **Heat** is a form of energy that can be measured as **temperature** and expressed in units called calories. A **calorie (cal)** is defined as the amount of heat required to raise the temperature of 1 gram of water 1°C. Because the calorie is a relatively small unit, the **kilocalorie (kcal)** or **Calorie (Cal),** spelled with an uppercase C, is often used to measure the body's metabolic rate and to express the energy content of foods. A kilocalorie equals 1000 calories. Thus, when we say that a particular food item contains 500 Calories, we are actually referring to kilocalories.

Metabolic Rate

The overall rate at which metabolic reactions use energy is termed the **metabolic rate.** Some of the energy is used to produce ATP, and some is released as heat. Because many factors affect metabolic rate, it is measured under standard conditions, with the body in a quiet, resting, and fasting condition called the **basal state.** The measurement obtained is the **basal metabolic rate (BMR).** The most common way to determine BMR is by measuring the amount of oxygen used per kilocalorie of food metabolized. When the body uses 1 liter of oxygen to oxidize a typical dietary mixture of triglycerides, carbohydrates, and proteins, about 4.8 Cal of energy is released. BMR is 1200–1800 Cal/day in adults, or about 24 Cal/kg of body mass in adult males and 22 Cal/kg in adult females. The added calories needed to support daily activities, such as digestion and walking, range from 500 Cal for a small, relatively sedentary person to over 300 Cal for a person in training for Olympic-level competitions or mountain climbing.

Body Temperature Homeostasis

Despite wide fluctuations in environmental temperature, homeostatic mechanisms can maintain a normal range for internal body temperature. If the rate of body heat production equals the rate of heat loss, the body maintains a constant core temperature near 37°C (98.6°F). **Core temperature** is the temperature in body structures deep to the skin and subcutaneous layer. **Shell temperature** is the temperature near the body surface—in the skin and subcutaneous layer. Depending on the environmental temperature, shell temperature is 1–6°C lower than core temperature. Too high a core temperature kills by denaturing body proteins, whereas too low a core temperature causes cardiac arrhythmias that result in death.

Heat Production

The production of body heat is proportional to metabolic rate. Several factors affect the metabolic rate and thus the rate of heat production:

- **Exercise.** During strenuous exercise, the metabolic rate may increase to as much as 15 times the basal rate. In well-trained athletes, the rate may increase up to 20 times.
- **Hormones.** Thyroid hormones (thyroxine and triiodothyronine) are the main regulators of BMR, which increases as the blood levels of thyroid hormones rise. The response to changing levels of thyroid hormones is slow, however, taking several days to appear. Thyroid hormones increase BMR in part by stimulating aerobic cellular respiration. As cells use more oxygen to produce ATP, more heat is given off, and body temperature rises. Other hormones have minor effects on BMR. Testosterone, insulin, and human growth hormone can increase the metabolic rate by 5–15%.
- **Nervous system.** During exercise or in a stressful situation, the sympathetic division of the autonomic nervous system is stimulated. Its postganglionic neurons release norepinephrine (NE), and it also stimulates release of the hormones epinephrine and norepinephrine by the adrenal medulla. Both epinephrine and norepinephrine increase the metabolic rate of body cells.
- **Body temperature.** The higher the body temperature, the higher the metabolic rate. Each 1°C rise in core temperature increases the rate of biochemical reactions by about 10%. As a result, metabolic rate may be substantially increased during a fever.
- **Ingestion of food.** The ingestion of food raises the metabolic rate 10–20% due to the energy "costs" of digesting, absorbing, and storing nutrients. This effect, *food-induced thermogenesis,* is greatest after eating a high-protein meal and is less after eating carbohydrates and lipids.
- **Age.** The metabolic rate of a child, in relation to its size, is about double that of an elderly person due to the high rates of reactions related to growth.
- **Other factors.** Other factors that affect metabolic rate are gender (lower in females, except during pregnancy and lactation), climate (lower in tropical regions), sleep (lower), and malnutrition (lower).

Mechanisms of Heat Transfer

Maintaining normal body temperature depends on the ability to lose heat to the environment at the same rate as it is produced by metabolic reactions. Heat can be transferred from the body to its surroundings in four ways: via conduction, convection, radiation, and evaporation.

1. **Conduction** is the heat exchange that occurs between molecules of two materials that are in direct contact with each other. At rest, about 3% of body heat is lost via conduction to solid materials in contact with the body, such as a chair, clothing, and jewelry. Heat can also be gained via conduction—for example, while soaking in a hot tub. Because water conducts heat 20 times more effectively than air, heat loss or heat gain via conduction is much greater when the body is submerged in cold or hot water.

2. **Convection** is the transfer of heat by the movement of a fluid (a gas or a liquid) between areas of different temperature. The contact of air or water with your body results in heat transfer by both conduction and convection. When cool air makes contact with the body, it becomes warmed and therefore less dense and is carried away by convection currents created as the less dense air rises. The faster the air moves—for example, by a breeze or a fan—the faster the rate of convection. At rest, about 15% of body heat is lost to the air via conduction and convection.

3. **Radiation** is the transfer of heat in the form of infrared rays between a warmer object and a cooler one without physical contact. Your body loses heat by radiating more infrared waves than it absorbs from cooler objects. If surrounding objects are warmer than you are, you absorb more heat than you lose by radiation. In a room at 21°C (70°F), about 60% of heat loss occurs via radiation in a resting person.

4. **Evaporation** is the conversion of a liquid to a vapor. Every milliliter of evaporating water takes with it a great deal of heat—about 0.58 Cal/mL. Under typical resting conditions, about 22% of heat loss occurs through evaporation of about 700 mL of water per day—300 mL in exhaled air and 400 mL from the skin surface. Because we are not normally aware of this water loss through the skin and mucous membranes of the mouth and respiratory system, it is termed **insensible water loss.** The rate of evaporation is inversely related to relative humidity, the ratio of the actual amount of moisture in the air to the maximum amount it can hold at a given temperature. The higher the relative humidity, the lower the rate of evaporation. At 100% humidity, heat is gained via condensation of water on the skin surface as fast as heat is lost via evaporation. Evaporation provides the main defense against overheating during exercise. Under extreme conditions, a maximum of about 3 liters of sweat can be produced each hour, removing more than 1700 Calories of heat if all of it evaporates. (Note: Sweat that drips off the body rather than evaporating removes very little heat.)

Hypothalamic Thermostat

The control center that functions as the body's thermostat is a group of neurons in the anterior part of the hypothalamus, the **preoptic area.** This area receives impulses from thermoreceptors in the skin and mucous membranes and in the hypothalamus. Neurons of the preoptic area generate nerve impulses at a higher frequency when blood temperature increases, and at a lower frequency when blood temperature decreases.

Nerve impulses from the preoptic area propagate to two other parts of the hypothalamus known as the **heat-losing center** and the **heat-promoting center,** which, when stimulated by the preoptic area, set into operation a series of responses that lower body temperature and raise body temperature, respectively.

Thermoregulation

If core temperature declines, mechanisms that help conserve heat and increase heat production act via several negative feedback loops to raise the body temperature to normal (Figure 25.19). Thermoreceptors in the skin and hypothalamus send nerve impulses to the preoptic area and the heat-promoting center in the hypothalamus, and to hypothalamic neurosecretory cells that produce thyrotropin-releasing hormone (TRH). In response, the hypothalamus discharges nerve impulses and secretes TRH, which in turn stimulates thyrotrophs in the anterior pituitary gland to release thyroid-stimulating hormone (TSH). Nerve impulses from the hypothalamus and TSH then activate several effectors.

Each effector responds in a way that helps increase core temperature to the normal value:

- Nerve impulses from the heat-promoting center stimulate sympathetic nerves that cause blood vessels of the skin to constrict. Vasoconstriction decreases the flow of warm blood, and thus the transfer of heat, from the internal organs to the skin. Slowing the rate of heat loss allows the internal body temperature to increase as metabolic reactions continue to produce heat.

- Nerve impulses in sympathetic nerves leading to the adrenal medulla stimulate the release of epinephrine and norepinephrine into the blood. The hormones, in turn, bring about an increase in cellular metabolism, which increases heat production.

- The heat-promoting center stimulates parts of the brain that increase muscle tone and hence heat production. As muscle tone increases in one muscle (the agonist), the small contractions stretch muscle spindles in its antagonist, initiating a stretch reflex. The resulting contraction in the antagonist stretches muscle spindles in the agonist, and it too develops a stretch reflex. This repetitive cycle—called **shivering**—greatly increases the rate of heat production. During maximal shivering, body heat production can rise to about four times the basal rate in just a few minutes.

- The thyroid gland responds to TSH by releasing more thyroid hormones into the blood. As increased levels of thyroid hormones slowly increase the metabolic rate, body temperature rises.

If core body temperature rises above normal, a negative feedback loop opposite to the one depicted in Figure 25.19 goes into action. The higher temperature of the blood stimulates thermoreceptors that send nerve impulses to the preoptic area, which, in turn, stimulate the heat-losing center and inhibit the heat-promoting center. Nerve impulses from the heat-losing center

Figure 25.19 Negative feedback mechanisms that conserve heat and increase heat production.

Core temperature is the temperature in body structures deep to the skin and subcutaneous layer; shell temperature is the temperature near the body surface.

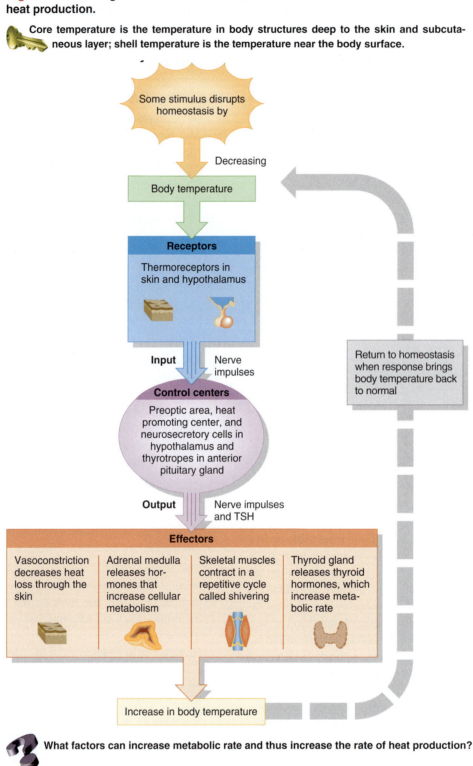

What factors can increase metabolic rate and thus increase the rate of heat production?

cause dilation of blood vessels in the skin. The skin becomes warm, and the excess heat is lost to the environment via radiation and conduction as an increased volume of blood flows from the warmer core of the body into the cooler skin. At the same time, metabolic rate decreases, and shivering does not occur. The high temperature of the blood stimulates sweat glands of the skin via hypothalamic activation of sympathetic nerves. As the water in perspiration evaporates from the surface of the skin, the skin is cooled. All these responses counteract heat-promoting effects and help return body temperature to normal.

Hypothermia

Hypothermia is a lowering of core body temperature to 35°C (95°F) or below. Causes of hypothermia include an overwhelming cold stress (immersion in icy water), metabolic diseases (hypoglycemia, adrenal insufficiency, or hypothyroidism), drugs (alcohol, antidepressants, sedatives, or tranquilizers), burns, and malnutrition. Hypothermia is characterized by the following as core body temperature falls: sensation of cold, shivering, confusion, vasoconstriction, muscle rigidity, bradycardia, acidosis, hypoventilation, hypotension, loss of spontaneous movement, coma, and death (usually caused by cardiac arrhythmias). Because the elderly have reduced metabolic protection against a cold environment coupled with a reduced perception of cold, they are at greater risk for developing hypothermia. ■

Energy Homeostasis and Regulation of Food Intake

Most mature animals and many men and women maintain **energy homeostasis,** the precise matching of energy intake (in food) to energy expenditure over time. When the energy content of food balances the energy used by all the cells of the body, body weight remains constant (unless there is a gain or loss of water). In many people, weight stability persists despite large day-to-day variations in activity and food intake. In the more affluent nations, however, a large fraction of the population is overweight. Easy access to tasty, high-calorie foods and a "couch-potato" lifestyle promote weight gain. Being overweight increases one's risk of dying from a variety of cardiovascular and metabolic disorders, such as diabetes mellitus.

Energy intake depends only on the amount of food consumed (and absorbed), but three components contribute to total energy expenditure.

1. The basal metabolic rate accounts for about 60% of energy expenditure.

2. Physical activity typically adds 30–35% but can be lower in sedentary people. The energy expenditure is partly from voluntary exercise, such as walking, and partly from **nonexercise thermogenesis (NEAT),** the energy costs for maintaining muscle tone, posture while sitting or standing, and involuntary fidgeting movements.

3. **Food-induced thermogenesis,** the heat produced while food is being digested, absorbed, and stored, represents 5–10% of total energy expenditure.

The major site of stored chemical energy in the body is adipose tissue. When energy use exceeds energy input, triglycerides in adipose tissue are catabolized to provide the extra energy, and when energy input exceeds energy expenditure, triglycerides are stored. Over time, the amount of stored triglycerides indicates the excess of energy intake over energy expenditure. Even small differences add up over time. A gain of 20 lb (9 kg) between ages 25 and 55 represents only a tiny imbalance, about 0.3% more energy intake in food than energy expenditure.

Clearly, negative feedback mechanisms are regulating both our energy intake and our energy expenditure. But no sensory receptors exist to monitor one's weight or size. How, then, is food intake regulated? The answer to this question is incomplete, but important advances in understanding regulation of food intake have occurred in the past decade. It depends on many factors, including neural and endocrine signals, levels of certain nutrients in the blood, psychological elements such as stress or depression, signals from the GI tract and the special senses, and neural connections between the hypothalamus and other parts of the brain.

Within the hypothalamus are clusters of neurons that play key roles in regulating food intake. These neurons receive signals that indicate hunger or satiety although they are not precisely organized into "feeding" and "satiety" centers, as was once thought. **Satiety** (sa-TĪ-i-tē) is a feeling of fullness accompanied by lack of desire to eat. Two hypothalamic areas involved in regulation of food intake are the *arcuate nucleus* and the *paraventricular nucleus* (see Figure 14.10 on page 466). In 1994, the first experiments were reported on a mouse gene, named *obese,* that causes overeating and severe obesity when it is mutated. The product of this gene is the hormone **leptin.** In both mice and humans, leptin helps decrease **adiposity,** total body-fat mass. Leptin is synthesized and secreted by adipocytes in proportion to adiposity; as more triglycerides are stored, more leptin is secreted into the bloodstream. Leptin acts on the hypothalamus to inhibit circuits that stimulate eating while also activating circuits that increase energy expenditure. The hormone insulin has a similar, but smaller, effect. Both leptin and insulin pass the blood–brain barrier.

When leptin and insulin levels are *low,* neurons that extend from the arcuate nucleus to the paraventricular nucleus release a neurotransmitter called **neuropeptide Y** that stimulates food intake. Other neurons that extend between the arcuate and paraventricular nuclei release a neurotransmitter called **melanocortin,** which is similar to melanocyte-stimulating hormone (MSH). Leptin stimulates release of melanocortin, which acts to inhibit food intake. Although leptin, neuropeptide Y, and melanocortin are key signaling molecules for maintaining energy homeostasis, several other hormones and neurotransmitters also contribute. An understanding of the brain circuits involved is still far from complete. Other areas of the hypothalamus plus nuclei in the brainstem, limbic system, and cerebral cortex take part.

Achieving energy homeostasis requires regulation of energy intake. Most increases and decreases in food intake are due to changes in meal size rather than changes in number of meals. Many experiments have demonstrated the presence of satiety signals, chemical or neural changes that help terminate eating when "fullness" is attained. For example, an increase in blood glucose level, as occurs after a meal, decreases appetite. Several hormones, such as glucagon, cholecystokinin, estrogens, and epinephrine (acting via beta receptors) act to signal satiety and to increase energy expenditure. Distention of the GI tract, particularly the stomach and duodenum, also contributes to termination of food intake. Other hormones increase appetite and decrease energy expenditure. These include growth hormone-releasing hormone (GHRH), androgens, glucocorticoids, epinephrine (acting via alpha receptors), and progesterone.

Emotional Eating

In addition to keeping us alive, eating serves countless psychological, social, and cultural purposes. We eat to celebrate, punish, comfort, defy, and deny. Eating in response to emotional drives, such as feeling stressed, bored, or tired is called **emotional eating.** Emotional eating is so common that, within limits, it is considered well within the range of normal behavior. Who hasn't at one time or another headed for the refrigerator after a bad day? Problems arise when emotional eating becomes so excessive that it interferes with health. Physical health problems include obesity and associated disorders such as hypertension and heart disease. Psychological health problems include poor self-esteem, an inability to cope effectively with feelings of stress, and in extreme cases, eating disorders, such as anorexia nervosa, bulimia, and obesity.

Eating provides comfort and solace, numbing pain and "feeding the hungry heart." Eating may provide a biochemical "fix" as well. Emotional eaters typically overeat carbohydrate foods (sweets and starches), which may raise brain serotonin levels and lead to feelings of relaxation. Food becomes a way to self-medicate when negative emotions arise. ■

▶ CHECKPOINT

25. Define a kilocalorie (kcal). How is the unit used? How does it relate to a calorie?

26. Distinguish between core temperature and shell temperature.

27. In what ways can a person lose heat to or gain heat from the surroundings? How is it possible for a person to lose heat on a sunny beach when the temperature is 40°C (104°F) and the humidity is 85%?

28. What does the term energy homeostasis mean?

29. How is food intake regulated?

NUTRITION

▶ OBJECTIVES

- **Describe how to select foods to maintain a healthy diet.**
- **Compare the sources, functions, and importance of minerals and vitamins in metabolism.**

Nutrients are chemical substances in food that body cells use for growth, maintenance, and repair. The six main types of nutrients are water, carbohydrates, lipids, proteins, minerals, and vitamins. Water is the nutrient needed in the largest amount—

about 2–3 liters per day. As the most abundant compound in the body, water provides the medium in which most metabolic reactions occur, and it also participates in some reactions (for example, hydrolysis reactions). The important roles of water in the body can be reviewed on pages 38–39. Three organic nutrients—carbohydrates, lipids, and proteins—provide the energy needed for metabolic reactions and serve as building blocks to make body structures. Some minerals and many vitamins are components of the enzyme systems that catalyze metabolic reactions. *Essential nutrients* are specific nutrient molecules that the body cannot make in sufficient quantity to meet its needs and thus must obtain from the diet. Some amino acids, some fatty acids, vitamins, and minerals are essential nutrients.

Next, we discuss some guidelines for healthy eating and the roles of minerals and vitamins in metabolism.

Guidelines for Healthy Eating

Each gram of protein or carbohydrate in food provides about 4 Calories, whereas a gram of fat (lipids) provides about 9 Calories. On a daily basis, many women and older people need about 1600 Calories; children, teenage girls, active women, and most men need about 2200 Calories; and teenage boys and active men need about 2800 Calories.

We do not know with certainty what levels and types of carbohydrate, fat, and protein are optimal in the diet. Different populations around the world eat radically different diets that are adapted to their particular lifestyles. However, many experts recommend the following distribution of calories: 50–60% from carbohydrates, with less than 15% from simple sugars; less than 30% from fats (triglycerides are the main type of dietary fat), with no more than 10% as saturated fats; and about 12–15% from proteins.

The guidelines for healthy eating are to:

- Eat a variety of foods.
- Maintain a healthy weight.
- Choose foods low in fat, saturated fat, and cholesterol.
- Eat plenty of vegetables, fruits, and grain products.
- Use sugars in moderation only.
- Use salt and sodium in moderation only (less than 6 grams daily).
- If you drink alcoholic beverages, do so in moderation (less than 1 ounce of the equivalent of pure alcohol per day).

To help people achieve a good balance of vitamins, minerals, carbohydrates, fats, and proteins in their food, the U.S. Department of Agriculture developed the food guide pyramid (Figure 25.20). The sections of the pyramid indicate how many servings of each of the five major food groups to eat each day. The smallest number of servings corresponds to a 1600 Cal/day diet, whereas the largest number of servings corresponds to a 2800 Cal/day diet. Because they should be consumed in largest

Figure 25.20 **The food guide pyramid.** The smallest number of servings corresponds to 1600 Calories per day, whereas the largest number of servings corresponds to 2800 Calories per day. Each example given equals one serving.

The sections of the pyramid show how many servings of five major food groups to eat each day.

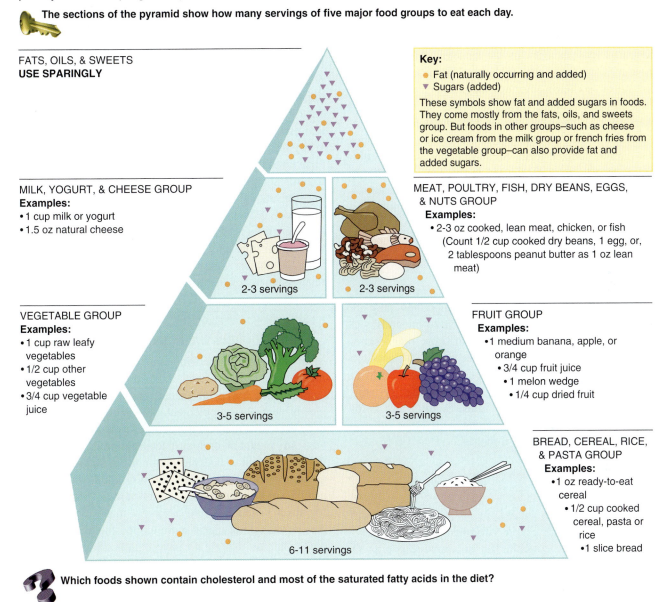

Key:
- Fat (naturally occurring and added)
▼ Sugars (added)

These symbols show fat and added sugars in foods. They come mostly from the fats, oils, and sweets group. But foods in other groups—such as cheese or ice cream from the milk group or french fries from the vegetable group—can also provide fat and added sugars.

FATS, OILS, & SWEETS
USE SPARINGLY

MILK, YOGURT, & CHEESE GROUP
Examples:
- 1 cup milk or yogurt
- 1.5 oz natural cheese

2-3 servings

MEAT, POULTRY, FISH, DRY BEANS, EGGS, & NUTS GROUP
Examples:
- 2-3 oz cooked, lean meat, chicken, or fish (Count 1/2 cup cooked dry beans, 1 egg, or, 2 tablespoons peanut butter as 1 oz lean meat)

2-3 servings

VEGETABLE GROUP
Examples:
- 1 cup raw leafy vegetables
- 1/2 cup other vegetables
- 3/4 cup vegetable juice

3-5 servings

FRUIT GROUP
Examples:
- 1 medium banana, apple, or orange
- 3/4 cup fruit juice
- 1 melon wedge
- 1/4 cup dried fruit

3-5 servings

BREAD, CEREAL, RICE, & PASTA GROUP
Examples:
- 1 oz ready-to-eat cereal
- 1/2 cup cooked cereal, pasta or rice
- 1 slice bread

6-11 servings

Which foods shown contain cholesterol and most of the saturated fatty acids in the diet?

quantity, foods rich in complex carbohydrates—the bread, cereal, rice, and pasta group—form the base of the pyramid. Vegetables and fruits form the next level. The health benefits of eating generous amounts of these foods are well documented. To be eaten in smaller quantities are foods on the next level up— the milk, yogurt, and cheese group and the meat, poultry, fish, dry beans, eggs, and nuts group. These two food groups have higher fat and protein content than the food groups below them. To lower your daily intake of fats, choose low-fat foods from these groups—nonfat milk and yogurt, low-fat cheese, fish, and poultry (remove the skin).

The apex of the pyramid is not a food group but rather a caution to use fats, oils, and sweets sparingly. The food guide pyramid does not distinguish among the different types of fatty acids—saturated, polyunsaturated, and monounsaturated—in dietary fats. However, atherosclerosis and coronary artery disease are prevalent in populations that consume large amounts of saturated fats and cholesterol. Populations living around the Mediterranean Sea, in contrast, have low rates of coronary artery disease despite eating a diet that contains up to 40% of the calories as fats. Most of their fat comes from olive oil, which is rich in monounsaturated fatty acids and has no cholesterol. Canola oil, avocados, nuts, and peanut oil are also rich in monounsaturated fatty acids.

Minerals

Minerals are inorganic elements that occur naturally in the Earth's crust. In the body they appear in combination with each other, in combination with organic compounds, or as ions in solution. Minerals constitute about 4% of total body mass and are concentrated most heavily in the skeleton. Minerals with known functions in the body include calcium, phosphorus, potassium, sulfur, sodium, chloride, magnesium, iron, iodide, manganese, copper, cobalt, zinc, fluoride, selenium, and chromium. Table 25.5 describes the vital functions of these minerals. Note that the body generally uses the ions of the minerals rather than the non-ionized form. Some minerals, such as chlorine, are toxic or even fatal if ingested in the non-ionized form. Other minerals—aluminum, boron, silicon, and molybdenum—are present but may have no functions. Typical diets supply adequate amounts of potassium, sodium, chloride, and magnesium. Some attention must be paid to eating foods that provide enough calcium, phosphorus, iron, and iodide. Excess amounts of most minerals are excreted in the urine and feces.

Calcium and phosphorus form part of the matrix of bone. Because minerals do not form long-chain compounds, they are otherwise poor building materials. A major role of minerals is to help regulate enzymatic reactions. Calcium, iron, magnesium, and manganese are constituents of some coenzymes. Magnesium also serves as a catalyst for the conversion of ADP to ATP. Minerals such as sodium and phosphorus work in buffer systems, which help control the pH of body fluids. Sodium also helps regulate the osmosis of water and, along with other ions, is involved in the generation of nerve impulses.

Vitamins

Organic nutrients required in small amounts to maintain growth and normal metabolism are called **vitamins.** Unlike carbohydrates, lipids, or proteins, vitamins do not provide energy or serve as the body's building materials. Most vitamins with known functions are coenzymes.

Most vitamins cannot be synthesized by the body and must be ingested in food. Other vitamins, such as vitamin K, are produced by bacteria in the GI tract and then absorbed. The body can assemble some vitamins if the raw materials, called **provitamins,** are provided. For example, vitamin A is produced by the body from the provitamin beta-carotene, a chemical present in yellow vegetables such as carrots and in dark green vegetables such as spinach. No single food contains all the required vitamins—one of the best reasons to eat a varied diet.

Vitamins are divided into two main groups: fat soluble and water soluble. The **fat-soluble vitamins** are vitamins A, D, E, and K. They are absorbed along with other dietary lipids in the small intestine and packaged into chylomicrons. They cannot be absorbed in adequate quantity unless they are ingested with other lipids. Fat-soluble vitamins may be stored in cells, particularly hepatocytes. The **water-soluble vitamins** include several B vitamins and vitamin C. They are dissolved in body fluids. Excess quantities of these vitamins are not stored but instead are excreted in the urine.

Besides their other functions, three vitamins—C, E, and beta-carotene (a provitamin)—are termed **antioxidant vitamins** because they inactivate oxygen free radicals. Recall that free radicals are highly reactive ions or molecules that carry an unpaired electron in their outermost electron shell (see Figure 2.3 on page 30). Free radicals damage cell membranes, DNA, and other cellular structures and contribute to the formation of artery-narrowing atherosclerotic plaques. Some free radicals arise naturally in the body, and others derive from environmental hazards such as tobacco smoke and radiation. Antioxidant vitamins are thought to play a role in protecting against some kinds of cancer, reducing the buildup of atherosclerotic plaque, delaying some effects of aging, and decreasing the chance of cataract formation in the lens of the eyes. Table 25.6 on pages 940–941 lists the major vitamins, their sources, their functions, and related deficiency disorders.

Vitamin and Mineral Supplements

Most nutritionists recommend eating a balanced diet that includes a variety of foods rather than taking vitamin or mineral supplements, except in special circumstances. Common examples of necessary supplementations include iron for women who have excessive menstrual bleeding; iron and calcium for women who are pregnant or breast-feeding; folic acid (folate) for all women who may become pregnant, to reduce the risk of fetal neural tube defects; calcium for most adults, because they do not receive the recommended amount in their diets; and vitamin B_{12} for strict vegetarians, who eat no meat. Because most North Americans do not ingest in their food the high levels of antioxidant vitamins thought to have beneficial effects, some experts recommend supplementing vitamins C and E. More is not always better, however; larger doses of vitamins or minerals can be very harmful. ■

► **CHECKPOINT**

30. What is a nutrient?

31. Describe the food guide pyramid and give examples of foods from each food group.

32. What is a mineral? Briefly describe the functions of the following minerals: calcium, phosphorus, potassium, sulfur, sodium, chloride, magnesium, iron, iodine, copper, zinc, fluoride, manganese, cobalt, chromium, and selenium.

33. Define a vitamin. Explain how we obtain vitamins. Distinguish between a fat-soluble vitamin and a water-soluble vitamin.

34. For each of the following vitamins, indicate its principal function and the effect(s) of deficiency: A, D, E, K, B_1, B_2, niacin, B_6, B_{12}, pantothenic acid, folic acid, biotin, and C.

Table 25.5 Minerals Vital to the Body

Mineral	Comments	Importance
Calcium	Most abundant mineral in body. Appears in combination with phosphates. About 99% is stored in bone and teeth. Blood Ca^{2+} level is controlled by parathyroid hormone(PTH). Calcitriol promotes absorption of dietary calcium. Excess is excreted in feces and urine. Sources are milk, egg yolk, shellfish, and leafy green vegetables.	Formation of bones and teeth, blood clotting, normal muscle and nerve activity, endocytosis and exocytosis, cellular motility, chromosome movement during cell division, glycogen metabolism, and release of neurotransmitters and hormones.
Phosphorus	About 80% is found in bones and teeth as phosphate salts. Blood phosphate level is controlled by parathyroid hormone (PTH). Excess is excreted in urine; small amount is eliminated in feces. Sources are dairy products, meat, fish, poultry, and nuts.	Formation of bones and teeth. Phosphates ($H_2PO_4^-$, HPO_4^-, and PO_4^{3-}) constitute a major buffer system of blood. Plays important role in muscle contraction and nerve activity. Component of many enzymes. Involved in energy transfer (ATP). Component of DNA and RNA.
Potassium	Major cation (K^+) in intracellular fluid. Excess excreted in urine. Present in most foods (meats, fish, poultry, fruits, and nuts).	Needed for generation and conduction of action potentials in neurons and muscle fibers.
Sulfur	Component of many proteins (such as insulin and chrondroitin sulfate), electron carriers in electron transport chain, and some vitamins (thiamine and biotin). Excreted in urine. Sources include beef, liver, lamb, fish, poultry, eggs, cheese, and beans.	As component of hormones and vitamins, regulates various body activities. Needed for ATP production by electron transport chain.
Sodium	Most abundant cation (Na^+) in extracellular fluids; some found in bones. Excreted in urine and perspiration. Normal intake of NaCl (table salt) supplies more than the required amounts.	Strongly affects distribution of water through osmosis. Part of bicarbonate buffer system. Functions in nerve and muscle action potential conduction.
Chloride	Major anion (Cl^-) in extracellular fluid. Excess excreted in urine. Sources include table salt (NaCl), soy sauce, and processed foods.	Plays role in acid–base balance of blood, water balance, and formation of HCl in stomach.
Magnesium	Important cation (Mg^{2+}) in intracellular fluid. Excreted in urine and feces. Widespread in various foods, such as green leafy vegetables, seafood, and whole-grain cereals.	Required for normal functioning of muscle and nervous tissue. Participates in bone formation. Constituent of many coenzymes.
Iron	About 66% found in hemoglobin of blood. Normal losses of iron occur by shedding of hair, epithelial cells, and mucosal cells, and in sweat, urine, feces, bile, and blood lost during menstruation. Sources are meat, liver, shellfish, egg yolk, beans, legumes, dried fruits, nuts, and cereals.	As component of hemoglobin, reversibly binds O_2. Component of cytochromes involved in electron transport chain.
Iodide	Essential component of thyroid hormones. Excreted in urine. Sources are seafood, iodized salt, and vegetables grown in iodine-rich soils.	Required by thyroid gland to synthesize thyroid hormones, which regulate metabolic rate.
Manganese	Some stored in liver and spleen. Most excreted in feces.	Activates several enzymes. Needed for hemoglobin synthesis, urea formation, growth, reproduction, lactation, bone formation, and possibly production and release of insulin, and inhibition of cell damage.
Copper	Some stored in liver and spleen. Most excreted in feces. Sources include eggs, whole-wheat flour, beans, beets, liver, fish, spinach, and asparagus.	Required with iron for synthesis of hemoglobin. Component of coenzymes in electron transport chain and enzyme necessary for melanin formation.
Cobalt	Constituent of vitamin B_{12}.	As part of vitamin B_{12}, required for erythropoiesis.
Zinc	Important component of certain enzymes. Widespread in many foods, especially meats.	As a component of carbonic anhydrase, important in carbon dioxide metabolism. Necessary for normal growth and wound healing, normal taste sensations and appetite, and normal sperm counts in males. As a component of peptidases, it is involved in protein digestion.
Fluoride	Components of bones, teeth, other tissues.	Appears to improve tooth structure and inhibit tooth decay.
Selenium	Important component of certain enzymes. Found in seafood, meat, chicken, tomatoes, egg yolk, milk, mushrooms, and garlic, and cereal grains grown in selenium-rich soil.	Needed for synthesis of thyroid hormones, sperm motility, and proper functioning of the immune system. Also functions as an antioxidant. Prevents chromosome breakage and may play a role in preventing certain birth defects, miscarriage, prostate cancer, and coronary artery disease.
Chromium	Found in high concentrations in brewer's yeast. Also found in wine and some brands of beer.	Needed for normal activity of insulin in carbohydrate and lipid metabolism.

Table 25.6   The Principal Vitamins

Vitamin	Comment and Source	Functions	Deficiency Symptoms and Disorders
Fat-soluble	All require bile salts and some dietary lipids for adequate absorption.		
A	Formed from provitamin beta-carotene (and other provitamins) in GI tract. Stored in liver. Sources of carotene and other provitamins include orange, yellow, and green vegetables; sources of vitamin A include liver and milk.	Maintains general health and vigor of epithelial cells. Beta-carotene acts as an antioxidant to inactivate free radicals.	Deficiency results in atrophy and keratinization of epithelium, leading to dry skin and hair; increased incidence of ear, sinus, respiratory, urinary, and digestive system infections; inability to gain weight; drying of cornea; and skin sores.
		Essential for formation of light-sensitive pigments in photoreceptors of retina.	**Night blindness** or decreased ability for dark adaptation.
		Aids in growth of bones and teeth by helping to regulate activity of osteoblasts and osteoclasts.	Slow and faulty development of bones and teeth.
D	Sunlight converts 7-dehydrocholesterol in the skin to cholecalciferol (vitamin D_3). A liver enzyme then converts cholecalciferol to 25-hydroxycholecalciferol. A second enzyme in the kidneys converts 25-hydroxycholecalciferol to calcitriol (1,25-dihydroxycalciferol), which is the active form of vitamin D. Most is excreted in bile. Dietary sources include fish-liver oils, egg yolk, and fortified milk.	Essential for absorption of calcium and phosphorus from GI tract. Works with parathyroid hormone (PTH) to maintain. Ca^{2+} homeostasis.	Defective utilization of calcium by bones leads to **rickets** in children and **osteomalacia** in adults. Possible loss of muscle tone.
E (tocopherols)	Stored in liver, adipose tissue, and muscles. Sources include fresh nuts and wheat germ, seed oils, and green leafy vegetables.	Inhibits catabolism of certain fatty acids that help form cell structures, especially membranes. Involved in formation of DNA, RNA, and red blood cells. May promote wound healing, contribute to the normal structure and functioning of the nervous system, and prevent scarring. May help protect liver from toxic chemicals such as carbon tetrachloride. Acts as an antioxidant to inactivate free radicals.	May cause oxidation of monounsaturated fats, resulting in abnormal structure and function of mitochondria, lysosomes, and plasma membranes. A possible consequence is hemolytic anemia.
K	Produced by intestinal bacteria. Stored in liver and spleen. Dietary sources include spinach, cauliflower, cabbage, and liver.	Coenzyme essential for synthesis of several clotting factors by liver, including prothrombin.	Delayed clotting time results in excessive bleeding.
Water-soluble	Dissolved in body fluids. Most are not stored in body. Excess intake is eliminated in urine.		
B_1 (thiamine)	Rapidly destroyed by heat. Sources include whole-grain products, eggs, pork, nuts, liver, and yeast.	Acts as coenzyme for many different enzymes that break carbon-to-carbon bonds and are involved in carbohydrate metabolism of pyruvic acid to CO_2 and H_2O. Essential for synthesis of the neurotransmitter acetylcholine.	Improper carbohydrate metabolism leads to buildup of pyruvic and lactic acids and insufficient production of ATP for muscle and nerve cells. Deficiency leads to: (1) **beriberi,** partial paralysis of smooth muscle of GI tract, causing digestive disturbances; skeletal muscle paralysis; and atrophy of limbs; (2) **polyneuritis,** due to degeneration of myelin sheaths; impaired reflexes, impaired sense of touch, stunted growth in children, and poor appetite.

Vitamin	Comment and Source	Functions	Deficiency Symptoms and Disorders
Water-soluble (continued)			
B$_2$ (riboflavin)	Small amounts supplied by bacteria of GI tract. Dietary sources include yeast, liver, beef, veal, lamb, eggs, whole-grain products, asparagus, peas, beets, and peanuts.	Component of certain coenzymes (for example, FAD and FMN) in carbohydrate and protein metabolism, especially in cells of eye, integument, mucosa of intestine, and blood.	Deficiency may lead to improper utilization of oxygen resulting in blurred vision, cataracts, and corneal ulcerations. Also dermatitis and cracking of skin, lesions of intestinal mucosa, and one type of anemia.
Niacin (nicotinamide)	Derived from amino acid tryptophan. Sources include yeast, meats, liver, fish, whole-grain products, peas, beans, and nuts.	Essential component of NAD and NADP, coenzymes in oxidation–reduction reactions. In lipid metabolism, inhibits production of cholesterol and assists in triglyceride breakdown.	Principal deficiency is **pellagra,** characterized by dermatitis, diarrhea, and psychological disturbances.
B$_6$ (pyridoxine)	Synthesized by bacteria of GI tract. Stored in liver, muscle, and brain. Other sources include salmon, yeast, tomatoes, yellow corn, spinach, wholegrain products, liver, and yogurt.	Essential coenzyme for normal amino acid metabolism. Assists production of circulating antibodies. May function as coenzyme in triglyceride metabolism.	Most common deficiency symptom is dermatitis of eyes, nose, and mouth. Other symptoms are retarded growth and nausea.
B$_{12}$ (cyanocobalamin)	Only B vitamin not found in vegetables; only vitamin containing cobalt. Absorption from GI tract depends on intrinsic factor secreted by gastric mucosa. Sources include liver, kidney, milk, eggs, cheese, and meat.	Coenzyme necessary for red blood cell formation, formation of the amino acid methionine, entrance of some amino acids into Krebs cycle, and manufacture of choline (used to synthesize acetylcholine).	Pernicious anemia, neuropsychiatric abnormalities (ataxia, memory loss, weakness, personality and mood changes, and abnormal sensations), and impaired activity of osteoblasts.
Pantothenic acid	Some produced by bacteria of GI tract. Stored primarily in liver and kidneys. Other sources include kidney, liver, yeast, green vegetables, and cereal.	Constituent of coenzyme A, which is essential for transfer of acetyl group from pyruvic acid into Krebs cycle, conversion of lipids and amino acids into glucose, and synthesis of cholesterol and steroidhormones.	Fatigue, muscle spasms, insufficient production of adrenal steroid hormones, vomiting, and insomnia.
Folic acid (folate, folacin)	Synthesized by bacteria of GI tract. Dietary sources include green leafy vegetables, broccoli, asparagus, breads, dried beans, and citrus fruits.	Component of enzyme systems synthesizing nitrogenous bases of DNA and RNA. Essential for normal production of red and white blood cells.	Production of abnormally large red blood cells (macrocytic anemia). Higher risk of neural tube defects in babies born to folate-deficient mothers.
Biotin	Synthesized by bacteria of GI tract. Dietary sources include yeast, liver, egg yolk, and kidneys.	Essential coenzyme for conversion of pyruvic acid to oxaloacetic acid and synthesis of fatty acids and purines.	Mental depression, muscular pain, dermatitis, fatigue, and nausea.
C (ascorbic acid)	Rapidly destroyed by heat. Some stored in glandular tissue and plasma. Sources include citrus fruits, tomatoes, and green vegetables.	Promotes protein synthesis including laying down of collagen in formation of connective tissue. As coenzyme, may combine with poisons, rendering them harmless until excreted. Works with antibodies, promotes wound healing, and functions as an antioxidant.	Scurvy; anemia; many symptoms related to poor collagen formation, including tender swollen gums, loosening of teeth (alveolar processes also deteriorate), poor wound healing, bleeding (vessel walls are fragile because of connective tissue degeneration), and retardation of growth.

DISORDERS: HOMEOSTATIC IMBALANCES

Fever

A **fever** is an elevation of core temperature caused by a resetting of the hypothalamic thermostat. The most common causes of fever are viral or bacterial infections and bacterial toxins; other causes are ovulation, excessive secretion of thyroid hormones, tumors, and reactions to vaccines. When phagocytes ingest certain bacteria, they are stimulated to secrete a **pyrogen** (PĪ-rō-gen; *pyro-* = fire; *-gen* = produce), a fever-producing substance. One pyrogen is interleukin-1. It circulates to the hypothalamus and induces neurons of the preoptic area to secrete prostaglandins. Some prostaglandins can reset the hypothalamic thermostat at a higher temperature, and temperature-regulating reflex mechanisms then act to bring the core body temperature up to this new setting. *Antipyretics* are agents that relieve or reduce fever. Examples include aspirin, acetaminophen (Tylenol), and ibuprofen (Advil), all of which reduce fever by inhibiting synthesis of certain prostaglandins.

Suppose that due to production of pyrogens the thermostat is reset at 39°C (103°F). Now the heat-promoting mechanisms (vasoconstriction, increased metabolism, shivering) are operating at full force. Thus, even though core temperature is climbing higher than normal—say, 38°C (101°F)—the skin remains cold, and shivering occurs. This condition, called a **chill,** is a definite sign that core temperature is rising. After several hours, core temperature reaches the setting of the thermostat, and the chills disappear. But now the body will continue to regulate temperature at 39°C (103°F). When the pyrogens disappear, the thermostat is reset at normal—37.0°C (98.6°F). Because core temperature is high in the beginning, the heat-losing mechanisms (vasodilation and sweating) go into operation to decrease core temperature. The skin becomes warm, and the person begins to sweat. This phase of the fever is called the **crisis,** and it indicates that core temperature is falling.

Although death results if core temperature rises above 44–46°C (112–114°F), up to a point, fever is beneficial. For example, a higher temperature intensifies the effects of interferons and the phagocytic activities of macrophages while hindering replication of some pathogens. Because fever increases heart rate, infection-fighting white blood cells are delivered to sites of infection more rapidly. In addition, antibody production and T cell proliferation increase. Moreover, heat speeds up the rate of chemical reactions, which may help body cells repair themselves more quickly.

Obesity

Obesity is body weight more than 20% above a desirable standard due to an excessive accumulation of adipose tissue. About one-third of the adult population in the United States is obese. (An athlete may be *overweight* due to higher-than-normal amounts of muscle tissue without being obese.) Even moderate obesity is hazardous to health; it is a risk factor in cardiovascular disease, hypertension, pulmonary disease, non-insulin-dependent diabetes mellitus, arthritis, certain cancers (breast, uterus, and colon), varicose veins, and gallbladder disease.

In a few cases, obesity may result from trauma of or tumors in the food-regulating centers in the hypothalamus. In most cases of obesity, no specific cause can be identified. Contributing factors include genetic factors, eating habits taught early in life, overeating to relieve tension, and social customs. Studies indicate that some obese people burn fewer calories during digestion and absorption of a meal, a smaller food-induced thermogenesis effect. Additionally, obese people who lose weight require about 15% fewer calories to maintain normal body weight than do people who have never been obese. Interestingly, people who gain weight easily when deliberately fed excess calories exhibit less NEAT (nonexercise activity thermogenesis, such as occurs with fidgeting) than people who resist weight gains in the face of excess calories. Although leptin suppresses appetite and produces satiety in experimental animals, it is not deficient in most obese people.

Most surplus calories in the diet are converted to triglycerides and stored in adipose cells. Initially, the adipocytes increase in size, but at a maximal size, they divide. As a result, proliferation of adipocytes occurs in extreme obesity. The enzyme endothelial lipoprotein lipase regulates triglyceride storage. The enzyme is very active in abdominal fat but less active in hip fat. Accumulation of fat in the abdomen is associated with higher blood cholesterol level and other cardiac risk factors.

Treatment of obesity is difficult because most people who are successful at losing weight gain it back within two years. Yet, even modest weight loss is associated with health benefits. Treatments for obesity include behavior modification programs, very-low-calorie diets, drugs, and surgery. Behavior modification programs, offered at many hospitals, strive to alter eating behaviors and increase exercise activity. The nutrition program includes a "heart-healthy" diet that includes abundant vegetables but is low in fats, especially saturated fats. A typical exercise program suggests walking for 30 minutes a day, five to seven times a week. Regular exercise enhances both weight loss and weight-loss maintenance. Very-low-calorie (VLC) diets include 400 to 800 kcal/day in a commercially made liquid mixture. The VLC diet is usually prescribed for 12 weeks, under close medical supervision. Two drugs are available to treat obesity. Sibutramine is an appetite suppressant that works by inhibiting reuptake of serotonin and norepinephrine in brain areas that govern eating behavior. Orlistat works by inhibiting the lipases released into the lumen of the GI tract. With less lipase activity, fewer dietary triglycerides are absorbed. For those with extreme obesity who have not responded to other treatments, a surgical procedure may be considered. The two operations most commonly performed—gastric bypass and gastroplasty—both greatly reduce the stomach size so that it can hold just a tiny quantity of food.

MEDICAL TERMINOLOGY

Heat cramps Cramps that result from profuse sweating. The salt lost in sweat causes painful contractions of muscles; such cramps tend to occur in muscles used while working but do not appear until the person relaxes once the work is done. Drinking salted liquids usually leads to rapid improvement.

Heat exhaustion (heat prostration) A condition in which the core temperature is generally normal, or a little below, and the skin is cool and moist due to profuse perspiration. Heat exhaustion is usually characterized by loss of fluid and electrolytes, especially salt (NaCl). The salt loss results in muscle cramps, dizziness, vomiting, and fainting; fluid loss may cause low blood pressure. Complete rest, rehydration, and electrolyte replacement are recommended.

Heatstroke (sunstroke) A severe and often fatal disorder caused by exposure to high temperatures, especially when the relative humidity is high, which makes it difficult for the body to lose heat. Blood flow to the skin is decreased, perspiration is greatly reduced, and body temperature rises sharply because of failure of the hypothalamic thermostat. Body temperature may reach 43°C (110°F). Treatment, which must be undertaken immediately, consists of cooling the body by immersing the victim in cool water and by administering fluids and electrolytes.

Kwashiorkor (kwash-ē-OR-kor) A disorder in which protein intake is deficient despite normal or nearly normal caloric intake, character-

ized by edema of the abdomen, enlarged liver, decreased blood pressure, low pulse rate, lower-than-normal body temperature, and sometimes mental retardation. Because the main protein in corn (zein) lacks two essential amino acids, which are needed for growth and tissue repair, many African children whose diet consists largely of cornmeal develop kwashiorkor.

Malnutrition (*mal-* = bad) An imbalance of total caloric intake or intake of specific nutrients, which can be either inadequate or excessive.

Marasmus (mar-AZ-mus) A type of protein–calorie undernutrition that results from inadequate intake of both protein and calories. Its characteristics include retarded growth, low weight, muscle wasting, emaciation, dry skin, and thin, dry, dull hair.

STUDY OUTLINE

INTRODUCTION (p. 907)

1. Our only source of energy for performing biological work is the food we eat. Food also provides essential substances that we cannot synthesize.
2. Most food molecules absorbed by the gastrointestinal tract are used to supply energy for life processes, serve as building blocks during synthesis of complex molecules, or are stored for future use.

METABOLIC REACTIONS (p. 907)

1. Metabolism refers to all chemical reactions of the body and is of two types: catabolism and anabolism.
2. Catabolism is the term for reactions that break down complex organic compounds into simple ones. Overall, catabolic reactions are exergonic; they produce more energy than they consume.
3. Chemical reactions that combine simple molecules into more complex ones that form the body's structural and functional components are collectively known as anabolism. Overall, anabolic reactions are endergonic; they consume more energy than they produce.
4. The coupling of anabolism and catabolism occurs via ATP.

ENERGY TRANSFER (p. 908)

1. Oxidation is the removal of electrons from a substance; reduction is the addition of electrons to a substance.
2. Two coenzymes that carry hydrogen atoms during coupled oxidation–reduction reactions are nicotinamide adenine dinucleotide (NAD⁺) and flavin adenine dinucleotide (FAD).
3. ATP can be generated via substrate-level phosphorylation, oxidative phosphorylation, and photophosphorylation.

CARBOHYDRATE METABOLISM (p. 909)

1. During digestion, polysaccharides and disaccharides are hydro-

lyzed into the monosaccharides glucose (about 80%), fructose, and galactose; the latter two are then converted to glucose.
2. Some glucose is oxidized by cells to provide ATP. Glucose also can be used to synthesize amino acids, glycogen, and triglycerides.
3. Glucose moves into most body cells via facilitated diffusion through glucose transporters (GluT) and becomes phosphorylated to glucose 6-phosphate. In muscle cells, this process is stimulated by insulin. Glucose entry into neurons and hepatocytes is always "turned on."
4. Cellular respiration, the complete oxidation of glucose to CO_2 and H_2O, involves glycolysis, the Krebs cycle, and the electron transport chain.
5. Glycolysis is the breakdown of glucose into two molecules of pyruvic acid; there is a net production of two molecules of ATP.
6. When oxygen is in short supply, pyruvic acid is reduced to lactic acid; under aerobic conditions, pyruvic acid enters the Krebs cycle.
7. Pyruvic acid is prepared for entrance into the Krebs cycle by conversion to a two-carbon acetyl group followed by the addition of coenzyme A to form acetyl coenzyme A.
8. The Krebs cycle involves decarboxylations, oxidations, and reductions of various organic acids.
9. Each molecule of pyruvic acid that is converted to acetyl coenzyme A and then enters the Krebs cycle produces three molecules of CO_2, four molecules of NADH and four H⁺, one molecule of $FADH_2$, and one molecule of ATP.
10. The energy originally stored in glucose and then in pyruvic acid is transferred primarily to the reduced coenzymes NADH and $FADH_2$.
11. The electron transport chain involves a series of oxidation–reduction reactions in which the energy in NADH and $FADH_2$ is liberated and transferred to ATP.
12. The electron carriers include FMN, cytochromes, iron–sulfur centers, copper atoms, and coenzyme Q.
13. The electron transport chain yields a maximum of 32 or 34 molecules of ATP and six molecules of H_2O.

14. Table 25.1 on page 918 summarizes the ATP yield during cellular respiration. The complete oxidation of glucose can be represented as follows:

$$C_6H_{12}O_6 + 6\,O_2 + 36 \text{ or } 38 \text{ ADPs} + 36 \text{ or } 38 \;\textcircled{P} \longrightarrow$$
$$6\,CO_2 + 6\,H_2O + 36 \text{ or } 38 \text{ ATPs}$$

15. The conversion of glucose to glycogen for storage in the liver and skeletal muscle is called glycogenesis. It is stimulated by insulin.

16. The conversion of glycogen to glucose is called glycogenolysis. It occurs between meals and is stimulated by glucagon and epinephrine.

17. Gluconeogenesis is the conversion of noncarbohydrate molecules into glucose. It is stimulated by cortisol and glucagon.

LIPID METABOLISM (p. 920)

1. Lipoproteins transport lipids in the bloodstream. Types of lipoproteins include chylomicrons, which carry dietary lipids to adipose tissue; very low-density lipoproteins (VLDLs), which carry triglycerides from the liver to adipose tissue; low-density lipoproteins (LDLs), which deliver cholesterol to body cells; and high-density lipoproteins (HDLs), which remove excess cholesterol from body cells and transport it to the liver for elimination.

2. Cholesterol in the blood comes from two sources: from food and from synthesis by the liver.

3. Lipids may be oxidized to produce ATP or stored as triglycerides in adipose tissue, mostly in the subcutaneous layer.

4. A few lipids are used as structural molecules or to synthesize essential molecules.

5. Adipose tissue contains lipases that catalyze the deposition of triglycerides from chylomicrons and hydrolyze triglycerides into fatty acids and glycerol.

6. In lipolysis, triglycerides are split into fatty acids and glycerol and released from adipose tissue under the influence of epinephrine, norepinephrine, cortisol, thyroid hormones, and insulinlike growth factors.

7. Glycerol can be converted into glucose by conversion into glyceraldehyde 3-phosphate.

8. In beta-oxidation of fatty acids, carbon atoms are removed in pairs from fatty acid chains; the resulting molecules of acetyl coenzyme A enter the Krebs cycle.

9. The conversion of glucose or amino acids into lipids is called lipogenesis; it is stimulated by insulin.

PROTEIN METABOLISM (p. 923)

1. During digestion, proteins are hydrolyzed into amino acids, which enter the liver via the hepatic portal vein.

2. Amino acids, under the influence of insulinlike growth factors and insulin, enter body cells via active transport.

3. Inside cells, amino acids are synthesized into proteins that function as enzymes, hormones, structural elements, and so forth; stored as fat or glycogen; or used for energy.

4. Before amino acids can be catabolized, they must be deaminated and converted to substances that can enter the Krebs cycle.

5. Amino acids may also be converted into glucose, fatty acids, and ketone bodies.

6. Protein synthesis is stimulated by insulinlike growth factors, thyroid hormones, insulin, estrogen, and testosterone.

7. Table 25.2 on page 927 summarizes carbohydrate, lipid, and protein metabolism.

KEY MOLECULES AT METABOLIC CROSSROADS (p. 925)

1. Three molecules play a key role in metabolism: glucose 6-phosphate, pyruvic acid, and acetyl coenzyme A.

2. Glucose 6-phosphate may be converted to glucose, glycogen, ribose 5-phosphate, and pyruvic acid.

3. When ATP is low and oxygen is plentiful, pyruvic acid is converted to acetyl coenzyme A; when oxygen supply is low, pyruvic acid is converted to lactic acid. One link between carbohydrate and protein metabolism occurs via pyruvic acid.

4. Acetyl coenzyme A is the molecule that enters the Krebs cycle; it is also used to synthesize fatty acids, ketone bodies, and cholesterol.

METABOLIC ADAPTATIONS (p. 927)

1. During the absorptive state, ingested nutrients enter the blood and lymph from the GI tract.

2. During the absorptive state, blood glucose is oxidized to form ATP, and glucose transported to the liver is converted to glycogen or triglycerides. Most triglycerides are stored in adipose tissue. Amino acids in hepatocytes are converted to carbohydrates, fats, and proteins. Table 25.3 on page 929 summarizes the hormonal regulation of metabolism during the absorptive state.

3. During the postabsorptive state, absorption is complete and the ATP needs of the body are satisfied by nutrients already present in the body. The major task is to maintain normal blood glucose level by converting glycogen in the liver and skeletal muscle into glucose, converting glycerol into glucose, and converting amino acids into glucose. Fatty acids, ketone bodies, and amino acids are oxidized to supply ATP. Table 25.4 on page 931 summarizes the hormonal regulation of metabolism during the postabsorptive state.

5. Fasting is going without food for a few days; starvation implies weeks or months of inadequate food intake. During fasting and starvation, fatty acids and ketone bodies are increasingly utilized for ATP production.

HEAT AND ENERGY BALANCE (p. 932)

1. Measurement of the metabolic rate under basal conditions is called the basal metabolic rate (BMR).

2. A kilocalorie (kcal) or Calorie is the amount of energy required to raise the temperature of 1000 g of water from 14°C to 15°C.

3. Normal core temperature is maintained by a delicate balance between heat-producing and heat-losing mechanisms.

4. Exercise, hormones, the nervous system, body temperature, ingestion of food, age, gender, climate, sleep, and malnutrition affect metabolic rate.

5. Mechanisms of heat transfer are conduction, convection, radiation, and evaporation. Conduction is the transfer of heat between two substances or objects in contact with each other. Convection is the transfer of heat by a liquid or gas between areas of different temperatures. Radiation is the transfer of heat from a warmer object to a cooler object without physical contact. Evaporation is the conversion of a liquid to a vapor; in the process, heat is lost.

6. The hypothalamic thermostat is in the preoptic area.

7. Responses that produce, conserve, or retain heat when core temperature falls are vasoconstriction; release of epinephrine, norepinephrine, and thyroid hormones; and shivering.

8. Responses that increase heat loss when core temperature increases

include vasodilation, decreased metabolic rate, and evaporation of perspiration.

9. Two nuclei in the hypothalamus that help regulate food intake are the arcuate and paraventricular nuclei. The hormone leptin, released by adipocytes, inhibits release of neuropeptide Y from the arcuate nucleus and thereby decreases food intake. Melanocortin also decreases food intake.

NUTRITION (p. 936)

1. Nutrients include water, carbohydrates, lipids, proteins, minerals, and vitamins.
2. Most teens and adults need between 1600 and 2800 Calories per day.
3. Nutrition experts suggest dietary calories be 50–60% from carbohydrates, 30% or less from fats, and 12–15% from proteins, although the optimal levels of these nutrients may vary.

4. The food guide pyramid indicates how many servings of five food groups are recommended each day to attain the number of calories and variety of nutrients needed for wellness.
5. Minerals known to perform essential functions include calcium, phosphorus, potassium, sulfur, sodium, chloride, magnesium, iron, iodide, manganese, copper, cobalt, zinc, fluoride, selenium, and chromium. Their functions are summarized in Table 25.5 on page 939.
6. Vitamins are organic nutrients that maintain growth and normal metabolism. Many function in enzyme systems.
7. Fat-soluble vitamins are absorbed with fats and include vitamins A, D, E, and K; water-soluble vitamins include the B vitamins and vitamin C.
8. The functions and deficiency disorders of the principal vitamins are summarized in Table 25.6 on pages 940–941.

Q SELF-QUIZ QUESTIONS

Fill in the blanks in the following statements.

1. The thermostat and food intake regulating center of the body is in the _____ of the brain.
2. The three key molecules of metabolism _____, _____, and _____.
3. _____ are the primary regulators of metabolism.

Indicate whether the following statements are true of false.

4. Glycogenesis occurs in the liver and cardiac muscle and is stimulated by insulin.
5. Most of the glucose in the body is used to produce ATP.

Choose the one best answer to the following questions.

6. Which of the following is *not* considered a nutrient? (a) water, (b) carbohydrates, (c) nucleic acids, (d) proteins, (e) lipids.
7. Which of the following factors affect metabolic rate and thus the production of body heat? (1) exercise, (2) hormones, (3) minerals, (4) food intake, (5) waste product removal, (6) age. (a) 1, 2, 3, and 4, (b) 2, 3, 4, and 5, (c) 3, 4, 5, and 6, (d) 1, 2, 4, and 6, (e) 1, 3, 5, and 6.
8. Which of the following are uses for glucose in the body? (1) ATP production, (2) amino acid synthesis, (3) glycogen synthesis, (4) triglyceride synthesis, (5) vitamin synthesis. (a) 1, 2, 3, and 4, (b) 2, 3, 4, and 5, (c) 1, 3, and 5, (d) 2, 4, and 5, (e) 1, 2, 4, and 5.
9. Which of the following is the *correct* sequence for the oxidation of glucose to produce ATP? (a) electron transport chain, Krebs cycle, glycolysis, formation of acetyl CoA, (b) Krebs cycle, formation of acetyl CoA, electron transport chain, glycolysis, (c) glycolysis, electron transport chain, Krebs cycle, formation of acetyl CoA, (d) glycolysis, formation of acetyl CoA, Krebs cycle, electron transport chain, (e) formation of acetyl CoA, Krebs cycle, glycolysis, electron transport chain.
10. Which of the following are absorptive state reactions? (1) aerobic cellular respiration, (2) glycogenesis, (3) glycogenolysis, (4) gluconeogenesis using lactic acid, (5) lipolysis. (a) 1 and 2, (b) 2 and 3, (c) 3 and 4, (d) 4 and 5, (e) 1 and 5.

11. Which of the following would you *not* expect to experience during fasting or starvation? (a) decrease in plasma fatty acid levels, (b) increase in ketone body formation, (c) lipolysis, (d) increased use of ketones for ATP production in the brain, (e) depletion of glycogen.
12. If core body temperature rises above normal, which of the following would occur to cool the body? (1) dilation of vessels in the skin, (2) increased radiation and conduction of heat to the environment, (3) increased metabolic rate, (4) evaporation of perspiration, (5) increased secretion of thyroid hormones. (a) 3, 4, and 5, (b) 1, 2, and 4, (c) 1, 2, and 5, (d) 1, 2, 3, 4, and 5, (e) 1, 2, 4, and 5.
13. Match the following:

    ____ (a) the breakdown of glycogen back to glucose
    ____ (b) the removal of the amino group from an amino acid
    ____ (c) the splitting of a triglyceride into glycerol and fatty acids
    ____ (d) the cleavage of one pair of carbon atoms at a time from a fatty acid
    ____ (e) the formation of ketone bodies
    ____ (f) the synthesis of lipids
    ____ (g) the conversion of glucose into glycogen
    ____ (h) the transfer of an amino group from an amino acid to a substance such as pyruvic acid
    ____ (i) the formation of glucose from noncarbohydrate sources
    ____ (j) the breakdown of glucose into two molecules of pyruvic acid

    (1) glycolysis
    (2) glycogenolysis
    (3) glycogenesis
    (4) gluconeogenesis
    (5) lipolysis
    (6) ketogenesis
    (7) lipogenesis
    (8) deamination
    (9) transamination
    (10) beta oxidation

14. Match the following:

_____ (a) deliver cholesterol to body cells for use in repair of cell membrances and synthesis of steroid hormones

_____ (b) remove excess cholesterol from body cells and transport it to the liver for elimination

_____ (c) organic nutrients required in tiny amounts to maintain growth and normal metabolism

_____ (d) the energy-transfering molecule of the body

_____ (e) nutrient molecules that can be oxidized to produce ATP or stored in adipose tissue

_____ (f) carriers of electrons in the electron transport chain

_____ (g) the body's preferred source for synthesizing ATP

_____ (h) composed of amino acids and are the primary regulatory molecules in the body

_____ (i) acetoacetic acid, betahydroxybutyric acid, and acetone

_____ (j) hormone that acts to decrease total body fat mass

_____ (k) neurotransmitter that stimulates food intake

_____ (j) inorganic substances that perform many vital functions in the body

(1) leptin
(2) minerals
(3) glucose
(4) lipids
(5) proteins
(6) neuropeptide Y
(7) cytochromes
(8) ketone bodies
(9) low-density lipoproteins
(10) ATP
(11) vitamins
(12) high-density lipoproteins

15. Match the following:

_____ (a) the mechanism of ATP generation that links chemical reactions with pumping of hydrogen ions

_____ (b) the removal of electrons from an atom or molecule resulting in a decrease in energy

_____ (c) refers to all the chemical reactions in the body

_____ (d) the oxidation of glucose to produce ATP

_____ (e) the addition of electrons to a molecule resulting in an increase in energy content of the molecule

_____ (f) chemical reactions that break down complex organic molecules into simpler ones

_____ (g) overall rate at which metabolic reactions use energy

_____ (h) removal of CO_2 from a molecule

_____ (i) chemical reactions that combine simple molecules and monomers to make more complex ones

_____ (j) the addition of a phosphate group to a molecule

(1) metabolism
(2) catabolism
(3) anabolism
(4) metabolic rate
(5) oxidation
(6) reduction
(7) phosphorylation
(8) chemiosmosis
(9) decarboxylation
(10) cellular respiration

CRITICAL THINKING QUESTIONS

1. Deb was training for a marathon. With only one day to go before the big event, she had shortened her morning training run and was loading up on bread and pasta for her last meal. Why is Deb eating lots of bread and spaghetti?
HINT *Bread and pasta are high in carbohydrates.*

2. Mr. Hernandez believed that the perfect lawn needed to be cut according to a strict schedule. As he was pushing the lawn mower around the yard on a scorching hot Saturday morning, he began to feel dizzy and nauseous. When he sat down, he was bothered by cramps in his legs, and he nearly fainted. What's wrong with Mr. Hernandez?
HINT *His shirt was soaked with sweat, but his forehead actually felt cool.*

3. Three-year-old Sara's mother was concerned about her child's eating habits. She complained "one day Sara will only eat a few bites, and then the next day she eats like a horse!" The pediatrician told Sara's mom that it sounded like Sara had perfectly normal control over her food intake. How is food intake controlled.
HINT *Sara is too young to count calories.*

? ANSWERS TO FIGURE QUESTIONS

25.1 In pancreatic acinar cells, anabolism predominates because they are synthesizing complex molecules (digestive enzymes).

25.2 Glycolysis is also called anaerobic cellular respiration.

25.3 The reactions of glycolysis consume two molecules of ATP but generate four molecules of ATP, for a net of two.

25.4 Kinases are enzymes that phosphorylate (add phosphate to) their substrate.

25.5 Glycolysis occurs in the cytosol.

25.6 CO_2 is given off during the production of acetyl coenzyme A and during the Krebs cycle. It diffuses into the blood, is transported by the blood to the lungs, and is exhaled.

25.7 The production of reduced coenzymes is important in the Krebs cycle because they will subsequently yield ATP in the electron transport chain.

25.8 The energy source that powers the proton pumps is electrons provided by $NADH + H^+$.

25.9 The concentration of H^+ is highest in the space between the inner and outer mitochondrial membranes.

25.10 During the complete oxidation of one glucose molecule, six molecules of O_2 are used and six molecules of CO_2 are produced.

25.11 Skeletal muscle fibers can synthesize glycogen, but they cannot release glucose into the blood because they lack the enzyme phosphatase.

25.12 Hepatocytes can carry out gluconeogenesis and glycogenesis.

25.13 LDLs deliver cholesterol to body cells.

25.14 Hepatocytes and adipose cells carry out lipogenesis, beta oxidation, and lipolysis; hepatocytes carry out ketogenesis.

25.15 An amino group is removed from an amino acid via deamination.

25.16 Acetyl coenzyme A is the gateway into the Krebs cycle.

25.17 Reactions of the absorptive state are mainly anabolic.

25.18 Processes that directly elevate blood glucose during the postabsorptive state include lipolysis (in adipocytes and hepatocytes), gluconeogenesis (in hepatocytes), and glycogenolysis (in hepatocytes).

25.19 Exercise, the sympathetic nervous system, hormones (epinephrine, norepinephrine, thyroxine, testosterone, human growth hormone), elevated body temperature, and ingestion of food increase metabolic rate.

25.20 Foods that contain cholesterol and most of the saturated fatty acids in the diet are milk, yogurt, cheeses, and meats.

The Urinary System

A Student's View

Reading this chapter through the eyes of a future nurse and midwife, the importance of the urinary system for the overall health and well being of a patient and developing fetus becomes evident. I knew that this system is vital in excreting wastes and foreign substances from the body, but discovered the critical role it plays in regulating blood chemistry and volume. I imagine myself as a health care professional with a patient whose kidneys are failing, and the importance of understanding this material and knowing the consequences to other organ systems becomes clear.

I found the sections that dealt with medical disorders most intriguing, specifically the box at the end of the chapter titled "Disorders: Homeostatic Imbalances", and the "stethoscope" sections encountered throughout the text. When I read explanations of disorders, I like to envision myself as a nurse encountering a patient with the symptoms described and think about how I would explain to the patient what exactly is going wrong with their urinary system. From a midwifery standpoint, I found "Development of the Urinary System" a very useful and fascinating section. And finally, the "Production of Dilute and Concentrated Urine" and the "Evaluation of Kidney Function" sections show the importance of urine as a diagnostic tool in assessing the overall health of a patient.

I thought all of the figures in this chapter were excellent and very helpful in explaining material that can sometimes get confusing.

Margaret Chambers
George Mason University

THE URINARY SYSTEM AND HOMEOSTASIS

The urinary system contributes to homeostasis by altering blood composition, forming urine, and regulating body fluid pH.

ENERGIZE YOUR STUDY
FOUNDATIONS CD

Anatomy Overview
- The Urinary System

Animations
- Systems Contributions to Homeostasis
- The Case of the Girl with the Fruity Breath
- The Case of the Man with the Yellow Eyes

Concepts and Connections and Exercises reinforce your understanding

www.wiley.com/college/apcentral

INSIGHTS AND EXPLORATIONS

Fortunately most of us do not have to worry about whether the food we eat, the drugs we take, or the water we drink will be eliminated from our bodies. We have healthy kidneys that do this for us, along with our digestive system. But many people must schedule their weeks around visits to dialysis centers to have waste and extra fluid removed from their blood by a process called dialysis. In this web-based activity we will explore why it is that some people lose the function of their kidneys and how dialysis works.

The **urinary system** consists of two kidneys, two ureters, one urinary bladder, and one urethra (Figure 26.1). After the kidneys filter blood plasma, they return most of the water and solutes to the bloodstream. The remaining water and solutes constitute **urine,** which passes through the ureters and is stored in the urinary bladder until it is excreted from the body through the urethra. **Nephrology** (nef-ROL-ō-jē; *nephro-* = kidney; *-logy* = study of) is the scientific study of the anatomy, physiology, and pathology of the kidneys. The branch of medicine that deals with the male and female urinary systems and the male reproductive system is **urology** (ū-ROL-ō-jē; *uro-* = urine).

OVERVIEW OF KIDNEY FUNCTIONS

► O B J E C T I V E

- **List the functions of the kidneys.**

The kidneys do the major work of the urinary system. The other parts of the system are mainly passageways and storage areas. Functions of the kidneys include:

- *Regulating blood ionic composition.* The kidneys help regulate the blood levels of several ions, most importantly sodium ions (Na^+), potassium ions (K^+), calcium ions (Ca^{2+}), chloride ions (Cl^-), and phosphate ions (HPO_4^{2-}).

- *Regulating blood pH.* The kidneys excrete a variable amount of hydrogen ions (H^+) into the urine and conserve bicarbonate ions (HCO_3^-), which are an important buffer of H^+ in the blood. Both of these activities help regulate blood pH.

Figure 26.1 **Organs of the urinary system in a female.** (See Tortora, *A Photographic Atlas of the Human Body,* Figure 13.2.)

Urine formed by the kidneys passes first into the ureters, then to the urinary bladder for storage, and finally through the urethra for elimination from the body.

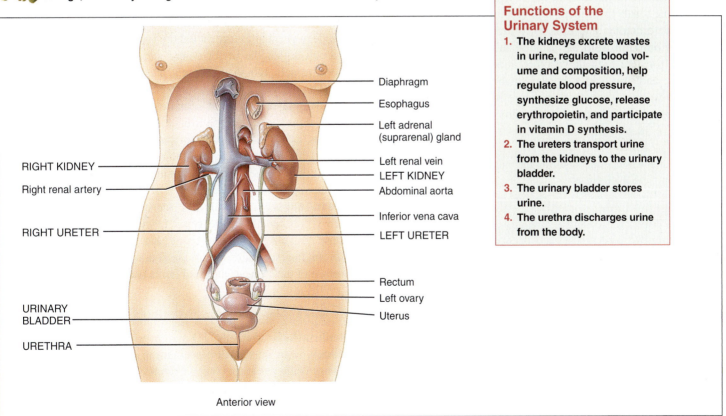

Diaphragm
Esophagus
Left adrenal (suprarenal) gland
Left renal vein
LEFT KIDNEY
Abdominal aorta
Inferior vena cava
LEFT URETER
Rectum
Left ovary
Uterus

RIGHT KIDNEY
Right renal artery
RIGHT URETER
URINARY BLADDER
URETHRA

Anterior view

Functions of the Urinary System

1. The kidneys excrete wastes in urine, regulate blood volume and composition, help regulate blood pressure, synthesize glucose, release erythropoietin, and participate in vitamin D synthesis.
2. The ureters transport urine from the kidneys to the urinary bladder.
3. The urinary bladder stores urine.
4. The urethra discharges urine from the body.

 Which organs constitute the urinary system?

- *Regulating blood volume.* The kidneys adjust blood volume by conserving or eliminating water in the urine. Also, an increase in blood volume increases blood pressure, whereas a decrease in blood volume decreases blood pressure.

- *Regulating blood pressure.* Besides adjusting blood volume, the kidneys help regulate blood pressure by secreting the enzyme renin, which activates the renin–angiotensin–aldosterone pathway (see Figure 18.16 on page 611). Increased renin causes an increase in blood pressure.

- *Maintaining blood osmolarity.* By separately regulating loss of water and loss of solutes in the urine, the kidneys maintain a relatively constant blood osmolarity close to 290 milliosmoles per liter (mOsm/liter).*

- *Producing hormones.* The kidneys produce two hormones. *Calcitriol,* the active form of vitamin D, helps regulate calcium homeostasis (see Figure 18.14 on page 608), and *erythropoietin* stimulates production of red blood cells (see Figure 19.5 on page 641).

- *Regulating blood glucose level.* Like the liver, the kidneys can use the amino acid glutamine in *gluconeogenesis,* the synthesis of new glucose molecules. They can then release glucose into the blood to help maintain a normal blood glucose level.

- *Excreting wastes and foreign substances.* By forming urine, the kidneys help excrete **wastes**—substances that have no useful function in the body. Some wastes excreted in urine result from metabolic reactions in the body. These include ammonia and urea from the deamination of amino acids; bilirubin from the catabolism of hemoglobin; creatinine from the breakdown of creatine phosphate in muscle fibers; and uric acid from the catabolism of nucleic acids. Other wastes excreted in urine are foreign substances from the diet, such as drugs and environmental toxins.

▶ CHECKPOINT

1. What are wastes, and how do the kidneys participate in their removal from the body?

---

*The **osmolarity** of a solution is a measure of the total number of dissolved particles per liter of solution. The particles may be molecules, ions, or a mixture of both. To calculate osmolarity, multiply molarity (see page 39) by the number of particles per molecule, once the molecule dissolves. A similar term, *osmolality,* is the number of particles of solute per *kilogram* of water. Because it is easier to measure volumes of solutions than to determine the mass of water they contain, osmolarity is used more commonly than osmolality. Most body fluids and solutions used clinically are dilute, in which case there is less than a 1% difference between the two measures.

ANATOMY AND HISTOLOGY OF THE KIDNEYS

▶ O B J E C T I V E S

- **Describe the external and internal gross anatomical features of the kidneys.**
- **Trace the path of blood flow through the kidneys.**
- **Describe the structure of renal corpuscles and renal tubules.**

The paired **kidneys** are reddish, kidney-bean-shaped organs located just above the waist between the peritoneum and the posterior wall of the abdomen. Because their position is posterior to the peritoneum of the abdominal cavity, they are said to be **retroperitoneal** (re'-trō-per-i-tō-NĒ-al; *retro-* = behind) organs (Figure 26.2). The kidneys are located between the levels of the last thoracic and third lumbar vertebrae, a position where they are partially protected by the eleventh and twelfth pairs of ribs. The right kidney is slightly lower than the left (see Figure 26.1) because the liver occupies considerable space on the right side superior to the kidney.

External Anatomy of the Kidneys

A typical kidney in an adult is 10–12 cm (4–5 in.) long, 5–7 cm (2–3 in.) wide, and 3 cm (1 in.) thick—about the size of a bar of bath soap—and has a mass of 135–150 g (4.5–5 oz). The concave medial border of each kidney faces the vertebral column (see Figure 26.1). Near the center of the concave border is a deep vertical fissure called the **renal hilus** (see Figure 26.3), through which the ureter emerges from the kidney along with blood vessels, lymphatic vessels, and nerves.

Three layers of tissue surround each kidney (Figure 26.2). The deep layer, the **renal capsule** (*ren-* = kidney), is a smooth, transparent sheet of dense irregular connective tissue that is continuous with the outer coat of the ureter. It serves as a barrier against trauma and helps maintain the shape of the kidney. The middle layer, the **adipose capsule,** is a mass of fatty tissue surrounding the renal capsule. It also protects the kidney from trauma and holds it firmly in place within the abdominal cavity. The superficial layer, the **renal fascia,** is another thin layer of dense irregular connective tissue that anchors the kidney to the surrounding structures and to the abdominal wall. On the anterior surface of the kidneys, the renal fascia is deep to the peritoneum.

Nephroptosis (Floating Kidney)

Nephroptosis (nef'-rōp-TŌ-sis; *ptosis* = falling), or **floating kidney,** is an inferior displacement or dropping of the kidney. It occurs when the kidney slips from its normal position because it is not securely held in place by adjacent organs or its covering of fat. Nephroptosis develops most often in very thin people whose adipose capsule or renal fascia is deficient. It is dangerous because the ureter may kink and block urine flow. The resulting backup of urine puts pressure on the kidney, which damages the tissue. Twisting of the ureter also causes pain. ■

Figure 26.2   Position and coverings of the kidneys.   (See Tortora, *A Photographic Atlas of the Human Body,* Figure 13.3.)

The kidneys are surrounded by a renal capsule, adipose capsule, and renal fascia.

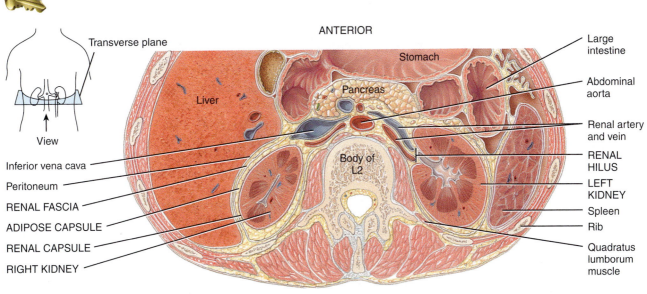

ANTERIOR

Transverse plane

View

Inferior vena cava

Peritoneum

RENAL FASCIA

ADIPOSE CAPSULE

RENAL CAPSULE

RIGHT KIDNEY

Liver

Stomach

Pancreas

Body of L2

Large intestine

Abdominal aorta

Renal artery and vein

RENAL HILUS

LEFT KIDNEY

Spleen

Rib

Quadratus lumborum muscle

POSTERIOR

(a) Inferior view of transverse section of abdomen (L2)

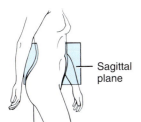

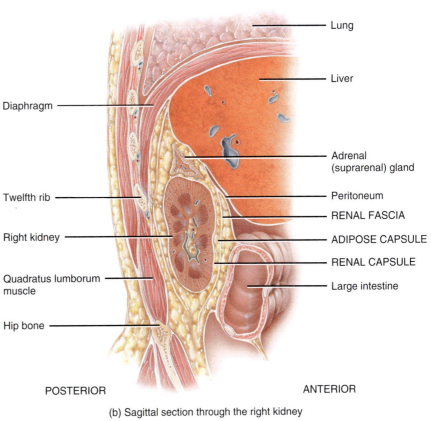

SUPERIOR

Sagittal plane

Diaphragm

Twelfth rib

Right kidney

Quadratus lumborum muscle

Hip bone

Lung

Liver

Adrenal (suprarenal) gland

Peritoneum

RENAL FASCIA

ADIPOSE CAPSULE

RENAL CAPSULE

Large intestine

POSTERIOR

ANTERIOR

(b) Sagittal section through the right kidney

Why are the kidneys said to be retroperitoneal?

Internal Anatomy of the Kidneys

A frontal section through the kidney reveals two distinct regions: a superficial, smooth-textured reddish area called the **renal cortex** (*cortex* = rind or bark) and a deep, reddish-brown inner region called the **renal medulla** (*medulla* = inner portion) (Figure 26.3). The medulla consists of 8 to 18 cone-shaped **renal pyramids.** The base (wider end) of each pyramid faces the renal cortex, and its apex (narrower end), called a **renal papilla,** points toward the renal hilus. The renal cortex is the smooth-textured area extending from the renal capsule to the bases of the renal pyramids and into the spaces between them. It is divided into an outer *cortical zone* and an inner *juxtamedullary zone.* Those portions of the renal cortex that extend between renal pyramids are called **renal columns.** A **renal lobe** consists of a renal pyramid, its overlying area of renal cortex, and one-half of each adjacent renal column.

Together, the renal cortex and renal pyramids of the renal medulla constitute the **parenchyma** (functional portion) of the kidney. Within the parenchyma are the functional units of the kidney—about 1 million microscopic structures called **nephrons** (NEF-rons). Urine formed by the nephrons drains into large **papillary ducts,** which extend through the renal papillae of the pyramids. The papillary ducts drain into cuplike structures called **minor** and **major calyces** (KĀ-li-sēz = cups; singular is *calyx*). Each kidney has 8 to 18 minor calyces and 2 to 3 major calyces. A minor calyx receives urine from the papillary ducts of one renal papilla and delivers it to a major calyx. From the major calyces, urine drains into a single large cavity called the **renal pelvis** (*pelv-* = basin) and then out through the ureter to the urinary bladder.

The hilus expands into a cavity within the kidney called the **renal sinus,** which contains part of the renal pelvis, the calyces, and branches of the renal blood vessels and nerves. Adipose tissue helps stabilize the position of these structures in the renal sinus.

Blood and Nerve Supply of the Kidneys

Because the kidneys remove wastes from the blood and regulate its volume and ionic composition, it is not surprising that they are abundantly supplied with blood vessels. Although the kidneys constitute less than 0.5% of total body mass, they receive

Figure 26.3 Internal anatomy of the kidneys. (See Tortora, *A Photographic Atlas of the Human Body,* Figures 13.4, 13.5.)

The two main regions of the kidney parenchyma are the renal cortex and the renal pyramids in the renal medulla.

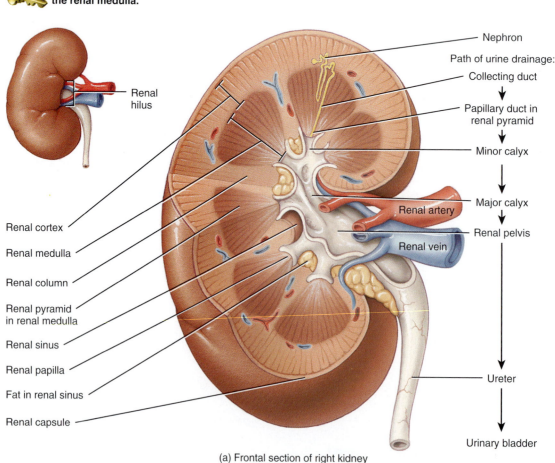

(a) Frontal section of right kidney

20–25% of the resting cardiac output via the right and left **renal arteries** (Figure 26.4).

Within the kidney, the renal artery divides into several **segmental arteries,** each of which gives off several branches that enter the parenchyma and pass through the renal columns between the renal pyramids as the interlobar arteries. At the bases of the renal pyramids, the **interlobar arteries** arch between the renal medulla and cortex; here they are known as the **arcuate arteries** (AR-kū-āt = shaped like a bow). Divisions of the arcuate arteries produce a series of **interlobular arteries,** which enter the renal cortex and give off branches called **afferent arterioles** (*af-* = toward; *-ferrent* = to carry).

Each nephron receives one afferent arteriole, which divides into a tangled, ball-shaped capillary network called the **glomerulus** (glō-MER-ū-lus = little ball; plural is *glomeruli*). The glomerular capillaries then reunite to form an **efferent arteriole** (*ef-* = out) that carries blood out of the glomerulus. Glomerular capillaries are unique among capillaries in the body because they are positioned between two arterioles, rather than between an arteriole and a venule. Because they are capillary networks and they also play an important role in urine formation, the glomeruli are part of both the cardiovascular and the urinary systems.

The efferent arterioles divide to form the **peritubular capillaries** (*peri-* = around), which surround tubular parts of the nephron in the renal cortex. Extending from some efferent arterioles are long loop-shaped capillaries called **vasa recta** (VĀ-sa REK-ta; *vasa* = vessels; *recta* = straight) that supply tubular portions of the nephron in the renal medulla (see Figure 26.5b).

The peritubular capillaries eventually reunite to form **peritubular venules** and then **interlobular veins,** which also receive blood from the vasa recta. Then the blood drains through the **arcuate veins** to the **interlobar veins** running between the renal pyramids. Blood leaves the kidney through a single **renal vein** that exits at the renal hilus and carries venous blood to the inferior vena cava.

Most renal nerves originate in the *celiac ganglion* and pass through the *renal plexus* into the kidneys along with the renal arteries. Renal nerves are part of the sympathetic division of the autonomic nervous system. Most are vasomotor nerves that regulate the flow of blood through the kidney by causing vasodilation or vasoconstriction of renal arterioles.

The Nephron

Parts of a Nephron

Nephrons are the functional units of the kidneys. Each nephron (Figure 26.5) consists of two parts: a **renal corpuscle** (KOR-pus-sul = tiny body), where blood plasma is filtered, and a **renal tubule** into which the filtered fluid passes. The two components

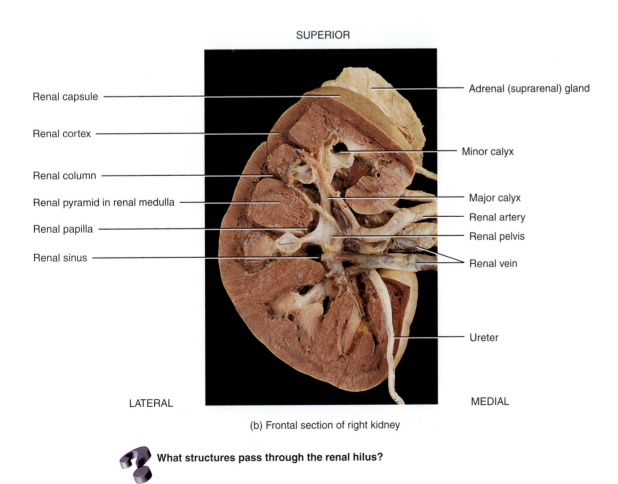

SUPERIOR

Renal capsule —
Renal cortex —
Renal column —
Renal pyramid in renal medulla —
Renal papilla —
Renal sinus —

— Adrenal (suprarenal) gland
— Minor calyx
— Major calyx
— Renal artery
— Renal pelvis
— Renal vein
— Ureter

LATERAL   MEDIAL

(b) Frontal section of right kidney

What structures pass through the renal hilus?

Figure 26.4   Blood supply of the kidneys.   (See Tortora, *A Photographic Atlas of the Human Body,* Figure 13.6.)

🔑 **The renal arteries deliver 20–25% of the resting cardiac output to the kidneys.**

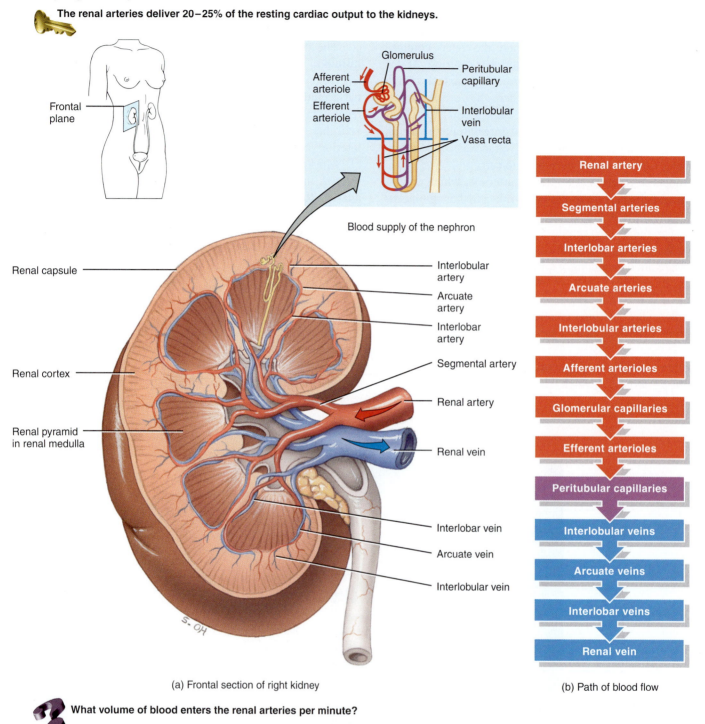

(a) Frontal section of right kidney

(b) Path of blood flow

❓ **What volume of blood enters the renal arteries per minute?**

of a renal corpuscle are the **glomerulus** (capillary network) and the **glomerular (Bowman's) capsule,** a double-walled epithelial cup that surrounds the glomerular capillaries. Blood plasma is first filtered in the glomerular capsule, and then the filtered fluid passes into the renal tubule, which has three main sections. In the order that fluid passes through them, the renal tubule consists of a (1) **proximal convoluted tubule,** (2) **loop of Henle (nephron loop),** and (3) **distal convoluted tubule.** *Proximal* denotes the part of the tubule attached to the glomerular capsule, and *distal* denotes the part that is farther away. *Convoluted* means the tubule is tightly coiled rather than straight. The renal corpuscle and both convoluted tubules lie within the renal cortex, whereas the loop of Henle extends into the renal medulla, makes a hairpin turn, and then returns to the renal cortex.

Figure 26.5 **The structure of nephrons (colored gold) and associated blood vessels.**
(a) A cortical nephron. (b) A juxtamedullary nephron.

🔑 **Nephrons are the functional units of the kidneys.**

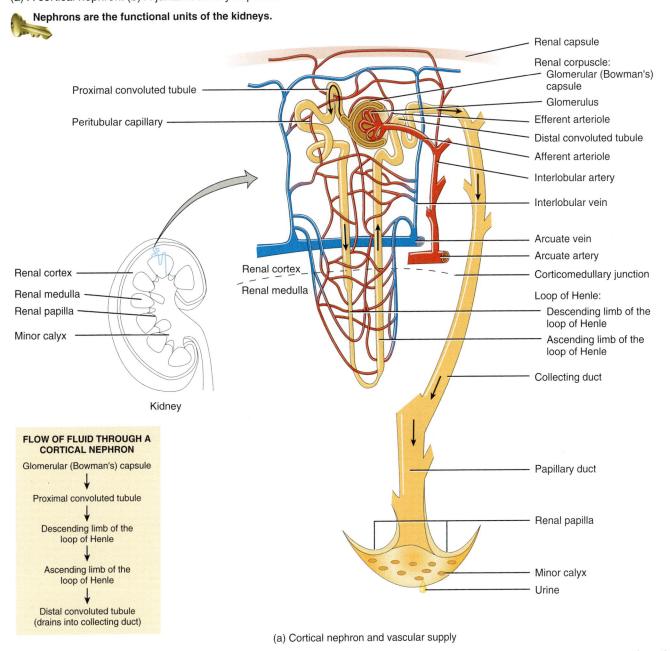

(a) Cortical nephron and vascular supply

FLOW OF FLUID THROUGH A CORTICAL NEPHRON

Glomerular (Bowman's) capsule
↓
Proximal convoluted tubule
↓
Descending limb of the loop of Henle
↓
Ascending limb of the loop of Henle
↓
Distal convoluted tubule (drains into collecting duct)

(continues)

The distal convoluted tubules of several nephrons empty into a single **collecting duct.** Collecting ducts then unite and converge until eventually there are only several hundred large **papillary ducts,** which drain into the minor calyces. The collecting ducts and papillary ducts extend from the renal cortex through the renal medulla to the renal pelvis. Although one kidney has about 1 million nephrons, it has a much smaller number of collecting ducts and even fewer papillary ducts.

In a nephron, the loop of Henle connects the proximal and distal convoluted tubules. The first part of the loop of Henle dips into the renal medulla, where it is called the **descending limb of the loop of Henle** (Figure 26.5). It then makes that hairpin turn and returns to the renal cortex as the **ascending limb of the loop of Henle.** About 80–85% of the nephrons are **cortical nephrons.** Their renal corpuscles lie in the outer portion of the renal cortex, and they have *short* loops of Henle that lie mainly in the cortex and penetrate only into the outer region of the renal medulla (Figure 26.5a). The short loops of Henle receive their blood supply from peritubular capillaries that arise from efferent arterioles. The other 15–20% of the nephrons are

Figure 26.5 *(continued)*

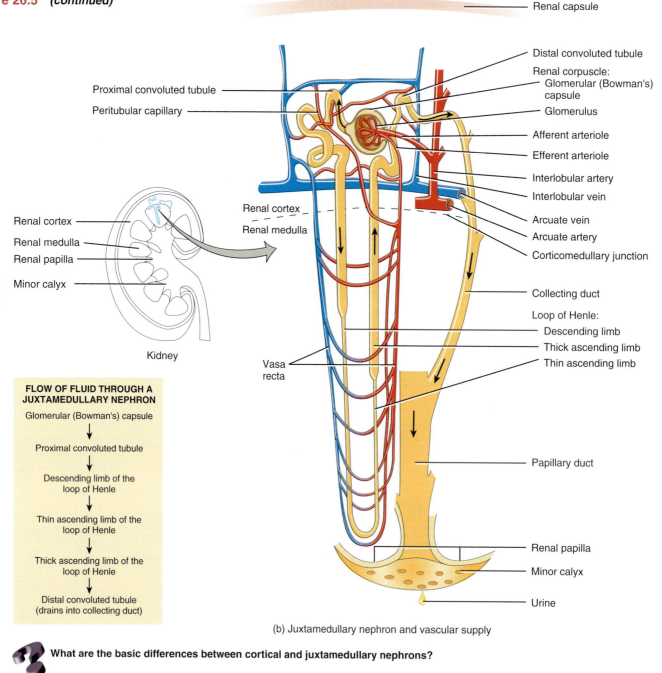

FLOW OF FLUID THROUGH A
JUXTAMEDULLARY NEPHRON

Glomerular (Bowman's) capsule
↓
Proximal convoluted tubule
↓
Descending limb of the
loop of Henle
↓
Thin ascending limb of the
loop of Henle
↓
Thick ascending limb of the
loop of Henle
↓
Distal convoluted tubule
(drains into collecting duct)

(b) Juxtamedullary nephron and vascular supply

What are the basic differences between cortical and juxtamedullary nephrons?

juxtamedullary nephrons (*juxta-* = near to). Their renal corpuscles lie deep in the cortex, close to the medulla, and they have a *long* loop of Henle that extends into the deepest region of the medulla (Figure 26.5b). Long loops of Henle receive their blood supply from peritubular capillaries and from the vasa recta that arise from efferent arterioles. In addition, the ascending limb of the loop of Henle of juxtamedullary nephrons consists of two portions: a **thin ascending limb** followed by a **thick ascending limb** (Figure 26.5b). The lumen of the thin ascending limb is the same as in other areas of the renal tubule; it is only the epithelium that is thinner. Nephrons with long loops of

Henle enable the kidneys to excrete very dilute or very concentrated urine (described on pages 972–973).

Histology of the Nephron and Collecting Duct

A single layer of epithelial cells forms the entire wall of the glomerular capsule, renal tubule, and ducts. Each part, however, has distinctive histological features that reflect its particular functions. In the order that fluid flows through them, the parts are the glomerular capsule, the renal tubule, and the collecting duct.

GLOMERULAR CAPSULE The glomerular (Bowman's) capsule consists of visceral and parietal layers (Figure 26.6a). The visceral layer consists of modified simple squamous epithelial cells called **podocytes** (PŌ-dō-cīts; *podo-* = foot; *-cytes* = cells). The many footlike projections of these cells (pedicels) wrap around the single layer of endothelial cells of the glomerular capillaries and form the inner wall of the capsule. The parietal layer of the glomerular capsule consists of simple squamous epithelium and forms the outer wall of the capsule. Fluid filtered from the glomerular capillaries enters the **capsular (Bowman's)**

Figure 26.6 Histology of a renal corpuscle.

🔑 A renal corpuscle consists of a glomerular (Bowman's) capsule and a glomerulus.

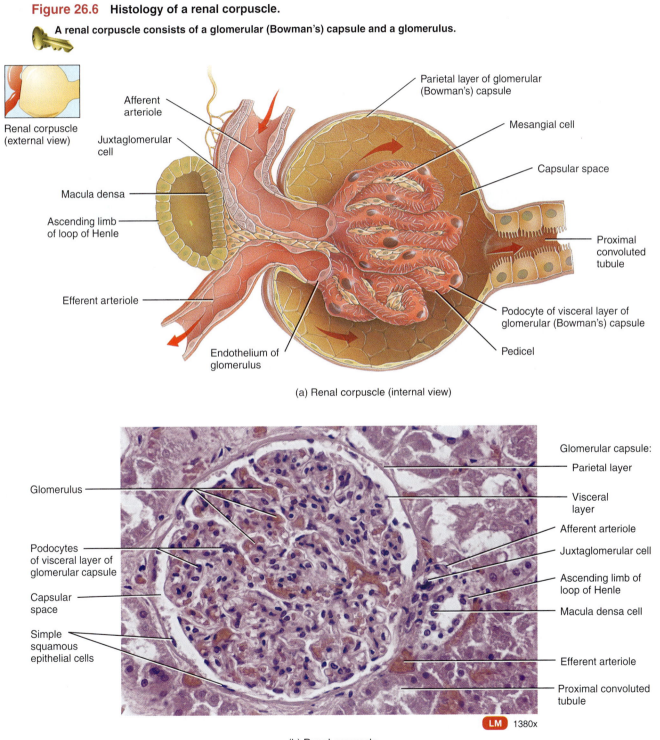

Renal corpuscle (external view)

Afferent arteriole
Juxtaglomerular cell
Macula densa
Ascending limb of loop of Henle
Efferent arteriole
Endothelium of glomerulus

Parietal layer of glomerular (Bowman's) capsule
Mesangial cell
Capsular space
Proximal convoluted tubule
Podocyte of visceral layer of glomerular (Bowman's) capsule
Pedicel

(a) Renal corpuscle (internal view)

Glomerulus
Podocytes of visceral layer of glomerular capsule
Capsular space
Simple squamous epithelial cells

Glomerular capsule:
Parietal layer
Visceral layer
Afferent arteriole
Juxtaglomerular cell
Ascending limb of loop of Henle
Macula densa cell
Efferent arteriole
Proximal convoluted tubule

LM 1380x

(b) Renal corpuscle

❓ Is the photomicrograph in (b) from a section through the renal cortex or renal medulla? How can you tell?

space, the space between the two layers of the glomerular capsule. Think of the relationship between the glomerulus and glomerular capsule in the following way. The glomerulus is a fist punched into a limp balloon (the glomerular capsule) until the fist is covered by two layers of balloon (visceral and parietal layers) with a space in between, the capsular space.

RENAL TUBULE AND COLLECTING DUCT Table 26.1 illustrates the histology of the cells that form the renal tubule and collecting duct. In the proximal convoluted tubule, the cells are simple cuboidal epithelial cells with a prominent brush border of microvilli on their apical surface (surface facing the lumen). These microvilli, like those of the small intestine, increase the surface area for reabsorption and secretion. The descending limb of the loop of Henle and the first part of the ascending limb of the loop of Henle (the thin ascending limb) are composed of simple squamous epithelium. (Recall that cortical or short-loop

nephrons lack the thin ascending limb.) The thick ascending limb of the loop of Henle is composed of simple cuboidal to low columnar epithelium.

In each nephron, the final part of the ascending limb of the loop of Henle makes contact with the afferent arteriole serving that renal corpuscle (Figure 26.6a). Because the columnar tubule cells in this region are crowded together, they are known as the **macula densa** (*macula* = spot; *densa* = dense). Alongside the macula densa, the wall of the afferent arteriole (and sometimes the efferent arteriole) contains modified smooth muscle fibers called **juxtaglomerular (JG) cells.** Together with the macula densa, they constitute the **juxtaglomerular apparatus (JGA).** As you will see later, the JGA helps regulate blood pressure within the kidneys. The distal convoluted tubule (DCT) begins a short distance past the macula densa. In the last part of the DCT and continuing into the collecting ducts, two different types of cells are present. Most are **principal cells,** which have receptors

Table 26.1	Histological Features of the Renal Tubule and Collecting Duct	
Region and Histology		**Description**
Proximal convoluted tubule (PCT)		Simple cuboidal epithelial cells with prominent brush borders of microvilli.
Loop of Henle: descending limb and thin ascending limb		Simple squamous epithelial cells.
Loop of Henle: thick ascending limb		Simple cuboidal to low columnar epithelial cells.
Most of distal convoluted tubule (DCT)		Simple cuboidal epithelial cells.
Last part of DCT and all of collecting duct (CD)		Simple cuboidal epithelium consisting of principal cells and intercalated cells.

for both antidiuretic hormone (ADH) and aldosterone, two hormones that regulate their functions. A smaller number are **intercalated cells,** which play a role in the homeostasis of blood pH. The collecting ducts drain into large papillary ducts, which are lined by simple columnar epithelium.

Number of Nephrons

The number of nephrons is constant from birth. Any increase in kidney size is due solely to the growth of individual nephrons. If nephrons are injured or become diseased, new ones do not form. Signs of kidney dysfunction usually do not become apparent until function declines to less than 25% of normal because the remaining functional nephrons adapt to handle a larger-than-normal load. Surgical removal of one kidney, for example, stimulates hypertrophy (enlargement) of the remaining kidney, which eventually is able to filter blood at 80% of the rate of two normal kidneys. ■

▶ CHECKPOINT

2. Why are the kidneys said to be retroperitoneal?

3. Which branch of the autonomic nervous system innervates renal blood vessels?

4. How do cortical nephrons and juxtamedullary nephrons differ structurally?

5. Where is the juxtaglomerular apparatus (JGA) located and what is its structure?

OVERVIEW OF RENAL PHYSIOLOGY

▶ OBJECTIVE

• **Identify the three basic functions performed by nephrons and collecting ducts, and indicate where each occurs.**

To produce urine, nephrons and collecting ducts perform three basic processes—glomerular filtration, tubular secretion, and tubular reabsorption (Figure 26.7):

❶ *Glomerular filtration.* In the first step of urine production, water and most solutes in blood plasma move across the wall of glomerular capillaries into the glomerular capsule and then into the renal tubule.

❷ *Tubular reabsorption.* As filtered fluid flows along the renal tubule and through the collecting duct, tubule cells reabsorb about 99% of the filtered water and many useful solutes. The water and solutes return to the blood as it flows through the peritubular capillaries and vasa recta. Note that the term *reabsorption* refers to the return of substances to the bloodstream. The term *absorption,* by contrast, means entry of new substances into the body, as occurs in the gastrointestinal tract.

❸ *Tubular secretion.* As fluid flows along the tubule and through the collecting duct, the tubule and duct cells secrete other materials, such as wastes, drugs, and excess ions, into the fluid. Notice that tubular secretion *removes* a substance from the blood. In other instances of secretion—for

Figure 26.7 Relation of a nephron's structure to its three basic functions: glomerular filtration, tubular reabsorption, and tubular secretion. Excreted substances remain in the urine and subsequently leave the body. For any substance S, excretion rate of S = filtration rate of S − reabsorption rate of S + secretion rate of S.

Glomerular filtration occurs in the renal corpuscle, whereas tubular reabsorption and tubular secretion occur all along the renal tubule and collecting duct.

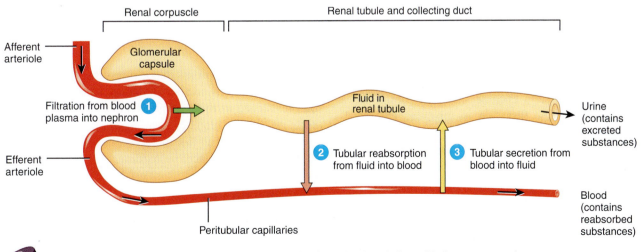

When cells of the renal tubules secrete the drug penicillin, is the drug being added to or removed from the bloodstream?

patient's peritoneal cavity and connected to a sterile dialysis solution. The dialysate flows into the peritoneal cavity from its plastic container by gravity. The solution remains in the cavity until metabolic waste products, excess electrolytes, and extracellular fluid diffuse into the dialysis solution. The solution is then drained from the cavity by gravity into a sterile bag that is discarded. ■

▶ **CHECKPOINT**

18. What are the characteristics of normal urine?

19. What chemical substances normally are present in urine?

20. How may kidney function be evaluated?

21. Why are the renal plasma clearances of glucose, urea, and creatinine different? How does each clearance compare to glomerular filtration rate?

URINE TRANSPORTATION, STORAGE, AND ELIMINATION

▶ **OBJECTIVE**

• **Describe the anatomy, histology, and physiology of the ureters, urinary bladder, and urethra.**

From collecting ducts, urine drains through papillary ducts into the minor calyces, which join to become major calyces that unite to form the renal pelvis (see Figure 26.3). From the renal pelvis, urine first drains into the ureters and then into the urinary bladder. Urine is then discharged from the body through the single urethra (see Figure 26.1).

Ureters

Each of the two **ureters** (Ū-rē-ters) transports urine from the renal pelvis of one kidney to the urinary bladder. Peristaltic contractions of the muscular walls of the ureters push urine toward the urinary bladder, but hydrostatic pressure and gravity also contribute. Peristaltic waves that pass from the renal pelvis to the urinary bladder vary in frequency from one to five per minute, depending on how fast urine is being formed.

The ureters are 25–30 cm (10–12 in.) long and are thick-walled, narrow tubes that vary in diameter from 1 mm to 10 mm along their course between the renal pelvis and the urinary bladder. Like the kidneys, the ureters are retroperitoneal. At the base of the urinary bladder, the ureters curve medially and pass obliquely through the wall of the posterior aspect of the urinary bladder (Figure 26.21).

Even though there is no anatomical valve at the opening of each ureter into the urinary bladder, a physiological one is quite effective. As the urinary bladder fills with urine, pressure within it compresses the oblique openings into the ureters and prevents the backflow of urine. When this physiological valve is not

Figure 26.21 Ureters, urinary bladder, and urethra in a female. (See Tortora, *A Photographic Atlas of the Human Body,* Figures 13.8, 13.10.)

🔑 **Urine is stored in the urinary bladder before being expelled by micturition.**

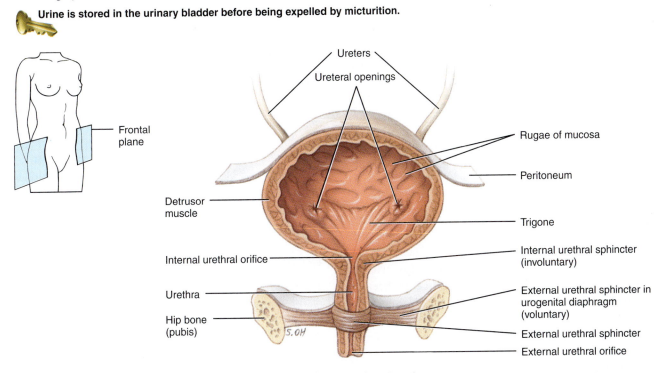

Anterior view of frontal section

❓ **What is a lack of voluntary control over micturition called?**

operating properly, it is possible for microbes to travel up the ureters from the urinary bladder to infect one or both kidneys.

Three coats of tissue form the wall of the ureters. The deepest coat, or **mucosa,** is a mucous membrane with **transitional epithelium** (see Table 4.1E on page 111) and an underlying **lamina propria** of areolar connective tissue with considerable collagen, elastic fibers, and lymphatic tissue. Transitional epithelium is able to stretch—a marked advantage for any organ that must accommodate a variable volume of fluid. Mucus secreted by the goblet cells of the mucosa prevents the cells from coming in contact with urine, the solute concentration and pH of which may differ drastically from the cytosol of cells that form the wall of the ureters. Throughout most of the length of the ureters, the intermediate coat, the **muscularis,** is composed of inner longitudinal and outer circular layers of smooth muscle fibers. This arrangement is opposite to that of the gastrointestinal tract, which contains inner circular and outer longitudinal layers. The muscularis of the distal third of the ureters also contains an outer layer of longitudinal muscle fibers. Thus, the muscularis in the distal third of the ureter is inner longitudinal, middle circular, and outer longitudinal. Peristalsis is the major function of the muscularis. The superficial coat of the ureters is the **adventitia,** a layer of areolar connective tissue containing blood vessels, lymphatic vessels, and nerves that serve the muscularis and mucosa. The adventitia blends in with surrounding connective tissue and anchors the ureters in place.

Urinary Bladder

The **urinary bladder** is a hollow, distensible muscular organ situated in the pelvic cavity posterior to the pubic symphysis. In males, it is directly anterior to the rectum; in females, it is anterior to the vagina and inferior to the uterus (see Figure 28.13b on page 1029). Folds of the peritoneum hold the urinary bladder in position. The shape of the urinary bladder depends on how much urine it contains. Empty, it is collapsed; when slightly distended it becomes spherical. As urine volume increases, it becomes pear-shaped and rises into the abdominal cavity. Urinary bladder capacity averages 700–800 mL. It is smaller in females because the uterus occupies the space just superior to the urinary bladder.

Anatomy and Histology of the Urinary Bladder

In the floor of the urinary bladder is a small triangular area called the **trigone** (TRĪ-gōn = triangle). The two posterior corners of the trigone contain the two ureteral openings, whereas the opening into the urethra, **the internal urethral orifice,** lies in the anterior corner (Figure 26.21). Because its mucosa is firmly bound to the muscularis, the trigone has a smooth appearance.

Three coats make up the wall of the urinary bladder. The deepest is the **mucosa,** a mucous membrane composed of **tran-**sitional epithelium and an underlying **lamina propria** similar to that of the ureters. Rugae (the folds in the mucosa) are also present to permit expansion of the urinary bladder. Surrounding the mucosa is the intermediate **muscularis,** also called the **detrusor muscle** (de-TROO-ser = to push down), which consists of three layers of smooth muscle fibers: the inner longitudinal, middle circular, and outer longitudinal layers. Around the opening to the urethra the circular fibers form an **internal urethral sphincter;** inferior to it is the **external urethral sphincter,** which is composed of skeletal muscle and is a modification of the urogenital diaphragm muscle (see Figure 11.12 on page 339). The most superficial coat of the urinary bladder on the posterior and inferior surfaces is the **adventitia,** a layer of areolar connective tissue that is continuous with that of the ureters. Over the superior surface of the urinary bladder is the **serosa,** a layer of visceral peritoneum.

The Micturition Reflex

Discharge of urine from the urinary bladder, called **micturition** (mik′-choo-RISH-un; *mictur-* = urinate), is also known as *urination* or *voiding*. Micturition occurs via a combination of involuntary and voluntary muscle contractions. When the volume of urine in the bladder exceeds 200–400 mL, pressure within the bladder increases considerably, and stretch receptors in its wall transmit nerve impulses into the spinal cord. These impulses propagate to the **micturition center** in sacral spinal cord segments S2 and S3 and trigger a spinal reflex called the **micturition reflex.** In this reflex arc, parasympathetic impulses from the micturition center propagate to the urinary bladder wall and internal urethral sphincter. The nerve impulses cause *contraction* of the detrusor muscle and *relaxation* of the internal urethral sphincter muscle. Simultaneously, the micturition center inhibits somatic motor neurons that innervate skeletal muscle in the external urethral sphincter. Upon contraction of the bladder wall and relaxation of the sphincters, urination takes place. Bladder filling causes a sensation of fullness that initiates a conscious desire to urinate before the micturition reflex actually occurs. Although emptying of the urinary bladder is a reflex, in early childhood we learn to initiate it and stop it voluntarily. Through learned control of the external urethral sphincter muscle and certain muscles of the pelvic floor, the cerebral cortex can initiate micturition or delay its occurrence for a limited period.

Urethra

The **urethra** (ū-RĒ-thra) is a small tube leading from the internal urethral orifice in the floor of the urinary bladder to the exterior of the body (see Figure 26.1). In both males and females, the urethra is the terminal portion of the urinary system and the passageway for discharging urine from the body. In males, it discharges semen (fluid that contains sperm) as well.

In females, the urethra lies directly posterior to the pubic symphysis, is directed obliquely inferiorly and anteriorly, and has a length of 4 cm (1.5 in.) (see Figure 28.13a on page 1028). The opening of the urethra to the exterior, the **external urethral orifice,** is located between the clitoris and the vaginal opening (see Figure 28.22 on page 1038). The wall of the female urethra consists of a deep **mucosa** and a superficial **muscularis.** The mucosa is a mucous membrane composed of **epithelium** and **lamina propria** (areolar connective tissue with elastic fibers and a plexus of veins). The muscularis consists of circularly arranged smooth muscle fibers and is continuous with that of the urinary bladder. Near the urinary bladder, the mucosa contains transitional epithelium that is continuous with that of the urinary bladder; near the external urethral orifice, the epithelium is nonkeratinized stratified squamous epithelium for protection. Between these areas, the mucosa contains stratified columnar or pseudostratified columnar epithelium.

In males, the urethra also extends from the internal urethral orifice to the exterior, but its length and passage through the body are considerably different than in females (see Figure 28.3 on page 1016 and Figure 28.12 on page 1028). The male urethra first passes through the prostate, then through the urogenital diaphragm, and finally through the penis, a distance of 15–20 cm (6–8 in.).

The male urethra, which also consists of a deep **mucosa** and a superficial **muscularis,** is subdivided into three anatomical regions: (1) The **prostatic urethra** passes through the prostate. (2) The **membranous urethra,** the shortest portion, passes through the urogenital diaphragm. (3) The **spongy urethra,** the longest portion, passes through the penis. The epithelium of the prostatic urethra is continuous with that of the urinary bladder and consists of transitional epithelium that becomes stratified columnar or pseudostratified columnar epithelium more distally. The mucosa of the membranous urethra contains stratified columnar or pseudostratified columnar epithelium. The epithelium of the spongy urethra is stratified columnar or pseudostratified columnar epithelium, except near the external urethral orifice. There it is nonkeratinized stratified squamous epithelium. The **lamina propria** of the male urethra is areolar connective tissue with elastic fibers and a plexus of veins.

The muscularis of the prostatic urethra is composed of mostly circular smooth muscle fibers superficial to the lamina propria; these circular fibers help form the internal urethral sphincter of the urinary bladder. The muscularis of the membranous urethra consists of circularly arranged skeletal muscle fibers of the urogenital diaphragm that help form the external urethral sphincter of the urinary bladder.

Several glands and other structures associated with reproduction deliver their contents into the male urethra. The prostatic urethra contains the openings from (1) ducts that transport secretions from the **prostate** and from (2) the **seminal vesicles** and **ductus (vas) deferens,** which deliver sperm into the urethra and provide secretions that both neutralize the acidity of the female reproductive tract and contribute to sperm motility and viability. The openings of the ducts of the **bulbourethral (Cowper's) glands** empty into the spongy urethra. They deliver an alkaline substance before ejaculation that neutralizes the acidity of the urethra. The glands also secrete mucus, which lubricates the end of the penis during sexual arousal. Throughout the urethra, but especially in the spongy urethra, the openings of the ducts of **urethral (Littré) glands** discharge mucus during sexual arousal or ejaculation.

Urinary Incontinence

A lack of voluntary control over micturition is called **urinary incontinence.** In infants and children under 2–3 years old, incontinence is normal because neurons to the external urethral sphincter muscle are not completely developed, and voiding occurs whenever the urinary bladder is sufficiently distended to stimulate the micturition reflex. Urinary incontinence also occurs in adults. Of the more than 10 million U.S. adults who suffer from urinary incontinence, 1–2 million have **stress incontinence.** In this condition, physical stresses that increase abdominal pressure, such as coughing, sneezing, laughing, exercising, pregnancy, or simply walking cause leakage of urine from the urinary bladder. Other causes of incontinence in adults are injury to the nerves controlling the urinary bladder, loss of bladder flexibility with age, disease or irritation of the urinary bladder or urethra, damage to the external urethral sphincter, or certain drugs. Also, compared with nonsmokers, those who smoke have twice the risk of developing incontinence. ■

► **CHECKPOINT**

22. What forces help propel urine from the renal pelvis to the urinary bladder?

23. What is micturition? How does the micturition reflex occur?

24. How do the location, length, and histology of the urethra compare in males and females?

WASTE MANAGEMENT IN OTHER BODY SYSTEMS

► **OBJECTIVE**

- **Describe the ways that body wastes are handled.**

As we have seen, just one of the many functions of the urinary system is to help rid the body of some kinds of waste materials. Besides the kidneys, several other tissues, organs, and processes contribute to the temporary confinement of wastes, the transport of waste materials for disposal, the recycling of materials, and the excretion of excess or toxic substances in the body. These waste management systems include the following:

- **Body buffers.** Buffers bind excess hydrogen ions (H^+), thereby preventing an increase in the acidity of body fluids. Buffers are like wastebaskets in that they have a limited capacity; eventually the H^+, like the paper in a wastebasket, must be eliminated from the body by excretion.

- **Blood.** The bloodstream provides pickup and delivery services for the transport of wastes, in much the same way that garbage trucks and sewer lines serve a community.

- **Liver.** The liver is the primary site for metabolic recycling, as occurs, for example, in the conversion of amino acids into glucose or of glucose into fatty acids. The liver also converts toxic substances into less toxic ones, such as ammonia into urea. These functions of the liver are described in Chapters 24 and 25.

- **Lungs.** With each exhalation, the lungs excrete CO_2, and expel heat and a little water vapor.

- **Sweat (sudoriferous) glands.** Especially during exercise, sweat glands in the skin help eliminate excess heat, water, and CO_2, plus small quantities of salts and urea as well.

- **Gastrointestinal tract.** Through defecation, the gastrointestinal tract excretes solid, undigested foods; wastes; some CO_2; water; salts; and heat.

At about the fifth week, a mesodermal outgrowth, called a **ureteric bud,** develops from the distal portion of the mesonephric duct near the cloaca. The **metanephros,** or ultimate kidney, develops from the ureteric bud and metanephric mesoderm. The ureteric bud forms the *collecting ducts, calyces, renal pelvis,* and *ureter.* The **metanephric mesoderm** forms the *nephrons* of the kidneys. By the third month, the fetal kidneys begin excreting urine into the surrounding amniotic fluid; indeed, fetal urine makes up most of the amniotic fluid.

During development, the cloaca divides into a **urogenital sinus,** into which urinary and genital ducts empty, and a *rectum* that discharges into the anal canal. The *urinary bladder* develops from the urogenital sinus. In females, the *urethra* develops as a result of lengthening of the short duct that extends from the urinary bladder to the urogenital sinus. In males, the urethra is considerably longer and more complicated, but it is also derived from the urogenital sinus.

▶ **CHECKPOINT**

25. Which tissue develops into nephrons?

26. Which tissue gives rise to collecting ducts, calyces, renal pelves, and ureters?

DEVELOPMENT OF THE URINARY SYSTEM

▶ **OBJECTIVE**

- **Describe the development of the urinary system.**

Starting in the third week of fetal development, a portion of the mesoderm along the posterior aspect of the embryo, the **intermediate mesoderm,** differentiates into the kidneys. Three pairs of kidneys form within the intermediate mesoderm in succession: the pronephros, the mesonephros, and the metanephros (Figure 26.22). Only the last pair remains as the functional kidneys of the newborn.

The first kidney to form, the **pronephros,** is the most superior of the three and has an associated **pronephric duct.** This duct empties into the **cloaca,** the expanded terminal part of the hindgut, which functions as a common outlet for the urinary, digestive, and reproductive ducts. The pronephros begins to degenerate during the fourth week and is completely gone by the sixth week.

The second kidney, the **mesonephros,** replaces the pronephros. The retained portion of the pronephric duct, which connects to the mesonephros, develops into the **mesonephric duct.** The mesonephros begins to degenerate by the sixth week and is almost gone by the eighth week.

AGING AND THE URINARY SYSTEM

▶ **OBJECTIVE**

- **Describe the effects of aging on the urinary system.**

With aging, the kidneys shrink in size, have a decreased blood flow, and filter less blood. The mass of the two kidneys decreases from an average of 260 g in 20-year-olds to less than 200 g by age 80. Likewise, renal blood flow and filtration rate decline by 50% between ages 40 and 70. Kidney diseases that become more common with age include acute and chronic kidney inflammations and renal calculi (kidney stones). Because the sensation of thirst diminishes with age, older individuals also are susceptible to dehydration. Urinary tract infections are more common among the elderly, as are polyuria (excessive urine production), nocturia (excessive urination at night), increased frequency of urination, dysuria (painful urination), urinary retention or incontinence, and hematuria (blood in the urine).

• • •

To appreciate the many ways that the urinary system contributes to homeostasis of other body systems, examine *Focus on Homeostasis: The Urinary System* on page 984. Next, in Chapter 27, we will see how the kidneys and lungs contribute to maintenance of homeostasis of body fluid volume, electrolyte levels in body fluids, and acid–base balance.

Figure 26.22 **Development of the urinary system.**

Three pairs of kidneys form within intermediate mesoderm in succession: pronephros, mesonephros, and metanephros.

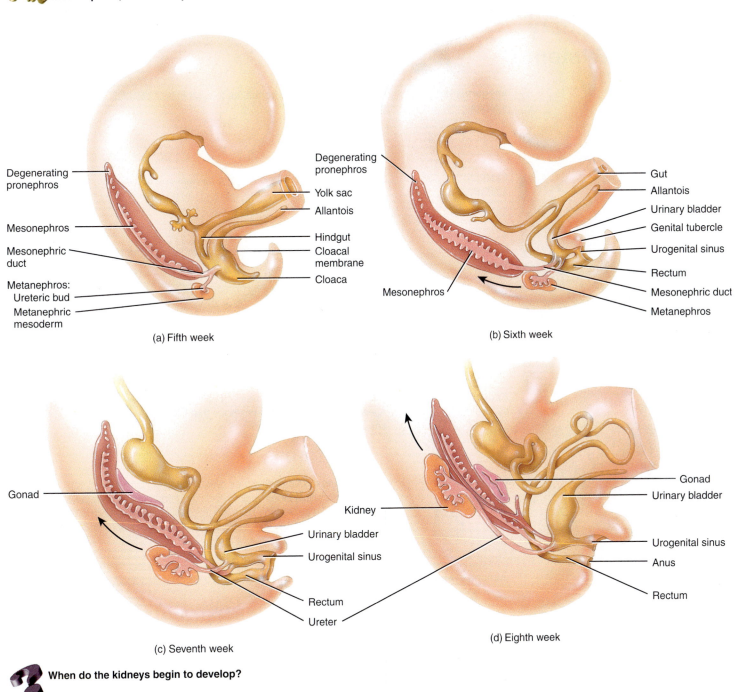

(a) Fifth week

Degenerating pronephros

Mesonephros

Mesonephric duct

Metanephros:
Ureteric bud
Metanephric mesoderm

Yolk sac
Allantois
Hindgut
Cloacal membrane
Cloaca

(b) Sixth week

Degenerating pronephros

Mesonephros

Gut
Allantois
Urinary bladder
Genital tubercle
Urogenital sinus
Rectum
Mesonephric duct
Metanephros

(c) Seventh week

Gonad

Urinary bladder
Urogenital sinus

Rectum
Ureter

(d) Eighth week

Kidney

Gonad
Urinary bladder

Urogenital sinus
Anus

Rectum

When do the kidneys begin to develop?

Focus on Homeostasis:

The Urinary System

Body System	Contribution of the Urinary System
For all body systems	Kidneys regulate volume, composition, and pH of body fluids by removing wastes and excess substances from blood and excreting them in urine; the ureters transport urine from the kidneys to the urinary bladder, which stores urine until it is eliminated through the urethra.
Integumentary system	Kidneys and skin both contribute to synthesis of calcitriol, the active form of vitamin D.
Skeletal system	Kidneys help adjust levels of blood calcium and phosphates, needed for building bone matrix.
Muscular system	Kidneys help adjust level of blood calcium, needed for contraction of muscle.
Nervous system	Kidneys perform gluconeogenesis, which provides glucose for ATP production in neurons, especially during fasting or starvation.
Endocrine system	Kidneys participate in synthesis of calcitriol, the active form of vitamin D, and release erythropoietin, the hormone that stimulates production of red blood cells.
Cardiovascular system	By increasing or decreasing their reabsorption of water filtered from blood, the kidneys help adjust blood volume and blood pressure; renin released by juxtaglomerular cells in the kidneys raises blood pressure; some bilirubin from hemoglobin breakdown is converted to a yellow pigment (urobilin), which is excreted in urine.
Lymphatic and Immune system	By increasing or decreasing their reabsorption of water filtered from blood, the kidneys help adjust the volume of interstitial fluid and lymph; urine flushes microbes out of urethra.
Respiratory system	Kidneys and lungs cooperate in adjusting pH of body fluids.
Digestive system	Kidneys help synthesize calcitriol, the active form of vitamin D, which is needed for absorption of dietary calcium.
Reproductive systems	In males, the portion of the urethra that extends through the prostate and penis is a passageway for semen as well as urine.

984

DISORDERS: HOMEOSTATIC IMBALANCES

Renal Calculi

The crystals of salts present in urine occasionally precipitate and solidify into insoluble stones called **renal calculi** (*calculi* = pebbles) or **kidney stones.** They commonly contain crystals of calcium oxalate, uric acid, or calcium phosphate. Conditions leading to calculus formation include the ingestion of excessive calcium, low water intake, abnormally alkaline or acidic urine, and overactivity of the parathyroid glands. When a stone lodges in a narrow passage, such as a ureter, the pain can be intense. **Shock-wave lithotripsy** (LITH-ō-trip′-sē; *litho* = stone) offers an alternative to surgical removal of kidney stones. A device, called a lithotripter, delivers brief, high-intensity sound waves through a water-filled cushion. Over a period of 30 to 60 minutes, 1000 or more hydraulic shock waves pulverize the stone, creating fragments that are small enough to wash out in the urine.

Urinary Tract Infections

The term **urinary tract infection (UTI)** is used to describe either an infection of a part of the urinary system or the presence of large numbers of microbes in urine. UTIs are more common in females due to their shorter urethra. Symptoms include painful or burning urination, urgent and frequent urination, low back pain, and bed-wetting. UTIs include *urethritis* (inflammation of the urethra), *cystitis* (inflammation of the urinary bladder), and *pyelonephritis* (inflammation of the kidneys). If pyelonephritis becomes chronic, scar tissue can form in the kidneys and severely impair their function.

Glomerular Diseases

A variety of conditions may damage the kidney glomeruli, either directly or indirectly because of disease elsewhere in the body. Typically, the filtration membrane sustains damage, and its permeability increases.

Glomerulonephritis is an inflammation of the kidney that involves the glomeruli. One of the most common causes is an allergic reaction to the toxins produced by streptococcal bacteria that have recently infected another part of the body, especially the throat. The glomeruli become so inflamed, swollen, and engorged with blood that the filtration membranes allow blood cells and plasma proteins to enter the filtrate. As a result, the urine contains many erythrocytes (hematuria) and a lot of protein. The glomeruli may be permanently damaged, leading to chronic renal failure.

Nephrotic syndrome is a condition characterized by *proteinuria* (protein in the urine) and *hyperlipidemia* (high blood levels of cholesterol, phospholipids, and triglycerides). The proteinuria is due to an increased permeability of the filtration membrane, which permits proteins, especially albumin, to escape from blood into urine. Loss of albumin results in *hypoalbuminemia* (low blood albumin level) once liver production of albumin fails to meet increased urinary losses. Edema, usually seen around the eyes, ankles, feet, and abdomen, occurs in nephrotic syndrome because loss of albumin from the blood decreases blood colloid osmotic pressure. Nephrotic syndrome is associated with several glomerular diseases of unknown cause, as well as with systemic disorders such as diabetes mellitus, systemic lupus erythematosus (SLE), a variety of cancers, and AIDS.

Renal Failure

Renal failure is a decrease or cessation of glomerular filtration. In acute renal failure (ARF), the kidneys abruptly stop working entirely (or almost entirely). The main feature of ARF is the suppression of urine flow, usually characterized either by *oliguria* (*olig-* = scanty; *-uria* = urine production), which is daily urine output less than 250 mL, or by *anuria*, daily urine output less than 50 mL. Causes include low blood volume (for example, due to hemorrhage), decreased cardiac output, damaged renal tubules, kidney stones, the dyes used to visualize blood vessels in angiograms, nonsteroidal anti-inflammatory drugs, and some antibiotic drugs.

Renal failure causes a multitude of problems. There is edema due to salt and water retention and acidosis due to an inability of the kidneys to excrete acidic substances. In the blood, urea builds up due to impaired renal excretion of metabolic waste products and potassium level rises, which can lead to cardiac arrest. Often, there is anemia because the kidneys no longer produce enough erythropoietin for adequate red blood cell production. Because the kidneys are no longer able to convert vitamin D to calcitriol, which is needed for adequate calcium absorption from the small intestine, osteomalacia also may occur.

Chronic renal failure (CRF) refers to a progressive and usually irreversible decline in glomerular filtration rate (GFR). CRF may result from chronic glomerulonephritis, pyelonephritis, polycystic kidney disease, or traumatic loss of kidney tissue. CRF develops in three stages. In the first stage, *diminished renal reserve*, nephrons are destroyed until about 75% of the functioning nephrons are lost. At this stage, a person may have no signs or symptoms because the remaining nephrons enlarge and take over the function of those that have been lost. Once 75% of the nephrons are lost, the person enters the second stage, called *renal insufficiency*, characterized by a decrease in GFR and increased blood levels of nitrogen-containing wastes and creatinine. Also, the kidneys cannot effectively concentrate or dilute the urine. The final stage, called *end-stage renal failure*, occurs when about 90% of the nephrons have been lost. At this stage, GFR diminishes to 10–15% of normal, oliguria is present, and blood levels of nitrogen-containing wastes and creatinine increase further. People with end-stage renal failure need dialysis therapy and are possible candidates for a kidney transplant operation.

Polycystic Kidney Disease

Polycystic kidney disease (PKD) is one of the most common inherited disorders. In PKD, the kidney tubules become riddled with hundreds or thousands of cysts (fluid-filled cavities). In addition, inappropriate apoptosis (programmed cell death) of cells in noncystic tubules leads to progressive impairment of renal function and eventually to end-stage renal failure.

People with PKD also may have cysts and apoptosis in the liver, pancreas, spleen, and gonads; increased risk of cerebral aneurysms; heart valve defects; and diverticuli in the colon. Typically, symptoms are not noticed until adulthood, when patients may have back pain, urinary tract infections, blood in the urine, hypertension, and large abdominal masses. Using drugs to restore normal blood pressure, restricting protein and salt in the diet, and controlling urinary tract infections may slow progression to renal failure.

Urinary Bladder Cancer

Each year, nearly 12,000 Americans die from **urinary bladder cancer.** It generally strikes people over 50 years of age and is three times more likely to develop in males than females. The disease is typically painless as it develops, but in most cases blood in the urine is a primary sign of the disease. Less often, people experience painful and/or frequent urination.

As long as the disease is identified early and treated promptly, the prognosis is favorable. Fortunately, about 75% of the urinary bladder cancers are confined to the epithelium of the urinary bladder and are easily removed by surgery. The lesions tend to be low grade, meaning that they have only a small potential for metastasis.

Urinary bladder cancer is frequently the result of a carcinogen. About half of all cases occur in people who smoke or have at some time smoked cigarettes. The cancer also tends to develop in people who are exposed to chemicals called aromatic amines. Workers in the leather, dye, rubber, and aluminum industries, as well as painters, are often exposed to these chemicals.

MEDICAL TERMINOLOGY

Azotemia (az-ō-TĒ-mē-a; *azot-* = nitrogen; *-emia* = condition of blood) Presence of urea or other nitrogen-containing substances in the blood.

Cystocele (SIS-tō-sēl; *cysto-* = bladder; *-cele* = hernia or rupture) Hernia of the urinary bladder.

Enuresis (en′-ū-RĒ-sis = to void urine) Involuntary voiding of urine after the age at which voluntary control has typically been attained.

Intravenous pyelogram (in′-tra-VĒ-nus PĪ-el-ō-gram′; *intra-* = within; *veno-* = vein; *pyelo-* = pelvis of kidney; *-gram* =record), or **IVP** Radiograph (x ray) of the kidneys after intravenous injection of a dye.

Nocturnal enuresis (nok-TUR-nal en′-ū-RĒ-sis) Discharge of urine during sleep, resulting in bed-wetting; occurs in about 15% of 5-year-old children and generally resolves spontaneously, afflicting only about 1% of adults. It may have a genetic basis, as bed-wetting occurs more often in identical twins than in fraternal twins and more often in children whose parents or siblings were bed-wetters. Possible causes include smaller-than-normal bladder capacity, failure to awaken in response to a full bladder, and above-normal production of urine at night. Also referred to as **nocturia.**

Polyuria (pol′-ē-Ū-rē-a; *poly-* = too much) Excessive formation of urine.

Stricture (STRIK-chur) Narrowing of the lumen of a canal or hollow organ, as may occur in the ureter, urethra, or any other tubular structure in the body.

Urinary retention A failure to completely or normally void urine; may be due to an obstruction in the urethra or neck of the urinary bladder, to nervous contraction of the urethra, or to lack of urge to urinate. In men, an enlarged prostate may constrict the urethra and cause urinary retention.

STUDY OUTLINE

INTRODUCTION (p. 949)

1. The organs of the urinary system are the kidneys, ureters, urinary bladder, and urethra.
2. After the kidneys filter blood and return most water and many solutes to the bloodstream, the remaining water and solutes constitute urine.

OVERVIEW OF KIDNEY FUNCTIONS (p. 949)

1. The kidneys regulate blood ionic composition, blood osmolarity, blood volume, blood pressure, and blood pH.
2. The kidneys also perform gluconeogenesis, release calcitriol and erythropoietin, and excrete wastes and foreign substances.

ANATOMY AND HISTOLOGY OF THE KIDNEYS (p. 950)

1. The kidneys are retroperitoneal organs attached to the posterior abdominal wall.

2. Three layers of tissue surround the kidneys: renal capsule, adipose capsule, and renal fascia.
3. Internally, the kidneys consist of a renal cortex, a renal medulla, renal pyramids, renal papillae, renal columns, calyces, and a renal pelvis.
4. Blood flows into the kidney through the renal artery and successively into segmental, interlobar, arcuate, and interlobular arteries; afferent arterioles; glomerular capillaries; efferent arterioles; peritubular capillaries and vasa recta; and interlobular, arcuate, and interlobar veins before flowing out of the kidney through the renal vein.
5. Vasomotor nerves from the sympathetic division of the autonomic nervous system supply kidney blood vessels; they help regulate flow of blood through the kidney.
6. The nephron is the functional unit of the kidneys. A nephron consists of a renal corpuscle (glomerulus and glomerular or Bowmanms capsule) and a renal tubule.

7. A renal tubule consists of a proximal convoluted tubule, a loop of Henle, and a distal convoluted tubule, which drains into a collecting duct (shared by several nephrons). The loop of Henle consists of a descending limb and an ascending limb.

8. A cortical nephron has a short loop that dips only into the superficial region of the renal medulla; a juxtamedullary nephron has a long loop of Henle that stretches through the renal medulla almost to the renal papilla.

9. The wall of the entire glomerular capsule, renal tubule, and ducts consists of a single layer of epithelial cells. The epithelium has distinctive histological features in different parts of the tubule. Table 26.1 on page 958 summarizes the histological features of the renal tubule and collecting duct.

10. The juxtaglomerular apparatus (JGA) consists of the juxtaglomerular cells of an afferent arteriole and the macula densa of the final portion of the ascending limb of the loop of Henle.

OVERVIEW OF RENAL PHYSIOLOGY (p. 959)

1. Nephrons perform three basic tasks: glomerular filtration, tubular secretion, and tubular reabsorption.

GLOMERULAR FILTRATION (p. 960)

1. Fluid that enters the capsular space is glomerular filtrate.
2. The filtration membrane consists of the glomerular endothelium, basal lamina, and filtration slits between pedicels of podocytes.
3. Most substances in blood plasma easily pass through the glomerular filter. However, blood cells and most proteins normally are not filtered.
4. Glomerular filtrate amounts to up to 180 liters of fluid per day. This large amount of fluid is filtered because the filter is porous and thin, the glomerular capillaries are long, and the capillary blood pressure is high.
5. Glomerular blood hydrostatic pressure (GBHP) promotes filtration, whereas capsular hydrostatic pressure (CHP) and blood colloid osmotic pressure (BCOP) oppose filtration. Net filtration pressure (NFP) = GBHP – CHP – BCOP. NFP is about 10 mmHg.
6. Glomerular filtration rate (GFR) is the amount of filtrate formed in both kidneys per minute; it is normally 105–125 mL/min.
7. Glomerular filtration rate depends on renal autoregulation, neural regulation, and hormonal regulation. Table 26.2 on page 964 summarizes regulation of GFR.

TUBULAR REABSORPTION AND TUBULAR SECRETION (p. 965)

1. Tubular reabsorption is a selective process that reclaims materials from tubular fluid and returns them to the bloodstream. Reabsorbed substances include water, glucose, amino acids, urea, and ions, such as sodium, chloride, potassium, bicarbonate, and phosphate (Table 26.3 on page 965).
2. Some substances not needed by the body are removed from the blood and discharged into the urine via tubular secretion. Included are ions (K^+, H^+, and NH_4^+), urea, creatinine, and certain drugs.
3. Reabsorption routes include both paracellular (between tubule cells) and transcellular (across tubule cells) routes.
4. The maximum amount of a substance that can be reabsorbed per unit time is called the transport maximum (T_m).
5. About 90% of water reabsorption is obligatory; it occurs via osmo-sis, together with reabsorption of solutes, and is not hormonally regulated. The remaining 10% is facultative water reabsorption, which varies according to body needs and is regulated by ADH.

6. Na^+ are reabsorbed throughout the basolateral membrane via primary active transport.
7. In the proximal convoluted tubule, sodium ions are reabsorbed through the apical membranes via Na^+-glucose symporters and Na^+/H^+ antiporters; water is reabsorbed via osmosis; Cl^-, K^+, Ca^{2+}, Mg^{2+}, and urea are reabsorbed via passive diffusion; and NH_3 and NH_4^+ are secreted.
8. The loop of Henle reabsorbs 20–30% of the filtered Na^+, K^+, Ca^{2+}, and HCO_3^-; 35% of the filtered Cl^-, and 15% of the filtered water.
9. The distal convoluted tubule reabsorbs sodium and chloride ions via Na^+-Cl^- symporters.
10. In the collecting duct, principal cells reabsorb Na^+ and secrete K^+, whereas intercalated cells reabsorb K^+ and HCO_3^- and secrete H^+.
11. Angiotensin II, aldosterone, antidiuretic hormone, and atrial natriuretic peptide regulate solute and water reabsorption, as summarized in Table 26.4 on page 972.

PRODUCTION OF DILUTE AND CONCENTRATED URINE (p. 972)

1. In the absence of ADH, the kidneys produce dilute urine; renal tubules absorb more solutes than water.
2. In the presence of ADH, the kidneys produce concentrated urine; large amounts of water are reabsorbed from the tubular fluid into interstitial fluid, increasing solute concentration of the urine.
3. The countercurrent mechanism establishes an osmotic gradient in the interstitial fluid of the renal medulla that enables production of concentrated urine when ADH is present.

EVALUATION OF KIDNEY FUNCTION (p. 977)

1. A urinalysis is an analysis of the volume and physical, chemical, and microscopic properties of a urine sample. Table 26.5 on page 977 summarizes the principal physical characteristics of urine.
2. Chemically, normal urine contains about 95% water and 5% solutes. The solutes normally include urea, creatinine, uric acid, urobilinogen, and various ions.
3. Table 26.6 on page 978 lists several abnormal components that can be detected in a urinalysis, including albumin, glucose, red and white blood cells, ketone bodies, bilirubin, excessive urobilinogen, casts, and microbes.
4. Renal clearance refers to the ability of the kidneys to clear (remove) a specific substance from blood.

URINE TRANSPORTATION, STORAGE, AND ELIMINATION (p. 979)

1. The ureters are retroperitoneal and consist of a mucosa, muscularis, and adventitia. They transport urine from the renal pelvis to the urinary bladder, primarily via peristalsis.
2. The urinary bladder is located in the pelvic cavity posterior to the pubic symphysis; its function is to store urine before micturition.
3. The urinary bladder consists of a mucosa with rugae, a muscularis (detrusor muscle), and an adventitia (serosa over the superior surface).
4. The micturition reflex discharges urine from the urinary bladder

via parasympathetic impulses that cause contraction of the detrusor muscle and relaxation of the internal urethral sphincter muscle and via inhibition of impulses in somatic motor neurons to the external urethral sphincter.

5. The urethra is a tube leading from the floor of the urinary bladder to the exterior. Its anatomy and histology differ in females and males. In both sexes, the urethra functions to discharge urine from the body; in males, it discharges semen as well.

WASTE MANAGEMENT IN OTHER BODY SYSTEMS (p. 981)

1. Besides the kidneys, several other tissues, organs, and processes temporarily confine wastes, transport waste materials for disposal, recycle materials, and excrete excess or toxic substances.
2. Buffers bind excess H^+, the blood transports wastes, the liver converts toxic substances into less toxic ones, the lungs exhale CO2, sweat glands help eliminate excess heat, and the gastrointestinal tract eliminates solid wastes.

DEVELOPMENT OF THE URINARY SYSTEM (p. 982)

1. The kidneys develop from intermediate mesoderm.
2. The kidneys develop in the following sequence: pronephros, mesonephros, and metanephros. Only the metanephros remains and develops into a functional kidney.

AGING AND THE URINARY SYSTEM (p. 982)

1. With aging, the kidneys shrink in size, have a decreased blood flow, and filter less blood.
2. Common problems related to aging include urinary tract infections, increased frequency of urination, urinary retention or incontinence, and renal calculi.

SELF-QUIZ QUESTIONS

Fill in the blanks in the following statements.

1. The _____ cells of the ureters and urinary bladder allow distension for urine flow and storage.
2. Micturition is both a _____ and an _____ reflex.
3. The three basic processes carried out by nephrons are, _____, _____, and _____.

Indicate whether the following statements are true or false.

4. When dilute urine is being formed, the osmolarity of the fluid in the tubular lumen increases as it flows down the descending limb of the loop of Henle, decreases as it flows up the ascending limb, and continues to decrease as it flows through the rest of the nephron and collecting duct.
5. The route of blood flow from the renal artery to the renal vein is renal artery, segmental arteries, interlobar arteries, arcuate arteries, interlobular arteries, afferent arterioles, glomeruli, efferent arterioles, peritubular capillaries, vasa recta, peritubular venules, interlobular veins, arcuate veins, interlobar veins, and renal vein.

Choose the one best answer to the following questions.

6. Which of the following are features of the renal corpuscle that enhance its filtering capacity? (1) large glomerular capillary surface area, (2) thick, selectively permeable filtration membrane, (3) low glomerular capillary pressure, (4) high glomerular capillary pressure, (5) thin and porous filtration membrane. (a) 1, 2, and 3, (b) 2, 4, and 5, (c) 1, 4, and 5, (d) 2, 3, and 4, (e) 2, 3, and 5.
7. The main pressure that promotes glomerular filtration is (a) glomerular blood hydrostatic pressure, (b) capsular hydrostatic pressure, (c) blood colloid osmotic pressure, (d) capsular osmotic pressure.
8. Which of the following statements are *correct*? (1) Glomerular filtration rate (GFR) is directly related to the pressures that determine net filtration pressure. (2) Angiotensin II and atrial natriuretic peptide help regulate GFR. (3) Mechanisms that regulate GFR work by adjusting blood flow into and out of the glomerulus and by altering the glomerular capillary surface area available for filtration. (4) GFR increases when blood flow into glomerular capillaries decreases. (5) Normally, GFR increases very little when systemic blood pressure rises. (a) 1, 2, and 3, (b) 2, 3, and 4, (c) 3, 4, and 5, (d) 1, 2, 3, and 5, (e) 2, 3, 4, and 5.

9. Which of the following are mechanisms that control GFR? (1) renal autoregulation, (2) neural regulation, (3) hormonal regulation, (4) chemical regulation of ions, (5) presence or absence of a transporter. (a) 1, 2, and 3, (b) 2, 3, and 4, (c) 3, 4, and 5, (d) 1, 3, and 5, (e) 1, 3, and 4.
10. Which of the following hormones affect Na^+, Cl^-, and water reabsorption and K^+ secretion by the renal tubules? (1) angiotensin II, (2) aldosterone, (3) ADH, (4) atrial natriuretic peptide, (5) thyroid hormone. (a) 1, 3, and 5, (b) 2, 3, and 4, (c) 2, 4, and 5, (d) 1, 2, 4, and 5, (e) 1, 2, 3, and 4.
11. Which of the following is the correct sequence of tubular fluid flow? (a) proximal convoluted tubule, descending limb of loop of Henle, ascending limb of loop of Henle, collecting duct, proximal convoluted tubule, papillary duct, (b) proximal convoluted tubule, descending limb of loop of Henle, ascending limb of loop of Henle, distal convoluted tubule, collecting duct, papillary duct, (c) papillary duct, distal convoluted tubule, proximal convoluted tubule, collecting duct, ascending limb of loop of Henle, descending limb of loop of Henle, (d) collecting duct, proximal convoluted tubule, ascending limb of loop of Henle, descending limb of loop of Henle, papillary duct, distal convoluted tubule, (e) distal convoluted tubule, ascending limb of loop of Henle, descending limb of loop of Henle, proximal convoluted tubule, papillary duct, collecting duct.
12. Which of the following are functions of the kidneys? (1) control of blood pH, (2) excretion of wastes, (3) regulation of blood ionic composition, (4) regulation of white blood cell production, (5) regulation of blood volume. (a) 1, 2, 3, and 4, (b) 1, 2, and 5, (c) 1, 2, 3, and 5, (d) 2, 3, and 5, (e) 2, 3, 4, and 5.

13. Match the following:

_____ (a) cells in the last portion of the distal convoluted tubule and in the collecting ducts; regulated by ADH and aldosterone

_____ (b) a capillary network lying in Bowman's capsule and functioning in filtration

_____ (c) the functional unit of the kidney

_____ (d) drains into a collecting duct

_____ (e) combined glomerulus and glomerular capsule; where plasma is filtered

_____ (f) the visceral layer of the glomerular capsule consisting of modified simple squamous epithelial cells

_____ (g) cells of the final portion of the ascending limb of the loop of Henle that make contact with the afferent arteriole

_____ (h) site of obligatory water reabsorption

_____ (i) can secrete H⁺ against a concentration gradient

_____ (j) modified smooth muscle cells in the wall of the afferent arteriole

(1) podocytes
(2) glomerulus
(3) renal corpuscle
(4) proximal convoluted tubule
(5) distal convoluted tubule
(6) juxtaglomerular cells
(7) macula densa
(8) principal cells
(9) intercalated cells
(10) nephron

14. Match the following:

_____ (a) measure of blood nitrogen resulting from the catabolism and deamination of amino acids

_____ (b) produced from the catabolism of creatine phosphate in skeletal muscle

_____ (c) volume of blood that is cleared of a substance per unit of time

_____ (d) can result from diabetes mellitus

_____ (e) insoluble stones of crystallized salts

_____ (f) indicates infection in the urinary system

_____ (g) can be caused by damage to the filtration membranes

(1) pyuria
(2) renal calculi
(3) plasma creatinine
(4) BUN test
(5) albuminuria
(6) glucosuria
(7) renal plasma clearance

15. Match the following:

_____ (a) membrane proteins that function as water channels

_____ (b) a secondary active transport process that achieves Na⁺ reabsorption, returns filtered HCO₃⁻ and water to the peritubular capillaries, and secretes H⁺

_____ (c) stimulates principal cells to secrete more K⁺ into tubular fluid

_____ (d) reduces glomerular filtration rate; increases blood volume and pressure

_____ (e) increases glomerular filtration rate

_____ (f) regulates facultative water reabsorption by increasing the water permeability of principal cells

_____ (g) reabsorb Na⁺ together with a variety of other solutes

(1) angiotensin II
(2) atrial natriuretic peptide
(3) Na⁺ symporters
(4) Na⁺/H⁺ antiporters
(5) aquaporins
(6) aldosterone
(7) ADH

CRITICAL THINKING QUESTIONS

1. Imagine the discovery of a new toxin that blocks renal tubule reabsorption but does not affect filtration. Predict the short-term effects of this toxin.

   HINT *Normally about 99% of glomerular filtrate is reabsorbed.*

2. While traveling down the interstate, little Caitlin told her father to stop the car "Now! I'm so full I'm gonna burst!" Her dad thinks she can make it to the next rest stop. What structures of the urinary bladder will help her "hold it"?

   HINT *Children often visit the emergency room with broken bones, but not with exploding bladders.*

3. Although urinary catheters come in one length only, the number of centimeters of the catheter that must be inserted to reach the bladder and release urine differs significantly in males and females. Why?

   HINT *Why do men and women usually assume different positions when they urinate?*

ANSWERS TO FIGURE QUESTIONS

26.1 The kidneys, ureters, urinary bladder, and urethra are the components of the urinary system.

26.2 The kidneys are retroperitoneal because they are posterior to the peritoneum.

26.3 Blood vessels, lymphatic vessels, nerves, and a ureter pass through the renal hilus.

26.4 About 1200 mL of blood enters the renal arteries each minute.

26.5 Cortical nephrons have glomeruli in the superficial renal cortex, and their short loops of Henle penetrate only into the superficial renal medulla. Juxtamedullary nephrons have glomeruli deep in the renal cortex, and their long loops of Henle extend through the renal medulla nearly to the renal papilla.

26.6 This section must pass through the renal cortex because there are no renal corpuscles in the renal medulla.

26.7 Secreted penicillin is being removed from the bloodstream.

26.8 Endothelial fenestrations (pores) in glomerular capillaries are too small for red blood cells to pass through.

26.9 Obstruction of the right ureter would increase CHP and thus decrease NEP in the right kidney; the obstruction would have no effect on the left kidney.

26.10 *Auto* means self; tubuloglomerular feedback is an example of autoregulation because it takes place entirely within the kidneys.

26.11 The tight junctions form a barrier that prevents diffusion of transporter, channel, and pump proteins between the apical and basolateral membranes.

26.12 Glucose enters a PCT cell via a Na^+-glucose symporter in the apical membrane and leaves via facilitated diffusion through the basolateral membrane.

26.13 The electrochemical gradient promotes movement of Na^+ into the tubule cell through the apical membrane antiporters.

26.14 Reabsorption of the solutes creates an osmotic gradient that promotes the reabsorption of water via osmosis.

26.15 This is secondary active transport because the symporter uses the energy stored in the concentration gradient of Na^+ between extracellular fluid and the cytosol. No water is reabsorbed here because the thick ascending limb of the loop of Henle is virtually impermeable to water.

26.16 In principal cells, aldosterone stimulates secretion of K^+ and reabsorption of Na^+ by increasing the activity of sodium-potassium pumps and number of leakage channels for Na^+ and K^+.

26.17 A carbonic anhydrase inhibitor reduces secretion of H^+ into the urine and reduces reabsorption of Na^+ and HCO_3^- into the blood. It has a diuretic effect, and it tends to cause acidosis (lowered pH of the blood) due to loss of HCO_3^- in the urine.

26.18 Aldosterone and atrial natriuretic peptide also influence renal water reabsorption.

26.19 Dilute urine is produced when the thick ascending limb of the loop of Henle, the distal convoluted tubule, and the collecting duct reabsorb more solutes than water.

26.20 The high osmolarity of interstitial fluid in the renal medulla is due mainly to Na^+, Cl^-, and urea.

26.21 Secretion occurs in the proximal convoluted tubule, the loop of Henle, and the collecting duct.

26.22 Lack of voluntary control over micturition is termed urinary incontinence.

26.23 The kidneys start to develop during the third week of development.

Fluid, Electrolyte, and
Acid–Base Homeostasis

FLUIDS, ELECTROLYTES, AND ACID-BASE HOMEOSTASIS

The regulation of the volume of water and solutes (body fluids), their distribution throughout the body, and balancing the pH of the body fluids is crucial to maintaining overall homeostasis and health.

ENERGIZE YOUR STUDY
FOUNDATIONS CD
Animation
  • Chemistry Folder: Water and Fluid Flow; Acids and Bases
Concepts and Connections and Exercises reinforce your understanding

www.wiley.com/college/apcentral
INSIGHTS AND EXPLORATIONS
Sports magazines and catalogues advertise thirst quenching drinks to rehydrate and restore the body's fluid balance. Health magazines warn of the need to drink plenty of fluids. Almost every summer there is a news report about a young athlete collapsing on the practice field from heat prostration. Sometimes death results. Then there are the victims of cholera who can die from dehydration in as little as 6 hours. And victims of anorexia and other eating disorders who by purging deplete the body of fluid. What fluids should you be drinking this summer while out enjoying nature, at a marching band camp, a sports camp or working? Explore the possibilities in this web-based activity.

Chapter 26 you learned how the kidneys form urine. One important function of the kidneys is to help maintain fluid balance in the body. The water and dissolved solutes in the body constitute the **body fluids.** Regulatory mechanisms involving the kidneys and other organs normally maintain homeostasis of the body fluids. Malfunction in any or all of them may seriously endanger the functioning of organs throughout the body. In this chapter, we will explore the mechanisms that regulate the volume and distribution of body fluids and examine the factors that determine the concentrations of solutes and the pH of body fluids.

FLUID COMPARTMENTS AND FLUID BALANCE

▶ **OBJECTIVES**

- **Compare the locations of intracellular fluid (ICF) and extracellular fluid (ECF), and describe the various fluid compartments of the body.**
- **Describe the sources of water and solute gain and loss, and explain how each is regulated.**

In lean adults, body fluids constitute between 55% and 60% of total body mass in females and males, respectively (Figure 27.1). Fluids are present in two main "compartments"—inside cells and outside cells. About two-thirds of body fluid is **intracellular fluid (ICF)** (*intra-* = within) or **cytosol,** the fluid within cells. The other third, called **extracellular fluid (ECF)** (*extra-* = outside), is outside cells and includes all other body fluids. About 80% of the ECF is **interstitial fluid** (*inter-* = between), which occupies the microscopic spaces between tissue cells, and 20% of the ECF is **plasma,** the liquid portion of the blood. Extracellular fluids also include lymph in lymphatic vessels; cerebrospinal fluid in the nervous system; synovial fluid in joints; aqueous humor and vitreous body in the eyes; endolymph and perilymph in the ears; and pleural, pericardial, and peritoneal fluids between serous membranes. Because most of these fluids are similar to interstitial fluid, they are grouped together with it.

Figure 27.1 **Body fluid compartments.**

The term body fluid refers to body water and its dissolved substances.

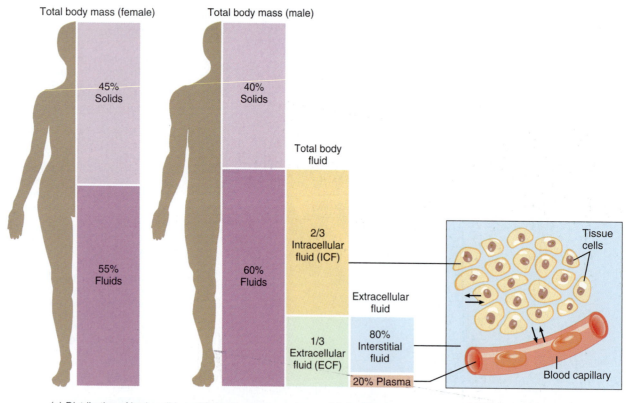

(a) Distribution of body solids and fluids in an average lean, adult female and male

(b) Exchange of water among body fluid compartments

What is the approximate volume of blood plasma in a lean 60-kg male? In a lean 60-kg female? *(NOTE: One liter of body fluid has a mass of 1 kilogram.)*

Two general "barriers" separate intracellular fluid, interstitial fluid, and blood plasma.

1. The *plasma membrane* of individual cells separates intracellular fluid from the surrounding interstitial fluid. You learned in Chapter 3 that the plasma membrane is a selectively permeable barrier: It allows some substances to cross but blocks the movement of other substances. In addition, active transport pumps work continuously to maintain different concentrations of certain ions in the cytosol and interstitial fluid.

2. *Blood vessel walls* divide the interstitial fluid from blood plasma. Only in capillaries, the smallest blood vessels, are the walls thin enough and leaky enough to permit the exchange of water and solutes between blood plasma and interstitial fluid.

The body is in **fluid balance** when the required amounts of water and solutes are present and are correctly proportioned among the various compartments. **Water** is by far the largest single component of the body, making up 45–75% of total body mass, depending on age and gender. Fatter people have proportionally less water than leaner people because water comprises less than 20% of the mass of adipose tissue. Skeletal muscle tissue, in contrast, is about 65% water. Infants have the highest percentage of water, up to 75% of body mass. The percentage of body mass that is water decreases until about 2 years of age. Until puberty, water accounts for about 60% of body mass in boys and girls. In lean adult males, water still accounts for about 60% of body mass. However, even lean females have more subcutaneous fat than do males. Thus, their percentage of total body water is lower, accounting for about 55% of body mass.

The processes of filtration, reabsorption, diffusion, and osmosis provide for continual exchange of water and solutes among body fluid compartments (Figure 27.1b). Yet, the volume of fluid in each compartment remains remarkably stable. The pressures that promote filtration of fluid from blood capillaries and reabsorption of fluid back into capillaries can be reviewed in Figure 21.7 on page 704. Because osmosis is the primary means of water movement between intracellular fluid and interstitial fluid, the concentration of solutes in these fluids determines the *direction* of water movement. Most solutes in body fluids are **electrolytes,** inorganic compounds that dissociate into ions. Fluid balance depends primarily on electrolyte balance so the two are closely interrelated. Because intake of water and electrolytes rarely occurs in exactly the same proportions as their presence in body fluids, the ability of the kidneys to excrete excess water by producing dilute urine, or to excrete excess electrolytes by producing concentrated urine, is of utmost importance.

Sources of Body Water Gain and Loss

The body can gain water by ingestion and by metabolic synthesis (Figure 27.2). The main sources of body water are ingested

Figure 27.2 **Sources of daily water gain and loss under normal conditions.** Numbers are average volumes for adults.

Normally, daily water loss equals daily water gain.

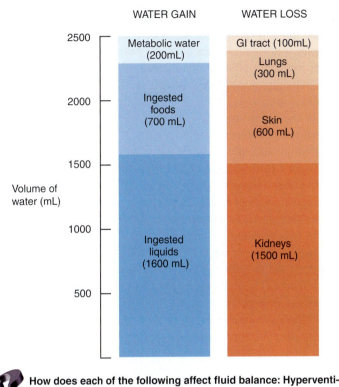

How does each of the following affect fluid balance: Hyperventilation? Vomiting? Fever? Diuretics?

liquids (about 1600 mL) and moist foods (about 700 mL) absorbed from the gastrointestinal (GI) tract, which total about 2300 mL/day. The other source of water is **metabolic water** that is produced in the body mainly when electrons are accepted by oxygen during aerobic cellular respiration (see Figure 25.2 on page 910) and to a smaller extent during dehydration synthesis reactions (see Figure 2.15 on page 44). Metabolic water gain accounts for only 200 mL/day. Daily water gain from these two sources totals about 2500 mL.

Normally, body fluid volume remains constant because water loss equals water gain. Water loss occurs in four ways (Figure 27.2). Each day the kidneys excrete about 1500 mL in urine, the skin evaporates about 600 mL (400 mL through insensible perspiration and 200 mL as sweat), the lungs exhale about 300 mL as water vapor, and the gastrointestinal tract eliminates about 100 mL in feces. In women of reproductive age, additional water is lost in menstrual flow. On average, daily water loss totals about 2500 mL. The amount of water lost by a given route can vary considerably over time. For example, water may literally pour from the skin in the form of sweat during strenuous exertion. In other cases, water may be lost in diarrhea during a GI tract infection.

Regulation of Body Water Gain

The rate of formation of metabolic water is not regulated to maintain homeostasis of body fluids. Rather, the volume of metabolic water formed depends on the level of aerobic cellular respiration, which reflects the demand for ATP in body cells. When more ATP is produced, more water is formed. Regulation of body water gain depends mainly on regulating the volume of water intake, in other words by drinking more or less fluid. An area in the hypothalamus known as the **thirst center** governs the urge to drink.

When water loss is greater than water gain, **dehydration**—a decrease in volume and an increase in osmolarity of body fluids—stimulates thirst (Figure 27.3). When body mass decreases by 2% due to fluid loss, mild dehydration exists. A decrease in blood volume causes blood pressure to fall. This change stimulates the kidneys to release renin, which promotes the formation of angiotensin II. Increased nerve impulses from osmoreceptors in the hypothalamus, triggered by increased blood osmolarity, and increased angiotensin II in the blood both stimulate the thirst center in the hypothalamus. Other signals that stimulate thirst come from (1) neurons in the mouth that detect dryness due to a decreased flow of saliva and (2) baroreceptors that detect lowered blood pressure in the heart and blood vessels. As a result, the sensation of thirst increases, which usually leads to increased fluid intake (if fluids are available) and restoration of normal fluid volume. Overall, fluid gain balances fluid loss. Sometimes, however, either the sensation of thirst does not occur quickly enough or access to fluids is restricted, and significant dehydration ensues. This happens most often in elderly people, in infants, and in those who are in a confused mental state. In situations where heavy sweating or fluid loss from diarrhea or vomiting occurs, it is wise to start replacing body fluids by drinking fluids even before the sensation of thirst appears.

Regulation of Water and Solute Loss

Even though the loss of water and solutes through sweating and exhalation increases during exercise, elimination of *excess* body water or solutes occurs mainly by controlling how much are lost in urine. As you will see, the extent of *urinary salt (NaCl) loss* is the main factor that determines body fluid *volume*. The reason is that in osmosis "water follows solutes," and the two main solutes in extracellular fluid (and in urine) are sodium ions (Na^+) and chloride ions (Cl^-). In a similar way, the main factor that determines body fluid *osmolarity* is the extent of *urinary water loss*.

Because our daily diet contains a highly variable amount of NaCl, urinary excretion of Na^+ and Cl^- must also vary to maintain homeostasis. Hormonal changes regulate the urinary loss of these ions, which in turn affects blood volume. Figure 27.4 depicts the sequence of changes that occur after a salty meal. The increased intake of NaCl produces an increase in plasma levels of Na^+ and Cl^- (the major contributors to osmolarity of extracellular fluid). As a result, the osmolarity of inter-

Figure 27.3 Pathways through which dehydration stimulates thirst.

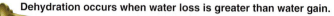

Dehydration occurs when water loss is greater than water gain.

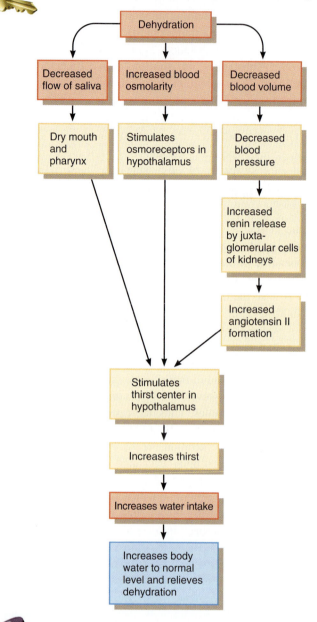

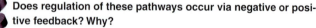

Does regulation of these pathways occur via negative or positive feedback? Why?

stitial fluid increases, which causes movement of water from intracellular fluid into interstitial fluid and then into plasma. Such water movement increases blood volume.

The three most important hormones that regulate the extent of renal Na^+ and Cl^- reabsorption (and thus how much is lost in the urine) are **angiotensin II, aldosterone,** and **atrial natriuretic peptide (ANP).** During dehydration angiotensin II and aldosterone promote urinary reabsorption of Na^+ and Cl^- (and water by osmosis with the electrolytes) and thereby conserve the volume of body fluids by reducing urinary loss. An increase in

Figure 27.4 Hormonal changes following increased NaCl intake. The resulting reduced blood volume is achieved by increasing urinary loss of Na⁺ and Cl⁻.

 The three main hormones that regulate renal Na⁺ and Cl⁻ reabsorption (and thus the amount lost in the urine) are angiotensin II, aldosterone, and atrial natriuretic peptide.

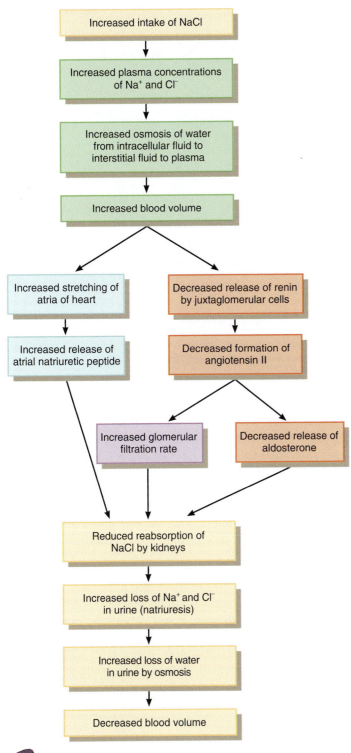

How does hyperaldosteronism (excessive aldosterone secretion) cause edema?

blood volume, as might occur after drinking a large volume of fluid, stretches the atria of the heart and promotes release of atrial natriuretic peptide. ANP promotes **natriuresis,** elevated urinary excretion of Na⁺ (and Cl⁻), which decreases blood volume. An increase in blood volume also slows release of renin from juxtaglomerular cells of the kidneys. When renin level declines, less angiotensin II is formed. Decline in angiotensin II from a moderate level to a low level increases glomerular filtration rate and reduces Na⁺, Cl⁻, and water reabsorption in the kidney tubules. In addition, less angiotensin II leads to less aldosterone, which causes reabsorption of filtered Na⁺ and Cl⁻ to slow in the renal collecting ducts. More filtered Na⁺ and Cl⁻ thus remain in the tubular fluid to be excreted in the urine. The osmotic consequence of excreting more Na⁺ and Cl⁻ is loss of more water in urine, which decreases blood volume and blood pressure.

The major hormone that regulates water loss is **antidiuretic hormone (ADH).** This hormone, also known as **vasopressin,** is produced by neurosecretory cells that extend from the hypothalamus to the posterior pituitary. Besides stimulating the thirst mechanism, an increase in the osmolarity of body fluids stimulates release of ADH (see Figure 26.17 on page 971). ADH promotes the insertion of water-channel proteins (aquaporin-2) into the apical membranes of principal cells in the collecting ducts of the kidneys. As a result, the permeability of these cells to water increases. Water molecules move by osmosis from the renal tubular fluid into the cells and then from the cells into the bloodstream. The result is production of a small volume of very concentrated urine. By contrast, intake of plain water decreases the osmolarity of blood and interstitial fluid. Within minutes, ADH secretion shuts down, and soon its blood level is close to zero. When the principal cells are not stimulated by ADH, aquaporin-2 molecules are removed from the apical membrane by endocytosis. As the number of water channels decreases, the water permeability of the principal cells' apical membrane falls, and more water is lost in the urine.

Under some conditions, factors other than blood osmolarity influence ADH secretion. A large decrease in blood volume, which is detected by baroreceptors (sensory neurons that respond to stretching) in the left atrium and in blood vessel walls, also stimulates ADH release. In severe dehydration, glomerular filtration rate decreases because blood pressure falls, and less water is lost in the urine. Conversely, the intake of too much water increases blood pressure, the rate of glomerular filtration rises, and more water is lost in the urine. Hyperventilation (abnormally fast and deep breathing) can increase fluid loss through the exhalation of more water vapor. Vomiting and diarrhea result in fluid loss from the GI tract. Finally, fever, heavy sweating, and destruction of extensive areas of the skin from burns can cause excessive water loss through the skin. In all of these conditions, an increase in ADH secretion will assist the conservation of body fluids.

Table 27.1 summarizes the factors that maintain body water balance.

Table 27.1	Summary of Factors that Maintain Body Water Balance	
Factor	**Mechanism**	**Effect**
Thirst center in hypothalamus	Stimulates desire to drink fluids.	Water gain if thirst is quenched.
Angiotensin II	Stimulates secretion of aldosterone.	Reduces loss of water in urine.
Aldosterone	By promoting urinary reabsorption of Na^+ and Cl^-, increases water reabsorption via osmosis.	Reduces loss of water in urine.
Atrial natriuretic peptide (ANP)	Promotes natriuresis, elevated urinary excretion of Na^+ (and Cl^-), accompanied by water.	Increases loss of water in urine.
Antidiuretic hormone (ADH), also known as **vasopressin**	Promotes insertion of water-channel proteins (aquaporin-2) into the apical membranes of principal cells in the collecting ducts of the kidneys. As a result, the water permeability of these cells increases and more water is reabsorbed.	Reduces loss of water in urine.

Figure 27.5   Series of events in water intoxication.

Water intoxication is a state in which excessive body water causes cells to become hypotonic and swell.

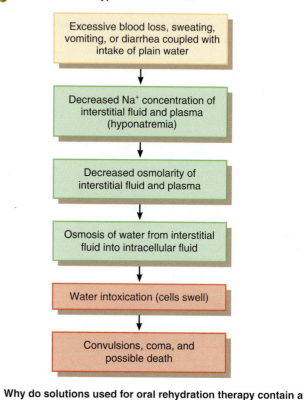

Why do solutions used for oral rehydration therapy contain a small amount of table salt (NaCl)?

Movement of Water Between Body Fluid Compartments

Normally, cells neither shrink nor swell because intracellular and interstitial fluids have the same osmolarity. Changes in the osmolarity of interstitial fluid, however, causes fluid imbalances. An increase in the osmolarity of interstitial fluid draws water out of cells, and they shrink slightly. A decrease in the osmolarity of interstitial fluid, by contrast, causes cells to swell. Changes in osmolarity most often result from changes in the concentration of Na^+.

A decrease in the osmolarity of interstitial fluid, as may occur after drinking a large volume of water, inhibits secretion of ADH. Normally, the kidneys then excrete a large volume of dilute urine, which restores the osmotic pressure of body fluids to normal. As a result, body cells swell only slightly, and only for a brief period. But when a person steadily consumes water faster than the kidneys can excrete it (the maximum urine flow rate is about 15 mL/min) or when renal function is poor, the result may be **water intoxication,** a state in which excessive body water causes cells to become hypotonic and to swell dangerously (Figure 27.5). If the body water and Na^+ lost during blood loss or excessive sweating, vomiting, or diarrhea is replaced by drinking plain water, then body fluids become more dilute. This dilution can cause the Na^+ concentration of plasma

and then of interstitial fluid to fall below the normal range (hyponatremia). When the Na^+ concentration of interstitial fluid decreases, its osmolarity also falls. The net result is osmosis of water from interstitial fluid into the cytosol. Water entering the cells causes them to become hypotonic and swell, producing convulsions, coma, and possibly death. To prevent this dire sequence of events in cases of severe electrolyte and water loss, solutions given for intravenous or oral rehydration therapy (ORT) include a small amount of table salt (NaCl).

Enemas and Fluid Balance

An **enema** (EN-e-ma) is the introduction of a solution into the rectum and colon to stimulate bowel activity and evacuate feces. Repeated enemas, especially in young children, increase the risk of fluid and electrolyte imbalances. Because tap water (which is hypotonic) and commercially prepared hypertonic solutions may cause rapid movement of fluid between compartments, an isotonic solution is preferred. ■

▶ CHECKPOINT

1. What is the approximate volume of each of *your* body fluid compartments?

2. Which routes of water gain and loss from the body are regulated?

3. How does thirst help regulate water intake?

4. How do angiotensin II, aldosterone, atrial natriuretic peptide, and antidiuretic hormone regulate the volume and osmolarity of body fluids?

5. What factors govern the movement of water between interstitial fluid and intracellular fluid?

ELECTROLYTES IN BODY FLUIDS

▶ **OBJECTIVES**

• **Compare the electrolyte composition of the three major fluid compartments: plasma, interstitial fluid, and intracellular fluid.**

• **Discuss the functions of sodium, chloride, potassium, bicarbonate, calcium, phosphate, and magnesium ions, and explain how their concentrations are regulated.**

The ions formed when electrolytes dissolve and dissociate serve four general functions in the body: (1) Because they are largely confined to particular fluid compartments and are more numerous than nonelectrolytes, certain ions *control the osmosis of water between fluid compartments.* (2) Ions *help maintain the acid–base balance* required for normal cellular activities. (3) Ions *carry electrical current,* which allows production of action potentials and graded potentials and controls secretion of some hormones and neurotransmitters. (4) Several ions *serve as cofactors* needed for optimal activity of enzymes.

Concentrations of Electrolytes in Body Fluids

To compare the charge carried by ions in different solutions, the concentration of ions is typically expressed in units of **milliequivalents per liter (mEq/liter).** These units give the concentration of cations or anions in a given volume of solution. One equivalent is the positive or negative charge equal to the amount of charge in one mole of H^+; a milliequivalent is one-thousandth of an equivalent. Recall that a mole of a substance is its molecular weight expressed in grams. For ions such as sodium (Na^+), potassium (K^+), and bicarbonate (HCO_3^-), which have a single positive or negative charge, the number of mEq/liter is equal to the number of mmol/liter. For ions such as calcium (Ca^{2+}) or phosphate (HPO_4^{2-}), which have two positive or negative charges, the number of mEq/liter is twice the number of mmol/liter.

Figure 27.6 compares the concentrations of the main electrolytes and protein anions in blood plasma, interstitial fluid, and intracellular fluid. The chief difference between the two extracellular fluids—blood plasma and interstitial fluid—is that

Figure 27.6 **Electrolyte and protein anion concentrations in plasma, interstitial fluid, and intracellular fluid.** The height of each column represents the milliequivalents per liter (mEq/liter).

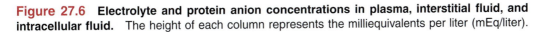

The electrolytes present in extracellular fluids are different from those present in intracellular fluid.

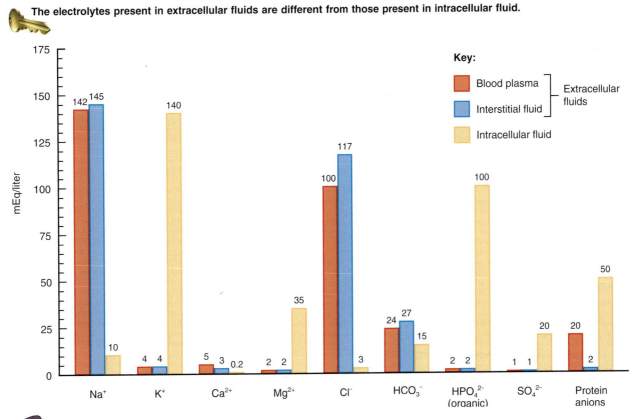

What are the major cation and the two major anions in ECF and ICF?

blood plasma contains many protein anions, whereas interstitial fluid has very few. Because normal capillary membranes are virtually impermeable to proteins, only a few plasma proteins leak out of blood vessels into the interstitial fluid. This difference in protein concentration is largely responsible for the blood colloid osmotic pressure exerted by blood plasma. In other respects, the two fluids are similar.

The electrolyte content of intracellular fluid, by contrast, differs considerably from that of extracellular fluid. In extracellular fluid, the most abundant cation is Na^+, and the most abundant anion is Cl^-. In intracellular fluid, the most abundant cation is K^+, and the most abundant anions are proteins and HPO_4^{2-}. By actively transporting Na^+ out of cells and K^+ into cells, sodium-potassium pumps (Na^+/K^+ ATPase) play a major role in maintaining the high intracellular concentration of K^+ and high extracellular concentration of Na^+.

Sodium

Sodium ions (Na^+) are the most abundant ions in extracellular fluid, accounting for 90% of the extracellular cations. The normal blood plasma Na^+ concentration is 136–148 mEq/liter. As we have already seen, Na^+ plays a pivotal role in fluid and electrolyte balance because it accounts for almost half of the osmolarity (142 of about 290 mOsm/liter) of extracellular fluid. The flow of Na^+ through voltage-gated channels in the plasma membrane also is necessary for the generation and conduction of action potentials in neurons and muscle fibers. The typical daily intake of Na^+ in North America often far exceeds the body's normal daily requirements due largely to excess dietary salt. The kidneys excrete excess Na^+, but they also can conserve it during periods of shortage.

The Na^+ level in the blood is controlled by aldosterone, antidiuretic hormone (ADH), and atrial natriuretic peptide (ANP). Aldosterone increases renal reabsorption of Na^+. When the blood plasma concentration of Na^+ drops below 135 mEq/liter, a condition called *hyponatremia*, ADH release ceases. The lack of ADH, in turn, permits greater excretion of water in urine and restoration of the normal Na^+ level in ECF. Atrial natriuretic peptide (ANP) increases Na^+ excretion by the kidneys when Na^+ level is above normal, a condition called *hypernatremia*.

Edema and Hypovolemia Indicate Na^+ Imbalance

If excess sodium ions remain in the body because the kidneys fail to excrete enough of them, water is also osmotically retained. The result is increased blood volume, increased blood pressure, and **edema,** an abnormal accumulation of interstitial fluid. Renal failure and hyperaldosteronism (excessive aldosterone secretion) are two causes of Na^+ retention. Excessive urinary loss of Na^+, by contrast, has the osmotic effect of causing excessive water loss, which results in **hypovolemia,** an abnormally low blood volume. Hypovolemia related to Na^+ loss is most often due to inadequate secretion of aldosterone associated with either adrenal insufficiency or overly vigorous therapy with diuretic drugs. ■

Chloride

Chloride ions (Cl^-) are the most prevalent anions in extracellular fluid. The normal blood plasma Cl^- concentration is 95–105 mEq/liter. Cl^- moves relatively easily between the extracellular and intracellular compartments because most plasma membranes contain many Cl^- leakage channels and antiporters. For this reason, Cl^- can help balance the level of anions in different fluid compartments. One example is the chloride shift that occurs between red blood cells and blood plasma as the blood level of carbon dioxide either increases or decreases (see Figure 23.24 on page 836). In this case, the antiporter exchange of Cl^- for HCO_3^- maintains the correct balance of anions between ECF and ICF. Chloride ions also are part of the hydrochloric acid secreted into gastric juice. ADH helps regulate Cl^- balance in body fluids because it governs the extent of water loss in urine. Processes that increase or decrease renal reabsorption of sodium ions also affect reabsorption of chloride ions. The negatively charged Cl^- follows the positively charged Na^+ due to the electrical attraction of oppositely charged particles.

Potassium

Potassium ions (K^+) are the most abundant cations in intracellular fluid (140 mEq/liter). K^+ plays a key role in establishing the resting membrane potential and in the repolarization phase of action potentials in neurons and muscle fibers; K^+ also helps maintain normal intracellular fluid volume. When K^+ moves into or out of cells, it often is exchanged for H^+ and thereby helps regulate the pH of body fluids.

The normal blood plasma K^+ concentration is 3.5–5.0 mEq/liter and is controlled mainly by aldosterone. When blood plasma K^+ concentration is high, more aldosterone is secreted into the blood. Aldosterone then stimulates principal cells of the renal collecting ducts to secrete more K^+ so excess K^+ is lost in the urine. Conversely, when blood plasma K^+ concentration is low, aldosterone secretion decreases and less K^+ is excreted in urine. Because K^+ is needed during the repolarization phase of action potentials, abnormal K^+ levels can be lethal. For instance, *hyperkalemia* (above normal concentration of K^+ in blood) can cause death due to ventricular fibrillation.

Bicarbonate

Bicarbonate ions (HCO_3^-) are the second most prevalent extracellular anions. Normal blood plasma HCO_3^- concentration is 22–26 mEq/liter in systemic arterial blood and 23–27 mEq/liter in systemic venous blood. HCO_3^- concentration increases as blood flows through systemic capillaries. This happens because the carbon dioxide released by metabolically active cells combines with water to form carbonic acid, which dissociates into H^+ and HCO_3^-. As blood flows through pulmonary capillaries, however, the concentration of HCO_3^- decreases again as carbon dioxide is exhaled. (Figure 23.24 on page 836 shows these reactions.) Intracellular fluid also contains a small amount of HCO_3^-. As previously noted, the exchange of Cl^- for HCO_3^- helps maintain the correct balance of anions in extracellular fluid and intracellular fluid.

The kidneys are the main regulators of blood HCO_3^- concentration. The intercalated cells of the renal tubule can either form HCO_3^- and release it into the blood when the blood level is low (see Figure 27.8) or excrete excess HCO_3^- in the urine when the level in blood is too high. Changes in the blood level of HCO_3^- are considered later in this chapter in the section on acid–base balance.

Calcium

Because such a large amount of calcium is stored in bone, it is the most abundant mineral in the body. About 98% of the calcium in adults is in the skeleton and teeth, where it is combined with phosphates to form a crystal lattice of mineral salts. In body fluids, calcium is mainly an extracellular cation (Ca^{2+}). The normal concentration of free or unattached Ca^{2+} in blood plasma is 4.5–5.5 mEq/liter. In addition, about the same amount of Ca^{2+} is attached to various plasma proteins. Besides contributing to the hardness of bones and teeth, Ca^{2+} plays important roles in blood clotting, neurotransmitter release, maintenance of muscle tone, and excitability of nervous and muscle tissue.

The two main regulators of Ca^{2+} concentration in blood plasma are parathyroid hormone (PTH) and calcitriol (1,25-dihydroxy vitamin D_3), the form of vitamin D that acts as a hormone (see Figure 18.14 on page 608). A low level of Ca^{2+} in blood plasma promotes release of more PTH, which stimulates osteoclasts in bone tissue to release calcium (and phosphate) from bone matrix. Thus, PTH increases bone *resorption*. Parathyroid hormone also enhances *reabsorption* of Ca^{2+} from glomerular filtrate through renal tubule cells and back into blood, and increases production of calcitriol, which in turn increases Ca^{2+} *absorption* from food in the gastrointestinal tract.

Phosphate

About 85% of the phosphate in adults is present as calcium phosphate salts, which are structural components of bone and teeth. The remaining 15% is ionized. Three phosphate ions ($H_2PO_4^-$, HPO_4^{2-}, and PO_4^{3-}) are important intracellular anions. At the normal pH of body fluids, HPO_4^{2-} is the most prevalent form. Phosphates contribute about 100 mEq/liter of anions to intracellular fluid. HPO_4^{2-} is an important buffer of H^+, both in body fluids and in the urine. Although some are "free," most phosphate ions are covalently bound to organic molecules such as lipids (phospholipids), proteins, carbohydrates, nucleic acids (DNA and RNA), and adenosine triphosphate (ATP).

The normal blood plasma concentration of ionized phosphate is only 1.7–2.6 mEq/liter. The same two hormones that govern calcium homeostasis—parathyroid hormone (PTH) and calcitriol—also regulate the level of HPO_4^{2-} in blood plasma. PTH stimulates resorption of bone matrix by osteoclasts, which releases both phosphate and calcium ions into the bloodstream. In the kidneys, however, PTH inhibits reabsorption of phosphate ions while stimulating reabsorption of calcium ions by renal tubular cells. Thus, PTH increases urinary excretion of phosphate and lowers blood phosphate level. Calcitriol promotes absorption of both phosphates and calcium from the digestive tract.

Magnesium

In adults, about 54% of the total body magnesium is part of bone matrix as magnesium salts. The remaining 46% occurs as magnesium ions (Mg^{2+}) in intracellular fluid (45%) and extracellular fluid (1%). Mg^{2+} is the second most common intracellular cation (35 mEq/liter). Functionally, Mg^{2+} is a cofactor for certain enzymes needed for the metabolism of carbohydrates and proteins and for Na^+/K^+ ATPase (the sodium-potassium pump). Mg^{2+} is essential for normal neuromuscular activity, synaptic transmission, and myocardial functioning. In addition, secretion of PTH depends on Mg^{2+}.

Normal blood plasma Mg^{2+} concentration is low, only 1.3–2.1 mEq/liter. Several factors regulate the blood plasma level of Mg^{2+} by varying the rate at which it is excreted in the urine. The kidneys increase urinary excretion of Mg^{2+} in hypercalcemia, hypermagnesemia, increases in extracellular fluid volume, decreases in parathyroid hormone, and acidosis. The opposite conditions decrease renal excretion of Mg^{2+}.

Table 27.2 describes the imbalances that result from the deficiency or excess of several electrolytes.

Those at Risk for Fluid and Electrolyte Imbalances

In a number of situations, people may be at risk for fluid and electrolyte imbalances. Those at risk include people who depend on others for fluid and food, such as infants, the elderly, and the hospitalized; and individuals undergoing medical treatment that involves intravenous infusions, drainages or suctions, and urinary catheters. People who receive diuretics experience excessive fluid losses and require increased fluid intake, or experience fluid retention and have fluid restrictions are also at risk. Finally at risk are postoperative individuals, severe burn or trauma cases, individuals with chronic diseases (congestive heart failure, diabetes, chronic obstructive lung disease, and cancer), people in confinement, and individuals with altered levels of consciousness who may be unable to communicate needs or respond to thirst. ∎

▶ **CHECKPOINT**

6. What are the functions of electrolytes in the body?

7. Name three important extracellular electrolytes and three important intracellular electrolytes.

Table 27.2 Blood Electrolyte Imbalances

Electrolyte*	Deficiency		Excess	
	Name and Causes	Signs and Symptoms	Name and Causes	Signs and Symptoms
Sodium (Na^+) 136–148 mEq/liter	**Hyponatremia** (hī-pō-na-TRĒ-mē-a) may be due to decreased sodium intake; increased sodium loss through vomiting, diarrhea, aldosterone deficiency, or taking certain diuretics; and excessive water intake.	Muscular weakness; dizziness, headache, and hypotension; tachycardia and shock; mental confusion, stupor, and coma.	**Hypernatremia** may occur with dehydration, water deprivation, or excessive sodium in diet or intravenous fluids; causes hypertonicity of ECF, which pulls water out of body cells into ECF, causing cellular dehydration.	Intense thirst, hypertension, edema, agitation, and convulsions.
Chloride (Cl^-) 95–105 mEq/liter	**Hypochloremia** (hī-pō-klō-RĒ-mē-a) may be due to excessive vomiting, overhydration, aldosterone deficiency, congestive heart failure, and therapy with certain diuretics such as furosemide (Lasix).	Muscle spasms, metabolic alkalosis, shallow respirations, hypotension, and tetany.	**Hyperchloremia** may result from dehydration due to water loss or water deprivation; excessive chloride intake; or severe renal failure, hyperaldosteronism, certain types of acidosis, and some drugs.	Lethargy, weakness, metabolic acidosis, and rapid, deep breathing.
Potassium (K^+) 3.5–5.0 mEq/liter	**Hypokalemia** (hī-pō-ka-LĒ-mē-a) may result from excessive loss due to vomiting or diarrhea, decreased potassium intake, hyperaldosteronism, kidney disease, and therapy with some diuretics.	Muscle fatigue, flaccid paralysis, mental confusion, increased urine output, shallow respirations, and changes in the electrocardiogram, including flattening of the T wave.	**Hyperkalemia** may be due to excessive intake, renal failure, aldosterone deficiency, crushing injuries to body tissues, or transfusion of hemolyzed blood.	Irritability, nausea, vomiting, diarrhea, muscular weakness; can cause death by inducing ventricular fibrillation.
Calcium (Ca^{2+}) Total = 9–10.5 mg/dL; ionized = 4.5–5.5 mEq/liter	**Hypocalcemia** (hī-pō-kal-SĒ-mē-a) may be due to increased calcium loss, reduced calcium intake, elevated levels of phosphate, or hypoparathyroidism.	Numbness and tingling of the fingers; hyperactive reflexes, muscle cramps, tetany, and convulsions; bone fractures; spasms of laryngeal muscles that can cause death by asphyxiation.	**Hypercalcemia** may result from hyperparathyroidism, some cancers, excessive intake of vitamin D, and Paget's disease of bone.	Lethargy, weakness, anorexia, nausea, vomiting, polyuria, itching, bone pain, depression, confusion, paresthesia, stupor, and coma.
Phosphate (HPO_4^{2-}) 1.7–2.6 mEq/liter	**Hypophosphatemia** (hī-pō-fos′-fa-TĒ-mē-a) may occur through increased urinary losses, decreased intestinal absorption, or increased utilization.	Confusion, seizures, coma, chest and muscle pain, numbness and tingling of the fingers, uncoordination, memory loss, and lethargy.	**Hyperphosphatemia** occurs when the kidneys fail to excrete excess phosphate, as happens in renal failure; can also result from increased intake of phosphates or destruction of body cells, which releases phosphates into the blood.	Anorexia, nausea, vomiting, muscular weakness, hyperactive reflexes, tetany, and tachycardia.
Magnesium (Mg^{2+}) 1.3–2.1 mEq/liter	**Hypomagnesemia** (hī′-pō-mag′-ne-SĒ-mē-a) may be due to inadequate intake or excessive loss in urine or feces; also occurs in alcoholism, malnutrition, diabetes mellitus, and diuretic therapy.	Weakness, irritability, tetany, delirium, convulsions, confusion, anorexia, nausea, vomiting, paresthesia, and cardiac arrhythmias.	**Hypermagnesemia** occurs in renal failure or due to increased intake of Mg^{2+}, such as Mg^{2+}-containing antacids; also occurs in aldosterone deficiency and hypothyroidism.	Hypotension, muscular weakness or paralysis, nausea, vomiting, and altered mental functioning.

*Values are normal ranges of blood plasma levels in adults.

ACID–BASE BALANCE

▶ **O B J E C T I V E S**

- **Compare the roles of buffers, exhalation of carbon dioxide, and kidney excretion of H⁺ in maintaining pH of body fluids.**
- **Define acid–base imbalances, describe their effects on the body, and explain how they are treated.**

From our discussion thus far, it is clear that various ions play different roles that help maintain homeostasis. A major homeostatic challenge is keeping the H^+ concentration (pH) of body fluids at an appropriate level. This task—the maintenance of acid–base balance—is of critical importance to normal cellular function. For example, the three-dimensional shape of all body proteins, which enables them to perform specific functions, is very sensitive to pH changes. When the diet contains a large amount of protein, as is typical in North America, cellular metabolism produces more acids than bases, which tends to acidify the blood. (You may wish to review the discussion of acids, bases, and pH on pages 40–41.)

In a healthy person, several mechanisms help maintain the pH of systemic arterial blood between 7.35 and 7.45. (A pH of 7.4 corresponds to a H^+ concentration of 0.00004 mEq/liter = 40 nEq/liter.) Because metabolic reactions often produce a huge excess of H^+, the lack of any mechanism for the disposal of H^+ would cause H^+ level in body fluids to quickly rise to a lethal level. Homeostasis of H^+ concentration within a narrow range is thus essential to survival. The removal of H^+ from body fluids and its subsequent elimination from the body depend on the following three major mechanisms:

1. **Buffer systems.** Buffers act quickly to temporarily bind H^+, removing the highly reactive, excess H^+ from solution. Buffers thus raise pH of body fluids but do not remove H^+ from the body.
2. **Exhalation of carbon dioxide.** By increasing the rate and depth of breathing, more carbon dioxide can be exhaled. Within minutes this reduces the level of carbonic acid in blood, which raises the blood pH (reduces blood H^+ level).
3. **Kidney excretion of H^+.** The slowest mechanism, but the only way to eliminate acids other than carbonic acid, is through their excretion in urine.

We next examine each of these mechanisms in turn.

The Actions of Buffer Systems

Most buffer systems in the body consist of a weak acid and the salt of that acid, which functions as a weak base. Buffers prevent rapid, drastic changes in the pH of body fluids by converting strong acids and bases into weak acids and bases within fractions of a second. Strong acids lower pH more than weak acids because strong acids release H^+ more readily and thus contribute more free hydrogen ions. Similarly, strong bases raise pH more than weak ones. The principal buffer systems of the body fluids

are the protein buffer system, the carbonic acid–bicarbonate buffer system, and the phosphate buffer system.

Protein Buffer System

The **protein buffer system** is the most abundant buffer in intracellular fluid and blood plasma. For instance, the protein hemoglobin is an especially good buffer within red blood cells, and albumin is the main protein buffer in blood plasma. Proteins are composed of amino acids, organic molecules that contain at least one carboxyl group ($-COOH$) and at least one amino group ($-NH_2$); these groups are the functional components of the protein buffer system. The free carboxyl group at one end of a protein acts like an acid by releasing H^+ when pH rises; it dissociates as follows:

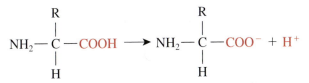

The H^+ is then able to react with any excess OH^- in the solution to form water. The free amino group at the other end of a protein can act as a base by combining with H^+ when pH falls, as follows:

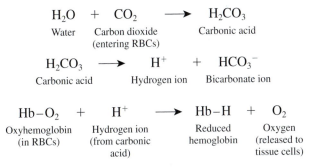

Thus, proteins can buffer both acids and bases. Besides the terminal carboxyl and amino groups, side chains that can buffer H^+ are present on seven of the 20 amino acids.

The protein hemoglobin is an important buffer of H^+ in red blood cells (see Figure 23.24 on page 836). As blood flows through the systemic capillaries, carbon dioxide (CO_2) passes from tissue cells into red blood cells, where it combines with water (H_2O) to form carbonic acid (H_2CO_3). Once formed, H_2CO_3 dissociates into H^+ and HCO_3^-. At the same time that CO_2 is entering red blood cells, oxyhemoglobin ($Hb-O_2$) is giving up its oxygen to tissue cells. Reduced hemoglobin (deoxyhemoglobin) is an excellent buffer of H^+, so it picks up most of the H^+. For this reason, reduced hemoglobin usually is written as $Hb-H$. The following reactions summarize these relations:

$$H_2O + CO_2 \longrightarrow H_2CO_3$$

Water   Carbon dioxide (entering RBCs)   Carbonic acid

$$H_2CO_3 \longrightarrow H^+ + HCO_3^-$$

Carbonic acid   Hydrogen ion   Bicarbonate ion

$$Hb-O_2 + H^+ \longrightarrow Hb-H + O_2$$

Oxyhemoglobin (in RBCs)   Hydrogen ion (from carbonic acid)   Reduced hemoglobin   Oxygen (released to tissue cells)

Carbonic Acid–Bicarbonate Buffer System

The **carbonic acid–bicarbonate buffer system** is based on the *bicarbonate ion* (HCO_3^-), which can act as a weak base, and *carbonic acid* (H_2CO_3), which can act as a weak acid. HCO_3^- is a significant anion in both intracellular and extracellular fluids (Figure 27.6), and its reaction with the CO_2 constantly released during cellular respiration produces H_2CO_3. Because the kidneys also synthesize new HCO_3^- and reabsorb filtered HCO_3^-, this important buffer is not lost in the urine. If there is an excess of H^+, the HCO_3^-, can function as a weak base and remove the excess H^+ as follows:

$$H^+ \quad + \quad HCO_3^- \quad \longrightarrow \quad H_2CO_3$$

Hydrogen ion   Bicarbonate ion   Carbonic acid
                          (weak base)

Then, H_2CO_3 dissociates into water and carbon dioxide, and the CO_2 is exhaled from the lungs.

Conversely, if there is a shortage of H^+, the H_2CO_3 can function as a weak acid and provide H^+ as follows:

$$H_2CO_3 \quad \longrightarrow \quad H^+ \quad + \quad HCO_3^-$$

Carbonic acid   Hydrogen ion   Bicarbonate ion
(weak acid)

At a pH of 7.4, HCO_3^- concentration is about 24 mEq/liter, whereas H_2CO_3 concentration is about 1.2 mmol/liter; thus, bicarbonate ions outnumber carbonic acid molecules by 20 to 1. Because CO_2 and H_2O combine to form H_2CO_3, this buffer system cannot protect against pH changes due to respiratory problems in which there is an excess or shortage of CO_2.

Phosphate Buffer System

The **phosphate buffer system** acts via a similar mechanism as the carbonic acid–bicarbonate buffer system. The components of the phosphate buffer system are the ions *dihydrogen phosphate* ($H_2PO_4^-$) and *monohydrogen phosphate* (HPO_4^{2-}). Recall that phosphates are major anions in intracellular fluid and minor ones in extracellular fluids (Figure 27.6). The dihydrogen phosphate ion acts as a weak acid and is capable of buffering strong bases such as OH^-, as follows:

$$OH^- \quad + \quad H_2PO_4^- \quad \longrightarrow \quad H_2O \, + \quad HPO_4^{2-}$$

Hydroxide ion   Dihydrogen   Water   Monohydrogen
(strong base)   phosphate   phosphate
                 (weak acid)   (weak base)

The monohydrogen phosphate ion, in contrast, acts as a weak base and is capable of buffering the H^+ released by a strong acid such as hydrochloric acid (HCl):

$$H^+ \quad + \quad HPO_4^{2-} \quad \longrightarrow \quad H_2PO_4^-$$

Hydrogen ion   Monohydrogen   Dihydrogen
(strong acid)   phosphate   phosphate
                 (weak acid)   (weak acid)

Because the concentration of phosphates is highest in intracellular fluid, the phosphate buffer system is an important regulator of pH in the cytosol. It also acts to a smaller degree in extracellular fluids, and it acts to buffer acids in urine. $H_2PO_4^{2-}$ is formed when excess H^+ in the kidney tubule fluid combine with HPO_4^{2-} (see Figure 27.8). The H^+ that becomes part of the $H_2PO_4^-$ passes into the urine. This reaction is one means whereby the kidneys help maintain blood pH by excreting H^+ in the urine.

Exhalation of Carbon Dioxide

Breathing also plays an important role in maintaining the pH of body fluids. An increase in the carbon dioxide (CO_2) concentration in body fluids increases H^+ concentration and thus lowers the pH (makes body fluids more acidic). Because H_2CO_3 can be eliminated by exhaling CO_2, it is called a **volatile acid.** Conversely, a decrease in the CO_2 concentration of body fluids raises the pH (makes body fluids more alkaline). This chemical interaction is illustrated by the following reversible reactions.

$$CO_2 \, + H_2O \, \rightleftharpoons \, H_2CO_3 \, \rightleftharpoons \, H^+ \, + \, HCO_3^-$$

Carbon   Water   Carbonic   Hydrogen   Bicarbonate
dioxide   acid   ion   ion

Changes in the rate and depth of breathing can alter the pH of body fluids within a couple of minutes. With increased ventilation, more CO_2 is exhaled, the preceding reaction is driven to the left, H^+ concentration falls, and blood pH rises. Doubling the ventilation increases pH by about 0.23 units, from 7.4 to 7.63. If ventilation is slower than normal, less carbon dioxide is exhaled, and the blood pH falls. Reducing ventilation to one-quarter of normal lowers the pH by 0.4 units, from 7.4 to 7.0. These examples show the powerful effect of alterations in breathing on the pH of body fluids.

The pH of body fluids and the rate and depth of breathing interact via a negative feedback loop (Figure 27.7). When the blood acidity increases, the decrease in pH (increase in concentration of H^+) is detected by central chemoreceptors in the medulla oblongata and peripheral chemoreceptors in the aortic and carotid bodies, both of which stimulate the inspiratory area in the medulla oblongata. As a result, the diaphragm and other respiratory muscles contract more forcefully and frequently, so more CO_2 is exhaled. As less H_2CO_3 forms and fewer H^+ are present, blood pH increases. When the response brings blood pH (H^+ concentration) back to normal, there is a return to acid–base homeostasis. The same negative feedback loop operates if the blood level of CO_2 increases. Ventilation increases, which removes more CO_2 from the blood, thus reducing its H^+ concentration so the blood's pH increases.

By contrast, if the pH of the blood increases, the respiratory center is inhibited and the rate and depth of breathing decreases. A decrease in the CO_2 concentration of the blood has the same effect. When breathing decreases, CO_2 accumulates in the blood so its H^+ concentration increases. This respiratory mechanism is a powerful eliminator of acid, but it can only get rid of the sole volatile acid—carbonic acid.

Figure 27.7 Negative feedback regulation of blood pH by the respiratory system.

Exhalation of carbon dioxide lowers the H^+ concentration of blood.

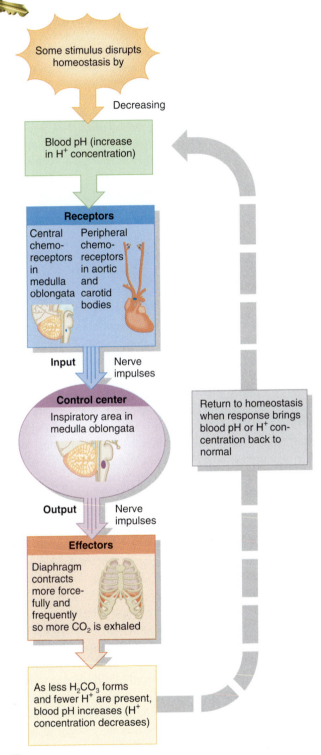

Some stimulus disrupts homeostasis by

Decreasing

Blood pH (increase in H^+ concentration)

Receptors

Central chemoreceptors in medulla oblongata

Peripheral chemoreceptors in aortic and carotid bodies

Input — Nerve impulses

Control center

Inspiratory area in medulla oblongata

Return to homeostasis when response brings blood pH or H^+ concentration back to normal

Output — Nerve impulses

Effectors

Diaphragm contracts more forcefully and frequently so more CO_2 is exhaled

As less H_2CO_3 forms and fewer H^+ are present, blood pH increases (H^+ concentration decreases)

If you hold your breath for 30 seconds, what is likely to happen to your blood pH?

Kidney Excretion of H^+

Metabolic reactions produce **nonvolatile acids** such as sulfuric acid at a rate of about 1 mEq of H^+ per day for every kilogram of body mass. The only way to eliminate this huge acid load is to excrete H^+ in the urine. Given the magnitude of these contributions to acid–base balance, it's not surprising that renal failure can quickly cause death.

Cells in both the proximal convoluted tubules (PCT) and the collecting ducts of the kidneys secrete hydrogen ions into the tubular fluid. In the PCT, Na^+/H^+ antiporters secrete H^+ as they reabsorb Na^+ (see Figure 26.13 on page 968). Even more important for regulation of pH of body fluids, however, are the intercalated cells of the collecting tubule. The *apical* membranes of some intercalated cells include **proton pumps (H^+ ATPases)** that secrete H^+ into the tubular fluid (Figure 27.8). Intercalated cells can secrete H^+ against a concentration gradient so effectively that urine can be up to 1000 times (3 pH units) more acidic than blood. HCO_3^- produced by dissociation of H_2CO_3 inside intercalated cells crosses the basolateral membrane by means of **Cl^-/HCO_3^- antiporters** and then diffuses into peritubular capillaries (Figure 27.8a). The HCO_3^- that enters the blood in this way is *new* (not filtered). For this reason, blood leaving the kidney in the renal vein may have a higher HCO_3^- concentration than blood entering the kidney in the renal artery.

Interestingly, a second type of intercalated cell has proton pumps in its *basolateral* membrane and Cl^-/HCO_3^- antiporters in its apical membrane. These intercalated cells secrete HCO_3^- and reabsorb H^+. Thus, the two types of intercalated cells help maintain the pH of body fluids in two ways—by excreting excess H^+ when pH of body fluids is too low or by excreting excess HCO_3^- when pH is too high.

Some H^+ secreted into the tubular fluid of the collecting duct are buffered. By now, most of the filtered HCO_3^- has been reabsorbed; few are left in the tubule lumen to combine with and buffer secreted H^+. Two other buffers, however, are present and available to combine with H^+ (Figure 27.8b). The most plentiful buffer in the tubular fluid of the collecting duct is HPO_4^{2-} (monohydrogen phosphate ion). In addition, a small amount of NH_3 (ammonia) also is present. H^+ combines with HPO_4^{2-} to form $H_2PO_4^-$ (dihydrogen phosphate ion) and with NH_3 to form NH_4^+ (ammonium ion). Because these ions cannot diffuse back into tubule cells, they are excreted in the urine.

Table 27.3 summarizes the mechanisms that maintain pH of body fluids.

Acid–Base Imbalances

The normal pH range of systemic arterial blood is between 7.35 (= 45 nEq of H^+/liter) and 7.45 (= 35 nEq of H^+/liter). **Acidosis** (or **acidemia**) is a condition in which blood pH is below 7.35; **alkalosis** (or **alkalemia**) is a condition in which blood pH is higher than 7.45.

Figure 27.8 Secretion of H⁺ by intercalated cells in the collecting duct. HCO_3^- = bicarbonate ion; CO_2 = carbon dioxide; H_2O = water; H_2CO_3 = carbonic acid; Cl^- = chloride ion; NH_3 = ammonia; NH_4^+ = ammonium ion; HPO_4^{2-} = monohydrogen phosphate ion; $H_2PO_4^-$ = dihydrogen phosphate ion.

 Urine can be up to 1000 times more acidic than blood due to the operation of the proton pumps in the collecting ducts of the kidneys.

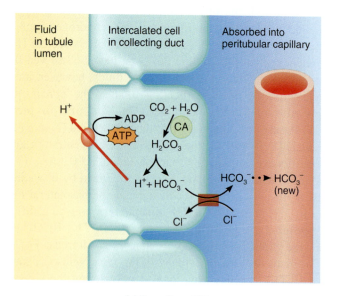

(a) Secretion of H⁺

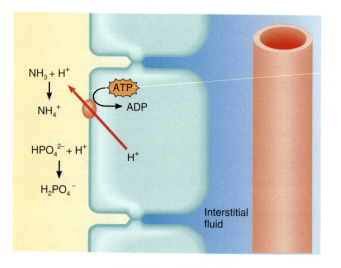

(b) Buffering of H⁺ in urine

Key:

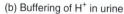 Proton pump (H⁺ ATPase) in apical membrane

HCO_3^-/Cl^- antiporter in basolateral membrane

•••► Diffusion

What would be the effects of a drug that blocks the activity of carbonic anhydrase?

Table 27.3	Mechanisms that Maintain pH of Body Fluids
Mechanism	**Comments**
Buffer Systems	Most consist of a weak acid and the salt of that acid, which functions as a weak base. They prevent drastic changes in body fluid pH.
Proteins	The most abundant buffers in body cells and blood. Histidine and cysteine are the two amino acids that contribute most of the buffering capacity of proteins. Hemoglobin inside red blood cells is a good buffer.
Carbonic acid–bicarbonate	Important regulator of blood pH. The most abundant buffers in extracellular fluid (ECF).
Phosphates	Important buffers in intracellular fluid and in urine.
Exhalation of CO₂	With increased exhalation of CO_2, pH rises (fewer H⁺). With decreased exhalation of CO_2, pH falls (more H⁺).
Kidneys	Renal tubules secrete H⁺ into the urine and reabsorb HCO_3^- so it is not lost in the urine.

The major physiological effect of acidosis is depression of the central nervous system through depression of synaptic transmission. If the systemic arterial blood pH falls below 7, depression of the nervous system is so severe that the individual becomes disoriented, then comatose, and may die. Patients with severe acidosis usually die while in a coma. A major physiological effect of alkalosis, by contrast, is overexcitability in both the central nervous system and peripheral nerves. Neurons conduct impulses repetitively, even when not stimulated by normal stimuli; the results are nervousness, muscle spasms, and even convulsions and death.

A change in blood pH that leads to acidosis or alkalosis may be countered by **compensation,** the physiological response to an acid–base imbalance that acts to normalize arterial blood pH. Compensation may be either *complete,* if pH indeed is brought within the normal range, or *partial,* if systemic arterial blood pH is still lower than 7.35 or higher than 7.45. If a person has altered blood pH due to metabolic causes, hyperventilation or hypoventilation can help bring blood pH back toward the normal range; this form of compensation, termed **respiratory compensation,** occurs within minutes and reaches its maximum within hours. If, however, a person has altered blood pH due to respiratory causes, then **renal compensation**—changes in secretion of H⁺ and reabsorption of HCO_3^- by the kidney tubules—can help reverse the change. Renal compensation may begin in minutes, but it takes days to reach maximum effectiveness.

In the discussion that follows, note that both respiratory acidosis and respiratory alkalosis are disorders resulting from changes in the partial pressure of CO_2 (P_{CO_2}) in systemic arterial blood (normal range is 35–45 mmHg). By contrast, both metabolic acidosis and metabolic alkalosis are disorders resulting from changes in HCO_3^- concentration (normal range is 22–26 mEq/liter in systemic arterial blood).

Respiratory Acidosis

The hallmark of **respiratory acidosis** is an abnormally high P_{CO_2} in systemic arterial blood—above 45 mmHg. Inadequate exhalation of CO_2 causes the blood pH to drop. Any condition that decreases the movement of CO_2 from the blood to the alveoli of the lungs to the atmosphere causes a buildup of CO_2, H_2CO_3, and H^+. Such conditions include emphysema, pulmonary edema, injury to the respiratory center of the medulla oblongata, airway obstruction, or disorders of the muscles involved in breathing. If the respiratory problem is not too severe, the kidneys can help raise the blood pH into the normal range by increasing excretion of H^+ and reabsorption of HCO_3^- (renal compensation). The goal in treatment of respiratory acidosis is to increase the exhalation of CO_2, as, for instance, by providing ventilation therapy. In addition, intravenous administration of HCO_3^- may be helpful.

Respiratory Alkalosis

In **respiratory alkalosis,** systemic arterial blood P_{CO_2} falls below 35 mmHg. The cause of the drop in P_{CO_2} and the resulting increase in pH is hyperventilation, which occurs in conditions that stimulate the inspiratory area in the brain stem. Such conditions include oxygen deficiency due to high altitude or pulmonary disease, cerebrovascular accident (stroke), or severe anxiety. Again, renal compensation may bring blood pH into the normal range if the kidneys decrease excretion of H^+ and reabsorption of HCO_3^-. Treatment of respiratory alkalosis is aimed at increasing the level of CO_2 in the body. One simple treatment is to have the person inhale and exhale into a paper bag for a short period; as a result, the person inhales air containing a higher-than-normal concentration of CO_2.

Metabolic Acidosis

In **metabolic acidosis,** the systemic arterial blood HCO_3^- level drops below 22 mEq/liter. Such a decline in this important buffer causes the blood pH to decrease. Three situations may lower the blood level of HCO_3^-: (1) actual loss of HCO_3^-, such as may occur with severe diarrhea or renal dysfunction; (2) accumulation of an acid other than carbonic acid, as may occur in ketosis (described on page 923); or (3) failure of the kidneys to excrete H^+ from metabolism of dietary proteins. If the problem is not too severe, hyperventilation can help bring blood pH into the normal range (respiratory compensation). Treatment of metabolic acidosis consists of administering intravenous solutions of sodium bicarbonate and correcting the cause of the acidosis.

Metabolic Alkalosis

In **metabolic alkalosis,** the systemic arterial blood HCO_3^- concentration is above 26 mEq/liter. A nonrespiratory loss of acid or excessive intake of alkaline drugs causes the blood pH to increase above 7.45. Excessive vomiting of gastric contents, which results in a substantial loss of hydrochloric acid, is probably the most frequent cause of metabolic alkalosis. Other causes include gastric suctioning, use of certain diuretics, endocrine disorders, excessive intake of alkaline drugs (antacids), and severe dehydration. Respiratory compensation through hypoventilation may bring blood pH into the normal range. Treatment of metabolic alkalosis consists of giving fluid solutions to correct Cl^-, K^+, and other electrolyte deficiencies plus correcting the cause of alkalosis.

Table 27.4 summarizes respiratory and metabolic acidosis and alkalosis.

Table 27.4	Summary of Acidosis and Alkalosis		
Condition	**Definition**	**Common Causes**	**Compensatory Mechanism**
Respiratory acidosis	Increased P_{CO_2} (above 45 mmHg) and decreased pH (below 7.35) if there is no compensation.	Hypoventilation due to emphysema, pulmonary edema, trauma to respiratory center, airway obstructions, or dysfunction of muscles of respiration.	*Renal:* increased excretion of H^+; increased reabsorption of HCO_3^-. If compensation is complete, pH will be within the normal range but P_{CO_2} will be high.
Respiratory alkalosis	Decreased P_{CO_2} (below 35 mmHg) and increased pH (above 7.45) if there is no compensation.	Hyperventilation due to oxygen deficiency, pulmonary disease, cerebrovascular accident (CVA), or severe anxiety.	*Renal:* decreased excretion of H^+; decreased reabsorption of HCO_3^-. If compensation is complete, pH will be within the normal range but P_{CO_2} will be low.
Metabolic acidosis	Decreased HCO_3^- (below 22 mEq/liter) and decreased pH (below 7.35) if there is no compensation.	Loss of bicarbonate ions due to diarrhea, accumulation of acid (ketosis), renal dysfunction.	*Respiratory:* hyperventilation, which increases loss of CO_2. If compensation is complete, pH will be within the normal range but HCO_3^- will be low.
Metabolic alkalosis	Increased HCO_3^- (above 26 mEq/liter) and increased pH (above 7.45) if there is no compensation.	Loss of acid due to vomiting, gastric suctioning, or use of certain diuretics; excessive intake of alkaline drugs.	*Respiratory:* hypoventilation, which slows loss of CO_2. If compensation is complete, pH will be within the normal range but HCO_3^- will be high.

Diagnosis of Acid–Base Imbalances

One can often pinpoint the cause of an acid–base imbalance by careful evaluation of three factors in a sample of systemic arterial blood: pH, concentration of HCO_3^-, and P_{CO_2}. These three blood chemistry values are examined in the following four-step sequence:

1. Note whether the pH is high (alkalosis) or low (acidosis).

2. Then decide which value—P_{CO_2} or HCO_3^-—is out of the normal range and could be the *cause* of the pH change. For example, *elevated pH* could be caused by *low P_{CO_2}* or *high HCO_3^-*.

3. If the cause is a *change in P_{CO_2}*, the problem is *respiratory;* if the cause is a *change in HCO_3^-*, the problem is *metabolic.*

4. Now look at the value that doesn't correspond with the observed pH change. If it is within its normal range, there is no compensation. If it is outside the normal range, compensation is occurring and partially correcting the pH imbalance. ■

▶ CHECKPOINT

8. Explain how each of the following buffer systems helps to maintain the pH of body fluids: proteins, carbonic acid–bicarbonate buffers, and phosphates.

9. Define acidosis and alkalosis. Distinguish among respiratory and metabolic acidosis and alkalosis.

10. What are the principal physiological effects of acidosis and alkalosis?

AGING AND FLUID, ELECTROLYTE, AND ACID–BASE BALANCE

▶ OBJECTIVE

• **Describe the changes in fluid, electrolyte, and acid–base balance that may occur with aging.**

There are significant differences between adults and infants, especially premature infants, with respect to fluid distribution, regulation of fluid and electrolyte balance, and acid–base homeostasis. Accordingly, infants experience more problems than adults in these areas. The differences are related to the following conditions:

• *Proportion and distribution of water.* Whereas a newborn's total body mass is about 75% water (and can be as high as 90% in a premature infant), an adult's is about 55–60% water. (The "adult" percentage is achieved at about 2 years of age.) Moreover, whereas adults have twice as much water in ICF as ECF, the opposite is true in premature infants. Because ECF is subject to more changes than ICF, rapid losses or gains of body water are much more critical in infants. Given that the rate of fluid intake and output is about seven times higher in infants than in adults, the slightest changes in fluid balance can result in severe abnormalities.

• *Metabolic rate.* The metabolic rate of infants is about double that of adults. This results in the production of more metabolic wastes and acids, which can lead to the development of acidosis in infants.

• *Functional development of the kidneys.* In infants, the kidneys are only about half as efficient in concentrating urine as those of adults. (Functional development is not complete until close to the end of the first month after birth.) As a result, the kidneys of newborns can neither concentrate urine nor rid the body of excess acids produced as a result of high metabolic rate as effectively as those of adults.

• *Body surface area.* The ratio of body surface area to body volume of infants is about three times greater than that of adults. This significantly increases water loss through the skin in infants.

• *Breathing rate.* The higher breathing rate of infants (about 30 to 80 times a minute) causes greater water loss from the lungs. Moreover, respiratory alkalosis may occur because greater ventilation eliminates more CO_2 and lowers the P_{CO_2}.

• *Ion concentrations.* Newborns have higher K^+ and Cl^- concentrations than adults. This creates a tendency toward metabolic acidosis.

By comparison with children and younger adults, older adults often have an impaired ability to maintain fluid, electrolyte, and acid–base balance. With increasing age, many people have a decreased volume of intracellular fluid and decreased total body K^+ due to declining skeletal muscle mass and increasing mass of adipose tissue (which contains very little water). Age-related decreases in respiratory and renal functioning may compromise acid–base balance by slowing the exhalation of CO_2 and the excretion of excess acids in urine. Other kidney changes, such as decreased blood flow, decreased glomerular filtration rate, and reduced sensitivity to antidiuretic hormone, have an adverse effect on the ability to maintain fluid and electrolyte balance. Due to a decrease in the number and efficiency of sweat glands, water loss from the skin declines with age. Because of these age-related changes, older adults are susceptible to several fluid and electrolyte disorders:

• *Dehydration* and *hypernatremia* often occur due to inadequate fluid intake or loss of more water than Na^+ in vomit, feces, or urine.

• *Hyponatremia* may occur due to inadequate intake of Na^+; elevated loss of Na^+ in urine, vomit, or diarrhea; or impaired ability of the kidneys to produce dilute urine.

• *Hypokalemia* often occurs in older adults who chronically use laxatives to relieve constipation or who take K^+-depleting diuretic drugs for treatment of hypertension or heart disease.

• *Acidosis* may occur due to impaired ability of the lungs and kidneys to compensate for acid–base imbalances. One cause of acidosis is that renal tubule cells produce less ammonia (NH_3), which then is not available to combine with H^+ and be excreted in urine as NH_4^+; another cause is reduced exhalation of CO_2.

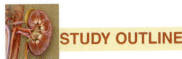

STUDY OUTLINE

FLUID COMPARTMENTS AND FLUID BALANCE (p. 992)

1. Body fluid includes water and dissolved solutes.

2. About two-thirds of the body's fluid is located within cells and is called intracellular fluid (ICF). The other one-third, called extracellular fluid (ECF), includes interstitial fluid; blood plasma and lymph; cerebrospinal fluid; gastrointestinal tract fluids; synovial fluid; fluids of the eyes and ears; pleural, pericardial, and peritoneal fluids; and glomerular filtrate.

3. Fluid balance means that the various body compartments contain the normal amount of water.

4. An inorganic substance that dissociates into ions in solution is called an electrolyte. Fluid balance and electrolyte balance are interrelated.

5. Water is the largest single constituent in the body. It makes up 45–75% of total body mass, depending on age, gender, and the amount of adipose tissue present.

6. Daily water gain and loss are each about 2500 mL. Sources of water gain are ingested liquids and foods, and water produced by cellular respiration and dehydration synthesis reactions (metabolic water). Water is lost from the body via urination, evaporation from the skin surface, exhalation of water vapor, and defecation. In women, menstrual flow is an additional route for loss of body water.

7. The main way to regulate body water gain is by adjusting the volume of water intake, mainly by drinking more or less fluid. The thirst center in the hypothalamus governs the urge to drink.

8. Although increased amounts of water and solutes are lost through sweating and exhalation during exercise, loss of excess body water or excess solutes depends mainly on regulating excretion in the urine. The extent of urinary NaCl loss is the main determinant of body fluid volume, whereas the extent of urinary water loss is the main determinant of body fluid osmolarity.

9. Table 27.1 on page 996 summarizes the factors that regulate water gain and water loss.

10. Angiotensin II and aldosterone reduce urinary loss of Na^+ and Cl^- and thereby increase the volume of body fluids. ANP promotes natriuresis, elevated excretion of Na^+ (and Cl^-), which decreases blood volume.

11. The major hormone that regulates water loss and thus body fluid osmolarity is antidiuretic hormone (ADH).

12. An increase in the osmolarity of interstitial fluid draws water out of cells, and they shrink slightly. A decrease in the osmolarity of interstitial fluid causes cells to swell. Most often a change in osmolarity is due to a change in the concentration of Na^+, the dominant solute in interstitial fluid.

13. When a person consumes water faster than the kidneys can excrete it or when renal function is poor, the result may be water intoxication, in which cells swell dangerously.

ELECTROLYTES IN BODY FLUIDS (p. 997)

1. Ions formed when electrolytes dissolve in body fluids control the osmosis of water between fluid compartments, help maintain acid–base balance, and carry electrical current.

2. The concentrations of cations and anions is expressed in units of milliequivalents/liter (mEq/liter).

3. Blood plasma, interstitial fluid, and intracellular fluid contain varying kinds and amounts of ions.

4. Sodium ions (Na^+) are the most abundant extracellular ions. They are involved in impulse transmission, muscle contraction, and fluid and electrolyte balance. Na^+ level is controlled by aldosterone, antidiuretic hormone, and atrial natriuretic peptide.

5. Chloride ions (Cl^-) are the major extracellular anions. They play a role in regulating osmotic pressure and forming HCl in gastric juice. Cl^- level is controlled indirectly by antidiuretic hormone and by processes that increase or decrease renal reabsorption of Na^+.

6. Potassium ions (K^+) are the most abundant cations in intracellular fluid. They play a key role in the resting membrane potential and action potential of neurons and muscle fibers; help maintain intracellular fluid volume; and contribute to regulation of pH. K^+ level is controlled by aldosterone.

7. Bicarbonate ions (HCO_3^-) are the second most abundant anions in extracellular fluid. They are the most important buffer in blood plasma.

8. Calcium is the most abundant mineral in the body. Calcium salts are structural components of bones and teeth. Ca^{2+}, which are principally extracellular cations, function in blood clotting, neurotransmitter release, and contraction of muscle. Ca^{2+} level is controlled mainly by parathyroid hormone and calcitriol.

9. Phosphate ions ($H_2PO_4^-$, HPO_4^{2-}, and PO_4^{3-}) are principally intracellular anions, and their salts are structural components of bones and teeth. They are also required for the synthesis of nucleic acids and ATP and participate in buffer reactions. Their level is controlled by parathyroid hormone and calcitriol.

10. Magnesium ions (Mg^{2+}) are primarily intracellular cations. They act as cofactors in several enzyme systems.

11. Table 27.2 on page 1000 describes the imbalances that result from deficiency or excess of important body electrolytes.

ACID–BASE BALANCE (p. 1001)

1. The overall acid–base balance of the body is maintained by controlling the H^+ concentration of body fluids, especially extracellular fluid.

2. The normal pH of systemic arterial blood is 7.35–7.45.

3. Homeostasis of pH is maintained by buffer systems, via exhalation of carbon dioxide, and via kidney excretion of H^+ and reabsorption of HCO_3^-.

4. The important buffer systems include proteins, carbonic acid–bicarbonate buffers, and phosphates.

5. An increase in exhalation of carbon dioxide increases blood pH; a decrease of CO_2 in exhalation decreases blood pH.

6. In the proximal convoluted tubules of the kidneys, Na^+/H^+ antiporters secrete H^+ as they reabsorb Na^+. In the collecting ducts of the kidneys, some intercalated cells reabsorb K^+ and HCO_3^- and secrete H^+, whereas other intercalated cells secrete HCO_3^-. In these ways, the kidneys can increase or decrease the pH of body fluids.

7. Table 27.3 on page 1004 summarizes the mechanisms that maintain pH of body fluids.

8. Acidosis is a systemic arterial blood pH below 7.35; its principal

effect is depression of the central nervous system (CNS). Alkalosis is a systemic arterial blood pH above 7.45; its principal effect is overexcitability of the CNS.

9. Respiratory acidosis and alkalosis are disorders due to changes in blood P_{CO_2}, whereas metabolic acidosis and alkalosis are disorders associated with changes in blood HCO_3^- concentration.

10. Metabolic acidosis or alkalosis can be compensated by respiratory mechanisms; respiratory acidosis or alkalosis can be compensated by renal mechanisms.

11. Table 27.4 on page 1005 summarizes respiratory and metabolic acidosis and alkalosis.

12. By examining systemic arterial blood pH, HCO_3^-, and P_{CO_2} values, it is possible to pinpoint the cause of an acid-base imbalance.

AGING AND FLUID, ELECTROLYTE, AND ACID–BASE BALANCE (p. 1006)

1. With increasing age, there is decreased intracellular fluid volume and decreased K^+ due to declining skeletal muscle mass.

2. Decreased kidney function with aging adversely affects fluid and electrolyte balance.

Q SELF-QUIZ QUESTIONS

Fill in the blanks in the following statements.

1. The two compartments in which water can be found are the _____ and the _____.

2. In the carbonic acid-bicarbonate buffer system, the _____ acts as a weak base, and _____ acts as a weak acid.

3. Most buffers consist of a _____ and a _____.

4. The principal buffer systems of the body are the _____ systems, the _____ systems, and the _____ systems.

Indicate whether the following statements are true or false.

5. The most abundant buffer in intracellular fluid and blood plasma is the protein buffer system.

6. Normally, water loss equals water gain, so body fluid volume is constant.

Choose the one best answer to the following questions.

7. The primary means of regulating body water gain is adjusting (a) the volume of water intake, (b) the rate of cellular respiration, (c) the formation of metabolic water, (d) the volume of metabolic water, (e) the metabolic use of water.

8. Which of the following are ways that dehydration stimulates thirst? (1) It decreases the production of saliva. (2) It increases the production of saliva. (3) It increases osmolarity of body fluids. (4) It decreases osmolarity of body fluids. (5) It decreases blood volume. (6) It increases blood volume. (a) 1, 2, 4, and 6, (b) 1, 3, 5, and 6, (c) 1, 3, and 5, (d) 2, 4, and 6, (e) 1, 4, 5, and 6.

9. Which of the following hormones regulate fluid loss? (1) antidiuretic hormone, (2) aldosterone, (3) atrial natriuretic peptide, (4) thyroxine, (5) cortisol. (a) 1, 3, and 5, (b) 1, 2, and 3, (c) 2, 4, and 5, (d) 2, 3, and 4, (e) 1, 3, and 4.

10. Which of the following are *true* concerning ions in the body? (1) They control osmosis of water between fluid compartments. (2) They help maintain acid–base balance. (3) They carry electrical current. (4) They serve as cofactors for enzyme activity. (5) They serve as neurotransmitters under special circumstances. (a) 1, 3, and 5, (b) 2, 4, and 5, (c) 1, 4, and 5, (d) 1, 2, and 4, (e) 1, 2, 3, and 4.

11. Which of the following statements are *true*? (1) Buffers prevent rapid, drastic changes in pH of a body fluid. (2) Buffers work slowly. (3) Strong acids lower pH more than weak acids because strong acids contribute fewer H^+. (4) Proteins can buffer both acids and bases. (5) Hemoglobin is an important buffer. (a) 1, 2, 3, and 5, (b) 1, 3, 4, and 5, (c) 1, 3, and 5, (d) 1, 4, and 5, (e) 2, 3, and 5.

12. Which of the following statements are *true*? (1) An increase in the carbon dioxide concentration in body fluids increases H^+ concentration and thus lowers pH. (2) Breath holding results in a decline in blood pH. (3) The respiratory buffer mechanism can eliminate a single volatile acid: carbonic acid. (4) The only way to eliminate fixed acids is to excrete H^+ in the urine. (5) When the diet contains a large amount of protein, normal metabolism produces more acids than bases. (a) 1, 2, 3, 4, and 5, (b) 1, 3, 4, and 5, (c) 1, 2, 3, and 4, (d) 1, 2, 4, and 5, (e) 1, 3, and 4.

13. Concerning acid–base imbalances: (1) Acidosis can cause depression of the central nervous system through depression of synaptic transmission. (2) Renal compensation can resolve respiratory alkalosis or acidosis. (3) A major physiological effect of alkalosis is lack of excitability in the central nervous system and peripheral nerves. (4) Resolution of metabolic acidosis and alkalosis occurs through renal compensation. (5) In adjusting blood pH, renal compensation occurs quickly whereas respiratory compensation takes days. (a) 1, 2, and 5, (b) 1 and 2, (c) 2, 3, and 4, (d) 2, 3, and 5, (e) 1, 2, 3, and 5.

14. Match the following:

____ (a) the most abundant cation in intracellular fluid; plays a key role in establishing the resting membrane potential

____ (b) the most abundant mineral in the body; plays important roles in blood clotting, neurotransmitter release, maintenance of muscle tone, and excitability of nervous and muscle tissue

____ (c) second most common intracellular cation; is a cofactor for enzymes involved in carbohydrate, protein, and Na^+/K^+ ATPase metabolism

____ (d) the most abundant extracellular cation; essential in fluid and electrolyte balance

____ (e) ions that are mostly combined with lipids, proteins, carbohydrates, nucleic acids, and ATP inside cells

____ (f) most prevalent extracellular anion; can help balance the level of anions in different fluid compartments

____ (g) second most prevalent extracellular anion; mainly regulated by the kidneys; important for acid–base balance

____ (h) substances that act to prevent rapid, drastic changes in the pH of a body fluid

____ (i) inorganic substances that dissociate into ions when in solution

(1) sodium
(2) chloride
(3) electrolytes
(4) bicarbonate
(5) buffers
(6) phosphate
(7) magnesium
(8) potassium
(9) calcium

15. Match the following:

____ (a) an abnormal increase in the volume of interstitial fluid

____ (b) the swelling of cells due to water moving from blood plasma into interstitial fluid and then into cells

____ (c) occurs when water loss is greater than water gain

____ (d) condition that can occur as water moves out of blood plasma into interstitial fluid and blood volume decreases

____ (e) can be caused by emphysema, pulmonary edema, injury to the respiratory center of the medulla oblongata, airway destruction, or disorders of the muscles involved in breathing

____ (f) can be caused by actual loss of bicarbonate ions, ketosis, or failure of kidneys to excrete H^+

____ (g) can be caused by excessive vomiting of gastric contents, gastric suctioning, use of certain diuretics, severe dehydration, or excessive intake of alkaline drugs

____ (h) can be caused by oxygen deficiency at high altitude, stroke, or severe anxiety

(1) respiratory acidosis
(2) respiratory alkalosis
(3) metabolic acidosis
(4) metabolic alkalosis
(5) dehydration
(6) hypo-volemia
(7) water intoxication
(8) edema

CRITICAL THINKING QUESTIONS

1. After a late-night party, Gary ended up in the emergency room. According to his friends, Gary had said he was so thirsty that he could drink a tub full of water, so someone dared him to try it. His friends called for an ambulance when Gary suffered a convulsion. What has happened to Gary?
   HINT *Gary was intoxicated, but not with alcohol.*

2. "I'm so mad at Jacques," Aimee told her girl friend. "We're about the same height and the same weight, but when we had our body fat measured in A & P lab, he had less fat than I did!" How do you explain this to Aimee?
   HINT *Fat and water don't mix.*

3. When Carlos ate some of his roommate's leftovers, he thought the food tasted a little odd, but he didn't feel like cooking so he ate it anyway. A while later, Carlos was vomiting the entire contents of his stomach. Carlos then tried to settle his queasy stomach by eating an entire bottle of antacids. What effect will this episode have on Carlos's fluid and electrolyte balance?
   HINT *What is the pH of gastric juice?*

4. Henry is in the intensive care unit because he suffered a severe myocardial infarction three days ago. The lab reports the following values from an arterial blood sample: pH = 7.30, $HCO_3^- = 20$ mEq/liter, $P_{CO_2} = 32$ mmHg. Diagnose Henry's acid–base status and decide whether or not compensation is occurring.
   HINT *Use the four-step method described on page 1006.*

ANSWERS TO FIGURE QUESTIONS

27.1 Plasma volume equals body mass × percent of body mass that is body fluid, × proportion of body fluid that is ECF × proportion of ECF that is plasma × a conversion factor (1 liter/kg). For male: blood plasma volume = 60kg × 0.60 × 1/3 × 0.20 × 1 liter/kg = 2.4 liters. For female, blood plasma volume is 2.2 liters.

27.2 Hyperventilation, vomiting, fever, and diuretics all increase fluid loss.

27.3 Negative feedback is in operation because the result (an increase in fluid intake) is opposite to the initiating stimulus (dehydration).

27.4 Elevated aldosterone promotes abnormally high renal reabsorption of NaCl and water, which expands blood volume and increases blood pressure. Increased blood pressure causes more fluid to filter out of capillaries and accumulate in the interstitial fluid; this is edema.

27.5 When both the salt and water are absorbed in the gastrointestinal tract, blood volume increases without a decrease in osmolarity, so water intoxication does not occur.

27.6 In ECF, the major cation is Na^+, and the major anions are Cl^- and HCO_3^-. In ICF, the major cation is K^+, and the major anions are proteins and organic phosphates (for example, ATP).

27.7 Holding one's breath causes blood pH to decrease slightly as CO_2 and H^+ accumulate.

27.8 A carbonic anhydrase inhibitor reduces secretion of H^+ into the urine and reduces reabsorption of Na^+ and HCO_3^- into the blood. It has a diuretic effect and can cause acidosis (lowered pH of the blood) due to loss of HCO_3^- in the urine.

The Reproductive Systems

THE REPRODUCTIVE SYSTEMS AND HOMEOSTASIS

The male and female reproductive organs work together to produce offspring. The female reproductive organs contribute to sustaining the growth of fetuses.

ENERGIZE YOUR STUDY

FOUNDATIONS CD

Anatomy Overview
• The Reproductive Systems

Animation
• Systems Contributions to Homeostasis

Concepts and Connections and Exercises reinforce your understanding

www.wiley.com/college/apcentral

INSIGHTS AND EXPLORATIONS

A young man of 23, planning to start a family, feels a painless lump in one of his testicles. A woman who feels fine is told, after her yearly examination at age 55, that she has ovarian cancer. An African American man of 56, who is without symptoms, is diagnosed with prostate cancer. Young or old, we can fall victim to cancers of the reproductive organs. What are the chances that you will develop one of these cancers? In this web-based activity we will explore who is at risk for developing these cancers and more.

Sexual reproduction is the process by which organisms produce offspring by making germ cells called **gametes** (GAM-ēts = spouses). After the male gamete (sperm cell) unites with the female gamete (secondary oocyte)—an event called **fertilization**—the resulting cell contains one set of chromosomes from each parent. Males and females have anatomically distinct reproductive organs that are adapted for producing gametes, facilitating fertilization, and in females sustaining the growth of the embryo and fetus.

The male and female reproductive organs can be grouped by function. The **gonads**—testes in males and ovaries in females—produce gametes and secrete sex hormones. Various **ducts** then store and transport the gametes, and **accessory sex glands** produce substances that protect the gametes and facilitate their movement. Finally, supporting structures, such as the penis and the uterus, assist the delivery and joining of gametes and, in females, the growth of the fetus during pregnancy.

Gynecology (gī-ne-KOL-ō-jē; *gyneco-* = woman; *-logy* = study of) is the specialized branch of medicine concerned with the diagnosis and treatment of diseases of the female reproductive system. As noted in Chapter 26, **urology** (ū-ROL-ō-jē) is the study of the urinary system. Urologists also diagnose and treat diseases and disorders of the male reproductive system.

THE CELL CYCLE IN THE GONADS

▶ **OBJECTIVE**

• **Describe the process of meiosis.**

Number of Chromosomes in Somatic Cells and Gametes

Before we can discuss the cell cycle in the gonads, we must first understand the distribution of genetic material in cells. Human somatic cells, such as those in the brain, stomach, and kidneys, contain 23 pairs of chromosomes, or a total of 46 chromosomes. One member of each pair was inherited from each parent. The two chromosomes that make up each pair are called **homologous chromosomes** (hō-MOL-ō-gus; *homo-* = same) or **homologs;** they contain similar genes arranged in the same (or almost the same) order. When examined under a light microscope (see Figure 29.24 on page 1096), homologous chromosomes generally look very similar. The exception to this rule is one pair of chromosomes called the **sex chromosomes,** designated X and Y. In females the homologous pair of sex chromosomes consists of two large X chromosomes; in males the pair consists of an X and a much smaller Y chromosome. The other 22 pairs of chromosomes are called **autosomes.** Because somatic cells contain two sets of chromosomes, they are called **diploid cells** (DIP-loid; *dipl-* = double; *-oid* = form). Geneticists use the symbol n to denote the number of different chromosomes in an organism; in humans, $n = 23$. Diploid cells are $2n$.

In sexual reproduction, each new organism is the result of the union of two different gametes, one produced by each parent. If gametes had the same number of chromosomes as somatic cells, the number of chromosomes would double at fertilization. Because **meiosis** (mī-Ō-sis; *mei-* = lessening; *-osis* = condition of), the reproductive cell division that occurs in the gonads, produces gametes in which the number of chromosomes is reduced by half, gametes contain a single set of 23 chromosomes and thus are **haploid (1n) cells** (HAP-loyd; *hapl-* = single). Fertilization then restores the diploid number of chromosomes.

Meiosis

Meiosis occurs in two successive stages: **meiosis I** and **meiosis II.** During the interphase that precedes meiosis I, the chromosomes of the diploid parent cell replicate. As a result of replication, each chromosome consists of two genetically identical (sister) chromatids, which are attached at their centromeres. This replication of chromosomes is similar to the one that precedes mitosis in somatic cell division.

Meiosis I

Meiosis I, which begins once chromosomal replication is complete, consists of four phases: prophase I, metaphase I, anaphase I, and telophase I (Figure 28.1a). Prophase I is an extended phase in which the chromosomes shorten and thicken, the nuclear envelope and nucleoli disappear, and the mitotic spindle forms. Two events that are not seen in prophase I of mitosis (or in prophase II of meiosis) occur during prophase I of meiosis. First, the two sister chromatids of each pair of homologous chromosomes pair off, an event called **synapsis** (Figure 28.1b). The resulting four chromatids form a structure called a **tetrad.** Second, parts of the chromatids of two homologous chromosomes may be exchanged with one another. Such an exchange between parts of genetically different (nonsister) chromatids is termed **crossing-over** (Figure 28.1b on page 1014). This process, among others, permits an exchange of genes between chromatids of homologous chromosomes. Due to crossing-over, the resulting daughter cells are genetically unlike each other and genetically unlike the parent cell that produced them. Crossing-over results in **genetic recombination**—that is, the formation of new combinations of genes—and accounts for part of the great genetic variation among humans and other organisms that form gametes via meiosis. The random assortment of maternally derived chromosomes and paternally derived chromosomes toward opposite poles is another reason for genetic variation among gametes.

In metaphase I, the homologous pairs of chromosomes line up along the metaphase plate of the cell, with homologous chromosomes side by side (see Figure 28.1a). (No such pairing of

Figure 28.1 **Meiosis, reproductive cell division.** Details of events are discussed in the text.

In reproductive cell division, a single diploid parent cell undergoes meiosis I and meiosis II to produce four haploid gametes that are genetically different from the parent cell.

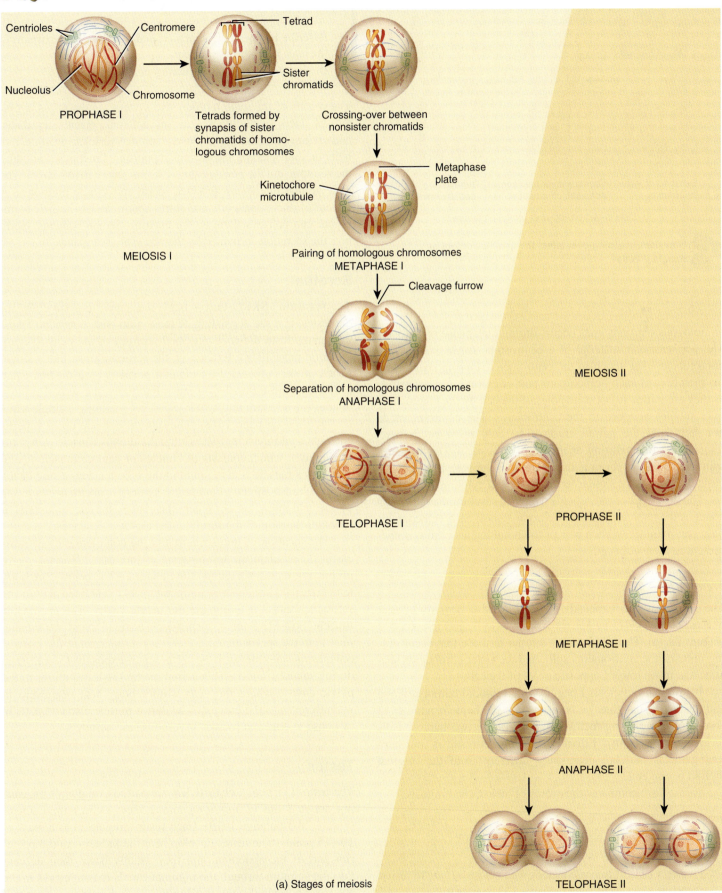

(a) Stages of meiosis

(continues)

Figure 28.1   *(continued)*

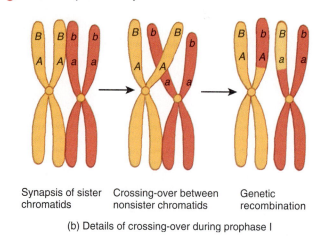

Synapsis of sister   Crossing-over between   Genetic
chromatids   nonsister chromatids   recombination

(b) Details of crossing-over during prophase I

How does crossing-over affect the genetic content of daughter cells?

homologous chromosomes occurs during metaphase of mitosis.) The pericentriolar material of a centrosome forms kinetochore microtubules that attach to the centromeres. During anaphase I, the members of each homologous pair of chromosomes separate as they are pulled to opposite poles of the cell by the kinetochore microtubules attached to the centromeres. The centromeres do not split, and the paired chromatids, held by a centromere, remain together. (During anaphase of mitosis, the centromeres split and the sister chromatids separate.) Telophase I and cytokinesis of meiosis are similar to telophase and cytokinesis of mitosis. The net effect of meiosis I is that each resulting daughter cell contains the haploid number of chromosomes; each cell contains only one member of each pair of the homologous chromosomes present in the parent cell.

Meiosis II

The second stage of meiosis, meiosis II, also consists of four phases: prophase II, metaphase II, anaphase II, and telophase II (Figure 28.1a). These phases are similar to those that occur during mitosis; the centromeres split, and the sister chromatids separate and move toward opposite poles of the cell.

In summary, meiosis I begins with a diploid parent cell and ends with two daughter cells, each with the haploid number of chromosomes. During meiosis II, each of the two haploid cells formed during meiosis I divides, and the net result is four haploid gametes that are genetically different from the original diploid parent cell.

Figure 28.2 compares the events of meiosis and mitosis.

▶ **CHECKPOINT**

1. How are haploid *(n)* and diploid *(2n)* cells different?

2. What are homologous chromosomes?

3. Prepare a table to compare meiosis with mitosis. (*Hint:* To review mitosis in more detail, refer to Figure 3.33 on page 92.)

MALE REPRODUCTIVE SYSTEM

▶ **O B J E C T I V E S**

*   **Describe the location, structure, and functions of the organs of the male reproductive system.**

*   **Discuss the process of spermatogenesis in the testes.**

The organs of the male reproductive system are the testes, a system of ducts (including the ductus deferens, ejaculatory ducts, and urethra), accessory sex glands (seminal vesicles, prostate, and bulbourethral gland), and several supporting structures, including the scrotum and the penis (Figure 28.3 on pages 1016–1017). The male gonads, the testes, produce sperm and secrete hormones. A system of ducts transports and stores sperm, assists in their maturation, and conveys them to the exterior. Semen contains sperm plus the secretions provided by the accessory sex glands.

Scrotum

The **scrotum** (SKRŌ-tum = bag), the supporting structure for the testes, is a sac consisting of loose skin and superficial fascia that hangs from the root (attached portion) of the penis (Figure 28.3a on page 1018). Externally, the scrotum looks like a single pouch of skin separated into lateral portions by a median ridge called the **raphe** (RĀ-fē = seam). Internally, the scrotal septum divides the scrotum into two sacs, each containing a single testis (Figure 28.4 on page 1018). The septum is made up of superficial fascia and muscle tissue called the **dartos muscle** (DAR-tōs = skinned), which consists of bundles of smooth muscle fibers. The dartos muscle is also found in the subcutaneous tissue of the scrotum and is directly continuous with the subcutaneous tissue of the abdominal wall. When the dartos muscle contracts, it wrinkles the skin of the scrotum and elevates the testes.

The location of the scrotum and the contraction of its muscle fibers regulate the temperature of the testes. Normal sperm production requires a temperature about 2–3°C below core body temperature. This lowered temperature is maintained within the scrotum because it is outside the pelvic cavity. The **cremaster muscle** (krē-MAS-ter = suspender), a small band of skeletal muscle in the spermatic cord that is a continuation of the internal oblique muscle, elevates the testes upon exposure to cold (and during sexual arousal). This action moves the testes closer to the pelvic cavity, where they can absorb body heat. Exposure to warmth reverses the process. The dartos muscle also contracts in response to cold and relaxes in response to warmth.

Testes

The **testes** (TES-tēz), or **testicles,** are paired oval glands measuring about 5 cm (2 in.) long and 2.5 cm (1 in.) in diameter (Figure 28.5 on page 1019). Each **testis** (singular) has a mass of 10–15 grams. The testes develop near the kidneys, in the posterior part of the abdomen, and they usually begin their descent into the scrotum through the inguinal canals (passageways in the anterior abdominal wall; see Figure 28.4) during the latter half

Figure 28.2 Comparison between mitosis (left) and meiosis (right) in which the parent cell has two pairs of homologous chromosomes.

The phases of meiosis II and mitosis are similar.

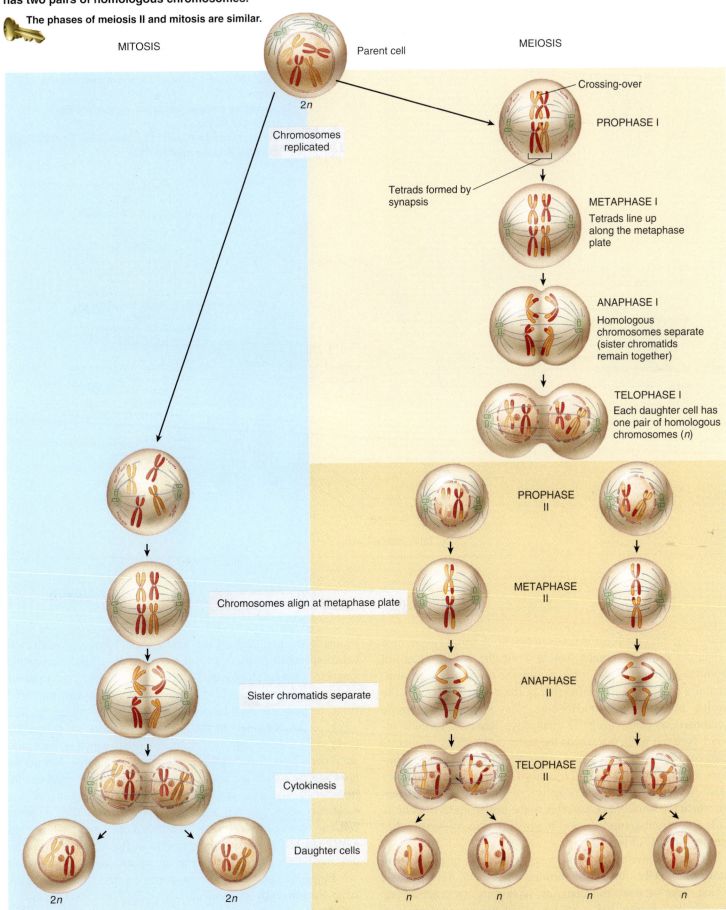

MITOSIS

MEIOSIS

Parent cell

2n

Chromosomes replicated

Crossing-over

PROPHASE I

Tetrads formed by synapsis

METAPHASE I

Tetrads line up along the metaphase plate

ANAPHASE I

Homologous chromosomes separate (sister chromatids remain together)

TELOPHASE I

Each daughter cell has one pair of homologous chromosomes (n)

PROPHASE II

Chromosomes align at metaphase plate

METAPHASE II

Sister chromatids separate

ANAPHASE II

Cytokinesis

TELOPHASE II

Daughter cells

2n   2n

n   n   n   n

How does anaphase I of meiosis differ from anaphase of mitosis and anaphase II of meiosis?

Figure 28.3 **Male organs of reproduction and surrounding structures.**
(See Tortora, *A Photographic Atlas of the Human Body*, Figure 14.2.)

Reproductive organs are adapted for producing new individuals and passing on genetic material from one generation to the next.

Functions of the Male Reproductive System
1. **Testes:** produce sperm and the male sex hormone testosterone.
2. **Ducts:** transport, store, and assist in maturation of sperm.
3. **Accessory sex glands:** secrete most of the liquid portion of semen.
4. **Penis:** contains the urethra, a passageway for ejaculation of semen and excretion of urine.

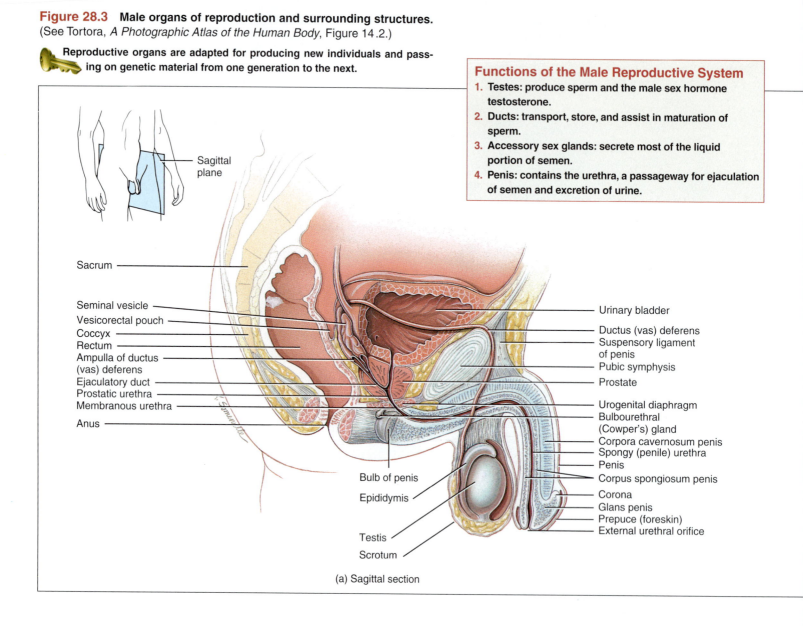

(a) Sagittal section

of the seventh month of fetal development. A serous membrane called the **tunica vaginalis** (*tunica* = sheath), which is derived from the peritoneum and forms during the descent of the testes, partially covers the testes. Internal to the tunica vaginalis is the **tunica albuginea** (al'-bū-JIN-ē-a; *albu-* = white), a capsule composed of dense irregular connective tissue. It extends inward, forming septa that divide the testis into a series of internal compartments called **lobules.** Each of the 200–300 lobules contains one to three tightly coiled **seminiferous tubules** (*semin-* = seed; *fer-* = to carry), where sperm are produced. The process by which the seminiferous tubules of the testes produce sperm is called **spermatogenesis** (sper'-ma-tō-JEN-e-sis; *genesis* = beginning process or production).

The seminiferous tubules contain two types of cells: **spermatogenic cells,** the sperm-forming cells, and **Sertoli cells,** which have several functions in supporting spermatogenesis (Figure 28.6 on page 1020). Starting at puberty, sperm production

begins at the periphery of the seminiferous tubules in stem cells called **spermatogonia** (sper'-ma-tō-GŌ-nē-a; *-gonia* = offspring; singular is **spermatogonium**). These cells develop from **primordial germ cells** (*primordi-* = primitive or early form) that arise from the yolk sac endoderm and enter the testes during the fifth week of development. In the embryonic testes, the primordial germ cells differentiate into spermatogonia, which remain dormant during childhood and become active at puberty. Toward the lumen of the tubule are layers of progressively more mature cells. In order of advancing maturity, these are primary spermatocytes, secondary spermatocytes, spermatids, and sperm. When a **sperm cell,** or **spermatozoon** (sper'-ma-tō-ZŌ-on; *-zoon* = life), has nearly reached maturity, it is released into the lumen of the seminiferous tubule. (The plural terms are **sperm** and **spermatozoa.**)

Embedded among the spermatogenic cells in the tubules are large **Sertoli cells** or *sustentacular cells* (sus'-ten-TAK-ū-lar), which extend from the basement membrane to the lumen of the

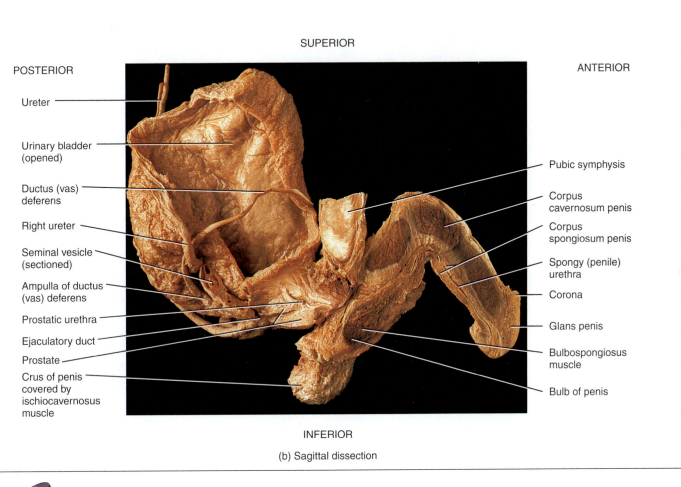

SUPERIOR

POSTERIOR

ANTERIOR

Ureter

Urinary bladder
(opened)

Ductus (vas)
deferens

Right ureter

Seminal vesicle
(sectioned)

Ampulla of ductus
(vas) deferens

Prostatic urethra

Ejaculatory duct

Prostate

Crus of penis
covered by
ischiocavernosus
muscle

Pubic symphysis

Corpus
cavernosum penis

Corpus
spongiosum penis

Spongy (penile)
urethra

Corona

Glans penis

Bulbospongiosus
muscle

Bulb of penis

INFERIOR

(b) Sagittal dissection

What are the groups of reproductive organs in males, and what are the functions of each group?

tubule. Internal to the basement membrane and spermatogonia, tight junctions join neighboring Sertoli cells to one another. These junctions form an obstruction known as the **blood–testis barrier** because substances must first pass through the Sertoli cells before they can reach the developing sperm. By isolating the developing gametes from the blood, the blood–testis barrier prevents an immune response against the spermatogenic cell's surface antigens, which are recognized as "foreign" by the immune system. The blood–testis barrier does not include spermatogonia.

Sertoli cells support and protect developing spermatogenic cells in several ways. They nourish spermatocytes, spermatids, and sperm; phagocytize excess spermatid cytoplasm as development proceeds; and control movements of spermatogenic cells and the release of sperm into the lumen of the seminiferous tubule. They also produce fluid for sperm transport, secrete androgen-binding protein and the hormone inhibin, and mediate

the effects of testosterone and FSH (follicle-stimulating hormone).

In the spaces between adjacent seminiferous tubules are clusters of cells called **Leydig cells** or *interstitial endocrinocytes* (Figure 28.6a). These cells secrete testosterone, the most prevalent androgen. Although androgens are hormones that promote development of masculine characteristics, they also have other functions, such as promoting libido (sexual desire) in both males and females.

Cryptorchidism

The condition in which the testes do not descend into the scrotum is called **cryptorchidism** (krip-TOR-ki-dizm; *crypt-* = hidden; *orchid* = testis). It occurs in about 3% of full-term infants and about 30% of premature infants. Untreated bilateral cryptorchidism results in sterility because the cells involved in the

Figure 28.4 **The scrotum, the supporting structure for the testes.**

The scrotum consists of loose skin and superficial fascia and supports the testes.

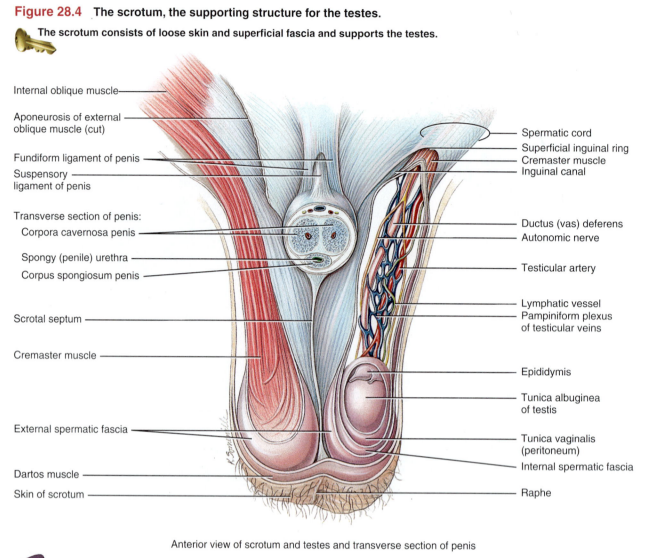

Internal oblique muscle

Aponeurosis of external oblique muscle (cut)

Fundiform ligament of penis

Suspensory ligament of penis

Transverse section of penis:

Corpora cavernosa penis

Spongy (penile) urethra

Corpus spongiosum penis

Scrotal septum

Cremaster muscle

External spermatic fascia

Dartos muscle

Skin of scrotum

Spermatic cord

Superficial inguinal ring

Cremaster muscle

Inguinal canal

Ductus (vas) deferens

Autonomic nerve

Testicular artery

Lymphatic vessel

Pampiniform plexus of testicular veins

Epididymis

Tunica albuginea of testis

Tunica vaginalis (peritoneum)

Internal spermatic fascia

Raphe

Anterior view of scrotum and testes and transverse section of penis

Which muscles help regulate the temperature of the testes?

initial stages of spermatogenesis are destroyed by the higher temperature of the pelvic cavity. The chance of testicular cancer is 30 to 50 times greater in cryptorchid testes, possibly due to abnormal division of germ cells due to the higher temperature of the pelvic cavity. The testes of about 80% of boys with cryptorchidism will descend spontaneously during the first year of life. When the testes remain undescended, the condition can be corrected surgically, ideally before 18 months of age. ■

Spermatogenesis

In humans, spermatogenesis takes 65 to 75 days. It begins in the spermatogonia, which contain the diploid (2n) chromosome number (Figure 28.7 on page 1021). Spermatogonia are *stem cells* because after they undergo mitosis, one daughter cell stays near the basement membrane of the seminiferous tubule in an undifferentiated state. As a result, stem cells remain for future cell divisions. The other daughter cell loses contact with the

basement membrane, squeezes through the tight junctions of the blood–testis barrier, undergoes developmental changes, and differentiates into a **primary spermatocyte** (SPER-ma-tō-sīt′). Primary spermatocytes, like spermatogonia, are diploid (2n); that is, they have 46 chromosomes.

Each primary spermatocyte enlarges and then begins meiosis (Figure 28.7). In meiosis I, DNA replicates, homologous pairs of chromosomes line up at the metaphase plate, and crossing-over occurs. Then, the meiotic spindle forms and pulls one (duplicated) chromosome of each pair to an opposite pole of the dividing cell. The two cells formed by meiosis I are called **secondary spermatocytes.** Each spermatocyte has 23 chromosomes, the haploid number. Each chromosome within a secondary spermatocyte, however, is made up of two chromatids (two copies of the DNA) still attached by a centromere.

In meiosis II, no further replication of DNA occurs. The chromosomes line up in single file along the metaphase plate, and the two chromatids of each chromosome separate. The four

Figure 28.5   Internal and external anatomy of a testis.

The testes are the male gonads, which produce haploid sperm.

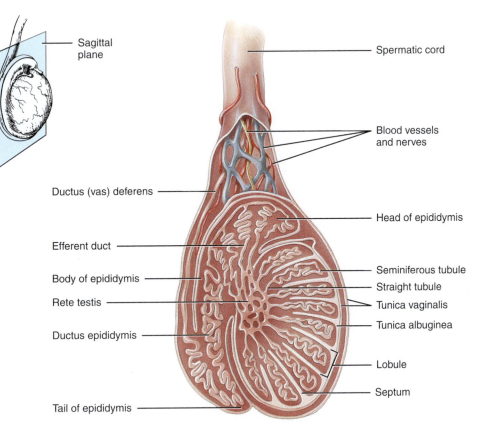

Sagittal plane

Spermatic cord

Blood vessels and nerves

Ductus (vas) deferens

Head of epididymis

Efferent duct

Body of epididymis

Seminiferous tubule

Straight tubule

Rete testis

Tunica vaginalis

Ductus epididymis

Tunica albuginea

Lobule

Septum

Tail of epididymis

(a) Sagittal section of a testis showing seminiferous tubules

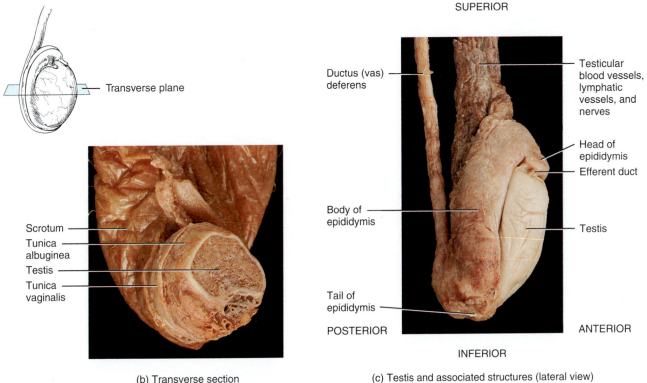

SUPERIOR

Transverse plane

Ductus (vas) deferens

Testicular blood vessels, lymphatic vessels, and nerves

Head of epididymis

Efferent duct

Body of epididymis

Testis

Scrotum

Tunica albuginea

Testis

Tunica vaginalis

Tail of epididymis

POSTERIOR

ANTERIOR

INFERIOR

(b) Transverse section

(c) Testis and associated structures (lateral view)

What tissue layers cover and protect the testes?

Figure 28.6 Microscopic anatomy of the seminiferous tubules and stages of sperm production (spermatogenesis).   Arrows in (b) indicate the progression of spermatogenic cells from least mature to most mature. The (*n*) and (*2n*) refer to haploid and diploid chromosome number, respectively.

Spermatogenesis occurs in the seminiferous tubules of the testes.

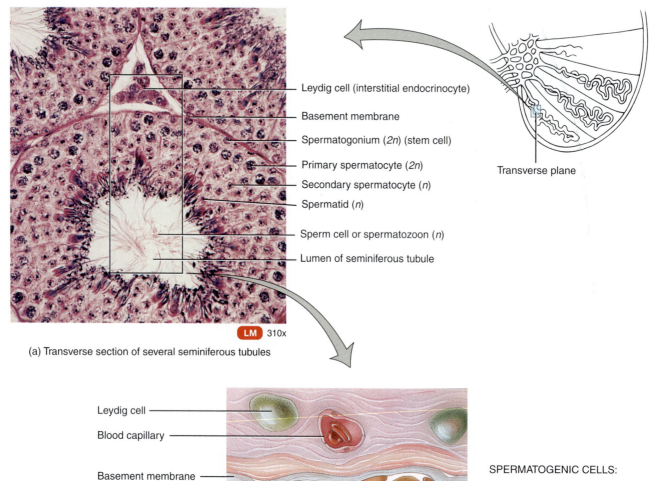

Leydig cell (interstitial endocrinocyte)

Basement membrane

Spermatogonium (*2n*) (stem cell)

Primary spermatocyte (*2n*)

Secondary spermatocyte (*n*)

Spermatid (*n*)

Sperm cell or spermatozoon (*n*)

Lumen of seminiferous tubule

Transverse plane

LM 310x

(a) Transverse section of several seminiferous tubules

Leydig cell

Blood capillary

Basement membrane

Sertoli cell nucleus

Blood–testis barrier (tight junction)

Cytoplasmic bridge

SPERMATOGENIC CELLS:

Spermatogonium (*2n*) (stem cell)

Primary spermatocyte (*2n*)

Secondary spermatocyte (*n*)

Early spermatid (*n*)

Late spermatid (*n*)

Sperm cell or spermatozoon (*n*)

(b) Transverse section of a portion of a seminiferous tubule

 Which cells produce testosterone?

Figure 28.7 Events in spermatogenesis. Diploid cells (*2n*) have 46 chromosomes; haploid cells (*n*) have 23 chromosomes.

🔑 **Spermiogenesis involves the maturation of spermatids into sperm.**

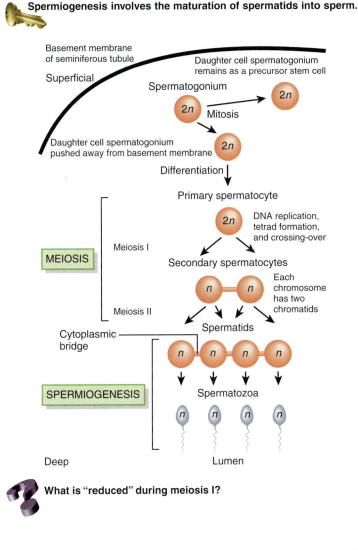

❓ **What is "reduced" during meiosis I?**

haploid cells resulting from meiosis II are called **spermatids.** A single primary spermatocyte therefore produces four spermatids via two rounds of cell division (meiosis I and meiosis II).

A unique process occurs during spermatogenesis. As the sperm cells proliferate, they fail to complete cytoplasmic separation (cytokinesis). The four daughter cells remain in contact via cytoplasmic bridges through their entire development (see Figures 28.6b and 28.7). This pattern of development most likely accounts for the synchronized production of sperm in any given area of seminiferous tubule. It may also have survival value in that half of the sperm contain an X chromosome and half contain a Y chromosome. The larger X chromosome may carry genes needed for spermatogenesis that are lacking on the smaller Y chromosome.

The final stage of spermatogenesis, **spermiogenesis** (sper′-mē-ō-JEN-e-sis), is the maturation of haploid spermatids into sperm. Because no cell division occurs in spermiogenesis, each spermatid develops into a single **sperm cell.** During this process, spherical spermatids transform into elongated, slender sperm. An acrosome (described shortly) forms atop the nucleus, which con-

denses and elongates, a flagellum develops, and mitochondria multiply. Sertoli cells dispose of the excess cytoplasm that sloughs off. Finally, sperm are released from their connections to Sertoli cells, an event known as **spermiation.** Sperm then enter the lumen of the seminiferous tubule. Fluid secreted by Sertoli cells pushes sperm along their way, toward the ducts of the testes.

Sperm

Spermatogenesis produces about 300 million sperm per day. Once ejaculated, most do not survive more than 48 hours within the female reproductive tract. A sperm cell consists of three structures highly adapted for reaching and penetrating a secondary oocyte: a head, a midpiece, and a tail (Figure 28.8). The **head** is 4–5 μm long. It contains an **acrosome** (*acro-* = atop; *-some* = body), a lysosomelike vesicle, and a nucleus that has the haploid number of chromosomes (23). Enzymes within the acrosome include hyaluronidase and proteases, which aid penetration of the sperm cell into a secondary oocyte. In the **midpiece** are many mitochondria, which provide ATP for locomotion. The **tail,** a typical flagellum, propels the sperm cell along its way. From head to tip of tail, human sperm are about 70 μm in length.

Hormonal Control of the Testes

Although the initiating factors are not known, at puberty certain hypothalamic neurosecretory cells increase their secretion of **gonadotropin-releasing hormone (GnRH).** This hormone, in

Figure 28.8 Parts of a sperm cell.

🔑 **About 300 million sperm mature each day.**

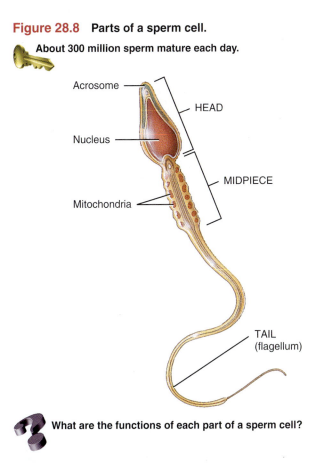

❓ **What are the functions of each part of a sperm cell?**

turn, stimulates gonadotrophs in the anterior pituitary to increase their secretion of the two gonadotropins, **luteinizing hormone (LH)** and **follicle-stimulating hormone (FSH).** Figure 28.9 shows the hormones and negative feedback loops that control secretion of testosterone and spermatogenesis.

LH stimulates Leydig cells, which are located between seminiferous tubules, to secrete the hormone **testosterone** (tes-TOS-te-rōn). This steroid hormone is synthesized from cholesterol in the testes and is the principal androgen. It is lipid soluble and readily diffuses out of Leydig cells into the interstitial fluid and then into blood. Testosterone acts in a negative feedback manner to suppress secretion of LH by anterior pituitary gonadotrophs and to suppress secretion of GnRH by hypothalamic neurosecretory cells. In some target cells, such as those in the prostate and seminal vesicles, the enzyme 5 alpha-reductase converts testosterone to another androgen called **dihydrotestosterone (DHT).**

FSH acts indirectly to stimulate spermatogenesis (Figure 28.9). FSH and testosterone act synergistically on the Sertoli cells to stimulate secretion of **androgen-binding protein (ABP)** into the lumen of the seminiferous tubules and into the interstitial fluid around the spermatogenic cells. ABP binds to testosterone, thereby keeping the concentration of testosterone high near the seminiferous tubules. Testosterone stimulates the final steps of spermatogenesis. Once the degree of spermatogenesis required for male reproductive functions has been achieved, Sertoli cells release **inhibin,** a protein hormone named for its inhibition of FSH secretion by the anterior pituitary (Figure 28.9). Inhibin thus inhibits the secretion of hormones needed for spermatogenesis. If spermatogenesis is proceeding too slowly, less inhibin is released, which permits more FSH secretion and an increased rate of spermatogenesis.

Testosterone and dihydrotestosterone both bind to the same androgen receptors, which are found within the nuclei of target cells. The hormone–receptor complex acts to regulate gene expression, turning some genes on and others off. Because of these changes, the androgens produce several effects:

- *Prenatal development.* Before birth, testosterone stimulates the male pattern of development of reproductive system ducts and the descent of the testes. Dihydrotestosterone, by contrast, stimulates development of the external genitals (described on page 1050). Testosterone also is converted in the brain to estrogens (feminizing hormones), which may play a role in the development of certain regions of the brain in males.

- *Development of male sexual characteristics.* At puberty, testosterone and dihydrotestosterone bring about development and enlargement of the male sex organs and the development of masculine secondary sexual characteristics. These include muscular and skeletal growth that results in wide shoulders and narrow hips; pubic, axillary, facial, and chest hair (within hereditary limits); thickening of the skin; increased sebaceous (oil) gland secretion; and enlargement of the larynx and consequent deepening of the voice.

Figure 28.9 Hormonal control of spermatogenesis and actions of testosterone and dihydrotestosterone (DHT). In response to stimulation by FSH and testosterone, Sertoli cells secrete androgen-binding protein (ABP). Dashed red lines indicate negative feedback inhibition.

🔑 Release of FSH is stimulated by GnRH and inhibited by inhibin; release of LH is stimulated by GnRH and inhibited by testosterone.

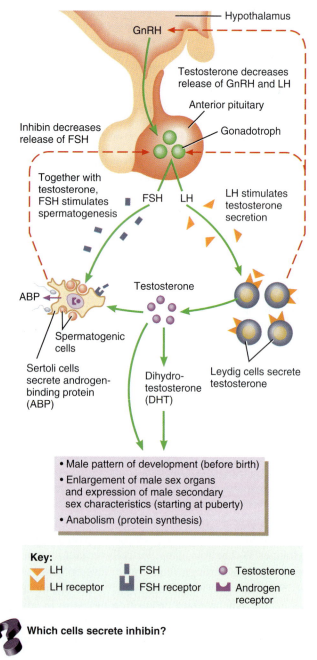

Which cells secrete inhibin?

- *Development of sexual function.* Androgens contribute to male sexual behavior and spermatogenesis and to sex drive (libido) in both males and females. Recall that the adrenal cortex is the main source of androgens in females.

- *Stimulation of anabolism.* Androgens are anabolic hormones; that is, they stimulate protein synthesis. This effect is obvious in the heavier muscle and bone mass of most men as compared to women.

A negative feedback system regulates testosterone production (Figure 28.10). When testosterone concentration in the blood increases to a certain level, it inhibits the release of GnRH by cells in the hypothalamus. As a result, there is less GnRH in the portal blood that flows from the hypothalamus to the anterior pituitary. Gonadotrophs in the anterior pituitary then release less LH, so the concentration of LH in systemic blood falls. With less stimulation by LH, the Leydig cells in the testes secrete less testosterone, and there is a return to homeostasis. If the testosterone concentration in the blood falls too low, however, GnRH is again released by the hypothalamus and stimulates secretion of LH by the anterior pituitary. LH, in turn, stimulates testosterone production by the testes.

▶ **CHECKPOINT**

4. How does the scrotum protect the testes?
5. Where are sperm cells produced?
6. What are the functions of Sertoli cells and Leydig cells?
7. What are the principal events of spermatogenesis?
8. Identify the main parts of a sperm cell, and list the functions of each.
9. What are the roles of FSH and LH in the male reproductive system? How is secretion of these hormones controlled?
10. What are the effects of testosterone and inhibin?
11. How is the blood level of testosterone controlled?

Reproductive System Ducts in Males

Ducts of the Testis

Pressure generated by the fluid secreted by Sertoli cells pushes sperm and fluid along the lumen of seminiferous tubules and then into a series of very short ducts called **straight tubules.** The straight tubules lead to a network of ducts in the testis called the **rete testis** (RĒ-tē = network) (see Figure 28.5a). From the rete testis, sperm move into a series of coiled **efferent ducts** in the epididymis that empty into a single tube called the **ductus epididymis.**

Epididymis

The **epididymis** (ep′-i-DID-i-mis; *epi-* = above or over; *-didymis* = testis) is a comma-shaped organ about 4 cm (1.5 in.) long that lies along the posterior border of each testis (see Figure 28.5a). The plural is **epididymides** (ep′-i-did-IM-i-dēs). Each epididymis consists mostly of the tightly coiled **ductus epididymis.** The larger, superior portion of the epididymis, the **head,** is where the efferent ducts from the testis join the ductus epididymis. The **body** is the narrow midportion of the epididymis, and the **tail** is the smaller, inferior portion. At its distal end, the tail of the

Figure 28.10 Negative feedback control of blood level of testosterone.

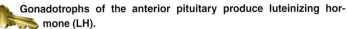 Gonadotrophs of the anterior pituitary produce luteinizing hormone (LH).

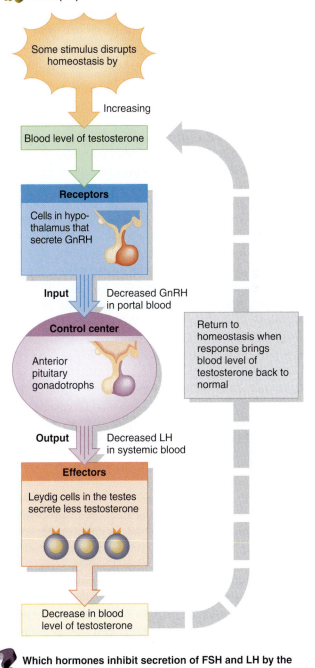

Which hormones inhibit secretion of FSH and LH by the anterior pituitary?

epididymis continues as the ductus (vas) deferens (discussed shortly).

The ductus epididymis would measure about 6 m (20 ft) in length if it were straightened out. It is lined with pseudostratified columnar epithelium and encircled by layers of smooth muscle. The free surfaces of the columnar cells contain **stereocilia,** which despite their name are long, branching microvilli that increase surface area for the reabsorption of degenerated sperm.

Functionally, the ductus epididymis is the site where sperm mature, that is, they acquire motility and the ability to fertilize an ovum. This occurs over a 10- to 14-day period. The ductus epididymis also stores sperm and helps propel them by peristaltic contraction of its smooth muscle into the ductus (vas) deferens. Sperm may remain in storage in the ductus epididymis for a month or more.

Ductus Deferens

Within the tail of the epididymis, the ductus epididymis becomes less convoluted, and its diameter increases. Beyond this point, the duct is known as the **ductus deferens** or **vas deferens** (see Figure 28.5a). The ductus deferens, which is about 45 cm (18 in.) long, ascends along the posterior border of the epididymis, passes through the inguinal canal (see Figure 28.4), and enters the pelvic cavity. There it loops over the ureter and passes over the side and down the posterior surface of the urinary bladder (see Figure 28.3a). The dilated terminal portion of the ductus deferens is the **ampulla** (am-PUL-la = little jar; see Figure 28.11). The ductus deferens is lined with pseudostratified columnar epithelium and contains a heavy coat of three layers of smooth muscle. The inner and outer layers are arranged longitudinally, whereas the middle layer encircles the ductus deferens.

Functionally, the ductus deferens stores sperm, which can remain viable here for up to several months. The ductus deferens also conveys sperm from the epididymis toward the urethra by peristaltic contractions of the muscular coat. Sperm that are not ejaculated are eventually reabsorbed.

Spermatic Cord

The **spermatic cord** is a supporting structure of the male reproductive system that ascends out of the scrotum (see Figure 28.4). It consists of the ductus (vas) deferens as it ascends through the scrotum, the testicular artery, autonomic nerves, veins that drain the testes and carry testosterone into circulation (the pampiniform plexus), lymphatic vessels, and the cremaster muscle. The spermatic cord and ilioinguinal nerve pass through the **inguinal canal** (IN-gwinal = groin), an oblique passageway in the anterior abdominal wall just superior and parallel to the medial half of the inguinal ligament. The canal, which is about 4–5 cm (about 2 in.) long, originates at the **deep (abdominal) inguinal ring,** a slitlike opening in the aponeurosis of the transversus abdominis muscle. The canal ends at the **superficial (subcutaneous) inguinal ring** (see Figure 28.4), a triangular opening in the aponeurosis of the external oblique muscle.

Inguinal Hernias

Because the inguinal region is a weak area in the abdominal wall, it is often the site of an **inguinal hernia**—a rupture or separation of a portion of the inguinal area of the abdominal wall

resulting in the protrusion of a part of the small intestine. In an *indirect inguinal hernia,* a part of the small intestine protrudes through the deep inguinal ring and enters the scrotum. In a *direct inguinal hernia,* a portion of the small intestine pushes into the posterior wall of the inguinal canal, usually causing a localized bulging in the wall of the canal. Inguinal hernias are much more common in males than in females because the larger inguinal canals in males represent larger weak points in the abdominal wall. ■

Ejaculatory Ducts

Each **ejaculatory duct** (e-JAK-ū-la-tō′-rē; *ejacul-* = to expel) is about 2 cm (1 in.) long and is formed by the union of the duct from the seminal vesicle and the ampulla of the ductus deferens (Figure 28.11). The ejaculatory ducts arise just superior to the base (superior portion) of the prostate and pass inferiorly and anteriorly through the prostate. They terminate in the prostatic urethra, where they eject sperm and seminal vesicle secretions just before **ejaculation,** the powerful propulsion of semen from the urethra to the exterior.

Urethra

In males, the **urethra** is the shared terminal duct of the reproductive and urinary systems; it serves as a passageway for both semen and urine. The urethra, which is about 20 cm (8 in.) long, has three parts that pass through the prostate, the urogenital diaphragm, and the penis (see Figures 28.3a and 28.11). The **prostatic urethra** is 2–3 cm (1 in.) long and passes through the prostate. As this part of the urethra continues inferiorly, it passes through the urogenital diaphragm (a muscular partition between the two ischial and pubic rami; see Figure 11.13 on page 341), where it is known as the **membranous urethra.** The membranous urethra is about 1 cm (0.5 in.) in length. As this duct passes into the corpus spongiosum of the penis, it becomes the **spongy (penile) urethra,** which is about 15–20 cm (6–8 in.) long, and ends at the **external urethral orifice.** The histology of the male urethra may be reviewed on page 981.

Accessory Sex Glands

Whereas the ducts of the male reproductive system store and transport sperm cells, the **accessory sex glands** secrete most of the liquid portion of semen. The accessory sex glands include the seminal vesicles, the prostate, and the bulbourethral glands.

Seminal Vesicles

The paired **seminal vesicles** (VES-i-kuls) or **seminal glands** are convoluted pouchlike structures, about 5 cm (2 in.) in length, lying posterior to the base of the urinary bladder and anterior to the rectum (Figure 28.11). They secrete an alkaline, viscous fluid that contains fructose (a monosaccharide sugar), prostaglandins, and clotting proteins that are different from those in blood. The alkaline nature of the seminal fluid helps to neutralize the acidic environment of the male urethra and female repro-

Figure 28.11 **Locations of several accessory reproductive organs in males.** The prostate, urethra, and penis have been sectioned to show internal details.

🔑 **The male urethra has three subdivisions: the prostatic, membranous, and spongy (penile) urethra.**

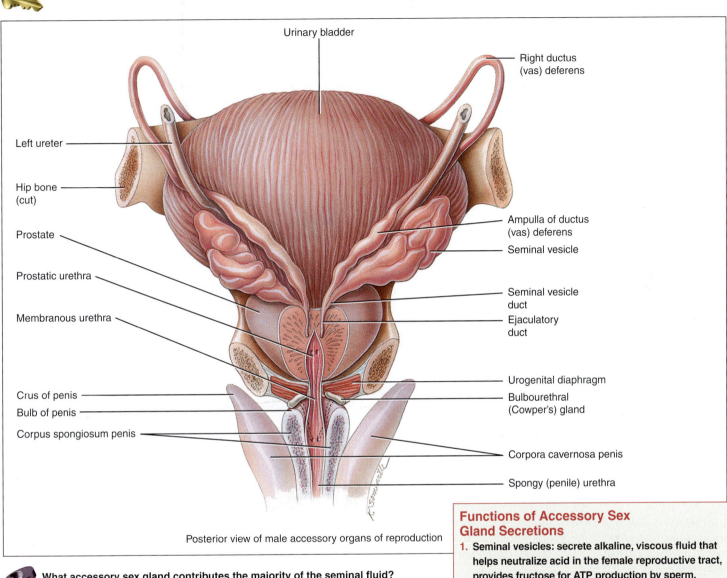

Posterior view of male accessory organs of reproduction

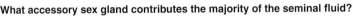 **What accessory sex gland contributes the majority of the seminal fluid?**

Functions of Accessory Sex Gland Secretions

1. **Seminal vesicles:** secrete alkaline, viscous fluid that helps neutralize acid in the female reproductive tract, provides fructose for ATP production by sperm, contributes to sperm motility and viability, and helps semen coagulate after ejaculation.
2. **Prostate:** secretes a milky, slightly acidic fluid that helps semen coagulate after ejaculation and subsequently breaks down the clot.
3. **Bulbourethral (Cowper's) glands:** secrete alkaline fluid that neutralizes the acidic environment of the urethra and mucus that lubricates the lining of the urethra and the tip of the penis during sexual intercourse.

---

ductive tract that otherwise would inactivate and kill sperm. The fructose is used for ATP production by sperm. Prostaglandins contribute to sperm motility and viability and may stimulate smooth muscle contractions within the female reproductive tract. The clotting proteins help semen coagulate after ejaculation. Fluid secreted by the seminal vesicles normally constitutes about 60% of the volume of semen.

Prostate

The **prostate** (PROS-tāt) is a single, doughnut-shaped gland about the size of a golf ball. It measures about 4 cm (1.6 in.) from side to side, about 3 cm (1.2 in.) from top to bottom, and about 2 cm (0.8 in.) from front to back. It is inferior to the urinary bladder and surrounds the prostatic urethra (Figure 28.11). The prostate slowly increases in size from birth to puberty, and then it expands rapidly. The size attained by age 30 typically remains stable until about age 45, when further enlargement may occur.

The prostate secretes a milky, slightly acidic fluid (pH about 6.5) that contains several important substances: (1) *Citric acid* in prostatic fluid is used by sperm for ATP production via the Krebs cycle. (2) Several *proteolytic enzymes,* such as *prostate-specific antigen (PSA),* pepsinogen, lysozyme, amylase, and hyaluronidase, eventually break down the clotting proteins from the seminal vesicles. (3) Acid phosphatase is secreted by the prostate, but its function is unknown. Secretions of the prostate enter the prostatic urethra through many prostatic ducts. Prostatic secretions make up about 25% of the volume of semen and contribute to sperm motility and viability.

Bulbourethral Glands

The paired **bulbourethral glands** (bul′-bō-ū-RĒ-thral), or **Cowper's glands,** are about the size of peas. They are located inferior to the prostate on either side of the membranous urethra within the urogenital diaphragm, and their ducts open into the spongy urethra (Figure 28.11). During sexual arousal, the bulbourethral glands secrete an alkaline fluid into the urethra that protects the passing sperm by neutralizing acids from urine in the urethra. They also secrete mucus that lubricates the end of the penis and the lining of the urethra, thereby decreasing the number of sperm damaged during ejaculation.

Semen

Semen (= seed) is a mixture of sperm and **seminal fluid,** a liquid that consists of the secretions of the seminiferous tubules, seminal vesicles, prostate, and bulbourethral glands. The volume of semen in a typical ejaculation is 2.5–5 milliliter (mL), with 50–150 million sperm per mL. When the number falls below 20 million/mL, the male is likely to be infertile. A very large number is required for successful fertilization because only a tiny fraction ever reaches the secondary oocyte.

Despite the slight acidity of prostatic fluid, semen has a slightly alkaline pH of 7.2–7.7 due to the higher pH and larger volume of fluid from the seminal vesicles. The prostatic secretion gives semen a milky appearance, whereas fluids from the seminal vesicles and bulbourethral glands give it a sticky consistency. Seminal fluid provides sperm with a transportation medium and nutrients, and it neutralizes the hostile acidic environment of the male's urethra and the female's vagina. Semen also contains *seminalplasmin,* an antibiotic that can destroy certain bacteria. Seminalplasmin may help decrease the number of naturally occurring bacteria in the semen and in the lower female reproductive tract.

Once ejaculated, liquid semen coagulates within 5 minutes due to the presence of clotting proteins from the seminal vesicles. The functional role of semen coagulation is not known, but the proteins involved are different from those that cause blood coagulation. After about 10 to 20 minutes, semen reliquefies because prostate-specific antigen (PSA) and other proteolytic enzymes produced by the prostate break down the clot. Abnormal or delayed liquefaction of clotted semen may cause complete or partial immobilization of sperm, thereby inhibiting their movement through the cervix of the uterus.

Penis

The **penis** contains the urethra and is a passageway for the ejaculation of semen and the excretion of urine (Figure 28.12). It is cylindrical in shape and consists of a root, body, and glans penis. The **root of the penis** is the attached portion (proximal portion). It consists of the **bulb of the penis,** the expanded portion of the base of the corpus spongiosum penis (described shortly), and the **crura of the penis** (singular is **crus** = resembling a leg), the two separated and tapered portions of the corpora cavernosa penis (Figure 28.12a). The bulb of the penis is attached to the inferior surface of the urogenital diaphragm and is enclosed by the bulbospongiosus muscle. Each crus of the penis is attached to the ischial and inferior pubic rami and is surrounded by the ischiocavernosus muscle (see Figure 11.13 on page 341). Contraction of these skeletal muscles aids ejaculation.

The **body of the penis** is composed of three cylindrical masses of tissue, each surrounded by fibrous tissue called the **tunica albuginea.** The two dorsolateral masses are called the **corpora cavernosa penis** (*corpora* = main bodies; *cavernosa* = hollow). The smaller midventral mass, the **corpus spongiosum penis,** contains the spongy urethra and functions in keeping the spongy urethra open during ejaculation. Fascia and skin enclose all three masses, which consist of erectile tissue permeated by blood sinuses.

Upon sexual stimulation, which may be visual, tactile, auditory, olfactory, or imagined, the arteries supplying the penis dilate, and large quantities of blood enter the blood sinuses. Expansion of these spaces compresses the veins draining the penis, so blood outflow is slowed. These vascular changes, due to local release of nitric oxide and a parasympathetic reflex, result in an **erection,** the enlargement and stiffening of the penis. The penis returns to its flaccid state when the arteries constrict and pressure on the veins is relieved.

Ejaculation is a sympathetic reflex. As part of the reflex, the smooth muscle sphincter at the base of the urinary bladder closes. Therefore, urine is not expelled during ejaculation, and semen does not enter the urinary bladder. Even before ejaculation occurs, peristaltic contractions in the ampulla of the ductus deferens, seminal vesicles, ejaculatory ducts, and prostate propel semen into the penile portion of the urethra (spongy urethra). Typically, this leads to **emission** (ē-MISH-un), the discharge of a small volume of semen before ejaculation. Emission may also occur during sleep (nocturnal emission).

The distal end of the corpus spongiosum penis is a slightly enlarged, acorn-shaped region called the **glans penis;** its margin is the **corona.** The distal urethra enlarges within the glans penis and forms a terminal slitlike opening, the **external urethral orifice.** Covering the glans in an uncircumcised penis is the loosely fitting **prepuce** (PRĒ-poos), or **foreskin.** The weight of the penis is supported by two ligaments that are continuous with the fascia of the penis. (1) The **fundiform ligament** arises

Figure 28.12 Internal structure of the penis. The inset in (b) shows details of the skin and fasciae. (See Tortora, *A Photographic Atlas of the Human Body*, Figure 14.6.)

🔑 **The penis contains the urethra, a pathway for the ejaculation of semen and the excretion of urine.**

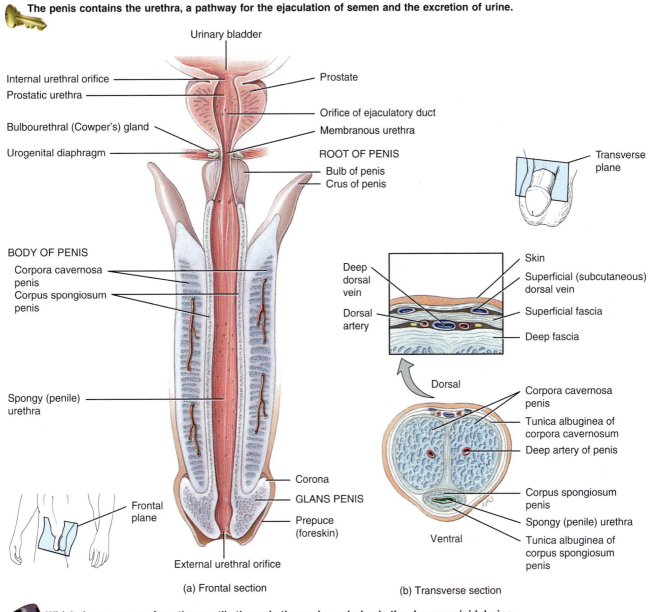

(a) Frontal section

(b) Transverse section

❓ **Which tissue masses form the erectile tissue in the penis, and why do they become rigid during sexual arousal?**

from the inferior part of the linea alba. (2) The **suspensory ligament of the penis** arises from the pubic symphysis.

🩺 Circumcision

Circumcision (= to cut around) is a surgical procedure in which part or the entire prepuce is removed. It is usually performed just after delivery, 3 to 4 days after birth, or on the eighth day as part of a Jewish religious rite. Although most health-care professionals find no medical justification for circumcision, some feel that it has benefits, such as a lower risk of urinary tract infections, protection against penile cancer, and possibly a lower risk for sexually trans-

mitted diseases. Indeed, studies in several African villages have found lower rates of HIV infection among circumcised men. ■

▶ **CHECKPOINT**

12. Which ducts transport sperm within the testes?

13. What structures comprise the spermatic cord?

14. Give the locations of the three subdivisions of the male urethra.

15. Trace the course of sperm through the system of ducts from the seminiferous tubules through the urethra.

16. What is semen? What is its function?

17. How does an erection occur?

FEMALE REPRODUCTIVE SYSTEM

▶ **O B J E C T I V E S**

- Describe the location, structure, and functions of the organs of the female reproductive system.
- Discuss the process of oogenesis in the ovaries.

The organs of the female reproductive system (Figure 28.13) include the ovaries; the uterine (Fallopian) tubes, or oviducts; the uterus; the vagina; and external organs, which are collectively called the vulva, or pudendum. The mammary glands are part of the integumentary system and also are considered part of the female reproductive system.

Ovaries

The female gonads are the two **ovaries,** one on either side of the uterus, that are the size and shape of unshelled almonds (Figure 28.14 on page 1030). The ovaries produce (1) gametes, secondary oocytes that develop into mature ova (eggs) after fertilization, and (2) hormones, including progesterone and estrogens (the female sex hormones), inhibin, and relaxin. During development, the ovaries arise from the same embryonic gonadal tissue as the testes. When two organs have the same embryonic origin, they are said to be *homologous.*

The ovaries descend to the brim of the superior portion of the pelvic cavity during the third month of development. A

Figure 28.13   Organs of reproduction and surrounding structures in females.

🔑 The organs of reproduction in females include the ovaries, uterine (Fallopian) tubes, uterus, vagina, vulva, and mammary glands.

> **Functions of the Female Reproductive System**
> 1. **Ovaries:** produce secondary oocytes and hormones, including progesterone and estrogens (female sex hormones), inhibin, and relaxin.
> 2. **Uterine tubes:** transport a secondary oocyte to the uterus and normally are the sites where fertilization occurs.
> 3. **Uterus:** site of implantation of a fertilized ovum, development of the fetus during pregnancy, and labor.
> 4. **Vagina:** receives the penis during sexual intercourse and is a passageway for childbirth.
> 5. **Mammary glands:** synthesize, secrete, and eject milk for nourishment of the newborn.

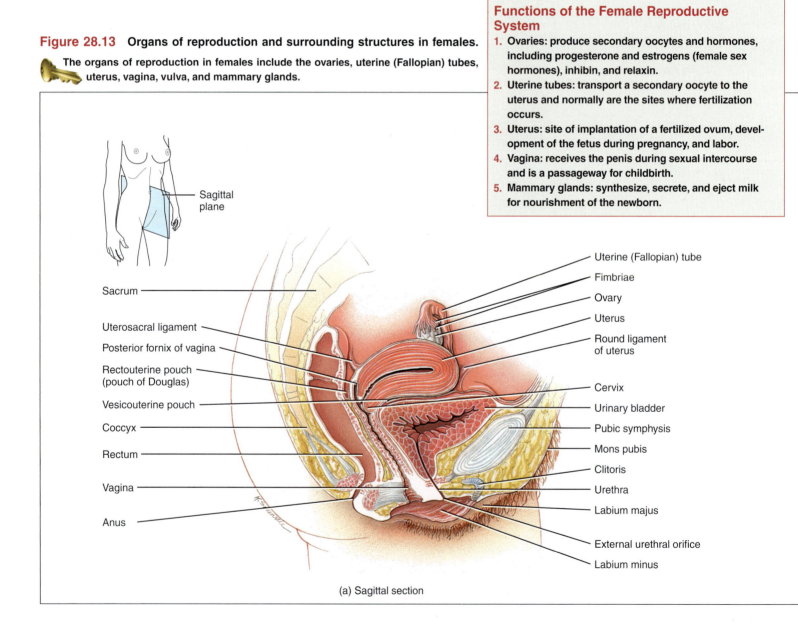

(a) Sagittal section

series of ligaments holds them in position (Figure 28.14). The **broad ligament** of the uterus (see also Figure 28.13b), which is itself part of the parietal peritoneum, attaches to the ovaries by a double-layered fold of peritoneum called the **mesovarium.** The **ovarian ligament** anchors the ovaries to the uterus, and the **suspensory ligament** attaches them to the pelvic wall. Each ovary contains a **hilus,** the point of entrance and exit for blood vessels and nerves and along which the mesovarium is attached.

Histology of the Ovary

Each ovary consists of the following parts (Figure 28.15 on page 1031):

- The **germinal epithelium** is a layer of simple epithelium (low cuboidal or squamous) that covers the surface of the ovary. It is continuous with the mesothelium that covers the mesovarium. The term *germinal epithelium* is a misnomer because it does not give rise to ova, although at one time

people believed that it did. Now we know that the progenitors of ova arise from the endoderm of the yolk sac and migrate to the ovaries during embryonic development.

- The **tunica albuginea** is a whitish capsule of dense irregular connective tissue located immediately deep to the germinal epithelium.

- The **ovarian cortex** is a region just deep to the tunica albuginea. It consists of ovarian follicles (described shortly) surrounded by dense irregular connective tissue that contains scattered smooth muscle cells.

- The **ovarian medulla** is deep to the ovarian cortex. The border between the cortex and medulla is indistinct, but the medulla consists of more loosely arranged connective tissue and contains blood vessels, lymphatic vessels, and nerves.

- **Ovarian follicles** (*folliculus* = little bag) are in the cortex and consist of **oocytes** in various stages of development, plus the cells surrounding them. When the surrounding cells form a single layer, they are called **follicular cells;** later in

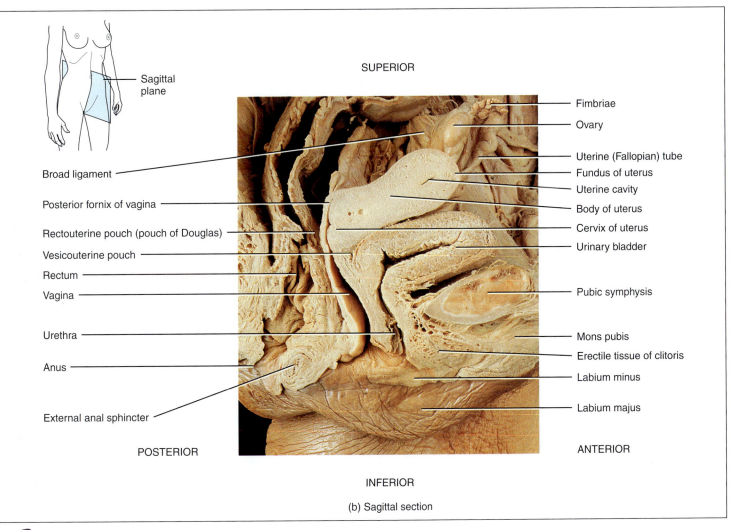

SUPERIOR

Sagittal plane

Fimbriae

Ovary

Uterine (Fallopian) tube

Broad ligament

Fundus of uterus

Uterine cavity

Posterior fornix of vagina

Body of uterus

Rectouterine pouch (pouch of Douglas)

Cervix of uterus

Vesicouterine pouch

Urinary bladder

Rectum

Vagina

Pubic symphysis

Urethra

Mons pubis

Erectile tissue of clitoris

Anus

Labium minus

Labium majus

External anal sphincter

POSTERIOR

ANTERIOR

INFERIOR

(b) Sagittal section

 Which structures in males are homologous to the ovaries, the clitoris, the paraurethral glands, and the greater vestibular glands?

Figure 28.14   Relative positions of the ovaries, the uterus, and the ligaments that support them.   (See Tortora, *A Photographic Atlas of the Human Body*, Figure 14.9.)

🔑 Ligaments holding the ovaries in position are the mesovarium, the ovarian ligament, and the suspensory ligament.

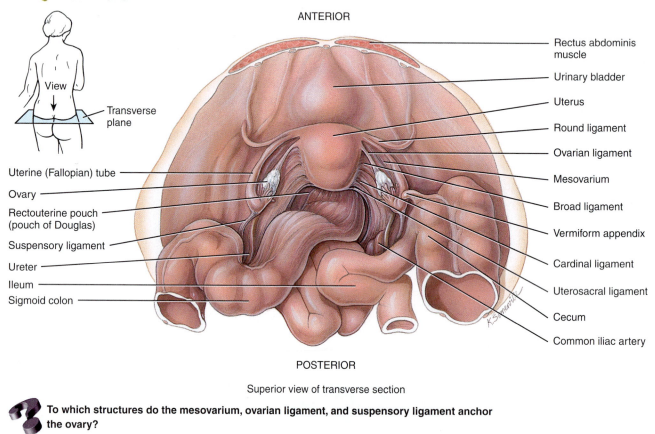

ANTERIOR

Rectus abdominis muscle

Urinary bladder

Uterus

Round ligament

Ovarian ligament

Mesovarium

Broad ligament

Vermiform appendix

Cardinal ligament

Uterosacral ligament

Cecum

Common iliac artery

Uterine (Fallopian) tube

Ovary

Rectouterine pouch (pouch of Douglas)

Suspensory ligament

Ureter

Ileum

Sigmoid colon

View

Transverse plane

POSTERIOR

Superior view of transverse section

❓ **To which structures do the mesovarium, ovarian ligament, and suspensory ligament anchor the ovary?**

development, when they form several layers, they are referred to as **granulosa cells.** The surrounding cells nourish the developing oocyte and begin to secrete estrogens as the follicle grows larger.

• A **mature (Graafian) follicle** is a large, fluid-filled follicle that is ready to rupture and expel its secondary oocyte, a process known as **ovulation.**

• A **corpus luteum** (= yellow body) contains the remnants of mature follicle after ovulation. The corpus luteum produces progesterone, estrogens, relaxin, and inhibin until it degenerates into fibrous scar tissue called the **corpus albicans** (= white body).

Oogenesis and Follicular Development

The formation of gametes in the ovaries is termed **oogenesis** (ō-ō-JEN-e-sis; *oo-* = egg). Whereas spermatogenesis begins in males at puberty, oogenesis begins in females before they are even born. Oogenesis occurs in essentially the same manner as spermatogenesis, with meiosis taking place and the resulting germ cells undergoing maturation.

During early fetal development, primordial (primitive) germ cells migrate from the endoderm of the yolk sac to the ovaries.

There, germ cells differentiate within the ovaries into **oogonia** (ō′-o-GŌ-nē-a; singular is **oogonium**). Oogonia are diploid (2n) stem cells that divide mitotically to produce millions of germ cells. Even before birth, most of these germ cells degenerate in a process known as **atresia** (a-TRĒ-zē-a). A few, however, develop into larger cells called **primary oocytes** (Ō-ō-sītz) that enter prophase of meiosis I during fetal development but do not complete that phase until after puberty. At birth, 200,000 to 2,000,000 oogonia and primary oocytes remain in each ovary. Of these, about 40,000 remain at puberty, and around 400 will mature and ovulate during a woman's reproductive lifetime. The rest undergo atresia.

At the same time oogenesis is occurring, the follicle cells surrounding the oocyte are also undergoing developmental changes. In the beginning, a single layer of follicular cells surrounds each primary oocyte, and the entire structure is called a **primordial follicle** (Figure 28.16a on page 1032). Although the stimulating mechanism is unclear, a few primordial follicles periodically start to grow, even during childhood. They become **primary follicles,** which are surrounded first by one layer of cuboidal follicular cells and then by six to seven layers of cuboidal and low-columnar cells called granulosa cells. As a fol-

Figure 28.15 Histology of the ovary. The arrows indicate the sequence of developmental stages that occur as part of the maturation of an ovum during the ovarian cycle.

🔑 **The ovaries are the female gonads; they produce haploid oocytes.**

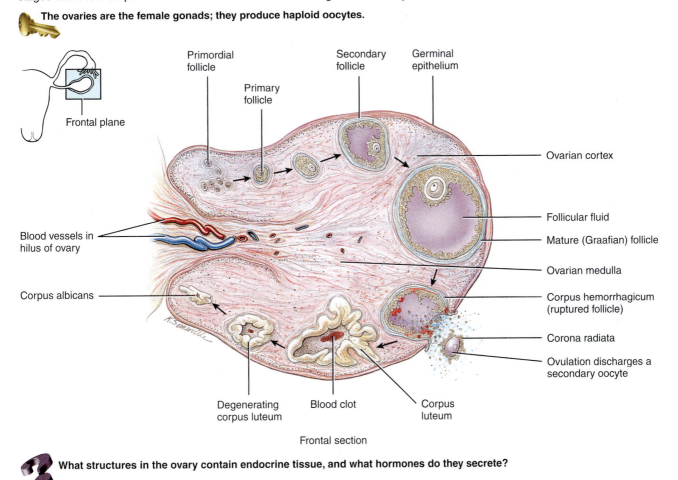

Frontal plane

Primordial follicle

Primary follicle

Secondary follicle

Germinal epithelium

Ovarian cortex

Follicular fluid

Mature (Graafian) follicle

Ovarian medulla

Blood vessels in hilus of ovary

Corpus albicans

Corpus hemorrhagicum (ruptured follicle)

Corona radiata

Ovulation discharges a secondary oocyte

Degenerating corpus luteum

Blood clot

Corpus luteum

Frontal section

❓ **What structures in the ovary contain endocrine tissue, and what hormones do they secrete?**

licle grows, it forms a clear glycoprotein layer, called the **zona pellucida** (pe-LOO-si-da), between the primary oocyte and the granulosa cells. The innermost layer of granulosa cells becomes firmly attached to the zona pellucida and is called the **corona radiata** (*corona* = crown; *radiata* = radiation) (Figure 28.16b).

The outermost granulosa cells rest on a basement membrane. Encircling the basement membrane is a region called the **theca folliculi.** Although the region of granulosa cells is avascular, many blood capillaries are located in the theca folliculi. As a primary follicle continues to grow, the theca differentiates into two layers: (1) the **theca interna,** a highly vascularized internal layer of cuboidal secretory cells, fibroblasts, and bundles of collagen fibers and (2) the **theca externa,** an outer layer of connective tissue cells, collagen fibers, and smooth muscle fibers. The granulosa cells begin to secrete follicular fluid, which builds up in a cavity called the **antrum** in the center of the follicle, which is now termed a **secondary follicle.**

Each month after puberty, gonadotropins secreted by the anterior pituitary stimulate the resumption of oogenesis (Figure 28.17). Meiosis I resumes in several secondary follicles, although only one will eventually reach the maturity needed for

ovulation. The diploid primary oocyte completes meiosis I, producing two haploid cells of unequal size—both with 23 chromosomes (n) of two chromatids each. The smaller cell produced by meiosis I, called the **first polar body,** is essentially a packet of discarded nuclear material. The larger cell, known as the **secondary oocyte,** receives most of the cytoplasm. Once a secondary oocyte is formed, it begins meiosis II but then stops in metaphase. The follicle in which these events are taking place—the **mature (Graafian) follicle**—soon ruptures and releases its secondary oocyte. It takes 90 days or longer for a primary follicle to develop into a secondary follicle and then a mature follicle that releases its secondary oocyte.

At ovulation, the secondary oocyte is expelled into the pelvic cavity together with the first polar body and corona radiata. Normally these cells are swept into the uterine tube. If fertilization does not occur, the cells degenerate. If sperm are present in the uterine tube and one penetrates the secondary oocyte, however, meiosis II resumes. The secondary oocyte splits into two haploid (n) cells, again of unequal size. The larger cell is the **ovum,** or mature egg; the smaller one is the **second polar body.** The nuclei of the sperm cell and the ovum then unite, forming a

An over-the-counter home test that detects a rising level of LH can be used to predict ovulation a day in advance. FSH also increases at this time, but not as dramatically as LH because FSH is stimulated only by the increase in gonadotropin-releasing hormone. The positive feedback effect of estrogens on the hypothalamus and anterior pituitary does not occur if progesterone is present at the same time.

Postovulatory Phase

The **postovulatory phase** of the female reproductive cycle is the time between ovulation and onset of the next menses. In duration, it is the most constant part of the female reproductive cycle. It lasts for 14 days in a 28-day cycle, from day 15 to day 28 (see Figure 28.26).

EVENTS IN ONE OVARY After ovulation, the mature follicle collapses, and the basement membrane between the granulosa cells and theca interna breaks down. Once a blood clot forms from minor bleeding of the ruptured follicle, the follicle becomes the **corpus hemorrhagicum** (*hemo-* = blood; *rrhagic-* = bursting forth) (see Figure 28.15). Theca interna cells mix with the granulosa cells as they all become transformed into corpus luteum cells under the influence of LH. Stimulated by LH, the corpus luteum secretes progesterone, estrogen, relaxin, and inhibin. The luteal cells also absorb the blood clot. With reference to the ovarian cycle, this phase is also called the **luteal phase.**

Later events in an ovary that has ovulated an oocyte depend on whether the oocyte is fertilized. If the oocyte *is not fertilized,* the corpus luteum has a lifespan of only 2 weeks. Then, its secretory activity declines, and it degenerates into a corpus albicans (see Figure 28.15). As the levels of progesterone, estrogens, and inhibin decrease, release of GnRH, FSH, and LH rise due to loss of negative feedback suppression by the ovarian hormones. Then, follicular growth resumes and a new ovarian cycle begins.

If the secondary oocyte *is fertilized* and begins to divide, the corpus luteum persists past its normal 2-week life span. It is "rescued" from degeneration by **human chorionic gonadotropin** (kō-rē-ON-ik) **(hCG).** This hormone is produced by the chorion of the embryo beginning about 8 days after fertilization. Like LH, hCG stimulates the secretory activity of the corpus luteum. The presence of hCG in maternal blood or urine is an indicator of pregnancy and is the hormone detected by home pregnancy tests.

EVENTS IN THE UTERUS Progesterone and estrogens produced by the corpus luteum promote growth and coiling of the endometrial glands, vascularization of the superficial endometrium, and thickening of the endometrium to 12–18 mm (0.48–0.72 in.). Because of the secretory activity of the endometrial glands, which begin to secrete glycogen, this period is termed the **secretory phase** of the uterine cycle. These preparatory changes peak about one week after ovulation, at the time a fertilized ovum might arrive in the uterus. If fertilization does not occur, the level of progesterone declines due to degeneration of the corpus luteum. Withdrawal of progesterone causes menstruation.

Figure 28.28 summarizes the hormonal interactions and cyclical changes in the ovaries and uterus during the ovarian and uterine cycles.

Female Athlete Triad: Disordered Eating, Amenorrhea, and Premature Osteoporosis

The female reproductive cycle can be disrupted by many factors, including weight loss, low body weight, disordered eating, and vigorous physical activity. The observation that three conditions—disordered eating, amenorrhea, and osteoporosis—often occur together in female athletes led researchers to coin the term **female athlete triad** to encompass all three.

Many athletes experience intense pressure from coaches, parents, peers, and themselves to lose weight to improve performance. Hence, they may develop disordered eating behaviors and engage in other harmful weight-loss practices in a struggle to maintain a very low body weight. **Amenorrhea** (ā-men'-ō-RĒ-a; *a-* = without; *men-* = month; *-rrhea* = a flow) is the absence of menstruation. The most common causes of amenorrhea are pregnancy and menopause. In female athletes, amenorrhea results from reduced secretion of gonadotropin-releasing hormone, which decreases the release of LH and FSH. As a result, ovarian follicles fail to develop, ovulation does not occur, synthesis of estrogens and progesterone wanes, and monthly menstrual bleeding ceases. Most cases of the female athlete triad occur in young women whose percentage of body fat is very low. Low levels of the hormone leptin, secreted by adipose cells, may be a contributing factor.

Because estrogens help bones retain calcium and other minerals, chronically low levels of estrogens are associated with loss of bone mineral density. The female athlete triad causes "old bones in young women." In one study, amenorrheic runners in their twenties had low bone mineral densities, similar to those of postmenopausal women 50 to 70 years old! Whereas short periods of amenorrhea in young athletes may cause no lasting harm, long-term cessation of the reproductive cycle may be accompanied by a loss of bone mass, and adolescent athletes may fail to achieve an adequate bone mass. Both situations can lead to premature osteoporosis and irreversible bone damage. ■

▶ **CHECKPOINT**

28. Describe the function of each of the following hormones in the uterine and ovarian cycles: GnRH, FSH, LH, estrogens, progesterone, and inhibin.

29. Briefly outline the major events of each phase of the uterine cycle, and correlate them with the events of the ovarian cycle.

30. Prepare a labeled diagram of the major hormonal changes that occur during the uterine and ovarian cycles.

Figure 28.28   Summary of hormonal interactions in the ovarian and uterine cycles.

Hormones from the anterior pituitary regulate ovarian function, and hormones from the ovaries regulate the changes in the endometrial lining of the uterus.

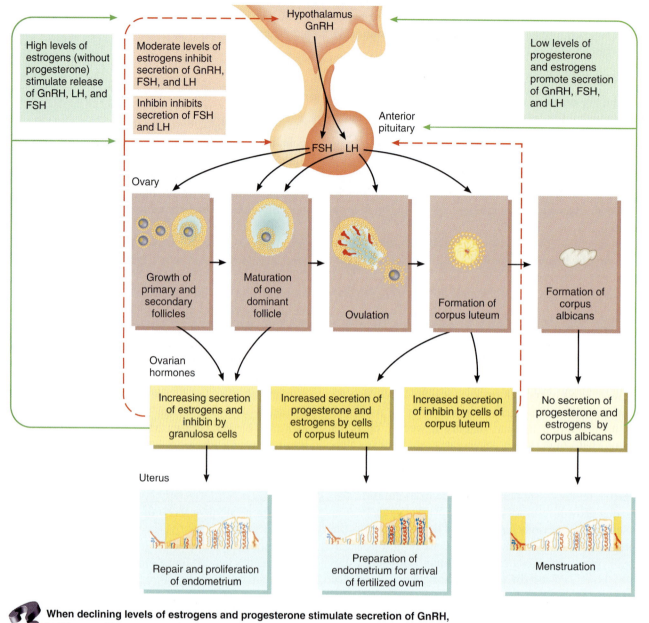

When declining levels of estrogens and progesterone stimulate secretion of GnRH, is this a positive or a negative feedback effect? Why?

THE HUMAN SEXUAL RESPONSE

▶ O B J E C T I V E

• **Describe the similarities and differences in the sexual responses of males and females.**

During heterosexual **sexual intercourse,** also called **coitus** (KŌ-i-tus = a coming together), sperm are ejaculated from the male urethra into the vagina. The similar sequence of physiological and emotional changes experienced by both males and females

before, during, and after intercourse is termed the **human sexual response.**

Stages of the Human Sexual Response

William Masters and Virginia Johnson, who began their pioneering research on human sexuality in the late 1950s, divided the human sexual response into four stages: excitement, plateau, orgasm, and resolution. During sexual **excitement,** also known as **arousal,** various physical and psychological stimuli trigger parasympathetic reflexes in which nerve impulses propagate from the gray matter of the second, third, and fourth sacral segments of the spinal cord along the pelvic splanchnic nerves. Some of these parasympathetic postganglionic fibers produce relaxation of vascular smooth muscle, which allows **vasocongestion**—engorgement with blood—of genital tissues. Parasympathetic impulses also stimulate the secretion of lubricating fluids, most of which are contributed by the female. Without satisfactory lubrication, sexual intercourse is difficult and painful for both partners and inhibits orgasm.

Other changes include increased heart rate and blood pressure, increased skeletal muscle tone throughout the body, and hyperventilation. Direct physical contact (as in kissing or touching), especially of the glans penis, clitoris, nipples of the breasts, and earlobes is a potent initiator of excitement. However, anticipation or fear, memories, visual, olfactory, and auditory sensations, and fantasies can enhance or diminish the likelihood that excitement will occur. Due to the tactile stimulation of the breast while nursing an infant, feelings of sexual arousal during breastfeeding are fairly common and should not be cause for concern.

The changes that begin during excitement are sustained at an intense level in the **plateau** stage, which may last for only a few seconds or for many minutes. At this stage, many females and some males display a sex flush, a rashlike redness of the face and chest. Generally, the briefest stage is **orgasm** *(climax),* during which the male ejaculates and both sexes experience several rhythmic muscular contractions about 0.8 sec apart, accompanied by intense, pleasurable sensations. The sex flush also is most prominent at this time. During orgasm, bursts of *sympathetic* nerve impulses leave the spinal cord at the first and second lumbar levels and cause rhythmic contractions of smooth muscle in genital organs. At the same time, *somatic motor neurons* from lumbar and sacral segments of the spinal cord trigger powerful, rhythmic contractions of perineal skeletal muscles, particularly the bulbocavernosus muscles, ischiocavernosus muscles, and anal sphincter (see Figure 11.13 on page 341 for males and Figure 28.23 for females).

In males, orgasm usually accompanies ejaculation. Reception of the ejaculate, however, provides little stimulus for a female, especially if she is not already at the plateau stage; this is why a female does not automatically experience orgasm simultaneously with her partner. In both males and females, orgasm is a total body response that may produce milder sensations on some occasions and more intense, explosive sensations at other times. Whereas females may experience two or more orgasms in rapid succession, males enter a **refractory period,** a recovery time during which a second ejaculation and orgasm is physiologically impossible. In some males the refractory period lasts only a few minutes; in others it lasts for several hours.

In the final stage—**resolution,** which begins with a sense of profound relaxation—genital tissues, heart rate, blood pressure, breathing, and muscle tone return to the unaroused state. If sexual excitement has been intense but orgasm has not occurred, resolution takes place more slowly. The four phases of the human sexual response are not always clearly separated from one another and may vary considerably among different people, and even in the same person at different times.

Changes in Males

Most of the time, the penis is flaccid (limp) because sympathetic impulses cause vasoconstriction of its arteries, which limits blood inflow. The first sign of sexual excitement is erection. Parasympathetic impulses cause release of neurotransmitters and local hormones, including the gas nitric oxide, which relaxes vascular smooth muscle in the penis. The arteries of the penis dilate, and blood fills the blood sinuses of the three corpora. Expansion of these erectile tissues compresses the superficial veins that normally drain the penis, resulting in engorgement and rigidity (erection). Parasympathetic impulses also cause the bulbourethral (Cowper's) glands to secrete mucus, which flows through the urethra and provides a small amount of lubrication for intercourse.

During the plateau stage, the head of the penis increases in diameter, and vasocongestion causes the testes to swell. As orgasm begins, rhythmic sympathetic impulses cause peristaltic contractions of smooth muscle in the ducts of each testis, epididymis, and ductus (vas) deferens, as well as in the walls of the seminal vesicles and prostate. These contractions propel sperm and fluid into the urethra (emission). Then, peristaltic contractions in the ducts and urethra combine with rhythmic contractions of skeletal muscles in the perineum and at the base of the penis to propel the semen from the urethra to the exterior.

Changes in Females

The first signs of sexual excitement in females are also due to vasocongestion. Within a few seconds to a minute, parasympathetic impulses stimulate release of fluids that lubricate the walls of the vagina. Although the vaginal mucosa lacks glands, engorgement of its connective tissue with blood during sexual excitement causes lubricating fluid to ooze from capillaries and seep through the epithelial layer via a process called **transudation** (tran′-soo-DĀ-shun; *trans-* = across or through; *sud-* = sweat). Glands within the cervical mucosa and the greater vestibular (Bartholin's) glands contribute a small quantity of lubricating mucus. During the excitement stage, parasympathetic impulses also trigger erection of the clitoris, vasocongestion of the labia, and relaxation of vaginal smooth muscle. Vasocongestion may also cause the breasts to swell and the nipples to become erect.

Late in the plateau stage, pronounced vasocongestion of the distal third of the vagina swells the tissue and narrows the opening. Because of this response, the vagina grips the penis more firmly. If effective sexual stimulation continues, orgasm may occur, associated with 3–15 rhythmic contractions of the vagina, uterus, and perineal muscles.

Erectile Dysfunction

Erectile dysfunction (ED), previously termed *impotence,* is the consistent inability of an adult male to ejaculate or to attain or hold an erection long enough for sexual intercourse. Many cases of impotence are caused by insufficient release of nitric oxide, which relaxes the smooth muscle of the penile arteries. The drug *Viagra* (sildenafil) enhances effects stimulated by nitric oxide in the penis. Other causes of erectile dysfunction include diabetes mellitus, physical abnormalities of the penis, systemic disorders such as syphilis, vascular disturbances (arterial or venous obstructions), neurological disorders, surgery, testosterone deficiency, and drugs (alcohol, antidepressants, antihistamines, antihypertensives, narcotics, nicotine, and tranquilizers). Psychological factors such as anxiety or depression, fear of causing pregnancy, fear of sexually transmitted diseases, religious inhibitions, and emotional immaturity may also cause ED. ■

▶ **CHECKPOINT**

31. What happens during each of the four stages of the human sexual response?

BIRTH CONTROL METHODS AND ABORTION

▶ **OBJECTIVE**

• **Compare the various kinds of birth control methods and their effectiveness.**

No single, ideal method of **birth control** exists. The only method of preventing pregnancy that is 100% reliable is total **abstinence,** the avoidance of sexual intercourse. Several other methods are available, each with advantages and disadvantages. These include surgical sterilization, hormonal methods, intrauterine devices, spermicides, barrier methods, periodic abstinence, and coitus interruptus. Table 28.3 gives the failure rates for each method of birth control. We will also discuss induced abortion, the intentional termination of pregnancy.

Surgical Sterilization

Sterilization is a procedure that renders an individual incapable of reproduction. The most common means of sterilization of males is **vasectomy** (vas-EK-tō-mē; *-ectomy* = cut out), in which a portion of each ductus deferens is removed. An incision is made in the posterior side of the scrotum, the ducts are located, each is tied in two places, and the part between the ties is removed. Even though sperm production continues in the testes, sperm can no longer reach the exterior. Instead, they degenerate and are destroyed by phagocytosis. Blood testosterone level is normal, so a vasectomy has no effect on sexual desire or performance. Sterilization in females most often is achieved by performing a **tubal ligation** (lī-GĀ-shun) in which both uterine tubes are tied closed and then cut. As a result, the secondary oocyte cannot pass

Table 28.3	Failure Rates of Several Birth Control Methods	
	Failure Rates*	
Method	**Perfect Use†**	**Typical Use**
None	85%	85%
Complete abstinence	0%	0%
Surgical sterilization		
Vasectomy	0.10%	0.15%
Tubal ligation	0.5%	0.5%
Hormonal methods		
Oral contraceptives	0.1%	3%‡
Norplant	0.3%	0.3%
Depo-provera	0.05%	0.05%
Intrauterine device		
Copper T 380A	0.6%	0.8%
Spermicides	6%	26%‡
Barrier methods		
Male condom	3%	14%‡
Vaginal pouch	5%	21%‡
Diaphragm	6%	20%‡
Periodic abstinence		
Rhythm	9%	25%‡
Sympto-thermal	2%	20%‡
Coitus interruptus	4%	18%‡

*Defined as percentage of women having an unintended pregnancy during the first year of use.
†Failure rate when the method is used correctly and consistently.
‡Includes couples who forgot to use the method.

through the uterine tubes, and sperm cannot reach the oocyte. Tubal ligation reduces the risk of pelvic inflammatory disease in women who are exposed to sexually transmitted infections; it may also reduce the risk of ovarian cancer.

Hormonal Methods

Aside from total abstinence or surgical sterilization, hormonal methods are the most effective means of birth control. Used by 50 million women worldwide, **oral contraceptives** ("the pill") contain various mixtures of synthetic estrogens and progestins (chemicals with actions similar to those of progesterone). They prevent pregnancy mainly by negative feedback inhibition of anterior pituitary secretion of the gonadotropins FSH and LH. The low levels of FSH and LH usually prevent development of a dominant follicle. As a result, estrogen level does not rise, the midcycle LH surge does not occur, and ovulation is not triggered. Thus, there is no secondary oocyte available for fertilization. Even if ovulation does occur, as it does in some cases, oral contraceptives also alter cervical mucus such that it is more hostile to sperm and block implantation in the uterus. If taken properly, the pill is close to 100% effective.

Among the noncontraceptive benefits of oral contraceptives are regulation of the length of menstrual cycles and decreased menstrual flow (and therefore decreased risk of anemia). The pill also provides protection against endometrial and ovarian cancers and reduces the risk of endometriosis. However, oral contraceptives may not be advised for women with a history of blood clotting disorders, cerebral blood vessel damage, migraine headaches, hypertension, liver malfunction, or heart disease. Women who take the pill and smoke face far higher odds of having a heart attack or stroke than do nonsmoking pill users. Smokers should quit smoking or use an alternative method of birth control.

Oral contraceptives also may be used for **emergency contraception (EC),** the so-called "morning-after pill." The relatively high levels of estrogen and progestin in EC pills provide negative feedback inhibition of FSH and LH secretion. Loss of the stimulating effects of these gonadotropic hormones causes the ovaries to cease secretion of their own estrogen and progesterone. In turn, declining levels of estrogen and progesterone induce shedding of the uterine lining, thereby blocking implantation. When two pills are taken within 72 hours after unprotected intercourse, and another two pills are taken 12 hours later, the chance of pregnancy is reduced by 75%.

Other hormonal methods of contraception are Norplant, Depo-provera, and the vaginal ring. **Norplant** consists of six slender hormone-containing capsules that are surgically implanted under the skin of the arm using local anesthesia. They slowly and continually release a progestin, which inhibits ovulation and thickens the cervical mucus. The effects last for 5 years, and Norplant is about as reliable as sterilization. Removing the Norplant capsules restores fertility. **Depo-provera,** which is given as an intramuscular injection once every 3 months, contains progestin, a hormone similar to progesterone, that prevents maturation of the ovum and causes changes in the uterine lining that make pregnancy less likely. The **vaginal ring** is a doughnut-shaped ring that fits in the vagina and releases either a progestin alone or a progestin and an estrogen. It is worn for 3 weeks and removed for 1 week to allow menstruation to occur.

Intrauterine Devices

An **intrauterine device (IUD)** is a small object made of plastic, copper, or stainless steel that is inserted into the cavity of the uterus. IUDs cause changes in the uterine lining that prevent implantation of a fertilized ovum. The IUD most commonly used in the United States today is the Copper T 380A, which is approved for up to 10 years of use and has long-term effectiveness comparable to that of tubal ligation. Some women cannot use IUDs because of expulsion, bleeding, or discomfort.

Spermicides

Various foams, creams, jellies, suppositories, and douches that contain sperm-killing agents, or **spermicides,** make the vagina and cervix unfavorable for sperm survival and are available without prescription. The most widely used spermicide is nonoxynol-9, which kills sperm by disrupting their plasma membrane. It also inactivates the AIDS virus and decreases the incidence of gonorrhea (described on page 1055). A spermicide is more effective when used together with a barrier method such as a diaphragm or a condom.

Barrier Methods

Barrier methods are designed to prevent sperm from gaining access to the uterine cavity and uterine tubes. In addition to preventing pregnancy, barrier methods may also provide some protection against sexually transmitted diseases (STDs) such as AIDS. In contrast, oral contraceptives and IUDs confer no such protection. Among the barrier methods are use of a condom, a vaginal pouch, or a diaphragm.

A **condom** is a nonporous, latex covering placed over the penis that prevents deposition of sperm in the female reproductive tract. A **vaginal pouch,** sometimes called a female condom, is made of two flexible rings connected by a polyurethane sheath. One ring lies inside the sheath and is inserted to fit over the cervix; the other ring remains outside the vagina and covers the female external genitals.

A **diaphragm** is a rubber, dome-shaped structure that fits over the cervix and is used in conjunction with a spermicide. It can be inserted up to 6 hours before intercourse. The diaphragm stops most sperm from passing into the cervix and the spermicide kills most sperm that do get by. Although diaphragm use does decrease the risk of some STDs, it does not fully protect against HIV infection.

Periodic Abstinence

A couple can use their knowledge of the physiological changes that occur during the female reproductive cycle to decide either to abstain from intercourse on those days when pregnancy is a likely result, or to plan intercourse on those days if they wish to conceive a child. In females with normal and regular menstrual cycles, these physiological events help to predict the day on which ovulation is likely to occur.

The first physiologically based method, developed in the 1930s, is known as the **rhythm method.** It takes advantage of the fact that a secondary oocyte is fertilizable for only 24 hours and is available for only 3 to 5 days in each reproductive cycle. During this time (3 days before ovulation, the day of ovulation, and 3 days after ovulation) the couple abstains from intercourse. The effectiveness of the rhythm method for birth control is poor in many women due to their irregular cycles.

Another system is the **sympto-thermal method,** in which couples are instructed to know and understand certain signs of fertility. The signs of ovulation include increased basal body temperature; the production of abundant clear, stretchy cervical mucus; and pain associated with ovulation (mittelschmerz). If a couple abstains from sexual intercourse when the signs of ovulation are present and for 3 days afterward, the chance of pregnancy is decreased. A big problem with this method is that fertilization is very likely if intercourse occurs one or two days *before* ovulation.

Coitus Interruptus

Coitus interruptus is withdrawal of the penis from the vagina just before ejaculation. Failures with this method are due either to failure to withdraw before ejaculation or to preejaculatory emission of sperm-containing fluid from the urethra. In addition, this method offers no protection against transmission of STDs.

Abortion

Abortion refers to the premature expulsion of the products of conception from the uterus, usually before the 20th week of pregnancy. An abortion may be spontaneous (naturally occurring; also called a miscarriage) or induced (intentionally performed). Induced abortions may be performed by vacuum aspiration (suction), infusion of a saline solution, or surgical evacuation (scraping).

Certain drugs, most notably RU 486, can induce a so-called nonsurgical abortion. **RU 486 (mifepristone)** is an antiprogestin; it blocks the action of progesterone by binding to and blocking progesterone receptors. Progesterone prepares the uterine endometrium for implantation and then maintains the uterine lining after implantation. If the level of progesterone falls during pregnancy or if the action of the hormone is blocked, menstruation occurs, and the embryo sloughs off along with the uterine lining. Within 12 hours after taking RU 486, the endometrium starts to degenerate, and within 72 hours, it begins to slough off. A form of prostaglandin E (misoprostol), which stimulates uterine contractions, is given after RU 486 to aid in expulsion of the endometrium. RU 486 can be taken up to 5 weeks after conception. One side effect of the drug is uterine bleeding.

▶ **CHECKPOINT**

32. How do oral contraceptives reduce the likelihood of pregnancy?

33. Why do some methods of birth control protect against sexually transmitted diseases, whereas others do not?

DEVELOPMENT OF THE REPRODUCTIVE SYSTEMS

▶ **OBJECTIVE**

- **Describe the development of the male and female reproductive systems.**

The *gonads* develop from the **intermediate mesoderm.** During the fifth week of development, they appear as bulges that protrude into the ventral body cavity (Figure 28.29). Adjacent to the gonads are the **mesonephric (Wolffian) ducts,** which eventually develop into structures of the reproductive system in males. A second pair of ducts, the **paramesonephric (Müllerian) ducts,** develop lateral to the mesonephric ducts and eventually form structures of the reproductive system in females. Both sets of ducts empty into the urogenital sinus. An early embryo has the potential to follow either the male or the female pattern of development because it contains both sets of ducts and primitive gonads that can differentiate into either testes or ovaries.

Cells of a male embryo have one X chromosome and one Y chromosome. The male pattern of development is initiated by a Y chromosome "master switch" gene named *SRY,* which stands for *S*ex-determining *R*egion of the *Y* chromosome. When the *SRY* gene is expressed during development, its protein product causes the primitive Sertoli cells to begin to differentiate in the gonadal tissues during the seventh week. The developing Sertoli cells secrete a hormone called **Müllerian-inhibiting substance (MIS),** which causes apoptosis of cells within the paramesonephric (Müllerian) ducts. As a result, those cells do not contribute any functional structures to the male reproductive system. Stimulated by human chorionic gonadotropin (hCG), primitive Leydig cells in the gonadal tissue begin to secrete the androgen **testosterone** during the eighth week. Testosterone then stimulates development of the mesonephric duct on each side into the *epididymis, ductus (vas) deferens, ejaculatory duct,* and *seminal vesicle.* The *testes* connect to the mesonephric duct through a series of tubules that eventually become the *seminiferous tubules.* The *prostate* and *bulbourethral glands* are **endodermal** outgrowths of the urethra.

Cells of a female embryo have two X chromosomes and no Y chromosome. Because *SRY* is absent, the gonads develop into *ovaries,* and because MIS is not produced, the paramesonephric ducts flourish. The distal ends of the paramesonephric ducts fuse to form the *uterus* and *vagina* whereas the unfused proximal portions become the *uterine (Fallopian) tubes.* The mesonephric ducts degenerate without contributing any functional structures to the female reproductive system because testosterone is absent. The *greater* and *lesser vestibular glands* develop from **endodermal** outgrowths of the vestibule.

The *external genitals* of both male and female embryos also remain undifferentiated until about the eighth week. Before differentiation, all embryos have an elevated midline swelling called the **genital tubercle** (Figure 28.30). The tubercle consists of the **urethral groove** (opening into the urogenital sinus), paired **urethral folds,** and paired **labioscrotal swellings.**

In male embryos, some testosterone is converted to a second androgen called **dihydrotestosterone (DHT).** DHT stimulates development of the urethra, prostate, and external genitals (scrotum and penis). Part of the genital tubercle elongates and develops into a penis. Fusion of the urethral folds forms the *spongy (penile) urethra* and leaves an opening to the exterior only at the distal end of the penis, the *external urethral orifice.* The labioscrotal swellings develop into the *scrotum.* In the absence of DHT, the genital tubercle gives rise to the *clitoris* in female embryos. The urethral folds remain open as the *labia minora,* and the labioscrotal swellings become the *labia majora.* The urethral groove becomes the *vestibule.* After birth, androgen levels

Figure 28.29 Development of the internal reproductive systems.

The gonads develop from intermediate mesoderm.

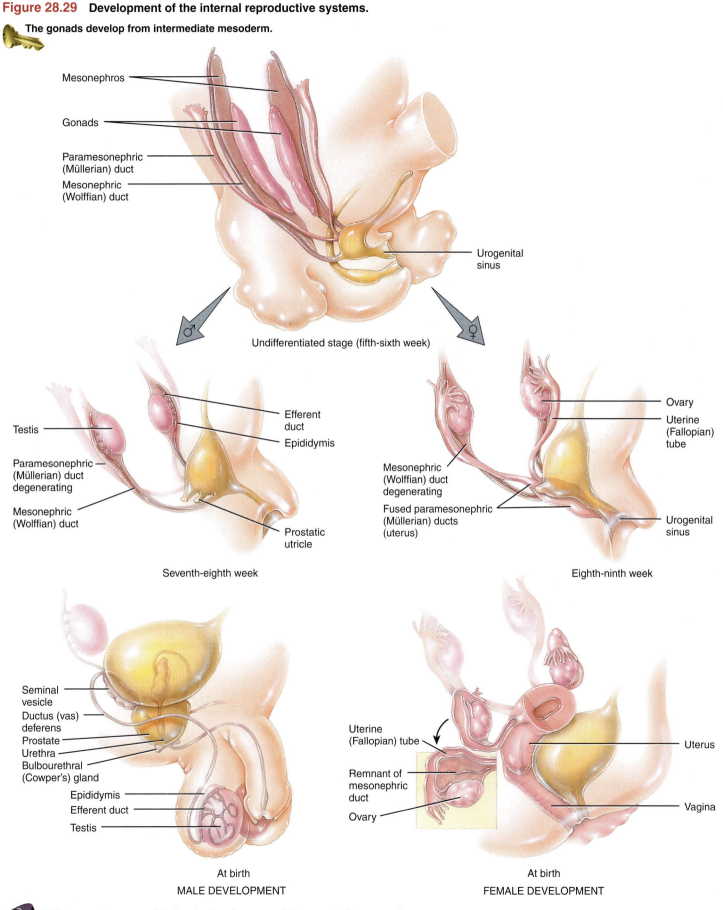

Mesonephros

Gonads

Paramesonephric (Müllerian) duct

Mesonephric (Wolffian) duct

Urogenital sinus

Undifferentiated stage (fifth-sixth week)

Testis

Efferent duct

Epididymis

Paramesonephric (Müllerian) duct degenerating

Mesonephric (Wolffian) duct

Prostatic utricle

Seventh-eighth week

Ovary

Uterine (Fallopian) tube

Mesonephric (Wolffian) duct degenerating

Fused paramesonephric (Müllerian) ducts (uterus)

Urogenital sinus

Eighth-ninth week

Seminal vesicle

Ductus (vas) deferens

Prostate

Urethra

Bulbourethral (Cowper's) gland

Epididymis

Efferent duct

Testis

At birth
MALE DEVELOPMENT

Uterine (Fallopian) tube

Remnant of mesonephric duct

Ovary

Uterus

Vagina

At birth
FEMALE DEVELOPMENT

Which gene is responsible for the development of the gonads into testes?

Figure 28.30   Development of the external genitals.

The external genitals of male and female embryos remain undifferentiated until about the eighth week.

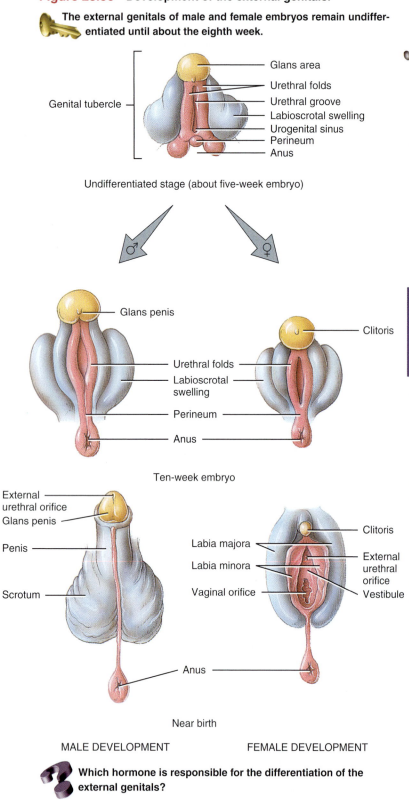

Undifferentiated stage (about five-week embryo)

Ten-week embryo

Near birth

MALE DEVELOPMENT   FEMALE DEVELOPMENT

Which hormone is responsible for the differentiation of the external genitals?

decline because hCG is no longer present to stimulate secretion of testosterone.

Deficiency of 5 Alpha-Reductase

A rare genetic mutation leads to a deficiency in 5 alpha-reductase, the enzyme that converts testosterone to dihydrotestosterone. At birth such a baby looks like a female externally because of the absence of dihydrotestosterone during development. At puberty, however, testosterone level rises, masculine characteristics start to appear; and the breasts fail to develop. Internal examination reveals testes and structures that normally develop from the mesonephric duct (epididymis, ductus deferens, seminal vesicle, and ejaculatory duct) rather than ovaries and a uterus. ■

AGING AND THE REPRODUCTIVE SYSTEMS

► **OBJECTIVE**

• Describe the effects of aging on the reproductive systems.

During the first decade of life, the reproductive system is in a juvenile state. At about age 10, hormone-directed changes start to occur in both sexes. **Puberty** (PŪ-ber-tē = a ripe age) is the period when secondary sexual characteristics begin to develop and the potential for sexual reproduction is reached. Onset of puberty is marked by pulses or bursts of LH and FSH secretion, each triggered by a pulse of GnRH. Most pulses occur during sleep. As puberty advances, the hormone pulses occur during the day as well as at night. The pulses increase in frequency during a three- to four-year period until the adult pattern is established. The stimuli that cause the GnRH pulses are still unclear, but a role for the hormone leptin is starting to unfold. Just before puberty, leptin levels rise in proportion to adipose tissue mass. Interestingly, leptin receptors are present in both the hypothalamus and anterior pituitary. Mice that lack a functional leptin gene from birth are sterile and remain in a prepubertal state. Giving leptin to such mice elicits secretion of gonadotropins, and they become fertile. Leptin may signal the hypothalamus that long-term energy stores (triglycerides in adipose tissue) are adequate for reproductive functions to begin.

In females, the reproductive cycle normally occurs once each month from **menarche** (me-NAR-kē), the first menses, to **menopause,** the permanent cessation of menses. Thus, the female reproductive system has a time-limited span of fertility between menarche and menopause. For the first 1 to 2 years after menarche, ovulation only occurs in about 10% of the cycles and the luteal phase is short. Gradually, the percentage of ovulatory cycles increases, and the luteal phase reaches its normal duration of 14 days. With age, fertility declines. Between the ages of 40 and 50 the pool of remaining ovarian follicles becomes exhausted. As a result, the ovaries become less responsive to hormonal stimulation. The production of estrogens declines, despite copious secretion of FSH and LH by the anterior pituitary. Many women experience hot flashes and heavy sweating, which coincide with bursts of GnRH release. Other symptoms of menopause are headache, hair loss, muscular pains, vaginal dryness, insomnia, depression, weight gain, and mood swings. Some atrophy of the ovaries, uterine tubes, uterus, vagina, external genitalia, and breasts occurs in postmenopausal women. Due to loss of estrogens, most women experience a decline in bone mineral density after menopause. Sexual desire (libido) does not show a parallel decline; it may be maintained by adrenal sex steroids. The risk of having uterine cancer peaks at about 65 years of age, but cervical cancer is more common in younger women.

In males, declining reproductive function is much more subtle than in females. Healthy men often retain reproductive capacity into their eighties or nineties. At about age 55 a decline in testosterone synthesis leads to reduced muscle strength, fewer viable sperm, and decreased sexual desire. However, abundant sperm may be present even in old age.

Enlargement of the prostate to two to four times its normal size occurs in approximately one-third of all males over age 60. This condition, called **benign prostatic hyperplasia (BPH),** decreases the size of the prostatic urethra and is characterized by frequent urination, nocturia (bed-wetting), hesitancy in urination, decreased force of urinary stream, postvoiding dribbling, and a sensation of incomplete emptying.

▶ **C H E C K P O I N T**

34. What are the roles of hormones in differentiation of the gonads, the mesonephric ducts, the paramesonephric ducts, and the external genitals?

35. What changes occur in males and females at puberty?

36. What do the terms menarche and menopause mean?

DISORDERS: HOMEOSTATIC IMBALANCES

Reproductive System Disorders in Males

Testicular Cancer

Testicular cancer is the most common cancer in males between the ages of 20 and 35. More than 95% of testicular cancers arise from spermatogenic cells within the seminiferous tubules. An early sign of testicular cancer is a mass in the testis, often associated with a sensation of testicular heaviness or a dull ache in the lower abdomen; pain usually does not occur. To increase the chance for early detection of a testicular cancer, all males should perform regular self-examinations of the testes.

Prostate Disorders

Because the prostate surrounds part of the urethra, any infection, enlargement, or tumor in it can obstruct the flow of urine. Acute and chronic infections of the prostate are common in postpubescent males, often in association with inflammation of the urethra. In **acute prostatitis,** the prostate becomes swollen and tender. **Chronic prostatitis** is one of the most common chronic infections in men of the middle and later years. On examination, the prostate feels enlarged, soft, and very tender, and its surface outline is irregular.

Prostate cancer is the leading cause of death from cancer in men in the United States, having surpassed lung cancer in 1991. Each year it is diagnosed in almost 200,000 U.S. men and causes nearly 40,000 deaths. A blood test can measure the level of prostate-specific antigen (PSA) in the blood. The amount of PSA, which is produced only by prostate epithelial cells, increases with enlargement of the prostate and may indicate infection, benign enlargement, or prostate cancer. Males over the age of 40 should have an annual examination of the prostate gland. In a **digital rectal exam,** a physician palpates the gland through the rectum with the fingers (digits). Many physicians also recommend an annual PSA test for males over age 50. Treatment for prostate cancer may involve surgery, cryotherapy, radiation, hormonal therapy, and chemotherapy. Because many prostate cancers grow very slowly, some urologists recommend "watchful waiting" before treating small tumors in men over age 70.

Reproductive System Disorders in Females

Premenstrual Syndrome and Premenstrual Dysphoric Disorder

Premenstrual syndrome (PMS) is a cyclical disorder of severe physical and emotional distress. It appears during the postovulatory (luteal) phase of the female reproductive cycle and dramatically disappears when menstruation begins. The signs and symptoms are highly variable from one woman to another. They may include edema, weight gain, breast swelling and tenderness, abdominal distension, backache, joint pain, constipation, skin eruptions, fatigue and lethargy, greater need for sleep, depression or anxiety, irritability, mood swings, headache, poor coordination and clumsiness, and cravings for sweet or salty foods. The cause of PMS is unknown. For some women, getting regular exercise; avoiding caffeine, salt, and alcohol; and eating a diet that is high in complex carbohydrates and lean proteins can bring considerable relief.

Premenstrual dysphoric disorder (PMDD) is a more severe syndrome in which PMS-like signs and symptoms do not resolve after the onset of menstruation. Clinical research studies have found that suppression of the reproductive cycle by a drug that interferes with GnRH (leuprolide) decreases symptoms significantly. Because symptoms reappear when estradiol or progesterone is given together with leuprolide, researchers propose that PMDD is caused by abnormal responses to normal levels of these ovarian hormones. **SSRIs** (selective serotonin receptor inhibitors) have shown promise in treating both PMS and PMDD.

Endometriosis

Endometriosis (en-dō-mē-trē-Ō-sis; *endo-* = within; *metri-* = uterus; *osis* = condition) is characterized by the growth of endometrial tissue outside the uterus. The tissue enters the pelvic cavity via the open uterine tubes and may be found in any of several sites—on the ovaries, the rectouterine pouch, the outer surface of the uterus, the sigmoid colon, pelvic and abdominal lymph nodes, the cervix, the abdominal wall, the kidneys, and the urinary bladder. Endometrial tissue responds to hormonal fluctuations, whether it is inside or outside the uterus. With each reproductive cycle, the tissue proliferates and then breaks down and bleeds. When this occurs outside the uterus, it can cause inflammation, pain, scarring, and infertility. Symptoms include premenstrual pain or unusually severe menstrual pain.

Breast Cancer

One in eight women in the United States faces the prospect of **breast cancer.** After lung cancer, it is the second-leading cause of death from cancer in U.S. women. Breast cancer can occur but is rare in males. In females, breast cancer is seldom seen before age 30, and its incidence rises rapidly after menopause. An estimated 5% of the 180,000 cases diagnosed each year in the United States, particularly those that arise in younger women, stem from inherited genetic mutations (changes in the DNA). Researchers have now identified two genes that increase susceptibility to breast cancer: *BRCA 1* (*breast cancer 1*) and *BRCA2*. Mutation of *BRCA1* also confers a high risk for ovarian cancer. In addition, mutations of the *p53* gene increase the risk of breast cancer in both males and females, and mutations of the androgen receptor gene are associated with the occurrence of breast cancer in some males. Despite the fact that breast cancer generally is not painful until it becomes quite advanced, any lump, no matter how small, should be reported to a physician at once. Early detection—by breast self-examination and mammograms—is the best way to increase the chance of survival.

The most effective technique for detecting tumors less than 1 cm (0.4 in.) in diameter is **mammography** (mam-OG-ra-fē; -*graphy* = to record), a type of radiography using very sensitive x-ray film. The image of the breast, called a **mammogram,** is best obtained by compressing the breasts, one at a time, using flat plates. A supplementary procedure for evaluating breast abnormalities is **ultrasound.** Although ultrasound cannot detect tumors smaller than 1 cm in diameter, it can be used to determine whether a lump is a benign, fluid-filled cyst or a solid (and therefore possibly malignant) tumor.

Among the factors that increase the risk of developing breast cancer are (1) a family history of breast cancer, especially in a mother or sister; (2) nulliparity (never having borne a child) or having a first child after age 35; (3) previous cancer in one breast; (4) exposure to ionizing radiation, such as x rays; (5) excessive alcohol intake; and (6) cigarette smoking.

The American Cancer Society recommends the following steps to help diagnose breast cancer as early as possible:

- All women over 20 should develop the habit of monthly breast self-examination.

- A physician should examine the breasts every 3 years when a woman is between the ages of 20 and 40, and every year after age 40.

- A mammogram should be taken in women between the ages of 35 and 39, to be used later for comparison (baseline mammogram).

- Women with no symptoms should have a mammogram every year or two between ages 40 and 49, and every year after age 50.

- Women of any age with a history of breast cancer, a strong family history of the disease, or other risk factors should consult a physician to determine a schedule for mammography.

Treatment for breast cancer may involve hormone therapy, chemotherapy, radiation therapy, **lumpectomy** (removal of the tumor and the immediate surrounding tissue), a modified or radical mastectomy, or a combination of these approaches. A **radical mastectomy** (*mast-* = breast) involves removal of the affected breast along with the underlying pectoral muscles and the axillary lymph nodes. (Lymph nodes are removed because metastasis of cancerous cells usually occurs through lymphatic or blood vessels.) Radiation treatment and chemotherapy may follow the surgery to ensure the destruction of any stray cancer cells. Several types of chemotherapeutic drugs are used to decrease the risk of relapse or disease progression. *Nolvadex (tamoxifen)*, is an estrogen antagonist that binds to and blocks estrogen receptors, thus decreasing the stimulating effect of estrogen itself on breast cancer cells. Used for 20 years now, tamoxifen greatly reduces the risk of cancer recurrence. *Herceptin*, a monoclonal antibody drug, targets an antigen on the surface of breast cancer cells. It is effective in causing regression of tumors and retarding progression of the disease. The early data from clinical trials of two new drugs, *Femara* and *Amimidex*, show relapse rates that are lower than those for tamoxifen. These drugs are inhibitors of aromatase, the enzyme needed for the final step in synthesis of estrogens. Finally, two drugs—tamoxifen and *Evista (raloxifene)*—are being marketed for breast cancer *prevention.* Interestingly, raloxifene blocks estrogen receptors in the breasts and uterus but activates estrogen receptors in bone. Thus, it can be used to treat osteoporosis without increasing a woman's risk of breast or endometrial (uterine) cancer.

Ovarian Cancer

Even though ovarian cancer is the sixth most common form of cancer in females, it is the leading cause of death from all gynecological malignancies (excluding breast cancer) because it is difficult to detect before it metastasizes (spreads) beyond the ovaries. Risk factors associated with ovarian cancer include age (usually over age 50); race (whites are at highest risk); family history of ovarian cancer; more than 40 years of active ovulation; nulliparity or first pregnancy after age 30; a high-fat, low-fiber, vitamin A–deficient diet; and prolonged exposure to asbestos or talc. Early ovarian cancer has no symptoms or only mild ones associated with other common problems, such as abdominal discomfort, heartburn, nausea, loss of appetite, bloating, and flatulence. Later-stage signs and symptoms include an enlarged abdomen, abdominal and/or pelvic pain, persistent gastrointestinal disturbances, urinary complications, menstrual irregularities, and heavy menstrual bleeding.

Cervical Cancer

Cervical cancer, carcinoma of the cervix of the uterus, starts with **cervical dysplasia** (dis-PLĀ-sē-a), a change in the shape, growth, and number of cervical cells. The cells may either return to normal or progress to cancer. In most cases, cervical cancer may be detected in its earliest stages by a Pap test (see page 108). Some evidence links cervical cancer to the virus that causes genital warts, human papillomavirus (HPV). Increased risk is associated with having a large number of sexual partners, having first intercourse at a young age, and smoking cigarettes.

Vulvovaginal Candidiasis

Candida albicans is a yeastlike fungus that commonly grows on mucous membranes of the gastrointestinal and genitourinary tracts. The organism is responsible for **vulvovaginal candidiasis** (vul-vō-VAJ-i-nal can-di-DĪ-a-sis), the most common form of **vaginitis** (vaj′-i′-NĪ-tis), inflammation of the vagina. Candidiasis is characterized by severe itching; a thick, yellow, cheesy discharge; a yeasty odor; and pain. The disorder, experienced at least once by about 75% of females, is usually a result of proliferation of the fungus following antibiotic therapy for another condition. Predisposing conditions include the use of oral contraceptives or cortisone-like medications, pregnancy, and diabetes.

Sexually Transmitted Diseases

A **sexually transmitted disease (STD)** is one that is spread by sexual contact. In most developed countries of the world, such as those of the European Community, Japan, Australia, and New Zealand, the incidence of STDs has declined markedly during the past 25 years. In the United States, by contrast, STDs have been rising to near-epidemic proportions, currently affecting more than 65 million people. AIDS and hepatitis B, which are sexually transmitted diseases that also may be contracted in other ways, are discussed in Chapters 22 and 24, respectively.

Chlamydia

Chlamydia (kla-MID-ē-a) is a sexually transmitted disease caused by the bacterium *Chlamydia trachomatis* (*chlamy-* = cloak). This unusual bacterium cannot reproduce outside body cells; it "cloaks" itself inside cells, where it divides. At present, chlamydia is the most prevalent sexually transmitted disease in the United States. In most cases, the initial infection is asymptomatic and thus difficult to recognize clinically. In males, urethritis is the principal result, causing a clear discharge, burning on urination, frequent urination, and painful urination. Without treatment, the epididymides may also become inflamed, leading to sterility. In 70% of females with chlamydia, symptoms are absent, but chlamydia is the leading cause of pelvic inflammatory disease. Moreover, the uterine tubes may also become inflamed, which increases the risk of ectopic pregnancy (implantation of a fertilized ovum outside the uterus) and infertility due to the formation of scar tissue in the tubes.

Gonorrhea

Gonorrhea (gon-ō-RĒ-a) or **"the clap"** is caused by the bacterium *Neisseria gonorrhoeae*. In the United States, 1–2 million new cases of gonorrhea appear each year, most among individuals aged 15–29 years. Discharges from infected mucus membranes are the source of transmission of the bacteria either during sexual contact or during the passage of a newborn through the birth canal. The infection site can be in the mouth and throat after oral-genital contact, in the vagina and penis after genital intercourse, or in the rectum after recto-genital contact.

Males usually experience urethritis with profuse pus drainage and painful urination. The prostate and epididymis may also become infected. In females, infection typically occurs in the vagina, often with a discharge of pus. Both infected males and females may harbor the disease without any symptoms, however, until it has progressed to a more advanced stage; about 5–10% of males and 50% of females are asymptomatic. In females, the infection and consequent inflammation can proceed from the vagina into the uterus, uterine tubes, and pelvic cavity. An estimated 50,000 to 80,000 women in the United States are made infertile by gonorrhea every year as a result of scar tissue formation that closes the uterine tubes. If bacteria in the birth canal are transmitted to the eyes of a newborn, blindness can result. Administration of a 1% silver nitrate solution in the infant's eyes prevents infection.

Syphilis

Syphilis, caused by the bacterium *Treponema pallidum*, is transmitted through sexual contact or exchange of blood, or through the placenta to a fetus. The disease progresses through several stages. During the *primary stage*, the chief sign is a painless open sore, called a **chancre** (SHANG-ker), at the point of contact. The chancre heals within 1 to 5 weeks. From 6 to 24 weeks later, signs and symptoms such as a skin rash, fever, and aches in the joints and muscles usher in the *secondary state*, which is systemic—the infection spreads to all major body systems. When signs of organ degeneration appear, the disease is said to be in the *tertiary stage*. If the nervous system is involved, the tertiary stage is called **neurosyphilis.** As motor areas become extensively damaged, victims may be unable to control urine and bowel movements. Eventually they may become bedridden and unable even to feed themselves. In addition, damage to the cerebral cortex produces memory loss and personality changes that range from irritability to hallucinations.

Genital Herpes

Genital herpes is an incurable STD. Type II herpes simplex virus (HSV-2) causes genital infections, producing painful blisters on the prepuce, glans penis, and penile shaft in males and on the vulva or sometimes high up in the vagina in females. The blisters disappear and reappear in most patients, but the virus itself remains in the body. A related virus, type I herpes simplex virus (HSV-1), causes cold sores on the mouth and lips. Infected individuals typically experience recurrences of symptoms several times a year.

Genital Warts

Warts are an infectious disease caused by viruses. *Human papillomavirus (HPV)* causes **genital warts,** which is commonly transmitted sexually. Nearly one million people a year develop genital warts in the United States. Patients with a history of genital warts may be at increased risk for cancers of the cervix, vagina, anus, vulva, and penis. There is no cure for genital warts.

MEDICAL TERMINOLOGY

Abnormal uterine bleeding Menstruation of excessive duration or excessive amount, diminished menstrual flow, too frequent menstruation, bleeding between menstrual periods, and postmenopausal bleeding. These abnormalities may be caused by disordered hormonal regulation, emotional factors, fibroid tumors of the uterus, and systemic diseases.

Castration (kas-TRĀ-shun = to prune) Removal, inactivation, or destruction of the gonads; commonly used in reference to removal of the testes only.

Culdoscopy (kul-DOS-kō-pē; *-scopy* = to examine) A procedure in which a culdoscope (endoscope) is inserted through the vagina to view the pelvic cavity.

Dysmenorrhea (dis'-men-oī-Ē-a; *dys-* = difficult or painful) Pain associated with menstruation; the term is usually reserved to describe menstrual symptoms that are severe enough to prevent a woman from functioning normally for one or more days each month. Some cases are caused by uterine tumors, ovarian cysts, pelvic inflammatory disease, or intrauterine devices.

Endocervical curettage (kū-re-TAHZH; *curette* = scraper) A procedure in which the cervix is dilated and the endometrium of the uterus is scraped with a spoon-shaped instrument called a curette; commonly called a D and C (dilation and curettage).

Hermaphroditism (her-MAF-rō-dīt-izm) Presence of both ovarian and testicular tissue in one individual.

Hypospadias (hī′-pō-SPĀ-dē-as; *hypo-* = below) A displaced urethral opening. In males, the displaced opening may be on the underside of the penis, at the penoscrotal junction, between the scrotal folds, or in the perineum; in females, the urethra opens into the vagina.

Leukorrhea (loo′-kō-RĒ-a; *leuko-* = white) A whitish (nonbloody) vaginal discharge containing mucus and pus cells that may occur at any age and affects most women at some time.

Oophorectomy (ō′-of-ō-REK-tō-mē; *oophor-* = bearing eggs) Removal of the ovaries.

Ovarian cyst The most common form of ovarian tumor, in which a fluid-filled follicle or corpus luteum persists and continues growing.

Pelvic inflammatory disease (PID) A collective term for any extensive bacterial infection of the pelvic organs, especially the uterus, uterine tubes, or ovaries, which is characterized by pelvic soreness, lower back pain, abdominal pain, and urethritis. Often the early symptoms of PID occur just after menstruation. As infection spreads, fever may develop, along with painful abscesses of the reproductive organs.

Salpingectomy (sal′-pin-JEK-tō-mē; *salpingo* = tube) Removal of a uterine tube (oviduct).

Smegma (SMEG-ma) The secretion, consisting principally of desquamated epithelial cells, found chiefly around the external genitalia and especially under the foreskin of the male.

STUDY OUTLINE

THE CELL CYCLE IN THE GONADS (p. 1012)

1. Reproduction is the process by which new individuals of a species are produced and the genetic material is passed from generation to generation.
2. The organs of reproduction are grouped as gonads (produce gametes), ducts (transport and store gametes), accessory sex glands (produce materials that support gametes), and supporting structures (have various roles in reproduction).
3. Gametes contain the haploid (*n*) chromosome number (23), and most somatic cells contain the diploid (*2n*) chromosome number (46).
4. Meiosis is the process that produces haploid gametes; it consists of two successive nuclear divisions called meiosis I and meiosis II.
5. During meiosis I, homologous chromosomes undergo synapsis (pairing) and crossing-over; the net result is two haploid daughter cells that are genetically unlike each other and unlike the parent cell that produced them.
6. During meiosis II, the two haploid daughter cells divide to form four haploid cells.

MALE REPRODUCTIVE SYSTEM (p. 1014)

1. The male structures of reproduction include the testes, ductus epididymis, ductus (vas) deferens, ejaculatory duct, urethra, seminal vesicles, prostate, bulbourethral (Cowper's) glands, and penis.
2. The scrotum is a sac that hangs from the root of the penis and consists of loose skin and superficial fascia; it supports the testes.
3. The temperature of the testes is regulated by contraction of the cremaster muscle and dartos muscle, which either elevates them and brings them closer to the pelvic cavity or relaxes and moves them farther from the pelvic cavity.
4. The testes are paired oval glands (gonads) in the scrotum containing seminiferous tubules, in which sperm cells are made; Sertoli cells (sustentacular cells), which nourish sperm cells and secrete

inhibin; and Leydig cells (interstitial endocrinocytes), which produce the male sex hormone testosterone.

5. The testes descend into the scrotum through the inguinal canals during the seventh month of fetal development. Failure of the testes to descend is called cryptorchidism.
6. Secondary oocytes and sperm, both of which are called gametes, are produced in the gonads.
7. Spermatogenesis, which occurs in the testes, is the process whereby immature spermatogonia develop into mature sperm. The spermatogenesis sequence, which includes meiosis I, meiosis II, and spermiogenesis, results in the formation of four haploid sperm (spermatozoa) from each primary spermatocyte.
8. Mature sperm consist of a head, a midpiece, and a tail. Their function is to fertilize a secondary oocyte.
9. At puberty, gonadotropin-releasing hormone (GnRH) stimulates anterior pituitary secretion of FSH and LH. LH stimulates production of testosterone; FSH and testosterone stimulate spermatogenesis. Sertoli cells secrete androgen-binding protein (ABP), which binds to testosterone and keeps its concentration high in the seminiferous tubule.
10. Testosterone controls the growth, development, and maintenance of sex organs; stimulates bone growth, protein anabolism, and sperm maturation; and stimulates development of masculine secondary sex characteristics.
11. Inhibin is produced by Sertoli cells; its inhibition of FSH helps regulate the rate of spermatogenesis.
12. The duct system of the testes includes the seminiferous tubules, straight tubules, and rete testis. Sperm flow out of the testes through the efferent ducts.
13. The ductus epididymis is the site of sperm maturation and storage.
14. The ductus (vas) deferens stores sperm and propels them toward the urethra during ejaculation.
15. Each ejaculatory duct, formed by the union of the duct from the seminal vesicle and ampulla of the ductus (vas) deferens, is the

passageway for ejection of sperm and secretions of the seminal vesicles into the first portion of the urethra, the prostatic urethra.

16. The urethra in males is subdivided into three portions: the prostatic, membranous, and spongy (penile) urethra.

17. The seminal vesicles secrete an alkaline, viscous fluid that contains fructose (used by sperm for ATP production). Seminal fluid constitutes about 60% of the volume of semen and contributes to sperm viability.

18. The prostate secretes a slightly acidic fluid that constitutes about 25% of the volume of semen and contributes to sperm motility.

19. The bulbourethral (Cowper's) glands secrete mucus for lubrication and an alkaline substance that neutralizes acid.

20. Semen is a mixture of sperm and seminal fluid; it provides the fluid in which sperm are transported, supplies nutrients, and neutralizes the acidity of the male urethra and the vagina.

21. The penis consists of a root, a body, and a glans penis.

22. Engorgement of the penile blood sinuses under the influence of sexual excitation is called erection.

FEMALE REPRODUCTIVE SYSTEM (p. 1028)

1. The female organs of reproduction include the ovaries (gonads), uterine (Fallopian) tubes or oviducts, uterus, vagina, and vulva.

2. The mammary glands are part of the integumentary system and also are considered part of the reproductive system in females.

3. The ovaries, the female gonads, are located in the superior portion of the pelvic cavity, lateral to the uterus.

4. Ovaries produce secondary oocytes, discharge secondary oocytes (the process of ovulation), and secrete estrogens, progesterone, relaxin, and inhibin.

5. Oogenesis (the production of haploid secondary oocytes) begins in the ovaries. The oogenesis sequence includes meiosis I and meiosis II, which goes to completion only after an ovulated secondary oocyte is fertilized by a sperm cell.

6. The uterine (Fallopian) tubes transport secondary oocytes from the ovaries to the uterus and are the normal sites of fertilization. Ciliated cells and peristaltic contractions help move a secondary oocyte or fertilized ovum toward the uterus.

7. The uterus is an organ the size and shape of an inverted pear that functions in menstruation, implantation of a fertilized ovum, development of a fetus during pregnancy, and labor. It also is part of the pathway for sperm to reach the uterine tubes to fertilize a secondary oocyte. Normally, the uterus is held in position by a series of ligaments.

8. Histologically, the layers of the uterus are an outer perimetrium (serosa), a middle myometrium, and an inner endometrium.

9. The vagina is a passageway for sperm and the menstrual flow, the receptacle of the penis during sexual intercourse, and the inferior portion of the birth canal. It is capable of considerable distension.

10. The vulva, a collective term for the external genitals of the female, consists of the mons pubis, labia majora, labia minora, clitoris, vestibule, vaginal and urethral orifices, hymen, bulb of the vestibule, and three sets of glands: the paraurethral (Skene's), greater vestibular (Bartholin's), and lesser vestibular glands.

11. The perineum is a diamond-shaped area at the inferior end of the trunk medial to the thighs and buttocks.

12. The mammary glands are modified sweat glands lying superficial to the pectoralis major muscles. Their function is to synthesize, secrete, and eject milk (lactation).

13. Mammary gland development depends on estrogens and progesterone.

14. Milk production is stimulated by prolactin, estrogens, and progesterone; milk ejection is stimulated by oxytocin.

THE FEMALE REPRODUCTIVE CYCLE (p. 1041)

1. The function of the ovarian cycle is to develop a secondary oocyte, whereas that of the uterine (menstrual) cycle is to prepare the endometrium each month to receive a fertilized egg. The female reproductive cycle includes both the ovarian and uterine cycles.

2. The uterine and ovarian cycles are controlled by GnRH from the hypothalamus, which stimulates the release of FSH and LH by the anterior pituitary.

3. FSH stimulates development of secondary follicles and initiates secretion of estrogens by the follicles. LH stimulates further development of the follicles, secretion of estrogens by follicular cells, ovulation, formation of the corpus luteum, and the secretion of progesterone and estrogens by the corpus luteum.

4. Estrogens stimulate the growth, development, and maintenance of female reproductive structures; stimulate the development of secondary sex characteristics; and stimulate protein synthesis.

5. Progesterone works with estrogens to prepare the endometrium for implantation and the mammary glands for milk synthesis.

6. Relaxin relaxes the myometrium at the time of possible implantation. At the end of a pregnancy, relaxin increases the flexibility of the pubic symphysis and helps dilate the uterine cervix to facilitate delivery.

7. During the menstrual phase, the stratum functionalis of the endometrium is shed, discharging blood, tissue fluid, mucus, and epithelial cells.

8. During the preovulatory phase, a group of follicles in the ovaries begins to undergo final maturation. One follicle outgrows the others and becomes dominant while the others degenerate. At the same time, endometrial repair occurs in the uterus. Estrogens are the dominant ovarian hormones during the preovulatory phase.

9. Ovulation is the rupture of the dominant mature (Graafian) follicle and the release of a secondary oocyte into the pelvic cavity. It is brought about by a surge of LH. Signs and symptoms of ovulation include increased basal body temperature; clear, stretchy cervical mucus; changes in the uterine cervix; and abdominal pain.

10. During the postovulatory phase, both progesterone and estrogens are secreted in large quantity by the corpus luteum of the ovary, and the uterine endometrium thickens in readiness for implantation.

11. If fertilization and implantation do not occur, the corpus luteum degenerates, and the resulting low level of progesterone allows discharge of the endometrium followed by the initiation of another reproductive cycle.

12. If fertilization and implantation do occur, the corpus luteum is maintained by placental hCG. The corpus luteum and later the placenta secrete progesterone and estrogens to support pregnancy and breast development for lactation.

THE HUMAN SEXUAL RESPONSE (p. 1046)

1. The similar sequence of changes experienced by both males and females before, during, and after intercourse is termed the human sexual response; it occurs in four stages: excitement (arousal), plateau, orgasm, and resolution.

2. During excitement and plateau, parasympathetic nerve impulses produce genital vasocongestion, engorgement of tissues with blood, and secretion of lubricating fluids. Heart rate, blood pressure, breathing rate, and muscle tone increase.

3. During orgasm, sympathetic and somatic motor nerve impulses cause rhythmical contractions of smooth and skeletal muscles.

4. During resolution, the body relaxes and returns to the unaroused state.

BIRTH CONTROL METHODS AND ABORTION (p. 1048)

1. Birth control methods include surgical sterilization (vasectomy, tubal ligation), hormonal methods, intrauterine devices, spermicides, barrier methods (condom, vaginal pouch, diaphragm), periodic abstinence (rhythm and sympto-thermal methods), coitus interruptus, and induced abortion. See Table 28.2 on page 1039 for failure rates for these methods.

2. Contraceptive pills of the combination type contain estrogens and progestins in concentrations that decrease the secretion of FSH and LH and thereby inhibit development of ovarian follicles and ovulation.

3. An abortion is a spontaneous or induced premature expulsion of the products of conception from the uterus. RU 486 can induce abortion by blocking the action of progesterone.

DEVELOPMENT OF THE REPRODUCTIVE SYSTEMS (p. 1051)

1. The gonads develop from intermediate mesoderm. In the presence of the *SRY* gene, the gonads begin to differentiate into testes during the seventh week. The gonads differentiate into ovaries when the *SRY* gene is absent.

2. In males, testosterone stimulates development of each mesonephric duct into an epididymis, ductus (vas) deferens, ejacu-

latory duct, and seminal vesicle, and Müllerian-inhibiting substance (MIS) causes the paramesonephric duct cells to die. In females, testosterone and MIS are absent; the paramesonephric ducts develop into the uterine tubes, uterus, and vagina and the mesonephric ducts degenerate.

3. The external genitals develop from the genital tubercle and are stimulated to develop into typical male structures by the hormone dihydrotestosterone (DHT). The external genitals develop into female structures when DHT is not produced, the normal situation in female embryos.

AGING AND THE REPRODUCTIVE SYSTEMS (p. 1052)

1. Puberty is the period when secondary sex characteristics begin to develop and the potential for sexual reproduction is reached.

2. Onset of puberty is marked by pulses or bursts of LH and FSH secretion, each triggered by a pulse of GnRH. The hormone leptin, released by adipose tissue, may signal the hypothalamus that long-term energy stores (triglycerides in adipose tissue) are adequate for reproductive functions to begin.

3. In females, the reproductive cycle normally occurs once each month from menarche, the first menses, to menopause, the permanent cessation of menses.

4. Between the ages of 40 and 50, the pool of remaining ovarian follicles becomes exhausted and levels of progesterone and estrogens decline. Most women experience a decline in bone mineral density after menopause, together with some atrophy of the ovaries, uterine tubes, uterus, vagina, external genitalia, and breasts. Uterine and breast cancer increase in incidence with age.

5. In older males, decreased levels of testosterone are associated with decreased muscle strength, waning sexual desire, and fewer viable sperm; prostate disorders are common.

Q SELF-QUIZ QUESTIONS

Fill in the blanks in the following statements.

1. The period of time when secondary sexual characteristics begin to develop and the potential for sexual reproduction is reached is called _____. The first menses is called _____, and the permanent cessation of menses is called _____.

2. The fluid produced by the male accessory sex glands is called _____.

Indicate whether the following statements are true or false.

3. Spermatogenesis does not occur at normal core body temperature.

4. The stages of the human sexual response are arousal, plateau, orgasm, and resolution.

Choose the one best answer to the following questions.

5. Which of the following are functions of Sertoli cells? (1) protection of developing spermatogenic cells, (2) nourishment of spermatocytes, spermatids, and sperm, (3) phagocytosis of excess sperm cytoplasm as development proceeds, (4) mediation of the effects of testosterone and FSH, (5) control of movements of sper-

matogenic cells and release of sperm into the lumen of seminiferous tubules. (a) 1, 2, 4, and 5, (b) 1, 2, 3, and 5, (c) 2, 3, 4, and 5, (d) 1, 2, 3, and 4, (e) 1, 2, 3, 4, and 5.

6. Which of the following are true concerning androgens? (1) They stimulate the male pattern of development. (2) They stimulate the male and female patterns of development. (3) They contribute to sex drive in males and females. (4) They contribute to male secondary sex characteristics. (5) They stimulate protein synthesis. (a) 1, 2, and 3, (b) 1, 3, 4, and 5, (c) 1, 2, 4, and 5, (d) 1, 2, 3, and 4, (e) 1, 3, and 5.

7. Which of the following are true concerning estrogens? (1) They promote development and maintenance of female reproductive structures and secondary sex characteristics. (2) They help control fluid and electrolyte balance. (3) They increase protein catabolism. (4) They lower blood cholesterol. (5) In moderate levels, they inhibit the release of GnRH and the secretion of LH and FSH. (a) 1, 4, and 5, (b) 1, 3, 4, and 5, (c) 1, 2, 3, and 5, (d) 1, 2, 3, and 4, (e) 1, 2, 3, 4, and 5.

8. Which of the following statements are correct? (1) A sperm head contains DNA and an acrosome. (2) An acrosome is a specialized lysosome that contains enzymes that enable sperm to produce the ATP needed to propel themselves out of the male reproductive tract. (3) Mitochondria in the midpiece of a sperm produce ATP for sperm motility. (4) A sperm's tail, a flagellum, propels it along its way. (5) Once ejaculated, sperm are viable and normally are able to fertilize a secondary oocyte for 5 days. (a) 1, 2, 3, and 4, (b) 2, 3, 4, and 5, (c) 1, 3, and 4, (d) 2, 4, and 5, (e) 2, 3, and 4.

9. Which of the following statements are correct? (1) Spermatogonia are stem cells because when they undergo mitosis, some of the daughter cells remain to serve as a reservoir of cells for future mitosis. (2) Meiosis I is a division of pairs of chromosomes resulting in daughter cells with only one member of each chromosome pair. (3) Meiosis II separates the chromatids of each chromosome. (4) Spermiogenesis involves the maturation of spermatids into sperm. (5) The process by which the seminiferous tubules produce haploid sperm is called spermatogenesis. (a) 1, 2, 3, and 5, (b) 1, 2, 3, 4, and 5, (c) 1, 3, 4, and 5, (d) 1, 2, 3, and 4, (e) 1, 3, and 5.

10. Which of the following statements are correct? (1) Cells from the endoderm of the yolk sac give rise to oogonia. (2) Ova arise from the germinal epithelium of the ovary. (3) Primary oocytes enter prophase of meiosis I during fetal development but do not complete it until after puberty. (4) Once a secondary oocyte is formed, it proceeds to metaphase of meiosis II and stops at this stage. (5) The secondary oocyte resumes meiosis II and forms the ovum and a polar body only if fertilization occurs. (6) A primary oocyte gives rise to an ovum and four polar bodies. (a) 1, 3, 4, and 5, (b) 1, 3, 4, and 6, (c) 1, 2, 4, and 6, (d) 1, 2, 4, and 5, (e) 1, 2, 5, and 6.

11. Which of the following statements are correct? (1) The female reproductive cycle consists of a menstrual phase, a preovulatory phase, ovulation, and a postovulatory phase. (2) During the menstrual phase, small secondary follicles in the ovary begin to enlarge while the uterus is shedding its lining. (3) During the preovulatory phase, a dominant follicle continues to grow and begins to secrete estrogens and inhibin while the uterine lining begins to rebuild. (4) Ovulation results in the release of an ovum and the shedding of the uterine lining to nourish and support the released ovum. (5) After ovulation, a corpus luteum forms from the ruptured follicle and begins to secrete progesterone and estrogens, which it will continue to do throughout pregnancy if the egg is fertilized. (6) If pregnancy does not occur, then the corpus luteum degenerates into a scar called the corpus albicans, and the uterine lining is prepared to be shed again. (a) 1, 2, 4, and 5, (b) 2, 4, 5, and 6, (c) 1, 4, 5, and 6, (d) 1, 3, 4, and 6, (e) 1, 2, 3, and 6.

12. Match the following:
____ (a) relaxes the uterus by inhibiting myometrial contractions during monthly cycles; increases flexibility of the pubic symphysis during childbirth
____ (b) stimulates Leydig cells to secrete testosterone
____ (c) inhibits production of FSH by the anterior pituitary
____ (d) stimulates male pattern of development; stimulates protein synthesis; contributes to sex drive
____ (e) maintains the corpus luteum during the first trimester of pregnancy
____ (f) promote development of female reproductive structures; lower blood cholesterol
____ (g) stimulates the initial secretion of estrogens by growing follicles; promotes follicle growth
____ (h) is secreted by the corpus luteum to maintain the uterine lining during the first trimester of pregnancy

(1) inhibin
(2) LH
(3) FSH
(4) testosterone
(5) estrogens
(6) progesterone
(7) relaxin
(8) human chorionic gonadotropin

13. Match the following:
____ (a) the process during meiosis when portions of homologous chromosomes may be exchanged with each other
____ (b) refers to cells containing one-half the chromosome number
____ (c) the cell produced by the union of an egg and a sperm
____ (d) all chromosomes except the sex chromosomes
____ (e) two chromosomes that belong to a pair (one maternal and one paternal)
____ (f) the degeneration of oogonia before and after birth
____ (g) a packet of discarded nuclear material from the first or second meiotic division of the egg
____ (h) refers to cells containing the full chromosome number

(1) autosomes
(2) zygote
(3) homologous chromosomes
(4) haploid
(5) diploid
(6) crossing-over
(7) polar body
(8) atresia

14. Match the following:

_____ (a) site of sperm maturation
_____ (b) the male copulatory organ; a passageway for ejaculation of sperm and excretion of urine
_____ (c) sperm-forming cells
_____ (d) during sexual arousal, produce an alkaline substance that protects sperm by neutralizing acids in the urethra
_____ (e) ejects sperm into the urethra just before ejaculation
_____ (f) the supporting structure for the testes
_____ (g) carries the sperm from the scrotum into the abdominopelvic cavity for release by ejaculation; is cut and tied as a means of sterilization
_____ (h) the shared terminal duct of the reproductive and urinary systems in the male
_____ (i) surrounds the urethra at the base of the urinary bladder; produces secretions that contribute to sperm motility and viability
_____ (j) produce testosterone
_____ (k) support and protect developing spermatogenic cells; secrete inhibin; form the blood-testis barrier
_____ (l) secrete an alkaline fluid to help neutralize acids in the female reproductive tract; secrete fructose for use in ATP production by sperm
_____ (m) site of spermatogenesis

(1) spermatogenic cells
(2) Sertoli cells
(3) Leydig cells
(4) penis
(5) scrotum
(6) epididymis
(7) ductus (vas) deferens
(8) ejaculatory duct
(9) seminiferous tubules
(10) seminal vesicles
(11) prostate
(12) bulbourethral glands
(13) urethra

15. Match the following:

_____ (a) a small, cylindrical mass of erectile tissue and nerves in the female; homologous to the male glans penis
_____ (b) produce mucus in the female during sexual arousal and intercourse; homologous to the male bulbourethral glands
_____ (c) the group of cells that nourish the developing oocyte and begin to secrete estrogens
_____ (d) a pathway for sperm to reach the uterine tubes; the site of menstruation; the site of implantation of a fertilized ovum; the womb
_____ (e) produces progesterone, estrogens, relaxin, and inhibin
_____ (f) draw the ovum into the uterine tube
_____ (g) the opening between the uterus and vagina
_____ (h) mucus-secreting glands in the female that are homologous to the prostate gland
_____ (i) the female copulatory organ; the birth canal
_____ (j) passageway for the ovum to the uterus; usual site of fertilization
_____ (k) refers to the external genitals of the female
_____ (l) the layer of the uterine lining that is partially shed during each monthly cycle

(1) follicle
(2) corpus luteum
(3) uterine tube
(4) fimbriae
(5) uterus
(6) cervix
(7) endometrium
(8) vagina
(9) vulva
(10) clitoris
(11) paraurethral glands
(12) greater vestibular glands

CRITICAL THINKING QUESTIONS

1. Melissa skipped A&P class when the reproductive system was covered because, as she said, "I already know all that stuff." On the exam, she answered "false" to the statement "One target tissue of LH is the testes." Melissa should have attended class. Explain her error.

   HINT *LH is named for its effect in females.*

2. Why is the germinal epithelium of the ovaries superficial to the tunica albuginea, whereas that of the testes is deep? Where is the tunica vaginalis located?

   HINT *Male and female structures develop from common embryonic origins.*

3. If two babies were produced following the fertilization of both a secondary oocyte and a polar body, would the babies be identical twins?

   HINT *Identical twins have an identical chromosomal makeup.*

?ANSWERS TO FIGURE QUESTIONS

28.1 The result of crossing-over is that daughter cells are genetically unlike each other and genetically unlike the parent cell that produced them.

28.2 During anaphase I of meiosis, the paired chromatids are held together by a centromere and do not separate. During anaphase II of meiosis, the centromeres do not split and the paired chromatids remain together.

28.3 The gonads (testes) produce gametes (sperm) and hormones; the ducts transport, store, and receive gametes; and the accessory sex glands secrete materials that support gametes.

28.4 The cremaster and dartos muscles help regulate the temperature of the testes.

28.5 The tunica vaginalis and tunica albuginea are tissue layers that cover and protect the testes.

28.6 Leydig cells secrete testosterone.

28.7 During meiosis I, the number of chromosomes in each cell is reduced by half.

28.8 The sperm head contains DNA and enzymes for penetration of a secondary oocyte; the midpiece contains mitochondria for ATP production; the tail consists of a flagellum that provides propulsion.

28.9 Sertoli cells secrete inhibin.

28.10 Testosterone inhibits secretion of LH, and inhibin inhibits secretion of FSH.

28.11 Seminal vesicles contribute the largest volume to seminal fluid.

28.12 During sexual arousal, blood enters the sinuses in the two corpora cavernosa penis faster than it can drain out because the veins become compressed. The trapped blood engorges and stiffens the penis, producing an erection. The corpus spongiosum penis keeps the spongy urethra open so that ejaculation can occur.

28.13 The testes are homologous to the ovaries; the glans penis is homologous to the clitoris; the prostate is homologous to the paraurethral glands; and the bulbourethral gland is homologous to the greater vestibular glands.

28.14 The mesovarium anchors the ovary to the broad ligament of the uterus and the uterine tube; the ovarian ligament anchors it to the uterus; the suspensory ligament anchors it to the pelvic wall.

28.15 Ovarian follicles secrete estrogens; the corpus luteum secretes progesterone, estrogens, relaxin, and inhibin.

28.16 Most ovarian follicles undergo atresia (degeneration).

28.17 Primary oocytes are present in the ovary at birth, so they are as old as the woman is. In males, primary spermatocytes are continually being formed from stem cells (spermatogonia) and thus are only a few days old.

28.18 Fertilization most often occurs in the ampulla of the uterine tube.

28.19 Ciliated columnar epithelial cells and secretory cells with microvilli line the uterine tubes.

28.20 The endometrium is a highly vascularized, secretory epithelium that provides the oxygen and nutrients needed to sustain a fertilized egg; the myometrium is a thick smooth muscle layer that supports the uterine wall during pregnancy and contracts to expel the fetus at birth.

28.21 The stratum basalis of the endometrium provides cells to replace those that are shed (the stratum functionalis) during each menstruation.

28.22 Anterior to the vaginal opening are the mons pubis, clitoris, and prepuce. Lateral to the vaginal opening are the labia minora and labia majora.

28.23 The anterior portion of the perineum is called the urogenital triangle because its borders form a triangle that encloses the urethral (uro-) and vaginal (-genital) orifices.

28.24 Prolactin, estrogens, and progesterone regulate the synthesis of milk. Oxytocin regulates the ejection of milk.

28.25 The principal estrogen is β-estradiol.

28.26 The hormones responsible for the proliferative phase of endometrial growth are estrogens; for ovulation, LH; for growth of the corpus luteum, LH; and for the midcycle surge of LH, estrogens.

28.27 The effect of rising but moderate levels of estrogens is negative feedback inhibition of the secretion of GnRH, LH, and FSH.

28.28 This is negative feedback, because the response is opposite to the stimulus. A reduced amount of negative feedback due to declining levels of estrogens and progesterone stimulates release of GnRH, which in turn increases the production and release of FSH and LH, which ultimately stimulate the secretion of estrogens.

28.29 The *SRY* gene on the Y chromosome is responsible for the development of the gonads into testes.

28.30 The presence of dihydrotestosterone (DHT) stimulates differentiation of the external genitals in males; its absence allows differentiation of the external genitals in females.

Development and Inheritance

DEVELOPMENT, INHERITANCE AND HOMEOSTASIS

The genetic material inherited from parents and normal embryonic and fetal development both play important roles in determining the birth of a healthy child.

ENERGIZE YOUR STUDY
FOUNDATIONS CD
Animation
  • **Positive Feedback Control of Labor**
Concepts and Connections and Exercises reinforce your understanding

www.wiley.com/college/apcentral
INSIGHTS AND EXPLORATIONS
Stem cells. What are they? Scientists claim that the study of stem cells holds the promise for new therapies for Parkinson's and Alzheimer's disease, spinal cord injury, stroke, burns, heart disease, diabetes, osteoarthritis, and a host of other diseases. Other scientists believe these cells could be a source of a multitude of cell types to test potential drugs on for safety. Yet there is great controversy and for many an ethical dilemma because some believe the best source for these cells are from human embryonic and fetal tissue from terminated pregnancies. We will explore answers to these questions and issues in this web-based exercise.

Developmental biology is the study of the sequence of events from the fertilization of a secondary oocyte to the formation of an adult organism. From fertilization through the eighth week of development, the developing human is called an **embryo** (-*bryo* = grow) and this is the period of **embryological development.** **Embryology** (em-brē-OL-ō-jē) is the study of development from the fertilized egg through the eighth week. **Fetal development** begins at week nine and continues until birth. During this time, the developing human is called a **fetus** (FĒ-tus = offspring).

Once sperm and a secondary oocyte have developed through meiosis and maturation, and the sperm have been deposited in the vagina, pregnancy can occur. **Pregnancy** is a sequence of events that begins with fertilization, proceeds to implantation, embryonic development, and fetal development, and normally ends with birth about 38 weeks later, or 40 weeks after the last menstrual period.

Obstetrics (ob-STET-riks; *obstetrix* = midwife) is the branch of medicine that deals with the management of pregnancy, labor, and the **neonatal period,** the first 28 days after birth. **Prenatal development** (prē-NĀ-tal; *pre-* = before; *natal* = birth) is the time from fertilization to birth and includes both embryological and fetal development. It is divided into three periods of three calendar months each, called **trimesters.** Accordingly, obstetricians divide the nine months of pregnancy into three trimesters.

1. The **first trimester** is the most critical stage of development during which the rudiments of all the major organs systems appear, and also during which the developing organism is the most vulnerable to the effects of drugs, radiation, and microbes.
2. The **second trimester** is characterized by the nearly complete development of organ systems. By the end of this stage, the fetus assumes distinctively human features.
3. The **third trimester** represents a period of rapid fetal growth. During the early stages of this period, most of the organ systems are becoming fully functional.

In this chapter, we focus on the developmental sequence from fertilization through implantation, embryonic and fetal development, labor, and birth.

EMBRYONIC PERIOD

▶ **O B J E C T I V E**

- **Explain the major developmental events that occur during the embryonic period.**

First Week of Development

The first week of development is characterized by several significant events including fertilization, cleavage of the zygote, blastocyst formation, and implantation.

Fertilization

During **fertilization** (fer-til-i-ZĀ-shun; *fertil-* = fruitful), the genetic material from a haploid sperm cell (spermatozoon) and a haploid secondary oocyte merges into a single diploid nucleus. Of the about 200 million sperm introduced into the vagina, fewer than 2 million (1%) reach the cervix of the uterus and only about 200 reach the secondary oocyte. Fertilization normally occurs in the uterine (Fallopian) tube within 12 to 24 hours after ovulation. Sperm can remain viable for about 48 hours after deposition in the vagina, although a secondary oocyte is viable for only about 24 hours after ovulation. Thus, pregnancy is *most likely* to occur if intercourse takes place during a 3-day "window"—from 2 days before ovulation to 1 day after ovulation.

The process leading to fertilization begins when sperm swim from the vagina into the cervical canal propelled by the whiplike movements of their tails (flagella). The passage of sperm through the rest of the uterus and then into the uterine tube results mainly from contractions of the walls of these organs. Prostaglandins in semen are believed to stimulate uterine motility at the time of intercourse and to aid in the movement of sperm through the uterus and into the uterine tube. Sperm that reach the vicinity of the oocyte within minutes after ejaculation *are not capable* of fertilizing it until about seven hours later. During this time in the female reproductive tract, mostly in the uterine tube, sperm undergo **capacitation** (ka-pas′-i-TĀ-shun; *capacit-* = capable of), a series of functional changes that cause the sperm's tail to beat even more vigorously and prepare its plasma membrane to fuse with the oocyte's plasma membrane. During capacitation, sperm are acted upon by secretions in the female reproductive tract that result in the removal of cholesterol, glycoproteins, and proteins from the plasma membrane around the acrosome of the sperm.

For fertilization to occur, a sperm cell first must penetrate the **corona radiata** (kō-RŌ-na = crown; rā-dē-A-ta = to shine), the granulosa cells that surround the secondary oocyte, and then the **zona pellucida** (ZŌ-na = zone; pe-LOO-si-da = allowing passage of light), the clear glycoprotein layer between the corona radiata and the oocyte's plasma membrane (Figure 29.1a). One of the glycoproteins in the zona pellucida called ZP3 acts as a sperm receptor. Its binding to specific membrane proteins in the sperm head triggers the **acrosomal**

reaction, the release of the contents of the acrosome. The acrosome is similar to a large lysosome and contains digestive enzymes. The acrosomal enzymes digest a path through the zona pellucida as the lashing sperm tail pushes the sperm cell onward. Many sperm bind to ZP3 molecules and undergo acrosomal reactions, but only the first sperm cell to penetrate the entire zona pellucida and reach the oocyte's plasma membrane fuses with the oocyte.

The fusion of a sperm with a secondary oocyte, called **syngamy** (*syn-* = coming together; *-gamy* = marriage), sets in motion events that block **polyspermy,** fertilization by more than one sperm cell. Within a few seconds, the cell membrane of the oocyte depolarizes, which acts as a *fast block to polyspermy*—a depolarized oocyte cannot fuse with another sperm. Depolarization also triggers the intracellular release of calcium ions, which stimulate exocytosis of secretory vesicles from the oocyte. The molecules released by exocytosis inactivate ZP3 and harden the entire zona pellucida, events that constitute the *slow block to polyspermy.*

Once a sperm cell enters a secondary oocyte, the oocyte first must complete meiosis II. It divides into a larger ovum (mature egg) and a smaller second polar body that fragments and disintegrates (see Figure 28.17 on page 1032). The nucleus in the head of the sperm develops into the **male pronucleus,** and the nucleus of the fertilized ovum develops into the **female pronucleus** (Figure 29.1c). After the male and female pronuclei form, they fuse, producing a single diploid nucleus that contains 23 chromosomes from each pronucleus. Thus, the fusion of the haploid (*n*) pronuclei restores the diploid number (*2n*) of 46 chromosomes. The fertilized ovum now is called a **zygote** (ZĪ-gōt = a joining).

Dizygotic (fraternal) twins are produced from the independent release of two secondary oocytes and the subsequent fertilization of each by different sperm. They are the same age and in the uterus at the same time, but they are genetically as dissimilar as are any other siblings. Dizygotic twins may or may not be the same sex. Because **monozygotic (identical) twins** develop from a single fertilized ovum, they contain exactly the same genetic material and are always the same sex. Monozygotic twins arise from separation of the developing cells into two embryos, which occurs before 8 days after fertilization 99% of the time. Separations that occur later than 8 days are likely to produce **conjoined twins,** a situation in which the twins are joined together and share some body structures.

Cleavage of the Zygote

After fertilization, rapid mitotic cell divisions of the zygote called **cleavage** (KLĒV-ij) take place (Figure 29.2). The first division of the zygote begins about 24 hours after fertilization and is completed about 6 hours later. Each succeeding division takes slightly less time. By the second day after fertilization, the second cleavage is completed and there are four cells (Figure 29.2b). By the end of the third day, there are 16 cells. The progressively smaller cells produced by cleavage are called **blas-**

Figure 29.1   Selected structures and events in fertilization. (a) A sperm cell penetrating the corona radiata and zona pellucida around a secondary oocyte. (b) A sperm cell in contact with a secondary oocyte. (c) Male and female pronuclei.

🔑 **During fertilization, genetic material from a sperm cell and a secondary oocyte merge to form a single diploid nucleus.**

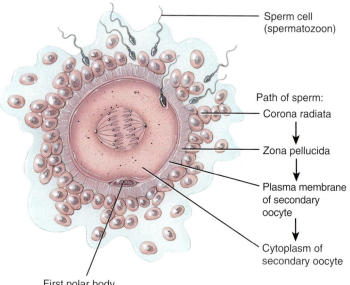

Sperm cell (spermatozoon)

Path of sperm:
Corona radiata
Zona pellucida
Plasma membrane of secondary oocyte
Cytoplasm of secondary oocyte

First polar body

(a) Sperm cell penetrating a secondary oocyte

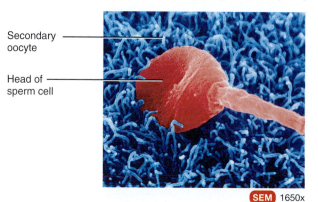

Secondary oocyte

Head of sperm cell

SEM 1650x

(b) Sperm cell in contact with a secondary oocyte

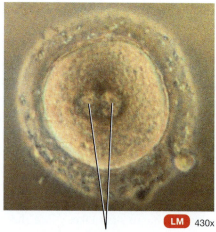

LM 430x

Pronuclei

(c) Male and female pronuclei

❓ **What is capacitation?**

tomeres (BLAS-tō-mērz; *blasto-* = germ or sprout; *-meres* = parts). Successive cleavages eventually produce a solid sphere of cells called the **morula** (MOR-ū-la; *morula* = mulberry). The morula is still surrounded by the zona pellucida and is about the same size as the original zygote (Figure 29.2c).

Blastocyst Formation

By the end of the fourth day, the number of cells in the morula increases as it continues to move through the uterine tube toward the uterine cavity. When the morula enters the uterine cavity on day 4 or 5, a glycogen-rich secretion from the glands of the endometrium of the uterus passes into the uterine cavity and enters the morula through the zona pellucida. This fluid, called **uterine milk,** along with nutrients stored in the cytoplasm of the blastomeres of the morula, provides nourishment for the developing morula. At the 32-cell stage, the fluid enters the morula, collects between the blastomeres, and reorganizes them around a large fluid-filled cavity called the **blastocyst cavity** (BLAS-tō-sist; *blasto-* = germ or sprout; *-cyst* = bag) (Figure 29.2e). With the formation of this cavity, the developing mass is then called the **blastocyst.** Though it now has hundreds of cells, the blastocyst is still about the same size as the original zygote.

Further rearrangement of the blastomeres results in the formation of two distinct structures: the inner cell mass and trophoblast (Figure 29.2e). The **inner cell mass** is located internally and eventually develops into the embryo. The **trophoblast** (TRŌF-ō-blast; *tropho-* = develop or nourish) is an outer superficial layer of cells that forms the wall of the blastocyst. It will ultimately develop into the fetal portion of the placenta, the site of exchange of nutrients and wastes between the mother and fetus. On about the fifth day after fertilization, the blastocyst hatches from the zona pellucida by digesting a hole in it with an enzyme, and then the blastocyst squeezes through the hole. Shedding of the zona pellucida is necessary to permit implantation.

Stem Cell Research and Therapeutic Cloning

Throughout the book, reference has been made to **stem cells,** the unspecialized cells that have the ability to divide for indefinite periods and give rise to specialized cells. In the context of human development, a zygote (fertilized ovum) is a stem cell. Because it has the potential to form an entire organism, a zygote is known as a *totipotent stem cell* (tō-TIP-ō-tent). Inner cell mass cells, by contrast, can give rise to many (but not all) different types of cells. They are called *pluripotent stem cells* (ploo-RIP-ō-tent). Later, pluripotent stem cells can undergo further specialization into cells that have a specific function. These cells are called *multipotent stem cells* (mul-TIP-ō-tent). Examples that we have already studied include keratinocytes that produce new skin cells, myeloid and lymphoid stem cells that develop into blood cells, and spermatogonia that give rise to sperm.

Figure 29.2 Cleavage and the formation of the morula and blastocyst.

Cleavage refers to the early, rapid mitotic divisions of a zygote.

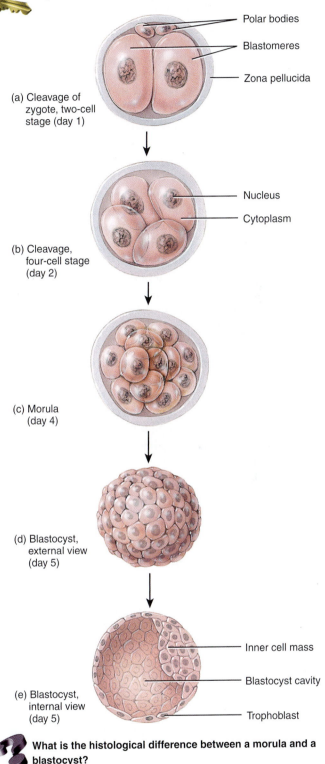

(a) Cleavage of zygote, two-cell stage (day 1)

Polar bodies
Blastomeres
Zona pellucida

(b) Cleavage, four-cell stage (day 2)

Nucleus
Cytoplasm

(c) Morula (day 4)

(d) Blastocyst, external view (day 5)

(e) Blastocyst, internal view (day 5)

Inner cell mass
Blastocyst cavity
Trophoblast

What is the histological difference between a morula and a blastocyst?

Pluripotent stem cells currently used in research are derived from (1) the inner cell mass of embryos in the blastocyst stage that were destined to be used for infertility treatments but were not needed and from (2) nonliving fetuses terminated during the first trimester of pregnancy. Because pluripotent stem cells give rise to almost all cell types in the body, they are extremely important in research and health care. For example, they might be used to generate cells and tissues for transplantation to treat conditions such as cancer, Parkinson and Alzheimer disease, spinal cord injury, diabetes, heart disease, stroke, burns, birth defects, osteoarthritis, and rheumatoid arthritis. Pluripotent stem cells may also be used to improve understanding of the underlying processes that occur during development to identify pathways that result in developmental defects and certain diseases. Additionally, pluripotent stem cells may be used to develop and test drugs by first testing them on cells developed from pluripotent stem cells and then using only the most promising drugs for further testing on animals and humans.

On October 13, 2001, researchers reported cloning of the first human embryo to grow cells to treat human diseases. **Therapeutic cloning** is envisioned as a procedure in which the genetic material of a patient with a particular disease is used to create pluripotent stem cells to treat the disease. In the procedure used to produce the first human cloned embryo the genetic material (nucleus) and polar body were removed from an unfertilized secondary oocyte. Then, the researchers injected the genetic material from a granulosa cell from the carona radiata of another secondary oocyte into the unfertilized secondary oocyte. Following injection, the secondary oocyte was exposed to chemicals that induced it to divide. After about 24 hours, the oocyte divided and the resulting cells contained the genetic material only from the injected granulosa cell. Researchers who produced the first human cloned embryos reported that two secondary oocytes divided to form early embryos of two cells each and one secondary oocyte divided to form an early embryo consisting of six cells before growth stopped.

Using the principles of therapeutic cloning, scientists hope to make an embryo clone of a patient, remove the pluripotent stem cells from the embryo, and then use them to grow tissues to treat particular diseases. Presumably, the tissues would not be rejected since they would contain the patient's own genetic material. ∎

Implantation

The blastocyst remains free within the uterine cavity for about 2 days before it attaches to the uterine wall. The endometrium is in its secretory phase. About 6 days after fertilization, the blastocyst loosely attaches to the endometrium, in a process called **implantation** (Figure 29.3). As the blastocyst implants, usually in either the posterior portion of the fundus or the body of the uterus, it orients with the inner cell mass toward the endometrium (Figure 29.3b). About seven days after fertilization, the blastocyst attaches to the endometrium more firmly, endome-

Figure 29.3 Relation of a blastocyst to the endometrium of the uterus at the time of implantation.

🔑 **Implantation, the attachment of a blastocyst to the endometrium, occurs about 6 days after fertilization.**

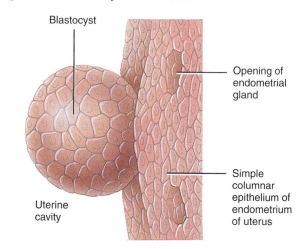

(a) External view of blastocyst, about 6 days after fertilization

trial glands in the vicinity enlarge, and the endometrium becomes more vascularized.

Following implantation, the endometrium is known as the **decidua** (dē-SID-ū-a = falling off). The decidua separates from the endometrium after the fetus is delivered much as it does in normal menstruation. Different regions of the decidua are named based on their positions relative to the site of the implanted blastocyst (Figure 29.4). The **decidua basalis** is the portion of the endometrium between the embryo and the stratum basalis of the uterus; it provides large amounts of glycogen and lipids for the developing embryo and fetus and later becomes the maternal part of the placenta. The **decidua capsularis** is the portion of the endometrium that is located between the embryo and the uterine cavity. The **decidua parietalis** (par-rī-e-TAL-is) is the remaining modified endometrium that lines the noninvolved areas of the rest of the uterus. As the embryo and later the fetus enlarges, the decidua capsularis bulges into the uterine cavity and fuses with the decidua parietalis, thereby obliterating the uterine cavity. By about 27 weeks, the decidua capsularis degenerates and disappears.

The major events associated with the first week of development are summarized in Figure 29.5.

Ectopic Pregnancy

Ectopic pregnancy (ek-TOP-ik; *ec-* = out of; *-topic* = place) is the development of an embryo or fetus outside the uterine cavity. An ectopic pregnancy usually occurs when movement of the fertilized ovum through the uterine tube is impaired. Situations that

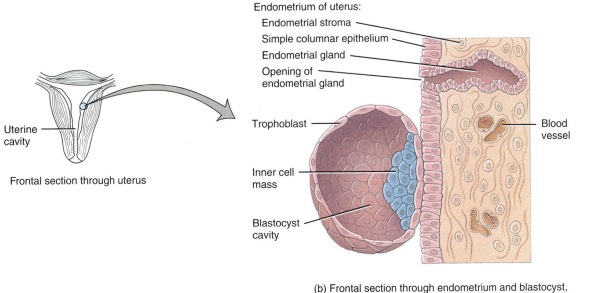

Endometrium of uterus:
Endometrial stroma
Simple columnar epithelium
Endometrial gland
Opening of
endometrial gland

Trophoblast

Blood
vessel

Inner cell
mass

Blastocyst
cavity

Uterine
cavity

Frontal section through uterus

(b) Frontal section through endometrium and blastocyst,
about 6 days after fertilization

How does the blastocyst merge with and burrow into the endometrium?

Figure 29.4   Regions of the decidua.

The decidua is a modified portion of the endometrium that develops after implantation.

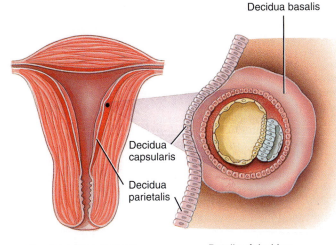

Decidua basalis

Decidua
capsularis

Decidua
parietalis

Frontal section of uterus   Details of decidua

Which part of the decidua helps form the maternal part of the placenta?

impair movement include scarring due to a prior tubal infection, decreased motility of the uterine tube smooth muscle, or abnormal tubal anatomy. Although the most common sites of ectopic pregnancies are the ampullar and infundibular portions of the uterine tube, ectopic pregnancies may also occur in the ovary, abdominal cavity, or uterine cervix. Compared to nonsmokers, women who smoke and become pregnant are twice as likely to have an ectopic pregnancy because nicotine in cigarette smoke paralyzes the cilia in the lining of the uterine tube (as it does those in the respiratory airways). Scars from pelvic inflammatory disease, previous uterine tube surgery, and previous ectopic pregnancy may also hinder movement of the fertilized ovum.

The signs and symptoms of ectopic pregnancy include one or two missed menstrual cycles followed by bleeding and acute abdominal and pelvic pain. Unless removed, the developing embryo can rupture the uterine tube, often resulting in death of the mother. ■

▶ **C H E C K P O I N T**

1. Where does fertilization normally occur?

2. How is polyspermy prevented?

3. What is a morula, and how is it formed?

4. Describe the components of a blastocyst.

5. When, where, and how does implantation occur?

Figure 29.5   Summary of events associated with the first week of development.

🔑 **Fertilization usually occurs in the uterine tube.**

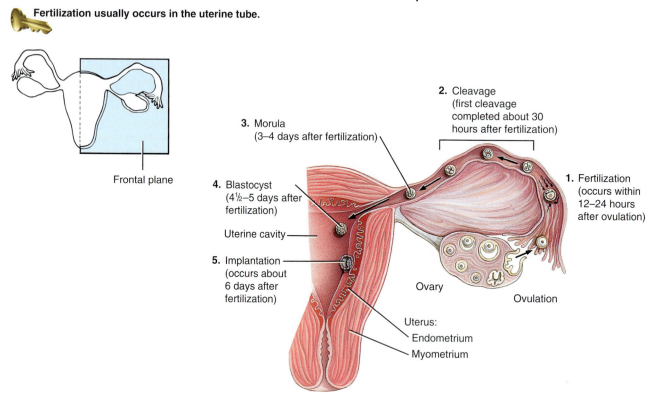

Frontal section through uterus, uterine tube, and ovary

❓ **In which phase of the uterine cycle does implantation occur?**

Second Week of Development

Development of the Trophoblast

About 8 days after fertilization, the trophoblast develops into two layers in the region of contact between the blastocyst and endometrium. These are a **syncytiotrophoblast** (sin-sīt′-ē-ō-TRŌF-ō-blast) that contains no distinct cell boundaries, and a **cytotrophoblast** (sī-tō-TRŌF-ō-blast) between the inner cell mass and syncytiotrophoblast that is composed of distinct cells (Figure 29.6a). The two layers of trophoblast become part of the chorion (one of the fetal membranes) as they undergo further growth (see Figure 29.11a inset). During implantation, the syn-cytiotrophoblast secretes enzymes that enable the blastocyst to penetrate the uterine lining by digesting and liquefying the endometrial cells. Eventually, the blastocyst becomes buried in the endometrium and inner one-third of the myometrium. Another secretion of the trophoblast is human chorionic gona-dotropin (hCG), which has actions similar to LH. Human chorionic gonadotropin rescues the corpus luteum from degener-ation and sustains its secretion of progesterone and estrogens. These hormones maintain the uterine lining in a secretory state and thereby prevent menstruation. Peak secretion of hGH occurs about the ninth week of pregnancy at which time the placenta is fully developed and produces the progesterone and estrogens that continue to sustain the pregnancy.

Development of the Bilaminar Embryonic Disc

Cells of the inner cell mass also differentiate into two layers around 8 days after fertilization: a hypoblast and epiblast (Figure 29.6a). The **hypoblast (primitive endoderm)** is a layer of columnar cells, whereas the **epiblast (primitive ectoderm)** is a layer of cuboidal cells (two-layered). Cells of the hypoblast and epiblast together form a flat disc referred to as the **bilaminar embryonic disc** (bī-LAM-in-ar = two-layered). In addition, a small cavity appears within the epiblast and eventually enlarges to form the **amniotic cavity** (am-nē-OT-ik; *amnio-* = lamb).

Development of the Amnion

As the amniotic cavity enlarges, a thin protective membrane called the **amnion** (AM-nē-on) develops from the epiblast (Figure 29.6a). Whereas the amnion forms the roof of the amniotic cavity, the epiblast forms the floor. Initially, the amnion overlies only the bilaminar embryonic disc. However, as the embryo grows, the amnion eventually surrounds the entire embryo (see Figure 29.11a inset), creating the amniotic cavity that becomes filled with **amniotic fluid.** Most amniotic fluid is initially derived from a filtrate of maternal blood. Later, the fetus contributes to the fluid by excreting urine into the amniotic cav-ity. Amniotic fluid serves as a shock absorber for the fetus, helps

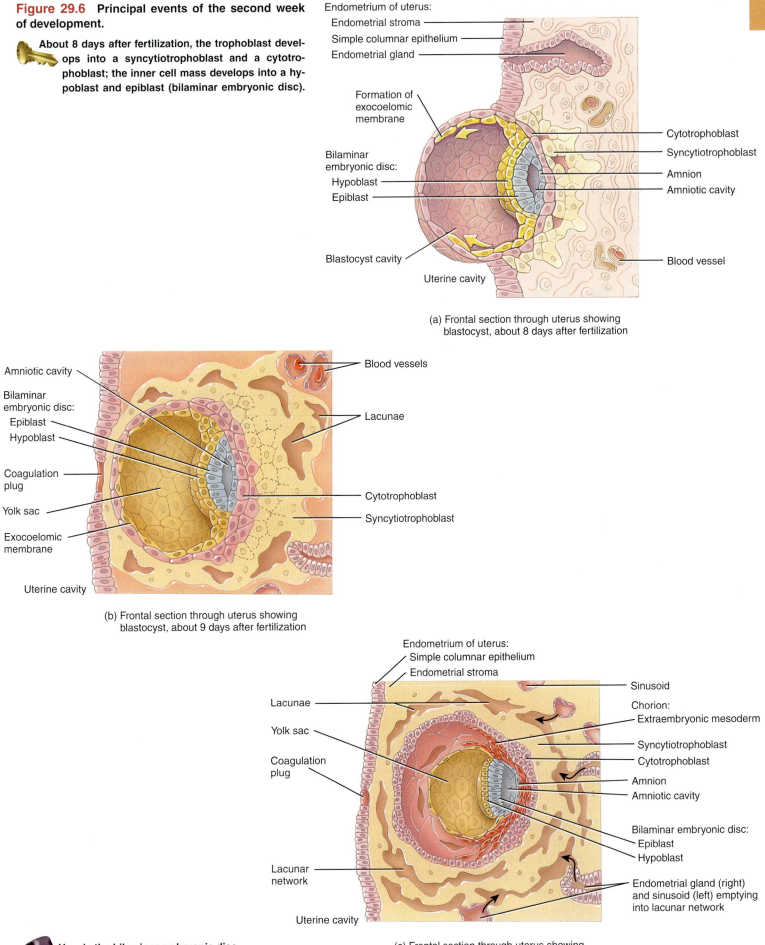

Figure 29.6 Principal events of the second week of development.

About 8 days after fertilization, the trophoblast develops into a syncytiotrophoblast and a cytotrophoblast; the inner cell mass develops into a hypoblast and epiblast (bilaminar embryonic disc).

Endometrium of uterus:
Endometrial stroma
Simple columnar epithelium
Endometrial gland

Formation of exocoelomic membrane

Bilaminar embryonic disc:
Hypoblast
Epiblast

Blastocyst cavity

Uterine cavity

Cytotrophoblast
Syncytiotrophoblast
Amnion
Amniotic cavity

Blood vessel

(a) Frontal section through uterus showing blastocyst, about 8 days after fertilization

Amniotic cavity

Bilaminar embryonic disc:
Epiblast
Hypoblast

Coagulation plug

Yolk sac

Exocoelomic membrane

Uterine cavity

Blood vessels

Lacunae

Cytotrophoblast
Syncytiotrophoblast

(b) Frontal section through uterus showing blastocyst, about 9 days after fertilization

Endometrium of uterus:
Simple columnar epithelium
Endometrial stroma

Lacunae

Yolk sac

Coagulation plug

Lacunar network

Uterine cavity

Sinusoid

Chorion:
Extraembryonic mesoderm

Syncytiotrophoblast
Cytotrophoblast

Amnion
Amniotic cavity

Bilaminar embryonic disc:
Epiblast
Hypoblast

Endometrial gland (right) and sinusoid (left) emptying into lacunar network

(c) Frontal section through uterus showing blastocyst, about 12 days after fertilization

How is the bilaminar embryonic disc connected to the trophoblast?

regulate fetal body temperature, helps prevent desiccation, and prevents adhesions between the skin of the fetus and surrounding tissues. Embryonic cells are normally sloughed off into amniotic fluid. They can be examined in a procedure called **amniocentesis** (am′-nē-ō-sen-TĒ-sis; *amnio-* = amnion; *-centesis* = puncture to remove fluid), which involves withdrawing some of the amniotic fluid that bathes the developing fetus and analyzing the fetal cells and dissolved substances (see page 1084). The amnion usually ruptures just before birth; it and its fluid constitute the "bag of waters."

Development of the Yolk Sac

Also on the eighth day after fertilization, cells at the edge of the hypoblast migrate and cover the inner surface of the blastocyst wall (Figure 29.6a). The migrating columnar cells become squamous (flat) and form a thin membrane called the **exocoelomic membrane** (ek′-sō-sē-LŌ-mik; *exo-* = outside; *-koilos* = space). On the ninth day after fertilization, the exocoelomic membrane is complete. Together with the hypoblast, the exocoelomic membrane forms the wall of the **yolk sac,** formerly called the blastocyst cavity (Figure 29.6b). As a result, the bilaminar embryonic disc is now positioned between the amniotic cavity and yolk sac. Human embryos receive their nutrients from the endometrium, so the yolk sac is relatively empty, small, and decreases in size as development progresses (see Figure 29.11a inset).

Nevertheless, the yolk sac has several important functions in humans. It transfers nutrients to the embryo during the second and third weeks while uteroplacental circulation (described shortly) is being established. The yolk sac also is the source of blood cells from the third through sixth weeks, at which time the liver takes over. Early in the third week, the yolk sac contains the first cells (primordial germ cells) that will eventually migrate into the developing gonads and there will differentiate into the primitive germ cells (spermatogonia and oogonia). Then during the fourth week, the yolk sac forms part of the gut (gastrointestinal tract). Finally, the yolk sac functions as a shock absorber and helps prevent desiccation of the embryo.

Development of Sinusoids

On the ninth day after fertilization, the blastocyst becomes completely embedded in the endometrium. The site in the endometrial epithelium where the blastocyst implants is sealed by a temporary mass of fibrin called the **coagulation plug** (Figure 29.6b). The ninth day is also when the syncytiotrophoblast expands into the endometrium and around the yolk sac. As the syncytiotrophoblast expands, small spaces called **lacunae** (la-KOO-nē = little lakes) develop within it (Figure 29.6b).

By the twelfth day of development, the lacunae fuse to form larger, interconnecting spaces called **lacunar networks** (Figure 29.6c). Endometrial capillaries around the developing embryo become dilated and are referred to as **sinusoids.** As the syncytiotrophoblast erodes some of the sinusoids and endometrial glands, maternal blood and secretions from the glands enter the lacunar networks and flow through them. Maternal blood is both a rich source of materials for embryonic nutrition and a disposal site for the embryo's wastes.

Development of the Extraembryonic Coelom

About the twelfth day after fertilization, the **extraembryonic mesoderm** develops. These mesodermal cells are derived from the yolk sac and form a connective tissue (mesenchyme) around the amnion and yolk sac (Figure 29.6c). Soon a number of large cavities develop in the extraembryonic mesoderm, which then fuse to form a single, larger cavity called the **extraembryonic coelom** (SĒ-lōm).

Development of the Chorion

The extraembryonic mesoderm, together with the two layers of the trophoblast (the cytotrophoblast and syncytiotrophoblast), constitutes the **chorion** (KOR-ē-on = membrane) (Figure 29.6c). It surrounds the embryo and, later, the fetus (see Figure 29.11a inset). Eventually the chorion becomes the principal embryonic part of the placenta, the structure for exchange of materials between mother and fetus. The chorion protects the embryo and fetus from the immune responses of the mother in two ways: (1) It secretes proteins that block antibody production by the mother. (2) It promotes the production of T lymphocytes that suppress the normal immune response in the uterus. The chorion also produces human chorionic gonadotropin (hCG), an important hormone of pregnancy (see Figure 29.16).

The inner layer of the chorion eventually fuses with the amnion. With the development of the chorion, the extraembryonic coelom is now referred to as the **chorionic cavity.** By the end of the second week of development, the bilaminar embryonic disc becomes connected to the trophoblast by a band of extraembryonic mesoderm called the **connecting (body) stalk** (see Figure 29.7, inset). The connecting stalk is the future umbilical cord.

▶ **CHECKPOINT**

6.   What are the functions of the trophoblast?

7.   How is the bilaminar embryonic disc formed?

8.   Describe the formation of the amnion, yolk sac, and chorion and explain their functions.

9.   Why are sinusoids important during embryonic development?

Third Week of Development

The third week of development begins a six-week period of very rapid embryonic development and differentiation. During the third week, the three primary germ layers are established and lay the groundwork for organ development in weeks four through eight.

Gastrulation

The first major event of the third week of development, **gastrulation** (gas′-troo-LĀ-shun), occurs about 15 days after fertilization. In this process, the two-dimensional bilaminar (two-layered) embryonic disc, consisting of epiblast and hypoblast, transforms into a two-dimensional trilaminar (three-layered) embryonic disc consisting of three primary germ layers, the ectoderm, mesoderm, and endoderm. The primary germ layers are the major embryonic tissues from which the various tissues and organs of the body develop.

Gastrulation involves the rearrangement and migration of cells from the epiblast. The first evidence of gastrulation is the formation of the **primitive streak,** a faint groove on the dorsal surface of the epiblast that elongates from the posterior to the anterior part of the embryo (Figure 29.7a). The primitive streak clearly establishes the head and tail ends of the embryo, as well

Figure 29.7 Gastrulation.

Gastrulation involves the rearrangement and migration of cells from the epiblast.

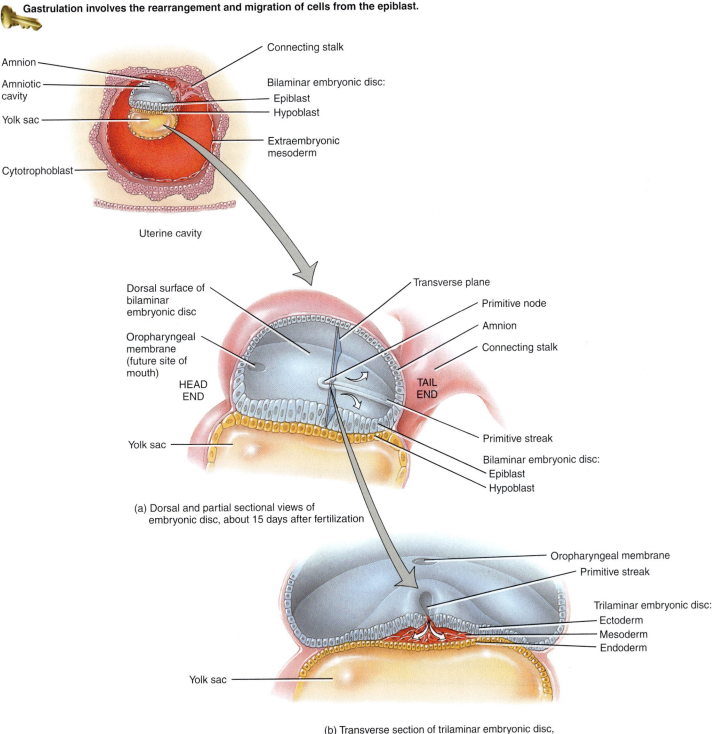

(a) Dorsal and partial sectional views of embryonic disc, about 15 days after fertilization

(b) Transverse section of trilaminar embryonic disc, about 16 days after fertilization

What is the significance of gastrulation?

as its right and left sides. At the head end of the primitive streak a small group of epiblastic cells forms a rounded structure called the **primitive node.**

Following formation of the primitive streak, cells of the epiblast move inward below the primitive streak and detach from the epiblast (Figure 29.7b). This inward movement is called **invagination** (in-vaj-i-NĀ-shun). Once the cells have invaginated, some of them displace the hypoblast, forming the **endoderm** (*endo-* = inside; *-derm* = skin). Other cells remain between the epiblast and newly formed endoderm to form the **mesoderm** (*meso-* = middle). Cells remaining in the epiblast then form the **ectoderm** (*ecto-* = outside). Whereas the ectoderm and endoderm are epithelia composed of tightly packed cells, the mesoderm is a loosely organized connective tissue (mesenchyme). As the embryo develops, the endoderm ultimately becomes the epithelial lining of the gastrointestinal tract, respiratory tract, and several other organs. The mesoderm gives rise to muscle, bone and other connective tissues, and the peritoneum. The ectoderm develops into the epidermis of the skin and the nervous system. Table 29.1 provides more details about the fates of these primary germ layers.

About 16 days after fertilization, mesodermal cells from the primitive node migrate toward the head end of the embryo and form a hollow tube of cells in the midline called the **notochordal process** (nō-tō-KOR-dal) (Figure 29.8). By days 22–24, the notochordal process becomes a solid cylinder of cells called the **notochord** (nō-tō-KORD; *noto-* = back; *-chord* = cord). This structure plays an extremely important role in **induction,** the process by which one tissue (*inducing tissue*) stimulates the development of an adjacent unspecialized tissue (*responding tissue*) into a specialized one. An inducing tissue usually produces a chemical substance that influences the responding tissue. The notochord induces certain mesodermal cells to develop into the vertebral bodies. It also forms the nucleus pulposus of the intervertebral discs (see Figure 7.24 on page 214).

During the third week of development, two faint depressions appear on the dorsal surface of the embryo. The structure closer to the head end is called the **oropharyngeal membrane** (or-ō-fa-RIN-jē-al; *oro-* = mouth; *-pharyngeal* = pertaining to the pharynx) (Figure 29.8a, b). It breaks down during the fourth week to connect the mouth cavity to the pharynx and the remainder of the gastrointestinal tract. The structure closer to the tail end is called the **cloacal membrane** (klō-Ā-kul = sewer), which degenerates in the seventh week to form the openings of the anus and urinary and reproductive tracts.

When the cloacal membrane appears, the wall of the yolk sac forms a small vascularized outpouching called the **allantois** (a-LAN-tō-is; *allant-* = sausage) that extends into the connecting stalk (Figure 29.8b). In most other mammals, the allantois is used for gas exchange and waste removal. Because of the role of the human placenta in these activities, the allantois is not a prominent structure in humans (see Figure 29.11a, inset). Nevertheless, it does function in early formation of blood and blood vessels and it is associated with the development of the urinary bladder.

Neurulation

In addition to inducing mesodermal cells to develop into vertebral bodies, the notochord also induces ectodermal cells over it to form the **neural plate** (Figure 29.9a on page 1074). (Also see Figure 14.25 on page 490.) By the end of the third week, the lateral edges of the neural plate become more elevated and form the **neural fold** (Figure 29.9b). The depressed midregion is called the **neural groove** (Figure 29.9c). Generally, the neural folds approach each other and fuse, thus converting the neural plate into a **neural tube** (Figure 29.9d). This occurs first near the middle of the embryo and then progresses toward the head and tail ends. Neural tube cells then develop into the brain and spinal cord. The process by which the neural plate, neural folds, and neural tube form is called **neurulation** (noor-oo-LĀ-shun).

Table 29.1	Structures Produced by the Three Primary Germ Layers	
Endoderm	**Mesoderm**	**Ectoderm**
Epithelial lining of gastrointestinal tract (except the oral cavity and anal canal) and the epithelium of its glands.	All skeletal and cardiac muscle tissue and most smooth muscle tissue.	All nervous tissue.
Epithelial lining of urinary bladder, gallbladder, and liver.	Cartilage, bone, and other connective tissues.	Epidermis of skin.
Epithelial lining of pharynx, auditory (Eustachian) tubes, tonsils, larynx, trachea, bronchi, and lungs.	Blood, red bone marrow, and lymphatic tissue.	Hair follicles, arrector pili muscles, nails, epithelium of skin glands (sebaceous and sudoriferous), and mammary glands.
Epithelium of thyroid gland, parathyroid glands, pancreas, and thymus.	Endothelium of blood vessels and lymphatic vessels.	Lens, cornea, and internal eye muscles.
Epithelial lining of prostate and bulbourethral (Cowper's) glands, vagina, vestibule, urethra, and associated glands such as the greater (Bartholin's) vestibular and lesser vestibular glands.	Dermis of skin.	Internal and external ear.
	Fibrous tunic and vascular tunic of eye.	Neuroepithelium of sense organs.
	Middle ear.	Epithelium of oral cavity, nasal cavity, paranasal sinuses, salivary glands, and anal canal.
	Mesothelium of ventral body cavity.	Epithelium of pineal gland, pituitary gland, and adrenal medullae.
	Epithelium of kidneys and ureters.	
	Epithelium of adrenal cortex.	
	Epithelium of gonads and genital ducts.	

Figure 29.8 Development of the notochordal process.

🔑 The notochordal process develops from the primitive node and later becomes the notochord.

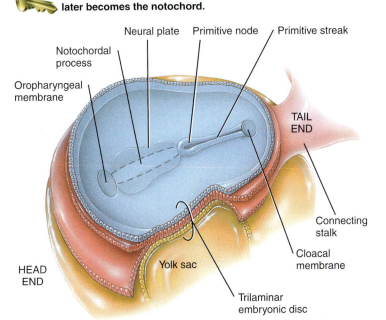

(a) Dorsal and partial sectional views of trilaminar embryonic disc, about 16 days after fertilization

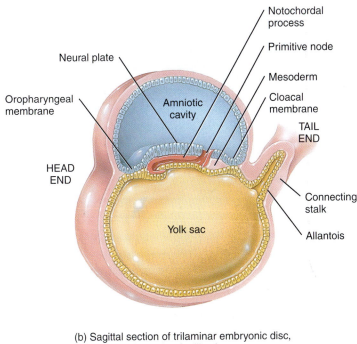

(b) Sagittal section of trilaminar embryonic disc, about 16 days after fertilization

❓ What is the significance of the notochord?

As the neural tube forms, some of the ectodermal cells from the tube migrate to form several layers of cells called the **neural crest** (see Figure 14.25b on page 490). Neural crest cells give rise to spinal and cranial nerves and their ganglia, autonomic nervous system ganglia, the meninges of the brain and spinal cord, the adrenal medullae, and several skeletal and muscular components of the head.

At about four weeks after fertilization, the head end of the neural tube develops into three enlarged areas called **primary brain vesicles** (see Figure 14.26 on page 491). These are the **prosencephalon (forebrain), mesencephalon (midbrain),** and **rhombencephalon (hindbrain).** Then, at about five weeks, the prosencephalon develops into **secondary brain vesicles** called the **telencephalon** and **diencephalon** and the rhombencephalon develops into secondary brain vesicles called the **metencephalon** and **myelencephalon.** The areas of the neural tube adjacent to the myelencephalon develop into the spinal cord. The parts of the brain that develop from the various brain vesicles are described on page 490 in Chapter 14.

Anencephaly

Neural tube defects (NTDs) are caused by arrest of the normal development and closure of the neural tube. These include spina bifida (discussed on page 214) and **anencephaly** (an′-en-SEPH-a-lē; *an-* = without; *encephal* = brain). In anencephaly, the prosencephalon (future cerebral hemispheres, thalamus, and hypothalamus) remains in contact with amniotic fluid and degenerates and the cranial bones fail to form. Usually, the brainstem also is abnormal, so vital functions such as breathing and regulation of the heart may not occur. Infants with anencephaly are stillborn or die within a few days after birth. The condition occurs about once in every 1000 births and is 2 to 4 times more common in females than males. ■

Development of Somites

By about the 17th day after fertilization, the mesoderm adjacent to the notochord and neural tube forms paired longitudinal columns of **paraxial mesoderm** (par-AK-sē-al; *para-* = near) (Figure 29.9b). The mesoderm lateral to the paraxial mesoderm forms paired cylindrical masses called **intermediate mesoderm.** The mesoderm lateral to the intermediate mesoderm consists of a pair of flattened sheets called **lateral plate mesoderm.** The paraxial mesoderm soon segments into a series of paired, cube-shaped structures called **somites** (SŌ-mīts = little bodies). By the end of the fifth week, 42–44 pairs of somites are present. The number of somites that develop over a given period can be correlated to the approximate age of the embryo.

Each somite differentiates into three regions: a **myotome,** a **dermatome,** and a **sclerotome** (see Figure 10.20b on page 301). The myotomes develop into the skeletal muscles of the neck, trunk, and limbs; the dermatomes form connective tissue, including the dermis of the skin; and the sclerotomes give rise to the vertebrae.

Figure 29.9 Neurulation and the development of somites.

Neurulation is the process by which the neural plate, neural folds, and neural tube form.

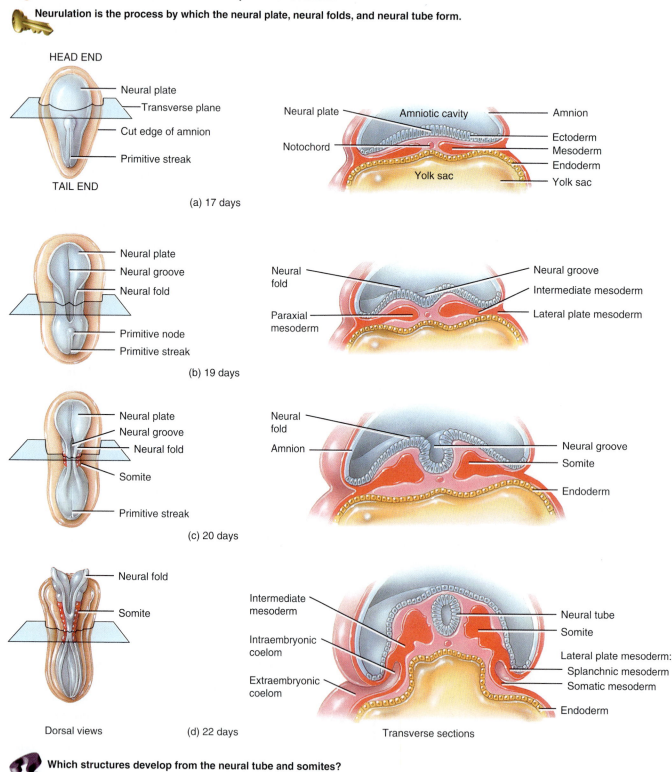

(a) 17 days

(b) 19 days

(c) 20 days

(d) 22 days

Dorsal views

Transverse sections

Which structures develop from the neural tube and somites?

Development of the Intraembryonic Coelom

In the third week of development, small spaces appear in the lateral plate mesoderm. These spaces soon merge to form a larger cavity called the **intraembryonic coelom.** This cavity splits the lateral plate mesoderm into two parts called the splanchnic mesoderm and somatic mesoderm (Figure 29.9d). During the second month of development, the intraembryonic coelom divides into the pericardial, pleural, and peritoneal cavities. **Splanchnic mesoderm** (SPLANGK-nik = visceral) forms the heart and the visceral layer of the serous pericardium, blood vessels, the smooth muscle and connective tissues of the respiratory and digestive organs, and the visceral layer of the serous mem-

The onset of labor is determined by complex interactions of several placental and fetal hormones. Because progesterone inhibits uterine contractions, labor cannot take place until its effects are diminished. Toward the end of gestation, the levels of estrogens in the mother's blood rise sharply, producing changes that overcome the inhibiting effects of progesterone. The rise in estrogens results from increasing secretion by the placenta of corticotropin-releasing hormone, which stimulates the fetal anterior pituitary gland to secrete ACTH (adrenocorticotropic hormone). In turn, ACTH stimulates the fetal adrenal gland to secrete cortisol and dehydroepiandrosterone (DHEA), the major adrenal androgen. The placenta then converts DHEA into an estrogen. High levels of estrogens cause uterine muscle fibers to display receptors for oxytocin and to form gap junctions with one another. Oxytocin released by the posterior pituitary stimulates uterine contractions, and relaxin from the placenta assists by increasing the flexibility of the pubic symphysis and helping dilate the uterine cervix. Estrogen also stimulates the placenta to release prostaglandins, which induce production of enzymes that digest collagen fibers in the cervix, causing it to soften.

Control of labor contractions during parturition occurs via a positive feedback loop (see Figure 1.4 on page 11). Contractions of the uterine myometrium force the baby's head or body into the cervix, thereby distending (stretching) the cervix. Stretch receptors in the cervix send nerve impulses to neurosecretory cells in the hypothalamus, causing them to release oxytocin into blood capillaries of the posterior pituitary gland. Oxytocin then is carried by the blood to the uterus, where it stimulates the myometrium to contract more forcefully. As the contractions intensify, the baby's body stretches the cervix still more, and the resulting nerve impulses stimulate the secretion of yet more oxytocin. With birth of the infant, the positive feedback cycle is broken because cervical distention suddenly lessens.

Uterine contractions occur in waves (quite similar to peristaltic waves) that start at the top of the uterus and move downward, eventually expelling the fetus. **True labor** begins when uterine contractions occur at regular intervals, usually producing pain. As the interval between contractions shortens, the contractions intensify. Another symptom of true labor in some women is localization of pain in the back that is intensified by walking. The reliable indicator of true labor is dilation of the cervix and the "show," a discharge of a blood-containing mucus that appears in the cervical canal during labor. In **false labor,** pain is felt in the abdomen at irregular intervals, but it does not intensify and walking does not alter it significantly. There is no "show" and no cervical dilation.

True labor can be divided into three stages (Figure 29.18):

1 *Stage of dilation.* The time from the onset of labor to the complete dilation of the cervix is the **stage of dilation.** This stage, which typically lasts 6–12 hours, features regular contractions of the uterus, usually a rupturing of the amniotic sac, and complete dilation (to 10 cm) of the cervix. If the amniotic sac does not rupture spontaneously, it is ruptured intentionally.

Figure 29.18 Stages of true labor.

The term *parturition* refers to birth.

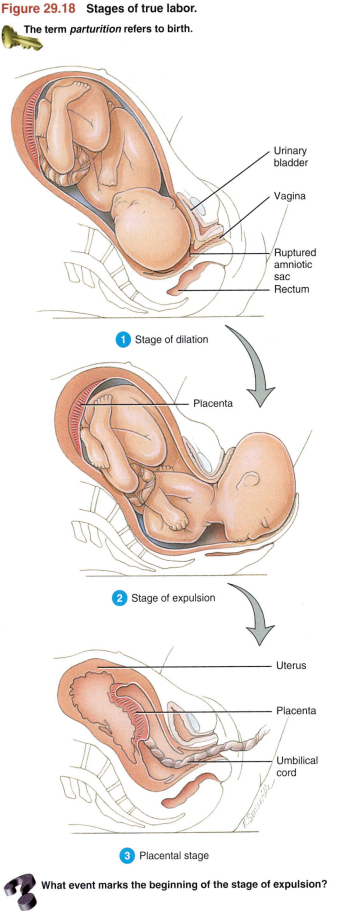

1 Stage of dilation

2 Stage of expulsion

3 Placental stage

What event marks the beginning of the stage of expulsion?

② *Stage of expulsion.* The time (10 minutes to several hours) from complete cervical dilation to delivery of the baby is the **stage of expulsion.**

③ *Placental stage.* The time (5–30 minutes or more) after delivery until the placenta or "afterbirth" is expelled by powerful uterine contractions is the **placental stage.** These contractions also constrict blood vessels that were torn during delivery, thereby reducing the likelihood of hemorrhage.

As a rule, labor lasts longer with first babies, typically about 14 hours. For women who have previously given birth, the average duration of labor is about 8 hours—although the time varies enormously among births. Because the fetus may be squeezed through the birth canal (cervix and vagina) for up to several hours, the fetus is stressed during childbirth: The fetal head is compressed, and the fetus undergoes some degree of intermittent hypoxia due to compression of the umbilical cord and the placenta during uterine contractions. In response to this stress, the fetal adrenal medullae secrete very high levels of epinephrine and norepinephrine, the "fight-or-flight" hormones. Much of the protection against the stresses of parturition, as well as preparation of the infant for surviving extrauterine life, is provided by these hormones. Among other functions, epinephrine and norepinephrine clear the lungs and alter their physiology in readiness for breathing air, mobilize readily usable nutrients for cellular metabolism, and promote an increased flow of blood to the brain and heart.

About 7% of pregnant women do not deliver by 2 weeks after their due date. Such cases carry an increased risk of brain damage to the fetus, and even fetal death, due to inadequate supplies of oxygen and nutrients from an aging placenta. Post-term deliveries may be facilitated by inducing labor, initiated by administration of oxytocin (Pitocin), or by surgical delivery (cesarean section).

Following the delivery of the baby and placenta is a 6-week period during which the maternal reproductive organs and physiology return to the prepregnancy state. This period is called the **puerperium** (pū′-er-PER-ē-um). Through a process of tissue catabolism, the uterus undergoes a remarkable reduction in size, called **involution** (in′-vō-LOO-shun), especially in lactating women. The cervix loses its elasticity and regains its prepregnancy firmness. For 2–4 weeks after delivery, women have a uterine discharge called **lochia** (LŌ-kē-a), which consists initially of blood and later of serous fluid derived from the former site of the placenta.

🩺 Dystocia and Cesarean Section

Dystocia (dis-TŌ-sē-a; *dys-* = painful or difficult; *toc-* = birth), or difficult labor, may result either from an abnormal position (presentation) of the fetus or a birth canal of inadequate size to permit vaginal delivery. In a **breech presentation,** for example, the fetal buttocks or lower limbs, rather than the head, enter the birth canal first; this occurs most often in premature births. If fetal or maternal distress prevents a vaginal birth, the baby may be delivered surgically through an abdominal incision. A low, horizontal cut is made through the abdominal wall and lower portion of the uterus, through which the baby and placenta are removed. Even though it is popularly associated with the birth of Julius Caesar, the true reason this procedure is termed a **cesarean section (C-section)** is because it was described in Roman Law, *lex cesarea,* about 600 years before Julius Caesar was born. Even a history of multiple C-sections need not exclude a pregnant woman from attempting a vaginal delivery. ■

▶ **CHECKPOINT**

27. What hormonal changes induce labor?

28. What is the difference between false labor and true labor?

29. What happens during the stage of dilation, the stage of expulsion, and the placental stage of true labor?

ADJUSTMENTS OF THE INFANT AT BIRTH

▶ **OBJECTIVE**

• **Explain the respiratory and cardiovascular adjustments that occur in an infant at birth.**

During pregnancy, the embryo (and later the fetus) is totally dependent on the mother for its existence. The mother supplies the fetus with oxygen and nutrients, eliminates its carbon dioxide and other wastes, protects it against shocks and temperature changes, and provides antibodies that confer protection against certain harmful microbes. At birth, a physiologically mature baby becomes much more self-supporting, and the newborn's body systems must make various adjustments. Next we examine some of the changes that occur in the respiratory and cardiovascular systems.

Respiratory Adjustments

The fetus depends entirely on the mother for obtaining oxygen and eliminating carbon dioxide. The fetal lungs are either collapsed or partially filled with amniotic fluid, which is absorbed at birth. The production of surfactant begins by the end of the sixth month of development. Because the respiratory system is fairly well developed at least 2 months before birth, premature babies delivered at 7 months are able to breathe and cry. After delivery, the baby's supply of oxygen from the mother ceases and carbon dioxide is no longer being removed. Circulation in the baby continues, but CO_2 builds up in the blood. A rising CO_2 level stimulates the respiratory center in the medulla oblongata, causing the respiratory muscles to contract, and the baby draws its first breath. Because the first inspiration is unusually deep, as the lungs contain no air, the baby exhales vigorously and naturally cries. A full-term baby may breathe 45 times a minute for the first 2 weeks after birth. Breathing rate gradually declines until it approaches a normal rate of 12 breaths per minute.

Cardiovascular Adjustments

After the baby's first inspiration, the cardiovascular system must make several adjustments (see Figure 21.31 on page 754). Closure of the foramen ovale between the atria of the fetal heart, which occurs at the moment of birth, diverts deoxygenated blood to the lungs for the first time. The foramen ovale is closed by two flaps of septal heart tissue that fold together and permanently fuse. The remnant of the foramen ovale is the fossa ovalis.

Once the lungs begin to function, the ductus arteriosus shuts off due to contractions of smooth muscle in its wall, and it becomes the ligamentum arteriosum. The muscle contraction is probably mediated by the polypeptide bradykinin, released from the lungs during their initial inflation. The ductus arteriosus generally does not close completely until about 3 months after birth. Prolonged incomplete closure results in a condition called **patent ductus arteriosus.**

After the umbilical cord is tied off and severed and blood no longer flows through the umbilical arteries, they fill with connective tissue, and their distal portions become the medial umbilical ligaments. The umbilical vein then becomes the ligamentum teres (round ligament) of the liver.

In the fetus, the ductus venosus connects the umbilical vein directly with the inferior vena cava, allowing blood from the placenta to bypass the fetal liver. When the umbilical cord is severed, the ductus venosus collapses, and venous blood from the viscera of the fetus flows into the hepatic portal vein to the liver and then via the hepatic vein to the inferior vena cava. The remnant of the ductus venosus becomes the ligamentum venosum.

At birth, an infant's pulse may range from 120 to 160 beats per minute and may go as high as 180 upon excitation. After birth, oxygen use increases, which stimulates an increase in the rate of red blood cell and hemoglobin production. Moreover, the white blood cell count at birth is very high—sometimes as much as 45,000 cells per microliter—but the count decreases rapidly by the seventh day.

Premature Infants

Delivery of a physiologically immature baby carries certain risks. A **premature infant** or "preemie" is generally considered a baby who weighs less than 2500 g (5.5 lb) at birth. Poor prenatal care, drug abuse, history of a previous premature delivery, and mother's age below 16 or above 35 increase the chance of premature delivery. The body of a premature infant is not yet ready to sustain some critical functions, and thus its survival is uncertain without medical intervention. The major problem after delivery of an infant under 36 weeks of gestation is respiratory distress syndrome (RDS) of the newborn due to insufficient surfactant. RDS can be eased by use of artificial surfactant and a ventilator that delivers oxygen until the lungs can operate on their own. ■

► C H E C K P O I N T

30. What is the importance of surfactant at birth and afterward?

31. What cardiovascular adjustments are made by an infant at birth?

THE PHYSIOLOGY OF LACTATION

► O B J E C T I V E

• Discuss the physiology and hormonal control of lactation.

Lactation (lak'-TĀ-shun) is the secretion and ejection of milk from the mammary glands. A principal hormone in promoting milk synthesis and secretion is **prolactin (PRL),** which is secreted from the anterior pituitary gland. Even though prolactin levels increase as the pregnancy progresses, no milk secretion occurs because progesterone inhibits the effects of prolactin. After delivery, the levels of estrogens and progesterone in the mother's blood decrease, and the inhibition is removed. The principal stimulus in maintaining prolactin secretion during lactation is the sucking action of the infant. Suckling initiates nerve impulses from stretch receptors in the nipples to the hypothalamus; the impulses decrease hypothalamic release of prolactin-inhibiting hormone (PIH) and increase release of prolactin-releasing hormone (PRH), so more prolactin is released by the anterior pituitary.

Oxytocin causes release of milk into the mammary ducts via the **milk ejection reflex** (Figure 29.19). Milk formed by the glandular cells of the breasts is stored until the baby begins active suckling. Stimulation of touch receptors in the nipple initiates sensory nerve impulses that are relayed to the hypothalamus. In response, secretion of oxytocin from the posterior pituitary increases. Carried by the bloodstream to the mammary glands, oxytocin stimulates contraction of myoepithelial (smooth-muscle-like) cells surrounding the glandular cells and ducts. The resulting compression moves the milk from the alveoli of the mammary glands into the mammary ducts, where it can be suckled. This process is termed **milk ejection (let-down).** Even though the actual ejection of milk does not occur until 30–60 seconds after nursing begins (the latent period), some milk stored in lactiferous sinuses near the nipple is available during the latent period. Stimuli other than suckling, such as hearing a baby's cry or touching the mother's genitals, also can trigger oxytocin release and milk ejection. The suckling stimulation that produces the release of oxytocin also inhibits the release of PIH; this results in increased secretion of prolactin, which maintains lactation.

During late pregnancy and the first few days after birth, the mammary glands secrete a cloudy fluid called **colostrum.** Although it is not as nutritious as milk—it contains less lactose and virtually no fat—colostrum serves adequately until the appearance of true milk on about the fourth day. Colostrum and maternal milk contain important antibodies that protect the infant during the first few months of life.

Following birth of the infant, the prolactin level starts to return to the nonpregnant level. However, each time the mother nurses the infant, nerve impulses from the nipples to the hypothalamus increase the release of PRH (and decrease the release of PIH), resulting in a tenfold increase in prolactin secretion by the anterior pituitary that lasts about an hour. Prolactin acts on the mammary glands to provide milk for the next nursing period.

Figure 29.19 The milk ejection reflex, a positive feedback cycle.

Oxytocin stimulates contraction of myoepithelial cells in the breasts, which squeezes the glandular and duct cells and causes milk ejection.

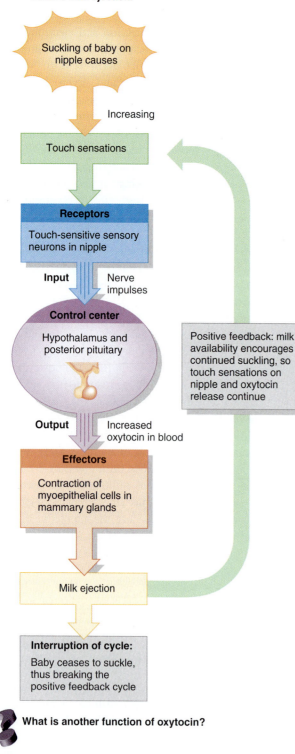

Suckling of baby on nipple causes

Increasing

Touch sensations

Receptors

Touch-sensitive sensory neurons in nipple

Input   Nerve impulses

Control center

Hypothalamus and posterior pituitary

Positive feedback: milk availability encourages continued suckling, so touch sensations on nipple and oxytocin release continue

Output   Increased oxytocin in blood

Effectors

Contraction of myoepithelial cells in mammary glands

Milk ejection

Interruption of cycle:

Baby ceases to suckle, thus breaking the positive feedback cycle

? What is another function of oxytocin?

If this surge of prolactin is blocked by injury or disease, or if nursing is discontinued, the mammary glands lose their ability to secrete milk in only a few days. Even though milk secretion normally decreases considerably within 7–9 months after birth, it can continue for several years if nursing continues.

Lactation often blocks ovarian cycles for the first few months following delivery, if the frequency of sucking is about 8–10 times a day. This effect is inconsistent, however, and ovulation commonly precedes the first menstrual period after delivery of a baby. As a result, the mother can never be certain she is not fertile. Breast-feeding is therefore not a very reliable birth control measure. The suppression of ovulation during lactation is believed to occur as follows: During breast-feeding, neural input from the nipple reaches the hypothalamus and causes it to produce neurotransmitters that suppress the release of gonadotropin-releasing hormone (GnRH). As a result, production of LH and FSH decreases, and ovulation is inhibited.

A primary benefit of **breast-feeding** is nutritional: Human milk is a sterile solution that contains amounts of fatty acids, lactose, amino acids, minerals, vitamins, and water that are ideal for the baby's digestion, brain development, and growth. Breast-feeding also benefits infants by providing the following:

- *Beneficial cells.* Several types of white blood cells are present in breast milk. Neutrophils and macrophages serve as phagocytes, ingesting microbes in the baby's gastrointestinal tract. Additionally macrophages produce lysozyme and other immune system components. Plasma cells, which develop from B lymphocytes, produce antibodies against specific microbes, and T lymphocytes kill microbes directly or help mobilize other defenses.

- *Beneficial molecules.* Breast milk also contains an abundance of beneficial molecules. Maternal IgA antibodies in breast milk bind to microbes in the baby's gastrointestinal tract and prevent their migration into other body tissues. Because a mother produces antibodies to whatever disease-causing microbes are present in her environment, her breast milk affords protection against the specific infectious agents to which her baby is also exposed. Additionally, two milk proteins bind to nutrients that many bacteria need to grow and survive, thereby making the nutrients unavailable: B_{12}-binding protein ties up vitamin B_{12}, and lactoferrin ties up iron. Some fatty acids can kill certain viruses by disrupting their membranes, and lysozyme kills bacteria by disrupting their cell walls. Finally, interferons enhance the antimicrobial activity of immune cells.

- *Decreased incidence of diseases later in life.* Breast-feeding provides children with a slight reduction in risk of lymphoma, heart disease, allergies, respiratory and gastrointestinal infections, ear infections, diarrhea, diabetes mellitus, and meningitis. Breast-feeding also protects the mother against osteoporosis and breast cancer.

- *Miscellaneous benefits.* Breast-feeding supports optimal infant growth, enhances intellectual and neurological development, and fosters mother–infant relations by establishing early and prolonged contact between them. Compared to cow's milk, the fats and iron in breast milk are more easily absorbed, the proteins in breast milk are more readily metabolized, and the lower sodium content of breast milk is more suited to an infant's needs. Premature infants benefit from breast-feeding because the milk produced by mothers of premature infants seems to be specially adapted to the infant's needs by having a higher protein content than the milk of mothers of full-term infants. Finally, a baby is less likely to have an allergic reaction to its mother's milk than to milk from another source.

Nursing and Childbirth

Years before oxytocin was discovered, it was common practice in midwifery to let a first-born twin nurse at the mother's breast to speed the birth of the second child. Now we know why this practice is helpful—it stimulates the release of oxytocin. Even after a single birth, nursing promotes expulsion of the placenta (afterbirth) and helps the uterus return to its normal size. Synthetic oxytocin (Pitocin) is often given to induce labor or to increase uterine tone and control hemorrhage just after parturition. ■

▶ CHECKPOINT

32. Which hormones contribute to lactation? What is the function of each?

33. What are the benefits of breast-feeding over bottle-feeding?

INHERITANCE

▶ OBJECTIVE

- **Define inheritance, and explain the inheritance of dominant, recessive, complex, and sex-linked traits.**

As previously indicated, the genetic material of a father and a mother unite when a sperm cell fuses with a secondary oocyte to form a zygote. Children resemble their parents because they inherit traits passed down from both parents. We now examine some of the principles involved in that process, called inheritance.

Inheritance is the passage of hereditary traits from one generation to the next. It is the process by which you acquired your characteristics from your parents and may transmit some of your traits to your children. The branch of biology that deals with inheritance is called **genetics** (je-NET-iks). The area of health care that offers advice on genetic problems (or potential problems) is called **genetic counseling.**

Genotype and Phenotype

The nuclei of all human cells except gametes contain 23 pairs of chromosomes—the diploid number (2n). One chromosome in each pair came from the mother, and the other came from the father. Each homologue—one of the two chromosomes that make up a pair—contains genes that control the same traits. If a chromosome contains a gene for body hair, for example, its homologue will also contain a gene for body hair in the same position on the chromosome. Such alternative forms of a gene

that code for the same trait and are at the same location on homologous chromosomes are called **alleles** (ah-LĒLZ). For example, one allele of a body hair gene might code for coarse hair, whereas another allele codes for fine hair. A **mutation** (mū-TĀ-shun; *muta-* = change) is a permanent heritable change in an allele that produces a different variant of the same trait.

The relationship of genes to heredity is illustrated by examining the alleles involved in a disorder called **phenylketonuria (PKU).** People with PKU (see page 925) are unable to manufacture the enzyme phenylalanine hydroxylase. The allele that codes for phenylalanine hydroxylase is symbolized as *P* whereas the mutated allele that fails to produce a functional enzyme is symbolized as *p*. The chart in Figure 29.20, which shows the possible combinations of gametes from two parents who each have one *P* and one *p* allele, is called a **Punnett square.** In constructing a Punnett square, the possible paternal alleles in sperm are written at the left side and the possible maternal alleles in ova (or secondary oocytes) are written at the top. The four spaces on the chart show how the alleles can combine in zygotes formed by the union of these sperm and ova to produce the three different genetic makeups, or **genotypes** (JĒ-nō-tīps): *PP, Pp,* or *pp.* Notice from the Punnett square that 25% of the offspring

Figure 29.20 Inheritance of phenylketonuria (PKU).

Whereas genotype refers to genetic makeup, phenotype refers to the physical or outward expression of a gene.

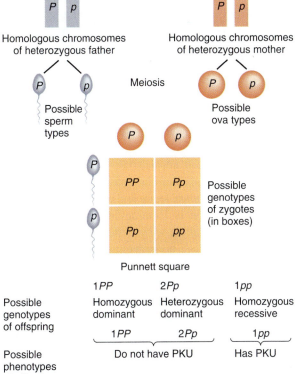

	Possible genotypes of offspring		
	1*PP*	2*Pp*	1*pp*
Possible genotypes of offspring	Homozygous dominant	Heterozygous dominant	Homozygous recessive
	1*PP*	2*Pp*	1*pp*
Possible phenotypes of offspring	Do not have PKU		Has PKU

If parents have the genotypes shown here, what is the percent chance that their first child will have PKU? What is the chance of PKU occurring in their second child?

will have the *PP* genotype, 50% will have the *Pp* genotype, and 25% will have the *pp* genotype. People who inherit *PP* or *Pp* genotypes do not have PKU; those with a *pp* genotype suffer from the disorder. Although people with a *Pp* genotype have one PKU allele (*p*), the allele that codes for the normal trait (*P*), is more dominant. An allele that dominates or masks the presence of another allele and is fully expressed (*P* in this example) is said to be a **dominant allele,** and the trait expressed is called a dominant trait. The allele whose presence is completely masked (*p* in this example) is said to be a **recessive allele,** and the trait it controls is called a recessive trait.

By tradition, the symbols for genes are written in italics, with dominant alleles written in capital letters and recessive alleles in lowercase letters. A person with the same alleles on homologous chromosomes (for example, *PP* or *pp*) is said to be **homozygous** for the trait. *PP* is homozygous dominant, and *pp* is homozygous recessive. An individual with different alleles on homologous chromosomes (for example, *Pp*) is said to be **heterozygous** for the trait.

Phenotype (FĒ-nō-tīp; *pheno-* = showing) refers to how the genetic makeup is expressed in the body; it is the physical or outward expression of a gene. A person with *Pp* (a heterozygote) has a different *genotype* from a person with *PP* (a homozygote), but both have the same *phenotype*—normal production of phenylalanine hydroxylase. Heterozygous individuals who carry a recessive gene but do not express it (*Pp*) can pass the gene on to their offspring. Such individuals are called **carriers** of the recessive gene.

Most genes give rise to the same phenotype whether they are inherited from the mother or the father. In a few cases, however, the phenotype is dramatically different, depending on the parental origin. This surprising phenomenon, first appreciated in the 1980s, is called **genomic imprinting.** In humans, the abnormalities most clearly associated with mutation of an imprinted gene are *Angelman syndrome,* which results when the gene for a particular abnormal trait is inherited from the mother, and *Prader-Willi syndrome,* which results when it is inherited from the father.

Alleles that code for normal traits do not always dominate over those that code for abnormal ones, but dominant alleles for severe disorders usually are lethal and cause death of the embryo or fetus. One exception is Huntington disease (HD), which is caused by a dominant allele whose effects do not become manifest until adulthood. Both homozygous dominant and heterozygous people exhibit the disease, whereas homozygous recessive people are normal. HD causes progressive degeneration of the nervous system and eventual death, but because symptoms typically do not appear until after age 30 or 40, many afflicted individuals have already passed the allele for the condition on to their children.

Occasionally an error in cell division, called **nondisjunction,** results in an abnormal number of chromosomes. In this situation, homologous chromosomes or sister chromatids fail to separate properly during anaphase of meiosis or mitosis. A cell that has one or more chromosomes of a set added or deleted is called an **aneuploid** (AN-ū-ployd). A monosomic cell ($2n - 1$) is missing a chromosome; a trisomic cell ($2n + 1$) has an extra chromosome. Most cases of Down syndrome (see page 1099) are aneuploid disorders in which there is trisomy of chromosome 21. Whereas nondisjunction occurs during gametogenesis (meiosis) in most instances, about 2% of Down syndrome cases result from nondisjunction during mitotic divisions in early embryonic development.

Another error in meiosis is a **translocation.** In this case, two chromosomes that are *not* homologous break and interchange portions of their chromosomes. The individual who has a translocation may be perfectly normal if no loss of genetic material took place when the rearrangement occurred. However, when the person makes gametes, some of the gametes may not contain the correct amount and type of genetic material. About 3% of Down syndrome cases result from a translocation of part of chromosome 21 to another chromosome, usually chromosome 14 or 15. The individual who has this translocation is normal and does not even know that he or she is a "carrier." When such a carrier produces gametes, however, some gametes end up with a whole chromosome 21 plus another chromosome with the translocated fragment of chromosome 21. Upon fertilization, the zygote then has three, rather than two, copies of that part of chromosome 21.

Table 29.3 lists some dominant and recessive inherited structural and functional traits in humans.

Variations on Dominant–Recessive Inheritance

Most patterns of inheritance are not simple **dominant–recessive inheritance** in which only dominant and recessive alleles interact. The phenotypic expression of a particular gene may be influenced not only by which alleles are present, but also by other genes and the environment. Moreover, most inherited traits are influenced by more than one gene, and most genes can influence

Table 29.3	Selected Hereditary Traits in Humans
Dominant	**Recessive**
Normal skin pigmentation	Albinism
Near- or farsightedness	Normal vision
PTC taster*	PTC nontaster
Polydactyly (extra digits)	Normal digits
Brachydactyly (short digits)	Normal digits
Syndactylism (webbed digits)	Normal digits
Diabetes insipidus	Normal urine excretion
Huntington disease	Normal nervous system
Widow's peak	Straight hairline
Curved (hyperextended) thumb	Straight thumb
Normal Cl⁻ transport	Cystic fibrosis
Hypercholesterolemia (familial)	Normal cholesterol level

*Ability to taste a chemical compound called phenylthiocarbamide (PTC).

more than a single trait. Next we discuss three variations on dominant–recessive inheritance.

Incomplete Dominance

In **incomplete dominance,** neither member of an allelic pair is dominant over the other, and the heterozygote has a phenotype intermediate between the homozygous dominant and the homozygous recessive phenotypes. An example of incomplete dominance in humans is the inheritance of **sickle-cell disease (SCD)** (Figure 29.21). People with the homozygous dominant genotype Hb^AHb^A form normal hemoglobin; those with the homozygous recessive genotype Hb^SHb^S have sickle-cell disease and severe anemia. Although they are usually healthy, those with the heterozygous genotype Hb^AHb^S have minor problems with anemia because half their hemoglobin is normal and half is not. Heterozygotes are carriers, and they are said to have *sickle-cell trait.*

Multiple-Allele Inheritance

Although a single individual inherits only two alleles for each gene, some genes may have more than two alternate forms, and this is the basis for **multiple-allele inheritance.** One example of multiple-allele inheritance is the inheritance of the ABO blood group. The four blood types (phenotypes) of the ABO group—A, B, AB, and O—result from the inheritance of six combinations of three different alleles of a single gene called the *I* gene: (1) allele I^A produces the A antigen, (2) allele I^B produces the B antigen, and (3) allele *i* produces neither A nor B antigen. Each person inherits two *I*-gene alleles, one from each parent, that

give rise to the various phenotypes. The six possible genotypes produce four blood types, as follows:

Genotype	Blood type (phenotype)
I^AI^A or I^Ai	A
I^BI^B or I^Bi	B
I^AI^B	AB
ii	O

Notice that both I^A and I^B are inherited as dominant traits, whereas *i* is inherited as a recessive trait. Because an individual with type AB blood has characteristics of both type A and type B red blood cells expressed in the phenotype, alleles I^A and I^B are said to be **codominant.** In other words, both genes are expressed equally in the heterozygote. Depending on the parental blood types, different offspring may have blood types different from each other. Figure 29.22 shows the blood types offspring could inherit, given the blood types of their parents.

Complex Inheritance

Most inherited traits are not controlled by one gene, but instead by the combined effects of two or more genes, a situation referred to as **polygenic inheritance** (*poly-* = many), or the combined effects of many genes and environmental factors, a situation referred to as **complex inheritance.** Examples of complex traits include skin color, hair color, eye color, height, metabolism rate, and body build. In complex inheritance, one genotype can have many possible phenotypes, depending on the environment, or one phenotype can include many possible genotypes. For

Figure 29.21 Inheritance of sickle-cell disease.

Sickle-cell disease is an example of incomplete dominance.

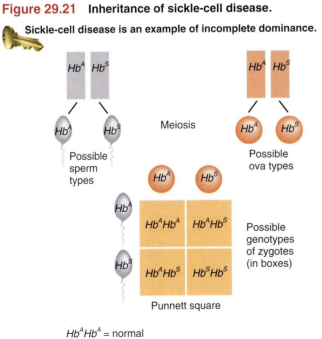

Hb^AHb^A = normal
Hb^AHb^S = carrier of sickle-cell disease
Hb^SHb^S = has sickle-cell disease

 What are the distinguishing features of incomplete dominance?

Figure 29.22 The ten possible combinations of parental ABO blood types and the blood types their offspring could inherit. For each possible set of parents, the blue letters represent the blood types their offspring could inherit.

Inheritance of ABO blood types is an example of multiple-allele inheritance.

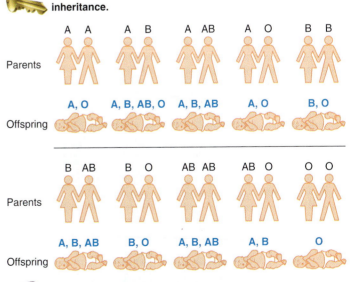

How is it possible for a baby to have type O blood if neither parent is type O?

example, even though a person inherits several genes for tallness, full height potential may not be reached due to environmental factors, such as disease or malnutrition during the growth years. Similarly, the risk of having a child with a neural tube defect is greater in pregnant women who lack adequate folic acid in their diet (an environmental factor). Because neural tube defects are more prevalent in some families than in others, however, one or more genes may also contribute.

Often, a complex trait shows a continuous gradation of small differences between extremes among individuals. Whereas it is relatively easy to predict the risk of passing on an undesirable trait that is due to a single dominant or recessive gene, it is very difficult to make this prediction when the trait is complex. Such traits are difficult to follow in a family because the range of variation is large, the number of different genes involved usually is not known, and the impact of environmental factors may be incompletely understood.

Skin color is a good example of a complex trait. It depends on environmental factors such as sun exposure and nutrition, as well as on several genes. Suppose that skin color is controlled by three separate genes, each having two alleles: *A, a; B, b;* and *C, c* (Figure 29.23). Whereas a person with the genotype *AABBCC* is very dark skinned, an individual with the genotype *aabbcc* is

very light skinned, and a person with the genotype *AaBbCc* has an intermediate skin color. Parents having an intermediate skin color may have children with very light, very dark, or intermediate skin color. Note that the **P generation** (parental generation) is the starting generation, the **F₁ generation** (first filial generation) is produced from the P generation, and the **F₂ generation** (second filial generation) is produced from the F₁ generation.

Autosomes, Sex Chromosomes, and Sex Determination

When viewed under a microscope, the 46 human chromosomes in a normal somatic cell can be identified by their size, shape, and staining pattern to be a member of 23 different pairs of chromosomes. In 22 of the pairs, the homologous chromosomes look alike and have the same appearance in both males and females; these 22 pairs are called **autosomes.** The two members of the 23rd pair are termed the **sex chromosomes;** they look different in males and females (Figure 29.24). In females, the pair consists of two chromosomes called X chromosomes. One X chromosome is also present in males, but its mate is a much smaller chromosome called a Y chromosome. The Y chromosome has only 231 genes, less than 10% of the 2968 genes present on chromosome 1, the largest chromosome.

When a spermatocyte undergoes meiosis to reduce its chromosome number, it gives rise to two sperm that contain an X chromosome and two sperm that contain a Y chromosome. Oocytes have no Y chromosomes and produce only X-containing gametes. If the secondary oocyte is fertilized by an X-bearing sperm, the offspring normally is female (XX). Fertilization by a Y-bearing sperm produces a male (XY). Thus, an individual's sex is determined by the father's chromosomes (Figure 29.25).

Both female and male embryos develop identically until about 7 weeks after fertilization. At that point, one or more genes set into motion a cascade of events that leads to the development of a male; in the absence of normal expression of the gene or genes, the female pattern of development occurs. It has been known since 1959 that the Y chromosome is needed to ini-

Figure 29.23   Polygenic inheritance of skin color.

🔑 In polygenic inheritance, a trait is controlled by the combined effects of several genes.

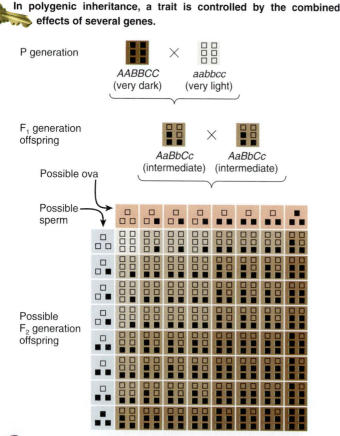

What other traits are transmitted by polygenic inheritance?

Figure 29.24   Autosomes and sex chromosomes.

🔑 Human somatic cells contain 23 different pairs of chromosomes.

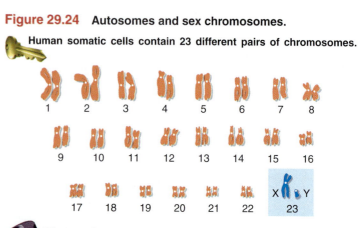

What are the two sex chromosomes in females and males?

Figure 29.25   Sex determination.

Sex is determined at the time of fertilization by the presence or absence of a Y chromosome in the sperm.

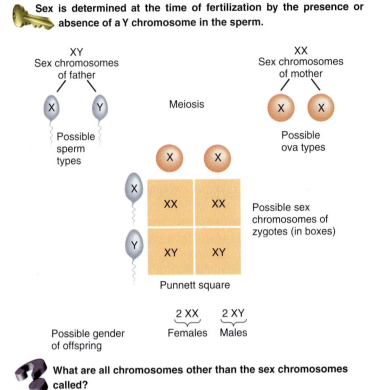

What are all chromosomes other than the sex chromosomes called?

tiate male development. Experiments published in 1991 established that the prime male-determining gene is one called **SRY (sex-determining region of the Y chromosome).** When a small DNA fragment containing this gene was inserted into 11 female mouse embryos, three of them developed as males. (The researchers suspected that the gene failed to be integrated into the genetic material in the other eight.) *SRY* acts as a molecular switch to turn on the male pattern of development. Only if the *SRY* gene is present and functional in a fertilized ovum will the fetus develop testes and differentiate into a male; in the absence of *SRY*, the fetus will develop ovaries and differentiate into a female.

Experiments of nature confirmed the key role of *SRY* in directing the male pattern of development in humans. In some cases, phenotypic females with an XY genotype were found to have mutated *SRY* genes. These individuals failed to develop normally as males because their *SRY* gene was defective. In other cases, phenotypic males with an XX genotype were found to have a small piece of the Y chromosome, including the *SRY* gene, inserted into one of their X chromosomes.

Sex-Linked Inheritance

The sex chromosomes also are responsible for the transmission of several nonsexual traits. Many of the genes for these traits are present on X chromosomes but are absent from Y chromosomes. This feature produces a pattern of heredity, termed **sex-linked inheritance,** that is different from the patterns already described.

Red–Green Color Blindness

One example of sex-linked inheritance is **red–green color blindness,** the most common type of color blindness. This condition is characterized by a deficiency in either red- or green-sensitive cones, so red and green are seen as the same color (either red or green, depending on which cone is present). The gene for red–green color blindness is a recessive one designated *c.* Normal color vision, designated *C,* dominates. The *C/c* genes are located only on the X chromosome, and thus the ability to see colors depends entirely on the X chromosomes. The possible combinations are as follows:

Genotype	Phenotype
$X^C X^C$	Normal female
$X^C X^c$	Normal female (but a carrier of the recessive gene)
$X^c X^c$	Red–green color-blind female
$X^C Y$	Normal male
$X^c Y$	Red–green color-blind male

Only females who have two X^c genes are red–green color blind. This rare situation can result only from the mating of a color-blind male and a color-blind or carrier female. (In $X^C X^c$ females the trait is masked by the normal, dominant gene.) Because males, in contrast, do not have a second X chromosome that could mask the trait, all males with an X^c gene will be red–green color blind. Figure 29.26 illustrates the inheritance of red–green color blindness in the offspring of a normal male and a carrier female.

Figure 29.26   An example of the inheritance of red–green color blindness.

Red–green color blindness and hemophilia are examples of sex-linked traits.

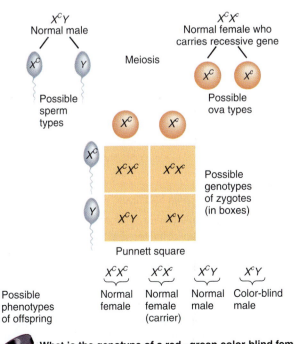

What is the genotype of a red–green color-blind female?

Traits inherited in the manner just described are called **sex-linked traits.** The most common type of **hemophilia**—a condition in which the blood fails to clot or clots very slowly after an injury—is also a sex-linked trait. Like the trait for red–green color blindness, hemophilia is caused by a recessive gene. Other sex-linked traits in humans are fragile X syndrome (described on page 1100), nonfunctional sweat glands, certain forms of diabetes, some types of deafness, uncontrollable rolling of the eyeballs, absence of central incisors, night blindness, one form of cataract, juvenile glaucoma, and juvenile muscular dystrophy.

X-Chromosome Inactivation

Whereas females have two X chromosomes in every cell (except developing oocytes), males have only one. Females thus have a double set of all genes on the X chromosome. A mechanism termed **X-chromosome inactivation (lyonization)** in effect reduces the X-chromosome genes to a single set in females. In each cell of a female's body, one X chromosome is randomly and permanently inactivated early in development, and most of the genes of the inactivated X chromosome are not expressed (transcribed and translated). The nuclei of cells in female mammals contain a dark-staining body, called a **Barr body,** that is not present in the nuclei of cells in males. Geneticist Mary Lyon correctly predicted in 1961 that the Barr body is the inactivated X chromosome. During inactivation, chemical groups that prevent transcription into RNA are added to the X chromosome's DNA. As a result, an inactivated X chromosome reacts differently to histological stains and has a different appearance than the rest of the DNA. In nondividing (interphase) cells, it remains tightly coiled and can be seen as a dark-staining body within the nucleus. In a blood smear, the Barr body of neutrophils is termed a "drumstick" because it looks like a tiny drumstick-shaped projection of the nucleus.

Teratogens

Exposure of a developing embryo or fetus to certain environmental factors can damage the developing organism or even cause death. A **teratogen** (TER-a-tō-jen; *terato-* = monster; *-gen* = creating) is any agent or influence that causes developmental defects in the embryo. In the following sections we briefly discuss several examples.

Chemicals and Drugs

Because the placenta is not an absolute barrier between the maternal and fetal circulations, any drug or chemical that is dangerous to an infant should be considered potentially dangerous to the fetus when given to the mother. Alcohol is by far the number-one fetal teratogen. Intrauterine exposure to even a small amount of alcohol may result in **fetal alcohol syndrome (FAS),** one of the most common causes of mental retardation and the most common preventable cause of birth defects in the United States. The symptoms of FAS may include slow growth

before and after birth, characteristic facial features (short palpebral fissures, a thin upper lip, and sunken nasal bridge), defective heart and other organs, malformed limbs, genital abnormalities, and central nervous system damage. Behavioral problems, such as hyperactivity, extreme nervousness, reduced ability to concentrate, and an inability to appreciate cause-and-effect relationships, are common.

Other teratogens include pesticides; defoliants (chemicals that cause plants to shed their leaves prematurely); industrial chemicals; some hormones; antibiotics; oral anticoagulants, anticonvulsants, antitumor agents, thyroid drugs, thalidomide, diethylstilbestrol (DES), and numerous other prescription drugs; LSD; and cocaine. A pregnant woman who uses cocaine, for example, subjects the fetus to higher risk of retarded growth, attention and orientation problems, hyperirritability, a tendency to stop breathing, malformed or missing organs, strokes, and seizures. The risks of spontaneous abortion, premature birth, and still-birth also increase with fetal exposure to cocaine.

Cigarette Smoking

Strong evidence implicates cigarette smoking during pregnancy as a cause of low infant birth weight and there is a strong association between smoking and a higher fetal and infant mortality rate. Women who smoke have a much higher risk of an ectopic pregnancy. Cigarette smoke may be teratogenic and may cause cardiac abnormalities and anencephaly (a developmental defect characterized by the absence of a cerebrum). Maternal smoking also is a significant factor in the development of cleft lip and palate and has been linked with sudden infant death syndrome (SIDS). Infants nursing from smoking mothers have also been found to have an increased incidence of gastrointestinal disturbances. Even a mother's exposure to secondhand cigarette smoke (breathing air containing tobacco smoke) during pregnancy or while nursing predisposes her baby to increased incidence of respiratory problems, including bronchitis and pneumonia, during the first year of life.

Irradiation

Ionizing radiation of various kinds is a potent teratogen. Exposure of pregnant mothers to x rays or radioactive isotopes during the embryo's susceptible period of development may cause microcephaly (small head size relative to the rest of the body), mental retardation, and skeletal malformations. Caution is advised, especially during the first trimester of pregnancy.

▶ **CHECKPOINT**

34. What do the terms genotype, phenotype, dominant, recessive, homozygous, and heterozygous mean?
35. What are genomic imprinting and nondisjunction?
36. Define incomplete dominance. Give an example.
37. What is multiple-allele inheritance? Give an example.
38. Define complex inheritance and give an example.
39. Why does X-chromosome inactivation occur?

DISORDERS: HOMEOSTATIC IMBALANCES

Infertility

Female infertility, or the inability to conceive, occurs in about 10% of all women of reproductive age in the United States. Female infertility may be caused by ovarian disease, obstruction of the uterine tubes, or conditions in which the uterus is not adequately prepared to receive a fertilized ovum. **Male infertility (sterility)** is an inability to fertilize a secondary oocyte; it does not imply erectile dysfunction (impotence). Male fertility requires production of adequate quantities of viable, normal sperm by the testes, unobstructed transport of sperm though the ducts, and satisfactory deposition in the vagina. The seminiferous tubules of the testes are sensitive to many factors—x rays, infections, toxins, malnutrition, and higher-than-normal scrotal temperatures—that may cause degenerative changes and produce male sterility.

One cause of infertility in females is inadequate body fat. To begin and maintain a normal reproductive cycle, a female must have a minimum amount of body fat. Even a moderate deficiency of fat—10% to 15% below normal weight for height—may delay the onset of menstruation (menarche), inhibit ovulation during the reproductive cycle, or cause amenorrhea (cessation of menstruation). Both dieting and intensive exercise may reduce body fat below the minimum amount and lead to infertility that is reversible, if weight gain or reduction of intensive exercise or both occurs. Studies of very obese women indicate that they, like very lean ones, experience problems with amenorrhea and infertility. Males also experience reproductive problems in response to undernutrition and weight loss. For example, they produce less prostatic fluid and reduced numbers of sperm having decreased motility.

Many fertility-expanding techniques now exist for assisting infertile couples to have a baby. The birth of Louise Joy Brown on July 12, 1978, near Manchester, England, was the first recorded case of **in vitro fertilization (IVF)**—fertilization in a laboratory dish. In the IVF procedure, the mother-to-be is given follicle-stimulating hormone (FSH) soon after menstruation, so that several secondary oocytes, rather than the typical single oocyte, will be produced (superovulation). When several follicles have reached the appropriate size, a small incision is made near the umbilicus, and the secondary oocytes are aspirated from the stimulated follicles and transferred to a solution containing sperm, where the oocytes undergo fertilization. Alternatively, an oocyte may be fertilized in vitro by suctioning a sperm or even a spermatid obtained from the testis into a tiny pipette and then injecting it into the oocyte's cytoplasm. This procedure, termed **intracytoplasmic sperm injection (ICSI),** has been used when infertility is due to impairments in sperm motility or to the failure of spermatids to develop into spermatozoa. When the zygote achieved by IVF reaches the 8-cell or 16-cell stage, it is introduced into the uterus for implantation and subsequent growth.

In **embryo transfer,** a man's semen is used to artificially inseminate a fertile secondary oocyte donor. After fertilization in the donor's uterine tube, the morula or blastocyst is transferred from the donor to the infertile woman, who then carries it (and subsequently the fetus) to term. Embryo transfer is indicated for women who are infertile or who do not want to pass on their own genes because they are carriers of a serious genetic disorder.

In **gamete intrafallopian transfer (GIFT)** the goal is to mimic the normal process of conception by uniting sperm and secondary oocyte in the prospective mother's uterine tubes. It is an attempt to bypass conditions in the female reproductive tract that might prevent fertilization, such as high acidity or inappropriate mucus. In this procedure, a woman is given FSH and LH to stimulate the production of several secondary oocytes, which are aspirated from the mature follicles, mixed outside the body with a solution containing sperm, and then immediately inserted into the uterine tubes.

Down Syndrome

Down syndrome (DS) is a disorder characterized by three, rather than two, copies of at least part of chromosome 21. Overall, one infant in 900 is born with Down syndrome. However, older women are more likely to have a DS baby. The chance of having a baby with this syndrome, which is less than 1 in 3000 for women under age 30, increases to 1 in 300 in the 35–39 age group and to 1 in 9 at age 48.

About 3% of cases are due to translocation of part of chromosome 21 to another chromosome. The rest are due to nondisjunction of chromosome 21. Nondisjunction most often occurs during meiosis, which causes an entire extra chromosome 21 to pass to one of the gametes. The extra chromosome typically comes from the mother, a not too surprising finding given that all her oocytes began meiosis when she herself was a fetus. They may have been exposed to chromosome-damaging chemicals and radiation for years. Undoubtedly, the kinetochore microtubules responsible for pulling sister chromatids to opposite poles of the cell (see Figure 28.1 on page 1013) sustain increasing damage with the passage of time. (Sperm, by contrast, usually are less than 10 weeks old when they fertilize a secondary oocyte.) In 1–2% of cases, nondisjunction of chromosome 21 occurs during an early mitotic division of embryonic development. The resulting individual is a *mosaic*, an organism composed of two genetically different types of cells. Some body cells have 46 chromosomes and other body cells have 47.

Down syndrome is characterized by mental retardation, retarded physical development (short stature and stubby fingers), distinctive facial structures (large tongue, flat profile, broad skull, slanting eyes, and round head), kidney defects, suppressed immune system, and malformations of the heart, ears, hands, and feet. Sexual maturity is rarely attained.

Trinucleotide Repeat Diseases

Repeating triplets of nucleotides in certain genes are known to cause more than a dozen different **trinucleotide repeat diseases.** In each case, a specific sequence of three DNA nucleotides, such as CAG (cytosine-adenine-guanine), that normally repeats several times within a gene becomes greatly expanded during gametogenesis. Sometimes, the number of repeats expands with each succeeding generation, causing a more severe disease that begins at an earlier age.

One trinucleotide repeat disease is **Huntington disease (HD).** Normal people have an average of 18 CAG repeats in each of their two *HD* genes, which codes for a protein named **huntingtin.** Because CAG codes for the amino acid glutamine, huntingtin normally has about 18 glutamines one after the other in a specific part of the protein. When the number of CAG repeats exceeds 35 to 40, the extra glutamines distort the shape of huntingtin, and Huntington disease slowly develops. The normal function of the huntingtin protein is not known, but the abnormal form appears to have a toxic effect on neurons. Because it is a dominant genetic disease, one mutant allele of the *HD* gene is sufficient to cause Huntington disease.

Another trinucleotide repeat disease is **fragile X syndrome,** in which the repeated trinucleotide is in a gene on the X chromosome. When the number of repeats exceeds 230, problems occur with both transcription of the gene and translation of the m-RNA. Effectively, the gene is inactive. Fragile X syndrome is second only to Down syndrome as a cause of inherited mental impairment. It causes learning difficulties, mental retardation, and physical abnormalities such as oversized ears, an elongated forehead, and enlarged testes. The syndrome may also be involved in autism, in which the individual exhibits extreme withdrawal and refusal to communicate. Because males have just one X chromosome and because the affected allele is recessive, fragile X syndrome occurs more often in males than in females. Both Huntington disease and fragile X syndrome can be diagnosed prenatally by measuring the number of repeats present in cells obtained via amniocentesis.

MEDICAL TERMINOLOGY

Breech presentation A malpresentation in which the fetal buttocks or lower limbs present into the maternal pelvis; the most common cause is prematurity.

Emesis gravidarum (EM-e-sis gra-VID-ar-um; *emeo* = to vomit; *gravida* = a pregnant woman) Episodes of nausea and possibly vomiting that are most likely to occur in the morning during the early weeks of pregnancy; also called **morning sickness.** Its cause is unknown, but the high levels of human chorionic gonadotropin (hCG) secreted by the placenta, and of progesterone secreted by the ovaries, have been implicated. In some women the severity of these symptoms requires hospitalization for intravenous feeding, and the condition is then known as **hyperemesis gravidarum.**

Karyotype (KAR-ē-ō-tīp; *karyo-* = nucleus) The chromosomal characteristics of an individual presented as a systematic arrangement of pairs of metaphase chromosomes arrayed in descending order of size and according to the position of the centromere; useful in judging whether chromosomes are normal in number and structure.

Klinefelter's syndrome A sex chromosome aneuploidy, usually due to trisomy XXY, that occurs once in every 500 births. Such individuals are somewhat mentally disadvantaged, sterile males with undeveloped testes, scant body hair, and enlarged breasts.

Lethal gene (LĒ-thal jēn; *lethum* = death) A gene that, when expressed, results in death either in the embryonic state or shortly after birth.

Metafemale syndrome A sex chromosome aneuploidy characterized by at least three X chromosomes (XXX) that occurs about once in every 700 births. These females have underdeveloped genital organs and limited fertility. Generally, they are mentally retarded.

Puerperal fever (pū-ER-per-al; *puer* = child) An infectious disease of childbirth, also called puerperal sepsis and childbed fever. The disease, which results from an infection originating in the birth canal, affects the endometrium. It may spread to other pelvic structures and lead to septicemia.

Turner's syndrome A sex chromosome aneuploidy caused by the presence of a single X chromosome (designated XO); occurs about once in every 5000 births and produces a sterile female with virtually no ovaries and limited development of secondary sex characteristics. Other features include short stature, webbed neck, underdeveloped breasts, and widely spaced nipples. Intelligence usually is normal.

STUDY OUTLINE

EMBRYONIC PERIOD (p. 1063)

1. Pregnancy is a sequence of events that begins with fertilization, proceeds to implantation, embryonic development, and fetal development. It normally ends in birth.
2. During fertilization a sperm cell penetrates a secondary oocyte and their pronuclei unite. Penetration of the zona pellucida is facilitated by enzymes in the sperm's acrosome. The resulting cell is a zygote.
3. Normally, only one sperm cell fertilizes a secondary oocyte because both fast and slow blocks to polyspermy exist.
4. Early rapid cell division of a zygote is called cleavage, and the cells produced by cleavage are called blastomeres. The solid sphere of cells produced by cleavage is a morula.
5. The morula develops into a blastocyst, a hollow ball of cells differentiated into a trophoblast and an inner cell mass.
6. The attachment of a blastocyst to the endometrium is termed implantation; it occurs as a result of enzymatic degradation of the endometrium.
7. After implantation, the endometrium becomes modified and is known as the decidua.
8. The trophoblast develops into the syncytiotrophoblast and cytotrophoblast, both of which become part of the chorion.
9. The inner cell mass differentiates into hypoblast and epiblast, the bilaminar embryonic disc.
10. The amnion is a thin protective membrane that develops from the cytotrophoblast.
11. The exocoelomic membrane and hypoblast form the yolk sac, which transfers nutrients to the embryo, forms blood cells, produces primordial germ cells, and forms part of the gut.
12. Erosion of sinusoids and endometrial glands provides blood and secretions, which enter lacunar networks to supply nutrition to and remove wastes from the embryo.
13. The extraembryonic coelom forms within extraembryonic mesoderm.
14. The extraembryonic mesoderm and trophoblast form the chorion, the principal embryonic part of the placenta.
15. The third week of development is characterized by gastrulation, the conversion of the bilaminar (two-layered) disc into a trilaminar

(three-layered) embryo consisting of ectoderm, mesoderm, and endoderm.

16. The first evidence of gastrulation is formation of the primitive streak and then the primitive node, notochordal process, and notochord.

17. The three primary germ layers form all tissues and organs of the developing organism. Table 29–1 on page 1072 summarizes the structures that develop from the primary germ layers.

18. Also during the third week, the oropharyngeal and cloacal membranes form. The wall of the yolk sac forms a small vascularized outpouching called the allantois, which functions in blood formation and development of the urinary bladder.

19. The process by which the neural plate, neural folds, and neural tube form is called neurulation. The brain and spinal cord develop from the neural tube.

20. Paraxial mesoderm segments to form somites from which skeletal muscles of the neck, trunk, and limbs develop. Somites also form connective tissues and vertebrae.

21. Blood vessel formation, called angiogenesis, begins in mesodermal cells called angioblasts.

22. The heart forms from mesodermal cells called the cardiogenic area. By the end of the third week, the primitive heart beats and circulates blood.

23. Chorionic villi, projections of the chorion, connect to the embryonic heart so that maternal and fetal blood vessels are brought into close proximity. Thus, nutrients and wastes are exchanged between maternal and fetal blood.

24. Placentation refers to formation of the placenta, the site of exchange of nutrients and wastes between the mother and fetus. The placenta also functions as a protective barrier, stores nutrients, and produces several hormones to maintain pregnancy.

25. The actual connection between the placenta and embryo (and later the fetus) is the umbilical cord.

26. Organogenesis refers to the formation of body organs and systems and occurs during the fourth week of development.

27. Conversion of the flat, two-dimensional trilaminar embryonic disc to a three-dimensional cylinder occurs by a process called embryonic folding.

28. Embryo folding brings various organs into their final adult positions and helps form the gastrointestinal tract.

29. Pharyngeal arches, clefts, and pouches give rise to the structures of the head and neck.

30. By the end of the fourth week, upper and lower limb buds develop and by the end of the eighth week the embryo has clearly human features.

FETAL PERIOD (p. 1081)

1. The fetal period is primarily concerned with the growth and differentiation of tissues and organs that developed during the embryonic period.

2. The rate of body growth is remarkable, especially during the ninth and sixteenth weeks.

3. The principal changes associated with embryonic and fetal growth are summarized in Table 29.2 on pages 1081–1082.

PRENATAL DIAGNOSTIC TESTS (p. 1084)

1. Several prenatal diagnostic tests are used to detect genetic disorders and to assess fetal well-being. These include fetal ultrasonography, in which an image of a fetus is displayed on a screen; amniocentesis, the withdrawal and analysis of amniotic fluid and the fetal cells within it; chorionic villi sampling (CVS), which involves withdrawal of chorionic villi tissue for chromosomal analysis.

2. CVS can be done earlier than amniocentesis, and the results are available more quickly, but it is also slightly riskier than amniocentesis.

3. Noninvasive prenatal tests include the maternal alpha-fetoprotein (AFP) test to detect neural tube defects and the Quad AFP Plus test to detect Down syndrome, trisomy 18, and neural tube defects.

MATERNAL CHANGES DURING PREGNANCY (p. 1085)

1. Pregnancy is maintained by human chorionic gonadotropin (hCG), estrogens, and progesterone.

2. Human chorionic somatomammotropin (hCS) contributes to breast development, protein anabolism, and catabolism of glucose and fatty acids.

3. Relaxin increases flexibility of the pubic symphysis and helps dilate the uterine cervix near the end of pregnancy.

4. Corticotropin-releasing hormone, produced by the placenta, is thought to establish the timing of birth, and stimulates the secretion of cortisol by the fetal adrenal gland.

5. During pregnancy, several anatomical and physiological changes occur in the mother.

EXERCISE AND PREGNANCY (p. 1088)

1. During pregnancy, some joints become less stable, and certain maneuvers are more difficult to execute.

2. Moderate physical activity does not endanger the fetus in a normal pregnancy.

LABOR (p. 1088)

1. Labor is the process by which the fetus is expelled from the uterus through the vagina to the outside. True labor involves dilation of the cervix, expulsion of the fetus, and delivery of the placenta.

2. Oxytocin stimulates uterine contractions via a positive feedback cycle.

ADJUSTMENTS OF THE INFANT AT BIRTH (p. 1090)

1. The fetus depends on the mother for oxygen and nutrients, the removal of wastes, and protection.

2. Following birth, an infant's respiratory and cardiovascular systems undergo changes in adjusting to becoming self-supporting during postnatal life.

THE PHYSIOLOGY OF LACTATION (p. 1091)

1. Lactation refers to the production and ejection of milk by the mammary glands.

2. Milk production is influenced by prolactin (PRL), estrogens, and progesterone.

3. Milk ejection is stimulated by oxytocin.

4. A few of the many benefits of breast-feeding include ideal nutrition for the infant, protection from disease, and decreased likelihood of developing allergies.

INHERITANCE (p. 1093)

1. Inheritance is the passage of hereditary traits from one generation to the next.

2. The genetic makeup of an organism is called its genotype; the traits expressed are called its phenotype.

3. Dominant genes control a particular trait; expression of recessive genes is masked by dominant genes.

4. Many patterns of inheritance do not conform to the simple dominant–recessive patterns.

5. In incomplete dominance, neither member of an allelic pair dominates; phenotypically, the heterozygote is intermediate between the homozygous dominant and the homozygous recessive. An example is sickle-cell disease.

6. In multiple-allele inheritance, genes have more than two alternate forms. An example is the inheritance of ABO blood groups.

7. In complex inheritance, a trait such as skin or eye color is controlled by the combined effects of two or more genes and may be influenced by environmental factors.

8. Each somatic cell has 46 chromosomes—22 pairs of autosomes and 1 pair of sex chromosomes.

9. In females, the sex chromosomes are two X chromosomes; in males, they are one X chromosome and a much smaller Y chromosome, which normally includes the prime male-determining gene, called *SRY.*

10. If the *SRY* gene is present and functional in a fertilized ovum, the fetus will develop testes and differentiate into a male. In the absence of *SRY,* the fetus will develop ovaries and differentiate into a female.

11. Red–green color blindness and hemophilia result from recessive genes located on the X chromosome. They are sex-linked traits that occur primarily in males because of the absence of any counterbalancing dominant genes on the Y chromosome.

12. A mechanism termed X-chromosome inactivation (lyonization) balances the difference in number of X chromosomes between males (one X) and females (two Xs). In each cell of a female's body, one X chromosome is randomly and permanently inactivated early in development and becomes a Barr body.

13. A given phenotype is the result of the interactions of genotype and the environment.

14. Teratogens, which are agents that cause physical defects in developing embryos, include chemicals and drugs, alcohol, nicotine, and ionizing radiation.

Q SELF-QUIZ QUESTIONS

Fill in the blanks in the following statements.

1. The three stages of true labor, in order of occurrence, are _____, _____, and _____.

2. Hormones produced by the _____, are responsible for maintaining the pregnancy during the first 3–4 months.

3. Indicate the germ layers responsible for development of the following structures: (a) muscle, bone, and peritoneum: _____; (b) nervous system and epidermis: _____; (c) epithelial linings of respiratory and gastrointestinal tracts: _____.

Indicate whether the following statements are true or false.

4. The principal stimulus in maintaining prolactin secretion during lactation is the sucking action of the infant.

Choose the one best answer to the following questions.

5. The group of cells that secrete enzymes that enable the blastocyst to penetrate the uterine lining is the (a) syncytiotrophoblast, (b) cytotrophoblast, (c) corona radiata, (d) zona pellucida, (e) trophoblast.

6. Which of the following statements are correct? (1) The embryonic period is the first four months of development. (2) The embryonic disc differentiates into the three primary germ layers. (3) The three primary germ layers are the endoderm, mesoderm, and ectoderm. (4) The embryonic membranes are the yolk sac, amnion, chorion, and allantois. (5) Placental development is accomplished during the fourth month of pregnancy. (a) 1, 2, and 3, (b) 2, 3, and 4, (c) 3, 4, and 5, (d) 1, 4, and 5, (e) 2, 3, and 5.

7. The hormone of pregnancy that mimics LH and acts to rescue the corpus luteum from degeneration and to stimulate its continued production of progesterone and estrogens is (a) relaxin, (b) human chorionic somatomammotropin, (c) human placental lactogen, (d) corticotropin-releasing hormone, (e) human chorionic gonadotropin.

8. Match the following:
____ (a) the penetration of a secondary oocyte by a single sperm cell
____ (b) fertilization of a secondary oocyte by more than one sperm
____ (c) the attachment of a blastocyst to the endometrium
____ (d) the fusion of the genetic material from a haploid sperm and a haploid secondary oocyte into a single diploid nucleus
____ (e) the induction by the female reproductive tract of functional changes in sperm that allow them to fertilize a secondary oocyte
____ (f) the examination of embryonic cells sloughed off into the amniotic fluid
____ (g) an abnormal condition of pregnancy characterized by sudden hypertension, large amounts of protein in urine, and generalized edema
____ (h) the process of giving birth
____ (i) the period of time (about 6 weeks) during which the maternal reproductive organs and physiology return to the pre-pregnancy state

(1) fertilization
(2) capacitation
(3) syngamy
(4) polyspermy
(5) implantation
(6) amniocentesis
(7) preeclampsia
(8) parturition
(9) puerperium

9. Which of the following are true concerning fertilization? (1) The sperm first penetrate the zona pellucida and then the corona radiata. (2) The binding of specific membrane proteins in the sperm head to ZP3 causes the release of acrosomal contents. (3) Sperm are able to fertilize the oocyte within minutes after ejaculation. (4) Depolarization of the cell membrane of the secondary oocyte inhibits fertilization by more than one sperm. (5) The oocyte completes meiosis II after fertilization. (a) 1, 2, 4, and 5, (b) 1, 3, and 5, (c) 1, 2, 3, and 4, (d) 1, 4, and 5, (e) 2, 4, and 5.

10. Amniotic fluid (1) is derived entirely from a filtrate of maternal blood, (2) acts as a fetal shock absorber, (3) provides nutrients to the fetus, (4) helps regulate fetal body temperature, (5) prevents adhesions between the skin of the fetus and surrounding tissues. (a) 1, 2, 3, 4, and 5, (b) 2, 4, and 5, (c) 2, 3, 4, and 5, (d) 1, 4, and 5, (e) 1, 2, 4, and 5.

11. Match the following:
_____ (a) the control of inherited traits by the combined effects of many genes
_____ (b) the two alternative forms of a gene that code for the same trait and are at the same location on homologous chromosomes
_____ (c) inheritance based on genes that have more than two alternate forms; an example is the inheritance of blood type
_____ (d) a cell in which one or more chromosomes of a set is added or deleted
_____ (e) refers to an individual with different alleles on homologous chromosomes
_____ (f) traits linked to the X chromosome
_____ (g) neither member of the allelic pair is dominant over the other, and the heterozygote has a phenotype intermediate between the homozygous dominant and homozygous recessive
_____ (h) refers to how the genetic makeup is expressed in the body
_____ (i) a homozygous dominant, homozygous recessive, or heterozygous genetic makeup; the actual gene arrangement
_____ (j) refers to a person with the same alleles on homologous chromosomes
_____ (k) individuals who possess a recessive gene (but do not express it) and can pass the gene on to their offspring
_____ (l) an allele that masks the presence of another allele and is fully expressed

(1) genotype
(2) phenotype
(3) alleles
(4) aneuploid
(5) incomplete dominance
(6) multiple-allele inheritance
(7) polygenic inheritance
(8) sex-linked inheritance
(9) homozygous
(10) heterozygous
(11) carriers
(12) dominant trait

12. Which of the following statements is correct? (a) Normal traits always dominate over abnormal traits. (b) Occasionally an error in meiosis called nondisjunction results in an abnormal number of chromosomes. (c) The mother always determines the sex of the child because she has either an X or a Y gene in her oocytes. (d) Most patterns of inheritance are simple dominant-recessive inheritances. (e) Genes are expressed normally regardless of any outside influence such as chemicals or radiation.

13. Match the following:
_____ (a) the embryonic membrane that entirely surrounds the embryo
_____ (b) functions as an early site of blood formation; contains cells that migrate into the gonads and differentiate into the primitive germ cells
_____ (c) becomes the principal part of the placenta; produces human chorionic gonadotropin
_____ (d) modified endometrium after implantation has occurred; separates from the endometrium after the fetus is delivered
_____ (e) contains the vascular connections between mother and fetus
_____ (f) allows oxygen and nutrients to diffuse from maternal blood into fetal blood
_____ (g) serves as an early site of blood vessel formation; its blood vessels serve as the umbilical connection in the placenta between mother and fetus
_____ (h) finger-like projections of the chorion that are bathed in maternal blood sinuses, thereby bringing maternal and fetal blood vessels into close proximity

(1) decidua
(2) placenta
(3) amnion
(4) chorion
(5) allantois
(6) yolk sac
(7) chorionic villi
(8) umbilical cord

14. Match the following:
_____ (a) a hollow sphere of cells that enters the uterine cavity
_____ (b) cells produced by cleavage
_____ (c) the developing individual from week nine of pregnancy until birth
_____ (d) the outer covering of cells of the blastocyst
_____ (e) membrane derived from trophoblast
_____ (f) early divisions of the zygote
_____ (g) a solid sphere of cells still surrounded by the zona pellucida
_____ (h) event in which differentiation into the three primary germ layers occurs

(1) cleavage
(2) blastomeres
(3) morula
(4) trophoblast
(5) blastocyst
(6) gastrulation
(7) chorion
(8) fetus

15. Which of the following are maternal changes that occur during pregnancy? (1) altered pulmonary function, (2) increased stroke volume, cardiac output, and heart rate, and decreased blood volume, (3) weight gain, (4) increased gastric motility, a delay in gastric emptying time, (5) edema and possible varicose veins. (a) 1, 2, 3, and 4, (b) 2, 3, 4, and 5, (c) 1, 3, 4, and 5, (d) 1, 3, and 5, (e) 2, 4, and 5.

CRITICAL THINKING QUESTIONS

1. The science class was studying genetics at school and Kendra came home in tears. She told her older sister, "We were doing our family tree and when I filled in our traits, I discovered that Mom and Dad can't be my real parents 'cause the traits don't match!" It turns out that Mom and Dad are PTC tasters but Kendra is a non-taster. Could Mom and Dad still be Kendra's parents?

   HINT   *Everyone says that Kendra is the "spitting image" of her mother.*

2. When pregnant Phoebe learned that Huntington disease (HD) was present in her husband's family, she said, "We won't have to worry about it because no one in my family has ever had it." Explain the inheritance of HD more clearly.

   HINT   *HD is a trinucleotide repeat disease that often worsens from one generation to the next.*

3. Usually, when you think about arteries, you think about oxygenated blood, but this textbook states that the umbilical arteries carry deoxygenated blood in the fetus. Could the authors, Tortora and Grabowski, be wrong about this? Please explain.

   HINT   *Think about the pulmonary arteries in an adult.*

ANSWERS TO FIGURE QUESTIONS

29.1 Capacitation entails the functional changes in sperm after they have been deposited in the female reproductive tract that enable them to fertilize a secondary oocyte.

29.2 A morula is a solid ball of cells, whereas a blastocyst consists of a rim of cells (trophoblast) surrounding a cavity (blastocele) and an inner cell mass.

29.3 The blastocyst secretes digestive enzymes that eat away the endometrial lining at the site of implantation.

29.4 The decidua basalis helps form the maternal part of the placenta.

29.5 Implantation occurs during the secretory phase of the uterine cycle.

29.6 The bilaminar embryonic disc is attached to the trophoblast by the connecting stalk.

29.7 Gastrulation converts a two-dimensional bilaminar embryonic disc into a two-dimensional trilaminar embryonic disc.

29.8 The notochord induces mesodermal cells to develop into vertebral bodies and forms the nucleus pulposus of intervertebral discs.

29.9 The neural tube forms the brain and spinal cord; somites develop into skeletal muscles, connective tissue, and the vertebrae.

29.10 Chorionic villi help to bring the fetal and maternal blood vessels into close proximity.

29.11 The placenta functions in exchange of materials between fetus and mother, as a protective barrier against many microbes, and stores nutrients.

29.12 Because of embryo folding, the embryo curves into a C-shape, various organs are brought into their eventual adult positions, and the primitive gut is formed.

29.13 They give rise to structures of the head and neck.

29.14 During this time, fetal weight quadruples.

29.15 Amniocentesis is used mainly to detect genetic disorders, but it also provides information concerning the maturity (and survivability) of the fetus.

29.16 Early pregnancy tests detect human chorionic gonadotropin (hCG).

29.17 Relaxin increases the flexibility of the pubic symphysis and helps dilate the cervix of the uterus to ease delivery.

29.18 Complete dilation of the cervix marks the onset of the stage of expulsion.

29.19 Oxytocin also stimulates contraction of the uterus during delivery of a baby.

29.20 The odds that a child will have PKU are the same for each child—25%.

29.21 In incomplete dominance, neither member of an allelic pair is dominant; the heterozygote has a phenotype intermediate between the homozygous dominant and the homozygous recessive phenotypes.

29.22 Yes, a baby can have blood type O if each parent has one i allele and passes it on to the offspring.

29.23 Hair color, height, and body build, among others, are polygenic traits.

29.24 The female sex chromosomes are XX, and the male sex chromosomes are XY.

29.25 The chromosomes that are not sex chromosomes are called autosomes.

29.26 A red–green color-blind female has an $X^c X^c$ genotype.

Appendix A
Measurements

U.S. Customary System

Parameter	Unit	Relation to Other U.S. Units	SI (Metric) Equivalent
Length	inch	1/12 foot	2.54 centimeters
foot	12 inches	0.305 meter	
yard	36 inches	9.144 meters	
mile	5,280 feet	1.609 kilometers	
Mass	grain	1/1000 pound	64.799 milligrams
dram	1/16 ounce	1.772 grams	
ounce	16 drams	28.350 grams	
pound	16 ounces	453.6 grams	
ton	2,000 pounds	907.18 kilograms	
Volume (Liquid)	ounce	1/16 pint	29.574 milliliters
pint	16 ounces	0.473 liter	
quart	2 pints	0.946 liter	
gallon	4 quarts	3.785 liters	
Volume (Dry)	pint	1/2 quart	0.551 liter
quart	2 pints	1.101 liters	
peck	8 quarts	8.810 liters	
bushel	4 pecks	35.239 liters	

International System (SI)

BASE UNITS

Unit	Quantity	Symbol
meter	length	M
kilogram	mass	Kg
second	time	S
liter	volume	L
mole	amount of matter	Mol

PREFIXES

Prefix	Multiplier	Symbol
tera-	$10^{12} = 1,000,000,000,000$	T
giga-	$10^{9} = 1,000,000,000$	G
mega-	$10^{6} = 1,000,000$	M
kilo-	$10^{3} = 1,000$	k
hecto-	$10^{2} = 100$	h
deca-	$10^{1} = 10$	da
deci-	$10^{-1} = 0.1$	d
centi-	$10^{-2} = 0.01$	c
milli-	$10^{-3} = 0.001$	m
micro-	$10^{-6} = 0.000,001$	μ
nano-	$10^{-9} = 0.000,000,001$	n
pico-	$10^{-12} = 0.000,000,000,001$	p

Temperature Conversion

Fahrenheit (F) To Celsius (C)

$$°C = (°F - 32) \div 1.8$$

Celsius (C) To Fahrenheit (F)

$$°F = (°C \times 1.8) + 32$$

U.S TO SI (Metric) Conversion

When you know	Multiply by	To find
inches	2.54	centimeters
feet	30.48	centimeters
yards	0.91	meters
miles	1.61	kilometers
ounces	28.35	grams
pounds	0.45	kilograms
tons	0.91	metric tons
fluid ounces	29.57	milliliters
pints	0.47	liters
quarts	0.95	liters
gallons	3.79	liters

SI (Metric) TO U.S. Conversion

When you know	Multiply by	To find
millimeters	0.04	inches
centimeters	0.39	inches
meters	3.28	feet
kilometers	0.62	miles
liters	1.06	quarts
cubic meters	35.32	cubic feet
grams	0.035	ounces
kilograms	2.21	pounds

Appendix B
Periodic Table

The periodic table lists the known **chemical elements,** the basic units of matter. The elements in the table are arranged left-to-right in rows in order of their **atomic number,** the number of protons in the nucleus. Each horizontal row, numbered from 1 to 7, is a **period.** All elements in a given period have the same number of electron shells as their period number. For example, an atom of hydrogen or helium each has one electron shell, while an atom of potassium or calcium each has four electron shells. The elements in each column, or **group,** share chemical properties. For example, the elements in column IA are very chemically reactive, whereas the elements in column VIIIA have full electron shells and thus are chemically inert.

Scientists now recognize 113 different elements; 92 occur naturally on Earth, and the rest are produced from the natural elements using particle accelerators or nuclear reactors. Elements are designated by **chemical symbols,** which are the first one or two letters of the element's name in English, Latin, or another language.

Twenty-six of the 92 naturally occurring elements normally are present in your body. Of these, just four elements—oxygen (O), carbon (C), hydrogen (H), and nitrogen (N) (coded blue)—constitute about 96% of the body's mass. Eight others—calcium (Ca), phosphorus (P), potassium (K), sulfur (S), sodium (Na), chlorine (Cl), magnesium (Mg), and iron (Fe) (coded pink)—contribute 3.8% of the body's mass. An additional 14 elements, called **trace elements** because they are present in tiny amounts, account for the remaining 0.2% of the body's mass. The trace elements are aluminum, boron, chromium, cobalt, copper, fluorine, iodine, manganese, molybdenum, selenium, silicon, tin, vanadium, and zinc (coded yellow). Table 2.1 on page 28 provides information about the main chemical elements in the body.

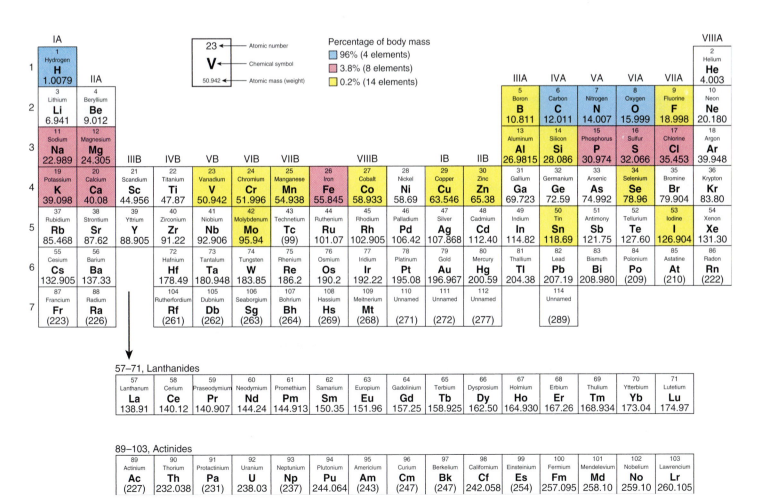

Appendix C
Normal Values For Selected Blood Tests

The system of international (SI) units (Système Internationale d'Unités) is used in most countries and in many medical and scientific journals. Clinical laboratories in the United States, by contrast, usually report values for blood and urine tests in conventional units. The laboratory values in this Appendix give conventional units first, followed by SI equivalents in parentheses. Values listed for various blood tests should be viewed as reference values rather than absolute "normal" values for all well people. Values may vary due to age, gender, diet, and environment of the subject or the equipment, methods, and standards of the lab performing the measurement.

Key To Symbols

g = gram
mg = milligram = 10^{-3} gram
μg = microgram = 10^{-6} gram
U = units
L = liter
dL = deciliter
mL = milliliter
μL = microliter
mEq/L = milliequivalents per liter
mmol/L = millimoles per liter
μmol/L = micromoles per liter
> = greater than; < = less than

Blood Tests

Test (Specimen)	U.S. Reference Values (SI Units)	Values Increase In	Values Decrease In
Aminotransferases (serum)			
Alanine aminotransferase (ALT)	0–35 U/L (same)	Liver disease or liver damage due to toxic drugs.	
Aspartate aminotransferase (AST)	0–35 U/L (same)	Myocardial infarction, liver disease, trauma to skeletal muscles, severe burns.	Beriberi, uncontrolled diabetes mellitus with acidosis, pregnancy.
Ammonia (plasma)	20–120 μg/dL (12–55 μmol/L)	Liver disease, heart failure, emphysema, pneumonia, hemolytic disease of newborn.	Hypertension.
Bilirubin (serum)	Conjugated: <0.5 mg/dL (<5.0 μmol/L)	Conjugated bilirubin: liver dysfunction or gallstones.	
	Unconjugated: 0.2–1.0 mg/dL (18–20 μmol/L) Newborn: 1.0–12.0 mg/dL (<200 μmol/L)	Unconjugated bilirubin: excessive hemolysis of red blood cells.	
Blood urea nitrogen (BUN) (serum)	8–26 mg/dL (2.9–9.3 mmol/L)	Kidney disease, urinary tract obstruction, shock, diabetes, burns, dehydration, myocardial infarction.	Liver failure, malnutrition, overhydration, pregnancy.

Blood Tests

Test (Specimen)	U.S. Reference Values (SI Units)	Values Increase In	Values Decrease In
Carbon dioxide content (bicarbonate + dissolved CO_2) (whole blood)	Arterial: 19–24 mEq/L (19–24 mmol/L) Venous: 22–26 mEq/L (22–26 mmol/L)	Severe diarrhea, severe vomiting, starvation, emphysema, aldosteronism.	Renal failure, diabetic ketoacidosis, shock.
Cholesterol, total (plasma)	<200 mg/dL (<5.2 mmol/L) is desirable	Hypercholesterolemia, uncontrolled diabetes mellitus, hypothyroidism, hypertension, atherosclerosis, nephrosis.	Liver disease, hyperthyroidism, fat malabsorption, pernicious or hemolytic anemia, severe infections.
HDL cholesterol (plasma)	>40 mg/dL (>1.0 mmol/L) is desirable		
LDL cholesterol (plasma)	<130 mg/dL (<3.2 mmol/L) is desirable		
Creatine (serum)	Males: 0.15–0.5 mg/dL (10–40 µmol/L) Females: 0.35–0.9 mg/dL (30–70 µmol/L)	Muscular dystrophy, damage to muscle tissue, electric shock, chronic alcoholism.	
Creatine Kinase (CK), also known as Creatine phosphokinase (CPK) (serum)	0–130 U/L (same)	Myocardial infarction, progressive muscular dystrophy, hypothyroidism, pulmonary edema.	
Creatinine (serum)	0.5–1.2 mg/dL (45–105 µmol/L)	Impaired renal function, urinary tract obstruction, giantism, acromegaly.	Decreased muscle mass, as occurs in muscular dystrophy or myasthenia gravis.
Electrolytes (plasma)	See Table 27.2 on page 1000.		
Gamma-glutamyl transferase (GGT) (serum)	0–30 U/L (same)	Bile duct obstruction, cirrhosis, alcoholism, metastatic liver cancer, congestive heart failure.	
Glucose (plasma)	70–110 mg/dL (3.9–6.1 mmol/L)	Diabetes mellitus, acute stress, hyperthyroidism, chronic liver disease, Cushing's disease.	Addison's disease, hypothyroidism, hyperinsulinism.
Hemoglobin (whole blood)	Males: 14–18 g/100 mL (140–180 g/L) Females: 12–16 g/100 mL (120–160 g/L) Newborns: 14–20 g/100 mL (140–200 g/L)	Polycythemia, congestive heart failure, chronic obstructive pulmonary disease, living at high altitude.	Anemia, severe hemorrhage, cancer, hemolysis, Hodgkin's disease, nutritional deficiency of vitamin B_{12}, systemic lupus erythematosus, kidney disease.
Iron, total (serum)	Males: 80–180 µg/dL (14–32 µmol/L) Females: 60–160 µg/dL (11–29 µmol/L)	Liver disease, hemolytic anemia, iron poisoning.	Iron-deficiency anemia, chronic blood loss, pregnancy (late), chronic heavy menstruation.
Lactic dehydrogenase (LDH) (serum)	71–207 U/L (same)	Myocardial infarction, liver disease, skeletal muscle necrosis, extensive cancer.	
Lipids (serum)		Hyperlipidemia, diabetes mellitus, hypothyroidism.	Fat malabsorption.
Total	400–850 mg/dL (4.0–8.5 g/L)		
Triglycerides	10–190 mg/dL (0.1–1.9 g/L)		
Platelet (thrombocyte) count (whole blood)	150,000–400,000/µL	Cancer, trauma, leukemia, cirrhosis.	Anemias, allergic conditions hemorrhage.
Protein (serum)		Dehydration, shock, chronic infections.	Liver disease, poor protein intake, hemorrhage, diarrhea, malabsorption, chronic renal failure, severe burns.
Total	6–8 g/dL (60–80 g/L)		
Albumin	4–6 g/dL (40–60 g/L)		
Globulin	2.3–3.5 g/dL (23–35 g/L)		

(continues)

Blood Tests

Test (Specimen)	U.S. Reference Values (SI Units)	Values Increase In	Values Decrease In
Red blood cell (erythrocyte) count (whole blood)	Males: 4.5–6.5 million/μL Females: 3.9–5.6 million/μL	Polycythemia, dehydration, living at high altitude.	Hemorrhage, hemolysis, anemias, cancer, overhydration.
Uric acid (urate) (serum)	2.0–7.0 mg/dL (120–420 μmol/L)	Impaired renal function, gout, metastatic cancer, shock, starvation.	
White blood cell (leukocyte) count, total (whole blood)	5,000–10,000/μL (See Table 19.3 on page 646 for relative percentages of different types of WBCs.)	Acute infections, trauma, malignant diseases, cardiovascular diseases. (See also Table 19.2 on page 645.)	Diabetes mellitus, anemia. (See also Table 19.2 on page 645.)

Appendix D
Normal Values For Selected Urine Tests

Urine Tests

Test (Specimen)	U.S. Reference Values (SI Units)	Clinical Implications
Amylase (2 hour)	35–260 Somogyi units/hr (6.5–48.1 units/hr)	Values increase in inflammation of the pancreas (pancreatitis) or salivary glands, obstruction of the pancreatic duct, and perforated peptic ulcer.
Bilirubin* (random)	Negative	Values increase in liver disease and obstructive biliary disease.
Blood* (random)	Negative	Values increase in renal disease, extensive burns, transfusion reactions, and hemolytic anemia.
Calcium (Ca^{2+}) (random)	10 mg/dL (2.5 mmol/liter); up to 300 mg/24 hr (7.5 mmol/24 hr)	Amount depends on dietary intake; values increase in hyperparathyroidism, metastatic malignancies, and primary cancer of breasts and lungs; values decrease in hypoparathyroidism and vitamin D deficiency.
Casts (24 hour)		
Epithelial	Occasional	Values increase in nephrosis and heavy metal poisoning.
Granular	Occasional	Values increase in nephritis and pyelonephritis.
Hyaline	Occasional	Values increase in kidney infections.
Red blood cell	Occasional	Values increase in glomerular membrane damage and fever.
White blood cell	Occasional	Values increase in pyelonephritis, kidney stones, and cystitis.
Chloride (Cl⁻) (24 hour)	140–250 mEq/24 hr (140–250 mmol/24 hr)	Amount depends on dietary salt intake; values increase in Addison's disease, dehydration, and starvation; values decrease in pyloric obstruction, diarrhea, and emphysema.
Color (random)	Yellow, straw, amber	Varies with many disease states, hydration, and diet.
Creatinine (24 hour)	Males: 1.0–2.0 g/24 hr (9–18 mmol/24 hr) Females: 0.8–1.8 g/24 hr (7–16 mmol/24 hr)	Values increase in infections; values decrease in muscular atrophy, anemia, and kidney diseases.
Glucose*	Negative	Values increase in diabetes mellitus, brain injury, and myocardial infarction.
Hydroxycortico-steroids (17-hydroxysteroids) (24 hour)	Males: 5–15 mg/24 hr (13–41 μmol/24 hr) Females: 2–13 mg/24 hr (5–36 μmol/24 hr)	Values increase in Cushing's syndrome, burns, and infections; values decrease in Addison's disease.
Ketone bodies* (random)	Negative	Values increase in diabetic acidosis, fever, anorexia, fasting, and starvation.
17-ketosteroids (24 hour)	Males: 8–25 mg/24 hr (28–87 μmol/24 hr) Females: 5–15 mg/24 hr (17–53 μmol/24 hr)	Values decrease in surgery, burns, infections, adrenogenital syndrome, and Cushing's syndrome.
Odor (random)	Aromatic	Becomes acetonelike in diabetic ketosis.
Osmolality (24 hour)	500–1400 mOsm/kg water (500–1400 mmol/kg water)	Values increase in cirrhosis, congestive heart failure (CHF), and high-protein diets; values decrease in aldosteronism, diabetes insipidus, and hypokalemia.
pH* (random)	4.6–8.0	Values increase in urinary tract infections and severe alkalosis; values decrease in acidosis, emphysema, starvation, and dehydration.

Test (Specimen)	U.S. Reference Values (SI Units)	Clinical Implications
Phenylpyruvic acid (random)	Negative	Values increase in phenylketonuria (PKU).
Potassium (K⁺) (24 hour)	40–80 mEq/24 hr (40–80 mmol/24 hr)	Values increase in chronic renal failure, dehydration, starvation, and Cushing's syndrome; values decrease in diarrhea, malabsorption syndrome, and adrenal cortical insufficiency
Protein* (albumin) (random)	Negative	Values increase in nephritis, fever, severe anemias, trauma, and hyperthyroidism.
Sodium (Na⁺) (24 hour)	75–200 mg/24 hr (75–200 mmol/24 hr)	Amount depends on dietary salt intake; values increase in dehydration, starvation, and diabetic acidosis; values decrease in diarrhea, acute renal failure, emphysema, and Cushing's syndrome.
Specific gravity* (random)	1.001–1.035 (same)	Values increase in diabetes mellitus and excessive water loss; values decrease in absence of antidiuretic hormone (ADH) and severe renal damage.
Urea (random)	25–35 g/24 hr (420–580 mmol/24 hr)	Values increase in response to increased protein intake; values decrease in impaired renal function.
Uric acid (24 hour)	0.4–1.0 g/24 hr (1.5–4.0 mmol/24 hr)	Values increase in gout, leukemia, and liver disease; values decrease in kidney disease.
Urobilinogen* (2 hour)	0.3–1.0 Ehrlich units (1.7–6.0 μmol/24 hr)	Values increase in anemias, hepatitis A (infectious), biliary disease, and cirrhosis; values decrease in cholelithiasis and renal insufficiency.
Volume, total (24 hour)	1000–2000 mL/24 hr (1.0–2.0 liters/24 hr)	Varies with many factors.

*Test often performed using a **dipstick,** a plastic strip impregnated with chemicals that is dipped into a urine specimen to detect particular substances. Certain colors indicate the presence or absence of a substance and sometimes give a rough estimate of the amount(s) present.

Appendix E
Answers

Answers to Self-Quiz Questions

Chapter 1
1. cell **2.** catabolism **3.** homeostasis **4.** false **5.** true **6.** true **7.** d
8. b **9.** b **10.** e **11.** e **12.** (a) 3, (b) 4, (c) 6, (d) 5, (e) 1, (f) 2
13. (a) 6, (b) 1, (c) 11, (d) 5, (e) 10, (f) 8, (g) 7, (h) 9, (i) 4, (j) 3, (k) 2
14. (a) 4, (b) 6, (c) 8, (d) 1, (e) 5, (f) 2, (g) 7, (h) 3 **15.** (a) 3, (b) 7,
(c) 1, (d) 5, (e) 6, (f) 4, (g) 8, (h) 2

Chapter 2
1. protons, electrons, neutrons **2.** 8 **3.** atomic number **4.** concentration, temperature **5.** true **6.** false **7.** b **8.** d **9.** a **10.** c **11.** b
12. (a) 1, (b) 2 (c) 1, (d) 4 **13.** a **14.** (a) 8, (b) 1, (c) 7, (d) 2, (e) 5, (f)
3, (g) 4, (h) 6 **15.** (a) 7, (b) 5, (c) 1, (d) 2, (e) 6, (f) 3, (g) 8, (h) 4

Chapter 3
1. plasma membrane, cytoplasm, nucleus **2.** apoptosis, necrosis
3. cytosol **4.** shortening and loss of protective telomeres on
chromosomes, cross-link formation between glucose and proteins, free
radical formation **5.** true **6.** true **7.** a **8.** d **9.** b **10.** a **11.** c
12. (a) 2, (b) 3, (c) 5, (d) 7, (e) 6, (f) 8, (g) 1, (h) 4 **13.** (a) 11, (b) 10,
(c) 1, (d) 2, (e) 5, (f) 9, (g) 12, (h) 7, (i) 6, (j) 3, (k) 4, (l) 8 **14.** (a) 2,
(b) 9, (c) 3, (d) 5, (e) 11, (f) 8, (g) 1, (h) 6, (i) 10, (j) 7, (k) 4 **15.** (a) 3,
(b) 9, (c) 1, (d) 5, (e) 11, (f) 4, (g) 8, (h) 7, (i) 2, (j) 10, (k) 6

Chapter 4
1. tissue **2.** endocrine **3.** cells, ground substance, fibers **4.** true **5.** false
6. a **7.** c **8.** b **9.** c **10.** b **11.** d **12.** (a) 4, (b) 8, (c) 5, (d) 2, (e) 3, (f)
6, (g) 1, (h) 7 **13.** (a) 3, (b) 5, (c) 8, (d) 12, (e) 9, (f) 7, (g) 11, (h) 6, (i) 2,
(j) 4, (k) 10, (l) 1 **14.** (a) 2, (b) 5, (c) 3, (d) 4, (e) 1 **15.** (a) 3, (b) 2, (c) 1

Chapter 5
1. palms, soles, fingertips **2.** strength, extensibility, elasticity
3. ceruminous **4.** melanin, carotene, hemoglobin **5.** a **6.** b **7.** a
8. c **9.** c **10.** e **11.** (a) 3, (b) 4, (c) 1, (d) 2 **12.** (a) 5, (b) 13, (c) 3,
(d) 7, (e) 4, (f) 2, (g) 6, (h) 8, (i) 10, (j) 12, (k) 11, (l) 1, (m) 9
13. (a) 3, (b) 5, (c) 4, (d) 1, (e) 2, (f) 6, (g) 7, (h) 8 **14.** (a) 3, (b) 1,
(c) 2, (d) 5, (e) 4 **15.** (a) 4, (b) 3, (c) 2, (d) 1, inflammatory, migratory, proliferative, maturation

Chapter 6
1. hardness, tensile strength **2.** osteons (Haversian systems), trabeculae **3.** hyaline cartilage, loose fibrous connective tissue membranes
4. true **5.** false **6.** b **7.** b **8.** a **9.** d **10.** a **11.** d **12.** (a) 3, (b) 9,
(c) 8, (d) 1, (e) 5, (f) 4, (g) 6, (h) 7, (i) 2, (j) 10 **13.** (a) 10, (b) 2, (c) 7,
(d) 1, (e) 8, (f) 3, (g) 6, (h) 4, (i) 9, (j) 5 **14.** (a) 9, (b) 7, (c) 6, (d) 1,
(e) 4, (f) 2, (g) 5, (h) 3, (i) 8, (j) 10 **15.** (a) 2, (b) 5, (c) 3, (d) 4, (e) 1

Chapter 7
1. fontanels **2.** thoracic **3.** intervertebral discs **4.** pituitary gland
5. false **6.** false **7.** a **8.** c **9.** b **10.** e **11.** d **12.** (a) 7, (b) 5, (c) 1,
(d) 6, (e) 2, (f) 4, (g) 8, (h) 9, (i) 3, (j) 10, (k) 11, (l) 13, (m) 12
13. (a) 2, (b) 3, (c) 5, (d) 6, (e) 4, (f) 1, (g) 5, (h) 4, (i) 2, (j) 4, (k) 3
14. (a) 4, (b) 7, (c) 6, (d) 5, (e) 3, (f) 1, (g) 2 **15.** (a) 3, (b) 1, (c) 6,
(d) 9, (e) 13, (f) 12, (g) 2, (h) 4, (i) 5, (j) 7, (k) 10, (l) 15, (m) 8, (n)
11, (o) 14

Chapter 8
1. protection, movement **2.** metacarpals **3.** true **4.** true **5.** b **6.** b
7. d **8.** e **9.** e **10.** c **11.** a **12.** b **13.** b **14.** (a) 4, (b) 3, (c) 3,
(d) 6, (e) 7, (f) 1, (g) 3, (h) 2, (i) 5, (j) 9, (k) 8, (l) 2, (m) 4, (n) 6, (o) 7
15. (a) 2, (b) 6, (c) 9, (d) 7, (e) 4, (f) 5, (g) 8, (h) 10, (i) 1, (j) 3

Chapter 9
1. joint, articulation, or arthrosis **2.** synovial fluid **3.** osteoarthritis
4. d **5.** b **6.** a **7.** c **8.** e **9.** e **10.** true **11.** false **12.** (a) 5, (b) 3,
(c) 6, (d) 1, (e) 4, (f) 7, (g) 2 **13.** (a) 4, (b) 3, (c) 6, (d) 2, (e) 5, (f) 1
14. (a) 6, (b) 4, (c) 5, (d) 1, (e) 3, (f) 2 **15.** (a) 6, (b) 7, (c) 9, (d) 4, (e)
8, (f) 3, (g) 10, (h) 2, (i) 11, (j) 1, (k) 5

Chapter 10
1. electrical excitability, contractility, extensibility, elasticity **2.** actin,
myosin **3.** motor unit **4.** false **5.** true **6.** e **7.** a **8.** b **9.** e **10.** d
11. c **12.** (a) 5, (b) 6, (c) 8, (d) 7, (e) 2, (f) 4, (g) 3, (h) 1 **13.** (a) 5, (b)
8, (c) 7, (d) 10, (e) 6, (f) 9, (g) 4, (h) 1, (i) 2, (j) 3 **14.** (a) 2, (b) 3, (c) 1,
(d) 1 and 2 (e) 2 and 3, (f) 2, (g) 1, (h) 3, (i) 1 and 2, (j) 3 **15.** (a) 10,
(b) 2, (c) 4, (d) 3, (e) 6, (f) 5, (g) 1, (h) 7, (i) 9, (j) 8

Chapter 11
1. origin, insertion **2.** buccinator **3.** true **4.** true **5.** d **6.** e **7.** a
8. e **9.** c **10.** a **11.** b **12.** (a) 12, (b) 9, (c) 8, (d) 6, (e) 3, (f) 11, (g)
10, (h) 1, (i) 2, (j) 7, (k) 4, (l) 5 **13.** (a) 3, (b) 1, (c) 2, (d) 1, (e) 2, (f) 3,
(g) 3 **14.** (a) 6, (b) 3, (c) 7, (d) 4, (e) 2, (f) 8, (g) 5, (h) 1 **15.** (a) 10,
(b) 1, (c) 9, (d) 8, (e) 12, (f) 17, (g) 2, (h) 6, (i) 8, (j) 14, (k) 5, (l) 4, (m)
2, (n) 15, (o) 1, (p) 11, (q) 13, (r) 12, (s) 7, (t) 16, (u) 11, (v) 17, (w) 16,
(x) 15, (y) 3

Chapter 12
1. somatic nervous system, autonomic nervous system, enteric nervous
system **2.** polarized **3.** true **4.** false **5.** c **6.** b **7.** a **8.** e **9.** c
10. e **11.** b **12.** a **13.** (a) 6, (b) 11, (c) 1, (d) 2, (e) 9, (f) 13, (g) 4, (h)
8, (i) 7, (j) 12, (k) 5, (l) 3, (m) 10 **14.** (a) 2, (b) 1, (c) 9, (d) 8, (e) 6, (f)
3, (g) 4, (h) 5, (i) 10, (j) 7 **15.** (a) 4, (b) 5, (c) 14, (d) 8, (e) 7, (f) 1,
(g) 2, (h) 10, (i) 13, (j) 6, (k) 3, (l) 12, (m) 9, (n) 11

Chapter 13

1. reflexes **2.** mixed **3.** true **4.** true **5.** d **6.** e **7.** d **8.** a **9.** c
10. (a) 1, (b) 8, (c) 4, (d) 2, (e) 10, (f) 1, (g) 6, (h) 5, (i) 3, (j) 9, (k) 1,
(l) 11, (m) 7, (n) 2 **11.** a **12.** b **13.** e **14.** (a) 10, (b) 8, (c) 9, (d) 1,
(e) 2, (f) 5, (g) 7, (h) 6, (i) 11, (j) 4, (k) 3 **15.** (a) 2, (b) 1, (c) 3, (d) 4,
(e) 1, (f) 2, (g) 4, (h) 3, (i) 5, (j) 1

Chapter 14

1. choroid plexuses **2.** longitudinal fissue **3.** true **4.** false **5.** c **6.** d
7. e **8.** d **9.** e **10.** a **11.** d **12.** (a) 3, (b) 5, (c) 6, (d) 8, (e) 11, (f) 10,
(g) 7, (h) 9, (i) 1, (j) 4, (k) 2, (l) 12, (m) 1, (n) 8, (o) 5, (p) 7, (q) 12,
(r) 10, (s) 9 **13.** (a) 5, (b) 9, (c) 11, (d) 6, (e) 3, (f) 1, (g) 10, (h) 8, (i) 2,
(j) 4, (k) 7 **14.** (a) 9, (b) 2, (c) 6, (d) 10, (e) 4, (f) 11, (g) 1, (h) 2, (i) 5,
(j) 8, (k) 12, (l) 7, (m) 3, (n) 6 and 8, (o) 13, (p) 7, (q) 1 **15.** (a) 9, (b)
2, (c) 6, (d) 8, (e) 7, (f) 5, (g) 3, (h) 10, (i) 12, (j) 11, (k) 4, (l) 1

Chapter 15

1. sensation, perception **2.** adaptation **3.** true **4.** true **5.** d **6.** b **7.** a
8. d **9.** e **10.** e **11.** d **12.** d **13.** (a) 9, (b) 8, (c) 4, (d) 7, (e) 10, (f) 2, (g)
3, (h) 1, (i) 5, (j) 6, (k) 11 **14.** (a) 10, (b) 8, (c) 7, (d) 1, (e) 4, (f) 3, (g) 5,
(h) 6, (i) 9, (j) 2 **15.** (a) 3, (b) 2, (c) 5, (d) 7, (e) 1, (f) 3, (g) 8, (h) 4, (i) 6

Chapter 16

1. sweet, sour, salty, bitter, umami **2.** static, dynamic **3.** true **4.** false
5. a **6.** d **7.** c **8.** c **9.** c **10.** c **11.** e **12.** a **13.** (a) 1, (b) 5, (c) 7,
(d) 6, (e) 1, (f) 8, (g) 2, (h) 4, (i) 3 **14.** (a) 3, (b) 6, (c) 8, (d) 12, (e) 1,
(f) 5, (g) 9, (h) 11, (i) 7, (j) 13, (k) 2, (l) 10, (m) 4 **15.** (a) 2, (b) 11, (c)
14, (d) 13, (e) 3, (f) 10, (g) 6, (h) 12, (i) 4, (j) 5, (k) 9, (l) 1, (m) 7, (n) 8

Chapter 17

1. acetylcholine, epinephrine, or norepinephrine **2.** thoracolumbar,
craniosacral **3.** true **4.** true **5.** b **6.** c **7.** d **8.** a **9.** a **10.** c **11.** d
12. e **13.** (a) 2, (b) 5, (c) 6, (d) 1, (e) 3, (f) 4 **14.** (a) 3, (b) 2, (c) 1, (d)
1, (e) 2, (f) 3, (g) 3 **15.** (a) 2, (b) 1, (c) 1, (d) 2, (e) 1, (f) 1, (g) 2, (h) 2

Chapter 18

1. fight-or-flight response, resistance reaction, exhaustion **2.** hypothalamus **3.** exocrine, endocrine **4.** false **5.** true **6.** b **7.** c **8.** d
9. a **10.** e **11.** a **12.** c **13.** (a) 8, (b) 2, (c) 7, (d) 1, (e) 12, (f) 18,
(g) 5, (h) 17, (i) 20, (j) 15, (k) 3, (l) 16, (m) 19, (n) 6, (o) 13, (p) 11,
(q) 4, (r) 10, (s) 14, (t) 9 **14.** (a) 10, (b) 8, (c) 2, (d) 4, (e) 1, (f) 6, (g)
9, (h) 7, (i) 5, (j) 3 **15.** (a) 12, (b) 1, (c) 11, (d) 7, (e) 3, (f) 10, (g) 2,
(h) 9, (i) 4, (j) 8, (k) 5, (l) 6

Chapter 19

1. blood plasma, formed elements, red blood cells, white blood cells,
platelets **2.** serum **3.** platelets, thrombocytes **4.** true **5.** false **6.** e
7. a **8.** e **9.** a **10.** b **11.** c **12.** d **13.** (a) 6, (b) 8, (c) 3, (d) 10, (e) 2,
(f) 4, (g) 1, (h) 7, (i) 9, (j) 5 **14.** (a) 5, (b) 1, (c) 2, (d) 3, (e) 6, (f) 8 (g)
9, (h) 4, (i) 7 **15.** (a) 4, (b) 6, (c) 8, (d) 1, (e) 7, (f) 3, (g) 5, (h) 2

Chapter 20

1. left ventricle **2.** systole, diastole **3.** stroke volume **4.** false
5. false **6.** true **7.** a **8.** c **9.** b **10.** a **11.** b **12.** (a) 3, (b) 2, (c) 9,
(d) 8, (e) 7, (f) 4, (g) 1, (h) 5, (i) 6 **13.** (a) 4, (b) 10, (c) 6, (d) 5, (e) 3,
(f) 9, (g) 2, (h) 8, (i) 1, (j) 7 **14.** (a) 3, (b) 7, (c) 2, (d) 5, (e) 1, (f) 6, (g)
4 and 7 **15.** (a) 3, (b) 6, (c) 1, (d) 5, (e) 2, (f) 4

Chapter 21

1. carotid sinus, aortic **2.** skeletal muscle pump, respiratory pump
3. true **4.** true **5.** a **6.** c **7.** e **8.** a **9.** d **10.** e **11.** c **12.** (a) 2, (b)
5, (c) 1, (d) 4, (e) 3 **13.** (a) 11, (b) 1, (c) 4, (d) 9, (e) 3, (f) 8, (g) 6, (h)
2, (i) 7, (j) 5, (k) 10, (l) 12 **14.** (a) 2, (b) 6, (c) 4, (d) 1, (e) 3, (f) 5
15. (a) 5, (b) 3, (c) 1, (d) 4, (e) 2, (f) 4, (g) 1, (h) 5

Chapter 22

1. skin, mucous membranes; antimicrobial proteins, NK cells, phagocytes **2.** specificity, memory **3.** true **4.** true **5.** e **6.** e **7.** d **8.** a
9. c **10.** e **11.** b **12.** c **13.** (a) 3, (b) 1, (c) 4, (d) 2, (e) 5 **14.** (a) 2,
(b) 3, (c) 4, (d) 7, (e) 1, (f) 6, (g) 5 **15.** (a) 11, (b) 11, (c) 8, (d) 1, (e) 2,
(f) 5, (g) 4, (h) 7, (i) 9, (j) 12, (k) 3, (l) 6, (m) 10

Chapter 23

1. pulmonary ventilation, external respiration, internal respiration
2. less, greater **3.** oxyhemoglogin; dissolved CO_2, carbamino compounds, and bicarbonate ion **4.** $CO_2 + H_2O \rightarrow H_2CO_3 \rightarrow H^+ + HCO_3^-$
5. false **6.** true **7.** b **8.** d **9.** a **10.** c **11.** e **12.** (a) 2, (b) 8, (c) 3, (d)
7, (e) 1, (f) 9, (g) 5, (h) 10, (i) 4, (j) 6 **13.** (a) 7, (b) 8, (c) 1, (d) 5, (e) 6,
(f) 2, (g) 3, (h) 4 **14.** (a) 3, (b) 8, (c) 5, (d) 2, (e) 7, (f) 1, (g) 4, (h) 6
15. (a) 7, (b) 3, (c) 5, (d) 1, (e) 6, (f) 4, (g) 2

Chapter 24

1. mucosa, submucosa, muscularis, serosa **2.** submucosal plexus
or plexus of Meissner; myenteric plexus or plexus of Auerbach
3. monosaccharides; amino acids; monoglycerides, fatty acids; pentoses, phosphates, nitrogenous bases **4.** false **5.** false **6.** a **7.** c
8. e **9.** b **10.** a **11.** b **12.** d **13.** (a) 3, (b) 10, (c) 5, (d) 8, (e) 6, (f)
1, (g) 4, (h) 9, (i) 11, (j) 2, (k) 7 **14.** (a) 4, (b) 6, (c) 7, (d) 1, (e) 5, (f)
3, (g) 8, (h) 2, (i) 10, (j) 9 **15.** (a) 4, (b) 8, (c) 2, (d) 10, (e) 7, (f) 11,
(g) 1, (h) 6, (i) 3, (j) 9, (k) 5

Chapter 25

1. hypothalamus **2.** glucose 6-phosphate, pyruvic acid, acetyl coenzyme A **3.** hormones **4.** false **5.** true **6.** c **7.** d **8.** a **9.** d **10.** a
11. a **12.** b **13.** (a) 2, (b) 8, (c) 5, (d) 10, (e) 6, (f) 7, (g) 3, (h) 9,
(i) 4, (j) 1 **14.** (a) 9, (b) 12, (c) 11, (d) 10, (e) 4, (f) 7, (g) 3, (h) 5,
(i) 8, (j) 1, (k) 6, (l) 2 **15.** (a) 8, (b) 5, (c) 1, (d) 10, (e) 6, (f) 2, (g)
4, (h) 9, (i) 3, (j) 7

Chapter 26

1. transitional epithelial **2.** voluntary, involuntary **3.** glomerular filtration, tubular reabsorption, tubular secretion **4.** true **5.** true **6.** c
7. a **8.** d **9.** a **10.** e **11.** b **12.** c **13.** (a) 8, (b) 2, (c) 10, (d) 5, (e) 3,
(f) 1, (g) 7, (h) 4, (i) 9, (j) 6 **14.** (a) 4, (b) 3, (c) 7, (d) 6, (e) 2, (f) 1, (g)
5 **15.** (a) 5, (b) 4, (c) 6, (d) 1, (e) 2, (f) 7, (g) 3

Chapter 27

1. intracellular fluid, extracellular fluid **2.** bicarbonate ion, carbonic
acid **3.** weak acid, weak base **4.** protein buffer, carbonic acid-
bicarbonate, phosphate **5.** true **6.** true **7.** a **8.** c **9.** b **10.** e **11.** d
12. a **13.** b **14.** (a) 8, (b) 9, (c) 7, (d) 1, (e) 6, (f) 2, (g) 4, (h) 5, (i) 3
15. (a) 8, (b) 7, (c) 5, (d) 6, (e) 1, (f) 3, (g) 4, (h) 2

Chapter 28

1. puberty, menarche, menopause **2.** semen **3.** true **4.** true **5.** e
6. b **7.** a **8.** c **9.** b **10.** a **11.** e **12.** (a) 7, (b) 2, (c) 1, (d) 4, (e) 8, (f)
5, (g) 3, (h) 6 **13.** (a) 6, (b) 4, (c) 2, (d) 1, (e) 3, (f) 8, (g) 7, (h) 5
14. (a) 6, (b) 4, (c) 1, (d) 12, (e) 8, (f) 5, (g) 7, (h) 13, (i) 11, (j) 3, (k)
2, (l) 10, (m) 9 **15.** (a) 10, (b) 12, (c) 1, (d) 5, (e) 2, (f) 4, (g) 6, (h) 11,
(i) 8, (j) 3, (k) 9, (l) 7

Chapter 29

1. dilation, expulsion, placental **2.** corpus luteum **3.** mesoderm; ectoderm; endoderm **4.** true **5.** a **6.** b **7.** e **8.** (a) 3, (b) 4, (c) 5, (d) 1,
(e) 2, (f) 6, (g) 7, (h) 8, (i) 9 **9.** e **10.** b **11.** (a) 7, (b) 3, (c) 6, (d) 4,
(e) 10, (f) 8, (g) 5, (h) 2, (i) 1, (j) 9, (k) 11, (l) 12 **12.** b **13.** (a) 3, (b)
6, (c) 4, (d) 1, (e) 8, (f) 2, (g) 5, (h) 7 **14.** (a) 5, (b) 2, (c) 8, (d) 4, (e) 7,
(f) 1, (g) 3, (h) 6 **15.** d

Answers to Critical Thinking Questions

Chapter 1

1. Homeostasis is the relative constancy of the body's internal environment. Body temperature should vary within a narrow range around normal body temperature (38°C or 98.6° F).
2. Bilateral means two sides or both arms in this case. The carpal region is the wrist area.
3. A x ray provides a good image of dense tissue such as bones. An MRI is used for imaging soft tissues, not bones. An MRI cannot be used when metal is present due to the magnetic field used.

Chapter 2

1. Fatty acids are in all lipids including plant oils and the phospholipids that compose cell membranes. Sugars (monosaccharides) are needed for ATP production, are a component of nucleotides, and are the basic units of disaccharides and polysaccharides including starch, glycogen, and cellulose.
2. DNA = deoxyribonucleic acid. DNA is composed of 4 different nucleotides. The order of the nucleotides is unique in every person. The 20 different amino acids form proteins.
3. One pH unit equals a tenfold change in H^+ concentration. Pure water has a pH of 7, which equals 1×10^{-7} moles of H^+ per liter. Blood has a pH of 7.4, which equals 0.4×10^{-7} moles H^+ per liter. Thus, blood has less than half as many H^+ as water.

Chapter 3

1. The tissues are destroyed due to autolysis of the cells caused by the release of acids and digestive enzymes from the lysosomes.
2. Synthesis of mucin by the ribosomes on rough endoplasmic reticulum, to transport vesicle, to entry face of Golgi, to transfer vesicle, to medial cisternae, where protein is modified, to transfer vesicle, to exit face, to secretory vesicle, to plasma membrane, where it undergoes exocytosis.
3. The sense strand is composed of DNA codons transcribed into mRNA. Only the exons will be found in the mRNA. The introns are clipped out.

Chapter 4

1. Many possible adaptations including: more adipose tissue for insulation, thicker bone for support, more red blood cells for oxygen transport, and so on.
2. The surface layer of the skin, the epidermis, is keratinized, stratified squamous epithelium. It is avascular. If the pin were stuck straight in, the vascularized connective tissue would be pierced and the finger would bleed.
3. The gelatin portion of the salad represents the ground substance of the connective tissue matrix. The grapes are cells such as fibroblasts. The shredded carrots and coconut are fibers embedded in the ground substance.

Chapter 5

1. The dust particles are mostly keratinocytes shed from the stratum corneum of the skin.
2. It would be neither wise nor feasible to remove the exocrine glands. Essential exocrine glands include the sudoriferous glands (sweat helps control body temperature), sebaceous glands (sebum lubricates the skin), and ceruminous glands (provide protective lubricant for ear canal).
3. The superficial layer of the epidermis of the skin is the stratum corneum. Its cells are full of intermediate filaments of keratin, kera-

tohyalin, and lipids from lamellar granules, making this layer a water-repellent barrier. The epidermis also contains an abundance of desmosomes.

Chapter 6

1. In greenstick fracture, which occurs only in children due to the greater flexibility of their bones, the bone breaks on one side but bends on the other side, resembling what happens when one tries to break a green (not dry) stick. Lynne Marie probably broke her fall with her extended arm, breaking the radius or ulna.
2. At Aunt Edith's age, the production of several hormones (such as estrogens and human growth hormone) necessary for bone remodeling is likely to be decreased. Her decreased height results, in part, from compression of her vertebrae due to flattening of the discs between them. She may also have osteoporosis and an increased susceptibility to fractures resulting in damage to the vertebrae and loss of height.
3. Exercise causes mechanical stress on bones, but because there is effectively zero gravity in space, the pull of gravity on bones is missing. The lack of stress from gravity results in bone demineralization and weakness.

Chapter 7

1. Fontanels, the soft spots between cranial bones, are fibrous connective tissue membranes that will be replaced by bone as the baby matures. They allow the infant's head to be molded during its passage through the birth canal and allow for brain and skull growth in the infant.
2. Barbara probably broke her coccyx or tailbone. The coccygeal vertebrae usually fuse by age 30. The coccyx points inferiorly in females.
3. Infants are born with a single concave curve in the vertebral column. Adults have four curves in their spinal cord-at the cervical (convex), thoracic (concave), lumbar (convex), and sacral regions. Buying a mattress from this company and sleeping on it would result in potentially severe back problems.

Chapter 8

1. Clawfoot is an abnormal elevation of the medial longitudinal arch of the foot.
2. There are 14 phalanges in each hand: two bones in the thumb and three in each of the other fingers. Farmer White has lost five phalanges on his left hand so he has nine remaining on his left and 14 remaining on his right for a total of 23.
3. Snakes don't have appendages, so they don't need shoulders and hips. Pelvic and pectoral girdles connect the upper and lower limbs (appendages) to the axial skeleton, and thus are considered part of the appendicular skeleton.

Chapter 9

1. Katie's vertebral column, head, thigh, lower leg, and lower arm are all flexed. Her lower arm and shoulder are medially rotated.
2. The elbow is a hinge-type synovial joint. The trochlea of the humerus articulates with the trochlear notch of the ulna. The elbow has monaxial (opening and closing) movement similar to a door.
3. Cartilaginous joints can be composed of hyaline cartilage, such as at the epiphyseal plate, or fibrocartilage, such as the intervertebral disc. Sutures are an example of fibrous joints, which are composed of dense fibrous tissue. Synovial (diarthrotic) joints are held together by an articular capsule, which has an inner synovial membrane layer.

Chapter 10

1. Smooth muscle contains both thick and thin filaments as well as intermediate filaments attached to dense bodies. The smooth muscle fiber contracts like a corkscrew turns; the fiber twists like a helix as it contracts and shortens.
2. Ming's muscles performed isometric contractions. Peng used primarily isotonic concentric contraction to lift, isometric to hold, and then isotonic eccentric contractions to lower the barbell.
3. Mr. Klopfer's students have muscle fatigue in their hands. Fatigue has many causes including an increase in lactic acid and ADP, and a decrease in Ca^{2+}, oxygen, creatine phosphate, and glycogen.

Chapter 11

1. Some of the antagonistic pairs for the regions are: upper arm-biceps brachii vs. triceps brachii; upper leg-quadriceps femoris vs. hamstrings; torso-rectus abdominus vs. erector spinae; lower leg-gastrocnemius vs. tibialis anterior.
2. The fulcrum is the knee joint, the load (resistance) is the weight of the upper body and the package, and the effort is the thigh muscles. This is a third-class lever.
3. Lifting the eyebrow uses the frontal belly of the occipitofrontalis; whistling uses the buccinator and orbicularis oris; closing the eyes uses the orbicularis oculi; shaking the head can use several muscles, including sternocleidomastoid, semispinalis capitis, splenius capitis, and longissimus capitis.

Chapter 12

1. The motor neuron is multipolar in structure with an axon and several dendrites projecting from the cell body. The converging circuit has several neurons that converge to form synapses with one common neuron. The bipolar neuron has one dendrite and an axon projecting from the cell body. In a simple circuit, each presynaptic neuron makes a single synapse with one postsynaptic neuron.
2. Gray matter appears gray in color due to the absence of myelin. It is composed of cell bodies, and unmyelinated axons and dendrites. White matter contains many myelinated axons. Lipofuscin, a yellowing pigment, collects in neurons with age.
3. Smelling coffee and hearing alarm are somatic sensory, stretching and yawning are somatic motor, salivating is autonomic (parasympathetic) motor, stomach rumble is enteric motor.

Chapter 13

1. A withdrawal/flexor reflex is ipsilateral and polysynaptic. The route is pain receptor of sensory neuron → spinal cord (integrating center) interneurons → motor neurons → flexor muscles in the leg. Also present is a crossed extensor reflex, which is contralateral and polysynaptic. The route diverges at the spinal cord: Interneurons cross to the other side → motor neurons → extensor muscles in the opposite leg.
2. The spinal cord is anchored in place by the filum terminale and the denticulate ligaments.
3. The needle will pierce the epidermis, the dermis, and the subcutaneous layer and then go between the vertebrae through the epidural space, the dura mater, the subdural space, the arachnoid mater, and into the CSF in the subarachnoid space. CSF is produced in the brain, and the spinal meninges are continuous with cranial meninges.

Chapter 14

1. Movement of the right arm is controlled by the left hemisphere's primary motor area, located in the precentral gyrus. Speech is controlled by Broca's area in the left hemisphere's frontal lobe just superior to the lateral cerebral sulcus.
2. The brain is enclosed by the cranial bones and meninges. The temporal bone houses the middle ear and inner ear, separating these from the brain.
3. The dentist has injected anesthetic into the inferior alveolar nerve, a branch of the mandibular nerve, which numbs the lower teeth and the lower lip. The tongue is numbed by blocking the lingual nerve. The upper teeth and lip are anesthetized by injecting the superior alveolar nerve, a branch of the maxillary nerve.

Chapter 15

1. Chemoreceptors in the nose detect odors. Proprioceptors detect body position and are involved in equilibrium. The receptors for smell are rapidly adapting (phasic), whereas proprioceptors are slowly adapting (tonic).
2. The tickle receptors are free nerve endings in the foot. The impulses travel along the first-order sensory neurons to the posterior gray horn, then along the second-order neurons, crossing to the other side of the spinal cord and up the anterior spinothalamic tract to the thalamus. The third-order neurons extend from the thalamus to the "foot" region of the somatosensory area of the cerebral cortex.
3. Yoshio's perception of feeling in his amputated foot is called phantom limb sensation. Impulses from the remaining proximal section of the sensory neuron are perceived by the brain as still coming from the amputated foot.

Chapter 16

1. Because smell and taste have ties to the cortex and limbic areas, Brenna may be recalling a memory of this or similar food. Pathway: olfactory receptors (cranial nerve I) → olfactory bulbs → olfactory tracts → lateral olfactory area in temporal lobe of cerebral cortex.
2. Fred may have cataracts, a loss of transparency in the lens of the eye. Cataracts are often associated with age, smoking, and exposure to UV light.
3. Auricle → external auditory meatus → tympanic membrane → maleus → incus → stapes → oval window → perilymph of scala vestibuli and tympani → vestibular membrane → endolymph of cochlear duct → basilar membrane → spiral organ.

Chapter 17

1. Digestion and relaxation are controlled by increased stimulation of the parasympathetic division of the ANS. The salivary glands, pancreas, and liver will show increased secretion; the stomach and intestines will have increased activity; the gallbladder will have increased contractions; heart contractions will have decreased force and rate.
2. Stretch receptors in the colon → autonomic sensory neurons → sacral region of spinal cord (integrating center) → parasympathetic preganglionic neuron → terminal ganglion → parasympathetic postganglionic neuron → smooth muscle in colon, rectum, and sphincter (effectors).
3. Nicotine from the cigarette smoke binds to nicotinic receptors on skeletal muscles, mimicking the effect of acetylcholine and causing increased contractions ("twitches"). Nicotine also binds to nicotinic receptors on cells in the adrenal medulla, stimulating release of epinephrine and norepinephrine, which mimics fight-or-flight responses.

Chapter 18

1. The beta cells are one of the cell types in the pancreatic islets of the pancreas. In Type 1 diabetes, only the beta cells are destroyed; the rest of the pancreas is not affected. A successful transplantation of

the beta cells would enable the recipient to produce the hormone insulin and would cure the diabetes.

2. Amanda has an enlarged thyroid gland or goiter. The goiter is probably due to hypothyroidism, which is causing the weight gain, fatigue, mental dullness, and other symptoms.

3. The two adrenal glands are located superior to the two kidneys. The adrenals are about 4 cm high, 2 cm wide, and 1 cm thick. The outer adrenal cortex is composed of three layers: zona glomerulosa, zona fasciculata, and zona reticularis. The inner adrenal medulla is composed of chromaffin cells.

Chapter 19

1. Blood is composed of liquid blood plasma and the formed elements: red blood cells (erythrocytes), white blood cells (leukocytes), and platelets. Blood plasma contains water, proteins, electrolytes, gases, and nutrients. There are about 5 million RBCs, 5000 WBCs, and 250,000 platelets per mL of blood. Leukocytes include neutrophils, eosinophils, basophils, monocytes, and lymphocytes.

2. To determine the blood type, a drop of each of three different antibody solutions (antisera) are added to three separate blood drops. The solutions are anti-A, anti-B, and anti-Rh. In Josef's test, anti-B caused agglutination (clump formation) when added to his blood, indicating the presence of antigen B on the RBCs. Anti-A and anti-Rh did not cause agglutination, indicating the absence of these antigens.

3. Hemostasis occurred. The stages involved are vascular spasm (if an arteriole was cut), platelet plug formation, and coagulation (clotting).

Chapter 20

1. With a normal resting CO of about 5.25 liters/min, and a heart rate of 55 beats per minute, Arian's stroke volume would be 95 mL/beat. His CO during strenuous exercise is 6 times his resting CO, about 31,500 mL/min.

2. Rheumatic fever is caused by inflammation of the bicuspid and the aortic valves following a streptococcal infection. Antibodies produced by the immune system to destroy the bacteria may also attack and damage the heart valves. Heart problems later in life may be related to this damage.

3. Mr. Perkins is suffering from angina pectoris and has several risk factors for coronary artery disease such as smoking, obesity, and male gender. Cardiac angiography involves the use of a cardiac catheter to inject a radiopaque medium into the heart and its vessels. The angiogram may reveal blockages such as atherosclerotic plaques in his coronary arteries.

Chapter 21

1. After birth, the foramen ovale and ductus arteriosus close to establish the pulmonary circulation. The umbilical artery and veins close because the placenta is no longer functioning. The ductus venosus closes so that the liver is no longer bypassed.

2. Right hand: left ventricle → ascending aorta → arch of aorta → brachiocephalic trunk → right subclavian artery → right axillary artery → right brachial artery → right radial and ulnar arteries → right superficial palmar arch. Left hand: left ventricle → ascending aorta → arch of aorta → left subclavian artery → left axillary artery → left brachial artery → left radial and ulnar arteries → left superficial palmar arch.

3. A vascular (venous) sinus is a vein that lacks smooth muscle in its tunica media. Dense connective tissue replaces the tunica media and tunica externa.

Chapter 22

1. Tariq had a hypersensitivity reaction to an insect sting; he was probably allergic to the venom. He had a localized anaphylactic or type I reaction.

2. The site of the embedded splinter had probably become infected with bacteria. The red streaks are the lymphatic vessels that drain the infected area; the swollen tender bumps are the axillary lymph nodes, which are swollen due to the immune response to the infection.

3. Influenza vaccination introduces a weakened or killed virus (which will not cause the disease) to the body. The immune system recognizes the antigen and mounts a primary immune response. Upon exposure to the same flu virus that was in the vaccine, the body will produce a secondary response, which will usually prevent a case of the flu. This is artificially acquired active immunity.

Chapter 23

1. Normal average male volumes are 500 mL for tidal volume during quiet breathing, 3100 mL for inspiratory reserve volume, and 1200 mL for expiratory reserve volume. Average female volumes are less than average male volumes, because females are generally smaller than males.

2. The bones of the external nose are the frontal bone, the two nasal bones (the location of the fracture), and the maxillae. The external nose is also composed of the septal, lateral nasal, and alar cartilages, as well as skin, muscle, and mucous membrane.

3. The cerebral cortex can temporarily allow voluntary breath holding. P_{CO_2} and H^+ levels in the blood and CSF increase with breath holding, strongly stimulating the inspiratory area, which stimulates inspiration to begin again. Breathing will begin again even if the person loses consciousness.

Chapter 24

1. Katie lost the two upper central permanent incisors. The remaining deciduous teeth include the lower central incisors, an upper and lower pair of lateral incisors, an upper and lower pair of cuspids, and upper and lower pair of first molars, and upper and lower pair of second molars.

2. CCK promotes ejection of bile, secretion of pancreatic juice, and contraction of the pyloric sphincter. It also promotes pancreatic growth and enhances the effects of secretin. CCK acts on the hypothalamus to induce satiety (the feeling of fullness) and should therefore decrease the appetite.

3. The smaller left lobe of the liver is separated from the larger right lobe by the falciform ligament. The left lobe is inferior to the diaphragm in the epigastric region of the abdominopelvic cavity.

Chapter 25

1. Deb is eating a diet high in carbohydrates in order to store maximum amounts of glycogen in her skeletal muscles and liver. This practice is called carbohydrate loading. Glycogenolysis, the breakdown of stored glycogen, supplies the muscles with the glucose needed for ATP production via cellular respiration.

2. Mr. Hernandez was suffering from heat exhaustion caused by loss of fluids and electrolytes. Lack of NaCl causes muscle cramps, nausea and vomiting, dizziness, and fainting. Low blood pressure may also result.

3. Sara's normal growth may be maintained despite day-to-day fluctuations in food intake. Many factors govern food intake, including neurons in the hypothalamus, blood glucose level, amount of adipose tissue, CCK, distention of the GI tract, and body temperature.

Chapter 26

1. Without reabsorption, initially 105–125 mL of filtrate would be lost per minute, assuming normal glomerular filtration rate. Fluid loss from the blood would cause a decrease in blood pressure, and therefore a decrease in GBHP. When GBHP dropped below 45mmHg, filtration would stop (assuming normal CHP and BCOP) because NFP would be zero.

2. The urinary bladder can stretch considerably due to the presence of transitional epithelium, rugae, and three layers of smooth muscle in the detrusor muscle.

3. In females the urethra is about 4 cm long; in males the urethra is 15–20 cm long, including its passage through the penis, urogenital diaphragm, and prostate.

Chapter 27

1. Gary has water intoxication. The Na^+ concentration of his plasma and interstitial fluid is below normal. Water moved by osmosis into the cells, resulting in hypotonic intracellular fluid and water intoxication. Decreased plasma volume, due to water movement into the interstitial fluid, caused hypovolemic shock.

2. The average female body contains about 55% water, versus 60% water in the average male. Due to the influence of male and female hormones, the average female has relatively more subcutaneous fat (which contains very little water) and relatively less muscle and other tissues (which have higher water content) than the average male.

3. Excessive vomiting causes a loss of hydrochloric acid in gastric juice, and intake of antacids increases the amount of alkali in the body fluids, resulting in metabolic alkalosis. Vomiting also causes a loss of fluid and may cause dehydration.

4. (Step 1) pH = 7.30 indicates slight acidosis, which could be caused by elevated P_{CO_2} or lowered HCO_3^-. (Step 2) The HCO_3^- is lower than normal (20 mEq/liter), so (step 3) the cause is metabolic. (Step 4) the P_{CO_2} is lower than normal (32 mmHg), so hyperventilation is providing some compensation. Diagnosis: Henry has partially compensated metabolic acidosis. A possible cause is kidney damage that resulted from interruption of blood flow during the heart attack.

Chapter 28

1. LH (luteinizing hormone) is an anterior pituitary hormone that acts on both the male and female reproductive systems. In males, LH stimulates the Leydig cells of the testes to secrete testosterone.

2. The germinal epithelium, located on the surface of the ovaries, is a misnomer. The oocytes are located in the ovarian cortex, deep to the tunica albuginea. The tunica vaginalis covers the testes, superficial to the tunica albuginea. It forms from the peritoneum during the descent of the testes from the abdomen into the scrotum during fetal development.

3. No. The first polar body, formed by meiosis I, would contain half the homologous chromosome pairs, the other half being in the secondary oocyte. The second polar body, formed by meiosis II, would contain chromosomes identical to the ovum; however, fertilization would be by two different sperm, so the babies would still not be identical.

Chapter 29

1. Being able to taste PCT is an autosomal dominant trait. Because Kendra is a PCT nontaster, her genotype is homozygous recessive for this trait. Her parents must both be heterozygous for the trait. They exhibit the dominant phenotype for PCT tasting and are carriers of the recessive trait.

2. Because Huntington disease (HD) is a trinucleotide repeat disease, in which the number of repeats expand with each succeeding generation, it can occur in a child whose parents did not suffer HD. Also, because it is a dominant gene, one mutant allele of the *HD* gene is sufficient to cause HD.

3. All arteries carry blood away from the heart. The umbilical arteries, located in the fetus, carry deoxygenated blood away from the fetal heart towards the placenta where the blood will become oxygenated as it passes through the placenta.

Glossary

Pronunciation Key

1. The most strongly accented syllable appears in capital letters, for example, bilateral (bī-LAT-er-al) and diagnosis (dī-ag-NŌ-sis).
2. If there is a secondary accent, it is noted by a prime (′), for example, constitution (kon′-sti-TOO-shun) and physiology (fiz′-ē-OL-ō-jē). Any additional secondary accents are also noted by a prime, for example, decarboxylation (dē′-kar-bok′-si-LĀ-shun).
3. Vowels marked by a line above the letter are pronounced with the long sound, as in the following common words:

   ā as in māke   ō as in pōle
   ē as in bē   ū as in cute
   ī as in īvy
4. Vowels not marked by a line above the letter are pronounced with the short sound, as in the following words:

   a as in above or at   o as in not
   e as in bet   u as in bud
   i as in sip

5. Other vowel sounds are indicated as follows:

   oy as in oil
   oo as in root
6. Consonant sounds are pronounced as in the following words:

b as in bat	m as in mother
ch as in chair	n as in no
d as in dog	p as in pick
f as in father	r as in rib
g as in get	s as in so
h as in hat	t as in tea
j as in jump	v as in very
k as in can	w as in welcome
ks as in tax	z as in zero
kw as in quit	zh as in lesion
l as in let	

A

Abdomen (ab-DŌ-men or AB-dō-men) The area between the diaphragm and pelvis.

Abdominal (ab-DŌM-i-nal) **cavity** Superior portion of the abdominopelvic cavity that contains the stomach, spleen, liver, gallbladder, most of the small intestine, and part of the large intestine.

Abdominal thrust maneuver A first-aid procedure for choking. Employs a quick, upward thrust against the diaphragm that forces air out of the lungs with sufficient force to eject any lodged material. Also called the **Heimlich** (HĪM-lik) **maneuver.**

Abdominopelvic (ab-dom′-i-nō-PEL-vic) **cavity** Inferior component of the ventral body cavity that is subdivided into a superior abdominal cavity and an inferior pelvic cavity.

Abduction (ab-DUK-shun) Movement away from the midline of the body.

Abortion (a-BOR-shun) The premature loss (spontaneous) or removal (induced) of the embryo or nonviable fetus; miscarriage due to a failure in the normal process of developing or maturing.

Abscess (AB-ses) A localized collection of pus and liquefied tissue in a cavity.

Absorption (ab-SORP-shun) Intake of fluids or other substances by cells of the skin or mucous membranes; the passage of digested foods from the gastrointestinal tract into blood or lymph.

Absorptive (fed) state Metabolic state during which ingested nutrients are absorbed into the blood or lymph from the gastrointestinal tract.

Accessory duct A duct of the pancreas that empties into the duodenum about 2.5 cm (1 in.) superior to the ampulla of Vater (hepatopancreatic ampulla). Also called the **duct of Santorini** (san′-tō-RĒ-nē).

Accommodation (a-kom-ō-DĀ-shun) An increase in the curvature of the lens of the eye to adjust for near vision.

Acetabulum (as′-e-TAB-ū-lum) The rounded cavity on the external surface of the hip bone that receives the head of the femur.

Acetylcholine (as′-e-til-KŌ-lēn) **(ACh)** A neurotransmitter liberated by many peripheral nervous system neurons and some central nervous system neurons. It is excitatory at neuromuscular junctions but inhibitory at some other synapses (for example, it slows heart rate).

Achalasia (ak′-a-LĀ-zē-a) A condition, caused by malfunction of the myenteric plexus, in which the lower esophageal sphincter fails to relax normally as food approaches. A whole meal may become lodged in the esophagus and enter the stomach very slowly. Distension of the esophagus results in chest pain that is often confused with pain originating from the heart.

Acid (AS-id) A proton donor, or a substance that dissociates into hydrogen ions (H^+) and anions; characterized by an excess of hydrogen ions and a pH less than 7.

Acidosis (as-i-DŌ-sis) A condition in which blood pH is below 7.35. Also known as **acidemia.**

Acini (AS-i-nē) Groups of cells in the pancreas that secrete digestive enzymes.

Acoustic (a-KOOS-tik) Pertaining to sound or the sense of hearing.

Acquired immunodeficiency syndrome (AIDS) A fatal disease caused by the human immunodeficiency virus (HIV). Characterized by a positive HIV-antibody test, low helper T cell count, and certain indicator diseases (for example Kaposi's sarcoma,

Pneumocystis carinii pneumonia, tuberculosis, fungal diseases). Other symptoms include fever or night sweats, coughing, sore throat, fatigue, body aches, weight loss, and enlarged lymph nodes.

Acrosome (AK-rō-sōm) A lysosomelike organelle in the head of a sperm cell containing enzymes that facilitate the penetration of a sperm cell into a secondary oocyte.

Actin (AK-tin) A contractile protein that is part of thin filaments in muscle fibers.

Action potential An electrical signal that propagates along the membrane of a neuron or muscle fiber (cell); a rapid change in membrane potential that involves a depolarization followed by a repolarization. Also called a **nerve action potential** or **nerve impulse** as it relates to a neuron, and a **muscle action potential** as it relates to a muscle fiber.

Activation (ak′-ti-VĀ-shun) **energy** The minimum amount of energy required for a chemical reaction to occur.

Active transport The movement of substances across cell membranes against a concentration gradient, requiring the expenditure of cellular energy (ATP).

Acute (a-KŪT) Having rapid onset, severe symptoms, and a short course; not chronic.

Adaptation (ad′-ap-TĀ-shun) The adjustment of the pupil of the eye to changes in light intensity. The property by which a sensory neuron relays a decreased frequency of action potentials from a receptor, even though the strength of the stimulus remains constant; the decrease in perception of a sensation over time while the stimulus is still present.

Adduction (ad-DUK-shun) Movement toward the midline of the body.

Adenoids (AD-e-noyds) The pharyngeal tonsils.

Adenosine triphosphate (a-DEN-ō-sēn trī-FOS-fāt) **(ATP)** The main energy currency in living cells; used to transfer the chemical energy needed for metabolic reactions. ATP consists of the purine base *adenine* and the five-carbon sugar *ribose,* to which are added, in linear array, three *phosphate* groups.

Adenylate cyclase (a-DEN-i-lāt SĪ-klās) An enzyme that is activated when certain neurotransmitters or hormones bind to their receptors; the enzyme that converts ATP into cyclic AMP, an important second messenger.

Adhesion (ad-HĒ-zhun) Abnormal joining of parts to each other.

Adipocyte (AD-i-pō-sīt) Fat cell, derived from a fibroblast.

Adipose (AD-i-pōz) **tissue** Tissue composed of adipocytes specialized for triglyceride storage and present in the form of soft pads between various organs for support, protection, and insulation.

Adrenal cortex (a-DRĒ-nal KOR-teks) The outer portion of an adrenal gland, divided into three zones; the zona glomerulosa secretes mineralocorticoids, the zona fasciculata secretes glucocorticoids, and the zona reticularis secretes androgens.

Adrenal glands Two glands located superior to each kidney. Also called the **suprarenal** (soo′-pra-RĒ-nal) **glands.**

Adrenal medulla (me-DUL-a) The inner part of an adrenal gland, consisting of cells that secrete epinephrine, norepinephrine, and a small amount of dopamine in response to stimulation by sympathetic preganglionic neurons.

Adrenergic (ad′-ren-ER-jik) **neuron** A neuron that releases epinephrine (adrenaline) or norepinephrine (noradrenaline) as its neurotransmitter.

Adrenocorticotropic (ad-rē′-nō-kor-ti-kō-TRŌP-ik) **hormone (ACTH)** A hormone produced by the anterior pituitary that influences the production and secretion of certain hormones of the adrenal cortex.

Adventitia (ad-ven-TISH-a) The outermost covering of a structure or organ.

Aerobic (air-Ō-bik) Requiring molecular oxygen.

Afferent arteriole (AF-er-ent ar-TĒ-rē-ōl) A blood vessel of a kidney that divides into the capillary network called a glomerulus; there is one afferent arteriole for each glomerulus.

Agglutination (a-gloo′-ti-NĀ-shun) Clumping of microorganisms or blood cells, typically due to an antigen–antibody reaction.

Aggregated lymphatic follicles Clusters of lymph nodules that are most numerous in the ileum. Also called **Peyer's** (PĪ-erz) **patches.**

Albinism (AL-bin-izm) Abnormal, nonpathological, partial, or total absence of pigment in skin, hair, and eyes.

Albumin (al-BŪ-min) The most abundant (60%) and smallest of the plasma proteins; it is the main contributor to blood colloid osmotic pressure (BCOP).

Aldosterone (al-DOS-ter-ōn) A mineralocorticoid produced by the adrenal cortex that promotes sodium and water reabsorption by the kidneys and potassium excretion in urine.

Alkaline (AL-ka-līn) Containing more hydroxide ions (OH^-) than hydrogen ions (H^+); a pH higher than 7.

Alkalosis (al-ka-LŌ-sis) A condition in which blood pH is higher than 7.45. Also known as **alkalemia.**

Allantois (a-LAN-tō-is) A small, vascularized outpouching of the yolk sac that serves as an early site for blood formation and development of the urinary bladder.

Alleles (a-LĒLZ) Alternate forms of a single gene that control the same inherited trait (such as type A blood) and are located at the same position on homologous chromosomes.

Allergen (AL-er-jen) An antigen that evokes a hypersensitivity reaction.

Alpha (AL-fa) **cell** A type of cell in the pancreatic islets (islets of Langerhans) in the pancreas that secretes the hormone glucagon. Also termed an **A cell.**

Alpha receptor A type of receptor for norepinephrine and epinephrine; present on visceral effectors innervated by sympathetic postganglionic neurons.

Alveolar duct Branch of a respiratory bronchiole around which alveoli and alveolar sacs are arranged.

Alveolar macrophage (MAK-rō-fāj) Highly phagocytic cell found in the alveolar walls of the lungs. Also called a **dust cell.**

Alveolar (al-VĒ-ō-lar) **pressure** Air pressure within the lungs. Also called **intrapulmonic pressure.**

Alveolar sac A cluster of alveoli that share a common opening.

Alveolus (al-VĒ-ō-lus) A small hollow or cavity; an air sac in the lungs; milk-secreting portion of a mammary gland. *Plural is* **alveoli** (al-VĒ-ol-ī).

Alzheimer (ALTZ-hī-mer) **disease (AD)** Disabling neurological disorder characterized by dysfunction and death of specific cerebral neurons, resulting in widespread intellectual impairment, personality changes, and fluctuations in alertness.

Amnesia (am-NĒ-zē-a) A lack or loss of memory.

Amenorrhea (ā-men-ō-RĒ-a) Absence of menstruation.

Amino (a-MĒ-nō) **acid** An organic acid, containing an acidic carboxyl group ($-COOH$) and a basic amino group ($-NH_2$); the monomer used to synthesize polypeptides and proteins.

Amnion (AM-nē-on) A thin, protective fetal membrane that develops from the epiblast; holds the fetus suspended in amniotic fluid. Also called the "**bag of waters.**"

Amniotic (am′-nē-OT-ik) **fluid** Fluid in the amniotic cavity, the space between the developing embryo (or fetus) and amnion; the fluid is initially produced as a filtrate from maternal blood and later includes fetal urine. It functions as a shock absorber, helps regulate fetal body temperature, and helps prevent desiccation.

Amphiarthrosis (am′-fē-ar-THRŌ-sis) A slightly movable joint, in which the articulating bony surfaces are separated by fibrous connective tissue or fibrocartilage to which both are attached; types are syndesmosis and symphysis.

Ampulla (am-PUL-la) A saclike dilation of a canal or duct.

Anabolism (a-NAB-ō-lizm) Synthetic, energy-requiring reactions whereby small molecules are built up into larger ones.

Anaerobic (an-ār-Ō-bik) Not requiring oxygen.

Anal (Ā-nal) **canal** The last 2 or 3 cm (1 in.) of the rectum; opens to the exterior through the anus.

Anal column A longitudinal fold in the mucous membrane of the anal canal that contains a network of arteries and veins.

Anal triangle The subdivision of the female or male perineum that contains the anus.

Analgesia (an-al-JĒ-zē-a) Pain relief; absence of the sensation of pain.

Anaphylaxis (an′-a-fi-LAK-sis) A hypersensitivity (allergic) reaction in which IgE antibodies attach to mast cells and basophils, causing them to produce mediators of anaphylaxis (histamine, leukotrienes, kinins, and prostaglandins) that bring about increased blood permeability, increased smooth muscle contraction, and increased mucus production. Examples are hay fever, hives, and anaphylactic shock.

Anaphase (AN-a-fāz) The third stage of mitosis in which the chromatids that have separated at the centromeres move to opposite poles of the cell.

Anastomosis (a-nas-tō-MŌ-sis) An end-to-end union or joining of blood vessels, lymphatic vessels, or nerves.

Anatomical (an′-a-TOM-i-kal) **position** A position of the body universally used in anatomical descriptions in which the body is erect, the head is level, the eyes face forward, the upper limbs are at the sides, the palms face forward, and the feet are flat on the floor.

Anatomic dead space Spaces of the nose, pharynx, larynx, trachea, bronchi, and bronchioles totaling about 150 mL of the 500 mL in a quiet breath (tidal volume); air in the anatomic dead space does not reach the alveoli to participate in gas exchange.

Anatomy (a-NAT-ō-mē) The structure or study of structure of the body and the relation of its parts to each other.

Androgens (AN-drō-jenz) Masculinizing sex hormones produced by the testes in males and the adrenal cortex in both sexes; responsible for libido (sexual desire); the two main androgens are testosterone and dihydrotestosterone.

Anemia (a-NĒ-mē-a) Condition of the blood in which the number of functional red blood cells or their hemoglobin content is below normal.

Anesthesia (an′-es-THĒ-zē-a) A total or partial loss of feeling or sensation; may be general or local.

Aneuploid (an′-ū-PLOYD) A cell that has one or more chromosomes of a set added or deleted.

Aneurysm (AN-ū-rizm) A saclike enlargement of a blood vessel caused by a weakening of its wall.

Angina pectoris (an-JI-na *or* AN-ji-na PEK-tō-ris) A pain in the chest related to reduced coronary circulation due to coronary artery disease (CAD) or spasms of vascular smooth muscle in coronary arteries.

Angiotensin (an-jē-ō-TEN-sin) Either of two forms of a protein associated with regulation of blood pressure. Angiotensin I is produced by the action of renin on angiotensinogen and is converted by the action of ACE (angiotensin-converting enzyme) into angiotensin II, which stimulates aldosterone secretion by the adrenal cortex, stimulates the sensation of thirst, and causes vasoconstriction with resulting increase in systemic vascular resistance.

Anion (AN-ī-on) A negatively charged ion. An example is the chloride ion (Cl⁻).

Ankylosis (ang′-ki-LŌ-sis) Severe or complete loss of movement at a joint as the result of a disease process.

Anoxia (an-OK-sē-a) Deficiency of oxygen.

Antagonist (an-TAG-ō-nist) A muscle that has an action opposite that of the prime mover (agonist) and yields to the movement of the prime mover.

Antagonistic (an-tag-ō-NIST-ik) **effect** A hormonal interaction in which the effect of one hormone on a target cell is opposed by another hormone. For example, calcitonin (CT) lowers blood calcium level, whereas parathyroid hormone (PTH) raises it.

Anterior (an-TĒR-ē-or) Nearer to or at the front of the body. Equivalent to **ventral** in bipeds.

Anterior pituitary Anterior lobe of the pituitary gland. Also called the **adenohypophysis** (ad′-e-nō-hī-POF-i-sis).

Anterior root The structure composed of axons of motor (efferent) neurons that emerges from the anterior aspect of the spinal cord and extends laterally to join a posterior root, forming a spinal nerve. Also called a **ventral root.**

Anterolateral (an′-ter-ō-LAT-er-al) **pathway** Sensory pathway that conveys information related to pain, temperature, crude touch, pressure, tickle, and itch.

Antibody (AN-ti-bod′-ē) A protein produced by plasma cells in response to a specific antigen; the antibody combines with that antigen to neutralize, inhibit, or destroy it. Also called an **immunoglobulin** (im-ū-nō-GLOB-ū-lin) or **Ig.**

Antibody-mediated immunity That component of immunity in which B lymphocytes (B cells) develop into plasma cells that produce antibodies that destroy antigens. Also called **humoral** (HŪ-mor-al) **immunity.**

Anticoagulant (an-tī-cō-AG-ū-lant) A substance that can delay, suppress, or prevent the clotting of blood.

Antidiuretic (an′-ti-dī-ū-RET-ik) Substance that inhibits urine formation.

Antidiuretic hormone (ADH) Hormone produced by neurosecretory cells in the paraventricular and supraoptic nuclei of the hypothalamus that stimulates water reabsorption from kidney tubule cells into the blood and vasoconstriction of arterioles. Also called **vasopressin** (vāz-ō-PRES-in).

Antigen (AN-ti-jen) A substance that has immunogenicity (the ability to provoke an immune response) and reactivity (the ability to react with the antibodies or cells that result from the immune response); contraction of *anti*body *gen*erator. Also termed a **complete antigen.**

Antigen-presenting cell (APC) Special class of migratory cell that processes and presents antigens to T cells during an immune response; APCs include macrophages, B cells, and dendritic cells, which are present in the skin, mucous membranes, and lymph nodes.

Antiporter A transmembrane transporter protein that moves two substances, often Na$^+$ and another substance, in opposite directions across a plasma membrane. Also called a **countertransporter.**

Anulus fibrosus (AN-ū-lus fī-BRŌ-sus) A ring of fibrous tissue and fibrocartilage that encircles the pulpy substance (nucleus pulposus) of an intervertebral disc.

Anuria (an-Ū-rē-a) Absence of urine formation or daily urine output of less than 50 mL.

Anus (Ā-nus) The distal end and outlet of the rectum.

Aorta (ā-OR-ta) The main systemic trunk of the arterial system of the body that emerges from the left ventricle.

Aortic (ā-OR-tik) **body** Cluster of chemoreceptors on or near the arch of the aorta that respond to changes in blood levels of oxygen, carbon dioxide, and hydrogen ions (H$^+$).

Aortic reflex A reflex that helps maintain normal systemic blood pressure; initiated by baroreceptors in the wall of the ascending aorta and arch of the aorta. Nerve impulses from aortic baroreceptors reach the cardiovascular center via sensory axons of the vagus nerves (cranial nerve X).

Aperture (AP-er-chur) An opening or orifice.

Apex (Ā-peks) The pointed end of a conical structure, such as the apex of the heart.

Aphasia (a-FA-zē-a) Loss of ability to express oneself properly through speech or loss of verbal comprehension.

Apnea (AP-nē-a) Temporary cessation of breathing.

Apneustic (ap-NOO-stik) **area** A part of the respiratory center in the pons that sends stimulatory nerve impulses to the inspiratory area that activate and prolong inhalation and inhibit exhalation.

Apocrine (AP-ō-krin) **gland** A type of gland in which the secretory products gather at the free end of the secreting cell and are pinched off, along with some of the cytoplasm, to become the secretion, as in mammary glands.

Aponeurosis (ap′-ō-noo-RŌ-sis) A sheetlike tendon joining one muscle with another or with bone.

Apoptosis (ap′-ō-TŌ-sis *or* ap′-ōp-TŌ-sis) Programmed cell death; a normal type of cell death that removes unneeded cells during embryological development, regulates the number of cells in tissues, and eliminates many potentially dangerous cells such as cancer cells. During apoptosis, the DNA fragments, the nucleus condenses, mitochondria cease to function, and the cytoplasm shrinks, but the plasma membrane remains intact. Phagocytes engulf and digest the apoptotic cells, and an inflammatory response does not occur.

Appositional (ap′-ō-ZISH-o-nal) **growth** Growth due to surface deposition of material, as in the growth in diameter of cartilage and bone. Also called **exogenous** (eks-OJ-e-nus) **growth.**

Aqueous humor (AK-wē-us HŪ-mer) The watery fluid, similar in composition to cerebrospinal fluid, that fills the anterior cavity of the eye.

Arachnoid (a-RAK-noyd) **mater** The middle of the three meninges (coverings) of the brain and spinal cord. Also termed the **arachnoid.**

Arachnoid villus (VIL-us) Berrylike tuft of the arachnoid mater that protrudes into the superior sagittal sinus and through which cerebrospinal fluid is reabsorbed into the bloodstream.

Arbor vitae (AR-bor VĪ-tē) The white matter tracts of the cerebellum, which have a treelike appearance when seen in midsagittal section.

Arch of the aorta The most superior portion of the aorta, lying between the ascending and descending segments of the aorta.

Areola (a-RĒ-ō-la) Any tiny space in a tissue. The pigmented ring around the nipple of the breast.

Arm The part of the upper limb from the shoulder to the elbow.

Arousal (a-ROW-zal) Awakening from sleep, a response due to stimulation of the reticular activating system (RAS).

Arrector pili (a-REK-tor PI-lē) Smooth muscles attached to hairs; contraction pulls the hairs into a vertical position, resulting in "goose bumps."

Arrhythmia (a-RITH-mē-a) An irregular heart rhythm. Also called a **dysrhythmia.**

Arteriole (ar-TĒ-rē-ōl) A small, almost microscopic, artery that delivers blood to a capillary.

Arteriosclerosis (ar-tē-rē-ō-skle-RŌ-sis) Group of diseases characterized by thickening of the walls of arteries and loss of elasticity.

Artery (AR-ter-ē) A blood vessel that carries blood away from the heart.

Arthritis (ar-THRI-tis) Inflammation of a joint.

Arthrology (ar-THROL-ō-jē) The study or description of joints.

Arthroscopy (ar-THROS-co-pē) A procedure for examining the interior of a joint, usually the knee, by inserting an arthroscope into a small incision; used to determine extent of damage, remove torn cartilage, repair cruciate ligaments, and obtain samples for analysis.

Arthrosis (ar-THRŌ-sis) A joint or articulation.

Articular (ar-TIK-ū-lar) **capsule** Sleevelike structure around a synovial joint composed of a fibrous capsule and a synovial membrane.

Articular cartilage (KAR-ti-lij) Hyaline cartilage attached to articular bone surfaces.

Articular disc Fibrocartilage pad between articular surfaces of bones of some synovial joints. Also called a **meniscus** (men-IS-kus).

Articulation (ar-tik′-ū-LĀ-shun) A joint; a point of contact between bones, cartilage and bones, or teeth and bones.

Arytenoid (ar′-i-TĒ-noyd) **cartilages** A pair of small, pyramidal cartilages of the larynx that attach to the vocal folds and intrinsic pharyngeal muscles and can move the vocal folds.

Ascending colon (KŌ-lon) The part of the large intestine that passes superiorly from the cecum to the inferior border of the liver, where it bends at the right colic (hepatic) flexure to become the transverse colon.

Ascites (as-SĪ-tēz) Abnormal accumulation of serous fluid in the peritoneal cavity.

Association areas Large cortical regions on the lateral surfaces of the occipital, parietal, and temporal lobes and on the frontal lobes anterior to the motor areas connected by many motor and sensory axons to other parts of the cortex. The association areas are concerned with motor patterns, memory, concepts of word-hearing and word-seeing, reasoning, will, judgment, and personality traits.

Asthma (AZ-ma) Usually allergic reaction characterized by smooth muscle spasms in bronchi resulting in wheezing and difficult breathing. Also called **bronchial asthma.**

Astigmatism (a-STIG-ma-tizm) An irregularity of the lens or cornea of the eye causing the image to be out of focus and producing faulty vision.

Astrocyte (AS-trō-sīt) A neuroglial cell having a star shape that participates in brain development and the metabolism of neurotransmitters, helps form the blood–brain barrier, helps maintain the proper balance of K$^+$ for generation of nerve impulses, and provides a link between neurons and blood vessels.

Ataxia (a-TAK-sē-a) A lack of muscular coordination, lack of precision.

Atherosclerotic (ath′-er-ō-skle-RO-tic) **plaque** (PLAK) A lesion that results from accumulated cholesterol and smooth muscle fibers (cells) of the tunica media of an artery; may become obstructive.

Atom Unit of matter that makes up a chemical element; consists of a nucleus (containing positively charged protons and uncharged neutrons) and negatively charged electrons that orbit the nucleus.

Atomic mass (weight) Average mass of all stable atoms of an element, reflecting the relative proportion of atoms with different mass numbers.

Atomic number Number of protons in an atom.

Atresia (a-TRĒ-zē-a) Degeneration and reabsorption of an ovarian follicle before it fully matures and ruptures; abnormal closure of a passage, or absence of a normal body opening.

Atrial fibrillation (Ā-trē-al fib-ri-LĀ-shun) Asynchronous contraction of cardiac muscle fibers in the atria that results in the cessation of atrial pumping.

Atrial natriuretic (na′-trē-ū-RET-ik) **peptide** (ANP) Peptide hormone, produced by the atria of the heart in response to stretching, that inhibits aldosterone production and thus lowers blood pressure; causes natriuresis, increased urinary excretion of sodium.

Atrioventricular (AV) (ā′-trē-ō-ven-TRIK-ū-lar) **bundle** The part of the conduction system of the heart that begins at the atrioventricular (AV) node, passes through the cardiac skeleton separating the atria and the ventricles, then extends a short distance down the interventricular septum before splitting into right and left bundle branches. Also called the **bundle of His** (HISS).

Atrioventricular (AV) node The part of the conduction system of the heart made up of a compact mass of conducting cells located in the septum between the two atria.

Atrioventricular (AV) valve A heart valve made up of membranous flaps or cusps that allows blood to flow in one direction only, from an atrium into a ventricle.

Atrium (Ā-trē-um) A superior chamber of the heart.

Atrophy (AT-rō-fē) Wasting away or decrease in size of a part, due to a failure, abnormality of nutrition, or lack of use.

Auditory ossicle (AW-di-tō-rē OS-si-kul) One of the three small bones of the middle ear called the **malleus, incus,** and **stapes.**

Auditory tube The tube that connects the middle ear with the nose and nasopharynx region of the throat. Also called the **Eustachian** (ū-STĀ-shun *or* ū-STĀ-kē-an) **tube** or **pharyngotympanic tube.**

Auscultation (aws-kul-TĀ-shun) Examination by listening to sounds in the body.

Autocrine (AW-tō-krin) Local hormone, such as interleukin-2, that acts on the same cell that secreted it.

Autoimmunity An immunological response against a person's own tissues.

Autolysis (aw-TOL-i-sis) Self-destruction of cells by their own lysosomal digestive enzymes after death or in a pathological process.

Autonomic ganglion (aw′-tō-NOM-ik GANG-lē-on) A cluster of cell bodies of sympathetic or parasympathetic neurons located outside the central nervous system.

Autonomic nervous system (ANS) Visceral sensory (afferent) and visceral motor (efferent) neurons. Autonomic motor neurons, both sympathetic and parasympathetic, conduct nerve impulses from the central nervous system to smooth muscle, cardiac muscle, and glands. So named because this part of the nervous system was thought to be self-governing or spontaneous.

Autonomic plexus (PLEK-sus) A network of sympathetic and parasympathetic axons; examples are the cardiac, celiac, and pelvic plexuses, which are located in the thorax, abdomen, and pelvis, respectively.

Autophagy (aw-TOF-a-jē) Process by which worn-out organelles are digested within lysosomes.

Autopsy (AW-top-sē) The examination of the body after death.

Autoregulation (aw-tō-reg-ū-LĀ-shun) A local, automatic adjustment of blood flow in a given region of the body in response to tissue needs.

Autorhythmic cells Cardiac or smooth muscle fibers that are self-excitable (generate impulses without an external stimulus); act as the heart's pacemaker and conduct the pacing impulse through the conduction system of the heart; self-excitable neurons in the central nervous system, as in the inspiratory area of the brain stem.

Autosome (AW-tō-sōm) Any chromosome other than the X and Y chromosomes (sex chromosomes).

Axilla (ak-SIL-a) The small hollow beneath the arm where it joins the body at the shoulders. Also called the **armpit.**

Axon (AK-son) The usually single, long process of a nerve cell that propagates a nerve impulse toward the axon terminals.

Axon terminal Terminal branch of an axon where synaptic vesicles undergo exocytosis to release neurotransmitter molecules.

Azygos (AZ-ī-gos) An anatomical structure that is not paired; occurring singly.

B

B cell A lymphocyte that can develop into a clone of antibody-producing plasma cells or memory cells when properly stimulated by a specific antigen.

Babinski (ba-BIN-skē) **sign** Extension of the great toe, with or without fanning of the other toes, in response to stimulation of the outer margin of the sole; normal up to 18 months of age and indicative of damage to descending motor pathways such as the corticospinal tracts after that.

Back The posterior part of the body; the dorsum.

Ball-and-socket joint A synovial joint in which the rounded surface of one bone moves within a cup-shaped depression or socket of another bone, as in the shoulder or hip joint. Also called a **spheroid** (SFĒ-royd) **joint.**

Baroreceptor (bar′-ō-re-SEP-tor) Neuron capable of responding to changes in blood, air, or fluid pressure. Also called a **pressoreceptor.**

Basal ganglia (GANG-glē-a) Paired clusters of gray matter deep in each cerebral hemisphere including the globus pallidus, putamen, and caudate nucleus. Together, the caudate nucleus and putamen are known as the **corpus striatum.** Nearby structures that are functionally linked to the basal ganglia are the substantia nigra of the midbrain and the subthalamic nuclei of the diencephalon.

Basal metabolic (BĀ-sal met′-a-BOL-ik) **rate** (BMR) The rate of metabolism measured under standard or basal conditions (awake, at rest, fasting).

Base A nonacid or a proton acceptor, characterized by excess of hydroxide ions (OH⁻) and a pH greater than 7. A ring-shaped, nitrogen-containing organic molecule that is one of the components of a nucleotide, namely, adenine, guanine, cytosine, thymine, and uracil; also known as a **nitrogenous base.**

Basement membrane Thin, extracellular layer between epithelium and connective tissue consisting of a basal lamina and a reticular lamina.

Basilar (BĀS-i-lar) **membrane** A membrane in the cochlea of the internal ear that separates the cochlear duct from the scala tympani and on which the spiral organ (organ of Corti) rests.

Basophil (BĀ-sō-fil) A type of white blood cell characterized by a pale nucleus and large granules that stain blue-purple with basic dyes.

Belly The abdomen. The gaster or prominent, fleshy part of a skeletal muscle.

Beta (BĀ-ta) **cell** A type of cell in the pancreatic islets (islets of Langerhans) in the pancreas that secretes the hormone insulin. Also termed a **B cell.**

Beta receptor A type of adrenergic receptor for epinephrine and norepinephrine; found on visceral effectors innervated by sympathetic postganglionic neurons.

Bicuspid (bī-KUS-pid) **valve** Atrioventricular (AV) valve on the left side of the heart. Also called the **mitral valve.**

Bilateral (bī-LAT-er-al) Pertaining to two sides of the body.

Bile (BĪL) A secretion of the liver consisting of water, bile salts, bile pigments, cholesterol, lecithin, and several ions; it emulsifies lipids prior to their digestion.

Bilirubin (bil-ē-ROO-bin) An orange pigment that is one of the end products of hemoglobin breakdown in the hepatocytes and is excreted as a waste material in bile.

Blastocele (BLAS-tō-sēl) The fluid-filled cavity within the blastocyst.

Blastocyst (BLAS-tō-sist) In the development of an embryo, a hollow ball of cells that consists of a blastocele (the internal cavity), trophoblast (outer cells), and inner cell mass.

Blastomere (BLAS-tō-mēr) One of the cells resulting from the cleavage of a fertilized ovum.

Blind spot Area in the retina at the end of the optic nerve (cranial nerve II) in which there are no photoreceptors.

Blood The fluid that circulates through the heart, arteries, capillaries, and veins and that constitutes the chief means of transport within the body.

Blood–brain barrier (BBB) A barrier consisting of specialized brain capillaries and astrocytes that prevents the passage of materials from the blood to the cerebrospinal fluid and brain.

Blood island Isolated mass of mesoderm derived from angioblasts and from which blood vessels develop.

Blood pressure (BP) Force exerted by blood against the walls of blood vessels due to contraction of the heart and influenced by the elasticity of the vessel walls; clinically, a measure of the pressure in arteries during ventricular systole and ventricular diastole. *See also* **mean arterial blood pressure.**

Blood reservoir (REZ-er-vwar) Systemic veins that contain large amounts of blood that can be moved quickly to parts of the body requiring the blood.

Blood–testis barrier (BTB) A barrier formed by Sertoli cells that prevents an immune response against antigens produced by spermatogenic cells by isolating the cells from the blood.

Body cavity A space within the body that contains various internal organs.

Body fluid Body water and its dissolved substances; constitutes about 60% of total body mass.

Bohr (BŌR) **effect** In an acidic environment, oxygen unloads more readily from hemoglobin because when hydrogen ions (H^+) bind to hemoglobin, they alter the structure of hemoglobin, thereby reducing its oxygen-carrying capacity.

Bolus (BŌ-lus) A soft, rounded mass, usually food, that is swallowed.

Bony labyrinth (LAB-i-rinth) A series of cavities within the petrous portion of the temporal bone forming the vestibule, cochlea, and semicircular canals of the inner ear.

Brachial plexus (BRĀ-kē-al PLEK-sus) A network of nerve axons of the ventral rami of spinal nerves C5, C6, C7, C8, and T1. The nerves that emerge from the brachial plexus supply the upper limb.

Bradycardia (brād′-i-KAR-dē-a) A slow resting heart or pulse rate (under 60 beats per minute).

Brain The part of the central nervous system contained within the cranial cavity.

Brain stem The portion of the brain immediately superior to the spinal cord, made up of the medulla oblongata, pons, and midbrain.

Brain waves Electrical signals that can be recorded from the skin of the head due to electrical activity of brain neurons.

Broad ligament A double fold of parietal peritoneum attaching the uterus to the side of the pelvic cavity.

Broca's (BRŌ-kaz) **area** Motor area of the brain in the frontal lobe that translates thoughts into speech. Also called the **motor speech area.**

Bohr effect Shifting of the oxygen–hemoglobin dissociation curve to the right when pH decreases; at any given partial pressure of oxygen, hemoglobin is less saturated with O_2 at lower pH.

Bronchi (BRONG-kē) Branches of the respiratory passageway including primary bronchi (the two divisions of the trachea), secondary or lobar bronchi (divisions of the primary bronchi that are distributed to the lobes of the lung), and tertiary or segmental bronchi (divisions of the secondary bronchi that are distributed to bronchopulmonary segments of the lung). *Singular is* **bronchus.**

Bronchial tree The trachea, bronchi, and their branching structures up to and including the terminal bronchioles.

Bronchiole (BRONG-kē-ōl) Branch of a tertiary bronchus further dividing into terminal bronchioles (distributed to lobules of the lung), which divide into respiratory bronchioles (distributed to alveolar sacs).

Bronchitis (brong-KĪ-tis) Inflammation of the mucous membrane of the bronchial tree; characterized by hypertrophy and hyperplasia of seromucous glands and goblet cells that line the bronchi and which results in a productive cough.

Bronchopulmonary (brong′-kō-PUL-mō-ner-ē) **segment** One of the smaller divisions of a lobe of a lung supplied by its own branches of a bronchus.

Buccal (BUK-al) Pertaining to the cheek or mouth.

Buffer (BUF-er) **system** A pair of chemicals—one a weak acid and the other the salt of the weak acid, which functions as a weak base—that resists changes in pH.

Bulbourethral (bul′-bō-ū-RĒ-thral) **gland** One of a pair of glands located inferior to the prostate on either side of the urethra that secretes an alkaline fluid into the cavernous urethra. Also called a **Cowper's** (KOW-perz) **gland.**

Bulimia (boo-LIM-ē-a *or* boo-LĒ-mē-a) A disorder characterized by overeating at least twice a week followed by purging by self-induced vomiting, strict dieting or fasting, vigorous exercise, or use of laxatives or diuretics. Also called **binge–purge syndrome.**

Bulk flow The movement of large numbers of ions, molecules, or particles in the same direction due to pressure differences (osmotic, hydrostatic, or air pressure).

Bundle branch One of the two branches of the atrioventricular (AV) bundle made up of specialized muscle fibers (cells) that transmit electrical impulses to the ventricles.

Bursa (BUR-sa) A sac or pouch of synovial fluid located at friction points, especially about joints.

Buttocks (BUT-oks) The two fleshy masses on the posterior aspect of the inferior trunk, formed by the gluteal muscles.

C

Calcaneal (kal-KĀ-ne-al) **tendon** The tendon of the soleus, gastrocnemius, and plantaris muscles at the back of the heel. Also called the **Achilles** (a-KIL-ēz) **tendon.**

Calcification (kal-si-fi-KĀ-shun) Deposition of mineral salts, primarily hydroxyapatite, in a framework formed by collagen fibers in which the tissue hardens. Also called **mineralization** (min′-e-ral-i-ZĀ-shun).

Calcitonin (kal-si-TŌ-nin) **(CT)** A hormone produced by the parafollicular cells of the thyroid gland that can lower the amount of blood calcium and phosphates by inhibiting bone resorption (breakdown of bone matrix) and by accelerating uptake of calcium and phosphates into bone matrix.

Calculus (KAL-kū-lus) A stone, or insoluble mass of crystallized salts or other material, formed within the body, as in the gallbladder, kidney, or urinary bladder.

Callus (KAL-lus) A growth of new bone tissue in and around a fractured area, ultimately replaced by mature bone. An acquired, localized thickening.

Calorie (KAL-ō-rē) A unit of heat. A calorie (cal) is the standard unit and is the amount of heat needed to raise the temperature of 1 g of water from 14°C to 15°C. The **kilocalorie (kcal)** or **Calorie** (spelled with an uppercase C), used to express the caloric value of foods and to measure metabolic rate, is equal to 1000 cal.

Calyx (KĀL-iks) Any cuplike division of the kidney pelvis. *Plural is* **calyces** (KĀ-li-sēz).

Canal (ka-NAL) A narrow tube, channel, or passageway.

Canaliculus (kan′-a-LIK-ū-lus) A small channel or canal, as in bones, where they connect lacunae. *Plural is* **canaliculi** (kan′-a-LIK-ū-lī).

Cancellous (KAN-sel-us) Having a reticular or latticework structure, as in spongy tissue of bone.

Capacitation (ka′-pas-i-TĀ-shun) The functional changes that sperm undergo in the female reproductive tract that allow them to fertilize a secondary oocyte.

Capillary (KAP-i-lar′-ē) A microscopic blood vessel located between an arteriole and venule through which materials are exchanged between blood and interstitial fluid.

Carbohydrate (kar′-bō-HĪ-drāt) An organic compound containing carbon, hydrogen, and oxygen in a particular amount and arrangement and composed of monosaccharide subunits; usually has the general formula $(CH_2O)n$.

Carcinogen (kar-SIN-ō-jen) A chemical substance or radiation that causes cancer.

Cardiac (KAR-dē-ak) **arrest** Cessation of an effective heartbeat in which the heart is completely stopped or in ventricular fibrillation.

Cardiac cycle A complete heartbeat consisting of systole (contraction) and diastole (relaxation) of both atria plus systole and diastole of both ventricles.

Cardiac muscle Striated muscle fibers (cells) that form the wall of the heart; stimulated by an intrinsic conduction system and autonomic motor neurons.

Cardiac notch An angular notch in the anterior border of the left lung into which part of the heart fits.

Cardiac output (CO) The volume of blood pumped from one ventricle of the heart (usually measured from the left ventricle) in 1 min; normally about 5.2 liters/min in an adult at rest.

Cardiac reserve The maximum percentage that cardiac output can increase above normal.

Cardinal ligament A ligament of the uterus, extending laterally from the cervix and vagina as a continuation of the broad ligament.

Cardiology (kar-dē-OL-ō-jē) The study of the heart and diseases associated with it.

Cardiovascular (kar-dē-ō-VAS-kū-lar) **center** Groups of neurons scattered within the medulla oblongata that regulate heart rate, force of contraction, and blood vessel diameter.

Carotene (KAR-o-tēn) Antioxidant precursor of vitamin A, which is needed for synthesis of photopigments; yellow-orange pigment present in the stratum corneum of the epidermis. Accounts for the yellowish coloration of skin. Also termed **beta-carotene.**

Carotid (ka-ROT-id) **body** Cluster of chemoreceptors on or near the carotid sinus that respond to changes in blood levels of oxygen, carbon dioxide, and hydrogen ions.

Carotid sinus A dilated region of the internal carotid artery just above the point where it branches from the common carotid artery; it contains baroreceptors that monitor blood pressure.

Carotid sinus reflex A reflex that helps maintain normal blood pressure in the brain. Nerve impulses propagate from the carotid sinus baroreceptors over sensory axons in the glossopharyngeal nerves (cranial nerve IX) to the cardiovascular center in the medulla oblongata.

Carpal bones The eight bones of the wrist. Also called **carpals.**

Carpus (KAR-pus) A collective term for the eight bones of the wrist.

Cartilage (KAR-ti-lij) A type of connective tissue consisting of chondrocytes in lacunae embedded in a dense network of collagen and elastic fibers and a matrix of chondroitin sulfate.

Cartilaginous (kar′-ti-LAJ-i-nus) **joint** A joint without a synovial (joint) cavity where the articulating bones are held tightly together by cartilage, allowing little or no movement.

Cast A small mass of hardened material formed within a cavity in the body and then discharged from the body; can originate in different areas and can be composed of various materials.

Catabolism (ka-TAB-ō-lizm) Chemical reactions that break down complex organic compounds into simple ones, with the net release of energy.

Catalyst (KAT-a-list) A substance that speeds up a chemical reaction without itself being altered; an enzyme.

Cataract (KAT-a-rakt) Loss of transparency of the lens of the eye or its capsule or both.

Cation (KAT-ī-on) A positively charged ion. An example is a sodium ion (Na^+).

Cauda equina (KAW-da ē-KWĪ-na) A tail-like array of roots of spinal nerves at the inferior end of the spinal cord.

Caudal (KAW-dal) Pertaining to any tail-like structure; inferior in position.

Cecum (SĒ-kum) A blind pouch at the proximal end of the large intestine that attaches to the ileum.

Celiac plexus (PLEK-sus) A large mass of autonomic ganglia and axons located at the level of the superior part of the first lumbar vertebra. Also called the **solar plexus.**

Cell The basic structural and functional unit of all organisms; the smallest structure capable of performing all the activities vital to life.

Cell cycle Growth and division of a single cell into two daughter cells; consists of interphase and cell division.

Cell division Process by which a cell reproduces itself that consists of a nuclear division (mitosis) and a cytoplasmic division (cytokinesis); types include somatic and reproductive cell division.

Cell-mediated immunity That component of immunity in which specially sensitized T lymphocytes (T cells) attach to antigens to destroy them. Also called **cellular immunity.**

Cementum (se-MEN-tum) Calcified tissue covering the root of a tooth.

Center of ossification (os′-i-fi-KĀ-shun) An area in the cartilage model of a future bone where the cartilage cells hypertrophy and then secrete enzymes that result in the calcification of their matrix, resulting in the death of the cartilage cells, followed by the invasion of the area by osteoblasts that then lay down bone.

Central canal A microscopic tube running the length of the spinal cord in the gray commissure. A circular channel running longitudinally in the center of an osteon (Haversian system) of mature compact bone, containing blood and lymphatic vessels and nerves. Also called a **Haversian** (ha-VER-shan) **canal.**

Central fovea (FŌ-vē-a) A depression in the center of the macula lutea of the retina, containing cones only and lacking blood vessels; the area of highest visual acuity (sharpness of vision).

Central nervous system (CNS) That portion of the nervous system that consists of the brain and spinal cord.

Centrioles (SEN-trē-ōlz) Paired, cylindrical structures of a centrosome, each consisting of a ring of microtubules and arranged at right angles to each other.

Centromere (SEN-trō-mēr) The constricted portion of a chromosome where the two chromatids are joined; serves as the point of attachment for the microtubules that pull chromatids during anaphase of cell division.

Centrosome (SEN-trō-sōm) A dense network of small protein fibers near the nucleus of a cell, containing a pair of centrioles and pericentriolar material.

Cephalic (se-FAL-ik) Pertaining to the head; superior in position.

Cerebellar peduncle (ser-e-BEL-ar pe-DUNG-kul) A bundle of nerve axons connecting the cerebellum with the brain stem.

Cerebellum (ser-e-BEL-um) The part of the brain lying posterior to the medulla oblongata and pons; governs balance and coordinates skilled movements.

Cerebral aqueduct (SER-ē-bral AK-we-dukt) A channel through the midbrain connecting the third and fourth ventricles and containing cerebrospinal fluid. Also termed the **aqueduct of Sylvius.**

Cerebral arterial circle A ring of arteries forming an anastomosis at the base of the brain between the internal carotid and basilar arteries and arteries supplying the cerebral cortex. Also called the **circle of Willis.**

Cerebral cortex The surface of the cerebral hemispheres, 2–4 mm thick, consisting of gray matter; arranged in six layers of neuronal cell bodies in most areas.

Cerebral peduncle One of a pair of nerve axon bundles located on the anterior surface of the midbrain, conducting nerve impulses between the pons and the cerebral hemispheres.

Cerebrospinal (se-rē′-brō-SPĪ-nal) **fluid (CSF)** A fluid produced by ependymal cells that cover choroid plexuses in the ventricles of the brain; the fluid circulates in the ventricles, the central canal, and the subarachnoid space around the brain and spinal cord.

Cerebrovascular (se rē′-brō-VAS-kū-lar) **accident (CVA)** Destruction of brain tissue (infarction) resulting from obstruction or rupture of blood vessels that supply the brain. Also called a **stroke** or **brain attack.**

Cerebrum (SER-e-brum *or* se-RĒ-brum) The two hemispheres of the forebrain (derived from the telencephalon), making up the largest part of the brain.

Cerumen (se-ROO-men) Waxlike secretion produced by ceruminous glands in the external auditory meatus (ear canal). Also termed **ear wax.**

Ceruminous (se-ROO-mi-nus) **gland** A modified sudoriferous (sweat) gland in the external auditory meatus that secretes cerumen (ear wax).

Cervical ganglion (SER-vi-kul GANG-glē-on) A cluster of cell bodies of postganglionic sympathetic neurons located in the neck, near the vertebral column.

Cervical plexus (PLEK-sus) A network formed by nerve axons from the ventral rami of the first four cervical nerves and receiving gray rami communicates from the superior cervical ganglion.

Cervix (SER-viks) Neck; any constricted portion of an organ, such as the inferior cylindrical part of the uterus.

Chemical bond Force of attraction in a molecule or compound that holds its atoms together. Examples include ionic and covalent bonds.

Chemical element Unit of matter that cannot be decomposed into a simpler substance by ordinary chemical reactions. Examples include hydrogen (H), carbon (C), and oxygen (O).

Chemically gated channel A channel in a membrane that opens and closes in response to a specific chemical stimulus, such as a neurotransmitter, hormone, or specific type of ion.

Chemical reaction The combination or separation of atoms in which chemical bonds are formed or broken and new products with different properties are produced.

Chemiosmosis (kem′-ē-oz-MŌ-sis) Mechanism for ATP generation that links chemical reactions (electrons passing along the electron transport chain) with pumping of H^+ out of the mitochondrial matrix. ATP synthesis occurs as H^+ diffuse back into the mitochondrial matrix through special H^+ channels in the membrane.

Chemoreceptor (kē′-mō-rē-SEP-tor) Sensory receptor that detects the presence of a specific chemical.

Chemotaxis (kē-mō-TAK-sis) Attraction of phagocytes to microbes by a chemical stimulus.

Chiasm (KĪ-azm) A crossing; especially the crossing of axons in the optic nerve (cranial nerve II).

Chief cell The secreting cell of a gastric gland that produces pepsinogen, the precursor of the enzyme pepsin, and the enzyme gastric lipase. Also called a **zymogenic** (zī′-mō-JEN-ik) **cell.** Cell in the parathyroid glands that secretes parathyroid hormone (PTH). Also called a **principal cell.**

Chiropractic (kī-rō-PRAK-tik) A system of treating disease by using one's hands to manipulate body parts, mostly the vertebral column.

Chloride shift Exchange of bicarbonate ions (HCO_3^-) for chloride ions (Cl^-) between red blood cells and plasma; maintains electrical balance inside red blood cells as bicarbonate ions are produced or eliminated during respiration.

Cholecystectomy (kō′-lē-sis-TEK-tō-mē) Surgical removal of the gallbladder.

Cholecystitis (kō′-lē-sis-TĪ-tis) Inflammation of the gallbladder.

Cholesterol (kō-LES-te-rol) Classified as a lipid, the most abundant steroid in animal tissues; located in cell membranes and used for the synthesis of steroid hormones and bile salts.

Cholinergic (kō′-lin-ER-jik) **neuron** A neuron that liberates acetylcholine as its neurotransmitter.

Chondrocyte (KON-drō-sīt) Cell of mature cartilage.

Chondroitin (kon-DROY-tin) **sulfate** An amorphous matrix material found outside connective tissue cells.

Chordae tendineae (KOR-dē TEN-di-nē-ē) Tendonlike, fibrous cords that connect atrioventricular valves of the heart with papillary muscles.

Chorion (KŌ-rē-on) The most superficial fetal membrane that becomes the principal embryonic portion of the placenta; serves a protective and nutritive function.

Chorionic villi (kō-rē-ON-ik VIL-lī) Fingerlike projections of the chorion that grow into the decidua basalis of the endometrium and contain fetal blood vessels.

Choroid (KŌ-royd) One of the vascular coats of the eyeball.

Choroid plexus (PLEK-sus) A network of capillaries located in the roof of each of the four ventricles of the brain; ependymal cells around choroid plexuses produce cerebrospinal fluid.

Chromaffin (KRŌ-maf-in) **cell** Cell that has an affinity for chrome salts, due in part to the presence of the precursors of the neurotransmitter epinephrine; found, among other places, in the adrenal medulla.

Chromatid (KRŌ-ma-tid) One of a pair of identical connected nucleoprotein strands that are joined at the centromere and separate during cell division, each becoming a chromosome of one of the two daughter cells.

Chromatin (KRŌ-ma-tin) The threadlike mass of genetic material, consisting of DNA and histone proteins, that is present in the nucleus of a nondividing or interphase cell.

Chromatolysis (krō-ma-TOL-i-sis) The breakdown of Nissl bodies into finely granular masses in the cell body of a neuron whose axon has been damaged.

Chromosome (KRŌ-mō-sōm) One of the small, threadlike structures in the nucleus of a cell, normally 46 in a human diploid cell, that bears the genetic material; composed of DNA and proteins (histones) that form a delicate chromatin thread during interphase; becomes packaged into compact rodlike structures that are visible under the light microscope during cell division.

Chronic (KRON-ik) Long term or frequently recurring; applied to a disease that is not acute.

Chronic obstructive pulmonary disease (COPD) A disease, such as bronchitis or emphysema, in which there is some degree of obstruction of airways and consequent increase in airway resistance.

Chyle (KĪL) The milky-appearing fluid found in the lacteals of the small intestine after absorption of lipids in food.

Chylomicron (kī-lō-MĪ-kron) Protein-coated spherical structure that contains triglycerides, phospholipids, and cholesterol and is absorbed into the lacteal of a villus in the small intestine.

Chyme (KĪM) The semifluid mixture of partly digested food and digestive secretions found in the stomach and small intestine during digestion of a meal.

Ciliary (SIL-ē-ar′-ē) **body** One of the three parts of the vascular tunic of the eyeball, the others being the choroid and the iris; includes the ciliary muscle and the ciliary processes.

Ciliary ganglion (GANG-glē-on) A very small parasympathetic ganglion whose preganglionic axons come from the oculomotor nerve (cranial nerve III) and whose postganglionic axons carry nerve impulses to the ciliary muscle and the sphincter muscle of the iris.

Cilium (SIL-ē-um) A hair or hairlike process projecting from a cell that may be used to move the entire cell or to move substances along the surface of the cell. *Plural is* **cilia.**

Circadian (ser-KĀ-dē-an) **rhythm** A cycle of active and nonactive periods in organisms determined by internal mechanisms and repeating about every 24 hours.

Circular folds Permanent, deep, transverse folds in the mucosa and submucosa of the small intestine that increase the surface area for absorption. Also called **plicae circulares** (PLĪ-kē SER-kū-lar-ēs).

Circulation time Time required for blood to pass from the right atrium, through pulmonary circulation, back to the left ventricle, through systemic circulation to the foot, and back again to the right atrium; normally about 1 min.

Circumduction (ser′-kum-DUK-shun) A movement at a synovial joint in which the distal end of a bone moves in a circle while the proximal end remains relatively stable.

Cirrhosis (si-RŌ-sis) A liver disorder in which the parenchymal cells are destroyed and replaced by connective tissue.

Cisterna chyli (sis-TER-na-KĪ-lē) The origin of the thoracic duct.

Cleavage The rapid mitotic divisions following the fertilization of a secondary oocyte, resulting in an increased number of progressively smaller cells, called blastomeres.

Climacteric (klī-mak-TER-ik) Cessation of the reproductive function in the female or diminution of testicular activity in the male.

Climax The peak period or moments of greatest intensity during sexual excitement.

Clitoris (KLI-to-ris) An erectile organ of the female, located at the anterior junction of the labia minora, that is homologous to the male penis.

Clone (KLŌN) A population of identical cells.

Clot The end result of a series of biochemical reactions that changes liquid plasma into a gelatinous mass; specifically, the conversion of fibrinogen into a tangle of polymerized fibrin molecules.

Clot retraction (rē-TRAK-shun) The consolidation of a fibrin clot to pull damaged tissue together.

Clotting Process by which a blood clot is formed. Also known as **coagulation** (cō-ag-ū-LĀ-shun).

Coccyx (KOK-six) The fused bones at the inferior end of the vertebral column.

Cochlea (KŌK-lē-a) A winding, cone-shaped tube forming a portion of the inner ear and containing the spiral organ (organ of Corti).

Cochlear duct The membranous cochlea consisting of a spirally arranged tube enclosed in the bony cochlea and lying along its outer wall. Also called the **scala media** (SCA-la MĒ-dē-a).

Coenzyme A nonprotein organic molecule that is associated with and activates an enzyme; many are derived from vitamins. An example is nicotinamide adenine dinucleotide (NAD), derived from the B vitamin niacin.

Coitus (KŌ-i-tus) Sexual intercourse.

Collagen (KOL-a-jen) A protein that is the main organic constituent of connective tissue.

Collateral circulation The alternate route taken by blood through an anastomosis.

Colliculus (ko-LIK-ū-lus) A small elevation.

Colloid (KOL-loyd) The material that accumulates in the center of thyroid follicles, consisting of thyroglobulin and stored thyroid hormones.

Colon The portion of the large intestine consisting of ascending, transverse, descending, and sigmoid portions.

Colony-stimulating factor (CSF) One of a group of molecules that stimulates development of white blood cells. Examples are macrophage CSF and granulocyte CSF.

Colostrum (kō-LOS-trum) A thin, cloudy fluid secreted by the mammary glands a few days prior to or after delivery before true milk is produced.

Column (KOL-um) Group of white matter tracts in the spinal cord.

Common bile duct A tube formed by the union of the common hepatic duct and the cystic duct that empties bile into the duodenum at the hepatopancreatic ampulla (ampulla of Vater).

Compact (dense) bone tissue Bone tissue that contains few spaces between osteons (Haversian systems); forms the external portion of all bones and the bulk of the diaphysis (shaft) of long bones; is found immediately deep to the periosteum and external to spongy bone.

Complement (KOM-ple-ment) A group of at least 20 normally inactive proteins found in plasma that forms a component of nonspecific resistance and immunity by bringing about cytolysis, inflammation, and opsonization.

Compliance The ease with which the lungs and thoracic wall or blood vessels can be expanded.

Compound A substance that can be broken down into two or more other substances by chemical means.

Concha (KONG-ka) A scroll-like bone found in the skull. *Plural is* **conchae** (KONG-kē).

Concussion (kon-KUSH-un) Traumatic injury to the brain that produces no visible bruising but may result in abrupt, temporary loss of consciousness.

Conduction system A group of autorhythmic cardiac muscle fibers that generates and distributes electrical impulses to stimulate coordinated contraction of the heart chambers; includes the sinoatrial (SA) node, the atrioventricular (AV) node, the atrioventricular (AV) bundle, the right and left bundle branches, and the Purkinje fibers.

Conductivity (kon′-duk-TIV-i-tē) The ability of a cell to propagate (conduct) action potentials along its plasma membrane; characteristic of neurons and muscle fibers (cells).

Condyloid (KON-di-loyd) **joint** A synovial joint structured so that an oval-shaped condyle of one bone fits into an elliptical cavity of another bone, permitting side-to-side and back-and-forth movements, such as the joint at the wrist between the radius and carpals. Also called an **ellipsoidal** (ē-lip-SOYD-al) **joint.**

Cone (KŌN) The type of photoreceptor in the retina that is specialized for highly acute color vision in bright light.

Congenital (kon-JEN-i-tal) Present at the time of birth.

Conjunctiva (kon′-junk-TĪ-va) The delicate membrane covering the eyeball and lining the eyes.

Connective tissue One of the most abundant of the four basic tissue types in the body, performing the functions of binding and supporting; consists of relatively few cells in a generous matrix (the ground substance and fibers between the cells).

Consciousness (KON-shus-nes) A state of wakefulness in which an individual is fully alert, aware, and oriented, partly as a result of feedback between the cerebral cortex and reticular activating system.

Continuous conduction (kon-DUK-shun) Propagation of an action potential (nerve impulse) in a step-by-step depolarization of each adjacent area of an axon membrane.

Contraception (kon′-tra-SEP-shun) The prevention of fertilization or impregnation without destroying fertility.

Contractility (kon′-trak-TIL-i-tē) The ability of cells or parts of cells to actively generate force to undergo shortening for movements. Muscle fibers (cells) exhibit a high degree of contractility.

Contralateral (kon′-tra-LAT-er-al) On the opposite side; affecting the opposite side of the body.

Control center The component of a feedback system, such as the brain, that determines the point at which a controlled condition, such as body temperature, is maintained.

Conus medullaris (KŌ-nus med-ū-LAR-is) The tapered portion of the spinal cord inferior to the lumbar enlargement.

Convergence (con-VER-jens) A synaptic arrangement in which the synaptic end bulbs of several presynaptic neurons terminate on one postsynaptic neuron. The medial movement of the two eyeballs so that both are directed toward a near object being viewed in order to produce a single image.

Convulsion (con-VUL-shun) Violent, involuntary contractions or spasms of an entire group of muscles.

Cornea (KOR-nē-a) The nonvascular, transparent fibrous coat through which the iris of the eye can be seen.

Corona radiata The innermost layer of granulosa cells that is firmly attached to the zona pellucida around a secondary oocyte.

Coronary artery disease (CAD) A condition such as atherosclerosis that causes narrowing of coronary arteries so that blood flow to the heart is reduced. The result is **coronary heart disease (CHD),** in which the heart muscle receives inadequate blood flow due to an interruption of its blood supply.

Coronary circulation The pathway followed by the blood from the ascending aorta through the blood vessels supplying the heart and returning to the right atrium. Also called **cardiac circulation.**

Coronary sinus (SĪ-nus) A wide venous channel on the posterior surface of the heart that collects the blood from the coronary circulation and returns it to the right atrium.

Corpus (KOR-pus) The principal part of any organ; any mass or body.

Corpus albicans (KOR-pus AL-bi-kanz) A white fibrous patch in the ovary that forms after the corpus luteum regresses.

Corpus callosum (kal-LŌ-sum) The great commissure of the brain between the cerebral hemispheres.

Corpus luteum (LOO-tē-um) A yellowish body in the ovary formed when a follicle has discharged its secondary oocyte; secretes estrogens, progesterone, relaxin, and inhibin.

Corpuscle of touch The sensory receptor for the sensation of touch; found in the dermal papillae, especially in palms and soles. Also called a **Meissner** (MĪZ-ner) **corpuscle.**

Cortex (KOR-teks) An outer layer of an organ. The convoluted layer of gray matter covering each cerebral hemisphere.

Costal (KOS-tal) Pertaining to a rib.

Costal cartilage (KAR-ti-lij) Hyaline cartilage that attaches a rib to the sternum.

Countercurrent mechanism One mechanism involved in the ability of the kidneys to produce hypertonic urine.

Cramp A spasmodic, usually painful contraction of a muscle.

Cranial (KRĀ-ne-al) **cavity** A subdivision of the dorsal body cavity formed by the cranial bones and containing the brain.

Cranial nerve One of 12 pairs of nerves that leave the brain; pass through foramina in the skull; and supply sensory and motor neurons to the head, neck, part of the trunk, and viscera of the thorax and abdomen. Each is designated by a Roman numeral and a name.

Craniosacral (krā-nē-ō-SĀ-kral) **outflow** The axons of parasympathetic preganglionic neurons, which have their cell bodies located in nuclei in the brain stem and in the lateral gray matter of the sacral portion of the spinal cord.

Cranium (KRĀ-nē-um) The skeleton of the skull that protects the brain and the organs of sight, hearing, and balance; includes the frontal, parietal, temporal, occipital, sphenoid, and ethmoid bones.

Creatine phosphate (KRĒ-a-tin FOS-fāt) Molecule in striated muscle fibers that contains high-energy phosphate bonds; used to generate ATP rapidly from ADP by transfer of a phosphate group. Also called **phosphocreatine** (fos'-fō-KRĒ-a-tin).

Crenation (krē-NĀ-shun) The shrinkage of red blood cells into knobbed, starry forms when they are placed in a hypertonic solution.

Crista (KRIS-ta) A crest or ridged structure. A small elevation in the ampulla of each semicircular duct that contains receptors for dynamic equilibrium.

Crossed extensor reflex A reflex in which extension of the joints in one limb occurs together with contraction of the flexor muscles of the opposite limb.

Crossing-over The exchange of a portion of one chromatid with another during meiosis. It permits an exchange of genes among chromatids and is one factor that results in genetic variation of progeny.

Crus (KRUS) **of penis** Separated, tapered portion of the corpora cavernosa penis. *Plural is* **crura** (KROO-ra).

Cryptorchidism (krip-TOR-ki-dizm) The condition of undescended testes.

Cuneate (KŪ-nē-āt) **nucleus** A group of neurons in the inferior part of the medulla oblongata in which axons of the cuneate fasciculus terminate.

Cupula (KUP-ū-la) A mass of gelatinous material covering the hair cells of a crista; a sensory receptor in the ampulla of a semicircular canal stimulated when the head moves.

Cushing's syndrome Condition caused by a hypersecretion of glucocorticoids characterized by spindly legs, "moon face," "buffalo hump," pendulous abdomen, flushed facial skin, poor wound healing, hyperglycemia, osteoporosis, hypertension, and increased susceptibility to disease.

Cutaneous (kū-TĀ-nē-us) Pertaining to the skin.

Cyanosis (sī-a-NŌ-sis) A blue or dark purple discoloration, most easily seen in nail beds and mucous membranes, that results from an increased concentration of deoxygenated (reduced) hemoglobin (more than 5 gm/dL).

Cyclic AMP (cyclic adenosine-3', 5'-monophosphate) Molecule formed from ATP by the action of the enzyme adenylate cyclase; serves as second messenger for some hormones and neurotransmitters.

Cyst (SIST) A sac with a distinct connective tissue wall, containing a fluid or other material.

Cystic (SIS-tik) **duct** The duct that carries bile from the gallbladder to the common bile duct.

Cystitis (sis-TĪ-tis) Inflammation of the urinary bladder.

Cytosol (SĪ-tō-sol) Fluid located within cells. Also called **intracellular** (in'-tra-SEL-ū-lar) **fluid (ICF).**

Cytolysis (sī-TOL-i-sis) The rupture of living cells in which the contents leak out.

Cytochrome (SĪ-tō-krōm) A protein with an iron-containing group (heme) capable of alternating between a reduced form (Fe^{2+}) and an oxidized form (Fe^{3+}).

Cytokines (SĪ-to-kīns) Small protein hormones produced by lymphocytes, fibroblasts, endothelial cells, and antigen-presenting cells that act as autocrine or paracrine substances to stimulate or inhibit cell growth and differentiation, regulate immune responses, or aid nonspecific defenses.

Cytokinesis (sī-tō-ki-NĒ-sis) Distribution of the cytoplasm into two separate cells during cell division; coordinated with nuclear division (mitosis).

Cytoplasm (SĪ-tō-plazm) Cytosol plus all organelles except the nucleus.

Cytoskeleton Complex internal structure of cytoplasm consisting of microfilaments, microtubules, and intermediate filaments.

Cytosol (SĪ-tō-sol) Semifluid portion of cytoplasm in which organelles and inclusions are suspended and solutes are dissolved. Also called **intracellular fluid.**

D

Dartos (DAR-tōs) The contractile tissue deep to the skin of the scrotum.

Decibel (DES-i-bel) **(dB)** A unit for expressing the relative intensity (loudness) of sound.

Decidua (dē-SID-ū-a) That portion of the endometrium of the uterus (all but the deepest layer) that is modified during pregnancy and shed after childbirth.

Deciduous (dē-SID-ū-us) Falling off or being shed seasonally or at a particular stage of development. In the body, referring to the first set of teeth.

Decussation (dē-ku-SĀ-shun) A crossing-over to the opposite (contralateral) side; an example is the crossing of 90% of the axons in the large motor tracts to opposite sides in the medullary pyramids.

Deep Away from the surface of the body or an organ.

Deep fascia (FASH-ē-a) A sheet of connective tissue wrapped around a muscle to hold it in place.

Deep inguinal (IN-gwi-nal) **ring** A slitlike opening in the aponeurosis of the transversus abdominis muscle that represents the origin of the inguinal canal.

Defecation (def-e-KĀ-shun) The discharge of feces from the rectum.

Deglutition (dē-gloo-TISH-un) The act of swallowing.

Dehydration (dē-hī-DRĀ-shun) Excessive loss of water from the body or its parts.

Delta cell A cell in the pancreatic islets (islets of Langerhans) in the pancreas that secretes somatostatin. Also termed a **D cell.**

Demineralization (de-min'-er-al-i-ZĀ-shun) Loss of calcium and phosphorus from bones.

Denaturation (de-nā-chur-Ā-shun) Disruption of the tertiary structure of a protein by heat, changes in pH, or other physical or chemical methods, in which the protein loses its physical properties and biological activity.

Dendrite (DEN-drīt) A neuronal process that carries electrical signals, usually graded potentials, toward the cell body.

Dendritic (den-DRIT-ik) **cell** One type of antigen-presenting cell with long branchlike projections that commonly is present in mucosal linings such as the vagina, in the skin (Langerhans cells in the epidermis), and in lymph nodes (follicular dendritic cells).

Dental caries (KA-rēz) Gradual demineralization of the enamel and dentin of a tooth that may invade the pulp and alveolar bone. Also called **tooth decay.**

Denticulate (den-TIK-ū-lāt) Finely toothed or serrated; characterized by a series of small, pointed projections.

Dentin (DEN-tin) The bony tissues of a tooth enclosing the pulp cavity.

Dentition (den-TI-shun) The eruption of teeth. The number, shape, and arrangement of teeth.

Deoxyribonucleic (dē-ok'-sē-rī'-bō-noo-KLĒ-ik) **acid (DNA)** A nucleic acid constructed of nucleotides consisting of one of four bases (adenine, cytosine, guanine, or thymine), deoxyribose, and a phosphate group; encoded in the nucleotides is genetic information.

Depolarization (dē-pō-lar-i-ZA-shun) A reduction of voltage across a plasma membrane; expressed as a change toward less negative (more positive) voltages on the interior surface of the plasma membrane.

Depression (de-PRESH-un) Movement in which a part of the body moves inferiorly.

Dermal papilla (pa-PILL-a) Fingerlike projection of the papillary region of the dermis that may contain blood capillaries or corpuscles of touch (Meissner corpuscles).

Dermatology (der-ma-TOL-ō-jē) The medical specialty dealing with diseases of the skin.

Dermatome (DER-ma-tōm) The cutaneous area developed from one embryonic spinal cord segment and receiving most of its sensory innervation from one spinal nerve. An instrument for incising the skin or cutting thin transplants of skin.

Dermis (DER-mis) A layer of dense irregular connective tissue lying deep to the epidermis.

Descending colon (KŌ-lon) The part of the large intestine descending from the left colic (splenic) flexure to the level of the left iliac crest.

Detritus (de-TRĪ-tus) Particulate matter produced by or remaining after the wearing away or disintegration of a substance or tissue; scales, crusts, or loosened skin.

Detrusor (de-TROO-ser) **muscle** Smooth muscle that forms the wall of the urinary bladder.

Developmental biology The study of development from the fertilized egg to the adult form.

Diagnosis (dī-ag-NŌ-sis) Distinguishing one disease from another or determining the nature of a disease from signs and symptoms by inspection, palpation, laboratory tests, and other means.

Dialysis (dī-AL-i-sis) The removal of waste products from blood by diffusion through a selectively permeable membrane.

Diaphragm (DĪ-a-fram) Any partition that separates one area from another, especially the dome-shaped skeletal muscle between the thoracic and abdominal cavities. Also a dome-shaped device that is placed over the cervix, usually with a spermicide, to prevent conception.

Diaphysis (dī-AF-i-sis) The shaft of a long bone.

Diarrhea (dī-a-RE-a) Frequent defecation of liquid feces caused by increased motility of the intestines.

Diarthrosis (dī-ar-THRŌ-sis) A freely movable joint; types are gliding, hinge, pivot, condyloid, saddle, and ball-and-socket.

Diastole (dī-AS-tō-lē) In the cardiac cycle, the phase of relaxation or dilation of the heart muscle, especially of the ventricles.

Diastolic (dī-as-TOL-ik) **blood pressure** The force exerted by blood on arterial walls during ventricular relaxation; the lowest blood pressure measured in the large arteries, normally about 80 mmHg in a young adult.

Diencephalon (dī′-en-SEF-a-lon) A part of the brain consisting of the thalamus, hypothalamus, epithalamus, and subthalamus.

Diffusion (dif-Ū-zhun) A passive process in which there is a net or greater movement of molecules or ions from a region of high concentration to a region of low concentration until equilibrium is reached.

Digestion (dī-JES-chun) The mechanical and chemical breakdown of food to simple molecules that can be absorbed and used by body cells.

Dilate (DĪ-lāt) To expand or swell.

Diploid (DIP-loyd) Having the number of chromosomes characteristically found in the somatic cells of an organism; having two haploid sets of chromosomes, one each from the mother and father. Symbolized 2*n*.

Direct motor pathways Collections of upper motor neurons with cell bodies in the motor cortex that project axons into the spinal cord, where they synapse with lower motor neurons or interneurons in the anterior horns. Also called the **pyramidal pathways.**

Disease Any change from a state of health.

Dislocation (dis-lō-KA-shun) Displacement of a bone from a joint with tearing of ligaments, tendons, and articular capsules. Also called **luxation** (luks-A-shun).

Dissect (di-SEKT) To separate tissues and parts of a cadaver or an organ for anatomical study.

Distal (DIS-tal) Farther from the attachment of a limb to the trunk; farther from the point of origin or attachment.

Diuretic (dī-ū-RET-ik) A chemical that increases urine volume by decreasing reabsorption of water, usually by inhibiting sodium reabsorption.

Divergence (dī-VER-jens) A synaptic arrangement in which the synaptic end bulbs of one presynaptic neuron terminate on several postsynaptic neurons.

Diverticulum (dī-ver-TIK-ū-lum) A sac or pouch in the wall of a canal or organ, especially in the colon.

Dominant allele An allele that overrides the influence of an alternate allele on the homologous chromosome; the allele that is expressed.

Dorsal body cavity Cavity near the dorsal (posterior) surface of the body that consists of a cranial cavity and vertebral canal.

Dorsal ramus (RA-mus) A branch of a spinal nerve containing motor and sensory axons supplying the muscles, skin, and bones of the posterior part of the head, neck, and trunk.

Dorsiflexion (dor′-si-FLEK-shun) Bending the foot in the direction of the dorsum (upper surface).

Down-regulation Phenomenon in which there is a decrease in the number of receptors in response to an excess of a hormone or neurotransmitter.

Ductus arteriosus (DUK-tus ar-tē-rē-Ō-sus) A small vessel connecting the pulmonary trunk with the aorta; found only in the fetus.

Ductus (vas) deferens (DEF-er-ens) The duct that carries sperm from the epididymis to the ejaculatory duct. Also called the **seminal duct.**

Ductus epididymis (ep′-i-DID-i-mis) A tightly coiled tube inside the epididymis, distinguished into a head, body, and tail, in which sperm undergo maturation.

Ductus venosus (ve-NŌ-sus) A small vessel in the fetus that helps the circulation bypass the liver.

Duodenal (doo-ō-DE-nal) **gland** Gland in the submucosa of the duodenum that secretes an alkaline mucus to protect the lining of the small intestine from the action of enzymes and to help neutralize the acid in chyme. Also called **Brunner's** (BRUN-erz) **gland.**

Duodenum (doo′-ō-DE-num *or* doo-OD-e-num) The first 25 cm (10 in.) of the small intestine, which connects the stomach and the ileum.

Dura mater (DOO-ra MA-ter) The outermost of the three meninges (coverings) of the brain and spinal cord.

Dynamic equilibrium (ē-kwi-LIB-rē-um) The maintenance of body position, mainly the head, in response to sudden movements such as rotation.

Dysfunction (dis-FUNK-shun) Absence of completely normal function.

Dysmenorrhea (dis′-men-ō-RE-a) Painful menstruation.

Dysplasia (dis-PLĀ-zē-a) Change in the size, shape, and organization of cells due to chronic irritation or inflammation; may either revert to normal if stress is removed or progress to neoplasia.

Dyspnea (DISP-nē-a) Shortness of breath.

E

Eardrum A thin, semitransparent partition of fibrous connective tissue between the external auditory meatus and the middle ear. Also called the **tympanic membrane.**

Ectoderm The primary germ layer that gives rise to the nervous system and the epidermis of skin and its derivatives.

Ectopic (ek-TOP-ik) Out of the normal location, as in ectopic pregnancy.

Edema (e-DĒ-ma) An abnormal accumulation of interstitial fluid.

Effector (e-FEK-tor) An organ of the body, either a muscle or a gland, that is innervated by somatic or autonomic motor neurons.

Efferent arteriole (EF-er-ent ar-TĒ-rē-ōl) A vessel of the renal vascular system that carries blood from a glomerulus to a peritubular capillary.

Efferent (EF-er-ent) **ducts** A series of coiled tubes that transport sperm from the rete testis to the epididymis.

Eicosanoids (ī-KŌ-sa-noyds) Local hormones derived from a 20-carbon fatty acid (arachidonic acid); two important types are prostaglandins and leukotrienes.

Ejaculation (e-jak-ū-LĀ-shun) The reflex ejection or expulsion of semen from the penis.

Ejaculatory (e-JAK-ū-la-tō-rē) **duct** A tube that transports sperm from the ductus (vas) deferens to the prostatic urethra.

Elasticity (e-las-TIS-i-tē) The ability of tissue to return to its original shape after contraction or extension.

Electrocardiogram (e-lek'-trō-KAR-dē-ō-gram) (**ECG** or **EKG**) A recording of the electrical changes that accompany the cardiac cycle that can be detected at the surface of the body; may be resting, stress, or ambulatory.

Electroencephalogram (e-lek'-trō-en-SEF-a-lō-gram) (**EEG**) A recording of the electrical activity of the brain from the scalp surface; used to diagnose certain diseases (such as epilepsy), furnish information regarding sleep and wakefulness, and confirm brain death.

Electrolyte (ē-LEK-trō-līt) Any compound that separates into ions when dissolved in water and that conducts electricity.

Electromyography (e-lek'-trō-mī-OG-ra-fē) Evaluation of the electrical activity of resting and contracting muscle to ascertain causes of muscular weakness, paralysis, involuntary twitching, and abnormal levels of muscle enzymes; also used as part of biofeedback studies.

Electron transport chain A sequence of electron carrier molecules on the inner mitochondrial membrane that undergo oxidation and reduction as they pump hydrogen ions (H^+) through the membrane. ATP synthesis then occurs as H^+ diffuse back into the mitochondrial matrix through special H^+ channels. *See also* **Chemiosmosis.**

Elevation (el-e-VĀ-shun) Movement in which a part of the body moves superiorly.

Embolism (EM-bō-lizm) Obstruction or closure of a vessel by an embolus.

Embolus (EM-bō-lus) A blood clot, bubble of air or fat from broken bones, mass of bacteria, or other debris or foreign material transported by the blood.

Embryo (EM-brē-ō) The young of any organism in an early stage of development; in humans, the developing organism from fertilization to the end of the eighth week of development.

Embryology (em'-brē-OL-ō-jē) The study of development from the fertilized egg to the end of the eighth week of development.

Emesis (EM-e-sis) Vomiting.

Emigration (em'-e-GRĀ-shun) Process whereby white blood cells (WBCs) leave the bloodstream by rolling along the endothelium, sticking to it, and squeezing between the endothelial cells. Adhesion molecules help WBCs stick to the endothelium. Also known as **migration** or **extravasation.**

Emission (ē-MISH-un) Propulsion of sperm into the urethra due to peristaltic contractions of the ducts of the testes, epididymides, and ductus (vas) deferens as a result of sympathetic stimulation.

Emmetropia (em'-e-TRŌ-pē-a) Normal vision in which light rays are focused exactly on the retina.

Emphysema (em'-fi'-SĒ-ma) A lung disorder in which alveolar walls disintegrate, producing abnormally large air spaces and loss of elasticity in the lungs; typically caused by exposure to cigarette smoke.

Emulsification (ē-mul'-si-fi-KĀ-shun) The dispersion of large lipid globules into smaller, uniformly distributed particles in the presence of bile.

Enamel (e-NAM-el) The hard, white substance covering the crown of a tooth.

End-diastolic (dī-as-TO-lik) **volume (EDV)** The volume of blood, about 130 mL, remaining in a ventricle at the end of its diastole (relaxation).

Endergonic (end'-er-GON-ik) **reaction** Type of chemical reaction in which the energy released as new bonds form is less than the energy needed to break apart old bonds; an energy-requiring reaction.

Endocardium (en-dō-KAR-dē-um) The layer of the heart wall, composed of endothelium and smooth muscle, that lines the inside of the heart and covers the valves and tendons that hold the valves open.

Endochondral ossification (en'-dō-KON-dral os'-i-fi-KĀ-shun) The replacement of cartilage by bone. Also called **intracartilaginous** (in'-tra-kar'-ti-LAJ-i-nus) **ossification.**

Endocrine (EN-dō-krin) **gland** A gland that secretes hormones into interstitial fluid and then the blood; a ductless gland.

Endocrinology (en'-dō-kri-NOL-ō-jē) The science concerned with the structure and functions of endocrine glands and the diagnosis and treatment of disorders of the endocrine system.

Endocytosis (en'-dō-sī-TŌ-sis) The uptake into a cell of large molecules and particles in which a segment of plasma membrane surrounds the substance, encloses it, and brings it in; includes phagocytosis, pinocytosis, and receptor-mediated endocytosis.

Endoderm (EN-dō-derm) A primary germ layer of the developing embryo; gives rise to the gastrointestinal tract, urinary bladder, urethra, and respiratory tract.

Endodontics (en'-dō-DON-tiks) The branch of dentistry concerned with the prevention, diagnosis, and treatment of diseases that affect the pulp, root, periodontal ligament, and alveolar bone.

Endogenous (en-DOJ-e-nus) Growing from or beginning within the organism.

Endolymph (EN-dō-limf') The fluid within the membranous labyrinth of the internal ear.

Endometrium (en'-dō-MĒ-trē-um) The mucous membrane lining the uterus.

Endomysium (en'-dō-MĪZ-ē-um) Invagination of the perimysium separating each individual muscle fiber (cell).

Endoneurium (en'-do-NOO-rē-um) Connective tissue wrapping around individual nerve axons (cells).

Endoplasmic reticulum (en'-do-PLAZ-mik re-TIK-ū-lum) **(ER)** A network of channels running through the cytoplasm of a cell that serves in intracellular transportation, support, storage, synthesis, and packaging of molecules. Portions of ER where ribosomes are attached to the outer surface are called **rough ER;** portions that have no ribosomes are called **smooth ER.**

Endorphin (en-DOR-fin) A neuropeptide in the central nervous system that acts as a painkiller.

Endosteum (en-DOS-tē-um) The membrane that lines the medullary (marrow) cavity of bones, consisting of osteogenic cells and scattered osteoclasts.

Endothelium (en'-dō-THĒ-lē-um) The layer of simple squamous epithelium that lines the cavities of the heart, blood vessels, and lymphatic vessels.

End-systolic (sis-TO-lik) **volume (ESV)** The volume of blood, about 60 mL, remaining in a ventricle after its systole (contraction).

Energy The capacity to do work.

Enkephalin (en-KEF-a-lin) A peptide found in the central nervous system that acts as a painkiller.

Enteric (EN-ter-ik) **nervous system** The part of the nervous system that is embedded in the submucosa and muscularis of the gastrointestinal (GI) tract; governs motility and secretions of the GI tract.

Enteroendocrine (en-ter-ō-EN-dō-krin) **cell** A cell of the mucosa of the gastrointestinal tract that secretes a hormone that governs function of the GI tract; hormones secreted include gastrin, cholecystokinin, glucose-dependent insulinotropic peptide (GIP), and secretin.

Enterogastric (en-te-rō-GAS-trik) **reflex** A reflex that inhibits gastric secretion; initiated by food in the small intestine.

Enzyme (EN-zīm) A substance that accelerates chemical reactions; an organic catalyst, usually a protein.

Eosinophil (ē'-ō-SIN-ō-fil) A type of white blood cell characterized by granules that stain red or pink with acid dyes.

Ependymal (e-PEN-de-mal) **cells** Neuroglial cells that cover choroid plexuses and produce cerebrospinal fluid (CSF); they also line the ventricles of the brain and probably assist in the circulation of CSF.

Epicardium (ep'-i-KAR-dē-um) The thin outer layer of the heart wall, composed of serous tissue and mesothelium. Also called the **visceral pericardium.**

Epidemiology (ep'-i-dē-mē-OL-ō-jē) Study of the occurrence and distribution of diseases and disorders in human populations.

Epidermis (ep-i-DERM-is) The superficial, thinner layer of skin, composed of keratinized stratified squamous epithelium.

Epididymis (ep'-i-DID-i-mis) A comma-shaped organ that lies along the posterior border of the testis and contains the ductus epididymis, in which sperm undergo maturation. *Plural is* **epididymides** (ep'-i-DID-i-mi-dēz).

Epidural (ep'-i-DOO-ral) **space** A space between the spinal dura mater and the vertebral canal, containing areolar connective tissue and a plexus of veins.

Epiglottis (ep'-i-GLOT-is) A large, leaf-shaped piece of cartilage lying on top of the larynx, attached to the thyroid cartilage and its unattached portion is free to move up and down to cover the glottis (vocal folds and rima glottidis) during swallowing.

Epimysium (ep'-i-MĪZ-ē-um) Fibrous connective tissue around muscles.

Epinephrine (ep-ē-NEF-rin) Hormone secreted by the adrenal medulla that produces actions similar to those that result from sympathetic stimulation. Also called **adrenaline** (a-DREN-a-lin).

Epineurium (ep'-i-NOO-rē-um) The superficial connective tissue covering around an entire nerve.

Epiphyseal (ep'-i-FIZ-ē-al) **line** The remnant of the epiphyseal plate in the metaphysis of a long bone.

Epiphyseal (ep'-i-FIZ-ē-al) **plate** The hyaline cartilage plate in the metaphysis of a long bone; site of lengthwise growth of long bones.

Epiphysis (ē-PIF-i-sis) The end of a long bone, usually larger in diameter than the shaft (diaphysis).

Episiotomy (e-piz'-ē-OT-ō-mē) A cut made with surgical scissors to avoid tearing of the perineum at the end of the second stage of labor.

Epistaxis (ep'-i-STAK-sis) Loss of blood from the nose due to trauma, infection, allergy, neoplasm, and bleeding disorders. Also called **nosebleed.**

Epithelial (ep'-i-THĒ-lē-al) **tissue** The tissue that forms innermost and outermost surfaces of body structures and forms glands.

Eponychium (ep'-o-NIK-ē-um) Narrow band of stratum corneum at the proximal border of a nail that extends from the margin of the nail wall. Also called the **cuticle.**

Erectile dysfunction Failure to maintain an erection long enough for sexual intercourse. Also known as **impotence** (IM-pō-tens).

Erection (ē-REK-shun) The enlarged and stiff state of the penis or clitoris resulting from the engorgement of the spongy erectile tissue with blood.

Eructation (e-ruk'-TĀ-shun) The forceful expulsion of gas from the stomach. Also called **belching.**

Erythema (er'-i-THĒ-ma) Skin redness usually caused by dilation of the capillaries.

Erythrocyte (e-RITH-rō-sīt) A mature red blood cell.

Erythropoiesis (e-rith'-rō-poy-Ē-sis) The process by which red blood cells are formed.

Erythropoietin (e-rith'-rō-POY-e-tin) A hormone released by the juxtaglomerular cells of the kidneys that stimulates red blood cell production.

Esophagus (e-SOF-a-gus) The hollow muscular tube that connects the pharynx and the stomach.

Essential amino acids Those 10 amino acids that cannot be synthesized by the human body at an adequate rate to meet its needs and therefore must be obtained from the diet.

Estrogens (ES-tro-jenz) Feminizing sex hormones produced by the ovaries; govern development of oocytes, maintenance of female reproductive structures, and appearance of secondary sex characteristics; also affect fluid and electrolyte balance, and protein anabolism. Examples are β-estradiol, estrone, and estriol.

Etiology (ē'-tē-OL-ō-jē) The study of the causes of disease, including theories of the origin and organisms (if any) involved.

Eupnea (ŪP-nē-a) Normal quiet breathing.

Eversion (ē-VER-zhun) The movement of the sole laterally at the ankle joint or of an atrioventricular valve into an atrium during ventricular contraction.

Excitability (ek-sīt'-a-BIL-i-tē) The ability of muscle fibers to receive and respond to stimuli; the ability of neurons to respond to stimuli and generate nerve impulses.

Excitatory postsynaptic potential (EPSP) A small depolarization of a postsynaptic membrane when it is stimulated by an excitatory neurotransmitter. An EPSP is both graded (variable size) and localized (decreases in size away from the point of excitation).

Excrement (EKS-kre-ment) Material eliminated from the body as waste, especially fecal matter.

Excretion (eks-KRĒ-shun) The process of eliminating waste products from the body; also the products excreted.

Exergonic (eks′-er-GON-ik) **reaction** Type of chemical reaction in which the energy released as new bonds form is greater than the energy needed to break apart old bonds; an energy-releasing reaction.

Exocrine (EK-sō-krin) **gland** A gland that secretes its products into ducts that carry the secretions into body cavities, into the lumen of an organ, or to the outer surface of the body.

Exocytosis (ex′-ō-sī-TŌ-sis) A process in which membrane-enclosed secretory vesicles form inside the cell, fuse with the plasma membrane, and release their contents into the interstitial fluid; achieves secretion of materials from a cell.

Exogenous (ex-SOJ-e-nus) Originating outside an organ or part.

Exhalation (eks-ha-LĀ-shun) Breathing out; expelling air from the lungs into the atmosphere. Also called **expiration.**

Exon (EX-on) A region of DNA that codes for synthesis of a protein.

Expiratory (eks-PĪ-ra-tō-rē) **reserve volume** The volume of air in excess of tidal volume that can be exhaled forcibly; about 1200 mL.

Extensibility (ek-sten′-si-BIL-i-tē) The ability of muscle tissue to stretch when it is pulled.

Extension (ek-STEN-shun) An increase in the angle between two bones; restoring a body part to its anatomical position after flexion.

External Located on or near the surface.

External auditory (AW-di-tōr-ē) **canal** or **meatus** (mē-Ā-tus) A curved tube in the temporal bone that leads to the middle ear.

External ear The outer ear, consisting of the pinna, external auditory canal, and tympanic membrane (eardrum).

External nares (NĀ-rez) The external nostrils, or the openings into the nasal cavity on the exterior of the body.

External respiration The exchange of respiratory gases between the lungs and blood. Also called **pulmonary respiration.**

Exteroceptor (eks′-ter-ō-SEP-tor) A sensory receptor adapted for the reception of stimuli from outside the body.

Extracellular fluid (ECF) Fluid outside body cells, such as interstitial fluid and plasma.

Extravasation (eks-trav-a-SĀ-shun) The escape of fluid, especially blood, lymph, or serum, from a vessel into the tissues.

Extrinsic (ek-STRIN-sik) Of external origin.

Extrinsic pathway (of blood clotting) Sequence of reactions leading to blood clotting that is initiated by the release of tissue factor (TF), also known as thromboplastin, that leaks into the blood from damaged cells outside the blood vessels.

Exudate (EKS-oo-dāt) Escaping fluid or semifluid material that oozes from a space and that may contain serum, pus, and cellular debris.

Eyebrow The hairy ridge superior to the eye.

F

Face The anterior aspect of the head.

Facilitated diffusion (fa-SIL-i-tā-ted dif-Ū-zhun) Diffusion in which a substance not soluble by itself in lipids diffuses across a selectively permeable membrane with the help of a transporter protein.

Falciform ligament (FAL-si-form LIG-a-ment) A sheet of parietal peritoneum between the two principal lobes of the liver. The ligamentum teres, or remnant of the umbilical vein, lies within its fold.

Falx cerebelli (FALKS ser′-e-BEL-lē) A small triangular process of the dura mater attached to the occipital bone in the posterior cranial fossa and projecting inward between the two cerebellar hemispheres.

Falx cerebri (FALKS SER-e-brē) A fold of the dura mater extending deep into the longitudinal fissure between the two cerebral hemispheres.

Fascia (FASH-ē-a) A fibrous membrane covering, supporting, and separating muscles.

Fascicle (FAS-i-kul) A small bundle or cluster, especially of nerve or muscle fibers (cells). Also called a **fasciculus** (fa-SIK-ū-lus). *Plural is* **fasciculi** (fa-SIK-yoo-lī).

Fasciculation (fa-sik′-ū-LĀ-shun) Abnormal, spontaneous twitch of all skeletal muscle fibers in one motor unit that is visible at the skin surface; not associated with movement of the affected muscle; present in progressive diseases of motor neurons, for example, poliomyelitis.

Fauces (FAW-sēz) The opening from the mouth into the pharynx.

F cell A cell in the pancreatic islets (islets of Langerhans) that secretes pancreatic polypeptide.

Feces (FĒ-sēz) Material discharged from the rectum and made up of bacteria, excretions, and food residue. Also called **stool.**

Feedback system A sequence of events in which information about the status of a situation is continually reported (fed back) to a control center.

Female reproductive cycle General term for the ovarian and uterine cycles, the hormonal changes that accompany them, and cyclic changes in the breasts and cervix; includes changes in the endometrium of a nonpregnant female that prepares the lining of the uterus to receive a fertilized ovum. Less correctly termed the **menstrual cycle.**

Fertilization (fer′-ti-li-ZĀ-shun) Penetration of a secondary oocyte by a sperm cell, meiotic division of secondary oocyte to form an ovum, and subsequent union of the nuclei of the gametes.

Fetal circulation The cardiovascular system of the fetus, including the placenta and special blood vessels involved in the exchange of materials between fetus and mother.

Fetus (FĒ-tus) In humans, the developing organism *in utero* from the beginning of the third month to birth.

Fever An elevation in body temperature above the normal temperature of 37°C (98.6°F) due to a resetting of the hypothalamic thermostat.

Fibrillation (fi-bri-LĀ-shun) Abnormal, spontaneous twitch of a single skeletal muscle fiber (cell) that can be detected with electromyography but is not visible at the skin surface; not associated with movement of the affected muscle; present in certain disorders of motor neurons, for example, amyotrophic lateral sclerosis (ALS). With reference to cardiac muscle, *see* **Atrial fibrillation** and **Ventricular fibrillation.**

Fibrin (FĪ-brin) An insoluble protein that is essential to blood clotting; formed from fibrinogen by the action of thrombin.

Fibrinogen (fī-BRIN-ō-jen) A clotting factor in blood plasma that by the action of thrombin is converted to fibrin.

Fibrinolysis (fī-bri-NOL-i-sis) Dissolution of a blood clot by the action of a proteolytic enzyme, such as plasmin (fibrinolysin), that dissolves fibrin threads and inactivates fibrinogen and other blood-clotting factors.

Fibroblast (FĪ-brō-blast) A large, flat cell that secretes most of the matrix (extracellular) material of areolar and dense connective tissues.

Fibrous (FĪ-brus) **joint** A joint that allows little or no movement, such as a suture or a syndesmosis.

Fibrous tunic (TOO-nik) The superficial coat of the eyeball, made up of the posterior sclera and the anterior cornea.

Fight-or-flight response The effects produced upon stimulation of the sympathetic division of the autonomic nervous system.

Filiform papilla (FIL-i-form pa-PIL-a) One of the conical projections that are distributed in parallel rows over the anterior two-thirds of the tongue and lack taste buds.

Filtration (fil-TRĀ-shun) The flow of a liquid through a filter (or membrane that acts like a filter) due to a hydrostatic pressure; occurs in capillaries due to blood pressure.

Filtration fraction The percentage of plasma entering the kidneys that becomes glomerular filtrate.

Filtration membrane Site of blood filtration in nephrons of the kidneys, consisting of the endothelium and basement membrane of the glomerulus and the epithelium of the visceral layer of the glomerular (Bowman's) capsule.

Filum terminale (FĪ-lum ter-mi-NAL-ē) Non-nervous fibrous tissue of the spinal cord that extends inferiorly from the conus medullaris to the coccyx.

Fimbriae (FIM-brē-ē) Fingerlike structures, especially the lateral ends of the uterine (Fallopian) tubes.

Fissure (FISH-ur) A groove, fold, or slit that may be normal or abnormal.

Fistula (FIS-tū-la) An abnormal passage between two organs or between an organ cavity and the outside.

Fixator A muscle that stabilizes the origin of the prime mover so that the prime mover can act more efficiently.

Fixed macrophage (MAK-rō-fāj) Stationary phagocytic cell found in the liver, lungs, brain, spleen, lymph nodes, subcutaneous tissue, and red bone marrow. Also called a **histiocyte** (HIS-tē-ō-sīt).

Flaccid (FLAS-sid) Relaxed, flabby, or soft; lacking muscle tone.

Flagellum (fla-JEL-um) A hairlike, motile process on the extremity of a bacterium, protozoan, or sperm cell. *Plural is* **flagella** (fla-JEL-a).

Flatus (FLĀ-tus) Gas in the stomach or intestines, commonly used to denote expulsion of gas through the anus.

Flexion (FLEK-shun) Movement in which there is a decrease in the angle between two bones.

Flexor reflex A protective reflex in which flexor muscles are stimulated while extensor muscles are inhibited.

Follicle (FOL-i-kul) A small secretory sac or cavity; the group of cells that contains a developing oocyte in the ovaries.

Follicle-stimulating hormone (FSH) Hormone secreted by the anterior pituitary that initiates development of ova and stimulates the ovaries to secrete estrogens in females, and initiates sperm production in males.

Fontanel (fon′-ta-NEL) A fibrous connective tissue membrane-filled space where bone formation is not yet complete, especially between the cranial bones of an infant's skull.

Foot The terminal part of the lower limb, from the ankle to the toes.

Foramen (fō-RĀ-men) A passage or opening; a communication between two cavities of an organ, or a hole in a bone for passage of vessels or nerves. *Plural is* **foramina** (fō-RAM-i-na).

Foramen ovale (fō-RĀ-men-ō-VAL-ē) An opening in the fetal heart in the septum between the right and left atria. A hole in the greater wing of the sphenoid bone that transmits the mandibular branch of the trigeminal nerve (cranial nerve V).

Forearm (FOR-arm) The part of the upper limb between the elbow and the wrist.

Fornix (FOR-niks) An arch or fold; a tract in the brain made up of association fibers, connecting the hippocampus with the mammillary bodies; a recess around the cervix of the uterus where it protrudes into the vagina.

Fossa (FOS-a) A furrow or shallow depression.

Fourth ventricle (VEN-tri-kul) A cavity filled with cerebrospinal fluid within the brain lying between the cerebellum and the medulla oblongata and pons.

Fracture (FRAK-choor) Any break in a bone.

Frenulum (FREN-ū-lum) A small fold of mucous membrane that connects two parts and limits movement.

Frontal plane A plane at a right angle to a midsagittal plane that divides the body or organs into anterior and posterior portions. Also called a **coronal** (kō-RŌ-nal) **plane.**

Functional residual (re-ZID-ū-al) **capacity** The sum of residual volume plus expiratory reserve volume; about 2400 mL.

Fundus (FUN-dus) The part of a hollow organ farthest from the opening.

Fungiform papilla (FUN-ji-form pa-PIL-a) A mushroomlike elevation on the upper surface of the tongue appearing as a red dot; most contain taste buds.

G

Gallbladder A small pouch, located inferior to the liver, that stores bile and empties by means of the cystic duct.

Gallstone A solid mass, usually containing cholesterol, in the gallbladder or a bile-containing duct; formed anywhere between bile canaliculi in the liver and the hepatopancreatic ampulla (ampulla of Vater), where bile enters the duodenum. Also called a **biliary calculus.**

Gamete (GAM-ēt) A male or female reproductive cell; a sperm cell or secondary oocyte.

Ganglion (GANG-glē-on) Usually, a group of neuronal cell bodies lying outside the central nervous system (CNS). *Plural is* **ganglia** (GANG-glē-a).

Gastric (GAS-trik) **glands** Glands in the mucosa of the stomach composed of cells that empty their secretions into narrow channels called gastric pits. Types of cells are chief cells (secrete pepsinogen), parietal cells (secrete hydrochloric acid and intrinsic factor), surface mucous and mucous neck cells (secrete mucus), and G cells (secrete gastrin).

Gastroenterology (gas′-trō-en′-ter-OL-ō-jē) The medical specialty that deals with the structure, function, diagnosis, and treatment of diseases of the stomach and intestines.

Gastrointestinal (gas-trō-in-TES-ti-nal) **(GI) tract** A continuous tube running through the ventral body cavity extending from the mouth to the anus. Also called the **alimentary** (al′-i-MEN-tar-ē) **canal.**

Gastrulation (gas′-troo-LĀ-shun) The migration of groups of cells from the epiblast that transform a bilaminar embryonic disc into a trilaminar embryonic disc that consists of the three primary germ layers; transformation of the blastula into the gastrula.

Gene (JĒN) Biological unit of heredity; a segment of DNA located in a definite position on a particular chromosome; a sequence of DNA that codes for a particular mRNA, rRNA, or tRNA.

Generator potential The graded depolarization that results in a change in the resting membrane potential in a receptor (specialized neuronal ending); may trigger a nerve action potential (nerve impulse) if depolarization reaches threshold.

Genetic engineering The manufacture and manipulation of genetic material.

Genetics The study of genes and heredity.

Genitalia (jen′-i-TĀ-lē-a) Reproductive organs.

Genome (JĒ-nōm) The complete set of genes of an organism.

Genotype (JĒ-nō-tīp) The genetic makeup of an individual; the combination of alleles present at one or more chromosomal locations, as distinguished from the appearance, or phenotype, that results from those alleles.

Geriatrics (jer′-ē-AT-riks) The branch of medicine devoted to the medical problems and care of elderly persons.

Gestation (jes-TĀ-shun) The period of development from fertilization to birth.

Gingivae (jin-JI-vē) Gums. They cover the alveolar processes of the mandible and maxilla and extend slightly into each socket.

Gland Specialized epithelial cell or cells that secrete substances; may be exocrine or endocrine.

Glans penis (glanz PĒ-nis) The slightly enlarged region at the distal end of the penis.

Glaucoma (glaw-KŌ-ma) An eye disorder in which there is increased intraocular pressure due to an excess of aqueous humor.

Gliding joint A synovial joint having articulating surfaces that are usually flat, permitting only side-to-side and back-and-forth movements, as between carpal bones, tarsal bones, and the scapula and clavicle. Also called an **arthrodial** (ar-THRŌ-dē-al) **joint.**

Glomerular (glō-MER-ū-lar) **capsule** A double-walled globe at the proximal end of a nephron that encloses the glomerular capillaries. Also called **Bowman's** (BŌ-manz) **capsule.**

Glomerular filtrate (glō-MER-ū-lar FIL-trāt) The fluid produced when blood is filtered by the filtration membrane in the glomeruli of the kidneys.

Glomerular filtration The first step in urine formation in which substances in blood pass through the filtration membrane and the filtrate enters the proximal convoluted tubule of a nephron.

Glomerular filtration rate (GFR) The total volume of fluid that enters all the glomerular (Bowman's) capsules of the kidneys in 1 min; about 100–125 mL/min.

Glomerulus (glō-MER-ū-lus) A rounded mass of nerves or blood vessels, especially the microscopic tuft of capillaries that is surrounded by the glomerular (Bowman's) capsule of each kidney tubule. *Plural is* **glomeruli.**

Glottis (GLOT-is) The vocal folds (true vocal cords) in the larynx plus the space between them (rima glottidis).

Glucagon (GLOO-ka-gon) A hormone produced by the alpha cells of the pancreatic islets (islets of Langerhans) that increases blood glucose level.

Glucocorticoids (gloo-kō-KOR-ti-koyds) Hormones secreted by the cortex of the adrenal gland, especially cortisol, that influence glucose metabolism.

Gluconeogenesis (gloo′-kō-nē-ō-JEN-e-sis) The synthesis of glucose from certain amino acids or lactic acid.

Glucose (GLOO-kōs) A hexose (six-carbon sugar), $C_6H_{12}O_6$, that is a major energy source for the production of ATP by body cells.

Glucosuria (gloo′-kō-SOO-rē-a) The presence of glucose in the urine; may be temporary or pathological. Also called **glycosuria.**

Glycogen (GLĪ-kō-jen) A highly branched polymer of glucose containing thousands of subunits; functions as a compact store of glucose molecules in liver and muscle fibers (cells).

Glycogenesis (glī′-kō-JEN-e-sis) The chemical reactions by which many molecules of glucose are used to synthesize glycogen.

Glycogenolysis (glī-kō-je-NOL-i-sis) The breakdown of glycogen into glucose.

Glycolysis (glī-KOL-i-sis) Series of chemical reactions in the cytosol of a cell in which a molecule of glucose is split into two molecules of pyruvic acid with net production of two ATPs.

Goblet cell A goblet-shaped unicellular gland that secretes mucus; present in epithelium of the airways and intestines.

Goiter (GOY-ter) An enlarged thyroid gland.

Golgi (GOL-jē) **complex** An organelle in the cytoplasm of cells consisting of four to six flattened sacs (cisternae), stacked on one another, with expanded areas at their ends; functions in processing, sorting, packaging, and delivering proteins and lipids to the plasma membrane, lysosomes, and secretory vesicles.

Gomphosis (gom-FŌ-sis) A fibrous joint in which a cone-shaped peg fits into a socket.

Gonad (GŌ-nad) A gland that produces gametes and hormones; the ovary in the female and the testis in the male.

Gonadotropic hormone Anterior pituitary hormone that affects the gonads.

Gracile (GRAS-il) **nucleus** A group of nerve cells in the inferior part of the medulla oblongata in which axons of the gracile fasciculus terminate.

Gray commissure (KOM-i-shur) A narrow strip of gray matter connecting the two lateral gray masses within the spinal cord.

Gray matter Areas in the central nervous system and ganglia containing neuronal cell bodies, dendrites, unmyelinated axons, axon terminals, and neuroglia; Nissl bodies impart a gray color and there is little or no myelin in gray matter.

Gray ramus communicans (RĀ-mus kō-MŪ-ni-kans) A short nerve containing axons of sympathetic postganglionic neurons; the cell bodies of the neurons are in a sympathetic chain ganglion, and the unmyelinated axons extend via the gray ramus to a spinal nerve and then to the periphery to supply smooth muscle in blood vessels, arrector pili muscles, and sweat glands. *Plural is* **rami communicantes** (RĀ-mē kō-mū-ni-KAN-tēz).

Greater omentum (ō-MEN-tum) A large fold in the serosa of the stomach that hangs down like an apron anterior to the intestines.

Greater vestibular (ves-TIB-ū-lar) **glands** A pair of glands on either side of the vaginal orifice that open by a duct into the space between the hymen and the labia minora. Also called **Bartholin's** (BAR-to-linz) **glands.**

Groin (GROYN) The depression between the thigh and the trunk; the inguinal region.

Gross anatomy The branch of anatomy that deals with structures that can be studied without using a microscope. Also called **macroscopic anatomy.**

Growth An increase in size due to an increase in (1) the number of cells, (2) the size of existing cells as internal components increase in size, or (3) the size of intercellular substances.

Gustatory (GUS-ta-tō′-rē) Pertaining to taste.

Gynecology (gī′-ne-KOL-ō-jē) The branch of medicine dealing with the study and treatment of disorders of the female reproductive system.

Gynecomastia (gīn′-e-kō-MAS-tē-a) Excessive growth (benign) of the male mammary glands due to secretion of estrogens by an adrenal gland tumor (feminizing adenoma).

Gyrus (JĪ-rus) One of the folds of the cerebral cortex of the brain. *Plural is* **gyri** (JĪ-rī). Also called a **convolution.**

H

Hair A threadlike structure produced by hair follicles that develops in the dermis. Also called a **pilus** (PĪ-lus).

Hair follicle (FOL-li-kul) Structure, composed of epithelium and surrounding the root of a hair, from which hair develops.

Hair root plexus (PLEK-sus) A network of dendrites arranged around the root of a hair as free or naked nerve endings that are stimulated when a hair shaft is moved.

Haldane effect Less carbon dioxide binds to hemoglobin when the partial pressure of oxygen is higher.

Hand The terminal portion of an upper limb, including the carpus, metacarpus, and phalanges.

Haploid (HAP-loyd) Having half the number of chromosomes characteristically found in the somatic cells of an organism; characteristic of mature gametes. Symbolized *n*.

Hard palate (PAL-at) The anterior portion of the roof of the mouth, formed by the maxillae and palatine bones and lined by mucous membrane.

Haustra (HAWS-tra) A series of pouches that characterize the colon; caused by tonic contractions of the teniae coli. *Singular is* **haustrum.**

Head The superior part of a human, cephalic to the neck. The superior or proximal part of a structure.

Heart A hollow muscular organ lying slightly to the left of the midline of the chest that pumps the blood through the cardiovascular system.

Heart block An arrhythmia (dysrhythmia) of the heart in which the atria and ventricles contract independently because of a blocking of electrical impulses through the heart at some point in the conduction system.

Heart murmur (MER-mer) An abnormal sound that consists of a flow noise that is heard before, between, or after the normal heart sounds, or that may mask normal heart sounds.

Heat exhaustion Condition characterized by cool, clammy skin, profuse perspiration, and fluid and electrolyte (especially sodium and chloride) loss that results in muscle cramps, dizziness, vomiting, and fainting. Also called **heat prostration.**

Heat stroke Condition produced when the body cannot easily lose heat and characterized by reduced perspiration and elevated body temperature. Also called **sunstroke.**

Hematocrit (hē-MAT-ō-krit) **(Hct)** The percentage of blood made up of red blood cells. Usually measured by centrifuging a blood sample in a graduated tube and then reading the volume of red blood cells and dividing it by the total volume of blood in the sample.

Hematology (hē′-ma-TOL-ō-jē) The study of blood.

Hemiplegia (hem-i-PLĒ-jē-a) Paralysis of the upper limb, trunk, and lower limb on one side of the body.

Hemodynamics (hē-mō-dī-NA-miks) The study of factors and forces that govern the flow of blood through blood vessels.

Hemoglobin (hē′-mō-GLŌ-bin) **(Hb)** A substance in red blood cells consisting of the protein globin and the iron-containing red pigment heme that transports most of the oxygen and some carbon dioxide in blood.

Hemolysis (hē-MOL-i-sis) The escape of hemoglobin from the interior of a red blood cell into the surrounding medium; results from disruption of the cell membrane by toxins or drugs, freezing or thawing, or hypotonic solutions.

Hemolytic disease of the newborn A hemolytic anemia of a newborn child that results from the destruction of the infant's erythrocytes (red blood cells) by antibodies produced by the mother; usually the antibodies are due to an Rh blood type incompatibility. Also called **erythroblastosis fetalis** (e-rith′-rō-blas-TŌ-sis fe-TAL-is).

Hemophilia (hē′-mō-FIL-ē-a) A hereditary blood disorder where there is a deficient production of certain factors involved in blood clotting, resulting in excessive bleeding into joints, deep tissues, and elsewhere.

Hemopoiesis (hē-mō-poy-Ē-sis) Blood cell production, which occurs in red bone marrow after birth. Also called **hematopoiesis** (hem′-a-tō-poy-Ē-sis).

Hemorrhage (HEM-or-rij) Bleeding; the escape of blood from blood vessels, especially when the loss is profuse.

Hemorrhoids (HEM-ō-royds) Dilated or varicosed blood vessels (usually veins) in the anal region. Also called **piles.**

Hemostasis (hē-MŌS-tā-sis) The stoppage of bleeding.

Heparin (HEP-a-rin) An anticoagulant given to slow the conversion of prothrombin to thrombin, thus reducing the risk of blood clot formation; found in basophils, mast cells, and various other tissues, especially the liver and lungs.

Hepatic (he-PAT-ik) Refers to the liver.

Hepatic duct A duct that receives bile from the bile capillaries. Small hepatic ducts merge to form the larger right and left hepatic ducts that unite to leave the liver as the common hepatic duct.

Hepatic portal circulation The flow of blood from the gastrointestinal organs to the liver before returning to the heart.

Hepatocyte (he-PAT-ō-cyte) A liver cell.

Hepatopancreatic (hep′-a-tō-pan′-krē-A-tik) **ampulla** A small, raised area in the duodenum where the combined common bile duct and main pancreatic duct empty into the duodenum. Also called the **ampulla of Vater** (VA-ter).

Hernia (HER-nē-a) The protrusion or projection of an organ or part of an organ through a membrane or cavity wall, usually the abdominal cavity.

Herniated (HER-nē-Ā′-ted) **disc** A rupture of an intervertebral disc so that the nucleus pulposus protrudes into the vertebral cavity. Also called a **slipped disc.**

Heterozygous (he-ter-ō-ZĪ-gus) Possessing different alleles on homologous chromosomes for a particular hereditary trait.

Hiatus (hī-Ā-tus) An opening; a foramen.

Hilus (HĪ-lus) An area, depression, or pit where blood vessels and nerves enter or leave an organ. Also called a **hilum.**

Hinge joint A synovial joint in which a convex surface of one bone fits into a concave surface of another bone, such as the elbow, knee, ankle, and interphalangeal joints. Also called a **ginglymus** (JIN-gli-mus) **joint.**

Hirsutism (HER-soot-izm) An excessive growth of hair in females and children, with a distribution similar to that in adult males, due to the conversion of vellus hairs into large terminal hairs in response to higher-than-normal levels of androgens.

Histamine (HISS-ta-mēn) Substance found in many cells, especially mast cells, basophils, and platelets, released when the cells are injured; results in vasodilation, increased permeability of blood vessels, and constriction of bronchioles.

Histology (hiss-TOL-ō-jē) Microscopic study of the structure of tissues.

Holocrine (HŌL-ō-krin) **gland** A type of gland in which entire secretory cells, along with their accumulated secretions, make up the secretory product of the gland, as in the sebaceous (oil) glands.

Homeostasis (hō′-mē-ō-STĀ-sis) The condition in which the body's internal environment remains relatively constant, within physiological limits.

Homologous chromosomes Two chromosomes that belong to a pair. Also called **homologues.**

Homozygous (hō-mō-ZĪ-gus) Possessing the same alleles on homologous chromosomes for a particular hereditary trait.

Hormone (HOR-mōn) A secretion of endocrine cells that alters the physiological activity of target cells of the body.

Horn An area of gray matter (anterior, lateral, or posterior) in the spinal cord.

Human chorionic gonadotropin (kō-rē-ON-ik gō-nad-ō-TRŌ-pin) **(hCG)** A hormone produced by the developing placenta that maintains the corpus luteum.

Human chorionic somatomammotropin (sō-mat-ō-mam-ō-TRŌ-pin) **(hCS)** Hormone produced by the chorion of the placenta that stimulates breast tissue for lactation, enhances body growth, and regulates metabolism. Also called **human placental lactogen (hPL).**

Human growth hormone (hGH) Hormone secreted by the anterior pituitary that stimulates growth of body tissues, especially skeletal and muscular tissues. Also known as **somatotropin** and **somatotropic hormone (STH).**

Hyaluronic (hī′-a-loo-RON-ik) **acid** A viscous, amorphous extracellular material that binds cells together, lubricates joints, and maintains the shape of the eyeballs.

Hyaluronidase (hī′-a-loo-RON-i-dās) An enzyme that breaks down hyaluronic acid, increasing the permeability of connective tissues by dissolving the substances that hold body cells together.

Hydride ion (HĪ-drīd Ī-on) A hydrogen nucleus with two orbiting electrons (H^-), in contrast to the more common **hydrogen ion,** which has no orbiting electrons (H^+).

Hymen (HĪ-men) A thin fold of vascularized mucous membrane at the vaginal orifice.

Hypercalcemia (hī′-per-kal-SĒ-mē-a) An excess of calcium in the blood.

Hypercapnia (hī′-per-KAP-nē-a) An abnormal increase in the amount of carbon dioxide in the blood.

Hyperextension (hī′-per-ek-STEN-shun) Continuation of extension beyond the anatomical position, as in bending the head backward.

Hyperglycemia (hī′-per-glī-SĒ-mē-a) An elevated blood glucose level.

Hyperkalemia (hī′-per-kā-LĒ-mē-a) An excess of potassium ions in the blood.

Hypermagnesemia (hī′-per-mag′-ne-SĒ-mē-a) An excess of magnesium ions in the blood.

Hypermetropia (hī′-per-mē-TRŌ-pē-a) A condition in which visual images are focused behind the retina, with resulting defective vision of near objects; farsightedness.

Hyperphosphatemia (hī-per-fos′-fa-TĒ-mē-a) An abnormally high level of phosphates in the blood.

Hyperplasia (hī′-per-PLĀ-zē-a) An abnormal increase in the number of normal cells in a tissue or organ, increasing its size.

Hyperpolarization (hī′-per-PŌL-a-ri-zā′-shun) Increase in the internal negativity across a cell membrane, thus increasing the voltage and moving it farther away from the threshold value.

Hypersecretion (hī′-per-se-KRĒ-shun) Overactivity of glands resulting in excessive secretion.

Hypersensitivity (hī′-per-sen-si-TI-vi-tē) Overreaction to an allergen that results in pathological changes in tissues. Also called **allergy.**

Hypertension (hī′-per-TEN-shun) High blood pressure.

Hyperthermia (hī′-per-THERM-ē-a) An elevated body temperature.

Hypertonia (hī′-per-TŌ-nē-a) Increased muscle tone that is expressed as spasticity or rigidity.

Hypertonic (hī′-per-TON-ik) Solution that causes cells to shrink due to loss of water by osmosis.

Hypertrophy (hī-PER-trō-fē) An excessive enlargement or overgrowth of tissue without cell division.

Hyperventilation (hī′-per-ven-ti-LĀ-shun) A rate of respiration higher than that required to maintain a normal partial pressure of carbon dioxide in the blood.

Hypocalcemia (hī′-pō-kal-SĒ-mē-a) A below-normal level of calcium in the blood.

Hypochloremia (hī′-pō-klō-RĒ-mē-a) Deficiency of chloride ions in the blood.

Hypoglycemia (hī′-pō-glī-SĒ-mē-a) An abnormally low concentration of glucose in the blood; can result from excess insulin (injected or secreted).

Hypokalemia (hī′-pō-ka-LĒ-mē-a) Deficiency of potassium ions in the blood.

Hypomagnesemia (hī′-pō-mag′-ne-SĒ-mē-a) Deficiency of magnesium ions in the blood.

Hyponatremia (hī′-pō-na-TRĒ-mē-a) Deficiency of sodium ions in the blood.

Hyponychium (hī′-pō-NIK-ē-um) Free edge of the fingernail.

Hypophosphatemia (hī-pō-fos′-fa-TĒ-mē-a) An abnormally low level of phosphates in the blood.

Hypophyseal fossa (hī′-pō-FIZ-ē-al FOS-a) A depression on the superior surface of the sphenoid bone that houses the pituitary gland.

Hypophyseal (hī′-pō-FIZ-ē-al) **pouch** An outgrowth of ectoderm from the roof of the mouth from which the anterior pituitary develops.

Hypophysis (hī-POF-i-sis) Pituitary gland.

Hyposecretion (hī′-pō-se-KRĒ-shun) Underactivity of glands resulting in diminished secretion.

Hypothalamohypophyseal (hī-pō-tha-lam′-ō-hī-pō-FIZ-ē-al) **tract** A bundle of axons containing secretory vesicles filled with oxytocin or antidiuretic hormone that extend from the hypothalamus to the posterior pituitary.

Hypothalamus (hī′-pō-THAL-a-mus) A portion of the diencephalon, lying beneath the thalamus and forming the floor and part of the wall of the third ventricle.

Hypothermia (hī′-pō-THER-mē-a) Lowering of body temperature below 35°C (95°F); in surgical procedures, it refers to deliberate cooling of the body to slow down metabolism and reduce oxygen needs of tissues.

Hypotonia (hī′-pō-TŌ-nē-a) Decreased or lost muscle tone in which muscles appear flaccid.

Hypotonic (hī′-pō-TON-ik) Solution that causes cells to swell and perhaps rupture due to gain of water by osmosis.

Hypoventilation (hī-pō-ven-ti-LĀ-shun) A rate of respiration lower than that required to maintain a normal partial pressure of carbon dioxide in plasma.

Hypovolemic (hī-pō-vō-LĒ-mik) **shock** A type of shock characterized by decreased blood volume; may be caused by acute hemorrhage or excessive loss of other body fluids, for example, by vomiting, diarrhea, or excessive sweating.

Hypoxia (hī-POKS-ē-a) Lack of adequate oxygen at the tissue level.

Hysterectomy (hiss-te-REK-tō-mē) The surgical removal of the uterus.

I

Ileocecal (il-ē-ō-SĒ-kal) **sphincter** A fold of mucous membrane that guards the opening from the ileum into the large intestine. Also called the **ileocecal valve.**

Ileum (IL-ē-um) The terminal part of the small intestine.

Immunity (im-Ū-ni-tē) The state of being resistant to injury, particularly by poisons, foreign proteins, and invading pathogens.

Immunogenicity (im′-ū-nō-je-NIS-i-tē) Ability of an antigen to provoke an immune response.

Immunoglobulin (im-ū-nō-GLOB-ū-lin) **(Ig)** An antibody synthesized by plasma cells derived from B lymphocytes in response to the introduction of an antigen. Immunoglobulins are divided into five kinds (IgG, IgM, IgA, IgD, IgE).

Immunology (im′-ū-NOL-ō-jē) The study of the responses of the body when challenged by antigens.

Implantation (im-plan-TĀ-shun) The insertion of a tissue or a part into the body. The attachment of the blastocyst to the stratum basalis of the endometrium about 6 days after fertilization.

Incontinence (in-KON-ti-nens) Inability to retain urine, semen, or feces through loss of sphincter control.

Indirect motor pathways Motor tracts that convey information from the brain down the spinal cord for automatic movements, coordination of body movements with visual stimuli, skeletal muscle tone and posture, and balance. Also known as **extrapyramidal pathways.**

Infarction (in-FARK-shun) A localized area of necrotic tissue, produced by inadequate oxygenation of the tissue.

Infection (in-FEK-shun) Invasion and multiplication of microorganisms in body tissues, which may be inapparent or characterized by cellular injury.

Inferior (in-FĒR-ē-or) Away from the head or toward the lower part of a structure. Also called **caudad** (KAW-dad).

Inferior vena cava (VĒ-na CĀ-va) **(IVC)** Large vein that collects blood from parts of the body inferior to the heart and returns it to the right atrium.

Infertility Inability to conceive or to cause conception. Also called **sterility.**

Inflammation (in′-fla-MĀ-shun) Localized, protective response to tissue injury designed to destroy, dilute, or wall off the infecting agent or injured tissue; characterized by redness, pain, heat, swelling, and sometimes loss of function.

Inflation reflex Reflex that prevents overinflation of the lungs. Also called **Hering–Breuer reflex.**

Infundibulum (in′-fun-DIB-ū-lum) The stalklike structure that attaches the pituitary gland to the hypothalamus of the brain. The funnel-shaped, open, distal end of the uterine (Fallopian) tube.

Ingestion (in-JES-chun) The taking in of food, liquids, or drugs, by mouth.

Inguinal (IN-gwi-nal) Pertaining to the groin.

Inguinal canal An oblique passageway in the anterior abdominal wall just superior and parallel to the medial half of the inguinal ligament that transmits the spermatic cord and ilioinguinal nerve in the male and round ligament of the uterus and ilioinguinal nerve in the female.

Inhalation (in-ha-LĀ-shun) The act of drawing air into the lungs. Also termed **inspiration.**

Inheritance The acquisition of body traits by transmission of genetic information from parents to offspring.

Inhibin A hormone secreted by the gonads that inhibits release of follicle-stimulating hormone (FSH) by the anterior pituitary.

Inhibiting hormone Hormone secreted by the hypothalamus that can suppress secretion of hormones by the anterior pituitary.

Inhibitory postsynaptic potential (IPSP) A small hyperpolarization in a neuron; caused by an inhibitory neurotransmitter at a synapse.

Inner cell mass A region of cells of a blastocyst that differentiates into the three primary germ layers—ectoderm, mesoderm, and endoderm—from which all tissues and organs develop; also called an **embryoblast.**

Inorganic (in′-or-GAN-ik) **compound** Compound that usually lacks carbon, usually is small, and often contains ionic bonds. Examples include water and many acids, bases, and salts.

Insertion (in-SER-shun) The attachment of a muscle tendon to a movable bone or the end opposite the origin.

Inspiratory (in-SPĪ-ra-tor-ē) **capacity** Total inspiratory capacity of the lungs; the total of tidal volume plus inspiratory reserve volume; averages 3600 mL.

Inspiratory (in-SPĪ-ra-tor-ē) **reserve volume** Additional inspired air over and above tidal volume; averages 3100 mL.

Insula (IN-soo-la) A triangular area of the cerebral cortex that lies deep within the lateral cerebral fissue, under the parietal, frontal, and temporal lobes.

Insulin (IN-soo-lin) A hormone produced by the beta cells of a pancreatic islet (islet of Langerhans) that decreases the blood glucose level.

Insulinlike growth factor (IGF) Small protein, produced by the liver and other tissues in response to stimulation by human growth hormone (hGH), that mediates most of the effects of human growth hormone. Previously called **somatomedin** (sō′-ma-tō-MĒ-din).

Integrins (IN-te-grinz) A family of transmembrane glycoproteins in plasma membranes that function in cell adhesion; they are present in hemidesmosomes, which anchor cells to a basement membrane, and they mediate adhesion of neutrophils to endothelial cells during emigration.

Integumentary (in-teg′-ū-MEN-tar-e) Relating to the skin.

Intercalated (in-TER-ka-lāt-ed) **disc** An irregular transverse thickening of sarcolemma that contains desmosomes, which hold cardiac muscle fibers (cells) together, and gap junctions, which aid in conduction of muscle action potentials from one fiber to the next.

Intercostal (in′-ter-KOS-tal) **nerve** A nerve supplying a muscle located between the ribs.

Interferons (in′-ter-FĒR-ons) **(IFNs)** Antiviral proteins produced by virus-infected host cells; induce uninfected host cells to synthesize proteins that inhibit viral replication and enhance phagocytic activity of macrophages; types include alpha interferon, beta interferon, and gamma interferon.

Intermediate Between two structures, one of which is medial and one of which is lateral.

Intermediate filament Protein filament, ranging from 8 to 12 nm in diameter, that may provide structural reinforcement, hold organelles in place, and give shape to a cell.

Internal Away from the surface of the body.

Internal capsule A large tract of projection fibers lateral to the thalamus that is the major connection between the cerebral cortex and the brain stem and spinal cord; contains axons of sensory neurons carrying auditory, visual, and somatic sensory signals to the cerebral cortex plus axons of motor neurons descending from the cerebral cortex to the thalamus, subthalamus, brain stem, and spinal cord.

Internal ear The inner ear or labyrinth, lying inside the temporal bone, containing the organs of hearing and balance.

Internal nares (NĀ-rez) The two openings posterior to the nasal cavities opening into the nasopharynx. Also called the **choanae** (kō-Ā-nē).

Internal respiration The exchange of respiratory gases between blood and body cells. Also called **tissue respiration.**

Interneurons (in′-ter-NOO-ronz) Neurons whose axons extend only for a short distance and contact nearby neurons in the brain, spinal cord, or a ganglion; they comprise the vast majority of neurons in the body.

Interoceptor (in′-ter-ō-SEP-tor) Sensory receptor located in blood vessels and viscera that provides information about the body's internal environment.

Interphase (IN-ter-fāz) The period of the cell cycle between cell divisions, consisting of the G_1-(gap or growth) phase, when the cell is engaged in growth, metabolism, and production of substances required for division; S-(synthesis) phase, during which chromosomes are replicated; and G_2-phase.

Interstitial (in′-ter-STISH-al) **fluid** The portion of extracellular fluid that fills the microscopic spaces between the cells of tissues; the internal environment of the body. Also called **intercellular** or **tissue fluid.**

Interstitial growth Growth from within, as in the growth of cartilage. Also called **endogenous** (en-DOJ-e-nus) **growth.**

Interventricular (in′-ter-ven-TRIK-ū-lar) **foramen** A narrow, oval opening through which the lateral ventricles of the brain communicate with the third ventricle. Also called the **foramen of Monro.**

Intervertebral (in′-ter-VER-te-bral) **disc** A pad of fibrocartilage located between the bodies of two vertebrae.

Intestinal gland A gland that opens onto the surface of the intestinal mucosa and secretes digestive enzymes. Also called a **crypt of Lieberkühn** (LĒ-ber-kūn).

Intrafusal (in′-tra-FŪ-sal) **fibers** Three to ten specialized muscle fibers (cells), partially enclosed in a spindle-shaped connective tissue capsule, that make up a muscle spindle.

Intramembranous ossification (in′-tra-MEM-bra-nus os′-i′-fi-KĀ-shun) The method of bone formation in which the bone is formed directly in membranous tissue.

Intraocular (in′-tra-OK-ū-lar) **pressure (IOP)** Pressure in the eyeball, produced mainly by aqueous humor.

Intrapleural pressure Air pressure between the two pleurae of the lungs, usually subatmospheric. Also called **intrathoracic pressure.**

Intrinsic (in-TRIN-sik) Of internal origin.

Intrinsic pathway (of blood clotting) Sequence of reactions leading to blood clotting that is initiated by damage to blood vessel endothelium or platelets; activators of this pathway are contained within blood itself or are in direct contact with blood.

Intrinsic factor (IF) A glycoprotein, synthesized and secreted by the parietal cells of the gastric mucosa, that facilitates vitamin B_{12} absorption in the small intestine.

Intron (IN-tron) A region of DNA that does not code for the synthesis of a protein.

In utero (Ū-ter-ō) Within the uterus.

Invagination (in-vaj′-i-NĀ-shun) The pushing of the wall of a cavity into the cavity itself.

Inversion (in-VER-zhun) The movement of the sole medially at the ankle joint.

In vitro (VĒ-trō) Literally, in glass; outside the living body and in an artificial environment such as a laboratory test tube.

In vivo (VĒ-vō) In the living body.

Ion (Ī-on) Any charged particle or group of particles; usually formed when a substance, such as a salt, dissolves and dissociates.

Ionization (ī′-on-i-ZĀ-shun) Separation of inorganic acids, bases, and salts into ions when dissolved in water. Also called **dissociation.**

Ipsilateral (ip′-si-LAT-er-al) On the same side, affecting the same side of the body.

Iris The colored portion of the vascular tunic of the eyeball seen through the cornea that contains circular and radial smooth muscle; the hole in the center of the iris is the pupil.

Irritable bowel syndrome (IBS) Disease of the entire gastrointestinal tract in which a person reacts to stress by developing symptoms (such as cramping and abdominal pain) associated with alternating patterns of diarrhea and constipation. Excessive amounts of mucus may appear in feces, and other symptoms include flatulence, nausea, and loss of appetite. Also known as **irritable colon** or **spastic colitis.**

Ischemia (is-KĒ-mē-a) A lack of sufficient blood to a body part due to obstruction or constriction of a blood vessel.

Isoantibody A specific antibody in blood plasma that reacts with specific isoantigens and causes the clumping of bacteria, red blood cells, or particles. Also called an **agglutinin.**

Isoantigen A genetically determined antigen located on the surface of red blood cells; basis for the ABO grouping and Rh system of blood classification. Also called an **agglutinogen.**

Isometric contraction A muscle contraction in which tension on the muscle increases, but there is only minimal muscle shortening so that no visible movement is produced.

Isotonic (Ī′-sō-TON-ik) Having equal tension or tone. A solution having the same concentration of impermeable solutes as cytosol.

Isotonic (i′-so-TON-ik) **contraction** Contraction in which the tension remains the same; occurs when a constant load is moved through the range of motions possible at a joint.

Isotopes (Ī-so-tōps′) Chemical elements that have the same number of protons but different numbers of neutrons. Radioactive isotopes change into other elements with the emission of alpha or beta particles or gamma rays.

Isovolumetric (ī-sō-vol-ū′-MET-rik) **contraction** The period of time, about 0.05 sec, between the start of ventricular systole and the opening of the semilunar valves; the ventricles contract and ventricular pressure rises rapidly, but ventricular volume does not change.

Isovolumetric relaxation The period of time, about 0.05 sec, between the closing of the semilunar valves and the opening of the atrioventricular (AV) valves; there is a drastic decrease in ventricular pressure without a change in ventricular volume.

Isthmus (IS-mus) A narrow strip of tissue or narrow passage connecting two larger parts.

J

Jaundice (JAWN-dis) A condition characterized by yellowness of the skin, the white of the eyes, mucous membranes, and body fluids because of a buildup of bilirubin.

Jejunum (je-JOO-num) The middle part of the small intestine.

Joint kinesthetic (kin′-es-THET-ik) **receptor** A proprioceptive receptor located in a joint, stimulated by joint movement.

Juxtaglomerular (juks-ta-glō-MER-ū-lar) **apparatus (JGA)** Consists of the macula densa (cells of the distal convoluted tubule adjacent to the afferent and efferent arteriole) and juxtaglomerular cells (modified cells of the afferent and sometimes efferent arteriole); secretes renin when blood pressure starts to fall.

K

Keratin (KER-a-tin) An insoluble protein found in the hair, nails, and other keratinized tissues of the epidermis.

Keratinocyte (ke-RAT-in′-ō-sīt) The most numerous of the epidermal cells; produces keratin.

Ketone (KĒ-tōn) **bodies** Substances produced primarily during excessive triglyceride catabolism, such as acetone, acetoacetic acid, and β-hydroxybutyric acid.

Ketosis (kē-TŌ-sis) Abnormal condition marked by excessive production of ketone bodies.

Kidney (KID-nē) One of the paired reddish organs located in the lumbar region that regulates the composition, volume, and pressure of blood and produces urine.

Kidney stone A solid mass, usually consisting of calcium oxalate, uric acid, or calcium phosphate crystals, that may form in any portion of the urinary tract. Also called a **renal calculus.**

Kinesiology (ki-nē′-sē-OL-ō-jē) The study of the movement of body parts.

Kinesthesia (kin-es-THĒ-zē-a) The perception of the extent and direction of movement of body parts; this sense is possible due to nerve impulses generated by proprioceptors.

Korotkoff (kō-ROT-kof) **sounds** The various sounds that are heard while taking blood pressure.

Krebs cycle A series of biochemical reactions that occurs in the matrix of mitochondria in which electrons are transferred to coenzymes and carbon dioxide is formed. The electrons carried by the coenzymes then enter the electron transport chain, which generates a large quantity of ATP. Also called the **citric acid cycle** or **tricarboxylic acid (TCA) cycle.**

Kyphosis (kī-FŌ-sis) An exaggeration of the thoracic curve of the vertebral column, resulting in a "round-shouldered" appearance. Also called **hunchback.**

L

Labia majora (LĀ-bē-a ma-JŌ-ra) Two longitudinal folds of skin extending downward and backward from the mons pubis of the female.

Labia minora (min-OR-a) Two small folds of mucous membrane lying medial to the labia majora of the female.

Labial frenulum (LĀ-bē-al FREN-ū-lum) A medial fold of mucous membrane between the inner surface of the lip and the gums.

Labium (LĀ-bē-um) A lip. A liplike structure. *Plural is* **labia** (LĀ-bē-a).

Labor The process of giving birth in which a fetus is expelled from the uterus through the vagina.

Lacrimal canal A duct, one on each eyelid, beginning at the punctum at the medial margin of an eyelid and conveying tears medially into the nasolacrimal sac.

Lacrimal gland Secretory cells, located at the superior anterolateral portion of each orbit, that secrete tears into excretory ducts that open onto the surface of the conjunctiva.

Lacrimal sac The superior expanded portion of the nasolacrimal duct that receives the tears from a lacrimal canal.

Lactation (lak-TĀ-shun) The secretion and ejection of milk by the mammary glands.

Lacteal (LAK-tē-al) One of many lymphatic vessels in villi of the intestines that absorb triglycerides and other lipids from digested food.

Lacuna (la-KOO-na) A small, hollow space, such as that found in bones in which the osteocytes lie. *Plural is* **lacunae** (la-KOO-nē).

Lambdoid (lam-DOYD) **suture** The joint in the skull between the parietal bones and the occipital bone; sometimes contains sutural (Wormian) bones.

Lamellae (la-MEL-ē) Concentric rings of hard, calcified matrix found in compact bone.

Lamellated corpuscle Oval-shaped pressure receptor located in the dermis or subcutaneous tissue and consisting of concentric layers of connective tissue wrapped around the dendrites of a sensory neuron. Also called a **Pacinian** (pa-SIN-ē-an) **corpuscle.**

Lamina (LAM-i-na) A thin, flat layer or membrane, as the flattened part of either side of the arch of a vertebra. *Plural is* **laminae** (LAM-i-nē).

Lamina propria (PRŌ-prē-a) The connective tissue layer of a mucosa.

Langerhans (LANG-er-hans) **cell** Epidermal dendritic cell that functions as an antigen-presenting cell (APC) during an immune response.

Lanugo (la-NOO-gō) Fine downy hairs that cover the fetus.

Large intestine The portion of the gastrointestinal tract extending from the ileum of the small intestine to the anus, divided structurally into the cecum, colon, rectum, and anal canal.

Laryngopharynx (la-rin′-gō-FAR-inks) The inferior portion of the pharynx, extending downward from the level of the hyoid bone that divides posteriorly into the esophagus and anteriorly into the larynx. Also called the **hypopharynx.**

Laryngotracheal (la-rin′-gō-TRĀ-ke-al) **bud** An outgrowth of endoderm of the foregut from which the respiratory system develops.

Larynx (LAR-inks) The voice box, a short passageway that connects the pharynx with the trachea.

Lateral (LAT-er-al) Farther from the midline of the body or a structure.

Lateral ventricle (VEN-tri-kul) A cavity within a cerebral hemisphere that communicates with the lateral ventricle in the other cerebral hemisphere and with the third ventricle by way of the interventricular foramen.

Leg The part of the lower limb between the knee and the ankle.

Lens A transparent organ constructed of proteins (crystallins) lying posterior to the pupil and iris of the eyeball and anterior to the vitreous body.

Lesion (LĒ-zhun) Any localized, abnormal change in a body tissue.

Lesser omentum (ō-MEN-tum) A fold of the peritoneum that extends from the liver to the lesser curvature of the stomach and the first part of the duodenum.

Lesser vestibular (ves-TIB-ū-lar) **gland** One of the paired mucus-secreting glands with ducts that open on either side of the urethral orifice in the vestibule of the female.

Leukemia (loo-KĒ-mē-a) A malignant disease of the blood-forming tissues characterized by either uncontrolled production and accumulation of immature leukocytes in which many cells fail to reach maturity (acute) or an accumulation of mature leukocytes in the blood because they do not die at the end of their normal life span (chronic).

Leukocyte (LOO-kō-sīt) A white blood cell.

Leukocytosis (loo′-kō-sī-TŌ-sis) An increase in the number of white blood cells, above 10,000 per μL, characteristic of many infections and other disorders.

Leukopenia (loo-kō-PĒ-nē-a) A decrease in the number of white blood cells below 5000 cells per μL.

Leukotriene (loo′-kō-TRĪ-ēn) A type of eicosanoid produced by basophils and mast cells; acts as a local hormone; produces increased vascular permeability and acts as a chemotactic agent for phagocytes in tissue inflammation.

Leydig (LĪ-dig) **cell** A type of cell that secretes testosterone; located in the connective tissue between seminiferous tubules in a mature testis. Also known as **interstitial cell of Leydig** or **interstitial endocrinocyte**.

Libido (li-BĒ-dō) Sexual desire.

Ligament (LIG-a-ment) Dense regular connective tissue that attaches bone to bone.

Ligand (LĪ-gand) A chemical substance that binds to a specific receptor.

Limbic system A part of the forebrain, sometimes termed the visceral brain, concerned with various aspects of emotion and behavior; includes the limbic lobe, dentate gyrus, amygdala, septal nuclei, mammillary bodies, anterior thalamic nucleus, olfactory bulbs, and bundles of myelinated axons.

Lingual frenulum (LIN-gwal FREN-ū-lum) A fold of mucous membrane that connects the tongue to the floor of the mouth.

Lipase An enzyme that splits fatty acids from triglycerides and phospholipids.

Lipid (LIP-id) An organic compound composed of carbon, hydrogen, and oxygen that is usually insoluble in water, but soluble in alcohol, ether, and chloroform; examples include triglycerides (fats and oils), phospholipids, steroids, and eicosanoids.

Lipid bilayer Arrangement of phospholipid, glycolipid, and cholesterol molecules in two parallel sheets in which the hydrophilic "heads" face outward and the hydrophobic "tails" face inward; found in cellular membranes.

Lipogenesis (li-pō-GEN-e-sis) The synthesis of triglycerides.

Lipolysis (lip-OL-i-sis) The splitting of fatty acids from a triglyceride or phospholipid.

Lipoprotein (lip′-ō-PRŌ-tēn) One of several types of particles containing lipids (cholesterol and triglycerides) and proteins that make it water soluble for transport in the blood; high levels of **low-density lipoproteins (LDLs)** are associated with increased risk of atherosclerosis, whereas high levels of **high-density lipoproteins (HDLs)** are associated with decreased risk of atherosclerosis.

Liver Large organ under the diaphragm that occupies most of the right hypochondriac region and part of the epigastric region. Functionally, it produces bile and synthesizes most plasma proteins; interconverts nutrients; detoxifies substances; stores glycogen, iron, and vitamins; carries on phagocytosis of worn-out blood cells and bacteria; and helps synthesize the active form of vitamin D.

Long-term potentiation (LTP) Prolonged, enhanced synaptic transmission that occurs at certain synapses within the hippocampus of the brain; believed to underlie some aspects of memory.

Lordosis (lor-DŌ-sis) An exaggeration of the lumbar curve of the vertebral column. Also called **swayback.**

Lower limb The appendage attached at the pelvic (hip) girdle, consisting of the thigh, knee, leg, ankle, foot, and toes. Also called **lower extremity.**

Lumbar (LUM-bar) Region of the back and side between the ribs and pelvis; loin.

Lumbar plexus (PLEK-sus) A network formed by the anterior (ventral) branches of spinal nerves L1 through L4.

Lumen (LOO-men) The space within an artery, vein, intestine, renal tubule, or other tubular structure.

Lungs Main organs of respiration that lie on either side of the heart in the thoracic cavity.

Lunula (LOO-noo-la) The moon-shaped white area at the base of a nail.

Luteinizing (LOO-tē-in′-īz-ing) **hormone (LH)** A hormone secreted by the anterior pituitary that stimulates ovulation, stimulates progesterone secretion by the corpus luteum, and readies the mammary glands for milk secretion in females; stimulates testosterone secretion by the testes in males.

Lymph (LIMF) Fluid confined in lymphatic vessels and flowing through the lymphatic system until it is returned to the blood.

Lymph node An oval or bean-shaped structure located along lymphatic vessels.

Lymphatic (lim-FAT-ik) **capillary** Closed-ended microscopic lymphatic vessel that begins in spaces between cells and converges with other lymphatic capillaries to form lymphatic vessels.

Lymphatic tissue A specialized form of reticular tissue that contains large numbers of lymphocytes.

Lymphatic vessel A large vessel that collects lymph from lymphatic capillaries and converges with other lymphatic vessels to form the thoracic and right lymphatic ducts.

Lymphocyte (LIM-fō-sīt) A type of white blood cell that helps carry out cell-mediated and antibody-mediated immune responses; found in blood and in lymphatic tissues.

Lysosome (LĪ-sō-sōm) An organelle in the cytoplasm of a cell, enclosed by a single membrane and containing powerful digestive enzymes.

Lysozyme (LĪ-sō-zīm) A bactericidal enzyme found in tears, saliva, and perspiration.

M

Macrophage (MAK-rō-fāj) Phagocytic cell derived from a monocyte; may be fixed or wandering.

Macula (MAK-ū-la) A discolored spot or a colored area. A small, thickened region on the wall of the utricle and saccule that contains receptors for static equilibrium.

Macula lutea (MAK-ū-la LOO-tē-a) The yellow spot in the center of the retina.

Major histocompatibility (MHC) antigens Surface proteins on white blood cells and other nucleated cells that are unique for each person (except for identical siblings); used to type tissues and help prevent rejection of transplanted tissues. Also known as **human leukocyte antigens (HLA).**

Malignant (ma-LIG-nant) Referring to diseases that tend to become worse and cause death, especially the invasion and spreading of cancer.

Mammary (MAM-ar-ē) **gland** Modified sudoriferous (sweat) gland of the female that produces milk for the nourishment of the young.

Mammillary (MAM-i-ler-ē) **bodies** Two small rounded bodies on the inferior aspect of the hypothalamus that are involved in reflexes related to the sense of smell.

Marrow (MAR-ō) Soft, spongelike material in the cavities of bone. Red bone marrow produces blood cells; yellow bone marrow contains adipose tissue that stores triglycerides.

Mast cell A cell found in areolar connective tissue that releases histamine, a dilator of small blood vessels, during inflammation.

Mastication (mas′-ti-KĀ-shun) Chewing.

Matrix (MĀ-triks) The ground substance and fibers between cells in a connective tissue.

Matter Anything that occupies space and has mass.

Mature follicle A large, fluid-filled follicle containing a secondary oocyte and surrounding granulosa cells that secrete estrogens. Also called a **Graafian** (GRAF-ē-an) **follicle.**

Maximal oxygen uptake Maximum rate of oxygen consumption during aerobic catabolism of pyruvic acid that is determined by age, sex, body size, and aerobic training.

Mean arterial blood pressure (MABP) The average force of blood pressure exerted against the walls of arteries; approximately equal to diastolic pressure plus one-third of pulse pressure; for example, 93 mmHg when systolic pressure is 120 mmHg and diastolic pressure is 80 mmHg.

Meatus (mē-Ā-tus) A passage or opening, especially the external portion of a canal.

Mechanoreceptor (me-KAN-ō-rē-sep-tor) Sensory receptor that detects mechanical deformation of the receptor itself or adjacent cells; stimuli so detected include those related to touch, pressure, vibration, proprioception, hearing, equilibrium, and blood pressure.

Medial (MĒ-dē-al) Nearer the midline of the body or a structure.

Medial lemniscus (lem-NIS-kus) A white matter tract that originates in the gracile and cuneate nuclei of the medulla oblongata and extends to the thalamus on the same side; sensory axons in this tract conduct nerve impulses for the sensations of proprioception, fine touch, vibration, hearing, and equilibrium.

Median aperture (AP-er-choor) One of the three openings in the roof of the fourth ventricle through which cerebrospinal fluid enters the subarachnoid space of the brain and cord. Also called the **foramen of Magendie.**

Median plane A vertical plane dividing the body into right and left halves. Situated in the middle.

Mediastinum (mē′-dē-as-TĪ-num) The broad, median partition between the pleurae of the lungs, that extends from the sternum to the vertebral column in the thoracic cavity.

Medulla (me-DUL-la) An inner layer of an organ, such as the medulla of the kidneys.

Medulla oblongata (me-DUL-la ob′-long-GA-ta) The most inferior part of the brain stem. Also termed the **medulla.**

Medullary (MED-ū-lar′-ē) **cavity** The space within the diaphysis of a bone that contains yellow bone marrow. Also called the **marrow cavity.**

Medullary rhythmicity (rith-MIS-i-tē) **area** The neurons of the respiratory center in the medulla oblongata that control the basic rhythm of respiration.

Meiosis (mē-Ō-sis) A type of cell division that occurs during production of gametes, involving two successive nuclear divisions that result in daughter cells with the haploid (*n*) number of chromosomes.

Melanin (MEL-a-nin) A dark black, brown, or yellow pigment found in some parts of the body such as the skin, hair, and pigmented layer of the retina.

Melanocyte (MEL-a-nō-sīt′) A pigmented cell, located between or beneath cells of the deepest layer of the epidermis, that synthesizes melanin.

Melanocyte-stimulating hormone (MSH) A hormone secreted by the anterior pituitary that stimulates the dispersion of melanin granules in melanocytes in amphibians; continued administration produces darkening of skin in humans.

Melatonin (mel-a-TŌN-in) A hormone secreted by the pineal gland that helps set the timing of the body's biological clock.

Membrane A thin, flexible sheet of tissue composed of an epithelial layer and an underlying connective tissue layer, as in an epithelial membrane, or of areolar connective tissue only, as in a synovial membrane.

Membranous labyrinth (mem-BRA-nus LAB-i-rinth) The part of the labyrinth of the internal ear that is located inside the bony labyrinth and separated from it by the perilymph; made up of the semicircular ducts, the saccule and utricle, and the cochlear duct.

Menarche (me-NAR-kē) The first menses (menstrual flow) and beginning of ovarian and uterine cycles.

Meninges (me-NIN-jēz) Three membranes covering the brain and spinal cord, called the dura mater, arachnoid mater, and pia mater. *Singular is* **meninx** (MEN-inks).

Menopause (MEN-ō-pawz) The termination of the menstrual cycles.

Menstruation (men′-stroo-Ā-shun) Periodic discharge of blood, tissue fluid, mucus, and epithelial cells that usually lasts for 5 days; caused by a sudden reduction in estrogens and progesterone. Also called the **menstrual phase** or **menses.**

Merkel (MER-kel) **cell** Type of cell in the epidermis of hairless skin that makes contact with a tactile (Merkel) disc, which functions in touch.

Merocrine (MER-ō-krin) **gland** Gland made up of secretory cells that remain intact throughout the process of formation and discharge of the secretory product, as in the salivary and pancreatic glands.

Mesenchyme (MEZ-en-kīm) An embryonic connective tissue from which all other connective tissues arise.

Mesentery (MEZ-en-ter′-ē) A fold of peritoneum attaching the small intestine to the posterior abdominal wall.

Mesocolon (mez′-ō-KŌ-lon) A fold of peritoneum attaching the colon to the posterior abdominal wall.

Mesoderm The middle primary germ layer that gives rise to connective tissues, blood and blood vessels, and muscles.

Mesothelium (mez′-ō-THĒ-lē-um) The layer of simple squamous epithelium that lines serous membranes.

Mesovarium (mez′-ō-VAR-ē-um) A short fold of peritoneum that attaches an ovary to the broad ligament of the uterus.

Metabolism (me-TAB-ō-lizm) All the biochemical reactions that occur within an organism, including the synthetic (anabolic) reactions and decomposition (catabolic) reactions.

Metacarpus (met′-a-KAR-pus) A collective term for the five bones that make up the palm.

Metaphase (MET-a-phāz) The second stage of mitosis, in which chromatid pairs line up on the metaphase plate of the cell.

Metaphysis (me-TAF-i-sis) Region of a long bone between the diaphysis and epiphysis that contains the epiphyseal plate in a growing bone.

Metarteriole (met′-ar-TĒ-rē-ōl) A blood vessel that emerges from an arteriole, traverses a capillary network, and empties into a venule.

Metastasis (me-TAS-ta-sis) The spread of cancer to surrounding tissues (local) or to other body sites (distant).

Metatarsus (met′-a-TAR-sus) A collective term for the five bones located in the foot between the tarsals and the phalanges.

Micelle (mī-SEL) A spherical aggregate of bile salts that dissolves fatty acids and monoglycerides so that they can be absorbed into small intestinal epithelial cells.

Microfilament (mī-krō-FIL-a-ment) Rodlike protein filament about 6 nm in diameter; constitutes contractile units in muscle fibers (cells) and provides support, shape, and movement in nonmuscle cells.

Microglia (mī-krō-GLĒ-a) Neuroglial cells that carry on phagocytosis.

Microtubule (mī-krō-TOO-būl′) Cylindrical protein filament, ranging in diameter from 18 to 30 nm, consisting of the protein tubulin; provides support, structure, and transportation.

Microvilli (mī′-krō-VIL-ē) Microscopic, fingerlike projections of the plasma membranes of cells that increase surface area for absorption, especially in the small intestine and proximal convoluted tubules of the kidneys.

Micturition (mik'-choo-RISH-un) The act of expelling urine from the urinary bladder. Also called **urination** (yoo-ri-NĀ-shun).

Midbrain The part of the brain between the pons and the diencephalon. Also called the **mesencephalon** (mes'-en-SEF-a-lon).

Middle ear A small, epithelial-lined cavity hollowed out of the temporal bone, separated from the external ear by the eardrum and from the internal ear by a thin bony partition containing the oval and round windows; extending across the middle ear are the three auditory ossicles. Also called the **tympanic** (tim-PAN-ik) **cavity.**

Midline An imaginary vertical line that divides the body into equal left and right sides.

Midsagittal plane A vertical plane through the midline of the body that divides the body or organs into *equal* right and left sides. Also called a **median plane.**

Milk ejection reflex Contraction of alveolar cells to force milk into ducts of mammary glands, stimulated by oxytocin (TO), which is released from the posterior pituitary in response to suckling action. Also called the **milk letdown reflex.**

Mineral Inorganic, homogeneous solid substance that may perform a function vital to life; examples include calcium and phosphorus.

Mineralocorticoids (min'-er-al-ō-KOR-ti-koyds) A group of hormones of the adrenal cortex that help regulate sodium and potassium balance.

Minimal volume The volume of air in the lungs even after the thoracic cavity has been opened, forcing out some of the residual volume.

Minute ventilation (MV) Total volume of air inhaled and exhaled per minute; about 6000 mL.

Miosis Constriction of the pupil of the eye.

Mitochondrion (mī'-tō-KON-drē-on) A double-membraned organelle that plays a central role in the production of ATP; known as the "powerhouse" of the cell.

Mitosis (mī-TŌ-sis) The orderly division of the nucleus of a cell that ensures that each new daughter nucleus has the same number and kind of chromosomes as the original parent nucleus. The process includes the replication of chromosomes and the distribution of the two sets of chromosomes into two separate and equal nuclei.

Mitotic spindle Collective term for a football-shaped assembly of microtubules (nonkinetochore, kinetochore, and aster) that is responsible for the movement of chromosomes during cell division.

Modality (mō-DAL-i-tē) Any of the specific sensory entities, such as vision, smell, taste, or touch.

Modiolus (mō-DĪ-ō'-lus) The central pillar or column of the cochlea.

Mole The weight, in grams, of the combined atomic weights of the atoms that make up a molecule of a substance.

Molecule (MOL-e-kūl) The chemical combination of two or more atoms covalently bonded together.

Monocyte (MON-ō-sīt') The largest type of white blood cell, characterized by agranular cytoplasm.

Monounsaturated fat A fatty acid that contains one double covalent bond between its carbon atoms; it is not completely saturated with hydrogen atoms. Plentiful in triglycerides of olive and peanut oils.

Mons pubis (monz PŪ-bis) The rounded, fatty prominence over the pubic symphysis, covered by coarse pubic hair.

Morula (MOR-ū-la) A solid sphere of cells produced by successive cleavages of a fertilized ovum about four days after fertilization.

Motor end plate Region of the sarcolemma of a muscle fiber (cell) that includes acetylcholine (ACh) receptors, which bind ACh released by synaptic end bulbs of somatic motor neurons.

Motor neurons (NOO-ronz) Neurons that conduct impulses from the brain toward the spinal cord or out of the brain and spinal cord into cranial or spinal nerves to effectors that may be either muscles or glands. Also called **efferent neurons.**

Motor unit A motor neuron together with the muscle fibers (cells) it stimulates.

Mucin (MŪ-sin) A protein found in mucus.

Mucosa-associated lymphatic tissue (MALT) Lymphatic nodules scattered throughout the lamina propria (connective tissue) of mucous membranes lining the gastrointestinal tract, respiratory airways, urinary tract, and reproductive tract.

Mucous (MŪ-kus) **cell** A unicellular gland that secretes mucus. Two types are mucous neck cells and surface mucous cells in the stomach.

Mucous membrane A membrane that lines a body cavity that opens to the exterior. Also called the **mucosa** (mū-KŌ-sa).

Mucus The thick fluid secretion of goblet cells, mucous cells, mucous glands, and mucous membranes.

Muscarinic (mus'-ka-RIN-ik) **receptor** Receptor for the neurotransmitter acetylcholine found on all effectors innervated by parasympathetic postganglionic axons and on sweat glands innervated by cholinergic sympathetic postganglionic axons; so named because muscarine activates these receptors but does not activate nicotinic receptors for acetylcholine.

Muscle An organ composed of one of three types of muscle tissue (skeletal, cardiac, or smooth), specialized for contraction to produce voluntary or involuntary movement of parts of the body.

Muscle action potential A stimulating impulse that propagates along the sarcolemma and transverse tubules; in skeletal muscle, it is generated by acetylcholine, which increases the permeability of the sarcolemma to cations, especially sodium ions (Na^+).

Muscle fatigue (fa-TĒG) Inability of a muscle to maintain its strength of contraction or tension; may be related to insufficient oxygen, depletion of glycogen, and/or lactic acid buildup.

Muscle spindle An encapsulated proprioceptor in a skeletal muscle, consisting of specialized intrafusal muscle fibers and nerve endings; stimulated by changes in length or tension of muscle fibers.

Muscle tissue A tissue specialized to produce motion in response to muscle action potentials by its qualities of contractility, extensibility, elasticity, and excitability; types include skeletal, cardiac, and smooth.

Muscle tone A sustained, partial contraction of portions of a skeletal or smooth muscle in response to activation of stretch receptors or a baseline level of action potentials in the innervating motor neurons.

Muscular dystrophies (DIS-trō-fēz') Inherited muscle-destroying diseases, characterized by degeneration of muscle fibers (cells), which causes progressive atrophy of the skeletal muscle.

Muscularis (MUS-kū-la'-ris) A muscular layer (coat or tunic) of an organ.

Muscularis mucosae (mū-KŌ-sē) A thin layer of smooth muscle fibers that underlie the lamina propria of the mucosa of the gastrointestinal tract.

Mutation (mū-TĀ-shun) Any change in the sequence of bases in a DNA molecule resulting in a permanent alteration in some inheritable trait.

Myasthenia (mī-as-THĒ-nē-a) **gravis** Weakness and fatigue of skeletal muscles caused by antibodies directed against acetylcholine receptors.

Myelin (MĪ-e-lin) **sheath** Multilayered lipid and protein covering, formed by Schwann cells and oligodendrocytes, around axons of many peripheral and central nervous system neurons.

Myenteric plexus A network of autonomic axons and postganglionic cell bodies located in the muscularis of the gastrointestinal tract. Also called the **plexus of Auerbach** (OW-er-bak).

Myocardial infarction (mī′-ō-KAR-dē-al in-FARK-shun) **(MI)** Gross necrosis of myocardial tissue due to interrupted blood supply. Also called a **heart attack.**

Myocardium (mī′-ō-KAR-dē-um) The middle layer of the heart wall, made up of cardiac muscle tissue, lying between the epicardium and the endocardium and constituting the bulk of the heart.

Myofibril (mī-ō-FĪ-bril) A threadlike structure, extending longitudinally through a muscle fiber (cell) consisting mainly of thick filaments (myosin) and thin filaments (actin, troponin, and tropomyosin).

Myoglobin (mī-ō-GLŌ-bin) The oxygen-binding, iron-containing protein present in the sarcoplasm of muscle fibers (cells); contributes the red color to muscle.

Myogram (MĪ-ō-gram) The record or tracing produced by a myograph, an apparatus that measures and records the force of muscular contractions.

Myology (mī-OL-ō-jē) The study of muscles.

Myometrium (mī′-ō-MĒ-trē-um) The smooth muscle layer of the uterus.

Myopathy (mī-OP-a-thē) Any abnormal condition or disease of muscle tissue.

Myopia (mī-Ō-pē-a) Defect in vision in which objects can be seen distinctly only when very close to the eyes; nearsightedness.

Myosin (MĪ-ō-sin) The contractile protein that makes up the thick filaments of muscle fibers.

Myotome (MĪ-ō-tōm) A group of muscles innervated by the motor neurons of a single spinal segment. In an embryo, the portion of a somite that develops into some skeletal muscles.

N

Nail A hard plate, composed largely of keratin, that develops from the epidermis of the skin to form a protective covering on the dorsal surface of the distal phalanges of the fingers and toes.

Nail matrix (MĀ-triks) The part of the nail beneath the body and root from which the nail is produced.

Nasal (NĀ-zal) **cavity** A mucosa-lined cavity on either side of the nasal septum that opens onto the face at the external nares and into the nasopharynx at the internal nares.

Nasal septum (SEP-tum) A vertical partition composed of bone (perpendicular plate of ethmoid and vomer) and cartilage, covered with a mucous membrane, separating the nasal cavity into left and right sides.

Nasolacrimal (nā′-zō-LAK-ri-mal) **duct** A canal that transports the lacrimal secretion (tears) from the nasolacrimal sac into the nose.

Nasopharynx (nā′-zō-FAR-inks) The superior portion of the pharynx, lying posterior to the nose and extending inferiorly to the soft palate.

Neck The part of the body connecting the head and the trunk. A constricted portion of an organ such as the neck of the femur or uterus.

Necrosis (ne-KRŌ-sis) A pathological type of cell death that results from disease, injury, or lack of blood supply in which many adjacent cells swell, burst, and spill their contents into the interstitial fluid, triggering an inflammatory response.

Negative feedback The principle governing most control systems; a mechanism of response in which a stimulus initiates actions that reverse or reduce the stimulus.

Neonatal (nē-ō-NĀ-tal) Pertaining to the first four weeks after birth.

Neoplasm (NĒ-ō-plazm) A new growth that may be benign or malignant.

Nephron (NEF-ron) The functional unit of the kidney.

Nerve A cordlike bundle of neuronal axons and/or dendrites and associated connective tissue coursing together outside the central nervous system.

Nerve fiber General term for any process (axon or dendrite) projecting from the cell body of a neuron.

Nerve impulse A wave of depolarization and repolarization that self-propagates along the plasma membrane of a neuron; also called a **nerve action potential.**

Nervous tissue Tissue containing neurons that initiate and conduct nerve impulses to coordinate homeostasis, and neuroglia that provide support and nourishment to neurons.

Net filtration pressure (NFP) Net pressure that promotes fluid outflow at the arterial end of a capillary, and fluid inflow at the venous end of a capillary; net pressure that promotes glomerular filtration in the kidneys.

Neural plate A thickening of ectoderm, induced by the notochord, that forms early in the third week of development and represents the beginning of the development of the nervous system.

Neuralgia (noo-RAL-jē-a) Attacks of pain along the entire course or branch of a peripheral sensory nerve.

Neuritis (noo-RĪ-tis) Inflammation of one or more nerves.

Neuroglia (noo-RŌG-lē-a) Cells of the nervous system that perform various supportive functions. The neuroglia of the central nervous system are the astrocytes, oligodendrocytes, microglia, and ependymal cells; neuroglia of the peripheral nervous system include Schwann cells and satellite cells. Also called **glial** (GLĒ-al) **cells.**

Neurohypophyseal (noo′-rō-hī′-pō-FIZ-ē-al) **bud** An outgrowth of ectoderm located on the floor of the hypothalamus that gives rise to the posterior pituitary.

Neurolemma (noo-rō-LEM-ma) The peripheral, nucleated cytoplasmic layer of the Schwann cell. Also called **sheath of Schwann** (SCHVON).

Neurology (noo-ROL-ō-jē) The study of the normal functioning and disorders of the nervous system.

Neuromuscular (noo-rō-MUS-kū-lar) **junction** A synapse between the axon terminals of a motor neuron and the sarcolemma of a muscle fiber (cell).

Neuron (NOO-ron) A nerve cell, consisting of a cell body, dendrites, and an axon.

Neuropeptide (noo-rō-PEP-tīd) Chain of three to about 40 amino acids that occurs naturally in the nervous system, and that acts primarily to modulate the response of or to a neurotransmitter. Examples are enkephalins and endorphins.

Neurosecretory (noo-rō-SĒC-re-tō-rē) **cell** A neuron that secretes a hypothalamic releasing hormone or inhibiting hormone into blood capillaries of the hypothalmus; a neuron that secretes oxytocin or antidiuretic hormone into blood capillaries of the posterior pituitary.

Neurotransmitter One of a variety of molecules within axon terminals that are released into the synaptic cleft in response to a nerve impulse, and that change the membrane potential of the postsynaptic neuron.

Neutrophil (NOO-trō-fil) A type of white blood cell characterized by granules that stain pale lilac with a combination of acidic and basic dyes.

Nicotinic (nik′-ō-TIN-ik) **receptor** Receptor for the neurotransmitter acetylcholine found on both sympathetic and parasympathetic postganglionic neurons and on skeletal muscle in the motor end plate; so named because nicotine activates these receptors but does not activate muscarinic receptors for acetylcholine.

Nipple A pigmented, wrinkled projection on the surface of the breast that is the location of the openings of the lactiferous ducts for milk release.

Nociceptor (nō′-sē-SEP-tor) A free (naked) nerve ending that detects painful stimuli.

Node of Ranvier (ron-vē-Ā) A space, along a myelinated axon, between the individual Schwann cells that form the myelin sheath and the neurolemma. Also called a **neurofibral node.**

Nondisjunction (non′-dis-JUNGK-shun) Failure of sister chromatids to separate properly during anaphase of mitosis (or meiosis II) or failure of homologous chromosomes to separate properly during meiosis I in which chromatids or chromosomes pass into the same daughter cell; the result is too many copies of that chromosome in the daughter cell or gamete.

Norepinephrine (nor′-ep-ē-NEF-rin) **(NE)** A hormone secreted by the adrenal medulla that produces actions similar to those that result from sympathetic stimulation. Also called **noradrenaline** (nor-a-DREN-a-lin).

Notochord (NŌ-tō-cord) A flexible rod of mesodermal tissue that lies where the future vertebral column will develop and plays a role in induction.

Nuclear medicine The branch of medicine concerned with the use of radioisotopes in the diagnosis and therapy of disease.

Nucleic (noo-KLĒ-ic) **acid** An organic compound that is a long polymer of nucleotides, with each nucleotide containing a pentose sugar, a phosphate group, and one of four possible nitrogenous bases (adenine, cytosine, guanine, and thymine or uracil).

Nucleolus (noo-KLĒ-ō-lus) Spherical body within a cell nucleus composed of protein, DNA, and RNA that is the site of the assembly of small and large ribosomal subunits.

Nucleosome (NOO-klē-ō-sōm) Structural subunit of a chromosome consisting of histones and DNA.

Nucleus (NOO-klē-us) A spherical or oval organelle of a cell that contains the hereditary factors of the cell, called genes. A cluster of unmyelinated nerve cell bodies in the central nervous system. The central part of an atom made up of protons and neutrons.

Nucleus pulposus (pul-PŌ-sus) A soft, pulpy, highly elastic substance in the center of an intervertebral disc; a remnant of the notochord.

Nutrient A chemical substance in food that provides energy, forms new body components, or assists in various body functions.

O

Obesity (ō-BĒS-i-tē) Body weight more than 20% above a desirable standard due to excessive accumulation of fat.

Oblique (ō-BLĒK) **plane** A plane that passes through the body or an organ at an angle between the transverse plane and either the mid-sagittal, parasagittal, or frontal plane.

Obstetrics (ob-STET-riks) The specialized branch of medicine that deals with pregnancy, labor, and the period of time immediately after delivery (about 6 weeks).

Olfactory (ōl-FAK-tō-rē) Pertaining to smell.

Olfactory bulb A mass of gray matter containing cell bodies of neurons that form synapses with neurons of the olfactory nerve (cranial nerve I), lying inferior to the frontal lobe of the cerebrum on either side of the crista galli of the ethmoid bone.

Olfactory receptor A bipolar neuron with its cell body lying between supporting cells located in the mucous membrane lining the superior portion of each nasal cavity; transduces odors into neural signals.

Olfactory tract A bundle of axons that extends from the olfactory bulb posteriorly to olfactory regions of the cerebral cortex.

Oligodendrocyte (ol′-i-gō-DEN-drō-sīt) A neuroglial cell that supports neurons and produces a myelin sheath around axons of neurons of the central nervous system.

Oligospermia (ol′-i-gō-SPER-mē-a) A deficiency of sperm cells in the semen.

Olive A prominent oval mass on each lateral surface of the superior part of the medulla oblongata.

Oncogenes (ONG-kō-jēnz) Cancer-causing genes; they derive from normal genes, termed **proto-oncogenes,** that encode proteins involved in cell growth or cell regulation but have the ability to transform a normal cell into a cancerous cell when they are mutated or inappropriately activated. One example is *p53.*

Oncology (ong-KOL-ō-jē) The study of tumors.

Oogenesis (ō′-ō-JEN-e-sis) Formation and development of female gametes (oocytes).

Oophorectomy (ō′-of-ō-REK-tō-me) Surgical removal of the ovaries.

Ophthalmic (of-THAL-mik) Pertaining to the eye.

Ophthalmologist (of′-thal-MOL-ō-jist) A physician who specializes in the diagnosis and treatment of eye disorders using drugs, surgery, and corrective lenses.

Ophthalmology (of′-thal-MOL-ō-jē) The study of the structure, function, and diseases of the eye.

Opsin (OP-sin) The glycoprotein portion of a photopigment.

Opsonization (op-sō-ni-ZĀ-shun) The action of some antibodies that renders bacteria and other foreign cells more susceptible to phagocytosis.

Optic (OP-tik) Refers to the eye, vision, or properties of light.

Optic chiasm (KĪ-azm) A crossing point of the optic nerves (cranial nerve II), anterior to the pituitary gland. Also called **optic chiasma.**

Optic disc A small area of the retina containing openings through which the axons of the ganglion cells emerge as the optic nerve (cranial nerve II). Also called the **blind spot.**

Optician (op-TISH-an) A technician who fits, adjusts, and dispenses corrective lenses on prescription of an ophthalmologist or optometrist.

Optic tract A bundle of axons that carry nerve impulses from the retina of the eye between the optic chiasm and the thalamus.

Optometrist (op-TOM-e-trist) Specialist with a doctorate degree in optometry who is licensed to examine and test the eyes and treat visual defects by prescribing corrective lenses.

Ora serrata (Ō-ra ser-RĀ-ta) The irregular margin of the retina lying internal and slightly posterior to the junction of the choroid and ciliary body.

Orbit (OR-bit) The bony, pyramidal-shaped cavity of the skull that holds the eyeball.

Organ A structure composed of two or more different kinds of tissues with a specific function and usually a recognizable shape.

Organelle (or-gan-EL) A permanent structure within a cell with characteristic morphology that is specialized to serve a specific function in cellular activities.

Organic (or-GAN-ik) **compound** Compound that always contains carbon in which the atoms are held together by covalent bonds. Examples include carbohydrates, lipids, proteins, and nucleic acids (DNA and RNA).

Organism (OR-ga-nizm) A total living form; one individual.

Orgasm (OR-gazm) Sensory and motor events involved in ejaculation for the male and involuntary contraction of the perineal muscles in the female at the climax of sexual intercourse.

Orifice (OR-i-fis) Any aperture or opening.

Origin (OR-i-jin) The attachment of a muscle tendon to a stationary bone or the end opposite the insertion.

Oropharynx (or'-ō-FAR-inks) The intermediate portion of the pharynx, lying posterior to the mouth and extending from the soft palate to the hyoid bone.

Orthopedics (or'-thō-PĒ-diks) The branch of medicine that deals with the preservation and restoration of the skeletal system, articulations, and associated structures.

Osmoreceptor (oz'-mō-re-CEP-tor) Receptor in the hypothalamus that is sensitive to changes in blood osmolarity and, in response to high osmolarity (low water concentration), stimulates synthesis and release of antidiuretic hormone (ADH).

Osmosis (os-MŌ-sis) The net movement of water molecules through a selectively permeable membrane from an area of higher water concentration to an area of lower water concentration until equilibrium is reached.

Osmotic pressure The pressure required to prevent the movement of pure water into a solution containing solutes when the solutions are separated by a selectively permeable membrane.

Osseous (OS-ē-us) Bony.

Ossicle (OS-si-kul) One of the small bones of the middle ear (malleus, incus, stapes).

Ossification (os'-i-fi-KĀ-shun) Formation of bone. Also called **osteogenesis.**

Osteoblast (OS-tē-ō-blast) Cell formed from an osteogenic cell that participates in bone formation by secreting some organic components and inorganic salts.

Osteoclast (OS-tē-ō-clast') A large, multinuclear cell that resorbs (destroys) bone matrix.

Osteocyte (OS-tē-ō-sīt') A mature bone cell that maintains the daily activities of bone tissue.

Osteogenic (os'-tē-ō-prō-JEN-i-tor) **cell** Stem cell derived from mesenchyme that has mitotic potential and the ability to differentiate into an osteoblast.

Osteogenic (os'-tē-ō-JEN-ik) **layer** The inner layer of the periosteum that contains cells responsible for forming new bone during growth and repair.

Osteology (os'-tē-OL-ō-jē) The study of bones.

Osteon (OS-tē-on) The basic unit of structure in adult compact bone, consisting of a central (Haversian) canal with its concentrically arranged lamellae, lacunae, osteocytes, and canaliculi. Also called a **Haversian** (ha-VER-shan) **system.**

Osteoporosis (os'-tē-ō-pō-RŌ-sis) Age-related disorder characterized by decreased bone mass and increased susceptibility to fractures, often as a result of decreased levels of estrogens.

Otic (Ō-tik) Pertaining to the ear.

Otolith (Ō-tō-lith) A particle of calcium carbonate embedded in the otolithic membrane that functions in maintaining static equilibrium.

Otolithic (ō-tō-LITH-ik) **membrane** Thick, gelatinous, glycoprotein layer located directly over hair cells of the macula in the saccule and utricle of the internal ear.

Otorhinolaryngology (ō'-tō-rī-nō-lar'-in-GOL-ō-jē) The branch of medicine that deals with the diagnosis and treatment of diseases of the ears, nose, and throat.

Oval window A small, membrane-covered opening between the middle ear and inner ear into which the footplate of the stapes fits.

Ovarian (ō-VAR-ē-an) **cycle** A monthly series of events in the ovary associated with the maturation of a secondary oocyte.

Ovarian follicle (FOL-i-kul) A general name for oocytes (immature ova) in any stage of development, along with their surrounding epithelial cells.

Ovarian ligament (LIG-a-ment) A rounded cord of connective tissue that attaches the ovary to the uterus.

Ovary (Ō-var-ē) Female gonad that produces oocytes and the estrogens, progesterone, inhibin, and relaxin hormones.

Ovulation (ov-ū-LĀ-shun) The rupture of a mature ovarian (Graafian) follicle with discharge of a secondary oocyte into the pelvic cavity.

Ovum (Ō-vum) The female reproductive or germ cell; an egg cell; arises through completion of meiosis in a secondary oocyte after penetration by a sperm.

Oxidation (ok-si-DĀ-shun) The removal of electrons from a molecule or, less commonly, the addition of oxygen to a molecule that results in a decrease in the energy content of the molecule. The oxidation of glucose in the body is called **cellular respiration.**

Oxyhemoglobin (ok'-sē-HĒ-mō-glō-bin) ($Hb-O_2$) Hemoglobin combined with oxygen.

Oxytocin (ok'-sē-TŌ-sin) **(OT)** A hormone secreted by neurosecretory cells in the paraventricular and supraoptic nuclei of the hypothalamus that stimulates contraction of smooth muscle in the pregnant uterus and myoepithelial cells around the ducts of mammary glands.

P

P wave The deflection wave of an electrocardiogram that signifies atrial depolarization.

Palate (PAL-at) The horizontal structure separating the oral and the nasal cavities; the roof of the mouth.

Palpate (PAL-pāt) To examine by touch; to feel.

Pancreas (PAN-krē-as) A soft, oblong organ lying along the greater curvature of the stomach and connected by a duct to the duodenum. It is both an exocrine gland (secreting pancreatic juice) and an endocrine gland (secreting insulin, glucagon, somatostatin, and pancreatic polypeptide).

Pancreatic (pan'-krē-AT-ik) **duct** A single large tube that unites with the common bile duct from the liver and gallbladder and drains pancreatic juice into the duodenum at the hepatopancreatic ampulla (ampulla of Vater). Also called the **duct of Wirsung.**

Pancreatic islet A cluster of endocrine gland cells in the pancreas that secretes insulin, glucagon, somatostatin, and pancreatic polypeptide. Also called an **islet of Langerhans** (LANG-er-hanz).

Pancreatic polypeptide Hormone secreted by the F cells of pancreatic islets (islets of Langerhans) that regulates release of pancreatic digestive enzymes.

Papanicolaou (pap´-a-NIK-ō-la-oo) **test** A cytological staining test for the detection and diagnosis of premalignant and malignant conditions of the female genital tract. Cells scraped from the epithelium of the cervix of the uterus are examined microscopically. Also called a **Pap test** or **Pap smear.**

Papilla (pa-PIL-a) A small nipple-shaped projection or elevation.

Paracrine (PAR-a-krin) Local hormone, such as histamine, that acts on neighboring cells without entering the bloodstream.

Paralysis (pa-RAL-a-sis) Loss or impairment of motor function due to a lesion of nervous or muscular origin.

Paranasal sinus (par´-a-NĀ-zal SĪ-nus) A mucus-lined air cavity in a skull bone that communicates with the nasal cavity. Paranasal sinuses are located in the frontal, maxillary, ethmoid, and sphenoid bones.

Paraplegia (par-a-PLĒ-jē-a) Paralysis of both lower limbs.

Parasagittal plane (par-a-SAJ-i-tal) A vertical plane that does not pass through the midline and that divides the body or organs into *unequal* left and right portions.

Parasympathetic (par´-a-sim-pa-THET-ik) **division** One of the two subdivisions of the autonomic nervous system, having cell bodies of preganglionic neurons in nuclei in the brain stem and in the lateral gray horn of the sacral portion of the spinal cord; primarily concerned with activities that conserve and restore body energy.

Parathyroid (par´-a-THĪ-royd) **gland** One of usually four small endocrine glands embedded in the posterior surfaces of the lateral lobes of the thyroid gland.

Parathyroid hormone (PTH) A hormone secreted by the chief (principal) cells of the parathyroid glands that increases blood calcium level and decreases blood phosphate level.

Paraurethral (par´-a-ū-RĒ-thral) **gland** Gland embedded in the wall of the urethra whose duct opens on either side of the urethral orifice and secretes mucus. Also called **Skene's** (SKĒNZ) **gland.**

Parenchyma (par-EN-ki-ma) The functional parts of any organ, as opposed to tissue that forms its stroma or framework.

Parietal (pa-RĪ-e-tal) Pertaining to or forming the outer wall of a body cavity.

Parietal cell A type of secretory cell in gastric glands that produces hydrochloric acid and intrinsic factor. Also called an **oxyntic cell.**

Parkinson disease (PD) Progressive degeneration of the basal ganglia and substantia nigra of the cerebrum resulting in decreased production of dopamine (DA) that leads to tremor, slowing of voluntary movements, and muscle weakness.

Parotid (pa-ROT-id) **gland** One of the paired salivary glands located inferior and anterior to the ears and connected to the oral cavity via a duct (Stensen's) that opens into the inside of the cheek opposite the maxillary (upper) second molar tooth.

Parturition (par´-too-RISH-un) Act of giving birth to young; childbirth, delivery.

Patellar (pa-TELL-ar) **reflex** Extension of the leg by contraction of the quadriceps femoris muscle in response to tapping the patellar ligament. Also called the **knee jerk reflex.**

Patent ductus arteriosus Congenital anatomical heart defect in which the fetal connection between the aorta and pulmonary trunk remains open instead of closing completely after birth.

Pathogen (PATH-ō-jen) A disease-producing microbe.

Pathological (path´-ō-LOJ-i-kal) **anatomy** The study of structural changes caused by disease.

Pectoral (PEK-tō-ral) Pertaining to the chest or breast.

Pediatrician (pē´-dē-a-TRISH-un) A physician who specializes in the care and treatment of children.

Pedicel (PED-i-sel) Footlike structure, as on podocytes of a glomerulus.

Pelvic (PEL-vik) **cavity** Inferior portion of the abdominopelvic cavity that contains the urinary bladder, sigmoid colon, rectum, and internal female and male reproductive structures.

Pelvic splanchnic (PEL-vik SPLANGK-nik) **nerves** Consist of preganglionic parasympathetic axons from the levels of S2, S3, and S4 that supply the urinary bladder, reproductive organs, and the descending and sigmoid colon and rectum.

Pelvis The basinlike structure formed by the two hip bones, the sacrum, and the coccyx. The expanded, proximal portion of the ureter, lying within the kidney and into which the major calyces open.

Penis (PĒ-nis) The organ of urination and copulation in males; used to deposit semen into the female vagina.

Pepsin Protein-digesting enzyme secreted by chief cells of the stomach in the inactive form pepsinogen, which is converted to active pepsin by hydrochloric acid.

Peptic ulcer An ulcer that develops in areas of the gastrointestinal tract exposed to hydrochloric acid; classified as a gastric ulcer if in the lesser curvature of the stomach and as a duodenal ulcer if in the first part of the duodenum.

Percussion (per-KUSH-un) The act of striking (percussing) an underlying part of the body with short, sharp blows as an aid in diagnosing the part by the quality of the sound produced.

Perforating canal A minute passageway by means of which blood vessels and nerves from the periosteum penetrate into compact bone. Also called **Volkmann's** (FŌLK-manz) **canal.**

Pericardial (per´-i-KAR-dē-al) **cavity** Small potential space between the visceral and parietal layers of the serous pericardium that contains pericardial fluid.

Pericardium (per´-i-KAR-dē-um) A loose-fitting membrane that encloses the heart, consisting of a superficial fibrous layer and a deep serous layer.

Perichondrium (per´-i-KON-drē-um) The membrane that covers cartilage.

Perilymph (PER-i-limf) The fluid contained between the bony and membranous labyrinths of the inner ear.

Perimetrium (per´-i-MĒ-trē-um) The serosa of the uterus.

Perimysium (per´-i-MĪZ-ē-um) Invagination of the epimysium that divides muscles into bundles.

Perineum (per´-i-NĒ-um) The pelvic floor; the space between the anus and the scrotum in the male and between the anus and the vulva in the female.

Perineurium (per´-i-NOO-rē-um) Connective tissue wrapping around fascicles in a nerve.

Periodontal (per-ē-ō-DON-tal) **disease** A collective term for conditions characterized by degeneration of gingivae, alveolar bone, periodontal ligament, and cementum.

Periodontal ligament The periosteum lining the alveoli (sockets) for the teeth in the alveolar processes of the mandible and maxillae.

Periosteum (per´-ē-OS-tē-um) The membrane that covers bone and consists of connective tissue, osteogenic cells, and osteoblasts; is essential for bone growth, repair, and nutrition.

Peripheral (pe-RIF-er-al) Located on the outer part or a surface of the body.

Peripheral nervous system (PNS) The part of the nervous system that lies outside the central nervous system, consisting of nerves and ganglia.

Peristalsis (per′-i-STAL-sis) Successive muscular contractions along the wall of a hollow muscular structure.

Peritoneum (per′-i-tō-NĒ-um) The largest serous membrane of the body that lines the abdominal cavity and covers the viscera.

Peritonitis (per′-i-tō-NĪ-tis) Inflammation of the peritoneum.

Permissive (per-MIS-sive) **effect** A hormonal interaction in which the effect of one hormone on a target cell requires previous or simultaneous exposure to another hormone(s) to enhance the response of a target cell or increase the activity of another hormone.

Peroxisome (per-OK-si-sōm) Organelle similar in structure to a lysosome that contains enzymes that use molecular oxygen to oxidize various organic compounds; such reactions produce hydrogen peroxide; abundant in liver cells.

Perspiration Sweat; produced by sudoriferous (sweat) glands and containing water, salts, urea, uric acid, amino acids, ammonia, sugar, lactic acid, and ascorbic acid. Helps maintain body temperature and eliminate wastes.

pH A measure of the concentration of hydrogen ions (H^+) in a solution. The pH scale extends from 0 to 14, with a value of 7 expressing neutrality, values lower than 7 expressing increasing acidity, and values higher than 7 expressing increasing alkalinity.

Phagocytosis (fag′-ō-sī-TŌ-sis) The process by which phagocytes ingest particulate matter; the ingestion and destruction of microbes, cell debris, and other foreign matter.

Phalanx (FĀ-lanks) The bone of a finger or toe. *Plural is* **phalanges** (fa-LAN-jēz).

Pharmacology (far′-ma-KOL-ō-jē) The science of the effects and uses of drugs in the treatment of disease.

Pharynx (FAR-inks) The throat; a tube that starts at the internal nares and runs partway down the neck, where it opens into the esophagus posteriorly and the larynx anteriorly.

Phenotype (FĒ-nō-tīp) The observable expression of genotype; physical characteristics of an organism determined by genetic makeup and influenced by interaction between genes and internal and external environmental factors.

Phlebitis (fle-BĪ-tis) Inflammation of a vein, usually in a lower limb.

Phosphorylation (fos′-for-i-LĀ-shun) The addition of a phosphate group to a chemical compound; types include substrate-level phosphorylation, oxidative phosphorylation, and photo-phosphorylation.

Photopigment A substance that can absorb light and undergo structural changes that can lead to the development of a receptor potential. An example is rhodopsin. In the eye, also called **visual pigment.**

Photoreceptor Receptor that detects light shining on the retina of the eye.

Physiology (fiz′-ē-OL-ō-jē) Science that deals with the functions of an organism or its parts.

Pia mater (PĪ-a-MĀ-ter *or* PĒ-a MA-ter) The innermost of the three meninges (coverings) of the brain and spinal cord.

Pineal (PĪN-ē-al) **gland** A cone-shaped gland located in the roof of the third ventricle that secretes melatonin. Also called the **epiphysis cerebri** (ē-PIF-i-sis se-RĒ-brē).

Pinealocyte (pin-ē-AL-ō-sīt) Secretory cell of the pineal gland that releases melatonin.

Pinna (PIN-na) The projecting part of the external ear composed of elastic cartilage and covered by skin and shaped like the flared end of a trumpet. Also called the **auricle** (OR-i-kul).

Pinocytosis (pi′-nō-sī-TŌ-sis) A process by which most body cells can ingest membrane-surrounded droplets of interstitial fluid.

Pituicyte (pi-TOO-i-sīt) Supporting cell of the posterior pituitary.

Pituitary (pi-TOO-i-tār-ē) **gland** A small endocrine gland occupying the hypophyseal fossa above the sella turcica of the sphenoid bone and attached to the hypothalamus by the infundibulum. Also called the **hypophysis** (hī-POF-i-sis).

Pivot joint A synovial joint in which a rounded, pointed, or conical surface of one bone articulates with a ring formed partly by another bone and partly by a ligament, as in the joint between the atlas and axis and between the proximal ends of the radius and ulna. Also called a **trochoid** (TRŌ-koyd) **joint.**

Placenta (pla-SEN-ta) The special structure through which the exchange of materials between fetal and maternal circulations occurs. Also called the **afterbirth.**

Plantar flexion (PLAN-tar FLEK-shun) Bending the foot in the direction of the plantar surface (sole).

Plaque (PLAK) A layer of dense proteins on the inside of a plasma membrane in adherens junctions and desmosomes. A mass of bacterial cells, dextran (polysaccharide), and other debris that adheres to teeth (dental plaque). See also atherosclerotic plaque.

Plasma (PLAZ-ma) The extracellular fluid found in blood vessels; blood minus the formed elements.

Plasma cell Cell that develops from a B cell (lymphocyte) and produces antibodies.

Plasma (cell) membrane Outer, limiting membrane that separates the cell's internal parts from extracellular fluid or the external environment.

Platelet (PLĀT-let) A fragment of cytoplasm enclosed in a cell membrane and lacking a nucleus; found in the circulating blood; plays a role in hemostasis. Also called a **thrombocyte** (THROM-bō-sīt).

Platelet plug Aggregation of platelets (thrombocytes) at a site where a blood vessel is damaged that helps stop or slow blood loss.

Pleura (PLOOR-a) The serous membrane that covers the lungs and lines the walls of the chest and the diaphragm.

Pleural cavity Small potential space between the visceral and parietal pleurae.

Plexus (PLEK-sus) A network of nerves, veins, or lymphatic vessels.

Pluripotent stem cell Immature stem cell in red bone marrow that gives rise to precursors of all the different mature blood cells.

Pneumotaxic (noo-mō-TAK-sik) **area** A part of the respiratory center in the pons that continually sends inhibitory nerve impulses to the inspiratory area, limiting inhalation and facilitating exhalation.

Podiatry (pō-DĪ-a-trē) The diagnosis and treatment of foot disorders.

Polar body The smaller cell resulting from the unequal division of primary and secondary oocytes during meiosis. The polar body has no function and degenerates.

Polycythemia (pol′-ē-sī-THĒ-mē-a) Disorder characterized by an above-normal hematocrit (above 55%) in which hypertension, thrombosis, and hemorrhage can occur.

Polysaccharide (pol′-ē-SAK-a-rīd) A carbohydrate in which three or more monosaccharides are joined chemically.

Polyunsaturated fat A fatty acid that contains more than one double covalent bond between its carbon atoms; abundant in triglycerides of corn oil, safflower oil, and cottonseed oil.

Polyuria (pol′-ē-Ū-rē-a) An excessive production of urine.

Pons (PONZ) The part of the brain stem that forms a "bridge" between the medulla oblongata and the midbrain, anterior to the cerebellum.

Positive feedback A feedback mechanism in which the response enhances the original stimulus.

Postabsorptive (fasting) state Metabolic state after absorption of food is complete and energy needs of the body must be satisfied using molecules stored previously.

Postcentral gyrus Gyrus of cerebral cortex located immediately posterior to the central sulcus; contains the primary somatosensory area.

Posterior (pos-TĒR-ē-or) Nearer to or at the back of the body. Equivalent to **dorsal** in bipeds.

Posterior column–medial lemniscus pathways Sensory pathways that carry information related to proprioception, fine touch, two-point discrimination, pressure, and vibration. First-order neurons project from the spinal cord to the ipsilateral medulla in the posterior columns (gracile fasciculus and cuneate fasciculus). Second-order neurons project from the medulla to the contralateral thalamus in the medial lemniscus. Third-order neurons project from the thalamus to the somatosensory cortex (postcentral gyrus) on the same side.

Posterior pituitary Posterior lobe of the pituitary gland. Also called the **neurohypophysis** (noo-rō-hī-POF-i-sis).

Posterior root The structure composed of sensory axons lying between a spinal nerve and the dorsolateral aspect of the spinal cord. Also called the **dorsal (sensory) root.**

Posterior root ganglion (GANG-glē-on) A group of cell bodies of sensory neurons and their supporting cells located along the posterior root of a spinal nerve. Also called a **dorsal (sensory) root ganglion.**

Postganglionic neuron (pōst′-gang-lē-ON-ik NOO-ron) The second autonomic motor neuron in an autonomic pathway, having its cell body and dendrites located in an autonomic ganglion and its unmyelinated axon ending at cardiac muscle, smooth muscle, or a gland.

Postsynaptic (pōst-sin-AP-tik) **neuron** The nerve cell that is activated by the release of a neurotransmitter from another neuron and carries nerve impulses away from the synapse.

Precapillary sphincter (SFINGK-ter) A ring of smooth muscle fibers (cells) at the site of origin of true capillaries that regulate blood flow into true capillaries.

Precentral gyrus Gyrus of cerebral cortex located immediately anterior to the central sulcus; contains the primary motor area.

Preganglionic (prē′-gang-lē-ON-ik) **neuron** The first autonomic motor neuron in an autonomic pathway, with its cell body and dendrites in the brain or spinal cord and its myelinated axon ending at an autonomic ganglion, where it synapses with a postganglionic neuron.

Pregnancy Sequence of events that normally includes fertilization, implantation, embryonic growth, and fetal growth and terminates in birth.

Premenstrual syndrome (PMS) Severe physical and emotional stress ocurring late in the postovulatory phase of the menstrual cycle and sometimes overlapping with menstruation.

Prepuce (PRĒ-poos) The loose-fitting skin covering the glans of the penis and clitoris. Also called the **foreskin.**

Presbyopia (prez-bē-Ō-pē-a) A loss of elasticity of the lens of the eye due to advancing age with resulting inability to focus clearly on near objects.

Presynaptic (prē-sin-AP-tik) **inhibition** Type of inhibition in which neurotransmitter released by an inhibitory neuron depresses the release of neurotransmitter by a presynaptic neuron.

Presynaptic (prē-sin-AP-tik) **neuron** A neuron that propagates nerve impulses toward a synapse.

Prevertebral ganglion (prē-VER-te-bral GANG-lē-on) A cluster of cell bodies of postganglionic sympathetic neurons anterior to the spinal column and close to large abdominal arteries. Also called a **collateral ganglion.**

Primary germ layer One of three layers of embryonic tissue, called ectoderm, mesoderm, and endoderm, that give rise to all tissues and organs of the body.

Primary motor area A region of the cerebral cortex in the precentral gyrus of the frontal lobe of the cerebrum that controls specific muscles or groups of muscles.

Primary somatosensory area A region of the cerebral cortex posterior to the central sulcus in the postcentral gyrus of the parietal lobe of the cerebrum that localizes exactly the points of the body where somatic sensations originate.

Prime mover The muscle directly responsible for producing a desired motion. Also called an **agonist** (AG-ō-nist).

Primitive gut Embryonic structure formed from the dorsal part of the yolk sac that gives rise to most of the gastrointestinal tract.

Principal cell Cell type in the distal convoluted tubules and collecting ducts of the kidneys that is stimulated by aldosterone and antidiuretic hormone.

Proctology (prok-TOL-ō-jē) The branch of medicine concerned with the rectum and its disorders.

Progeny (PROJ-e-nē) Offspring or descendants.

Progesterone (prō-JES-te-rōn) A female sex hormone produced by the ovaries that helps prepare the endometrium of the uterus for implantation of a fertilized ovum and the mammary glands for milk secretion.

Prognosis (prog-NŌ-sis) A forecast of the probable results of a disorder; the outlook for recovery.

Prolactin (prō-LAK-tin) **(PRL)** A hormone secreted by the anterior pituitary that initiates and maintains milk secretion by the mammary glands.

Prolapse (PRŌ-laps) A dropping or falling down of an organ, especially the uterus or rectum.

Proliferation (prō-lif′-er-Ā-shun) Rapid and repeated reproduction of new parts, especially cells.

Pronation (prō-NĀ-shun) A movement of the forearm in which the palm is turned posteriorly or inferiorly.

Prophase (PRŌ-fāz) The first stage of mitosis during which chromatid pairs are formed and aggregate around the metaphase plate of the cell.

Proprioception (prō-prē-ō-SEP-shun) The perception of the position of body parts, especially the limbs, independent of vision; this sense is possible due to nerve impulses generated by proprioceptors.

Proprioceptor (prō′-prē-ō-SEP-tor) A receptor located in muscles, tendons, joints, or the internal ear (muscle spindles, tendon organs, joint kinesthetic receptors, and hair cells of the vestibular apparatus) that provides information about body position and movements.

Prostaglandin (pros′-ta-GLAN-din) **(PG)** A membrane-associated lipid; released in small quantities and acts as a local hormone.

Prostate (PROS-tāt) A doughnut-shaped gland inferior to the urinary bladder that surrounds the superior portion of the male urethra and secretes a slightly acidic solution that contributes to sperm motility and viability.

Protein An organic compound consisting of carbon, hydrogen, oxygen, nitrogen, and sometimes sulfur and phosphorus; synthesized on ribosomes and made up of amino acids linked by peptide bonds.

Proteasome Tiny cellular organelle in cytosol and nucleus containing proteases that destroy unneeded, damaged, or faulty proteins.

Prothrombin (prō-THROM-bin) An inactive blood-clotting factor synthesized by the liver, released into the blood, and converted to active thrombin in the process of blood clotting by the activated enzyme prothrombinase.

Protraction (prō-TRAK-shun) The movement of the mandible or shoulder girdle forward on a plane parallel with the ground.

Proximal (PROK-si-mal) Nearer the attachment of a limb to the trunk; nearer to the point of origin or attachment.

Pseudopods (SOO-dō-pods) Temporary protrusions of the leading edge of a migrating cell; cellular projections that surround a particle undergoing phagocytosis.

Pterygopalatine ganglion (ter′-i-gō-PAL-a-tīn GANG-glē-on) A cluster of cell bodies of parasympathetic postganglionic neurons ending at the lacrimal and nasal glands.

Ptosis (TŌ-sis) Drooping, as of the eyelid or the kidney.

Puberty (PŪ-ber-tē) The time of life during which the secondary sex characteristics begin to appear and the capability for sexual reproduction is possible; usually occurs between the ages of 10 and 17.

Pubic symphysis A slightly movable cartilaginous joint between the anterior surfaces of the hip bones.

Puerperium (pū′-er-PER-ē-um) The period immediately after childbirth, usually 4–6 weeks.

Pulmonary (PUL-mo-ner′-ē) Concerning or affected by the lungs.

Pulmonary circulation The flow of deoxygenated blood from the right ventricle to the lungs and the return of oxygenated blood from the lungs to the left atrium.

Pulmonary edema (e-DĒ-ma) An abnormal accumulation of interstitial fluid in the tissue spaces and alveoli of the lungs due to increased pulmonary capillary permeability or increased pulmonary capillary pressure.

Pulmonary embolism (EM-bō-lizm) **(PE)** The presence of a blood clot or a foreign substance in a pulmonary arterial blood vessel that obstructs circulation to lung tissue.

Pulmonary ventilation The inflow (inhalation) and outflow (exhalation) of air between the atmosphere and the lungs. Also called **breathing.**

Pulp cavity A cavity within the crown and neck of a tooth, which is filled with pulp, a connective tissue containing blood vessels, nerves, and lymphatic vessels.

Pulse (PULS) The rhythmic expansion and elastic recoil of a systemic artery after each contraction of the left ventricle.

Pulse pressure The difference between the maximum (systolic) and minimum (diastolic) pressures; normally about 40 mmHg.

Pupil The hole in the center of the iris, the area through which light enters the posterior cavity of the eyeball.

Purkinje (pur-KIN-jē) **fiber** Muscle fiber (cell) in the ventricular tissue of the heart specialized for conducting an action potential to the myocardium; part of the conduction system of the heart.

Pus The liquid product of inflammation containing leukocytes or their remains and debris of dead cells.

Pyloric (pī-LOR-ik) **sphincter** A thickened ring of smooth muscle through which the pylorus of the stomach communicates with the duodenum. Also called the **pyloric valve.**

Pyogenesis (pī′-ō-JEN-e-sis) Formation of pus.

Pyramid (PIR-a-mid) A pointed or cone-shaped structure. One of two roughly triangular structures on the anterior aspect of the medulla oblongata composed of the largest motor tracts that run from the cerebral cortex to the spinal cord. A triangular structure in the renal medulla.

Pyramidal (pi-RAM-i-dal) **tracts (pathways)** *See* **Direct motor pathways**.

Q

QRS wave The deflection waves of an electrocardiogram that represent onset of ventricular depolarization.

Quadriplegia (kwod′-ri-PLĒ-jē-a) Paralysis of four limbs: two upper and two lower.

R

Radiographic (rā′-dē-ō-GRAF-ic) **anatomy** Diagnostic branch of anatomy that includes the use of x rays.

Rami communicantes (RĀ-mē kō-mū-ni-KAN-tēz) Branches of a spinal nerve. *Singular is* **ramus communicans** (RĀ-mus kō-MŪ-ni-kans).

Rapid eye movement (REM) sleep Stage of sleep in which dreaming occurs, lasting for 5 to 10 minutes several times during a sleep cycle; characterized by rapid movements of the eyes beneath the eyelids.

Reactivity (rē-ak-TI-vi-tē) Ability of an antigen to react specifically with the antibody whose formation it induced.

Receptor A specialized cell or a distal portion of a neuron that responds to a specific sensory modality, such as touch, pressure, cold, light, or sound, and converts it to an electrical signal (generator or receptor potential). A specific molecule or cluster of molecules that recognizes and binds a particular ligand.

Receptor-mediated endocytosis A highly selective process whereby cells take up specific ligands, which usually are large molecules or particles, by enveloping them within a sac of plasma membrane. Ligands are eventually broken down by enzymes in lysosomes.

Receptor potential Depolarization or hyperpolarization of the plasma membrane of a receptor that alters release of neurotransmitter from the cell; if the neuron that synapses with the receptor cell becomes depolarized to threshold, a nerve impulse is triggered.

Recessive allele An allele whose presence is masked in the presence of a dominant allele on the homologous chromosome.

Reciprocal innervation (re-SIP-rō-kal in-ner-VĀ-shun) The phenomenon by which action potentials stimulate contraction of one muscle and simultaneously inhibit contraction of antagonistic muscles.

Recombinant DNA Synthetic DNA, formed by joining a fragment of DNA from one source to a portion of DNA from another.

Recovery oxygen consumption Elevated oxygen use after exercise ends due to metabolic changes that start during exercise and continue after exercise. Previously called **oxygen debt.**

Recruitment (rē-KROOT-ment) The process of increasing the number of active motor units. Also called **motor unit summation.**

Rectouterine pouch A pocket formed by the parietal peritoneum as it moves posteriorly from the surface of the uterus and is reflected onto the rectum; the most inferior point in the pelvic cavity. Also called the **pouch** or **cul de sac of Douglas.**

Rectum (REK-tum) The last 20 cm (8 in.) of the gastrointestinal tract, from the sigmoid colon to the anus.

Red nucleus A cluster of cell bodies in the midbrain, occupying a large part of the tectum from which axons extend into the rubroreticular and rubrospinal tracts.

Red pulp That portion of the spleen that consists of venous sinuses filled with blood and thin plates of splenic tissue called splenic (Billroth's) cords.

Reduction The addition of electrons to a molecule or, less commonly, the removal of oxygen from a molecule that results in an increase in the energy content of the molecule.

Referred pain Pain that is felt at a site remote from the place of origin.

Reflex Fast response to a change (stimulus) in the internal or external environment that attempts to restore homeostasis.

Reflex arc The most basic conduction pathway through the nervous system, connecting a receptor and an effector and consisting of a receptor, a sensory neuron, an integrating center in the central nervous system, a motor neuron, and an effector.

Refraction (rē-FRAK-shun) The bending of light as it passes from one medium to another.

Refractory (re-FRAK-to-rē) **period** A time period during which an excitable cell (neuron or muscle fiber) cannot respond to a stimulus that is usually adequate to evoke an action potential.

Regional anatomy The division of anatomy dealing with a specific region of the body, such as the head, neck, chest, or abdomen.

Regurgitation (rē-gur′-ji-TĀ-shun) Return of solids or fluids to the mouth from the stomach; backward flow of blood through incompletely closed heart valves.

Relaxin (RLX) A female hormone produced by the ovaries and placenta that increases flexibility of the pubic symphysis and helps dilate the uterine cervix to ease delivery of a baby.

Releasing hormone Hormone secreted by the hypothalamus that can stimulate secretion of hormones of the anterior pituitary.

Remodeling Replacement of old bone by new bone tissue.

Renal (RĒ-nal) Pertaining to the kidneys.

Renal corpuscle (KOR-pus-l) A glomerular (Bowman's) capsule and its enclosed glomerulus.

Renal pelvis A cavity in the center of the kidney formed by the expanded, proximal portion of the ureter, lying within the kidney, and into which the major calyces open.

Renal pyramid A triangular structure in the renal medulla containing the straight segments of renal tubules and the vasa recta.

Renin (RĒ-nin) An enzyme released by the kidney into the plasma, where it converts angiotensinogen into angiotensin I.

Renin–angiotensin–aldosterone (RAA) pathway A mechanism for the control of blood pressure, initiated by the secretion of renin by juxtaglomerular cells of the kidney in response to low blood pressure; renin catalyzes formation of angiotensin I, which is converted to angiotensin II by angiotensin-converting enzyme (ACE), and angiotensin II stimulates secretion of aldosterone.

Repolarization (rē-pō-lar-i-ZĀ-shun) Restoration of a resting membrane potential after depolarization.

Reproduction (rē-prō-DUK-shun) The formation of new cells for growth, repair, or replacement; the production of a new individual.

Reproductive cell division Type of cell division in which gametes (sperm and oocytes) are produced; consists of meiosis and cytokinesis.

Residual (re-ZID-ū-al) **volume** The volume of air still contained in the lungs after a maximal exhalation; about 1200 mL.

Resistance (re-ZIS-tans) Hindrance (impedance) to blood flow as a result of higher viscosity, longer total blood vessel length, and smaller blood vessel radius. Ability to ward off disease. The hindrance encountered by electrical charges as they move from one point to another. The hindrance encountered by air as it moves through the respiratory passageways.

Respiration (res-pi-RĀ-shun) Overall exchange of gases between the atmosphere, blood, and body cells consisting of pulmonary ventilation, external respiration, and internal respiration.

Respiratory center Neurons in the pons and medulla oblongata of the brain stem that regulate the rate and depth of pulmonary ventilation.

Respiratory membrane Structure in the lungs consisting of the alveolar wall and its basement membrane and a capillary endothelium and its basement membrane through which the diffusion of respiratory gases occurs.

Resting membrane potential The voltage difference between the inside and outside of a cell membrane when the cell is not responding to a stimulus; in many neurons and muscle fibers it is −70 to −90 mV, with the inside of the cell negative relative to the outside.

Retention (rē-TEN-shun) A failure to void urine due to obstruction, nervous contraction of the urethra, or absence of sensation of desire to urinate.

Rete (RĒ-tē) **testis** The network of ducts in the testes.

Reticular (re-TIK-ū-lar) **activating system (RAS)** A portion of the reticular formation that has many ascending connections with the cerebral cortex; when this area of the brain stem is active, nerve impulses pass to the thalamus and widespread areas of the cerebral cortex, resulting in generalized alertness or arousal from sleep.

Reticular formation A network of small groups of neuronal cell bodies scattered among bundles of axons (mixed gray and white matter) beginning in the medulla oblongata and extending superiorly through the central part of the brain stem.

Reticulocyte (re-TIK-ū-lō-sīt) An immature red blood cell.

Reticulum (re-TIK-ū-lum) A network.

Retina (RET-i-na) The deep coat of the posterior portion of the eyeball consisting of nervous tissue (where the process of vision begins) and a pigmented layer of epithelial cells that contact the choroid.

Retinal (RE-ti-nal) A derivative of vitamin A that functions as the light-absorbing portion of the photopigment rhodopsin.

Retraction (rē-TRAK-shun) The movement of a protracted part of the body posteriorly on a plane parallel to the ground, as in pulling the lower jaw back in line with the upper jaw.

Retroflexion (re-trō-FLEK-shun) A malposition of the uterus in which it is tilted posteriorly.

Retrograde degeneration (RE-trō-grād dē-jen-er-Ā-shun) Changes that occur in the proximal portion of a damaged axon only as far as the first node of Ranvier; similar to changes that occur during Wallerian degeneration.

Retroperitoneal (re′-trō-per-i-tō-NĒ-al) External to the peritoneal lining of the abdominal cavity.

Rh factor An inherited antigen on the surface of red blood cells in Rh$^+$ individuals; not present in Rh$^-$ individuals.

Rhinology (rī-NOL-ō-jē) The study of the nose and its disorders.

Rhodopsin (rō-DOP-sin) The photopigment in rods of the retina, consisting of a glycoprotein called opsin and a derivative of vitamin A called retinal.

Ribonucleic (rī-bō-noo-KLĒ-ik) **acid (RNA)** A single-stranded nucleic acid made up of nucleotides, each consisting of a nitrogenous base (adenine, cytosine, guanine, or uracil), ribose, and a phosphate group; three types are messenger RNA (mRNA), transfer RNA (tRNA), and ribosomal RNA (rRNA), each of which has a specific role during protein synthesis.

Ribosome (RĪ-bō-sōm) An organelle in the cytoplasm of cells, composed of a small subunit and a large subunit that contain ribosomal RNA and ribosomal proteins; the site of protein synthesis.

Right heart (atrial) reflex A reflex concerned with maintaining normal venous blood pressure.

Right lymphatic (lim-FAT-ik) **duct** A vessel of the lymphatic system that drains lymph from the upper right side of the body and empties it into the right subclavian vein.

Rigidity (ri-JID-i-tē) Hypertonia characterized by increased muscle tone, but reflexes are not affected.

Rigor mortis State of partial contraction of muscles after death due to lack of ATP; myosin heads (cross bridges) remain attached to actin, thus preventing relaxation.

Rod One of two types of photoreceptor in the retina of the eye; specialized for vision in dim light.

Root canal A narrow extension of the pulp cavity lying within the root of a tooth.

Root of penis Attached portion of penis that consists of the bulb and crura.

Rotation (rō-TĀ-shun) Moving a bone around its own axis, with no other movement.

Round ligament (LIG-a-ment) A band of fibrous connective tissue enclosed between the folds of the broad ligament of the uterus, emerging from the uterus just inferior to the uterine tube, extending laterally along the pelvic wall and through the deep inguinal ring to end in the labia majora.

Round window A small opening between the middle and internal ear, directly inferior to the oval window, covered by the secondary tympanic membrane.

Rugae (ROO-gē) Large folds in the mucosa of an empty hollow organ, such as the stomach and vagina.

S

Saccule (SAK-ūl) The inferior and smaller of the two chambers in the membranous labyrinth inside the vestibule of the internal ear containing a receptor organ for static equilibrium.

Sacral plexus (SĀ-kral PLEK-sus) A network formed by the ventral branches of spinal nerves L4 through S3.

Sacral promontory (PROM-on-tor'-ē) The superior surface of the body of the first sacral vertebra that projects anteriorly into the pelvic cavity; a line from the sacral promontory to the superior border of the pubic symphysis divides the abdominal and pelvic cavities.

Saddle joint A synovial joint in which the articular surface of one bone is saddle shaped and the articular surface of the other bone is shaped like the legs of the rider sitting in the saddle, as in the joint between the trapezium and the metacarpal of the thumb.

Sagittal (SAJ-i-tal) **plane** A plane that divides the body or organs into left and right portions. Such a plane may be **midsagittal (median),** in which the divisions are equal, or **parasagittal,** in which the divisions are unequal.

Saliva (sa-LĪ-va) A clear, alkaline, somewhat viscous secretion produced mostly by the three pairs of salivary glands; contains various salts, mucin, lysozyme, salivary amylase, and lingual lipase (produced by glands in the tongue).

Salivary amylase (SAL-i-ver-ē AM-i-lās) An enzyme in saliva that initiates the chemical breakdown of starch.

Salivary gland One of three pairs of glands that lie external to the mouth and pour their secretory product (saliva) into ducts that empty into the oral cavity; the parotid, submandibular, and sublingual glands.

Salt A substance that, when dissolved in water, ionizes into cations and anions, neither of which are hydrogen ions (H^+) or hydroxide ions (OH^-).

Saltatory (sal-ta-TŌ-rē) **conduction** The propagation of an action potential (nerve impulse) along the exposed parts of a myelinated axon. The action potential appears at successive nodes of Ranvier and therefore seems to leap from node to node.

Sarcolemma (sar'-kō-LEM-ma) The cell membrane of a muscle fiber (cell), especially of a skeletal muscle fiber.

Sarcomere (SAR-kō-mēr) A contractile unit in a striated muscle fiber (cell) extending from one Z disc to the next Z disc.

Sarcoplasm (SAR-kō-plazm) The cytoplasm of a muscle fiber (cell).

Sarcoplasmic reticulum (sar'-kō-PLAZ-mik re-TIK-ū-lum) A network of saccules and tubes surrounding myofibrils of a muscle fiber (cell), comparable to endoplasmic reticulum; functions to reabsorb calcium ions during relaxation and to release them to cause contraction.

Saturated fat A fatty acid that contains only single bonds (no double bonds) between its carbon atoms; all carbon atoms are bonded to the maximum number of hydrogen atoms; prevalent in triglycerides of animal products such as meat, milk, milk products, and eggs.

Scala tympani (SKA-la TIM-pan-ē) The inferior spiral-shaped channel of the bony cochlea, filled with perilymph.

Scala vestibuli (ves-TIB-ū-lē) The superior spiral-shaped channel of the bony cochlea, filled with perilymph.

Schwann (SCHVON) **cell** A neuroglial cell of the peripheral nervous system that forms the myelin sheath and neurolemma around a nerve axon by wrapping around the axon in a jelly-roll fashion.

Sciatica (sī-AT-i-ka) Inflammation and pain along the sciatic nerve; felt along the posterior aspect of the thigh extending down the inside of the leg.

Sclera (SKLE-ra) The white coat of fibrous tissue that forms the superficial protective covering over the eyeball except in the most anterior portion; the posterior portion of the fibrous tunic.

Scleral venous sinus A circular venous sinus located at the junction of the sclera and the cornea through which aqueous humor drains from the anterior chamber of the eyeball into the blood. Also called the **canal of Schlemm** (SHLEM).

Sclerosis (skle-RŌ-sis) A hardening with loss of elasticity of tissues.

Scoliosis (skō'-lē-Ō-sis) An abnormal lateral curvature from the normal vertical line of the backbone.

Scrotum (SKRŌ-tum) A skin-covered pouch that contains the testes and their accessory structures.

Sebaceous (se-BĀ-shus) **gland** An exocrine gland in the dermis of the skin, almost always associated with a hair follicle, that secretes sebum. Also called an **oil gland.**

Sebum (SĒ-bum) Secretion of sebaceous (oil) glands.

Secondary response Accelerated, more intense cell-mediated or antibody-mediated immune response upon a subsequent exposure to an antigen after the primary response.

Secondary sex characteristic A characteristic of the male or female body that develops at puberty under the influence of sex hormones but is not directly involved in sexual reproduction; examples are distribution of body hair, voice pitch, body shape, and muscle development.

Second messenger An intracellular mediator molecule that is produced in response to a first messenger (hormone or neurotransmitter) binding to its receptor in the plasma membrane of a target cell. Initiates a cascade of chemical reactions that produce characteristic effects for that particular target cell.

Secretion (se-KRĒ-shun) Production and release from a cell or a gland of a physiologically active substance.

Selective permeability (per'-mē-a-BIL-i-tē) The property of a membrane by which it permits the passage of certain substances but restricts the passage of others.

Semen (SĒ-men) A fluid discharged at ejaculation by a male that consists of a mixture of sperm and the secretions of the seminiferous tubules, seminal vesicles, prostate, and bulbourethral (Cowper's) glands.

Semicircular canals Three bony channels (anterior, posterior, lateral), filled with perilymph, in which lie the membranous semicircular canals filled with endolymph. They contain receptors for equilibrium.

Semicircular ducts The membranous semicircular canals filled with endolymph and floating in the perilymph of the bony semicircular canals; they contain cristae that are concerned with dynamic equilibrium.

Semilunar (sem'-ē-LOO-nar) **valve** A valve between the aorta or the pulmonary trunk and a ventricle of the heart.

Seminal vesicle (SEM-i-nal VES-i-kul) One of a pair of convoluted, pouchlike structures, lying posterior and inferior to the urinary bladder and anterior to the rectum, that secrete a component of semen into the ejaculatory ducts. Also termed **seminal gland.**

Seminiferous tubule (sem'-i-NI-fer-us TOO-būl) A tightly coiled duct, located in the testis, where sperm are produced.

Senescence (se-NES-ens) The process of growing old.

Sensation A state of awareness of external or internal conditions of the body.

Sensory neurons (NOO-ronz) Neurons that carry sensory information from cranial and spinal nerves into the brain and spinal cord or from a lower to a higher level in the spinal cord and brain. Also called **afferent neurons.**

Septal defect An opening in the septum (interatrial or interventricular) between the left and right sides of the heart.

Septum (SEP-tum) A wall dividing two cavities.

Serous (SIR-us) **membrane** A membrane that lines a body cavity that does not open to the exterior. The external layer of an organ formed by a serous membrane. The membrane that lines the pleural, pericardial, and peritoneal cavities. Also called a **serosa** (se-RŌ-sa).

Sertoli (ser-TŌ-lē) **cell** A supporting cell in the seminiferous tubules that secretes fluid for supplying nutrients to sperm and the hormone inhibin, removes excess cytoplasm from spermatogenic cells, and mediates the effects of FSH and testosterone on spermatogenesis. Also called a **sustentacular** (sus'-ten-TAK-ū-lar) **cell.**

Serum Blood plasma minus its clotting proteins.

Sesamoid (SES-a-moyd) **bones** Small bones usually found in tendons.

Sex chromosomes The twenty-third pair of chromosomes, designated X and Y, which determine the genetic sex of an individual; in males, the pair is XY; in females, XX.

Sexual intercourse The insertion of the erect penis of a male into the vagina of a female. Also called **coitus** (KŌ-i-tus).

Shivering Involuntary contraction of skeletal muscles that generates heat. Also called **involuntary thermogenesis.**

Shock Failure of the cardiovascular system to deliver adequate amounts of oxygen and nutrients to meet the metabolic needs of the body due to inadequate cardiac output. It is characterized by hypotension; clammy, cool, and pale skin; sweating; reduced urine formation; altered mental state; acidosis; tachycardia; weak, rapid pulse; and thirst. Types include hypovolemic, cardiogenic, vascular, and obstructive.

Shoulder joint A synovial joint where the humerus articulates with the scapula.

Sigmoid colon (SIG-moyd KŌ-lon) The S-shaped part of the large intestine that begins at the level of the left iliac crest, projects medially, and terminates at the rectum at about the level of the third sacral vertebra.

Sign Any objective evidence of disease that can be observed or measured such as a lesion, swelling, or fever.

Sinoatrial (si-nō-Ā-trē-al) **(SA) node** A small mass of cardiac muscle fibers (cells) located in the right atrium inferior to the opening of the superior vena cava that spontaneously depolarize and generate a cardiac action potential about 100 times per minute. Also called the **pacemaker.**

Sinus (SĪ-nus) A hollow in a bone (paranasal sinus) or other tissue; a channel for blood (vascular sinus); any cavity having a narrow opening.

Sinusoid (SĪ-nū-soyd) A large, thin-walled, and leaky type of capillary, having large intercellular clefts that may allow proteins and blood cells to pass from a tissue into the bloodstream; present in the liver, spleen, anterior pituitary, parathyroid glands, and red bone marrow.

Skeletal muscle An organ specialized for contraction, composed of striated muscle fibers (cells), supported by connective tissue, attached to a bone by a tendon or an aponeurosis, and stimulated by somatic motor neurons.

Skin The external covering of the body that consists of a superficial, thinner epidermis (epithelial tissue) and a deep, thicker dermis (connective tissue) that is anchored to the subcutaneous layer.

Skull The skeleton of the head consisting of the cranial and facial bones.

Sleep A state of partial unconsciousness from which a person can be aroused; associated with a low level of activity in the reticular activating system.

Sliding-filament mechanism The explanation of how thick and thin filaments slide relative to one another during striated muscle contraction to decrease sarcomere length.

Small intestine A long tube of the gastrointestinal tract that begins at the pyloric sphincter of the stomach, coils through the central and inferior part of the abdominal cavity, and ends at the large intestine; divided into three segments: duodenum, jejunum, and ileum.

Smooth muscle A tissue specialized for contraction, composed of smooth muscle fibers (cells), located in the walls of hollow internal organs, and innervated by autonomic motor neurons.

Sodium-potassium ATPase An active transport pump located in the plasma membrane that transports sodium ions out of the cell and potassium ions into the cell at the expense of cellular ATP. It functions to keep the ionic concentrations of these ions at physiological levels. Also called the **sodium-potassium pump.**

Soft palate (PAL-at) The posterior portion of the roof of the mouth, extending from the palatine bones to the uvula. It is a muscular partition lined with mucous membrane.

Solution A homogeneous molecular or ionic dispersion of one or more substances (solutes) in a dissolving medium (solvent) that is usually liquid.

Somatic (sō-MAT-ik) **cell division** Type of cell division in which a single starting parent cell duplicates itself to produce two identical cells; consists of mitosis and cytokinesis.

Somatic nervous system (SNS) The portion of the peripheral nervous system consisting of somatic sensory (afferent) neurons and somatic motor (efferent) neurons.

Somite (SŌ-mīt) Block of mesodermal cells in a developing embryo that is distinguished into a myotome (which forms most of the skeletal muscles), dermatome (which forms connective tissues), and sclerotome (which forms the vertebrae).

Spasm (SPAZM) A sudden, involuntary contraction of large groups of muscles.

Spasticity (spas-TIS-i-tē) Hypertonia characterized by increased muscle tone, increased tendon reflexes, and pathological reflexes (Babinski sign).

Spermatic (sper-MAT-ik) **cord** A supporting structure of the male reproductive system, extending from a testis to the deep inguinal ring, that includes the ductus (vas) deferens, arteries, veins, lymphatic vessels, nerves, cremaster muscle, and connective tissue.

Spermatogenesis (sper′-ma-tō-JEN-e-sis) The formation and development of sperm in the seminiferous tubules of the testes.

Sperm cell A mature male gamete. Also termed **spermatozoon** (sper′-ma-tō-ZŌ-on).

Spermiogenesis (sper′-mē-ō-JEN-e-sis) The maturation of spermatids into sperm.

Sphincter (SFINGK-ter) A circular muscle that constricts an opening.

Sphincter of the hepatopancreatic ampulla A circular muscle at the opening of the common bile and main pancreatic ducts in the duodenum. Also called the **sphincter of Oddi** (OD-ē).

Sphygmomanometer (sfig′-mō-ma-NOM-e-ter) An instrument for measuring arterial blood pressure.

Spinal (SPĪ-nal) **cord** A mass of nerve tissue located in the vertebral canal from which 31 pairs of spinal nerves originate.

Spinal nerve One of the 31 pairs of nerves that originate on the spinal cord from posterior and anterior roots.

Spinal shock A period from several days to several weeks following transection of the spinal cord and characterized by the abolition of all reflex activity.

Spinothalamic (spī-nō-tha-LAM-ik) **tracts** Sensory (ascending) tracts that convey information up the spinal cord to the thalamus for sensations of pain, temperature, crude touch, and deep pressure.

Spinous (SPĪ-nus) **process** A sharp or thornlike process or projection. Also called a **spine.** A sharp ridge running diagonally across the posterior surface of the scapula.

Spiral organ The organ of hearing, consisting of supporting cells and hair cells that rest on the basilar membrane and extend into the endolymph of the cochlear duct. Also called the **organ of Corti** (KOR-tē).

Spirometer (spī-ROM-e-ter) An apparatus used to measure lung volumes and capacities.

Splanchnic (SPLANK-nik) Pertaining to the viscera.

Spleen (SPLĒN) Large mass of lymphatic tissue between the fundus of the stomach and the diaphragm that functions in formation of blood cells during early fetal development, phagocytosis of ruptured blood cells, and proliferation of B cells during immune responses.

Spongy (cancellous) bone tissue Bone tissue that consists of an irregular latticework of thin plates of bone called trabeculae; spaces between trabeculae of some bones are filled with red bone marrow; found inside short, flat, and irregular bones and in the epiphyses (ends) of long bones.

Sprain Forcible wrenching or twisting of a joint with partial rupture or other injury to its attachments without dislocation.

Squamous (SKWĀ-mus) Flat or scalelike.

Starling's law of the capillaries The movement of fluid between plasma and interstitial fluid is in a state of near equilibrium at the arterial and venous ends of a capillary; that is, filtered fluid and absorbed fluid plus that returned to the lymphatic system are nearly equal.

Starling's law of the heart The force of muscular contraction is determined by the length of the cardiac muscle fibers just before they contract; within limits, the greater the length of stretched fibers, the stronger the contraction.

Starvation (star-VĀ-shun) The loss of energy stores in the form of glycogen, triglycerides, and proteins due to inadequate intake of nutrients or inability to digest, absorb, or metabolize ingested nutrients.

Stasis (STĀ-sis) Stagnation or halt of normal flow of fluids, as blood or urine, or of the intestinal contents.

Static equilibrium (ē-kwi-LIB-rē-um) The maintenance of posture in response to changes in the orientation of the body, mainly the head, relative to the ground.

Stellate reticuloendothelial (STEL-āt re-tik′-ū-lō-en′-dō-THĒ-lē-al) **cell** Phagocytic cell bordering a sinusoid of the liver. Also called a **Kupffer** (KOOP-fer) **cell.**

Stenosis (sten-Ō-sis) An abnormal narrowing or constriction of a duct or opening.

Stereocilia (ste′-rē-ō-SIL-ē-a) Groups of extremely long, slender, nonmotile microvilli projecting from epithelial cells lining the epididymis.

Stereognosis (ste′-rē-og-NŌ-sis) The ability to recognize the size, shape, and texture of an object by touching it.

Sterile (STE-ril) Free from any living microorganisms. Unable to conceive or produce offspring.

Sterilization (ster′-i-li-ZĀ-shun) Elimination of all living microorganisms. Any procedure that renders an individual incapable of reproduction (for example, castration, vasectomy, hysterectomy, or oophorectomy).

Stimulus Any stress that changes a controlled condition; any change in the internal or external environment that excites a sensory receptor, a neuron, or a muscle fiber.

Stomach The J-shaped enlargement of the gastrointestinal tract directly inferior to the diaphragm in the epigastric, umbilical, and left hypochondriac regions of the abdomen, between the esophagus and small intestine.

Straight tubule (TOO-būl) A duct in a testis leading from a convoluted seminiferous tubule to the rete testis.

Stratum (STRĀ-tum) A layer.

Stratum basalis (ba-SAL-is) The layer of the endometrium next to the myometrium that is maintained during menstruation and gestation and produces a new stratum functionalis following menstruation or parturition.

Stratum functionalis (funk′-shun-AL-is) The layer of the endometrium next to the uterine cavity that is shed during menstruation and that forms the maternal portion of the placenta during gestation.

Stressor A stress that is extreme, unusual, or long-lasting and triggers the stress response.

Stress response Wide-ranging set of bodily changes, triggered by a stressor, that gears the body to meet an emergency. Also known as **general adaptation syndrome (GAS).**

Stretch receptor Receptor in the walls of blood vessels, airways, or organs that monitors the amount of stretching. Also termed **baroreceptor.**

Stretch reflex A monosynaptic reflex triggered by sudden stretching of muscle spindles within a muscle that elicits contraction of that same muscle. Also called a **tendon jerk.**

Stroke volume The volume of blood ejected by either ventricle in one systole; about 70 mL in an adult at rest.

Stroma (STRŌ-ma) The tissue that forms the ground substance, foundation, or framework of an organ, as opposed to its functional parts (parenchyma).

Subarachnoid (sub′-a-RAK-noyd) **space** A space between the arachnoid mater and the pia mater that surrounds the brain and spinal cord and through which cerebrospinal fluid circulates.

Subcutaneous (sub′-kū-TA-nē-us) Beneath the skin. Also called **hypodermic** (hi-pō-DER-mik).

Subcutaneous layer A continuous sheet of areolar connective tissue and adipose tissue between the dermis of the skin and the deep fascia of the muscles. Also called the **superficial fascia** (FASH-ē-a).

Subdural (sub-DOO-ral) **space** A space between the dura mater and the arachnoid mater of the brain and spinal cord that contains a small amount of fluid.

Sublingual (sub-LING-gwal) **gland** One of a pair of salivary glands situated in the floor of the mouth deep to the mucous membrane and to the side of the lingual frenulum, with a duct (Rivinus') that opens into the floor of the mouth.

Submandibular (sub′-man-DIB-ū-lar) **gland** One of a pair of salivary glands found inferior to the base of the tongue deep to the mucous membrane in the posterior part of the floor of the mouth, posterior to the sublingual glands, with a duct (Wharton's) situated to the side of the lingual frenulum. Also called the **submaxillary** (sub′-MAK-si-ler-ē) **gland.**

Submucosa (sub-mū-KŌ-sa) A layer of connective tissue located deep to a mucous membrane, as in the gastrointestinal tract or the urinary bladder; the submucosa connects the mucosa to the muscularis layer.

Submucosal plexus A network of autonomic nerve fibers located in the superficial part of the submucous layer of the small intestine. Also called the **plexus of Meissner** (MĪZ-ner).

Substrate A molecule upon which an enzyme acts.

Subthreshold stimulus A stimulus of such weak intensity that it cannot initiate an action potential (nerve impulse).

Sudoriferous (soo′-dor-IF-er-us) **gland** An apocrine or eccrine exocrine gland in the dermis or subcutaneous layer that produces perspiration. Also called a **sweat gland.**

Sulcus (SUL-kus) A groove or depression between parts, especially between the convolutions of the brain. *Plural is* **sulci** (SUL-sī).

Summation (sum-MA-shun) The addition of the excitatory and inhibitory effects of many stimuli applied to a neuron. The increased strength of muscle contraction that results when stimuli follow one another in rapid succession.

Superficial (soo′-per-FISH-al) Located on or near the surface of the body or an organ.

Superficial fascia (FASH-ē-a) A continuous sheet of fibrous connective tissue between the dermis of the skin and the deep fascia of the muscles. Also called **subcutaneous** (sub′-kū-TA-nē-us) **layer.**

Superficial inguinal (IN-gwi-nal) **ring** A triangular opening in the aponeurosis of the external oblique muscle that represents the termination of the inguinal canal.

Superior (soo-PĒR-ē-or) Toward the head or upper part of a structure. Also called **cephalad** (SEF-a-lad) or **craniad.**

Superior vena cava (VĒ-na CĀ-va) **(SVC)** Large vein that collects blood from parts of the body superior to the heart and returns it to the right atrium.

Supination (soo-pi-NĀ-shun) A movement of the forearm in which the palm is turned anteriorly or superiorly.

Surface anatomy The study of the structures that can be identified from the outside of the body.

Surfactant (sur-FAK-tant) Complex mixture of phospholipids and lipoproteins, produced by type II alveolar (septal) cells in the lungs, that decreases surface tension.

Susceptibility (sus-sep′-ti-BIL-i-tē) Lack of resistance to the damaging effects of an agent such as a pathogen.

Suspensory ligament (sus-PEN-so-rē LIG-a-ment) A fold of peritoneum extending laterally from the surface of the ovary to the pelvic wall.

Sutural (SOO-cher-al) **bone** A small bone located within a suture between certain cranial bones. Also called **Wormian** (WER-mē-an) **bone.**

Suture (SOO-cher) An immovable fibrous joint that joins skull bones.

Sympathetic (sim′-pa-THET-ik) **division** One of the two subdivisions of the autonomic nervous system, having cell bodies of preganglionic neurons in the lateral gray columns of the thoracic segment and the first two or three lumbar segments of the spinal cord; primarily concerned with process involving the expenditure of energy.

Sympathetic trunk ganglion (GANG-glē-on) A cluster of cell bodies of sympathetic postganglionic neurons lateral to the vertebral column, close to the body of a vertebra. These ganglia extend inferiorly through the neck, thorax, and abdomen to the coccyx on both sides of the vertebral column and are connected to one another to form a chain on each side of the vertebral column. Also called **sympathetic chain** or **vertebral chain ganglia.**

Sympathomimetic (sim′-pa-thō-mi-MET-ik) Producing effects that mimic those brought about by the sympathetic division of the autonomic nervous system.

Symphysis (SIM-fi-sis) A line of union. A slightly movable cartilaginous joint such as the pubic symphysis.

Symporter A transmembrane transporter protein that moves two substances, often Na^+ and another substance, in the same direction across a plasma membrane. Also called a **cotransporter.**

Symptom (SIMP-tum) A subjective change in body function not apparent to an observer, such as pain or nausea, that indicates the presence of a disease or disorder of the body.

Synapse (SYN-aps) The functional junction between two neurons or between a neuron and an effector, such as a muscle or gland; may be electrical or chemical.

Synapsis (sin-AP-sis) The pairing of homologous chromosomes during prophase I of meiosis.

Synaptic (sin-AP-tik) **cleft** The narrow gap at a chemical synapse that separates the axon terminal of one neuron from another neuron or muscle fiber (cell) and across which a neurotransmitter diffuses to affect the postsynaptic cell.

Synaptic delay The length of time between the arrival of the action potential at a presynaptic axon terminal and the membrane potential

(IPSP or EPSP) change on the postsynaptic membrane; usually about 0.5 msec.

Synaptic end bulb Expanded distal end of an axon terminal that contains synaptic vesicles. Also called a **synaptic knob.**

Synaptic vesicle Membrane-enclosed sac in a synaptic end bulb that stores neurotransmitters.

Synarthrosis (sin'-ar-THRŌ-sis) An immovable joint such as a suture, gomphosis, and synchondrosis.

Synchondrosis (sin'-kon-DRŌ-sis) A cartilaginous joint in which the connecting material is hyaline cartilage.

Syndesmosis (sin'-dez-MŌ-sis) A slightly movable joint in which articulating bones are united by fibrous connective tissue.

Syndrome (SIN-drōm) A group of signs and symptoms that occur together in a pattern that is characteristic of a particular disease or abnormal condition.

Synergist (SIN-er-jist) A muscle that assists the prime mover by reducing undesired action or unnecessary movement.

Synergistic (syn-er-JIS-tik) **effect** A hormonal interaction in which the effects of two or more hormones acting together is greater or more extensive than the sum of each hormone acting alone.

Synostosis (sin'-os-TŌ-sis) A joint in which the dense fibrous connective tissue that unites bones at a suture has been replaced by bone, resulting in a complete fusion across the suture line.

Synovial (si-NŌ-vē-al) **cavity** The space between the articulating bones of a synovial joint, filled with synovial fluid. Also called a **joint cavity.**

Synovial fluid Secretion of synovial membranes that lubricates joints and nourishes articular cartilage.

Synovial joint A fully movable or diarthrotic joint in which a synovial (joint) cavity is present between the two articulating bones.

Synovial membrane The deeper of the two layers of the articular capsule of a synovial joint, composed of areolar connective tissue that secretes synovial fluid into the synovial (joint) cavity.

System An association of organs that have a common function.

Systemic (sis-TEM-ik) Affecting the whole body; generalized.

Systemic anatomy The anatomic study of particular systems of the body, such as the skeletal, muscular, nervous, cardiovascular, or urinary systems.

Systemic circulation The routes through which oxygenated blood flows from the left ventricle through the aorta to all the organs of the body and deoxygenated blood returns to the right atrium.

Systemic vascular resistance (SVR) All the vascular resistance offered by systemic blood vessels. Also called **total peripheral resistance.**

Systole (SIS-tō-lē) In the cardiac cycle, the phase of contraction of the heart muscle, especially of the ventricles.

Systolic (sis-TOL-ik) **blood pressure** The force exerted by blood on arterial walls during ventricular contraction; the highest pressure measured in the large arteries, about 120 mmHg under normal conditions for a young adult.

T

T cell A lymphocyte that becomes immunocompetent in the thymus and can differentiate into a helper T cell or a cytotoxic T cell, both of which function in cell-mediated immunity.

T wave The deflection wave of an electrocardiogram that represents ventricular repolarization.

Tachycardia (tak'-i-KAR-dē-a) An abnormally rapid resting heartbeat or pulse rate (over 100 beats per minute).

Tactile (TAK-tīl) Pertaining to the sense of touch.

Tactile disc Modified epidermal cell in the stratum basale of hairless skin that functions as a cutaneous receptor for discriminative touch. Also called a **Merkel** (MER-kel) **disc.**

Teniae coli (TĒ-nē-ē KŌ-lī) The three flat bands of thickened, longitudinal smooth muscle running the length of the large intestine, except in the rectum. *Singular is* **tenia coli.**

Target cell A cell whose activity is affected by a particular hormone.

Tarsal bones The seven bones of the ankle. Also called **tarsals.**

Tarsal gland Sebaceous (oil) gland that opens on the edge of each eyelid. Also called a **Meibomian** (mī-BŌ-mē-an) **gland.**

Tarsal plate A thin, elongated sheet of connective tissue, one in each eyelid, giving the eyelid form and support. The aponeurosis of the levator palpebrae superioris is attached to the tarsal plate of the superior eyelid.

Tarsus (TAR-sus) A collective term for the seven bones of the ankle.

Tectorial (tek-TŌ-rē-al) **membrane** A gelatinous membrane projecting over and in contact with the hair cells of the spiral organ (organ of Corti) in the cochlear duct.

Teeth (TĒTH) Accessory structures of digestion, composed of calcified connective tissue and embedded in bony sockets of the mandible and maxilla, that cut, shred, crush, and grind food. Also called **dentes** (DEN-tēz).

Telophase (TEL-ō-fāz) The final stage of mitosis in which the daughter nuclei become established.

Tendon (TEN-don) A white fibrous cord of dense regular connective tissue that attaches muscle to bone.

Tendon organ A proprioceptive receptor, sensitive to changes in muscle tension and force of contraction, found chiefly near the junctions of tendons and muscles. Also called a **Golgi** (GOL-jē) **tendon organ.**

Tendon reflex A polysynaptic, ipsilateral reflex that protects tendons and their associated muscles from damage that might be brought about by excessive tension. The receptors involved are called tendon organs (Golgi tendon organs).

Tentorium cerebelli (ten-TŌ-rē-um ser'-e-BEL-ē) A transverse shelf of dura mater that forms a partition between the occipital lobe of the cerebral hemispheres and the cerebellum and that covers the cerebellum.

Teratogen (TER-a-tō-jen) Any agent or factor that causes physical defects in a developing embryo.

Terminal ganglion (TER-min-al GANG-glē-on) A cluster of cell bodies of parasympathetic postganglionic neurons either lying very close to the visceral effectors or located within the walls of the visceral effectors supplied by the postganglionic neurons.

Testis (TES-tis) Male gonad that produces sperm and the hormones testosterone and inhibin. Also called a **testicle.**

Testosterone (tes-TOS-te-rōn) A male sex hormone (androgen) secreted by interstitial endocrinocytes (Leydig cells) of a mature testis; needed for development of sperm; together with a second androgen termed **dihydrotestosterone (DHT),** controls the growth and development of male reproductive organs, secondary sex characteristics, and body growth.

Tetany (TET-a-nē) Hyperexcitability of neurons and muscle fibers (cells) caused by hypocalcemia and characterized by intermittent or continuous tonic muscular contractions; may be due to hypoparathyroidism.

Thalamus (THAL-a-mus) A large, oval structure located bilaterally on either side of the third ventricle, consisting of two masses of gray matter organized into nuclei; main relay center for sensory impulses ascending to the cerebral cortex.

Thermoreceptor (THER-mō-rē-sep-tor) Sensory receptor that detects changes in temperature.

Thigh The portion of the lower limb between the hip and the knee.

Third ventricle (VEN-tri-kul) A slitlike cavity between the right and left halves of the thalamus and between the lateral ventricles of the brain.

Thirst center A cluster of neurons in the hypothalamus that is sensitive to the osmotic pressure of extracellular fluid and brings about the sensation of thirst.

Thoracic (thō-RAS-ik) **cavity** Superior portion of the ventral body cavity that contains two pleural cavities, the mediastinum, and the pericardial cavity.

Thoracic duct A lymphatic vessel that begins as a dilation called the cisterna chyli, receives lymph from the left side of the head, neck, and chest, the left arm, and the entire body below the ribs, and empties into the left subclavian vein. Also called the **left lymphatic** (lim-FAT-ik) **duct.**

Thoracolumbar (thō′-ra-kō-LUM-bar) **outflow** The axons of sympathetic preganglionic neurons, which have their cell bodies in the lateral gray columns of the thoracic segments and first two or three lumbar segments of the spinal cord.

Thorax (THŌ-raks) The chest.

Threshold potential The membrane voltage that must be reached to trigger an action potential.

Threshold stimulus Any stimulus strong enough to initiate an action potential or activate a sensory receptor.

Thrombin (THROM-bin) The active enzyme formed from prothrombin that converts fibrinogen to fibrin during formation of a blood clot.

Thrombolytic (throm-bō-LIT-ik) **agent** Chemical substance injected into the body to dissolve blood clots and restore circulation; mechanism of action is direct or indirect activation of plasminogen; examples include tissue plasminogen activator (t-PA), streptokinase, and urokinase.

Thrombosis (throm-BŌ-sis) The formation of a clot in an unbroken blood vessel, usually a vein.

Thrombus A stationary clot formed in an unbroken blood vessel, usually a vein.

Thymus (THĪ-mus) A bilobed organ, located in the superior mediastinum posterior to the sternum and between the lungs, in which T cells develop immunocompetence.

Thyroglobulin (thī-rō-GLŌ-bū-lin) **(TGB)** A large glycoprotein molecule produced by follicular cells of the thyroid gland in which some tyrosines are iodinated and coupled to form thyroid hormones.

Thyroid cartilage (THĪ-royd KAR-ti-lij) The largest single cartilage of the larynx, consisting of two fused plates that form the anterior wall of the larynx.

Thyroid follicle (FOL-i-kul) Spherical sac that forms the parenchyma of the thyroid gland and consists of follicular cells that produce thyroxine (T_4) and triiodothyronine (T_3).

Thyroid gland An endocrine gland with right and left lateral lobes on either side of the trachea connected by an isthmus; located anterior to the trachea just inferior to the cricoid cartilage; secretes thyroxine (T_4), triiodothyronine (T_3), and calcitonin.

Thyroid-stimulating hormone (TSH) A hormone secreted by the anterior pituitary that stimulates the synthesis and secretion of thyroxine (T_4) and triiodothyronine (T_3).

Thyroxine (thī-ROK-sēn) **(T_4)** A hormone secreted by the thyroid gland that regulates metabolism, growth and development, and the activity of the nervous system.

Tic Spasmodic, involuntary twitching of muscles that are normally under voluntary control.

Tidal volume The volume of air breathed in and out in any one breath; about 500 mL in quiet, resting conditions.

Tissue A group of similar cells and their intercellular substance joined together to perform a specific function.

Tissue factor (TF) A factor, or collection of factors, whose appearance initiates the blood clotting process. Also called **thromboplastin** (throm-bō-PLAS-tin).

Tissue plasminogen activator (t-PA) An enzyme that dissolves small blood clots by initiating a process that converts plasminogen to plasmin, which degrades the fibrin of a clot.

Tongue A large skeletal muscle covered by a mucous membrane located on the floor of the oral cavity.

Tonicity (tō-NIS-i-tē) A measure of the concentration of impermeable solute particles in a solution relative to cytosol. When cells are bathed in an **isotonic solution,** they neither shrink nor swell.

Tonsil (TON-sil) An aggregation of large lymphatic nodules embedded in the mucous membrane of the throat.

Torn cartilage A tearing of an articular disc (meniscus) in the knee.

Total lung capacity The sum of tidal volume, inspiratory reserve volume, expiratory reserve volume, and residual volume; about 6000 mL in an average adult.

Trabecula (tra-BEK-ū-la) Irregular latticework of thin plates of spongy bone. Fibrous cord of connective tissue serving as supporting fiber by forming a septum extending into an organ from its wall or capsule. *Plural is* **trabeculae** (tra-BEK-ū-lē).

Trabeculae carneae (KAR-nē-ē) Ridges and folds of the myocardium in the ventricles.

Trachea (TRĀ-kē-a) Tubular air passageway extending from the larynx to the fifth thoracic vertebra. Also called the **windpipe.**

Tract A bundle of nerve axons in the central nervous system.

Transcription (trans-KRIP-shun) The first step in the expression of genetic information in which a single strand of DNA serves as a template for the formation of an RNA molecule.

Translation (trans-LĀ-shun) The synthesis of a new protein on a ribosome as dictated by the sequence of codons in messenger RNA.

Transverse colon (trans-VERS KŌ-lon) The portion of the large intestine extending across the abdomen from the right colic (hepatic) flexure to the left colic (splenic) flexure.

Transverse fissure (FISH-er) The deep cleft that separates the cerebrum from the cerebellum.

Transverse plane A plane that divides the body or organs into superior and inferior portions. Also called a **horizontal plane.**

Transverse tubules (TOO-būls) **(T tubules)** Small, cylindrical invaginations of the sarcolemma of striated muscle fibers (cells) that conduct muscle action potentials toward the center of the muscle fiber.

Trauma (TRAW-ma) An injury, either a physical wound or psychic disorder, caused by an external agent or force, such as a physical blow or emotional shock; the agent or force that causes the injury.

Tremor (TREM-or) Rhythmic, involuntary, purposeless contraction of opposing muscle groups.

Triad (TRĪ-ad) A complex of three units in a muscle fiber composed of a transverse tubule and the sarcoplasmic reticulum terminal cisterns on both sides of it.

Tricuspid (trī-KUS-pid) **valve** Atrioventricular (AV) valve on the right side of the heart.

Triglyceride (trī-GLI-cer-īd) A lipid formed from one molecule of glycerol and three molecules of fatty acids that may be either solid (fats) or liquid (oils) at room temperature; the body's most highly concentrated source of chemical potential energy. Found mainly within adipocytes. Also called a **neutral fat** or a **triacylglycerol.**

Trigone (TRĪ-gon) A triangular region at the base of the urinary bladder.

Triiodothyronine (trī-ī-ō-dō-THĪ-rō-nēn) **(T₃)** A hormone produced by the thyroid gland that regulates metabolism, growth and development, and the activity of the nervous system.

Trophoblast (TRŌF-ō-blast) The superficial covering of cells of the blastocyst.

Tropic (TRŌ-pik) **hormone** A hormone whose target is another endocrine gland.

Trunk The part of the body to which the upper and lower limbs are attached.

Tubal ligation (lī-GĀ-shun) A sterilization procedure in which the uterine (Fallopian) tubes are tied and cut.

Tubular reabsorption The movement of filtrate from renal tubules back into blood in response to the body's specific needs.

Tubular secretion The movement of substances in blood into renal tubular fluid in response to the body's specific needs.

Tumor suppressor gene A gene coding for a protein that normally inhibits cell division; loss or alteration of a tumor suppressor gene called *p53* is the most common genetic change in a wide variety of cancer cells.

Tunica albuginea (TOO-ni-ka al′-bū-JIN-ē-a) A dense white fibrous capsule covering a testis or deep to the surface of an ovary.

Tunica externa (eks-TER-na) The superficial coat of an artery or vein, composed mostly of elastic and collagen fibers. Also called the **adventitia.**

Tunica interna (in-TER-na) The deep coat of an artery or vein, consisting of a lining of endothelium, basement membrane, and internal elastic lamina. Also called the **tunica intima** (IN-ti-ma).

Tunica media (MĒ-dē-a) The intermediate coat of an artery or vein, composed of smooth muscle and elastic fibers.

Twitch contraction Brief contraction of all muscle fibers (cells) in a motor unit triggered by a single action potential in its motor neuron.

Tympanic antrum (tim-PAN-ik AN-trum) An air space in the middle ear that leads into the mastoid air cells or sinus.

Type II cutaneous mechanoreceptor A sensory receptor embedded deeply in the dermis and deeper tissues that detects stretching of skin. Also called a **Ruffini corpuscle.**

U

Umbilical cord The long, ropelike structure containing the umbilical arteries and vein that connect the fetus to the placenta.

Umbilicus (um-BIL-i-kus *or* um-bil-Ī-kus) A small scar on the abdomen that marks the former attachment of the umbilical cord to the fetus. Also called the **navel.**

Upper limb The appendage attached at the shoulder girdle, consisting of the arm, forearm, wrist, hand, and fingers. Also called **upper extremity.**

Up-regulation Phenomenon in which there is an increase in the number of receptors in response to a deficiency of a hormone or neurotransmitter.

Uremia (ū-RĒ-mē-a) Accumulation of toxic levels of urea and other nitrogenous waste products in the blood, usually resulting from severe kidney malfunction.

Ureter (Ū-rē-ter) One of two tubes that connect the kidney with the urinary bladder.

Urethra (ū-RĒ-thra) The duct from the urinary bladder to the exterior of the body that conveys urine in females and urine and semen in males.

Urinary (Ū-ri-ner-ē) **bladder** A hollow, muscular organ situated in the pelvic cavity posterior to the pubic symphysis; receives urine via two ureters and stores urine until it is excreted through the urethra.

Urine The fluid produced by the kidneys that contains wastes and excess materials; excreted from the body through the urethra.

Urogenital (ū′-ter-ō-JEN-i-tal) **triangle** The region of the pelvic floor inferior to the pubic symphysis, bounded by the pubic symphysis and the ischial tuberosities, and containing the external genitalia.

Urology (ū-ROL-ō-jē) The specialized branch of medicine that deals with the structure, function, and diseases of the male and female urinary systems and the male reproductive system.

Uterine (Ū-ter-in) **tube** Duct that transports ova from the ovary to the uterus. Also called the **Fallopian** (fal-LŌ-pē-an) **tube** or **oviduct.**

Uterosacral ligament (ū′-ter-ō-SĀ-kral LIG-a-ment) A fibrous band of tissue extending from the cervix of the uterus laterally to the sacrum.

Uterus (Ū-te-rus) The hollow, muscular organ in females that is the site of menstruation, implantation, development of the fetus, and labor. Also called the **womb.**

Utricle (Ū-tri-kul) The larger of the two divisions of the membranous labyrinth located inside the vestibule of the inner ear, containing a receptor organ for static equilibrium.

Uvea (Ū-vē-a) The three structures that together make up the vascular tunic of the eye.

Uvula (Ū-vū-la) A soft, fleshy mass, especially the V-shaped pendant part, descending from the soft palate.

V

Vagina (va-JĪ-na) A muscular, tubular organ that leads from the uterus to the vestibule, situated between the urinary bladder and the rectum of the female.

Vallate papilla (VAL-āt pa-PIL-a) One of the circular projections that is arranged in an inverted V-shaped row at the back of the tongue; the largest of the elevations on the upper surface of the tongue containing taste buds. Also called **circumvallate papilla.**

Valence (VĀ-lens) The combining capacity of an atom; the number of deficit or extra electrons in the outermost electron shell of an atom.

Varicocele (VAR-i-kō-sēl) A twisted vein; especially, the accumulation of blood in the veins of the spermatic cord.

Varicose (VAR-i-kōs) Pertaining to an unnatural swelling, as in the case of a varicose vein.

Vasa recta (VĀ-sa REK-ta) Extensions of the efferent arteriole of a juxtamedullary nephron that run alongside the loop of the nephron (Henle) in the medullary region of the kidney.

Vasa vasorum (va-SŌ-rum) Blood vessels that supply nutrients to the larger arteries and veins.

Vascular (VAS-kū-lar) Pertaining to or containing many blood vessels.

Vascular spasm Contraction of the smooth muscle in the wall of a damaged blood vessel to prevent blood loss.

Vascular (venous) sinus A vein with a thin endothelial wall that lacks a tunica media and externa and is supported by surrounding tissue.

Vascular tunic (TOO-nik) The middle layer of the eyeball, composed of the choroid, ciliary body, and iris. Also called the **uvea** (Ū-ve-a).

Vasectomy (va-SEK-tō-mē) A means of sterilization of males in which a portion of each ductus (vas) deferens is removed.

Vasoconstriction (vāz-ō-kon-STRIK-shun) A decrease in the size of the lumen of a blood vessel caused by contraction of the smooth muscle in the wall of the vessel.

Vasodilation (vāz′-ō-DĪ-lā-shun) An increase in the size of the lumen of a blood vessel caused by relaxation of the smooth muscle in the wall of the vessel.

Vasomotion (vāz-ō-MŌ-shun) Intermittent contraction and relaxation of the smooth muscle of the metarterioles and precapillary sphincters that result in an intermittent blood flow.

Vein A blood vessel that conveys blood from tissues back to the heart.

Vena cava (VĒ-na KĀ-va) One of two large veins that open into the right atrium, returning to the heart all of the deoxygenated blood from the systemic circulation except from the coronary circulation.

Ventral (VEN-tral) Pertaining to the anterior or front side of the body; opposite of dorsal.

Ventral body cavity Cavity near the ventral aspect of the body that contains viscera and consists of a superior thoracic cavity and an inferior abdominopelvic cavity.

Ventral ramus (RĀ-mus) The anterior branch of a spinal nerve, containing sensory and motor fibers to the muscles and skin of the anterior surface of the head, neck, trunk, and the limbs.

Ventricle (VEN-tri-kul) A cavity in the brain filled with cerebrospinal fluid. An inferior chamber of the heart.

Ventricular fibrillation (ven-TRIK-ū-lar fib-ri-LĀ-shun) Asynchronous ventricular contractions; unless reversed by defibrillation, results in heart failure.

Venule (VEN-ūl) A small vein that collects blood from capillaries and delivers it to a vein.

Vermiform appendix (VER-mi-form a-PEN-diks) A twisted, coiled tube attached to the cecum.

Vermilion (ver-MIL-yon) The area of the mouth where the skin on the outside meets the mucous membrane on the inside.

Vermis (VER-mis) The central constricted area of the cerebellum that separates the two cerebellar hemispheres.

Vertebral (VER-te-bral) **canal** A cavity within the vertebral column formed by the vertebral foramina of all the vertebrae and containing the spinal cord. Also called the **spinal canal.**

Vertebral column The 26 vertebrae of an adult and 33 vertebrae of a child; encloses and protects the spinal cord and serves as a point of attachment for the ribs and back muscles. Also called the **backbone, spine,** or **spinal column.**

Vesicle (VES-i-kul) A small bladder or sac containing liquid.

Vesicouterine (ves′-ik-ō-Ū-ter-in) **pouch** A shallow pouch formed by the reflection of the peritoneum from the anterior surface of the uterus, at the junction of the cervix and the body, to the posterior surface of the urinary bladder.

Vestibular (ves-TIB-ū-lar) **apparatus** Collective term for the organs of equilibrium, which includes the saccule, utricle, and semicircular ducts.

Vestibular membrane The membrane that separates the cochlear duct from the scala vestibuli.

Vestibule (VES-ti-būl) A small space or cavity at the beginning of a canal, especially the inner ear, larynx, mouth, nose, and vagina.

Villus (VIL-lus) A projection of the intestinal mucosal cells containing connective tissue, blood vessels, and a lymphatic vessel; functions in the absorption of the end products of digestion. *Plural is* **villi** (VIL-ī).

Viscera (VIS-er-a) The organs inside the ventral body cavity. *Singular is* **viscus** (VIS-kus).

Visceral (VIS-er-al) Pertaining to the organs or to the covering of an organ.

Visceral effectors (e-FEK-torz) Organs of the ventral body cavity that respond to neural stimulation, including cardiac muscle, smooth muscle, and glands.

Vital capacity The sum of inspiratory reserve volume, tidal volume, and expiratory reserve volume; about 4800 mL.

Vital signs Signs necessary to life that include temperature (T), pulse (P), respiratory rate (RR), and blood pressure (BP).

Vitamin An organic molecule necessary in trace amounts that acts as a catalyst in normal metabolic processes in the body.

Vitreous (VIT-rē-us) **body** A soft, jellylike substance that fills the vitreous chamber of the eyeball, lying between the lens and the retina.

Vocal folds Pair of mucous membrane folds below the ventricular folds that function in voice production. Also called **true vocal cords.**

Voltage-gated channel An ion channel in a plasma membrane composed of integral proteins that functions like a gate to permit or restrict the movement of ions across the membrane in response to changes in the voltage.

Vulva (VUL-va) Collective designation for the external genitalia of the female. Also called the **pudendum** (poo-DEN-dum).

W

Wallerian (wal-LE-rē-an) **degeneration** Degeneration of the portion of the axon and myelin sheath of a neuron distal to the site of injury.

Wandering macrophage (MAK-rō-fāj) Phagocytic cell that develops from a monocyte, leaves the blood, and migrates to infected tissues.

Wave summation (sum-MĀ-shun) The increased strength of muscle contraction that results when muscle action potentials occur one after another in rapid succession.

White matter Aggregations or bundles of myelinated and unmyelinated axons located in the brain and spinal cord.

White pulp The regions of the spleen composed of lymphatic tissue, mostly B lymphocytes.

White ramus communicans (RĀ-mus kō-MŪ-ni-kans) The portion of a preganglionic sympathetic axon that branches from the anterior ramus of a spinal nerve to enter the nearest sympathetic trunk ganglion.

X

X-chromosome inactivation The random and permanent inactivation of one X chromosome in each cell of a developing female embryo. Also called **lyonization.**

Xiphoid (ZĪ-foyd) Sword-shaped. The inferior portion of the sternum is the **xiphoid process.**

Y

Yolk sac An extraembryonic membrane composed of the exocoelomic membrane and hypoblast. It transfers nutrients to the embryo, is a source of blood cells, contains primordial germ cells that migrate into the gonads to form primitive germ cells, forms part of the gut, and helps prevent desiccation of the embryo.

Z

Zona fasciculata (ZŌ-na fa-sik′-ū-LA-ta) The middle zone of the adrenal cortex consisting of cells arranged in long, straight cords that secrete glucocorticoid hormones, mainly cortisol.

Zona glomerulosa (glo-mer′-ū-LŌ-sa) The outer zone of the adrenal cortex, directly under the connective tissue covering, consisting of cells arranged in arched loops or round balls that secrete mineralocorticoid hormones, mainly aldosterone.

Zona pellucida (pe-LOO-si-da) Clear glycoprotein layer between a secondary oocyte and the surrounding granulosa cells of the corona radiata.

Zona reticularis (ret-ik′-ū-LAR-is) The inner zone of the adrenal cortex, consisting of cords of branching cells that secrete sex hormones, chiefly androgens.

Zygote (ZĪ-gōt) The single cell resulting from the union of male and female gametes; the fertilized ovum.

EPONYMS USED IN THIS TEXT

In the life sciences, an eponym is the name of a structure, drug, or disease that is based on the name of a person. For example, you may be more familiar with the Achilles tendon than you are with its more anatomically descriptive term, the calcaneal tendon. Because eponyms remain in frequent use, this listing correlates common eponyms with their anatomical terms.

EPONYM	ANATOMICAL TERM	EPONYM	ANATOMICAL TERM
Achilles tendon	calcaneal tendon	Kupffer (KOOP-fer) cell	stellate reticuloendothelial cell
Adam's apple	thyroid cartilage	Leydig (LĪ-dig) cell	interstitial endocrinocyte
ampulla of Vater (VA-ter)	hepatopancreatic ampulla	loop of Henle (HEN-lē)	loop of the nephron
		Luschka's (LUSH-kaz) aperture	lateral aperture
Bartholin's (BAR-tō-linz) gland	greater vestibular gland		
Billroth's (BIL-rōtz) cord	splenic cord	Magendie's (ma-JEN-dēz) aperture	median aperture
Bowman's (BŌ-manz) capsule	glomerular capsule	Meibomian (mi-BŌ-mē-an) gland	tarsal gland
Bowman's (BŌ-manz) gland	olfactory gland	Meissner (MĪS-ner) corpuscle	corpuscle of touch
Broca's (BRŌ-kaz) area	motor speech area	Merkel (MER-kel) disc	tactile disc
Brunner's (BRUN-erz) gland	duodenal gland	Müllerian (mil-E rē-an) duct	paramesonephric duct
bundle of His (HISS)	artrioventricular (AV) bundle	organ of Corti (KOR-tē)	spiral organ
canal of Schlemm (SHLEM)	scleral venous sinus	Pacinian (pa-SIN-ē-an) corpuscle	lamellated corpuscle
circle of Willis (WIL-is)	cerebral arterial circle	Peyer's (PĪ-erz) patch	aggregated lymphatic follicle
Cooper's (KOO-perz) ligament	suspensory ligament of the breast	plexus of Auerbach (OW-er-bak)	myenteric plexus
		plexus of Meissner (MĪS-ner)	submucosal plexus
Cowper's (KOW-perz) gland	bulbourethral gland	pouch of Douglas	rectouterine pouch
crypt of Lieberkühn (LE-ber-kyūn)	intestinal gland	Purkinje (pur-KIN-jē) fiber	conduction myofiber
duct of Santorini (san'-tō-RĒ-nē)	accessory duct	Rathke's (rath-KĒZ) pouch	hypophyseal pouch
duct of Wirsung (VĒR-sung)	pancreatic duct	Ruffini (roo-FĒ-nē) corpuscle	type II cutaneous mechanoreceptor
Eustachian (yoo-STĀ-kē-an)	auditory tube		
		Sertoli (ser-TŌ-lē) cell	sustentacular cell
Fallopian (fal-LŌ-pē-an) tube	uterine tube	Skene's (SKĒNZ) gland	paraurethral gland
		sphincter of Oddi (OD-dē)	sphincter of the hepatopancreatic ampulla
gland of Littré (LĒ-tra)	urethral gland		
Golgi (GOL-jē) tendon organ	tendon organ	Volkmann's (FŌLK-manz) canal	perforating canal
Graafian (GRAF-ē-an) follicle	mature ovarian follicle		
		Wernicke's (VER-ni-kēz) area	auditory association area
Hassall's (HAS-alz) corpuscle	thymic corpuscle	Wharton's (HWAR-tunz) jelly	mucous connective tissue
Haversian (ha-VĒR-shun) canal	central canal	Wolffian duct	mesonephric duct
Haversian (ha-VĒR-shun) system	osteon	Wormian (WER-mē-an) bone	sutural bone
Heimlich (HĪM-lik) maneuver	abdomial thrust maneuver		
islet of Langerhans (LANG-er-hanz)	pancreatic islet		

COMBINING FORMS, WORD ROOTS, PREFIXES, AND SUFFIXES

Many of the terms used in anatomy and phsiology are compound words; that is, they are made up of word roots and one or more prefixes or suffixes. For example, *leukocyte* is formed from the word roots *leuk-* meaning "white", a connecting vowel (o), and *cyte* meaning "cell." Thus, a leukocyte is a white blood cell. The following list includes some of the most commonly used combining forms, word roots, prefixes, ad suffixes used in the study of anatomy and physiology. Each entry includes a usage example. Learning the meanings of these fundamental word parts will help you remember terms that, at first glance, may seem long or complicated.

COMBINING FORMS AND WORD ROOTS

Acous-, Acu- hearing Acoustics.
Acr- extremity Acromegaly.
Aden- gland Adenoma.
Alg-, Algia- pain Neuralgia.
Angi- vessel Angiocardiography.
Anthr- joint Arthropathy.
Aut-, Auto- self Autolysis.
Audit- hearing Auditory canal.

Bio- life, living Biopsy.
Blast- germ, bud Blastula.
Blephar- eyelid Blepharitis.
Brachi- arm Brachial plexus.
Bronch- trachea, windpipe Bronchoscopy.
Bucc- cheek Buccal.

Capit- head Decapitate.
Carcin- cancer Carcinogenic.
Cardi-, Cardia-, Cardio- heart Cardiogram.
Cephal- head Hydrocephalus.
Cerebro- brain Cerebrospinal fluid.
Chole- bile, gall Cholecystogram.
Chondr-, cartilage Chondrocyte.
Cor-, Coron- heart Coronary.
Cost- rib Costal.
Crani- skull Craniotomy.
Cut- skin Subcutaneous.
Cyst- sac, bladder Cystoscope.

Derma-, Dermato- skin Dermatosis.
Dura- hard Dura mater.

Enter- intestine Enteritis.
Erythr- red Erythrocyte.

Gastr- stomach Gastrointestinal.
Gloss- tongue Hypoglossal.
Glyco- sugar Glycogen.
Gyn-, Gynec- female, woman Gynecology.

Hem-, Hemat- blood Hematoma.
Hepar-, Hepat- liver Hepatitis.
Hist-, Histio- tissue Histology.
Hydr- water Dehydration.
Hyster- uterus Hysterectomy.

Ischi- hip, hip joint Ischium.

Kines- motion Kinesiology.

Labi- lip Labial.
Lacri- tears Lacrimal glands.
Laparo- loin, flank, abdomen Laparoscopy.
Leuko- white Leukocyte.
Lingu- tongue Sublingual glands.
Lip- fat Lipid.
Lumb- lower back, loin Lumbar.

Macul- spot, blotch Macula.
Malign- bad, harmful Malignant.
Mamm-, Mast- breast Mammography, Mastitis.
Meningo- membrane Meningitis.
Myel- marrow, spinal cord Myeloblast.
My-, Myo- muscle Myocardium.

Necro- corpse, dead Necrosis.
Nephro- kidney Nephron.
Neuro- nerve Neurotransmitter.

Ocul- eye Binocular.
Odont- tooth Orthodontic
Onco- mass, tumor Oncology.
Oo- egg Oocyte.
Opthalm- eye Ophthalmology.
Or- mouth Oral.
Osm- odor, sense of small Anosmia.
Os-, Osseo-, Osteo- bone Osteocyte.
Ot- ear Otitus media.

Palpebr- eyelid Palpebra.
Patho- disease Pathogen.
Pelv- basin Renal pelvis.
Phag- to eat Phagocytosis.
Phleb- vein Phlebitis.
Phren- diaphragm Phrenic.
Pilo- hair Depilatory.
Pneumo- lung, air Pneumothorax.
Pod- foot Podocyte.
Procto- anus, rectum Proctology.
Pulmon- lung Pulmonary.

Ren- kidneys Renal artery.
Rhin- nose Rhinitis.

Scler-, Sclero- hard Atherosclerosis.
Sep-, Spetic- toxic condition due to micoorganisms Septicemia.
Soma-, Somato- body Somatotropin.
Sten- narrow Stenosis.
Stasis-, Stat- stand still Homeostasis.

Tegument- skin, covering Integumentary.
Therm- heat Thermogenesis.
Thromb- clot, lump Thrombus.

Vas- vessel, duct Vasoconstriction.

Zyg- joined Zygote.

PREFIXES

A-, An- without, lack of, deficient Anesthesia.
Ab- away from, from Abnormal.
Ad-, Af- to, toward Adduction, Afferent neuron.
Alb- white Albino.
Alveol- cavity, socket Alveolus.
Andro- male, masculine Androgen.
Ante- before Antebrachial vein.
Anti- against Anticoagulant.

Bas- base, foundation Basal ganglia.
Bi- two, double Biceps.
Brady- slow Bradycardia.

Cata- down, lower, under Catabolism.
Circum- around Circumduction.
Cirrh- yellow Cirrhosis of the liver.
Co-, Con-, Com with, together Congenital.
Contra- against, opposite Contraception.
Crypt- hidden, concealed Cryptorchidism.
Cyano- blue Cyanosis.

De- down, from Deciduous.
Demi-, hemi- half Hemiplegia.
Di-, Diplo- two Diploid.
Dis- separation, apart, away from Dissection.
Dys- painful, difficult Dyspnea.

E-, Ec-, Ef- out from, out of Efferent neuron.
Ecto-, Exo- outside Ectopic pregnancy.
Em-, En- in, on Emmetropia.
End-, Endo- within, inside Endocardium.
Epi- upon, on, above Epidermis.
Eu- good, easy, normal Eupnea.
Ex-, Exo- outside, beyond Exocrine gland.
Extra- outside, beyond, in addition to Extracellular fluid.

Fore- before, in front of Forehead.

Gen- originate, produce, form Genitalia.
Gingiv- gum Gingivitis.

Hemi- half Hemiplegia.
Heter-, Hetero- other, different Heterozygous.
Homeo-, Homo- unchanging, the same, steady Homeostasis.
Hyper- over, above, excessive Hyperglycemia.
Hypo- under, beneath, deficient Hypothalamus.

In-, Im- in, inside, not Incontinent.
Infra- beneath Infraorbital.
Inter- among, between Intercostal.
Intra- within, inside Intracellular fluid.
Ipsi- same Ipsilateral.
Iso- equal, like Isotonic.

Juxta- near to Juxtaglomerular apparatus.

Later- side Lateral.

Macro- large, great Macrophage.
Mal- bad, abnormal Malnutrition.
Medi-, Meso- middle Medial.
Mega-, Megalo- great, large Magakaryocyte.
Melan- black Melanin.
Meta- after, beyond Metacarpus.
Micro- small Microfilament.
Mono- one Monounsaturated fat.

Neo- new Neonatal.

Oligo- small, few Oliguria.
Ortho- straight, normal Orthopedics.

Para- near, beyond, beside Paranasal sinus.
Peri- around Pericardium.
Poly- much, many, too much Polycythemia.
Post- after, beyond Postnatal.
Pre-, Pro- before, in front of Presynaptic.
Pseudo- false Pseudostratified.

Retro- backward, behind Retroperitoneal.

Semi- half Semicircular canals.
Sub- under, beneath, below Submucosa.
Super- above, beyond Superficial.
Supra- above, over Suprarenal.
Sym-, Syn- with, together Symphysis.

Tachy- rapid Tachycardia.
Trans- across, through, beyond Transudation.
Tri- three Trigone.

SUFFIXES

-able capable of, having ability to Viable.
-ac, -al pertaining to Cardiac.
-algia painful condition Myalgia.
-an, -ian pertaining to Circadian.
-ant having the characteristic of Malignant.
-ary connected with Ciliary.
-asis, -asia, -esis, -osis condition or state of Hemostasis.

-asthenia weakness Myasthenia.
-ation process, action, condition Inhalation.
-centesis puncture, usually for drainage Amniocentesis.
-cid, -cide, -cis, cut, kill destroy Spermicide.
-ectomize, -ectomy excision of, removal of Thyroidectomy.
-emia condition of blood Anemia.
-esthesia sensation, feeling Anesthesia.

-fer carry Efferent arteriole.

-gen agent that produces or originates Pathogen.
-genic producing Pyogenic.
-gram record Electrocardiogram.
-graph instrument for recording Electroencephalograph.

-ia state, condition Hypermetropia.
-ician person associated with Pediatrician.
-ics art of, science of Optics.
-ism condition, state Rheumatism.
-itis inflammation Neuritis.

-logy the study or science of Physiology.
-lysis dissolution, loosening, destruction Hemolysis.

-malacia softening Osteomalacia.
-megaly enlarged Cardiomegaly.
-mers, -meres parts Polymers.

-oma tumor Fibroma.
-osis condition, disease Necrosis.
-ostomy create an opening Colostomy.
-otomy surgical incision Tracheotomy.

-pathy disease Myopathy.
-penia deficiency Thrombocytopenia
-philic to like, have an affinity for Hydrophilic.
-phobe, -phobia fear of, aversion to Photophobia.
-plasia, -plasty forming, molding Rhinoplasty.
-pnea breath Apnea.
-poiesis making Hemopoiesis.
-ptosis falling, sagging Blepharoptosis.

-rrhage bursting forth, abnormal discharge Hemorrhage.
-rrhea flow, discharge Diarrhea.

-scope instrument for viewing Bronchoscope.
-stomy creation of a mouth or artificial opening Tracheostomy.

-tomy cutting into, incision into Laparotomy.
-tripsy crushing Lithotripsy.
-trophy relating to nutrition or growth Atrophy.

-uria urine Polyuria.